BLACKWELL'S FIVE-MINUTE VETERINARY CONSULT: AVIAN

BLACKWELL'S FIVE-MINUTE VETERINARY CONSULT

AVIAN

Jennifer E. Graham, DVM, DABVP (Avian/ECM), DACZM
Assistant Professor of Zoological Companion Animal Medicine,
Department of Clinical Sciences,
Tufts Cummings School of Veterinary Medicine,
North Grafton, Massachusetts, USA

This edition first published 2016 © 2016 by John Wiley & Sons, Inc.

Editorial offices: 1606 Golden Aspen Drive, Suites 103 and 104, Ames, Iowa 50010, USA
The Atrium, Southern Gate, Chichester, West Sussex, PO19 8SQ, UK
9600 Garsington Road, Oxford, OX4 2DQ, UK

For details of our global editorial offices, for customer services and for information about how to apply for permission to reuse the copyright material in this book please see our website at www.wiley.com/wiley-blackwell.

Authorization to photocopy items for internal or personal use, or the internal or personal use of specific clients, is granted by Blackwell Publishing, provided that the base fee is paid directly to the Copyright Clearance Center, 222 Rosewood Drive, Danvers, MA 01923. For those organizations that have been granted a photocopy license by CCC, a separate system of payments has been arranged. The fee codes for users of the Transactional Reporting Service are ISBN-13: 978-1-1189-3459-3 / 2016

Designations used by companies to distinguish their products are often claimed as trademarks. All brand names and product names used in this book are trade names, service marks, trademarks or registered trademarks of their respective owners. The publisher is not associated with any product or vendor mentioned in this book.

The contents of this work are intended to further general scientific research, understanding, and discussion only and are not intended and should not be relied upon as recommending or promoting a specific method, diagnosis, or treatment by health science practitioners for any particular patient. The publisher and the author make no representations or warranties with respect to the accuracy or completeness of the contents of this work and specifically disclaim all warranties, including without limitation any implied warranties of fitness for a particular purpose. In view of ongoing research, equipment modifications, changes in governmental regulations, and the constant flow of information relating to the use of medicines, equipment, and devices, the reader is urged to review and evaluate the information provided in the package insert or instructions for each medicine, equipment, or device for, among other things, any changes in the instructions or indication of usage and for added warnings and precautions. Readers should consult with a specialist where appropriate. The fact that an organization or Website is referred to in this work as a citation and/or a potential source of further information does not mean that the author or the publisher endorses the information the organization or Website may provide or recommendations it may make. Further, readers should be aware that Internet Websites listed in this work may have changed or disappeared between when this work was written and when it is read. No warranty may be created or extended by any promotional statements for this work. Neither the publisher nor the author shall be liable for any damages arising herefrom.

Library of Congress Cataloging-in-Publication Data

Names: Graham, Jennifer E. (Jennifer Erin), 1974- , editor.
Title: Blackwell's five-minute veterinary consult. Avian / [edited by] Jennifer E. Graham.
Other titles: Five-minute veterinary consult. Avian | Avian
Description: Ames, Iowa : John Wiley and Sons, Inc., 2016. | Includes bibliographical references and index.
Identifiers: LCCN 2015039615 | ISBN 9781118934593 (cloth)
Subjects: LCSH: Birds–Diseases. | MESH: Bird Diseases. | Veterinary Medicine–methods.
Classification: LCC SF994 .B63 2016 | NLM SF 994 | DDC 636.5/0896–dc23
LC record available at http://lccn.loc.gov/2015039615

A catalogue record for this book is available from the British Library.

Wiley also publishes its books in a variety of electronic formats. Some content that appears in print may not be available in electronic books.

Set in 9/10pt GaramondPro by Aptara Inc., New Delhi, India

Printed in Singapore by C.O.S. Printers Pte Ltd

1 2016

This book is dedicated to my heroes: my dear 'ol dad, Lewis, and my sister, Amy.

My father had me convinced at one point that he was Dr. Bob, with a side job as a brain surgeon (this was a total lie). This resulted in me telling an entire busload of children about his craniotomy adventures on the ride home from school and making the bus driver wait outside the house so he could wave to them all. He turned bright red and didn't tell so many stories after that day as I recall.

Amy-sis—thanks for always being there. I am so lucky to have you as "my person".

Preface

As a veterinary student and new graduate, I remember reaching for my *Blackwell's Five-Minute Veterinary Consult: Canine and Feline* text on a regular basis. The concise and thorough coverage organized by topic was exactly what I needed to refresh memories from lectures and make sure I was not forgetting an important aspect of a case. Although I have since left canine and feline practice behind, I still find myself reaching for this invaluable text when collaborating with colleagues, consulting on a case, or if I am in need of a quick overview of a particular disease process. It probably comes as no surprise that I am also a fan of the *Blackwell Five-Minute Veterinary Consult: Ferret and Rabbit* text.

When Wiley contacted me about working on a Five-minute Veterinary Consult: Avian text, I had two thoughts. The first thought was surprise at the realization that this had not already been done. The second was the fact that it was a complete no-brainer, that I would love to be the one to head up such a worthy project. The other no-brainer was who I would be inviting to write chapters. I have been fortunate to get an amazing group of talented contributors. All are leaders in the field of avian medicine and eminently qualified to address the topics included in this text.

This book is divided into 123 topics covering a wide range of diseases and syndromes in avian patients, seven appendices and accompanying algorithms, and a companion website that includes client handouts and descriptions and pictures of common clinical procedures. The approach on topics is focused on practical clinical knowledge and organized to offer fast access to essential information.

A helpful aspect of the table of contents is multiple listings of topics based on common terminology. For example, the topic of "viral neoplasms" can also be found under the headings: Marek's disease, lymphoid leukosis, and reticuloendotheliosis. Topics have been chosen to encompass the majority of syndromes and diseases seen in avian practice. Species covered in the text include psittacine birds, passerines, poultry, raptors, ratites, and waterfowl. The template design of topic layout ensures quick access to information without the need to read the entire section—you can jump to any section on an as-needed basis. At the end of each section is a "see also" section which lists similar diseases or syndromes.

The appendices are broken down into user-friendly tables. The formulary, Appendix 1, includes drugs mentioned throughout the book and lists dosing recommendations along with indications for use. Hematology and biochemistry reference ranges are outlined in Appendix 2. Appendix 3 lists common laboratory tests available for avian species and laboratory contact information. Viral diseases are outlined in Appendix 4 with information supplied on species affected, common lesions, and transmission routes. Zoonotic diseases of concern and personal protective guidelines are listed in Appendix 5. Appendix 6 outlines common plant toxins along with systems affected and treatment recommendations. Appendix 7 includes clinical algorithms for 15 common clinical presentations ranging from non-specific signs like "sick bird syndrome" and to specific presentations like "feather damaging behavior" and "lameness".

The companion website will be a helpful resource for practitioners. The website includes client education handouts that can be downloaded and edited for distribution to clients. Common procedures with accompanying descriptions and pictures are also available on the website.

We are hopeful this text will be a useful addition to your library—whether you see birds occasionally or as a regular part of your practice caseload.

Jennifer E. Graham

Acknowledgments

There are many individuals whose contributions helped make this book happen. To Hugues Beaufrère, who helped with modification and improvement of the table of contents along with some fabulous topic summaries—thank you for always being willing to lend a hand and for sharing your brilliance. To Erika Cervasio—thanks so much for the extra help with contributions to the online procedures and helping create some last-minute topic summaries that were originally omitted. To Lauren Powers—your images for online reference material are much appreciated. George Messenger—even though you are supposedly "retired", all the more time for you to help with future endeavors—thanks for your willingness to help with this project. To all my amazing colleagues who have contributed to the topics in this text—thank you for the time and effort you dedicated to this book. To Nancy, Erica, Shalini, Gayle, and the rest of the wonderful team at Wiley—working with you is a pleasure. And last but not least—to my husband James—thank you for your continued support, encouragement, and attendance at all the "boring vet events"; may our love outlast even our tattoos.

Contributors

ANDREW D. BEAN, DVM, MPH, CPH
Veterinarian
Pet Care Veterinary Hospital
Virginia Beach, Virginia, USA

HUGUES BEAUFRÈRE, Dr.Med.Vet, PhD,
ECZM (Avian), DABVP (Avian), DACZM
Service Chief, Avian and Exotics Service
Health Sciences Centre
Ontario Veterinary College
University of Guelph
Guelph, Ontario, Canada

JOAO BRANDAO, LMV, MS
Assistant Professor, Zoological Medicine
Service
Department of Veterinary Clinical Sciences
Center for Veterinary Health Sciences
Oklahoma State University
Stillwater, Oklahoma, USA

JAMES W. CARPENTER, MS, DVM, DACZM
Professor, Zoological Medicine
Veterinary Medical Teaching Hospital
Kansas State University
Manhattan, Kansas, USA

ERIKA L. CERVASIO, DVM, DABVP (Avian)
North Paws Veterinary Center
Mansfield, Massachusetts, USA

SUE CHEN, DVM, DABVP (Avian)
Gulf Coast Avian & Exotics
Gulf Coast Veterinary Specialists
Houston, Texas, USA

LEIGH ANN CLAYTON, DVM, DABPV
(Avian, Reptile-Amphibian)
Director Animal Health
National Aquarium
Baltimore, Maryland, USA

SUSAN L. CLUBB, DVM, DABVP (Avian)
Rainforest Clinic For Birds and Exotics, Inc.
Hurricane Aviaries, Inc.
Loxahatchee, Florida, USA

ROB L. COKE, DVM, DACZM, DABVP
(Reptile & Amphibian), CVA
Senior Staff Veterinarian
San Antonio Zoo
San Antonio, Texas, USA

GREGORY J. COSTANZO, DVM
Stahl Exotic Animal Veterinary Services
Fairfax, Virginia, USA

LORENZO CROSTA, DVM, GP CERT (Exotic
Animal Practice), PhD
FNOVI Accredited Veterinarian in Avian
Medicine and Surgery and Zoo Animal
Medicine and Surgery and Zoo Management
ECZM
Veterinari Montevecchia
Montevecchia (LC), Italy

JULIE DECUBELLIS, DVM, MS
Calgary Avian and Exotic Pet Clinic
Calgary, Alberta, Canada

MARION DESMARCHELIER, DMV, MSc,
DACZM
Clinical Instructor
Zoological Medicine Service
Department of Clinical Sciences
Faculté de medicine vétérinaire
Université de Montréal
Saint-Hyacinthe, Quebec, Canada

RYAN DEVOE, DVM, MSpVM, DACZM,
DABVP (Reptile & Amphibian)
Clinical Veterinarian
Department of Animal Health
Disney's Animal Kingdom
Bay Lake, Florida, USA

STEPHEN M. DYER, DVM, DABVP (Avian)
NorthPaws Veterinary Center
Greenville, Rhode Island, USA

THOMAS M. EDLING, DVM, MSpVM, MPH
Vice President, Veterinary Medicine
Petco Animal Supplies, Inc.
San Diego, California, USA

SHANNON FERRELL, DVM, DABVP (Avian),
DACZM
Veterinarian Zoo de Granby 525
Granby, Quebec, Canada

SUSAN G. FRIEDMAN, PhD
Department of Psychology
Utah State University
Logan, Utah, USA

ALAN M. FUDGE, DVM, DABVP (Avian)
Mobile Veterinary Services, Birds & Fish
Greenville, South Carolina, USA

JENNIFER E. GRAHAM, DVM, DABVP
(Avian/ECM), DACZM
Assistant Professor of Zoological Companion
Animal Medicine
Department of Clinical Sciences
Cummings School of Veterinary Medicine
Tufts University
North Grafton, Massachusetts, USA

DAVID E. HANNON, DVM, DABVP (Avian)
Avian and Exotic Animal Veterinary Services
Memphis Veterinary Specialists
Memphis, Tennessee, USA

KENDAL E. HARR, DVM, MS, DACVP
Pathologist
Florida Veterinary Pathology Consultants Inc.
Bushnell, Florida, USA

LISA HARRENSTIEN, DVM, DACZM
Zoological Medicine Consultant
Portland, Oregon, USA

J. JILL HEATLEY, DVM, MS, DABVP (Avian,
Reptile & Amphibian), DACZM
Associate Professor, Zoological Medicine
College of Veterinary Medicine
Texas A&M University
College Station, Texas, USA

CHRISTINE T. HIGBIE, DVM
Department of Clinical Sciences
College of Veterinary Medicine
Kansas State University
Manhattan, Kansas, USA

HEIDI L. HOEFER, DVM, DABVP (Avian)
Island Exotic Veterinary Care
Huntington Station, New York, USA

MICHAEL P. JONES, DVM, DABVP (Avian)
Associate Professor of Avian & Zoological
Medicine
Department of Small Animal Clinical
Sciences
College of Veterinary Medicine
University of Tennessee
Knoxville, Tennessee, USA

ERIC KLAPHAKE, DVM, DABVP (Avian), DACZM, DABVP (Reptile-Amphibian)
Associate Veterinarian
Cheyenne Mountain Zoo
Colorado Springs, Colorado, USA

LAURA M. KLEINSCHMIDT, BS, DVM
Department of Small Animal Clinical Sciences, Zoological Medicine Service
College of Veterinary Medicine and Biomedical Science, Texas A&M University
College Station, Texas, USA

ISABELLE LANGLOIS, DMV, DABVP (Avian)
Clinical Instructor, Zoological Medicine Service
Department of Clinical Sciences
Faculty of Veterinary Medicine
Université de Montréal
Saint-Hyacinthe, Québec, Canada

DELPHINE LANIESSE, Dr. Med. Vet, IPSAV
Ontario Veterinary College
University of Guelph
Guelph, Ontario, Canada

ANGELA M. LENNOX, DVM, DABVP (Avian, Exotic Companion Mammal)
Avian and Exotic Animal Clinic
Indianapolis, Indiana, USA

SHACHAR MALKA, DVM, DABVP (Avian)
The Avian and Exotics Service
The Humane Society of New York
New York, New York, USA

RUTH MARRION, DVM, DACVO, PhD
Ophthalmologist
Bulger Veterinary Hospital
North Andover, Massachusetts, USA

KEMBA L. MARSHALL, DVM, DABVP (Avian)
Director of Merchandising Pet Quality and Education
PetSmart SSG
Phoenix, Arizona, USA

GEORGE MESSENGER, DVM, DABVP (Avian)
Concord, New Hampshire, USA

MAUREEN MURRAY, DVM, DABVP (Avian)
Clinical Assistant Professor
Wildlife Clinic
Department of Infectious Disease and Global Health
Cummings School of Veterinary Medicine
Tufts University
North Grafton, Massachusetts, USA

HEIDE M. NEWTON, DVM, DACVD
Southern Arizona Veterinary Specialty & Emergency Center
Dermatology for Animals
Tucson, Arizona, USA

GEOFFREY P. OLSEN, DVM, DABVP (Avian)
Associate
Medical Center for Birds
Oakley, California, USA

DAVID N. PHALEN, DVM, PhD, DABVP (Avian)
Associate Professor
Faculty of Veterinary Science
University of Sydney
Sydney, New South Wales, Australia

ANTHONY A. PILNY, DVM, DABVP (Avian)
Medical Director
The Center for Avian and Exotic Medicine
New York City, New York, USA

CHRISTAL POLLOCK, DVM, DABVP (Avian)
Lafeber Company Veterinary Consultant
Cornell, Illinois, USA

LAUREN V. POWERS, DVM, DABVP (Avian/ECM)
Service Head, Avian and Exotic Pet Service
Carolina Veterinary Specialists
Adjunct Assistant Professor
School of Veterinary Medicine
North Carolina State University
North Carolina, USA

DRURY R. REAVILL, DVM, DABVP (Avian/Reptile & Amphibian Practice), DACVP
Zoo/Exotic Pathology Service
Carmichael, California, USA

GREGORY RICH, DVM, BS Medical Technology
Owner
West Esplanade Veterinary Clinic
Metairie, Louisiana, USA

SHANNON M. RIGGS, DVM
Director of Animal Care
Pacific Wildlife Care
Morro Bay, California, USA

VANESSA ROLFE, DVM, ABVP (Avian)
The Bird & Exotic Hospital, Inc.
Greenacres, Florida, USA

KAREN ROSENTHAL, DVM, MS
Dean, School of Veterinary Medicine
St. Matthew's University
Grand Cayman, Cayman Islands, British West Indies

PETRA SCHNITZER, DVM, GP Cert (Exotic Animal Practice)
ECZM Resident (Avian)
Veterinari Montevecchia
Montevecchia (LC), Italy

NICO J. SCHOEMAKER, DVM, PhD, DECZM (Small mammal, Avian), DABVP (Avian)
Associate Professor
Division of Zoological Medicine
Department of Clinical Sciences of Companion Animals
Faculty of Veterinary Medicine
Utrecht University
Utrecht, The Netherlands

ELISABETH SIMONE-FREILICHER, DVM, DABVP (Avian)
Senior Clinician
Avian and Exotic Animal Medicine Department
Angell Animal Medical Center
Boston, Massachusetts, USA

KRISTIN M. SINCLAIR, DVM, DABVP (Avian)
Staff Veterinarian
Kensington Bird and Animal Hospital
Kensington, Connecticut, USA

KURT K. SLADKY, MS, DVM, DACZM, DECZM (Herpetology)
Clinical Associate Professor
Zoological Medicine
Department of Surgical Sciences
School of Veterinary Medicine
University of Wisconsin
Madison, Wisconsin, USA

DALE A. SMITH, DVM, DVSc
Professor
Department of Pathobiology
Ontario Veterinary College
University of Guelph
Guelph, Ontario, Canada

MARCY J. SOUZA, DVM, MPH, DABVP (Avian), DACVPM
Assistant Professor
Director of Veterinary Public Health
Department of Biomedical & Diagnostic Sciences
College of Veterinary Medicine
University of Tennessee
Knoxville, Tennessee, USA

BRIAN SPEER, DVM, DABVP (Avian), DECZM (Avian)
Medical Center for Birds
Oakley, California, USA

JONATHAN STOCKMAN, DVM, DACVN
Senior Research Scientist
Effem Foods Co., Ltd.
Wyong, New South Wales, Australia

ANNELIESE STRUNK, DVM, DABVP (Avian)
Medical Director
The Center for Bird and Exotic Animal Medicine
Bothell, Washington, USA

JOHN H. TEGZES, MA, VMD, DABVT
Professor of Toxicology
College of Veterinary Medicine
Western University of Health Sciences
Pomona, California, USA

FLORINA S. TSENG, DVM
Associate Professor
Department of Infectious Disease and Global Health
Director, Tufts Wildlife Clinic
Cummings School of Veterinary Medicine
Tufts University
North Grafton, Massachusetts, USA

YVONNE VAN ZEELAND, DVM, MVR, PhD, DECZM (Avian, Small Mammal)
Associate Professor
Division of Zoological Medicine
Department of Clinical Sciences of Companion Animals
Faculty of Veterinary Medicine
Utrecht University
Utrecht, The Netherlands

KENNETH R. WELLE, DVM, DABVP (Avian)
College of Veterinary Medicine
University of Illinois
Urbana, Illinois, USA

AMY B. WORELL, DVM, ABVP (Avian)
Avian Veterinary Services
Wet Hills, California, USA

NICOLE R. WYRE, DVM, DABVP (Avian)
Attending Clinician – Exotic Companion Animal Medicine and Surgery
Matthew J. Ryan Veterinary Hospital
University of Pennsylvania
Philadelphia, Pennsylvania, USA

ASHLEY ZEHNDER, DVM, PhD, DABVP (Avian)
Postdoctoral Fellow, Program in Epithelial Biology
Stanford University
Stanford, California, USA

About the Companion Website

This book is accompanied by a companion website:

www.fiveminutevet.com/avian

The website includes:

- Client Education Handouts
- Procedures
- Algorithms

The password for the companion website is the first word on page 72 (first word in the bulleted list at the start of the page). Please use all lowercase.

Client Education Handouts
- Airborne Toxins
- Angel Wing
- Arthritis
- Aspergillosis
- Atherosclerosis
- Avocado/Plant Toxins
- Cardiac Disease
- Chlamydiosis
- Chronic Egg Laying
- Cloacal Diseases
- Cystic Ovarian Disease
- Feather Damaging Behavior
- Heavy Metal Toxicity
- Hypocalcemia (and Hypomagnesemia)
- Liver Disease
- Macrorhabdosis (Avian Gastric Yeast)
- Nutritional Imbalances
- Obesity
- Ovarian Neoplasia
- Overgrown Beak and Nails
- Pododermatitis
- Polyomavirus Infection in Psittacines
- Problem Behavior: Ten Things Your Parrot Wants You to Know about Behavior
- Problem Behavior: Top Ten List of Behavior Tips
- Proventricular Dilatation Disease
- Regurgitation and Vomiting
- Renal Disease
- Rhinitis and Sinusitis
- Sick Bird Syndrome
- Trauma

Procedures
- Air Sac Cannula
- Beak and Nail Trimming
- Blood Transfusion
- Bone Marrow Aspiration
- Choanal Swab
- Cloacal Swab
- Celiocentesis
- Conjunctival Swab
- Deslorelin Implant
- E-Collars
- Fecal Wet Mount and Gram's Stain
- Figure-of-Eight Bandage
- Handling and Restraint
- Indirect Blood Pressure Monitoring
- Ingluvial Gavage
- Intramuscular Injection
- Intraosseous Catheter
- Intravenous Catheter
- Microchip Placement
- Nasal Flush
- Oral Medications
- Sinus Aspiration
- Subcutaneous Injection
- Tracheal Swab or Wash
- Venipuncture
- Ventplasty
- Wing Trimming

Contents

Topic	
Adenoviruses	1
Aggression (in Problem Behaviors: Aggression, Biting and Screaming)	(244)
Airborne Toxins	3
Air Sac Mites	5
Air Sac Rupture	7
Anemia	9
Angel Wing	11
Anorexia	13
Anticoagulant Rodenticide Toxicosis	15
Arthritis	18
Ascites	21
Aspergillosis	24
Aspiration	27
Atherosclerosis	29
Avian Influenza	33
Avocado Toxins (in Plant and Avocado Toxins)	(225)
Beak Injuries	36
Beak Malocclusion (Scissors Beak)	38
Bite Wounds	41
Biting (in Problem behaviors: Aggression, Biting and Screaming)	(244)
Bordetellosis	44
Botulism	46
Brain Tumors (in CNS Tumors, Brain Tumors, Pituitary Tumors)	(68)
Campylobacteriosis	48
Candidiasis	50
Carbamate Toxicity (in Organophosphate and Carbamate Toxicity)	(200)
Cardiac Disease	52
Cere and Skin, Color Changes	54
Chlamydiosis	56
Chronic Egg Laying	59
Circoviruses	61
Cloacal Disease	63
Clostridiosis	66
CNS Tumors, Brain Tumors, Pituitary Tumors	68
Coagulation (in Coagulopathies and Coagulation)	(71)
Coagulopathies and Coagulation	71
Coccidiosis, Intestinal	74

 Client education handouts are available at www.fiveminutevet.com/avian for you to download and use in practice

Coccidiosis, Systemic	76
Coelomic Distention	79
Colibacillosis	82
Crop Stasis (in Ingluvial Hypomotility, Crop Stasis and Ileus)	(153)
Cryptosporidiosis	85
Cystic Ovaries (in Ovarian Cysts, Neoplasia and Cystic Ovaries)	(206)
Dehydration	87
Dermatitis	89
Diabetes Insipidus	91
Diabetes Mellitus (in Hyperglycemia and Diabetes Mellitus)	(143)
Diarrhea	93
Dyslipidemia/Hyperlipidemia	95
Dystocia and Egg Binding	98
Ectoparasites	101
Egg Binding (in Dystocia and Egg Binding)	(98)
Egg Yolk and Reproductive Coelomitis	103
Emaciation	105
Enteritis and Gastritis	108
Feather Cyst	111
Feather Damaging Behavior and Self-Injurious Behavior	113
Feather Disorders	117
Fecal Discoloration (in Urate and Fecal Discoloration)	(295)
Flagellate Enteritis	119
Fractures and Luxations	121
Gastritis (in Enteritis and Gastritis)	(108)
Gastrointestinal Foreign Bodies	124
Gastrointestinal Helminthiasis	126
Heavy Metal Toxicity	129
Hemoparasites	131
Hemorrhage	134
Hepatic Lipidosis	137
Herpesviruses	140
Hyperglycemia and Diabetes Mellitus	143
Hyperlipidemia (in Dyslipidemia/Hyperlipidemia)	(95)
Hyperuricemia	145
Hypocalcemia and Hypomagnesemia	147
Hypomagnesemia (in Hypocalcemia and Hypomagnesemia)	(147)
Ileus (in Ingluvial Hypomotility, Crop Stasis and Ileus)	(153)
Infertility	150
Ingluvial Hypomotility, Crop Stasis and Ileus	153
Intestinal Helminthiasis (in Gastrointestinal Helminthiasis)	(126)
Iron Storage Disease	156
Lameness	158
Lipomas	161
Liver Disease	163
Luxations (in Fractures and Luxations)	(121)

 Client education handouts are available at www.fiveminutevet.com/avian for you to download and use in practice

Lymphoid Leukosis (in Viral Neoplasms: Marek's Disease, Lymphoid Leukosis and Reticuloendotheliosis)	(302)
Lymphoid Neoplasia	166
Macrorhabdus Ornithogaster	169
Marek's Disease (in Viral Neoplasms: Marek's Disease, Lymphoid Leukosis and Reticuloendotheliosis)	(302)
Metabolic Bone Disease	171
Mycobacteriosis	174
Mycoplasmosis	177
Mycotoxicosis	179
Neurologic Conditions	182
Nutritional Imbalances	185
Obesity	188
Ocular Lesions	190
Oil Exposure	193
Oral Plaques	197
Organophosphate and Carbamate Toxicity	200
Osteomyelosclerosis and Polyostotic Hyperostosis	202
Otitis	204
Ovarian Cysts, Neoplasia and Cystic Ovaries	206
Ovarian Neoplasia (in Ovarian Cysts, Neoplasia and Cystic Ovaries)	(206)
Overgrown Beak and Nails	210
Overgrown Nails (in Overgrown Beak and Nails)	(210)
Oviductal Disease (in Salpingitis, Oviductal Disease and Uterine Disorders)	(262)
Pancreatic Diseases	213
Papillomas, Cutaneous	216
Paramyxoviruses	218
Pasteurellosis	221
Phallus Prolapse	223
Pituitary Tumors (in CNS Tumors, Brain Tumors, Pituitary Tumors)	(68)
Plant and Avocado Toxins	225
Pneumonia	228
Pododermatitis	231
Polydipsia	234
Polyomavirus	237
Polyostotic Hyperostosis (in Osteomyelosclerosis and Polyostotic Hyperostosis)	(202)
Polyuria	239
Poxvirus	241
Problem Behaviors: Aggression, Biting and Screaming	244
Proventricular Dilatation Disease (PDD)	247
Regurgitation and Vomiting	249
Renal Disease	251
Respiratory Distress	254
Reproductive Coelomitis (in Egg Yolk and Reproductive Coelomitis)	(103)
Reticuloendotheliosis (in Viral Neoplasms: Marek's Disease, Lymphoid Leukosis and Reticuloendotheliosis)	(302)
Rhinitis and Sinusitis	257

 Client education handouts are available at www.fiveminutevet.com/avian for you to download and use in practice

Salmonellosis	260
Salpingitis, Oviductal Disease and Uterine Disorders	262
Sarcocystis	265
Scissors Beak (in Beak Malocclusion)	(38)
Screaming (in Problem Behaviors: Aggression, Biting and Screaming)	(244)
Seizures	267
Self-Injurious Behavior (in Feather Damaging Behavior and Self-Injurious Behavior)	(113)
Sick Bird Syndrome	270
Sinusitis (in Rhinitis and Sinusitis)	(257)
Slipped Tendon (in Splay Leg and Slipped Tendon)	(272)
Splay Leg and Slipped Tendon	272
Squamous Cell Carcinoma	274
Syringeal Disease (in Tracheal Disease and Syringeal Disease)	(281)
Thyroid Diseases	277
Toxoplasmosis	279
Tracheal Disease and Syringeal Disease	281
Trauma	284
Trichomoniasis	287
Tumors	289
Undigested Food in Droppings	292
Urate and Fecal Discoloration	295
Uropygial Gland Disease	297
Uterine Disorders (in Salpingitis, Oviductal Disease and Uterine Disorders)	(262)
Viral Disease	299
Viral Neoplasms: Marek's Disease, Lymphoid Leukosis and Reticuloendotheliosis	302
Vitamin D Toxicosis	305
Vomiting (in Regurgitation and Vomiting)	(249)
West Nile Virus	307
Xanthomas	310
Appendix 1: Common Dosages for Birds	312
Appendix 2: Avian Hematology Reference Values	318
Appendix 3: Laboratory Testing (USA)	320
Appendix 4: Viral Diseases of Concern	330
Appendix 5: Selected Zoonotic Diseases of Concern and Personal Protection	335
Appendix 6: Common Avian Toxins and their Clinical Signs	337
Appendix 7: Clinical Algorithms	340
Algorithm 1: Sick Bird Syndrome	340
Algorithm 2: Diarrhea	341
Algorithm 3: Regurgitation/Vomiting	342
Algorithm 4: Respiratory Distress	343
Algorithm 5: Coelomic Distention	344
Algorithm 6: Egg Binding	345
Algorithm 7: Anemia	346
Algorithm 8: Oropharyngeal Lesions	347

 Client education handouts are available at www.fiveminutevet.com/avian for you to download and use in practice

Algorithm 9: Neurologic Signs	348
Algorithm 10: Polydipsia	349
Algorithm 11: Polyuria	350
Algorithm 12: Feather Damaging Behavior	351
Algorithm 13: Lameness	352
Algorithm 14: Hyperuricemia	353
Algorithm 15: Hepatopathy	354
Index	355

ADENOVIRUSES

BASICS

DEFINITION
Adenoviruses are double-stranded nonenveloped DNA viruses. Specific adenoviruses are known to infect and cause disease in passerine birds, psittacine birds, pigeons (aka rock doves), falcons, hawks and owls, and gallinaceous birds.

PATHOPHYSIOLOGY
• Not all adenovirus infections result in disease. • When they do, most cause a systemic infection and may cause considerable morbidity and mortality. • Diseased birds that survive and subclinically infected birds may remain infected for life and be a persistent source of infection. • The Egg Drop Syndrome Virus in chickens replicates extensively in the oviduct causing abnormalities in egg shells and the production of unshelled eggs.

SYSTEMS AFFECTED
• Gastrointestinal—falcons, finches, hawks, owls, pigeons, psittacine birds, turkeys.
• Hemic/Lymphatic/Immune—falcons, finches, pheasants, pigeon, psittacine birds, owls, pigeon, turkeys. • Hepatobiliary: Necrosis—pigeons, psittacine birds, falcons, hawks, owls. • Renal/Urologic—finches, pigeons, psittacine birds. • Reproductive—chickens. • Respiratory—pheasants.
• Respiratory—quail.

GENETICS
• At least one falcon adenovirus is believed to subclinically infect peregrine falcons and is more likely to cause disease in other species.
• The adenoviruses causing Marbled Spleen Disease in pheasants and Hemorrhagic Enteritis in turkeys are asymptomatically carried by waterfowl. • It is likely that many outbreaks of adenoviruses in mixed collections of birds are caused by cross species infection.

INCIDENCE/PREVALENCE
• Falcons: Rare outbreaks have been described. Prevalence of infection in peregrine falcons may be high. • Hawks and owls: Two outbreaks have been described. Prevalence is unknown. • Pigeons: Outbreaks occur sporadically, prevalence is unknown, but subclinical infections are likely to be common. • Psittacine birds: Prevalence is variable. There have been extensive outbreaks in Europe in Budgerigars. Individual infections and outbreaks in other psittacine birds are sporadic. • Chickens: Prevalence is variable, but can be high. • Turkeys: Prevalence is variable, but can be high.
• Pheasants: Prevalence is variable. • Quail: High prevalence of infection.

GEOGRAPHIC DISTRIBUTION
• Finches: Described in North America.
• Falcons: Described in North America.
• Hawks and owls: Described in the United Kingdom. • Pigeons: Worldwide. • Psittacine birds: Outbreaks have occurred on multiple continents, not all adenoviruses have been adequately characterized. The distribution of each adenovirus therefore is not fully known.
• Chickens: Worldwide, but not in North America. • Turkeys: Worldwide. • Pheasants: Worldwide. • Quail: Worldwide.

SIGNALMENT
• Finches: Adult finches, multiple species, both sexes. • Falcons: Nestling northern aplomado falcon, peregrine falcon, Taita falcon, and orange-breasted falcon, and adult American kestrel, both sexes. • Hawks and owls: Harris hawk, Bengal eagle owl, Verreaux's eagle owl, both sexes various ages.
• Pigeons: Less than one year old, both sexes.
• Psittacine birds: Most common in budgerigars, lovebirds and Poicephalus species; occurs sporadically in other parrot species. • Chickens: Laying hens. • Turkeys: Growing birds of both sexes 6–12 weeks old.
• Pheasants: Three to eight months old, both sexes. • Quail: One to six weeks old, both sexes.

SIGNS

Historical Findings
• Finches: Unexpected deaths in a flock.
• Falcons: Death after a short duration of nonspecific signs. • Hawks and owls: Death without premonitory signs or a short duration of nonspecific signs. • Pigeons:
• Type 1: Vomiting, watery diarrhea and depression, rapid spread through the loft, increased mortality; • Type 2: Multiple unexpected deaths.
• Psittacine birds: Unexpected mortality in nestling parrots. • Chickens: Sudden drop in egg production, abnormally colored eggs, shell-less eggs. • Turkeys: Sudden onset of hemorrhagic enteritis and depression.
• Pheasants: Dyspnea and death. • Quail: Sudden and dramatic increase in mortality, nonspecific signs of illness, increased respiratory effort and increased respiratory sounds.

Physical Examination Findings
• Finches: N/A • Falcons: N/A • Hawks and owls: Birds die before they can be presented for examination. • Pigeons: ○ Type 1: Vomiting, watery diarrhea, depression, weight loss; ○ Type 2: N/A • Psittacine birds: N/A
• Chickens: Abnormally colored eggs, shell-less eggs. Chickens appear normal.
• Turkeys: Bloody diarrhea and depression.
• Pheasants: Dyspnea, cyanosis. • Quail: Nasal discharge, open-mouthed breathing, respiratory sounds.

CAUSES
Three genera of adenoviruses (*Aviadenovirus*, *Siadenovirus*, and *Atadenovirus*) have been shown to cause disease in birds. The signs associated with infection depend on the organ targeted by the virus and the host's immune response.

RISK FACTORS
• Failure to quarantine new birds. • Housing multiple species together in the same collection. • High stocking densities.
• Pheasants, turkeys, chickens: exposure to waterfowl. • Pigeons: Concurrent infection with pigeon circovirus. • Quail: Exposure to infected birds.

DIAGNOSIS

DIFFERENTIAL DIAGNOSIS
All species: Other systemic viral infections, septicemia, gross management errors.

CBC/BIOCHEMISTRY/URINALYSIS
In birds experiencing hepatitis, elevations in the aspartate aminotransferase are expected.

OTHER LABORATORY TESTS
• Falcons: A virus neutralization assay has been developed that can detect serological evidence of virus infection. • Chickens: Antibodies can be detected by hemagglutination inhibition and enzyme-linked immunoassays.

IMAGING
Pigeons: Hepatomegaly and splenomegaly would be expected.

DIAGNOSTIC PROCEDURES
N/A

PATHOLOGIC FINDINGS
• Finches: Grossly, liver and spleen enlargement. Microscopically, multiple round-to-irregular pale tan (necrotic) foci. Hepatic, splenic, and intestinal mucosal necrosis with varying numbers of large intranuclear basophilic to amphophilic inclusion bodies. • Falcons: Grossly, liver and spleen enlargement. Microscopically, hepatic and splenic necrosis with a mild-to-moderate lymphoplasmacytic inflammatory response. Varying numbers of large intranuclear basophilic to amphophilic inclusion bodies are present. • Hawks and owls: Grossly, liver and spleen enlargement. Microscopically, hepatic necrosis and mild-to-moderate inflammatory response, splenic necrosis, and proventricular and ventricular and necrosis resulting in ulceration. Varying numbers of large intranuclear basophilic to amphophilic inclusion bodies are present in all affected tissues. • Pigeons: ○ Type 1: Grossly, fibrinous and hemorrhagic enteritis, variable liver enlargement with necrotic foci. Microscopically, villus atrophy of the duodenum, characteristic inclusion bodies are found in intestinal epithelial cells. Hepatic necrosis may occur, but it is infrequent. Inclusion bodies are infrequently found in the liver ○ Type 2: Grossly, hepatic and possibly splenic enlargement are seen. There may be

multifocal discoloration of the liver. Microscopically there is a moderate to massive necrosis of the liver with intranuclear eosinophilic inclusion bodies. • Psittacine birds: Lesions depend on the virus and species of bird. Grossly there may be evidence of one or more of: conjunctivitis, hepatitis, pancreatitis, enteritis, and splenic enlargement. The virus causes necrosis of the affected tissues, which will be accompanied by inflammation depending on how long the bird lives after the lesions develop. Intranuclear inclusions are generally common, but may be difficult to find. Inclusions in the tubular epithelial cells of the kidneys may be incidental findings in birds dying of other causes. • Chickens: Grossly, inactive ovaries and atrophied oviducts. Microscopically, severe chronic active inflammation of the shell gland with intranuclear inclusion bodies in the epithelial cells. Microscopically there is expansion of the histiocytic population surrounding the sheathed arteries of the spleen with lymphoid necrosis with pannuclear inclusion bodies. Digestive tract lesions include epithelial sloughing, hemorrhage within the villi and the submucosa, a variable degree of inflammation which can include heterophils and mononuclear cells and the presence of intranuclear inclusion bodies. Lesions are most severe in the duodenum. • Turkeys: Grossly, well muscled but pale, may have still been eating, hemorrhage into the intestine, hepatomegaly and splenomegaly. Lesions resemble those seen in the chicken, but do not involve the digestive tract. • Pheasants: Pulmonary edema and enlarged mottled spleens. Lesions resemble those seen in the chicken, but do not involve the digestive tract. • Quail: Exudate in the nasal passages and in the trachea with tracheal mucosal thickening. Exudate may extend into the mainstem bronchi. Microscopically there is necrosis and sloughing of the tracheal epithelium and the presence of intranuclear inclusion bodies and nuclear enlargement. There will be varying degrees of inflammation, which may be complicated by secondary bacterial infections. Multifocal hepatic necrosis may also occur.

TREATMENT
NURSING CARE
• Pigeons: ○ Type 1: Supportive care with fluids, supplemental heat and assist feeding of easily digested food. Broad spectrum antibiotics to prevent secondary *E. coli* enteritis and sepsis. • Quail: Supportive care. • All other species: N/A

ACTIVITY
Pigeons who survive Type 1 infections may take months to return to racing condition.

DIET
N/A

CLIENT EDUCATION
N/A

SURGICAL CONSIDERATIONS
N/A

MEDICATIONS
DRUG(S) OF CHOICE
N/A
CONTRAINDICATIONS
N/A
PRECAUTIONS
N/A
POSSIBLE INTERACTIONS
N/A
ALTERNATIVE DRUGS
N/A

FOLLOW-UP
PATIENT MONITORING
N/A
PREVENTION/AVOIDANCE
• Falcons: Do not raise other species of falcons with peregrine falcons. • Chickens, turkeys, pheasants: Avoid contact with waterfowl. • Chickens: Disease has been eradicated from laying stock. Infection is prevented by strict quarantine and hygiene methods. Inactivated vaccines have been developed and used effectively. • Turkeys: Vaccination by water administration. • Quail: Strict biosecurity measures.
POSSIBLE COMPLICATIONS
Pigeons, turkeys: Secondary *E. coli* infections.
EXPECTED COURSE AND PROGNOSIS
• Finches: Diseased birds die, low level or sporadic mortality. • Falcons: Most of the diseased birds will die. • Hawks and owls: The only known infected birds died. • Pigeons: ○ Type 1: High levels of morbidity (up to 100%), low mortality unless secondary *E. coli* infections occur. ○ Type 2: Sporadic mortality, most birds that develop the disease die. • Psittacine birds: Birds with disease die. It is likely that there are many subclinically infected birds. Nestling deaths may occur in subsequent clutches. • Chickens: A 10–40% reduction in egg production. • Turkeys: Average mortality of 10–15%, but may be higher. Secondary *E. coli* infections may increase the morbidity and mortality. • Pheasants: Flock mortality ranges from 2 to 15%. • Quail: Mortality rates may exceed 50% of susceptible birds.

MISCELLANEOUS
ASSOCIATED CONDITIONS
N/A
AGE-RELATED FACTORS
N/A
ZOONOTIC POTENTIAL
N/A
FERTILITY/BREEDING
N/A
SYNONYMS
N/A
SEE ALSO
Appendix 3: Laboratory Testing
Colibacillosis
Herpesviruses
Liver disease
Viral disease
ABBREVIATIONS
N/A
INTERNET RESOURCES
N/A

Suggested Reading
Marlier, D., Vindevogel, H. (2006). Viral infections in pigeons. *The Veterinary Journal*, 172:40–51.
Oaks, J.L., Schrenzel, M., Rideout, B., Sandfort, C. (2005). Isolation and epidemiology of Falcon Adenovirus. *Journal of Clinical Microbiology*, 43:3414–3420.
Saif, Y.M. (ed.) (2008). *Diseases of Poultry*, 12th edn. Oxford, UK: Blackwell Publishing.
Schmidt, R., Reavell, D., Phalen, D.N. (2003). *Pathology of Exotic Birds*. Ames, IA: Iowa State University Press.
Zsivanovits, P., Monks, D.J., Forbes, N.A., Ursu, K., Raue, R., Benko, M. (2006). Presumptive identification of a novel adenovirus in a Harris hawk (*Parabuteo unicinctus*), a Bengal eagle owl (*Bubo bengalensis*), and a Verreaux's eagle owl (*Bubo lacteus*). *Journal of Avian Medicine and Surgery*, 20:105–112.

Author David N. Phalen, DVM, PhD, DABVP (Avian)

Airborne Toxins

BASICS

DEFINITION
Airborne toxins are defined as particles or chemicals that are inspired and cause damage to various tissues of the body.

PATHOPHYSIOLOGY
The avian respiratory tract is particularly sensitive to airborne toxins because of specific anatomic and physiologic features that allow them to absorb oxygen more efficiently than can mammals. These include a cross-current flow of air and blood that allows the potential for blood oxygen levels to be higher than the oxygen levels in the expired breath. With this ability also comes the risk of absorbing higher amounts of toxins from the air, causing them to reach toxic levels sooner than would mammals.

SYSTEMS AFFECTED
• Respiratory system—direct exposure to the toxin. • Nervous system—secondary to hypoxia. • Cardiovascular system—secondary to a compromised respiratory system. • Ocular and upper gastrointestinal—inflammation and irritation.

GENETICS
N/A

INCIDENC/PREVALENCE
N/A

GEOGRAPHIC DISTRIBUTION
N/A

SIGNALMENT
• No sex or age predilections have been described. • PTFE—smaller birds like budgerigars may be more susceptible than larger birds. • COPD (hypersensitivity syndrome)—macaws are more susceptible than other species.

SIGNS
• Increased respiratory effort, open mouth breathing, exercise intolerance, cyanosis of facial skin, depression, ataxia, weakness, tail bobbing. • CO toxicity—cherry red mm. • Acute death or coma. • Weight loss, sneezing and coughing may occur with COPD of macaws. • The onset clinical signs of acute smoke inhalation may be delayed several hours after the exposure.

Historical Findings
• Recent toxin exposure. • Acute death or coma may occur after PTFE or carbon monoxide poisoning. • Sneezing and nasal discharge. • Smoking habit of the owner. • Multiple species of birds housed nearby (COPD).

Physical Examination Findings
• Open mouth breathing. • Cyanosis. • Sneezing and/or coughing. • Increased respiratory effort. • Ataxia, incoordination. • Cherry red mm. • Dyspnea. • Lethargy, depression. • Nasal discharge on nares and feathers of the face. • Weight loss.

CAUSES
Many types of toxins may be encountered and include the following:
• PTFE found on the surface of nonstick cookware, irons and ironing boards, heat lamps and self-cleaning ovens produce acidic fluorinated gases and particles. • Feather dander and dust from powder-down-producing birds like cockatoos (*Cacatua* spp), cockatiels (*Nymphicus hollandicus*), and African grey parrots (*Psittacus erithacus*) that can cause hypersensitivity reactions known as COPD of macaws or macaw hypersensitivity syndrome. • Smoke—solid or liquid material released into the air by pyrolysis (combustion). • CO, CO_2. • Nicotine, butadiene and other chemicals released in cigarette smoke. • Many other airborne toxins can have variable effects on birds, including air fresheners, scented candles, aerosols, methane, gasoline fumes, glues, paint fumes, self-cleaning ovens, solvents, bleach, ammonia, propellants and grooming products (nail polish, hair products).

RISK FACTORS
• Presence and use of nonstick cookware or other source of PTFE. • Presence of powder-down-producing birds in immediate environment of a macaw. • Cigarette smoking by the owner. • Housework involving painting or cleaning with aerosol producing chemicals. • Recent fire or other event releasing smoke into the environment.

DIAGNOSIS

DIFFERENTIAL DIAGNOSIS
• Respiratory compromise caused by trauma and secondary air sac rupture, bacterial, fungal or viral infections, neoplasia, ascites or hypovitaminosis A with secondary sinusitis. • Primary heart disease, artherosclerosis causing left heart failure, congenital heart disease. • Avocado toxicity. • Ataxia and weakness secondary to other neurologic disease (see neurologic conditions), metabolic derangements or systemic disease.

CBC/BIOCHEMISTRY/URINALYSIS
• Hemogram—in most cases, the hemogram will not show any consistent changes, except with polycythemia of COPD. The PCV can be as high as 80%. • Biochemistry profile—varied based on the systems affected.

IMAGING
• Radiographs may be useful in ruling out causes of respiratory disease and to evaluate the heart and lungs for secondary complications. Radiographic changes are often not apparent until the disease is advanced. • COPD–Often unremarkable. Occasionally right sided heart failure is seen due to chronic polycythemia. • CT scan may show smaller lesions not readily identifiable on radiographs.

DIAGNOSTIC PROCEDURES
Coelomic endoscopic examination and lung biopsy may reveal consistent histologic changes associated with damage to the lungs caused by airborne toxin (see pathologic findings). It may also help elucidate the presence of other secondary diseases such as aspergillosis or bacterial infections that may require specific treatment.

PATHOLOGIC FINDINGS
• PTFE toxicity—Grossly, red, wet lungs, eosinophilic fluid filled bronchi, and multifocal to confluent hemorrhage. Microscopic changes include air capillary collapse, congestion, hemorrhage and edema. • Chronic smoke inhalation may cause tertiary bronchi obliterans. • COPD of macaws—Grossly, firm and "rubbery" lungs. Microscopic changes include eosinophilic infiltration of the interstitium, proliferative fibrous connective tissue, and a mixed cellular infiltrate. Tertiary bronchi may be obstructed due to hypertrophy of smooth muscle. These lesions are usually well advanced by the time polycythemia has occurred. • Artherosclerotic plaques may results from chronic exposure to butadiene in cigarette smoke.

TREATMENT

APPROPRIATE HEALTH CARE
• Inpatient intensive care management is often required. • Administer bronchodilators and antianxiety analgesic, then place in oxygen. • Administer diuretics if heart failure is present and antimicrobials for potential secondary infections.

NURSING CARE
• Oxygen therapy—78–85% O_2 at a flow rate of 5 L/min. • HEPA filtration. • Fluid therapy to maintain hydration or correct dehydration

ACTIVITY
• Acute—Exercise restriction until symptoms have resolved. • Chronic—Lifelong exercise restriction due to permanent respiratory system damage.

DIET
Ingluvial gavage for anorectic patients.

CLIENT EDUCATION
• Prognosis varies based on the level of exposure and chronicity of disease. • Educate owners on the sources of airborne toxins and their role in removing these toxins from the environment. • Separate macaws from powder-down-producing bird species. HEPA filtration can be helpful.

AIRBORNE TOXINS (CONTINUED)

SURGICAL CONSIDERATIONS
Caution in patients with chronic respiratory system damage.

MEDICATIONS

DRUGS OF CHOICE
• Antianxiety analgesic—Butorphanol at 0.5–2 mg/kg IM. • Terbutaline 0.01 mg/kg IM q6–12h or 0.1 mg/kg PO q12–24h. • Eye ointment if ocular irritation. • Nonsteroidal anti-inflammatories (NSAIDs)–Meloxicam at 0.5 mg/kg PO q12–24h. • Short acting corticosteroids—Use is controversial; but some will use for smoke inhalation and COPD: ○ dexamethasone may be considered at a dose of 0.2–1.0 mg/kg IM once or q12–24h; ○ dexamethasone sodium phosphate at 2 mg/kg once or q6–12h during the acute phase.

CONTRAINDICATIONS
Housing New World and Old World species in the same space without adequate ventilation/filtration.

PRECAUTIONS
Use caution if using corticosteroids in birds; consider concurrent antibiotic and antifungal therapy.

POSSIBLE INTERACTIONS
N/A

ALTERNATIVE DRUGS
• Midazolam: 0.5–1.0 mg/kg IM may be used to reduce anxiety if butorphanol is not sufficient. • Other bronchodilators include theophylline or aminophylline. These may be less effective at bronchodilation in birds, but clinical improvement has been noted with their use: ○ theophylline: 2 mg/kg PO q12h; ○ aminophylline: 10 mg/kg IV q3h, 4 mg/kg IM q12h, or 5 mg/kg PO q12h; ○ albuterol nebulization: 2.5 mg in 3 cc saline q4–6h during acute clinical signs.

FOLLOW-UP

PATIENT MONITORING
• COPD—frequent monitoring of PCV to assess treatment effectiveness. • Radiographs to evaluate lungs and air sacs, and heart size and to check for the presence of atherosclerotic plaques.

PREVENTION/AVOIDANCE
Airborne toxicosis a complication of captivity. Elimination of the potential toxins before exposure is often possible and carries the best prognosis.

POSSIBLE COMPLICATIONS
Heart failure can result if polycythemia or pulmonary fibrosis is significant.

EXPECTED COURSE AND PROGNOSIS
• In the case of exposure to PTFE and clinical signs are present, the prognosis is usually very poor. • In the case of COPD, the condition can be improved with medication, HEPA filtration, and elimination of allergens from the environment, but even with good control, the condition will often shorten the normal lifespan of the patient.

MISCELLANEOUS

ASSOCIATED CONDITIONS
N/A

AGE-RELATED FACTORS
N/A

ZOONOTIC POTENTIAL
N/A

FERTILITY/BREEDING
Birds with COPD may have decreased breeding success.

SYNONYMS
N/A

SEE ALSO
Air sac rupture
Appendix 6: Common Avian Toxins

Aspergillosis
Avocado/plant toxins
Hemorrhage
Pneumonia
Respiratory distress
Tracheal/syringeal diseases

ABBREVIATIONS
PCV—packed cell volume
COPD—chronic obstructive pulmonary disease
PTFE—polytetrafluoroethylene
CO—carbon monoxide
CO_2—carbon dioxide
MM—mucus membranes
O_2—oxygen
HEPA—high-efficiency particulate air

INTERNET RESOURCES
N/A

Suggested Reading
Lichtenberger, M. (2006). Emergency case approach to hypotension, hypertension, and acute respiratory distress. *Proceedings of the Association of Avian Veterinarians Annual Conference*, San Antonio, TX, pp. 281–290.
Lightfoot, T.L., Yeager, J.M. (2008). Pet bird toxicity and related environmental concerns. *Vetinary Clinics of North America. Exotic Animal Practice*, 11(2):229–259.
Orosz, S.E., Lichtenberger, M. (2011). Avian respiratory distress: Etiology, diagnosis and treatment. *Vetinary Clinics of North America. Exotic Animal Practice*, 14(2):241–255.
Phalen, D.N. (2000). Respiratory medicine in cage and aviary birds. *Vetinary Clinics of North America. Exotic Animal Practice*, 3(2):423–452
Schmidt, R.E. (2013). The Avian Respiratory System. *Proceedings of the Western Veterinary Conference*, Las Vegas, NV.
Author Stephen M. Dyer, DVM, DABVP (Avian)
Acknowledgement Erika L. Cervasio, DVM, DABVP (Avian)

Client Education Handout available online

AIR SAC MITES

BASICS

DEFINITION
Infection with mites (including *Sternostoma tracheacolum* and less commonly *Cytodites nudus*, *Ptilonyssus spp*.) in the upper and lower respiratory systems (including nasal passages, trachea, smaller air passages, and air sacs). Some mite species (especially *Cytodites*) have also invaded other visceral areas including the coelom.

PATHOPHYSIOLOGY
Mites transmitted from infected birds travel throughout the respiratory system, restricting air flow by being present within narrowed spaces as well as causing inflammation and increased mucous production.

SYSTEMS AFFECTED
- Respiratory (because of a foreign body response, inflammation, and increased fluid/mucous production). • Behavioral (bird's reaction to the respiratory system being affected).

GENETICS
None known other than species predilections.

INCIDENCE/PREVALENCE
Common.

GEOGRAPHIC DISTRIBUTION
Worldwide both in captive and wild individuals.

SIGNALMENT
- **Species:** Finches (especially Australian species including Gouldian finches), canaries, pigeons, small psittacine birds (budgerigars and cockatiels), poultry and waterfowl. Society finches may be resistant to infection.
- **Mean age and range:** N/A • **Predominant sex:** N/A

SIGNS

Historical Findings
- No signs may have been noted by the owner. • Vocalization changes (cessation or tonal change of singing, clicking sounds). • Nonspecific signs of illness (lethargy, fluffed feathers, reduced appetite), • Head shaking and frequent swallowing, coughing or sneezing, • Open-mouth breathing and tail bobbing,

Physical Examination Findings
- Normal in mild cases, • Dyspnea (open-mouth breathing, tail bobbing, increased respiratory rate and effort). • Click (may be variously loud; may require bird being close to ear to hear this sound).
- Frequent swallowing motions and increased oral mucous. • Beak rubbing (upper and lower together or both on perch), head shaking. • Nasal discharge. • Moist breathing sounds auscultated. • Weight loss. • Various degrees of general lethargy/reduced activity/"fluffing" of feathers. • Death.

CAUSES
- Mites or mite eggs eliminated in sneezes, coughs, and feces of infected birds. • Waste, food, and environment may become contaminated. • Infected parents feeding chicks. • No intermediate host required.

RISK FACTOR
Poor quarantine and flock management.

DIAGNOSIS

DIFFERENTIAL DIAGNOSIS
Other causes of respiratory distress:
- bacterial respiratory infection (*Enterococcus fecalis*); • fungal infection (aspergillosis);
- other parasitic infection (*Syngamus* or *Trichomonas* infection); • viral infection (poxvirus); • space-occupying lesions in or around the respiratory system (including obesity, dystocia); • airborne irritants and toxicants (PTFE).

Susceptible species have higher infection rate and mites are more likely with respiratory signs.

CBC/BIOCHEMISTRY/URINALYSIS
There are usually no changes in biochemical or urinalysis tests; reported hematological changes include eosinophilia and/or a basophilia with infection.

OTHER LABORATORY TESTS
The eggs of the mites, or the mites themselves, may be seen in oral swab samples or in fecal samples (via direct wet mount microscopy). Test results may be false negative.

IMAGING
A respiratory mite infection may have generalized nonspecific radiopacity changes to the pulmonary and air sac fields with radiology.

DIAGNOSTIC PROCEDURES
- Transilluminate the neck/trachea with bright light source after moistening skin with alcohol. • Dark specks can be seen in tracheal lumen. • The mites may be in the lower respiratory areas instead and not seen in trachea. • Tracheal endoscopy (using 1.2-mm endoscope) may be useful.

PATHOLOGIC FINDINGS
At necropsy dark specks may be seen in the mucous at any location of the respiratory system. Pneumonia, thickened and opaque air sac membranes and tracheitis can be seen. Mites embed their legs into the tissue and live in the mucous layer. Histopathology may show mucous epithelial necrosis, mucosal hyperplasia, and inflammation.

TREATMENT

APPROPRIATE HEALTH CARE
- Mild-to-moderate signs: Home care is usually sufficient. • Very dyspneic, lethargic, inappetant, or thin/weak birds: Hospital care may be required.

Infection with concurrent primary or secondary infectious agents may need additional therapy.

NURSING CARE
If severe dyspnea is present, oxygen and humidity supplementation might be useful. If weight loss and/or reduced self-feeding is present, the bird may require supplemental/assisted feeding and crystalloid fluid administration (usually delivered subcutaneously).

ACTIVITY
If respiratory distress is present, the bird's exercise range should be limited and agitation/stress should be carefully limited/monitored.

DIET
Although nutritional needs must be met in all species, caloric intake must be aggressively maintained in the species with high metabolic rate that may have decreased intake with illness, as well as addressing any longer-term nutritional deficiencies that might be concurrent.

CLIENT EDUCATION
The owner should be advised to expect possible exaggerated symptoms after therapy as it has been reported with heavy infections; the massive die-off of the mites may cause symptoms to worsen shortly after treatment, before improving.

SURGICAL CONSIDERATIONS
It is best to fully treat these mites before surgery; in the case of emergency surgical needs, there is a higher risk of airway maintenance difficulties because of the increased mucous production and mite blockage of the respiratory passageways. The species affected tend to be small, which may affect endotracheal tube use.

MEDICATIONS

DRUG(S) OF CHOICE
Drugs of the avermectin class, especially Ivermectin at 0.2 mg/kg treated topically, orally, or parenterally repeated as often as weekly, for as long as several months. Other avermectins used include moxidectin and doramectin of unknown regimen.

CONTRAINDICATIONS
None known.

AIR SAC MITES (CONTINUED)

PRECAUTIONS
Drugs that depress respiration should be used with caution such as sedatives and opioids.

POSSIBLE INTERACTIONS
None likely.

ALTERNATIVE DRUGS
Some texts describe the use of a dichlorvos or a "no pest" strip near the affected birds, or an aerosol of rotenone or pyrethrin sprays. These pesticides have risks of toxicity, as precise dosing is impossible.

 FOLLOW-UP

PATIENT MONITORING
Recheck examination performed within weeks for mild symptoms; sooner or later as symptom degree necessitates. Hemogram changes can be followed.

PREVENTION/AVOIDANCE
Quarantine, examine and treat new arrivals into the flock to prevent spread to birds already treated.

POSSIBLE COMPLICATIONS
• Acaricide may worsen symptoms due to mite die-off. • Insufficient therapy or resistance may allow recurrence of symptoms if all mites are not killed. • Secondary infections may progress even with resolution of primary issue. • Anesthesia and tracheoscopy higher risk in very symptomatic birds.

EXPECTED COURSE AND PROGNOSIS
Resolution of symptoms with therapy; prognosis is good unless symptoms are severe.

 MISCELLANEOUS

ASSOCIATED CONDITIONS
N/A

AGE-RELATED FACTORS
N/A

ZOONOTIC POTENTIAL
N/A

FERTILITY/BREEDING
Heavy air sac mite infections can decrease breeding success. Some finch breeders will use surrogate parents of less-susceptible species to foster the chicks of more-susceptible species to limit the spread of this organism to new chicks.

SYNONYMS
Tracheal mite, respiratory mite, visceral mite, respiratory acariasis

SEE ALSO
Airborne toxicosis
Aspiration
Respiratory distress
Sick-bird syndrome
Tracheal/syringeal disease

ABBREVIATIONS
N/A

INTERNET RESOURCES
www.petmd.com/bird/conditions/respiratory/c_bd_respiratory_parasites-air_sac_mites

Suggested Reading
Ritchie, B., Harrison, G., Harrison, L. (1994). *Avian Medicine: Principles and Application.* Lake Worth, FL: Wingers Publishing.
Rosskopf, W., Woerpel, R. (1996). *Diseases of Cage and Aviary Birds*, 3rd edn. Baltimore, MD: Williams and Wilkins.
Samour, J. (ed.) (2000). *Avian Medicine.* London, UK: Mosby Publishing.
Altman, R., Clubb, S., Dorrestein, G., Quesenberry, K. (1997). *Avian Medicine and Surgery.* Philadelphia, PA: WB Saunders.
Author Vanessa Rolfe, DVM, ABVP (Avian)

AIR SAC RUPTURE

BASICS
DEFINITION
The cervicocephalic, abdominal, or caudal thoracic air sacs can contribute to subcutaneous air accumulation when ruptured. Affected birds show emphysematous enlargement of various body parts depending on which air sac is leaking.

PATHOPHYSIOLOGY
This condition occurs when the air sac lining is disrupted secondary to a traumatic event, allowing air to accumulate under the skin. The location of the rupture is generally not identifiable. The cervicocephalic air sacs appear most commonly involved, with subcutaneous emphysema affecting the head, neck and extending over the dorsum/ventrum in severe cases. The cervicocephalic air sacs are the only air sacs that do not communicate directly with the pulmonary system. They communicate with the infraorbital sinuses. No oxygen exchange occurs within the cervicocephalic air sacs.

SYSTEMS AFFECTED
- Respiratory—rupture of the cervicocephalic air sacs and sometimes the abdominal/caudal thoracic air sacs. • Skin—subcutaneous emphysema leading to skin expansion over the affected area; the degree of skin tension varies with the quantity of air present in the subcutaneous space. • Musculoskeletal—air accumulation under the skin may restrict body movement.

GENETICS
None.

INCIDENCE/PREVALENCE
Unknown, but relatively common.

GEOGRAPHIC DISTRIBUTION
N/A

SIGNALMENT
- No specific species, age or sex predilection.
- More commonly reported in Amazon parrots, macaws and cockatiels.

SIGNS
General Comments
- Rupture of an air sac is not life-threatening in most cases. However, fatalities may occur.
- Subcutaneous emphysema causes discomfort to the avian patient and likely affects the bird's quality of life.

Historical Findings
- Traumatic event reported by the owner.
- "Ballooning" under the skin of various body parts, most often affecting the head, neck, ventrum, and dorsum.

Physical Examination Findings
- Subcutaneous emphysema – in large species, the accumulation of air is most often confined to the dorsal aspect of the neck, whereas generalized subcutaneous emphysema may be seen more commonly in small bird species.

CAUSES
- Traumatic—air sacs generally rupture secondary to a traumatic event. Fractures of pneumatized bones may cause or contribute to the emphysema. • Infectious—chronic upper respiratory infection may also be involved. A pathologic process in the vicinity of the narrow connecting passage between the infraorbital sinus and the cervicocephalic air sac can act as a one-way valve, trapping air in the lumen of the air sac. • Nutritional—nutritional deficiencies such as hypovitaminosis A may predispose birds to respiratory infection.

RISK FACTORS
- Environmental—birds free-flying outside their cage present a higher risk of traumatic events, especially if the environment presents some dangers (ceiling fan, large mirror, wide unprotected windows, etc.). • Medical conditions—malnutrition may predispose to upper respiratory disease.

DIAGNOSIS
DIFFERENTIAL DIAGNOSIS
- Fracture of a pneumatized bone. • Luxation or subluxation of the humero–scapular joint causing disruption of the clavicular air sac.
- Distension of the cervicocephalic air sacs without rupture—with upper respiratory disease involving the infraorbital sinuses, air may become entrapped within the cervicocephalic air sacs leading to their distension. • Infection with gas-producing bacteria.

CBC/BIOCHEMISTRY/URINALYSIS
- No specific abnormalities are seen with air sac rupture. • CBC/biochemistry—indicated to rule out concurrent disease.

OTHER LABORATORY TESTS
- Bacterial/fungal culture—to evaluate the presence of respiratory disease. The rostral aspect of the choana may be sampled. Nasal or sinus flush may be considered to obtain a more representative sample of the upper respiratory tract microbial flora. Skin culture if subcutaneous infection is suspected.
- *Chlamydia* testing—may be considered if respiratory signs are present. • *Aspergillus* testing—may be considered if respiratory signs are present.

IMAGING
- Whole body radiographs—to identify musculoskeletal abnormalities. • Computed tomographic examination—to assess the infraorbital sinuses for disease processes; to identify musculoskeletal abnormalities.

DIAGNOSTIC PROCEDURES
No additional diagnostic procedures are indicated.

PATHOLOGIC FINDINGS
Air sac distention with concurrent signs of inflammation.

TREATMENT
APPROPRIATE HEALTH CARE
- Outpatient medical management—patient otherwise normal; diagnostic approach may require brief hospitalization. • Inpatient medical management—patient presenting with severe subcutaneous emphysema with depression/lethargy or concurrent disease that requires close monitoring. • Surgical management if relapse occurs with air aspiration

NURSING CARE
- Air aspiration—air can easily be removed using a syringe and hypodermic needle to decrease skin tension. However, the space often quickly refills as the patient breathes since the breach within the respiratory system is still patent. The procedure may be repeated multiple times. This technique may be used initially, especially if severe emphysema is present to decrease skin tension and improve patient's comfort. • Elizabethan collar—may be considered if a Teflon dermal stent is used in an area that the bird can reach with its beak.

ACTIVITY
Exercise may exacerbate the subcutaneous emphysema. Activity level recommendations should be adjusted to each individual avian patient.

DIET
Suboptimal diets should be improved.

CLIENT EDUCATION
Spontaneous resolution is possible but problem may be chronic and recurrent.

SURGICAL CONSIDERATIONS
- Fistula: Using a 2–6 mm skin biopsy punch or a scapel blade, a piece of skin is removed over the inflated area and the wound is left to heal by second intention. This opening will allow the skin to lie against the traumatized tissue and allow the rupture site to heal properly. In many cases, there is no reinflation of the subcutaneous space by the time the skin wound has healed. • Teflon® dermal stent (McAllister Technical Services, Coeur d'Alene, ID): Nonabsorbable sutures are preplaced in the four pairs of holes found around the stent. Then, a skin incision is performed over the distended skin ideally in an area that the bird is unable to reach with its beak. The bird's skin will deflate once the skin is incised and skin tension will decrease. The incision size should be just large enough to

insert the stent into the subcutaneous space. A hypodermic needle is used to retrieve the sutures through the skin and tie the stent in place. The four sutures should be placed two on each side of the incision and two at both ends of the incision. A purse-string suture may be placed around the rim of the stent that remains above the skin. The stent is generally left in place permanently.
• Cervicocephalic–clavicular air sac shunt: Possible when one of these two air sacs is ruptured. A skin incision is performed in the left lateral thoracic inlet area to avoid the esophagus. A small endotracheal tube is inserted in the hyperinflated air sac. The tube is then directed caudally along the esophagus to the cranial aspect of the clavicular air sac. The tube is sutured to the *longus coli* muscle to prevent migration before skin closure. The shunt is generally left in place permanently.

MEDICATIONS

DRUG(S) OF CHOICE
• Broad-spectrum antibiotic—consider perioperatively and until cleaning of the stent is minimal postoperatively. • Nonsteroidal anti-inflammatory medication (meloxicam 0.5 mg/kg PO/IM q12h) may be considered to alleviate inflammation at the surgical site.
• Opioids (butorphanol 1–4 mg/kg q4–8h) is recommended perioperatively.

CONTRAINDICATIONS
None.

PRECAUTIONS
None.

POSSIBLE INTERACTIONS
N/A

ALTERNATIVE DRUGS
Doxycycline: Consider to attempt pleurodesis. Air sac lining of birds is similar to the pleural lining of mammals. Medication instilled in the air sac will cause irritation between the air sac surfaces, closing off the space between them and preventing further air from accumulating. This procedure may be considered for cervicocephalic air sac rupture only to avoid dissemination of the instilled product to the lower respiratory system. Since pleurodesis has only been described in one Amazon parrot thus far, this technique should be used in last resort.

FOLLOW-UP

PATIENT MONITORING
• Aspiration: Recurrence of emphysema indicates the rupture site is not healed. The procedure may be repeated or surgical approach may be considered. • Fistula: Recurrence of emphysema indicates the rupture site is not healed. Depending on the location, placement of a dermal stent or cervicocephalic–clavicular air sac shunt may be considered. • Teflon® dermal stent: Postoperative care requires cleaning of the stent opening to prevent obstruction by debris and tissue fluids. Sterile swabs or needles are recommended to decrease bacterial contamination. Initially, cleaning twice daily is recommended. The cleaning frequency should be adjusted to each patient and decrease progressively. Recurrence of emphysema indicates the stent is no longer patent. • Cervicocephalic–clavicular air sac shunt: Recurrence of emphysema may indicate the shunt is no longer patent and surgical exploration is indicated.

PREVENTION/AVOIDANCE
Environment—ensure the bird's environment is safe to decrease the likelihood of trauma.

POSSIBLE COMPLICATIONS
• Teflon® dermal stent: Occlusion of the stent opening by debris and tissue fluids is common initially. On rare occasions, bird may pick at the stent. • Cervicocephalic-clavicular air sac shunt: Occlusion of the shunt. • Unable to achieve complete resolution of the emphysema despite all therapeutic measures.

EXPECTED COURSE AND PROGNOSIS
• Aspiration: If performed rapidly following the traumatic event, chances that the air will continue to accumulate in the subcutaneous space may be decreased. This technique is expected to provide only a temporary relief in chronic cases of air sac rupture. • Fistula: If performed rapidly following the traumatic event, chances that the air will continue to accumulate in the subcutaneous space are decreased. This technique is expected to provide only a temporary relief in chronic cases of air sac rupture. • Teflon® dermal stent: Complete resolution of the subcutaneous emphysema in most birds, partial resolution of the subcutaneous emphysema in some birds, minimally invasive procedure associated with a good prognosis.
• Cervicocephalic–clavicular air sac shunt: Complete resolution of the subcutaneous emphysema in most birds based on scant literature, moderately invasive procedure associated with a good prognosis.

MISCELLANEOUS

ASSOCIATED CONDITIONS
None.

AGE-RELATED FACTORS
None.

ZOONOTIC POTENTIAL
None.

FERTILITY/BREEDING
Avoid teratogenic antibiotics in laying hen.

SYNONYMS
N/A

SEE ALSO
Bite wounds
Fracture/luxation
Respiratory distress
Rhinitis and sinusitis
Tracheal/syringeal diseases
Trauma

ABBREVIATIONS
None.

INTERNET RESOURCES
http://avianmedicine.net/content/uploads/2013/03/41.pdf

Suggested Reading
Antinoff, N. (2008). Attempted pleurodesis for an air sac rupture in an Amazon parrot. *Proceedings of the Association of Avian Veterinarians Annual Conference*, August 11–14, Savannah, GA, p. 437.
Bennett, R.A., Harrison, G.J. (1994). Soft tissue surgery. In: Ritchie, B.W., Harrison, G.J., Harrison, L.R. (eds), *Avian Medicine: Principles and Application*. Lake Worth, FL: Wingers Publishing, pp.1096–1136.
Harris, J.M. (1991). Teflon dermal stent for the correction of subcutaneous emphysema. *Proceedings of the Association of Avian Veterinarians Annual Conference*, September 23–28, Chicago, IL, pp. 20–21.
Levine, B.S. (2005). Cervicocephalic-clavicular air sac shunts to correct cervicocephalic-clavicular air sac emphysema. *Proceedings of the Association of Avian Veterinarians Annual Conference*, August 9–11, Monterey, CA, pp. 59–60.
Petevinos, H. (2006). A method for resolving subcutaneous emphysema in a griffon vulture chick (*Gyps fulvus*). *Journal of Exotic Pet Medicine*, **15**(2):132–137.
Author Isabelle Langlois, DMV, DABVP (Avian)

Avian

ANEMIA

BASICS

DEFINITION
The literal translation of the word "anemia" is lack of blood. In practice, anemia is more commonly defined by a decrease in the red blood cell population or in a decrease in the normal quantity or quality of hemoglobin in the red blood cells. In other words, anemia can also be defined as a decrease in the oxygen-carrying capacity of the blood. Anemia can be further delineated by the cause of the condition. This can be grouped into three large categories:
1. Impairment of red blood cell production leading to decreased red blood cells or improperly functioning red blood cells.
2. Increased destruction of red blood cells (hemolytic anemia).
3. Blood loss (both acute and chronic).
Finally, one further way of classifying anemia is based on the number of circulating immature red blood cells (reticulocytes). In this scheme, anemia is either regenerative (excess of reticulocytes) or non-regenerative (lack of reticulocytes) anemia.

PATHOPHYSIOLOGY
The pathophysiology depends on the type of anemia that is present. The pathophysiology of blood loss is self-explanatory. Hemolytic anemia occurs when healthy red blood cells are being destroyed at a faster rate than normal and red blood cell production cannot keep pace with destruction. Hemolysis can occur for a variety of reasons including if there are antibodies directed against the red blood cells. Red blood cells can be phagocytized by macrophages (extravascular hemolysis) or destroyed in the blood vessels (intravascular hemolysis). Impairment of red blood cell production can occur for a number of reasons. Intrinsic causes of decreased red blood cell production are due to a defect in the bone marrow and erythrocyte stem cells are not produced. Extrinsic causes also cause disease in the bone marrow leading to decreased production of or defective red blood cells but are due to conditions outside of the bone marrow. In either case, not enough functioning erythrocytes are produced. Without enough competent, hemoglobin-rich cells, the tissues of the avian body lack sufficient oxygen to properly carry out necessary functions leading to an overall weakness and lethargy of the patient. Depending on the chronicity and severity of the anemia, the lack of oxygen to tissues can lead to eventual organ failure and death of the patient.

SYSTEMS AFFECTED
• Behavioral—anemia may not change the behavior, *per se*, of the patient, but if the patient is weakened from anemia, the typical behaviors of the pet may be dampened or non-existent due to lack of energy.
• Cardiovascular—chronic anemia can lead to increased cardiac output, increased blood flow, and cardiac murmurs. • Endocrine/Metabolic—metabolic functions may be affected if anemia is severe and/or chronic due to the lack of oxygen • Hepatobiliary—the liver and/or spleen may enlarge in cases of hemolytic anemia.
• Musculoskeletal—skeletal tissue may appear pale due to decreased erythrocyte population. Weakness may also be present.
• Ophthalmic—conjunctiva will appear pale with anemia. Retinal hemorrhage can occur with some forms of anemia.
• Renal/Urologic—hemoglobinuria can occur with hemolytic anemia and urine and urates can have a darker color.
• Reproductive—breeding hens may have a disruption in breeding cycle or decreased production of chicks if anemia is present.
• Respiratory—with severe anemia, respiratory rate and depth may be increased significantly. • Skin—the skin and mucous membranes can take on a very pale appearance with severe anemia.

GENETICS
In the avian literature, there is a disease condition termed, "conure bleeding syndrome" but no etiology was ever discovered although vitamin (e.g., vitamin D or K) or mineral (i.e., Ca) deficiency was suspected in certain lines of conures.

INCIDENCE/PREVALENCE
Anemia of chronic disease is common in birds. Blood loss due to trauma or parasites is also common. Hemolytic anemias are the least common types of anemia in birds.

GEOGRAPHIC DISTRIBUTION
There is no geographic distribution of anemia.

SIGNALMENT
• **Species:** All species of birds can have anemia. • **Mean age and range:** All ages are susceptible to anemia. • **Predominant sex:** Both sexes are equally affected.

SIGNS

General Comments
It is possible that no signs of anemia will be apparent to the owner, there will be no significant aspect of the history, and anemia may be found as an incidental finding during veterinary visit for an annual physical examination and minimum database.

Historical Findings
If anemia is due to blood loss, owners may report trauma and blood loss due to trauma. If there is hemolytic anemia, owners may report darkened urine and urates in the droppings. Although unusual, frank blood or melena can be seen in the fecal portion of the droppings. Or there may be frank blood not associated with the droppings but seen emanating from the cloaca as could be the case with ulcerated lesions associated with cloacal papilloma disease.

Physical Examination Findings
• Pale mucous membranes including conjunctiva. • Overall body weakness.
• Increased respiratory rate and effort.
• Evidence of trauma and subsequent blood loss. • Puncture wounds if trauma. • Blood in or around the cloaca. • Color change or blood in feces, urine and urates. • Petechia and ecchymosis.

CAUSES

Blood Loss Anemia
• Trauma • Gastrointestinal bleeding: neoplasia, bacterial infection, viral infection, parasitic infection, toxins • Reproductive disease in hens • External and internal parasites • Anticoagulant toxicosis.

Hemolytic Anemia
• Zinc toxicosis • Copper toxicosis
• Aflatoxicosis • Transfusion reaction
• *Plasmodium* infection • Viral diseases
• Immune mediated hemolytic anemia-idiopathic • Other toxins • Septicemias.

Impaired Erythrocyte Production
• Heavy metal toxicosis, especially lead
• Chronic disease: common diseases include: Chlamydia, aspergillosis, mycobacteria, egg yolk coelomitis, circovirus • Neoplasia
• Corticosteroid administration • Iron deficiency • Nutritional deficiencies
• Chronic organ disease: Renal disease, liver disease • Ingestion of petroleum products.

RISK FACTORS
Exposure to toxins, exposure to predators, chronic disease, red blood cell parasites, iron or folic acid deficiency, neoplasia.

DIAGNOSIS

DIFFERENTIAL DIAGNOSIS
Iatrogenic hemolysis of the red blood cells due to poor venipuncture technique or sample handing before analysis.

CBC/BIOCHEMISTRY/URINALYSIS
• PCV below the normal reference interval.
• Hemoglobin below the normal reference interval. • Increased polychromasia and anisocytosis in regenerative anemia.
• Increased reticulocytes in regenerative anemia.

OTHER LABORATORY TESTS
• Specific tests for infectious diseases. • Heavy metal concentration in blood samples.
• Examination of droppings for red blood cells. • Examination of blood drop on slide for evidence of agglutination. • Fecal parasitic examination.

IMAGING
Whole body radiographs may be necessary depending on the differential diagnoses such

ANEMIA (CONTINUED)

as: heavy metal ingestion, coelomitis, reproductive disease, splenic disease, liver disease.

DIAGNOSTIC PROCEDURES
Additional diagnostics may be necessary depending on the rule out list.

PATHOLOGIC FINDINGS
Dependent on the cause of the anemia. Unless there is erythrophagocytosis in the bone marrow, liver or spleen, there may be no pathologic findings present with anemia except for the lack of red blood cells.

TREATMENT

APPROPRIATE HEALTH CARE
Management will depend on the cause of the anemia.

NURSING CARE
Despite the cause, the patient may need supportive care until the cause of the anemia can be determined and treated. Nursing care may consist of:
- Fluids, especially if due to blood loss. Fluids may consist of a physiologically balanced replacement and/or a plasma expanding solution such as one containing hydroxyethyl starch. • Blood transfusion. Blood should only be transfused from a member of the same species. • Assisted feedings may be necessary.

ACTIVITY
In severe cases, activity should be limited until oxygen carrying capacity is restored to normal.

DIET
If dietary deficiencies led to anemia, the diet should be corrected.

CLIENT EDUCATION
N/A

SURGICAL CONSIDERATIONS
If anemia is due to heavy metal ingestion, in some cases, surgical or endoscopic retrieval of the metal objects can be performed. Some cases of trauma, especially causing internal bleeding, may require surgical intervention.

MEDICATIONS

DRUG(S) OF CHOICE
• Erythropoietin. • Iron dextran 10 mg/kg IM once, repeat if necessary.

CONTRAINDICATIONS
Iron supplementation in species prone to hemochromatosis.

PRECAUTIONS
Any drug that has the potential side effect of causing further anemia should be avoided.

POSSIBLE INTERACTIONS
N/A

ALTERNATIVE DRUGS
N/A

FOLLOW-UP

PATIENT MONITORING
PCV should be closely monitored and can be checked on a daily basis during the acute phase of treatment. The entire CBC should be evaluated on a regular basis until the patient is stable.

PREVENTION /AVOIDANCE
This is dependent on the cause of anemia.

POSSIBLE COMPLICATIONS
Severe anemia can lead to death.

EXPECTED COURSE AND PROGNOSIS
Depends on the cause of the anemia.

MISCELLANEOUS

ASSOCIATED CONDITIONS
N/A

AGE-RELATED FACTORS
Young chicks, geriatric birds will be more susceptible to the severe complications of anemia than healthy adult birds.

ZOONOTIC POTENTIAL
Depends on the cause of anemia.

FERTILITY/BREEDING
Anemia can disrupt egg production.

SYNONYMS
N/A

SEE ALSO
Anticoagulant rodenticide
Circoviruses
Coagulopathies
Heavy metal toxicity
Hemoparasites
Hemorrhage
Liver disease
Nutritional deficiencies
Oil exposure
Polyomavirus
Renal disease
Sick-bird syndrome
Urate/fecal discoloration
Trauma
Viral disease
Appendix 7, Algorithm 7. Anemia

ABBREVIATIONS
N/A

INTERNET RESOURCES
N/A

Suggested Reading
Campbell, T. (2004). Hematology of birds. In: Thrall, M., *Veterinary Hematology and Clinical Chemistry*. Philadelphia, PA: Lippincott Williams, and Wilkins, pp. 225–258.
Fudge, A. (2000). *Laboratory Medicine: Avian and Exotic Pets*. Philadelphia, PA: WB Saunders.
Tully, T., Lawton, M., Dorrestein, G. (2000). *Avian Medicine*. Woburn, MA: Butterworth-Heinemann.
Author Karen Rosenthal, DVM, MS

Angel Wing

BASICS

DEFINITION
Angel wing is a commonly-used term to describe the condition of carpal valgus in avian patients resulting from malnutrition and captive mismanagement.

PATHOPHYSIOLOGY
Angel wing is the result of the plumage developing at a faster rate than the musculoskeletal structures of the wing. The immature musculoskeletal structures of the wing are not strong enough to support the weight of the blood-filled quills of the rapidly developing plumage. The weight of the developing feathers increasingly pulls the wing into a deformed position.

SYSTEMS AFFECTED
• Musculoskeletal. • Integument.

GENETICS
There is information to suggest a genetic predisposition to angel wing in certain lines of birds. In a study of white Roman geese, angel wing severity was worse in certain lines of birds regardless of diet.

INCIDENCE/PREVALENCE
Angel wing is typically a disorder of captive birds or those being fed artificial diets (such as geese and ducks in parks that receive significant amounts of bread from well-meaning bird lovers). Incidence is rare in well-managed captive animals. Occasionally, outbreaks of angel wing are noted in the nestlings of certain wild populations.

GEOGRAPHIC DISTRIBUTION
N/A

SIGNALMENT
• **Species:** All species. Birds belonging to the orders anseriformes and otidiformes are most frequently affected. Slower growing species from temperate regions of the world are especially susceptible. • **Mean age and range:** Varies according to species. Occurs when initial set of primary wing feathers develop. • **Predominant sex:** No sex predilection.

SIGNS
Angel wing presents as unilateral or bilateral drooping of the wing at the carpi and elbows with outward (valgus) rotation of the wing distal to the carpi. The primary remiges are often deviated dorsally and laterally with the wing in a relaxed, flexed position. In young birds with immature musculoskeletal structures, the wing(s) can usually be manually manipulated into normal conformation. In older birds, bone mineralization and maturation of soft tissue structures result in permanent deformities that cannot be corrected manually.

CAUSES
Carpal valgus occurs when the weight of the developing primary feathers exceeds the musculoskeletal structure's ability to hold the wing in a normal position. Diets containing excessive protein and carbohydrates result in inappropriately rapid feather development leading to angel wing. This often occurs in populations of birds in parks that are fed large amounts of high-energy foods such as bread. Lack of exercise and musculoskeletal fitness can also result in angel wing even when the diet is appropriate.

RISK FACTORS
There are a number of other factors thought to be involved in the development of angel wing. Excessive dietary protein and energy are most commonly associated with angel wing. However, lack of adequate exercise, genetic predisposition, interruption of egg incubation, excessive heat during early development, vitamin D, E and manganese deficiency are also implicated in some cases.

DIAGNOSIS

DIFFERENTIAL DIAGNOSES
Occasionally fractures (traumatic or pathologic) of the distal wing(s) will mimic angel wing.

CBC/BIOCHEMISTRY/URINALYSIS
N/A

OTHER LABORATORY TESTS
N/A

IMAGING
Radiography of the affected wing(s) may help rule out traumatic injury, osteopenia, etc.

DIAGNOSTIC PROCEDURES
The diagnosis of angel wing in birds is usually quite straightforward and is made via signalment, history, physical examination and sometimes radiography. Rarely will other diagnostic modalities be necessary.

PATHOLOGIC FINDINGS
Abnormalities are limited to the gross anatomic deformities. Histologic lesions are not typically appreciated.

TREATMENT

APPROPRIATE HEALTH CARE
• Early intervention is key to correction of angel wing in young, growing birds. • In cases where clinical signs are limited to a mild wing droop and valgus has not begun to develop, trimming of the primary feathers to relieve weight on the distal wing can be corrective. • In more severe cases where valgus of the distal wing has begun to develop, intervention usually requires fixing the wing in a normal resting position with or without weight relief by primary feather trimming. The wing is placed into a normal, resting position and a light figure-of-eight bandage applied. In some cases taping the humerus, radius/ulna, and phalanges in line is sufficient; other cases require a more substantial bandage such as a figure-of-eight. A body wrap is not usually required. Holding the wing in a normal position for 3–5 days is typically sufficient to correct the deformity. • If not caught early and musculoskeletal maturation has progressed to the point where the wing cannot be manually manipulated easily back into normal position, treatment via wing taping is usually not effective.

NURSING CARE
N/A

ACTIVITY
Activity and exercise are encouraged in young birds to stimulate development of strong wings capable of supporting mature plumage. Typically exercise can be encouraged simply by providing spacious quarters for the young birds to move about in.

DIET
• A balanced diet with appropriate protein and energy levels for the species is paramount to preventing angel wing and other development deformities in young, growing birds • Relatively slow-growing avian species should not be fed high-energy and high-protein diets as it can create mis-matches between musculoskeletal and plumage development leading to angel wing. In waterfowl, northern/arctic species are adapted to essentially feed around the clock on high quality foods in order to maximize growth in the small window of time available. Conversely, temperate/tropical species typically have a much longer window of time in which to achieve the necessary growth, therefore they do not have need for a constant intake of high energy diet. • Dietary protein levels between 8 and 15% are recommended during the first three weeks of life in slow-growing waterfowl.

CLIENT EDUCATION
Clients who are interested in breeding bird species predisposed to developing angel wing should be well versed in their dietary and husbandry requirements. Exact diet and husbandry requirements will, of course, depend on the species in question, but in general protein and energy levels should be kept at the minimum acceptable level and opportunity for exercise maximized.

SURGICAL CONSIDERATIONS
• Mature birds with angel wing typically require surgical correction. Phalangeal amputation ("pinioning") or osteotomy with rotation and fixation are typically required to create a comfortable and cosmetic wing.

ANGEL WING (CONTINUED)

- Pinioning of older birds can result in significant complications as the resultant stumps are often prone to injury.

MEDICATIONS

DRUG(S) OF CHOICE
Although typically not indicated, analgesic and/or anti-inflammatory medications may be prescribed for relief of discomfort at the discretion of the clinician.

CONTRAINDICATIONS
Birds affected by angel wing are usually young, but otherwise there are usually no unique contraindications for analgesic/anti-inflammatory medications.

PRECAUTIONS
N/A

POSSIBLE INTERACTIONS
N/A

ALTERNATIVE DRUGS
N/A

FOLLOW-UP

PATIENT MONITORING
- Wing bandages/tape should not be left in place beyond 72 hours if possible. If bandaging is going to be successful, it will only need to be in place for 48–72 hours prior to removal. • In adult birds with mature deformities, primary wing feathers can be periodically trimmed to make the abnormal conformation less obvious.

PREVENTION/AVOIDANCE
Most cases of angel wing can be easily prevented with appropriate diet and husbandry. Proper nutrition during early development is key to prevention of angel wing. Neonatal birds should be provided with adequate opportunity to exercise and strengthen the developing musculoskeletal system. Removal from the breeding population of affected birds or birds with significant occurrence of the deformity in offspring may be warranted.

POSSIBLE COMPLICATIONS
- If intervention is not initiated early enough, usually within 3–5 days of first noticing clinical signs, bandaging will fail to correct carpal valgus. • In cases of permanent deformity, the wing can sometimes become traumatized if the bird cannot manipulate it appropriately.

EXPECTED COURSE AND PROGNOSIS
- Birds with angel wing typically are incapable of normal flight as adults despite intervention. • Angel wing deformities are not life-threatening and many birds will survive and thrive if the deformities are not corrected. • The birds will obviously continue to display the abnormal conformation and will be unable to fly. Complications can occur if the bird is not able to manipulate the wing to keep it from being traumatized.

MISCELLANEOUS

ASSOCIATED CONDITIONS
Angular limb deformities of the pelvic limbs are sometime encountered in conjunction with angel wing deformities.

AGE-RELATED FACTORS
Angel wing only occurs in young birds with developing plumage.

ZOONOTIC POTENTIAL
N/A

FERTILITY/BREEDING
As the role genetics play in the occurrence of angel wing is not understood, in some cases it may be prudent to avoid breeding affected birds.

SYNONYMS
Carpal valgus, airplane wing, dropped wing, crooked wing.

SEE ALSO
Arthritis
Feather disorders
Fracture/luxation
Lameness
Metabolic bone disease
Nutritional deficiencies
Splay leg/slipped tendon
Trauma

ABBREVIATIONS
N/A

INTERNET RESOURCES
N/A

Suggested Reading
Olsen, J.H. (1994). Anseriformes. In: Ritchie, B.W., Harrison, G.J., Harrison, L.R. (eds), *Avian Medicine: Principles and Application*. Lake Worth, FL: Wingers Publishing, pp. 1237–1275.

Author Ryan S. De Voe, DVM, DABVP (Avian, Reptile/Amphibian), DACZM

 Client Education Handout available online

ANOREXIA

BASICS

DEFINITION
Anorexia literally means lack of appetite. It is used in avian medicine to refer to a patient that is not eating or cannot eat. The term is very general and does not define the cause of the lack of appetite. Almost any disease condition in a bird can lead to secondary anorexia. When a bird is anorexic, two things need to be addressed. First, the cause of anorexia must be investigated using history, signalment, physical examination findings, and associated diagnostics. The second aspect that should be addressed in a bird with anorexia is to provide nutrition and fluids to the patient.

PATHOPHYSIOLOGY
In general, it is unknown exactly why a disease condition, not associated with the gastrointestinal system can cause a lack of appetite in a bird. It is a well-known phenomenon, in all veterinary patients, that a sick animal may not eat. It is more common in birds, though, than may be seen in other species. Anorexia can be caused by any condition that is perceived as painful. Anorexia can be caused by any disease affecting the gastrointestinal tract. This can be due to mechanical interference that causes the patient not to be able to swallow. This could be a foreign body in the upper gastrointestinal system, especially an object in the crop. This can also be due to an infection in the gastrointestinal system causing an overgrowth of bacterial or fungal organisms. Anorexia can lead to a deficiency of calories, vitamins, minerals, and other nutrients. As anorexia progresses, birds show signs typical of caloric deficiency. Typically, birds that are anorexic also have decreased fluid intake and can become dehydrated.

SYSTEMS AFFECTED
• Endocrine/Metabolic—theoretically, it has been suggested that long-term anorexia may decrease plasma glucose concentration and increase uric acid concentration. These changes are unlikely to be seen during periods of anorexia induced due to disease. In fact, during the first few days of anorexia, hyperglycemia may be present.
• Gastrointestinal—with continued anorexia, there will be smaller droppings as fecal production is reduced and if dehydration is also present, urine production will also be reduced. • Hepatobiliary—during periods of anorexia, one of the most significant and serious consequences is hepatic lipidosis. As the bird mobilizes fat stores to maintain homeostasis, fat becomes stored in the liver. The result is a yellow tinged, enlarged liver. Liver dysfunction may result with prolonged disease. • Musculoskeletal—loss of fat stores and eventually muscle mass will be present as anorexia continues. • Ophthalmic—for birds that store fat within the globe, there can be the appearance of "sunken" eyes. This can also be due to dehydration. • Reproductive—long-term anorexic hens will have disruption and possibly cessation of egg production.

GENETICS
N/A

INCIDENCE/PREVALENCE
This is a very common reaction to behavioral disturbances, disease processes, and pain. Anorexia can be secondary to any disease process and should be considered in all sick birds.

GEOGRAPHIC DISTRIBUTION
There is no geographic distribution for anorexia.

SIGNALMENT
• **Species:** Seen in all species of birds. This is a normal physiologic response in some wild birds, especially birds that migrate. • **Mean age and range:** Can be seen in birds of any age. • **Predominant sex:** Seen in both males and females.

SIGNS

General Comments
Anorexia should be suspected in any sick bird, no matter what the initial disease diagnosis is determined to be.

Historical Findings
Owners will report the obvious signs of anorexia such as uneaten food in the cage. Owners may also notice signs of weight loss and decreased size and frequency of droppings.

Physical Examination Findings
• Weight loss can be best seen with loss of pectoral muscle mass. In severe cases, the keel is easily palpated and can be termed a "butterknife" keel. • Evidence of dehydration may also be seen with eyelid tenting. • If there is a foreign object in the crop or esophagus causing a mechanical reason for anorexia, this may be found while palpating the crop.

CAUSES
In most cases of anorexia, the primary disease will be obvious. With birds that have anorexia as the only sign of disease, then some causes to consider are:
• Organic disease: liver disease, hepatic lipidosis (can be secondary to anorexia), renal disease, reproductive disease. • Toxin: lead toxicosis, zinc toxicosis. • Infectious disease: bacterial or fungal infection of the gastrointestinal tract. • Neoplasia. • Trauma.

RISK FACTORS
Any medication can be a possible cause of anorexia. Mood-altering medications used for behavioral issues such as feather damaging behavior, can cause anorexia. Haloperidol administration is well known for causing anorexia. Some antifungal medications can induce anorexia. Potentially, birds that are on an improper diet and are being transitioned to a more balanced diet may not initially eat the new food. These birds are at risk for developing anorexia. Any disease condition should be considered a risk factor for anorexia.

DIAGNOSIS

DIFFERENTIAL DIAGNOSIS
Anorexia is defined as the loss of appetite. Any other disease or condition that mechanically interferes with eating could be considered a differential diagnosis, as, in those cases, theoretically, the patient still has an appetite but cannot eat. This could include conditions such as trauma to the lower and upper beak and associated bony structures.

CBC/BIOCHEMISTRY/URINALYSIS
• CBC: Anorexia, if secondary, should cause no changes to the complete blood count. Primary disease may alter the CBC.
• Biochemistry: As with the CBC, anorexia should cause no changes to the biochemistry panel. The exception to this occurs if there is secondary hepatic lipidosis due to the anorexia. Liver analytes, including AST and bile acids, may be elevated.

OTHER LABORATORY TESTS
Consider performing other laboratory tests to diagnosis the primary cause of disease. If hepatic lipidosis is suspected, possible further testing includes a fine needle aspirate of the liver or biopsy of the liver can confirm the diagnosis.

IMAGING
Diagnostic imaging may be necessary to diagnose the primary disease process. If hepatic lipidosis is present the liver is usually increased in size. On the lateral projection, if the liver is enlarged, there may be a displacement of the proventriculus and ventriculus. The proventriculus will appear to be more dorsal in positioning and the ventriculus will appear more caudal. On the ventro-dorsal projection, the cardiac-hepatic silhouette will be abnormal with possible loss of the waist and an increase in the size of the hepatic portion of the silhouette.

DIAGNOSTIC PROCEDURES
Diagnostic procedures may be necessary to discern the primary cause of disease.

PATHOLOGIC FINDINGS
N/A

TREATMENT

APPROPRIATE HEALTH CARE
Health care is aimed at the primary cause of disease. As the primary cause is treated, it is expected that the patient will begin eating.

ANOREXIA (CONTINUED)

NURSING CARE
Nursing care will also be aimed at the primary problem. Birds may not begin eating immediately and likely assisted feeding will be necessary. The type and quantity of food that is to be used for assisted feeding depends on the species and age of the patient. In most cases, food will be deposited into the crop using a crop needle 2–4 times a day. In unusual cases where disease involves the crop or lower or upper beak, an esophageal feeding tube may need to be surgically placed. Since many birds that are anorexic also are deficient in fluid intake, patients that are receiving assisted feedings may also need fluid support.

ACTIVITY
N/A

DIET
It is recommended that the patient be given access to foods with high palatability, even if the diet is not balanced, for a short period of time until the patient is no longer anorexic.

CLIENT EDUCATION
N/A

SURGICAL CONSIDERATIONS
N/A

MEDICATIONS

DRUG(S) OF CHOICE
Commercial assisted feeding diets are available that are specifically formulated for the sick or debilitated patient. There are also diets that are specifically designed for hand feeding chicks. Diets should be formulated based on the caloric needs of the patient. Most quality commercial assisted feed diets are well-balanced and additional vitamins and minerals are not necessary if caloric needs are being met.

CONTRAINDICATIONS
N/A

PRECAUTIONS
When assist feeding patients, before each episode, the crop should be palpated to be certain that no material remains from the previous feeding.

POSSIBLE INTERACTIONS
N/A

ALTERNATIVE DRUGS
N/A

FOLLOW-UP

PATIENT MONITORING
The patient should be monitored for improvement on the primary disease process but the bird may remain anorexic even as the primary disease is being treated. Assisted feeds should continue until the patient is eating enough to satisfy its caloric requirements.

PREVENTION/AVOIDANCE
N/A

POSSIBLE COMPLICATIONS
If anorexia continues, hepatic lipidosis is a possible complication. If the patient develops hepatic lipidosis, anorexia may continue long past the treatment of the initial disease process.

EXPECTED COURSE AND PROGNOSIS
Prognosis depends on the treatment of the primary disease process.

MISCELLANEOUS

ASSOCIATED CONDITIONS
Hepatic lipidosis if anorexia is prolonged and not treated.

AGE-RELATED FACTORS
N/A

ZOONOTIC POTENTIAL
N/A

FERTILITY/BREEDING
Reproductively active hens that are anorexic may be poor breeders or may not be able to normally oviposit. Developing eggs in the oviduct may be affected if anorexia occurs during that episode.

SYNONYMS
N/A

SEE ALSO
Beak fracture
Crop stasis/ileus
Dehydration
Emaciation
Gastrointestinal foreign bodies
Heavy metal toxicosis
Hepatic lipidosis
Liver disease
Regurgitation/vomiting
Renal disease
Sick-bird syndrome

ABBREVIATIONS
N/A

INTERNET RESOURCES
N/A

Suggested Reading
Tully, T., Lawton, M., Dorrestein, G. (2000). *Avian Medicine*. Woburn, MA: Butterworth-Heinemann.

Author Karen Rosenthal, DVM, MS

Anticoagulant Rodenticide Toxicosis

BASICS

DEFINITION
• An impairment of coagulation resulting in life-threatening hemorrhage caused by ingestion of rodent poison bait (primary poisoning) or ingestion of a poisoned prey item (secondary poisoning). • Anticoagulant rodenticides (ARs) are divided into two groups that share the same mechanism of action but differ in their potencies and half-lives. • First-generation anticoagulant rodenticides (FGARs) include warfarin, diphacinone, and chlorphacinone. The FGARs are less potent and have shorter half-lives. • Second-generation anticoagulant rodenticides (SGARs) include brodifacoum, difethialone, and bromadiolone. The SGARs are more potent and have longer half-lives.

PATHOPHYSIOLOGY
ARs interfere with blood clotting by inhibiting the enzyme vitamin K epoxide reductase. This inhibition results in the accumulation of an inactive (oxidized) form of vitamin K, which in turn is unable to activate the vitamin-K dependent clotting factors (II, VII, IX, and X). The depletion of these activated clotting factors causes a coagulopathy and hemorrhage. Although, it was previously thought that birds lacked factors VII and X, these conclusions were drawn from studies that used nonavian tissue thromboplastin in their laboratory assays; this is now recognized to produce inaccurate results. Clinical signs resulting from AR toxicosis will not be apparent for 2–5 days following ingestion of a toxic dose due to the presence of activated clotting factors in the circulation at the time the AR is consumed. When these activated clotting factors are depleted, which is dependent upon the half-lives of the factors, coagulopathy occurs.

SYSTEMS AFFECTED
• Cardiovascular—anemia, hypoproteinemia, hypovolemic shock (pallor of mucous membranes, poor capillary refill time).
• Musculoskeletal—intramuscular bleeding, weakness, massive swelling in the absence of fractures. • Skin—bruising, subcutaneous hemorrhage, severe bleeding from minor lacerations. • Nervous—dull mentation secondary to hypovolemic shock, neurologic signs possible if hemorrhage occurs within the CNS or surrounding spinal cord.
• Respiratory—dyspnea possible if pulmonary hemorrhage occurs. • Urologic/Reproductive/Gastrointestinal—bleeding from the vent possible if hemorrhage from renal, reproductive, or gastrointestinal systems occurs.

GENETICS
N/A

INCIDENCE/PREVALENCE
• The avian species most commonly affected by AR exposure and toxicosis are free-living birds of prey. Studies in the United States published in 2003 and 2011 found exposure, primarily to SGARs, among multiple species of free-living birds of prey in 49% and 86% of tested populations, respectively (see SUGGESTED READING). • Poisonings have occurred in avian species housed in zoos.
• Poisoning in psittacine birds kept as pets is not commonly reported.

GEOGRAPHIC DISTRIBUTION
Mortalities and exposures have been documented in free-living birds of prey in multiple regions worldwide where ARs are used, including the United States, Canada, and Europe.

SIGNALMENT
• **Species:** There appear to be differences in sensitivity to ARs among avian species with evidence that birds of prey may be more sensitive than other species that have been studied. • **Mean age and range:** N/A
• **Predominant sex:** N/A

SIGNS

General Comments
Bleeding can be spontaneous or may be induced by traumatic injury. If the bleeding is initiated by a traumatic event, however, the amount of hemorrhage will be in excess of what would be expected with uncomplicated trauma. Signs are in part dependent on the location of the hemorrhage.

Historical Findings
Free-living birds of prey are often found on the ground, unable to fly due to weakness and hypovolemic shock. Captive or pet birds may have a known exposure to AR bait.

Physical Examination Findings
• Dull mentation. • Weakness. • Profound pallor of the mucous membranes. • Collapse of the cutaneous ulnar vein. • Bleeding may or may not be obvious. In many cases extensive subcutaneous hemorrhage and intramuscular bruising will be present but can only be appreciated if the clinician parts the feathers with isopropyl alcohol and looks for these signs. In other instances, a traumatically induced injury may be present but the extent of hemorrhage will be disproportionate to the injury; for example, severe external hemorrhage from a minor laceration or formation of an excessively large hematoma at a point of possible traumatic impact.
• Increased respiratory rate and/or effort can indicate hemorrhage involving the respiratory system. • Bleeding from the vent can indicate hemorrhage involving the gastrointestinal, renal, or reproductive systems. • Neurologic abnormalities can be associated with hemorrhage involving the central nervous system or spinal cord.

CAUSES
• Primary poisoning—ingestion of AR bait.
• Secondary poisoning—ingestion of poisoned prey.

RISK FACTORS
• Free-living birds hunting in areas of AR use.
• Accessibility of AR baits or contaminated food to captive or pet birds.

DIAGNOSIS

DIFFERENTIAL DIAGNOSIS
• Blood loss anemia secondary to trauma: Traumatic blood loss will be accompanied by other obvious signs of trauma, such as fractures, ocular injury (which may only affect the posterior segment of the eye with the anterior chamber appearing normal), or large wounds. Traumatic blood loss in birds will generally not result in anemia as severe as that seen with AR toxicosis. Profound anemia with or without visible massive bruising or with an externally bleeding wound that does not clot should raise the index of suspicion for AR toxicosis. • Anemia secondary to malnutrition: Malnourished birds will lack subcutaneous fat deposits, will show muscle wasting with a prominent keel, and will have normal coagulation. • Disseminated intravascular coagulation (DIC): Birds may show other signs of systemic illness. • Other causes of coagulopathy. Not well described in birds; liver dysfunction may result in coagulopathy.

CBC/BIOCHEMISTRY/URINALYSIS
• Anemia—often profound (PCV may be <10%). • Hypoproteinemia.

OTHER LABORATORY TESTS
• Assessment of coagulation: Commercial coagulation profiles are not available for avian species and tests run with mammalian thromboplastin will provide inaccurate results. However, the lack of clotting in birds suffering AR toxicosis is usually quite evident if 0.1–0.2 ml of blood are placed in a serum collection tube, which is then inverted occasionally and observed for clotting. In normal birds, signs of clot formation in the tube should be obvious within five minutes. In many AR poisoned birds, no signs of clotting are seen after several hours. • AR testing: Identification of the rodenticide from a plasma sample can be done by many diagnostic laboratories. False negatives are possible, however, as the half-lives of ARs in blood are short. Also, the need to pursue AR identification should be weighed against the drawbacks of taking additional blood for testing from an anemic and shocky bird. AR screens can be performed on liver tissue post-mortem.

ANTICOAGULANT RODENTICIDE TOXICOSIS (CONTINUED)

IMAGING
Radiographs may help rule in or rule out trauma. Radiography and/or ultrasonography may aid in visualizing blood within the coelom, if present. However, these procedures should be performed cautiously in critical patients.

DIAGNOSTIC PROCEDURES
N/A

PATHOLOGIC FINDINGS
Findings on gross necropsy vary with the location of hemorrhage. Common lesions include extensive subcutaneous and intramuscular bruising, coelomic hemorrhage, generalized bruising of the sternum, and pallor of internal organs. Focal, limited hemorrhage or bruising is more consistent with trauma. Histopathology may corroborate a gross diagnosis of AR toxicosis with findings of marked, acute hemorrhage of multiple tissues, hypoxic insult of organs such as liver, or possibly extramedullary hematopoiesis.

TREATMENT

APPROPRIATE HEALTH CARE
• Emergency inpatient intensive care management: For birds with active hemorrhage and severe blood loss. • Inpatient medical management: Free-living birds of prey require admission to a wildlife clinic or rehabilitation center for the duration of medical management; may be required for captive or pet birds if treatment cannot be reliably administered by caretaker or owner. • Outpatient medical management: Can be considered for captive or pet birds once condition is stable.

NURSING CARE
• Fluid therapy. Repeated subcutaneous crystalloid fluid therapy can be a very effective means of volume replacement in hypovolemic birds. Although conspecific whole blood or plasma transfusions can provide red blood cells and/or clotting factors, suitable blood donors and appropriate volumes for transfusion are often unavailable. The stress to the patient of intravenous or intraosseous catheter placement should be carefully considered. • Thermal support. Anemic, hypovolemic birds will benefit from supplemental heat. • Oxygen therapy. Beneficial for anemic birds and important therapy for birds in respiratory distress.

ACTIVITY
Activity should be restricted until medical management restores normal coagulation.

DIET
Once patient condition is stable, proper nutritional intake should be ensured. If patient will not self-feed, nutritional support in the form of hand feeding or gavage feeding should be considered.

CLIENT EDUCATION
For captive or pet birds, caretakers or owners should be counseled regarding preventing future exposure to AR products.

SURGICAL CONSIDERATIONS
N/A

MEDICATIONS

DRUG(S) OF CHOICE
Vitamin K_1: 2.5–5 mg/kg SC q8h while active hemorrhage is occurring. Once patient is stable, 2.5 mg/kg PO once daily. Bioavailablity is enhanced if given with food. Duration of treatment depends on type of AR due to the difference in the half-lives of FGARs and SGARs. Four weeks of vitamin K_1 therapy has been successful in the treatment of birds of prey with SGAR toxicosis. Shorter treatment may be adequate for FGAR toxicosis but has not been documented in birds. If the type of AR is unknown, treatment should be continued for four weeks due to the difficulty of accurately assessing coagulation.

CONTRAINDICATIONS
None.

PRECAUTIONS
• Intravenous administration of vitamin K_1 may result in anaphylaxis. This route should not be used. • Intramuscular administration in actively bleeding birds can result in the formation of a large hematoma and is best avoided. • Jugular venipuncture can result in life-threatening hemorrhage.

POSSIBLE INTERACTIONS
None.

ALTERNATIVE DRUGS
None. Vitamin K_3 is not an effective antidote for AR toxicosis.

FOLLOW-UP

PATIENT MONITORING
• Normal coagulation is restored following the synthesis and activation of new clotting factors, which occurs following administration of vitamin K_1. Response to vitamin K_1 therapy may be seen within 12–36 hours, depending on the extent of factor depletion. • PCV may show marked increase within 3–5 days as birds regenerate red blood cells quickly. • Re-assessment of coagulation via the method described under DIAGNOSIS 3–4 days after discontinuation of vitamin K_1 should be performed to ensure normal clotting factor activation in the absence of vitamin K_1 therapy. Because of the lack of accurate avian coagulation tests, patients should be closely observed during this time.

PREVENTION/AVOIDANCE
• Consumer education regarding the risks of ARs to free-living birds of prey is an important factor in decreasing exposure in these species. Avian clinicians that treat free-living birds can play a role in public education on this subject. • In the United States, confirmed or suspected cases of AR toxicosis can be reported to the U.S. Environmental Protection Agency via the National Pesticide Information Center at http://npic.orst.edu/eco. • For captive or pet birds, preventing access to AR baits or contaminated food.

POSSIBLE COMPLICATIONS
• Hemorrhage can recur if vitamin K_1 therapy is interrupted or discontinued too soon. Vitamin K_1 must be administered for the duration of the toxic effects of the AR, which is dependent upon the type of AR. • If the patient survives the initial acute crisis and appropriate therapy is maintained for the proper duration, adverse sequelae should not be expected.

EXPECTED COURSE AND PROGNOSIS
Successful treatment, rehabilitation, and release of even severely anemic free-living birds of prey with AR toxicosis is possible. Birds appear able to recover from severe blood loss, quickly regenerate red blood cells, and respond well to treatment consisting of crystalloid or other fluid replacement and vitamin K_1 therapy. Prognosis largely depends on the location of hemorrhage. Birds with external, intramuscular, or subcutaneous hemorrhage have a better prognosis for survival than birds that bleed internally, such as into the respiratory tract, the pericardial sac, or the central nervous system.

MISCELLANEOUS

ASSOCIATED CONDITIONS
None.

AGE-RELATED FACTORS
None.

ZOONOTIC POTENTIAL
None.

FERTILITY/BREEDING
None.

SYNONYMS
None.

SEE ALSO
Anemia
Appendix 3: Laboratory Testing
Appendix 6: Common Avian Toxins

Anticoagulant Rodenticide Toxicosis (Continued)

Coagulopathies
Heavy metal toxicity
Hemorrhage
Trauma
Sick-bird syndrome

ABBREVIATIONS
AR—anticoagulant rodenticide
FGAR—first generation anticoagulant rodenticide
SGAR—second generation anticoagulant rodenticide

INTERNET RESOURCES
- http://www2.epa.gov/rodenticides
- http://www.lafebervet.com/emergency-medicine/birds/presenting-problem-anticoagulant-rodenticide-toxicosis-in-free-living-birds-of-prey/
- http://npic.orst.edu/eco

Suggested Reading
Murray, M. (2011). Anticoagulant rodenticide exposure and toxicosis in four species of birds of prey presented to a wildlife clinic in Massachusetts, 2006–2010. *Journal of Zoo and Wildlife Medicine*, **42**(1):88–97.
Murray, M, Tseng, F. (2008). Diagnosis and treatment of secondary anticoagulant rodenticide toxicosis in a red-tailed hawk (*Buteo jamaicensis*). *Journal of Avian Medicine and Surgery*, **22**(1):41–46.
Nevill, H. (2009). Diagnosis of nontraumatic blood loss in birds and reptiles. *Journal of Exotic Pet Medicine*, **18**(2):140–145.
Stone, W.B., Okoniewski, J.C., Stedelin, J.R. (2003). Anticoagulant rodenticides and raptors: recent findings from New York, 1998–2001. *Bulletin of Environmental Contamination and Toxicology*, **70**(1):34–40.
Wernick, M.B., Steinmetz, H.W., Martin-Jurado, O., *et al.* (2013). Comparison of fluid types for resuscitation in acute hemorrhagic shock and evaluation of gastric luminal and transcutaneous P_{CO_2} in leghorn chickens. *Journal of Avian Medicine and Surgery*, **27**(2):109–119.

Author Maureen Murray, DVM, DABVP (Avian)

Portions Adapted From
Murray, M. Presenting problem: Anticoagulant rodenticide toxicosis in free-living birds of prey. LafeberVet.com website. [http://www.lafebervet.com/emergency-medicine/birds/presenting-problem-anticoagulant-rodenticide-toxicosis-in-free-living-birds-of-prey/; last accessed September 14, 2015].

ARTHRITIS

BASICS

DEFINITION
Arthritis is relatively common among many avian species and is often characterized by joint pain and discomfort associated with deterioration of the articular cartilage of diarthoidial (synovial) joints. The most common forms of arthritis are osteoarthritis, articular gout, and infectious arthritis, but immune-mediated arthritis and neoplasia have also been reported. Septic arthritis is defined as the presence of pathogenic organisms within a synovial joint. Even with aggressive treatment, arthritis is typically progressive and it can be challenging to maintain an adequate quality of life for the patient over the long term.

PATHOPHYSIOLOGY
• Arthritis is characterized by deterioration of articular cartilage and formation of new bone at the joint surfaces and margins. • This damage can have a number of causes, including, but not limited to, trauma, infection, and degeneration. As the cartilage becomes damaged, its ability to transmit and resist loads is decreased. Inflammation is not a factor in the initiation of osteoarthritis, but it does play a significant role in the progression of the disease. • Immune-mediated arthritis is characterized by inflammation of the synovium due to deposition of immune complexes in the joint. • Articular gout arises secondary to hyperuricemia and is characterized by precipitation and accumulation of monosodium urate crystals within the synovial capsule and tendon sheaths, resulting in inflammation. Hyperuricemia develops subsequent to reduced excretion of urates due to primary renal disease or prerenal or postrenal causes. • Once initiated, arthritis is typically progressive, resulting in continued degradation of the joint and increasing patient discomfort. • Neoplasia—pathophysiology is poorly understood.

SYSTEMS AFFECTED
• Musculoskeletal—lameness and falling; postural abnormalities; joint swelling; white, nodular swellings of the joint or tendon sheaths visible through the skin (articular gout); reduced range of motion of affected joints; disuse muscle atrophy.
• Behavioral—changes in activity; some birds may demonstrate increasing learned fear and escape/avoidance behaviors if painful.
• Gastrointestinal—decreased appetite due to pain and unwillingness to maneuver for access to food and water; hyporexia more common in obese individuals.
• Respiratory—tachypnea as a response to pain and discomfort.

GENETICS
No proven genetic component, but breeds that are selectively bred for large body mass (broiler chickens, Pekin ducks, turkeys, etc.) are at higher risk.

INCIDENCE/PREVALENCE
• Common in middle-aged to geriatric and in obese birds. • Young, large-bodied birds on growing formulas are predisposed to developing osteoarthritis. • In one study, 33% of waterfowl necropsy specimens were found to have arthritis of at least one pelvic limb joint. No arthritic changes of the thoracic limb joints and no septic arthritis was identified. • Arthritis of pelvic limb joints is most common, but arthritis of thoracic limb joints and intervertebral joints does occur.

GEOGRAPHIC DISTRIBUTION
None.

SIGNALMENT
• Middle-aged to geriatric birds of all species and both sexes. • Heavy breed chickens, waterfowl, turkeys, and peafowl are commonly affected.

SIGNS

General Comments
• Severity of clinical signs depends on the location and severity of the lesion. • There may be a history of a high-calorie diet.

Historical Findings
• Intermittent, persistent, and/or progressive lameness. • Falling. • Lethargy/reduced activity level; unwillingness to walk, step up, or fly. • Septic arthritis commonly associated with a history of a prior traumatic incident. • White nodules associated with any joint (owners are most likely to identify these in the feet and digits).

Physical Examination Findings
• Lameness: Limping, stilted gait, nonweight bearing, falling, unwillingness or inability to fly. • Abnormal standing or wing posture. • Soft tissue swelling and/or hyperemia associated with the affected joint, distention of the joint capsule, pain elicited upon manipulation of the joint. • Firmly enlarged and/or irregularly contoured joint. • Crepitus, joint instability, reduced range of motion. • Articular gout: White nodular enlargement of any joint or tendon sheath (most frequently recognized in the stifles, hock joints, and interphalangeal joints). • Muscle atrophy of the affected limbs (chronic disease).
• Lethargy/depression. • Asymmetrical wear pattern on the plantar surface of the feet.

CAUSES
• Noninfectious: Previous trauma or excessive wear to the joints results in damage to the chondrocytes; also, low-grade wear and tear, instability, abnormal weight bearing.
• Infectious: Typically either by direct inoculation or hematogenous spread but many infectious agents reported, including *Staphylococcus* spp., *Streptococcus* spp., *Pseudomonas* spp., *Salmonella* spp., *Actinobacillus* spp., *Mycoplasma* spp., *Escherichia coli, Pasteurella multocida, Erysipelothrix rhusiopathiae, Mycobacterium avium, Chlamydia psittaci, Reovirus,* and microfilaria. • Gout: Hyperuricemia subsequent to primary renal disease of any etiology or prerenal or postrenal causes.
• Immune-mediated: Polyarthritis associated with granulomatous vasculitis reported in a Mississippi sandhill crane. • Neoplasia: Unknown cause.

RISK FACTORS
• Clinically obese birds and heavy body breeds at high risk. • Previous trauma or fractures.
• Wounds adjacent to or involving the joint.

DIAGNOSIS

DIFFERENTIAL DIAGNOSIS
• Acute trauma (including fracture or joint luxation): Differentiated by history, physical examination findings, radiographic imaging, response to therapy, and outcome over the long term. • Bone cyst: Cyst identified by a combination of radiographs, ultrasound, and surgical biopsy/excision. • Amyloidosis: Presence of amyloid within the joint—identified by biopsy. • Numerous disease processes producing nonspecific signs of lethargy, weakness, reduced activity level, or impaired mobility. • Neurologic deficits, sensory or motor dysfunction (ataxia, paresis/paralysis, impaired proprioception, neurogenic pain). • Vascular disease (intermittent claudication). • Neoplasia: Synovial cell carcinoma; bronchial carcinoma; chondrosarcoma; mesothelioma; diagnosed primarily by biopsy of the joint.
• Subcutaneous filariasis: Rare condition in wild cockatoos confirmed by identification of microfilaria within the joint. • Tenosynovitis.

CBC/BIOCHEMISTRY/URINALYSIS
• AST and/or CK may be elevated due to muscle damage associated with trauma.
• Plasma uric acid—hyperuricemia should be considered a risk factor for development of articular gout. • Leukocytosis may be present in some cases of severe infectious arthritis or neoplasia.

OTHER LABORATORY TESTS
Serologic evaluation for possible infectious causes.

IMAGING

Radiographic findings
• Narrowing of the joint space, misalignment of the joint, sclerosis of the subchondral bone, or osteophyte formation. • Changes to the renal silhouette and postrenal obstructions including ureteroliths may be present in cases of articular gout. • Bony destruction/lysis in cases of septic arthritis with osteomyelitis.
• Radiographic changes may not directly

ARTHRITIS (CONTINUED)

correlate with clinical severity. • Changes can be challenging to identify in small patients.

Ultrasonography
• Identification of joint effusion and synovial abnormalities in larger patients.

Advanced imaging
• Computed tomography (CT). • Magnetic resonance imaging (MRI). • Micro-CT shows promise for improved definition of pathologic changes of the affected joints, but access to this imaging modality is currently limited.

DIAGNOSTIC PROCEDURES
• Arthrocentesis for cytology, culture, or pathogen-specific PCR of synovial fluid when septic arthritis is suspected. • Biopsy of synovial tissue and bone to rule out neoplasia or bone cyst.

PATHOLOGIC FINDINGS
• Degeneration or erosion of the articular cartilage, bony remodeling, and osteophyte formation. • Fibrosis of the joint capsule. • Presence of urate crystals within the joint(s). • Inflammatory infiltrates and/or pathogenic organisms.

TREATMENT
APPROPRIATE HEALTH CARE
Patients that are self-sustaining and ambulatory are typically treated as outpatients, but brief hospitalization may be appropriate for initiation of analgesic therapy and collection of baseline diagnostic data (including screening radiographs). Articular gout should be considered severely painful and the primary cause should be determined. Debilitated patients may need extended periods of hospitalization for supportive care until stable.

NURSING CARE
• Supportive care may include provision of padded substrates or supportive cushioning for nonweight-bearing patients, multimodal analgesia, fluid therapy (subcutaneous fluids can be administered at 5–10% of bodyweight), and nutritional support (gavage feeding). • Fluid therapy should be considered if articular gout is present, especially when there is current hyperuricemia.

ACTIVITY
Cage rest and modifications to the cage set-up can be used to minimize arthritis pain, reduce the risk of falling and continued trauma to the joints, facilitate greater ease of access to food and water dishes, and help patients compensate for reduced mobility. Wire cages tend to result in increased trauma to the legs, feet, and wings and permit climbing and, thereby, greater risk of falls. Plastic, glass, or plexiglass enclosures may be more appropriate with horizontal space emphasized over vertical space. Lowered perches, padded substrates, soft perches, platforms, and ramps can be used to aid mobility, improve patient comfort, and reduce risk of injury from falls. Food and water dishes should be easily accessible and placed in multiple locations. Some birds will require permanently altered caging.

DIET
When obesity is a contributing factor, conversion to a lower-calorie or restricted-calorie diet is appropriate to minimize fat stores and continued degradation and erosion of the articular cartilage.

CLIENT EDUCATION
• It is important to review with clients the long-term need to limit falling and repeated trauma, as these will exacerbate existing arthritis. • Handling methods, such as holding the bird's digits while it is perched on the hand, may become increasingly painful and should be avoided. • Nails should be trimmed short enough to reduce the risk of entanglement while ambulating but not so short that grip is compromised.

SURGICAL CONSIDERATIONS
• Bimodal analgesia should be considered for all procedures requiring restraint and positioning. • Arthrodesis of severely affected or unstable joints. • Amputation of affected limbs if the condition is nonresponsive to medical management and cannot be otherwise resolved surgically. • Surgical removal of urate crystals from gout lesions can be therapeutic, but does not address the primary underlying cause.

MEDICATIONS
DRUG(S) OF CHOICE
• Nonsteroid antiinflammatory drugs (NSAIDS)—inhibit prostaglandin synthesis by the COX enzyme: meloxicam 0.5–2 mg/kg PO q12–24h, carprofen 1–5 mg/kg PO q12–24h, ketoprofen 1–5 mg/kg IM q12–24h. • Analgesics—primarily function as kappa antagonists: butorphanol 0.5–2 mg/kg IM q1–4h; synthetic mu-receptor opiate-like agonist that also inhibits reuptake of serotonin and nor-epinephrine: tramadol 20–30 mg/kg PO q8–12h. • Gabapentin—10–30 mg/kg PO q12h. • Omega 3 fatty acids—0.22 mL/kg PO q24h. • Antibiotic therapy—systemic and/or local; ideally based on culture and sensitivity results; can include placement of antibiotic-impregnated poly(methyl methacrylate) (PMMA) beads within affected joint(s). • Articular gout—allopurinol 10–30 mg/kg PO q12h; colchicine 0.04 mg/kg PO q12–24h (may not be beneficial and could worsen disease).

CONTRAINDICATIONS
NSAIDS if articular gout or renal disease is present.

PRECAUTIONS
• NSAIDS should not be administered if renal pathology is known to be present or discontinued immediately if clinical signs suggestive of renal dysfunction (such as polyuria/polydipsia) are observed. • Adequan (polysulfated glycosaminoglycan) (PSGAG) has been used, but fatal hemorrhage has been reported.

POSSIBLE INTERACTIONS
Mild sedative effect if tramadol and gabapentin are used in combination.

ALTERNATIVE DRUGS
Glucosamine/chondroitin sulfate: 20 mg/kg PO BID.

FOLLOW-UP
PATIENT MONITORING
• Initial follow-up primarily consists of physical examination and observation to evaluate for resolving lameness, improved mobility, and reduction in pain and discomfort. • Annual examinations should be performed to review for changes in nutrition and potential increases in body weight, and to review husbandry (including cage set-up). • If NSAIDS are being used, serial biochemical evaluation of renal function should be performed.

PREVENTION/AVOIDANCE
Increases in body weight will result in increased wear of pelvic limb joints; body weight should be monitored for stability and dietary changes made as needed.

POSSIBLE COMPLICATIONS
• Arthritic changes commonly develop in contralateral pelvic limb joints as well as pododermatitis of the plantar foot/digits owing to increased, compensatory weight bearing. • Repeated and progressive falling which may result in additional injuries • Difficulty posturing preventing regular passage of droppings. • Difficulty accessing dishes/bowls to adequately meet food and water needs.

EXPECTED COURSE AND PROGNOSIS
Arthritis is a disease that is managed over the life of the bird. There may be periodic events in which clinical signs worsen, requiring at least temporary hospitalization for supportive care.

MISCELLANEOUS
ASSOCIATED CONDITIONS
Metabolic bone disease, pododermatitis.

AGE-RELATED FACTORS
More common in older birds.

ARTHRITIS (CONTINUED)

ZOONOTIC POTENTIAL
Some potential causes of infectious arthritis have zoonotic potential (*Chlamydia psittaci*, *Mycobacterium* spp, and *Erysipelothrix rhusiopathiae*).

FERTILITY/BREEDING
Birds with arthritis may be unable to copulate successfully.

SYNONYMS
Degenerative joint disease (DJD); osteoarthritis.

SEE ALSO
Fracture/luxation
Hyperuricemia
Lameness
Metabolic bone disease
Obesity
Pododermatitis
Renal disease
Trauma

ABBREVIATIONS
N/A

INTERNET RESOURCES
N/A

Suggested Reading

Baine, K. (2012) Management of the geriatric psittacine patient. *Journal of Exotic Pet Medicine*, **21**:140–148.

Alderson, M., Nade, S. (1987) Natural history of acute septic arthritis in an avian model. *Journal of Orthopedic Research*, **5**:261–274.

Degernes, L.A., Lynch, P.S., Shivaprassad, H.L. (2011) Degenerative joint disease in captive waterfowl. *Avian Pathology*, **40**:103–110.

Author Geoffrey P. Olsen, DVM, Dip ABVP (Avian)

 **Client Education Handout available online**

ASCITES

BASICS

DEFINITION
Accumulation of fluid within the peritoneal cavity.

PATHOPHYSIOLOGY
• Increased hydrostatic pressure in the vessels, decreased oncotic pressure, leakage of vessels, or the combination of these factors lead to accumulation of fluid in the coelomic cavity. • Increased capillary and lymphatic hydrostatic pressure may be due to portal hypertension from a chronic hepatopathy or pulmonary hypertension from congestive heart failure. • Hypoproteinemia can be due to loss a loss of proteins (e.g. protein-losing enteritis or nephropathy) or from decreased production (e.g., hepatopathy). The decreased protein levels lead to the inability to maintain oncotic pressure to keep fluid in the circulatory system. • Compromise of vasculature from inflammation or infection can also lead to leakage of fluid into the coelom. • The coelom is divided into the intestinal peritoneal cavity, right and left ventral hepatic peritoneal cavities, and the right and left dorsal hepatic peritoneal cavities; fluid can accumulate within any one or more of these five cavities.

SYSTEMS AFFECTED
• Respiratory—increased respiratory effort due to compression of the air sacs by the accumulated fluid. • Integument—increased visualization of skin around the caudal coelom, either from stretching of skin or localized plucking of feathers in response to discomfort from the distension.

GENETICS
• Direct and maternal genetic effects play a role in the development of broiler ascites syndrome. • Heritable cardiac conditions have been described in neonatal and juvenile parrots.

INCIDENCE/PREVALENCE
• Common clinical sign for cardiomyopathies, hepatopathies, and ovarian neoplasms.

GEOGRAPHIC DISTRIBUTION
N/A

SIGNALMENT
• **Species predilections** ○ Cockatiels—egg yolk peritonitis, ovarian neoplasms, cystic ovarian disease. ○ Toucans and mynah birds—hemochromatosis (iron storage disease). ○ Broiler chickens—pulmonary arterial hypertension (ascites syndrome). ○ Turkeys—spontaneous cardiomyopathy (round heart disease). • **Mean age and range and predominant sex** ○ Cardiomyopathies—atherosclerosis and associated CHF is usually noted in geriatric birds, however congenital heart conditions have been reported in neonates and juvenile birds. ○ Egg-yolk peritonitis, cystic ovarian disease, and ovarian neoplasms—occurs in sexually mature hens. ○ Hepatopathies—neoplasms and cirrhosis usually seen in older birds; toxins and infectious causes of hepatitis can affect any age.

SIGNS

Historical Findings
• Distended abdomen. • Tail bobbing and/or open-mouth breathing. • Generalized lethargy and depression; may be on bottom of the cage. • Decreased appetite. • Decreased ability to fly.

Physical Examination Findings
• Swollen coelom that is compressible. • Dyspnea and tachypnea. • Wheezing or crackles auscultated over lungs and caudal air sacs. • Lethargic and depressed. • Anorexic. • Fecal and/or urate accumulation over vent and tail feathers. • Exercise intolerance. • Holosystolic heart murmur or arrhythmia if cardiovascular disease is present. • Yellow or green urates if a hepatopathy is present.

CAUSES
• Reproductive disease—egg yolk coelomitis, cystic ovarian disease, ovarian neoplasia, oophoritis. • Carcinomatosis, neoplastic effusion. • Hepatic cirrhosis (aflatoxins, chlamydiosis, immune-mediated disease). • Protein losing nephropathy or enteropathy. • Congestive heart failure. • Iron storage disease (ISD)/hemochromatosis. • Aspergillosis (granulomatous pneumonia leading to pulmonary hypertension). • Mycobacteriosis (liver failure and protein-losing enteropathy). • Polyomavirus.

RISK FACTORS
• The lungs of bird are more rigid and noncompliant than that of mammals, thus possibly making them more predisposed to pulmonary hypertension. • Ingestion of moldy peanuts and grain may lead to aflatoxicosis, which can result in hepatic cirrhosis and ascites. • Birds in the ramphastidae (toucan) and sturnidae (mynah) families are at greater risk for ISD.

DIAGNOSIS

DIFFERENTIAL DIAGNOSIS
• Coelomic distension—egg binding or dystocia (common); body wall hernias; neoplasia, abscesses, or granulomas of any of the coelomic organs; obesity; diagnostic imaging and/or celiocentesis used to differentiate. • Dyspnea—cardiac disease; organomegaly or space-occupying masses in the coelomic cavity; primary respiratory condition such as pneumonia, sinusitis, or tracheal obstruction; pain; heat exhaustion; diagnostic imaging such as ultrasound or radiographs used to differentiate.

CBC/BIOCHEMISTRY/URINALYSIS
• Leukocytosis, with or without a monocytosis, may be noted in patients with systemic infection; may have severe leukocytosis if infectious disease process such as mycobacteria or chlamydia is involved. • Moderate to severe hypoalbuminemia and decreased A:G ratio. • Elevated bile acids if a hepatopathy is present. • Damage to skeletal or smooth musculature may lead to elevated aspartate aminotransferase and creatine phosphokinase levels.

OTHER LABORATORY TESTS
• Bacterial culture and sensitivity of aspirated fluid should be performed. • Fluid analysis and cytology of aspirated fluid allows for characterization of the fluid and aids in the diagnosis of the underlying cause. Neoplastic cells may be noted in cases of neoplastic effusions or carcinomatosis: ○ Transudate: Clear and odorless with low cellularity ($<1 \times 10^9$/L), low specific gravity < 1.020, low protein <30 g/L; may contain macrophages and mesothelial cells; due to hepatic cirrhosis, cardiac insufficiency, or hypoproteinemia. ○ Modified transudate: Serosanguineous, may be clear to slightly cloudy with moderate cellularity ($1-5 \times 10^9$/L), specific gravity usually <1.020, low protein usually <30 g/L; mesothelial cells and rare heterophils present, mesothelial cells may be reactive, neoplastic cells may be noted; may be associated with vascular leakage from increase capillary hydrostatic pressure or neoplasms. ○ Exudates: Frequently viscous and tend to clot, have increased cellularity ($>5 \times 10^9$/L), high specific gravity >1.020, and high protein >30 g/L; mixed cellularity that is predominately heterophils and macrophages if acute, lymphocytes and plasma cells are present if more chronic; can be characterized as septic or nonseptic; yolk or fat globules may be present in cases of egg yolk peritonitis; may be associated with inflammatory, necrotizing, infectious, or neoplastic disorders. ○ Hemorrhagic effusion: Bloody effusion with measurable hematocrit representing 10–25% of systemic blood PCV, erythrophagocytosis present and no platelets noted; should be differentiated from peripheral blood contamination.

IMAGING

Radiographic findings
• Poor serosal detail of the distended coelomic cavity noted on survey films. • Upper gastrointestinal contrast study may help evaluate if a mass effect is also present in the coelomic cavity. • Cardiomegaly and enlarged hepatic silhouette may be present in cases of CHF.

Ultrasonographic findings
• Coelomic ultrasound is the preferred method of diagnostic imaging for the evaluation of ascites. • Fluid distension enhances visualization of a bird's coelomic

cavity. • Multiple cystic structures of varying size or a single large cystic structure may be noted in the left craniodorsal region of the coelom if cystic ovarian disease is present. • A large heart and large hypoechoic liver are noted in cases of CHF. • Hepatic changes associated with hepatic cirrhosis include small hyperechoic liver with nodular margins.

DIAGNOSTIC PROCEDURES
Celiocentesis is both diagnostic and therapeutic. Removal of coelomic fluid before performing other diagnostic procedures may improve the bird's ability to breathe.

PATHOLOGIC FINDINGS
• Pathologic lesions will vary depending on the cause of the ascites. • Fluid may fill the entire coelomic cavity or may surround a specific liver lobe; fluid may be transudative, exudative, or serosanguineous. • The membranes of the hepatic and peritoneal cavities may be thickened, especially in cases of chronic ascites. • CHF: Cardiomegaly with right-sided, left-sided, or bi-ventricular enlargement; enlarged and congested liver with rounded edges; moist pulmonary parenchyma. • Hepatic cirrhosis: Small fibrotic hepatic parenchyma; may be pale tan, firm, and nodular. • Ovarian neoplasia: Cystic and/or soft tissue mass may be noted in the region of the ovary. • Carcinomatosis: Neoplastic lesions may be disseminated throughout the coelomic cavity. • Polyomavirus: Cardiomegaly, myocardial/epicardial hemorrhage, hepatomegaly with yellow-white foci, GI hemorrhage, splenomegaly, and pale swollen kidneys; large, clear, basophilic or amphophilic intranuclear inclusion bodies on histology. • Mycobacteriosis: Enlarged and mottled pale liver and spleen, thickening of the small intestinal wall with numerous pale miliary nodules on the mucosal surface; on histology, macrophages and giant cells containing acid-fast bacteria noted through multiple organs.

TREATMENT

APPROPRIATE HEALTH CARE
• Measures should be undertaken to improve the patient's ability to breathe. Treatments include diuretics, celiocentesis, and oxygen supplementation. • Centesis of excessive fluid can be repeated as necessary on an outpatient basis. • Control of infection and inflammation needs to be addressed when medically managing cases of egg yolk peritonitis; surgical intervention may be required once the patient is stabilized. • Cystic ovarian disease is rarely treated surgically due to the vascular supply to the ovary; medical management with hormone therapy may provide symptomatic treatment. • Surgical excision of some neoplasms may alleviate symptoms if the neoplasm has not already metastasized.

NURSING CARE
• Measures to improve respiratory function including oxygen supplementation and therapeutic celiocentesis should be performed prior to undertaking procedures that may compromise the bird. • In some cases, administering midazolam may be beneficial to help calm the bird down to avoid hyperventilation. • During centesis of the coelom, the needle must be directed midline to avoid puncturing through an air sac. • Intravenous fluid therapy may be beneficial, especially hetastarch if patient is hypoproteinemic. • Nutritional support is necessary if the patient is not eating.

ACTIVITY
Minimize stress and activity since the patient has limited ability to breathe.

DIET
• Gavage feed if patient is not eating.
• Species susceptible to ISD should be on a specially formulated low iron diet.

CLIENT EDUCATION
Celiocentesis may need to be repeated as needed to provide relief for dyspnea.

SURGICAL CONSIDERATIONS
• Compromised ability to breathe due to compression of air sacs. Intubation with positive pressure ventilation strongly recommended. • Extra care must be taken to prevent damaging the air sacs during the surgical approach to minimize risk of fluid aspiration through the air sacs.

MEDICATIONS

DRUG(S) OF CHOICE
• Broad-spectrum antibiotics should be administered if a septic exudate is noted and/or an infectious process is suspected until culture and sensitivity results of the aspirated fluid are available. Antibiotics often used include amoxicillin/clavulanate (125 mg/kg PO q8–12h) and enrofloxacin (10–20 mg/kg PO q12–24h). • Meloxicam (0.5–2 mg/kg PO, IM q12–24h)—administration of an anti-inflammatory medication is indicated if an inflammatory or painful condition is present. • Furosemide (0.1–2 mg/kg PO, SC, IM, IV q6–24h)—diuretic to remove excess fluid in cases of CHF or liver insufficiency.

For reproductive related disorders
• Leuprolide acetate (1500–3000 μg/kg IM q 2–4weeks)—gonadotropin-releasing hormone (GnRH) agonist; can be administered to birds with ovarian cysts or reproductive neoplasms.
• Deslorelin (4.7 or 9.4 mg SC implant)—gonadotropin-releasing hormone (GnRH) agonist; long-term implant that can be administered to birds (use of this implant in birds is considered off-label use) with ovarian cysts or reproductive neoplasms.

For congestive heart failure
• Benazepril HCl (0.25–0.5 mg/kg PO q24h)—ACE inhibitor used in the management of CHF and hypertension.
• Pimobendan (0.25–0.35 mg/kg PO q12h)—inodilator (has inotropic and vasodilator effects) used in the management of CHF. • Sildenafil citrate (1–3 mg/kg PO q12h)—phosphodiesterase-5 inhibitor used for the treatment of pulmonary hypertension.

For iron storage disease
• Deferoxamine (100 mg/kg SC/IM q24h); iron chelator for birds with ISD.

For hepatic cirrhosis/fibrosis
• Colchicine (0.05–0.1 mg/kg PO q12–24h); anti-inflammatory used in the treatment of cirrhosis of the liver.

CONTRAINDICATIONS
• Furosemide is contraindicated in patients with anuria. • Colchicine is contraindicated in patients with serious renal, GI, or cardiac dysfunction. • Deferoxamine should not be used in patients with severe renal failure.

PRECAUTIONS
Furosemide should be used with caution in patients with hepatic dysfunction or preexisting electrolyte or water imbalances.

POSSIBLE INTERACTIONS
Colchicine may cause additive myelosuppression when used in conjunction with medications with bone marrow depressant effects (i.e., antineoplastics, immunosuppressants, chloramphenicol, amphotericin B).

ALTERNATIVE DRUGS
Hepatoprotective medication such as silymarin may be beneficial if a hepatopathy is suspected.

FOLLOW-UP

PATIENT MONITORING
• Frequency of monitoring will vary depending on how quickly fluid re-accumulates in the coelomic cavity. Daily to weekly weight checks (either at home or at the clinic) evaluate for increases in body weight which may indicate that fluid has refilled the coelomic cavity. • Repeating coelomic ultrasound can be helpful in confirming and evaluating the amount of fluid in the coelomic cavity.

PREVENTION/AVOIDANCE
Birds that are prone to developing ISD should be placed on a low-iron diet.

POSSIBLE COMPLICATIONS
Lateral puncture of the coelomic cavity when performing celiocentesis may result in leakage

ASCITES (CONTINUED)

of fluid directly into the air sac and can result in drowning of the patient.

EXPECTED COURSE AND PROGNOSIS
- Depends on the cause of the ascites; in general prognosis is guarded, especially if the respiratory system is severely compromised.
- Ascites from congestive heart failure can be managed with medical treatment; however, this condition is usually progressive over time.
- Yolk coelomitis, cystic ovarian disease, and ovarian neoplasia have a guarded to grave prognosis. • Hepatic insufficiency from hepatic cirrhosis generally carries a guarded long-term prognosis. • Cases of mycobacteriosis require prolonged treatment; treatment of this disease in avian species is controversial.

MISCELLANEOUS

ASSOCIATED CONDITIONS
- Edema may also be noted if ascites is due to hypoalbuminemia. • Unilateral or bilateral hind limb lameness may be noted if the ascites is due to a neoplasm that is compressing the ischiatic nerve.

AGE-RELATED FACTORS
N/A

ZOONOTIC POTENTIAL
Though no cases of bird to human mycobacteriosis are reported, owners must be advised of the potential zoonotic risk, especially to immunocompromised individuals.

FERTILITY/BREEDING
Birds with ascites associated with reproductive disease may be infertile or produce abnormal eggs.

SYNONYMS
Coelomic effusion
Hydroperitoneum

SEE ALSO
Atherosclerosis
Cardiac disease
Coelomic distention
Cystic ovaries
Egg yolk coelomitis / reproductive coelomitis
Liver disease
Mycobacteriosis
Polyomavirus
Salpingitis / uterine disorders
Tumors
Viral disease

ABBREVIATIONS
CHF—congestive heart failure
ACE—angiotensin-converting enzyme
ISD—iron storage disease
GnRH—gonadotropin-releasing hormone

INTERNET RESOURCES
N/A

Suggested Reading

Antinoff, N. (2011). Reproductive Disorders and Abdominal Distension in the Bird: Medical Management. Proceedings of the International Veterinary Emergency and Critical Care Symposium, September 14–18, Nashville, TN.

Juan-Sallés, C., Soto, S., Garner, M.M., et al. (2011). Congestive heart failure in 6 African grey parrots (Psittacus e erithacus). *Veterinary Pathology*, **48**(3):691–697.

Keller, K.A., Beaufrère, H., Brandão, J., et al. (2013). Long-term management of ovarian neoplasia in two cockatiels (*Nymphicus hollandicus*). *Journal of Avian Medicine and Surgery*, **27**(1):44–52.

Sedacca, C.D., Campbell, T.W., Bright, J.M., et al. (2009). Chronic cor pulmonale secondary to pulmonary atherosclerosis in an African grey parrot. *Journal of the American Veterinary Medical Association*, **234**:1055–1059.

Wheeler, C.L., Webber, R.A. (2002). Localized ascites in a cockatiel (*Nymphicus hollandicus*) with hepatic cirrhosis. *Journal of Avian Medicine and Surgery*, **16**(4):300–305.

Author Sue Chen, DVM, Dipl. ABVP (Avian)

ASPERGILLOSIS

BASICS

DEFINITION
Aspergillosis – caused by members of the fungal genus *Aspergillus* – is a noncontagious, opportunistic infection that can produce disease varying in severity, chronicity, and affected body systems.

PATHOPHYSIOLOGY
- The fungal spores of *Aspergillus* spp. are ubiquitous and widespread within a given environment. • Many birds may "carry" spores within their respiratory tract for prolonged periods and it is not until they become immunocompromised in some way that aspergillosis may manifest as an active infection within an animal.

SYSTEMS AFFECTED
- Respiratory—increased respiratory rate and effort, respiratory distress; mycotic tracheitis, air sacculitis, rhinitis, and sinusitis.
- Behavioral—usually the first system that owners note changes to in early infection; depression and lethargy. • Cardiovascular. • Gastrointestinal—regurgitation, diarrhea, and abnormal droppings. • Hemic/Lymphatic/Immune. • Hepatobiliary—biliverdinuria and hepatomegaly. • Central nervous system—encephalitic and meningoencephalic lesions may occur with disseminated disease. • Ophthalmic—rare, but ocular fungal granulomas have been reported. • Renal/Urologic—polyuria, polydipsia, and renomegaly. • Skin/Exocrine.

GENETICS
N/A

INCIDENCE/PREVALENCE
- Most frequent cause of respiratory disease and most commonly diagnosed fungal disease in pet birds. • Most common, nontraumatically-induced medical problem in free-ranging birds of prey.

GEOGRAPHIC DISTRIBUTION
No specific geographic distribution, but a combination of humid and dry dusty environmental conditions increases disease prevalence.

SIGNALMENT
Specific species known to be more at risk for the development of aspergillosis include African grey parrots (*Psittacus erithacus*), Pionus parrots (*Pionus* spp.), Amazon parrots (*Amazona* spp.), goshawks (*Accipiter gentilis*), immature red-tailed hawks (*Buteo jamaicensis*), golden eagles (*Aquila chrysaetos*), snowy owls (*Nyctea scandiaca*), gyrfalcons (*Falco rusticolus*), rough-legged hawks (*Buteo lagopus*), swans (*Cygnus* spp.), and penguins (*Spheniscus* spp.).

SIGNS

General Comments
Unfortunately, many signs are nonspecific or have an exhaustive list of other differential diagnoses that may need to be explored through diagnostics.

Historical Findings
- Use of particulate bedding (corn cob or walnut-shell bedding, hay, wood shavings, etc.) within enclosure/mew. • Inappropriate sanitation, ventilation. • Change in voice/call, with or without breathing abnormalities. • Depression/weakness—reluctance to fly or perch, drooped wings. • Inappetance.

Physical Examination Findings
- Lethargy. • Increased respiratory rate and/or effort—tail bobbing, dyspnea, tachypnea, cyanosis, vocalization when breathing, harsh breathing sounds on auscultation. • Weight loss. • Polyuria/polydipsia.

CAUSES
- Husbandry practices—poor hygiene and particulate bedding in cages/enclosures increase exposure to fungal spores.
- Anatomic (congenital)—any alterations to respiratory tract may predispose a bird to disease. • Nutritional—poor diet, especially Vitamin A deficiencies. • Infectious—primary bacterial or viral infections. • Traumatic and Toxic—previous exposure to respiratory irritant (e.g., PTFE) or trauma of the nasal passageways can predispose a bird to aspergillosis.

RISK FACTORS
- Environmental factors: Poor husbandry/ventilation, use of particulate bedding. • Nutritional disorders/deficiencies (e.g., hypovitaminosis A). • Stress – in both captivity in birds of prey and Psittacine spp. • Impaired immune function. • Prolonged antibiotic use and steroids. • Preexisting disease condition.

DIAGNOSIS

DIFFERENTIAL DIAGNOSIS
- Avian chlamydiosis. • Environmental toxicosis or chronic exposure to respiratory irritant. • Allergic pneumonitis or noninfectious pneumonitis. • Fowl cholera. • Mycobacteriosis. • Mycoplasmosis. • Air sacculitis. • Foreign body obstruction within respiratory tract. • Neoplasia.

CBC/BIOCHEMISTRY/URINALYSIS
- In both chronic and acute disease, aspergillosis commonly causes profound heterophilic leukocytosis (25,000 cells/µl to 100,000 cells/µl). A left shift may be present, and quite often white blood cells are very reactive. • Chronic inflammation elicited by *Aspergillus* infection may also produce a non-regenerative anemia. • Serum biochemistry analysis may reveal elevations in aspartate aminotransferase (AST), lactate dehydrogenase (LDH), and serum bile acids if liver involvement is present. Hypoalbuminemia and hypergammaglobulinemia can also be characteristic of the disease.

OTHER LABORATORY TESTS
- Serology: often of limited value, especially in psittacine birds, due to ubiquitous nature of *Aspergillus* spp. ○ A positive titer of an indirect ELISA assay may suggest active infection, long-term exposure, or previous infection antibody. ○ A negative titer in an infected bird can either be a result of a lack of reactivity between the test conjugate and patient immunoglobulins, or by a lack of patient humoral response (i.e., immunosuppression). • Plasma protein electrophoresis: may help support the diagnosis of aspergillosis in psittacine birds but is not specific for the disease since changes are usually only indicative of a non-specific inflammatory response. • Galactomannan assay: Galactomannan is an *Aspergillus* species antigen. The assay appears to be useful for detecting *Aspergillus* antibody when coupled with plasma protein electrophoresis.

IMAGING

Radiography
- May reveal the distribution and severity of fungal lesions in the lungs and air sacs, but of limited value until advanced stages of disease.
- One of the more common radiologic findings is a bronchopneumonia with marked parabronchial patterns. Asymmetry, hyperinflation or consolidation of the air sacs as well as soft-tissue densities may also be observed. • Even after a successful resolution of clinical *Aspergillus* infection, the respiratory tract may remain thickened and irregular.

DIAGNOSTIC PROCEDURES

Endoscopy (rigid or flexible)
- Best antemortem diagnostic tool to confirm diagnosis of aspergillosis. • An effective way for visualization of the trachea, syrinx, and lower respiratory tract (air sacs and lungs) and the presence of fungal plaques. Endoscopy also allows the user to obtain samples to submit for culture and histopathologic analysis.

PATHOLOGIC FINDINGS
- *Aspergillus* air sacculitis is the most common form of the disease, sometimes extending into the lungs. • Thickened air sac membranes infiltrated with large numbers of inflammatory cells and germinating conidia are usually observed. Germinating conidia may also be observed within macrophages.

ASPERGILLOSIS (CONTINUED)

TREATMENT

APPROPRIATE HEALTH CARE
The patient's overall stability will determine in-hospital management versus outpatient care. The main goals of treatment are to stabilize, reduce stress, and improve patient's physical condition until complete resolution of infection.

NURSING CARE
• Maintain environmental temperature to 85–90°F (29–32°C) whilst providing a humidified environment. • Supplemental oxygen support may be indicated in some advanced stages of disease. • Fluid therapy: Warmed crystalloid fluids SC, IV, IO (50–150 mL/kg/day maintenance plus dehydration deficit if needed) at a rate of 10–25 mL/kg over a five-minute period or at a continuous rate of 100 mL/kg/q 24h. • Nutritional support may be required if patient remains anorexic throughout treatment.

ACTIVITY
Restricted activity should be maintained during this time to allow for complete resolution.

DIET
Appropriate caloric intake should be closely monitored in the patient to maintain the overall health during treatment.

CLIENT EDUCATION
• Aspergillosis is generally a preventable disease when appropriate husbandry and dietary needs are met • Definitive diagnosis and successful treatment requires patience and persistence. Prolonged antifungal therapy for periods up to 4–6 months is often necessary and, despite various treatment approaches, failure to respond is not uncommon.

SURGICAL CONSIDERATIONS
• In an emergency situation, where tracheal and syringeal granulomas create life-threatening respiratory obstructions, air sac cannulas may be placed within the clavicular, caudal thoracic, or abdominal air sacs. • Large granulomatous lesions that occlude air flow within the respiratory tract or are resistant to medical therapy (due to poor drug penetration) may require surgical/endosurgical debulking or, when possible, complete resection. • Infections involving the sinus cavities may require trephination of the frontal sinuses permitting direct access for topical treatments and debridement as needed.

MEDICATIONS

DRUG(S) OF CHOICE
• Amphotericin B—can be administered intratracheally, intravenously, in sinus flushes, and through nebulization: ○ Systemic therapy: 1.5 mg/kg IV q8h for 3–5 days. ○ Topical therapy: intratracheal use at 1 mg/kg q8–12h. ○ Nebulization therapy: 1 mg/mL sterile water for 15 minutes q12h. • Itraconazole—effective when administered with nebulized clotrimazole and/or amphotericin B (intravenous or nebulized): ○ 5–10 mg/kg PO q12h for five days, then daily until resolution. ○ Lower dose (2.5–5 mg/kg PO q24h) is recommended for African grey parrots. • Fluconazole—useful for treating mycoses of the eye and CNS; less effective against aspergillosis than itraconazole systemically: ○ 5–15 mg/kg PO q12h. • Clotrimazole—used as topical/nebulization therapy in conjunction with other antifungals: ○ Used as a 1% aqueous solution for 30 minutes every 24 hours. • Terbinafine hydrochloride—excellent ability to penetrate mycotic granulomas; usually used in combination with itraconazole; ○ Systemic therapy: 10–15 mg/kg PO q12–24h. ○ Nebulization therapy: 1 mg/mL (500 mg terbinafine plus 1 mL acetylcysteine plus 500 mL of distilled water).

CONTRAINDICATIONS
N/A

PRECAUTIONS
• Amphotericin B is potentially nephrotoxic and should be used only for a short duration. When nebulizing Amphotericin B, a well-ventilated area should be used for the safety of staff members. • Amphotericin B can be irritating when applied topically and should be diluted in water (saline inactivates amphotericin B) to reduce iatrogenic inflammation. • African grey parrots are known to potentially be sensitive to itraconazole; sometimes exhibiting anorexia and depression. If used with this species, lower dosing and close monitoring of the patient is required. • Itraconazole should be used with extreme caution in patients with liver impairment. • Administration of fluconazole at 10 mg/kg q12 h has been known to cause death in budgerigars.

POSSIBLE INTERACTIONS
Itraconazole: coadministration with other drugs primarily metabolized by cytochrome enzyme system may lead to increased plasma concentrations of itraconazole and could increase or prolong both therapeutic and adverse effects.

ALTERNATIVE DRUGS
Voriconazole—further studies need to be completed but proving to be a promising substitution to itraconazole—2–18 mg/kg PO q12h.

FOLLOW-UP

PATIENT MONITORING
• Repeated survey radiographs and endoscopy may be used to assess treatment response. • Serial biochemistry profiles to monitor patient's condition during administration of antifungal medications are important. • Treatment is usually extended one month after a normal complete blood count is achieved and resolution of clinical signs are observed.

PREVENTION/AVOIDANCE
Proper diet and good husbandry practices can drastically reduce the incidence of aspergillosis.

POSSIBLE COMPLICATIONS
Long-term respiratory impairment or sensitivity may be observed after treatment, dependent upon severity of initial clinical disease. These changes may be noted on radiographs for the life of the bird.

EXPECTED COURSE AND PROGNOSIS
• Uncomplicated cases have a good prognosis; however, the prognosis is generally dependent upon the immune status, species being treated, and chronicity of the illness in the patient. • Infections affecting the nasal passageways may cause anatomical deformities to nares.

MISCELLANEOUS

ASSOCIATED CONDITIONS
N/A

AGE-RELATED FACTORS
N/A

ASPERGILLOSIS (CONTINUED)

ZOONOTIC POTENTIAL
Aspergillosis has been reported in humans but only acquired from the environment, not from contact with infected birds.

FERTILITY/BREEDING
N/A

SYNONYMS
Aspergillus, aspergillosis

SEE ALSO
Aspiration
Mycobacteriosis
Oil exposure
Oral plaques
Pneumonia
Poxvirus
Respiratory distress

Rhinitis and sinusitis
Sick-bird syndrome
Tracheal/syringeal diseases
Trichomoniasis

ABBREVIATIONS
PTFE—polytetrafluoroethylene

INTERNET RESOURCES
N/A

Suggested Reading
Carpenter, J.W. (2013). *Exotic Animal Formulary*, 4th edn. St. Louis, MO: Elsevier Saunders.
Cray, C., Reavill, D., Romagnano, A., *et al.* (2009). Galactomannan assay and plasma protein electrophoresis findings in psittacine birds with aspergillosis. *Journal of Avian Medicine and Surgery*, **23**(2):125–135.
Flammer, K. (2003). Antifungal therapy in avian medicine. *Proceedings of the Western Veterinary Conference*. Accessed via: Veterinary Information Network.
Guzman, D.S.-M., Flammer, K., Papich, M.G., *et al.* (2010). Pharmacokinetics of voriconazole after oral administration of single and multiple doses in Hispaniolan Amazon parrots (*Amazona ventralis*). *American Journal of Veterinary Research*, **71**(4):460–467.
Harrison, G.J., Lightfoot, T.L. (eds) (2006). *Clinical Avian Medicine*. Palm Beach, FL, Spix Publishing Inc.

Author Gregory J. Costanzo, DVM

 **Client Education Handout available online**

ASPIRATION

BASICS

DEFINITION
Injury to the lungs causes by inhalation of fluid, food material, barium or vomitus.

PATHOPHYSIOLOGY
Inhalation of foreign material into the larynx, air sacs or lungs that triggers inflammatory response in the lower respiratory tract, physical obstruction of the airways, secondary infection and damage to the respiratory epithelium.

SYSTEMS AFFECTED
• Respiratory. • Cardiac-secondary to respiratory changes. • Upper gastrointestinal tract—as a primary cause of aspiration.

GENETICS
N/A

INCIDENCE/PREVALENCE
Not reported.

GEOGRAPHIC DISTRIBUTION
N/A

SIGNALMENT
• Leading cause of respiratory disease in handfed psittacines that are near weaning. • Common occurrence in ducks and geese in the foie gras industry. • Aspiration occurs in budgerigars on a seed diet or those fed a diet low in iodine. • Proventricular dilatation disease (PDD) has been found in over 50 species of birds and can be associated with gastrointestinal hypomotility leading to secondary aspiration.

SIGNS
General Comments
The more chronic form of aspiration pneumonia is more likely to result in weight loss.

Historical Findings
• Death shortly after feeding. • Labored breathing. • Cough-like sound shortly after feeding in weanlings. • Poor weight gain of chicks. • Chronic weight loss. • Recent barium administration, oral medication administration or ingluvial gavage feeding.

Physical Examination Findings
• Dyspnea • Cyanosis • Loud inspiratory sound • Anorexia • Depression • Lethargy • Tail bobbing • Open-mouthed breathing • Intermittent dyspnea • Cough after feeding • Regurgitation/vomiting • Vomitus crusted around nares or face • Acute death • Thin body condition • Crop distention • Mass effect in the neck • Passage of undigested food in feces (PDD).

CAUSES
• Hand feeding due to resisted attempts at feeding or by inexperienced individuals. • Overflow of barium, supplemental feeding or oral medications at administration. • Any gastrointestinal disease which causes vomiting, regurgitation or an esophageal motility disorder (heavy metal toxicosis, foreign body ingestion, infectious gastrointestinal disease, ingestion of oral irritants or a rancid diet, pancreatitis, renal failure, hepatopathy, septicemia, PDD). • Courtship behavior. • Periglottal disease. • Intraluminal or extraluminal neoplasia of the esophagus or neck. • Goiter in budgerigars secondary to a diet low in iodine. • Proventricular dilatation disease. • Aspiration of a seed or foreign body. • Oral administration of medications in birds with a motility disorder or that is vomiting or regurgitating.

RISK FACTORS
• Gastrointestinal disease causing vomiting or regurgitation. • Hand fed psittacines near weaning. • Budgerigars fed a low iodine diet.

DIAGNOSIS

DIFFERENTIAL DIAGNOSIS
• Infectious respiratory disease: fungal (aspergillosis), bacterial, viral, parasitic. • Neoplasia. • Anemia. • Respiratory irritants (PTFE, smoke, aerosols, cigarette smoke, candles). • Hypersensitivity syndrome of blue and gold macaws. • Tachypnea from pain or metabolic disease. • Ascites. • Normal growl of African grey parrots. • Normal hiss of Cockatoo species. • Normal moist respiratory sounds in Pionus species. • Mimicry of human cough or sneeze.

CBC/BIOCHEMISTRY
Hemogram—leukocytosis with heterophilia and monocytosis may be seen.

OTHER LABORATORY TESTS
• Blood culture may be positive for a secondary bacterial infection. • Tracheal wash, air sac wash, bronchoscopy, endoscopy and tissue biopsy can also be used to identify the cause and obtain cultures. However, these can only be performed on stable patients.

IMAGING
Radiography
• Appropriately positioned and exposed ventrodorsal and lateral radiographic projections recommended to evaluate for abnormalities of the lungs and air sacs.

Thoracic radiography
• May identify a mass effect in the neck, crop distention, a foreign body or show increase radiodensity or accentuation of the honeycomb pattern of the lungs.

Coelomic radiography
• Increased diffuse to focal radiodensity of the abdominal air sacs. • Loss of definition of the air sacs. • Hyperinflation in the air sacs. • Dilated proventriculus or generalized ileus in birds with PDD or other motility disorders.

Additional imaging
• CT scan will proved more detailed information especial lesions that are too small to identify radiographically.

PATHOLOGIC FINDINGS
• Lesions are more likely to form in the abdominal air sacs due to gravity. • With aspiration pneumonia focally extensive areas of the lungs are generally edematous, firm, and congested. There is loss of the sharp honeycomb pattern and irregular areas appear opaque red to brown. On cut surface it may be possible to see the aspirated material within larger airways. • On histology of the lung it is generally severely congested especially in areas supporting the intraluminal aspirated materials. Plant material and other foreign material will be admixed with hemorrhage and variable numbers of bacteria, cell debris, and fibrin filling the lumen of many tertiary and even the secondary bronchi. Some of these foci can be associated with mild acute heterophilic inflammation.

TREATMENT

APPROPRIATE HEALTH CARE
• Inpatient vs. outpatient care depends on whether it is acute or chronic. The acute presentation or those with multisystemic signs (anorexia, lethargy, weight loss) are more likely to require hospitalization. • For large airway obstruction, immediate air sac cannulation. • Nonsurgical resolution of tracheal foreign bodies in cockatiels or small birds can be performed (after air sac cannulation). Transilluminate the trachea and identify the foreign body. Position the patient with the head down and the tail up. Place a 27 or 25-guage needle between the tracheal rings just proximal to the seed. Either use gentle force to dislodge the seed, inject air forcefully to dislodge the seed or forcefully instill 0.1 cc of sterile saline into the trachea to move the seed rostrally and out the glottis. • More stable patients can be given a bronchodilator and anti-anxiety analgesic before being placed in oxygen.

NURSING CARE
• Fluid therapy to maintain or correct dehydration. • Oxygen therapy. Place in 78–85% O_2 concentration at a flow rate of 5 L/min. • Nebulization with 0.9% saline, bronchodilators or antibiotics. • Supplemental heat at 85–90°F.

ACTIVITY
Activity should be restricted until symptoms resolve.

DIET
• If the patient is vomiting, regurgitating or has an esophageal motility disorder feed multiple small meals of easily digestible diet.

ASPIRATION (CONTINUED)

• Complete the weaning process in psittacine chicks. • If anorexic gavage feed. • Warn owners that birds in severe respiratory distress or barium aspiration have a poor prognosis.

CLIENT EDUCATION
• Discourage hand feeding in novice aviculturists. • Discourage mating behavior. • Iodine deficiency: change to an appropriate diet with the proper amount of iodine.

SURGICAL CONSIDERATIONS
Be cautious with anesthesia due to respiratory compromise and the difficulty of measuring oxygen saturation in avian species.

MEDICATIONS
DRUG(S) OF CHOICE
• Broad spectrum antibiotics: Fluoroquinolone, TMS, Clavamox. • Bronchodilator: Terbutaline 0.01 mg/kg IM q 6–8h or 0.1 mg/kg PO in larger birds. Followed by nebulization at 0.01 mg/kg in 9 cc of saline. • Anti-anxiety analgesic: Butorphanol 1–2 mg/kg IM q 2-4h. • PDD treatment (see PDD).

CONTRAINDICATIONS
• Steroids. • Careful use of oral medications if the patient has vomiting, regurgitation or motility disorder.

ALTERNATIVE DRUGS
• Other bronchodilators include theophylline or aminophylline. These may be less effective at bronchodilation in birds, but clinical improvement has been noted with their use. • Theophylline has been used at a dose of 2 mg/kg PO q12h. • Aminophylline are 10 mg/kg IV q3h, 4 mg/kg IM q12h and 5 mg/kg PO q12h. • Albuterol nebulization or orally 0.5 mg/kg PO q1h.

FOLLOW-UP
PATIENT MONITORING
• CBC: Resolution of leukocytosis.
• Radiologic resolution of signs.

PREVENTION/AVOIDANCE
• Avoid hand feeding chicks by inexperienced individuals. • Feed budgerigars an appropriate diet. • Discourage mating behavior.

POSSIBLE COMPLICATIONS
• Granuloma formation in lungs or air sacs.
• Poor weight gain of chicks.

EXPECTED COURSE AND PROGNOSIS
• Depends on the severity of the disease. Prognosis is more guarded in young or immune compromised individual. • PDD carries a grave prognosis.

MISCELLANEOUS
ASSOCIATED CONDITIONS
N/A
AGE-RELATED FACTORS
Condition more common in young (weaning) birds.
ZOONOTIC POTENTIAL
N/A
FERTILITY/BREEDING
N/A
SYNONYMS
Aspiration pneumonia
SEE ALSO
Aspergillosis
Crop stasis/ileus
Gastrointestinal foreign body
Heavy metal toxicosis
Macrorhabdus ornithogaster
Pneumonia
Proventricular Dilatation Disease (PDD)
Regurgitation/vomiting
Sick-bird syndrome
Thyroid diseases
Undigested food in droppings

ABBREVIATIONS
PTFE—polytetrafluoroethylene
PDD—proventricular dilatation disease
SC—subcutaneous
IV—intravenous
IO—intraosseous
IM—intramuscular
O_2—oxygen

INTERNET RESOURCES
N/A

Suggested Reading
AAV (2005). Avicultural and Pediatric Medicine. *Proceedings of the Association of Avian Veterinarians.*
Clubb, S.L. (1997). Psittacine pediatric husbandry and medicine. In: Altman, R.B., Clubb, S.L., Dorrestein, G.M., Quesenberry, K. (eds), *Avian Medicine and Surgery.* Philadelphia, PA: WB Saunders, pp. 73–95.
Flammer, K., Clubb, S.L. (1994). *Neonatology. Avian Medicine: Principles and Applications.* West Palm Beach, FL: Zoological Education Network, pp. 805–838.
Orosz, S.E., Lichtenberger, M. (2011). Avian respiratory distress: Etiology, diagnosis and treatment. *Vet Clinics of North America Exotic Animal Practice,* **14**(2):241–255.
Phalen, D.N., (2000). Respiratory medicine in cage and aviary birds. *Vet Clinics of North America Exotic Animal Practice,* **3**(2):423–452.

Author Erika Cervasio, DVM, DABVP (Avian)
Acknowledgements Stephen Dyer, DVM, DABVP (Avian)
Drury Reavill, DVM DABVP (Avian), DACVP
Barbara Oglesbee, DVM

Atherosclerosis

BASICS

DEFINITION
Atherosclerosis is an inflammatory and degenerative disease of the arterial wall characterized by the disorganization of the arterial wall due to the accumulation of fat, cholesterol, calcium, cellular debris, and inflammatory cells and potentially leading to complications such as stenosis, ischemia, thrombosis, hemorrhage, and aneurysm.

Atherosclerotic diseases include a spectrum of disease processes caused by atherosclerotic lesions such as coronary arterial disease and peripheral arterial disease.

PATHOPHYSIOLOGY
• Response-to-injury hypothesis: Oxidative damage of the vascular endothelium and endothelial dysfunction promoted by risk factors set the stage for atherogenesis.
• Subendothelial retention of oxidized lipoproteins and other compounds, proliferation and accumulation of macrophages and other inflammatory cells, proliferation of vascular smooth muscle cells and formation of fibrous tissue. • Histologic lesions progressed from subendothelial and medial accumulation of lipid vacuoles and foam cells to atheromatous plaque formation further expanding in a large lipidonecrotic core that may be covered by a fibrous cap (fibroatheromatous lesion). • Progressive flow-limiting stenosis of the arterial lumen (advanced lesion types) is the main mechanism of atherosclerotic diseases in birds. It leads to chronic hypoperfusion and chronic ischemia of the supplied anatomical areas such as the heart, neurological system, and locomotory muscles. • The increase vascular resistance at the great vessels may lead to increase afterload and cardiac failure. • The chronic cardiac ischemia may lead to cardiac arrhythmias. • Atherothrombosis (rupture of the fibrous cap and exposure of the thrombotic material to the circulation) and emboli may also occur but are rare in birds in comparison to mammals, particularly humans. This seems to be associated with the decreased ability of avian thrombocytes to form shear-resistant three-dimensional aggregates compared to mammalian platelets.
• Ruptured vascular aneurysm may be due to atherosclerotic lesions. • In most birds the brachiocephalic arteries, ascending aorta, and carotid arteries are most frequently affected but lesions in the abdominal aorta, pulmonary arteries (*cor pulmonale*), coronary arteries, and peripheral arteries are not uncommon. Distribution of lesions may be different between avian species.

SYSTEMS AFFECTED
• Cardiovascular: ◦ Atheromatous, fibroatheromatous, and atherothrombotic lesions of the arteries. ◦ Chronic or acute ischemia of the myocardium potentially leading to cardiac arrhythmias, ischemic cardiomyopathy, coronary arterial disease, increased afterload. ◦ Peripheral arterial disease. • Endocrine/Metabolic—associated with multiple endocrine factors (metabolic syndrome) and dyslipidemia.
• Hepatobiliary—frequently associated with liver lipid disorders. • Musculoskeletal—may be associated with ischemic muscle lesions and resulting muscular signs. • Nervous—may cause central neurological signs due to central ischemia, hypoperfusion, or strokes.
• Respiratory—respiratory signs are common

GENETICS
A genetic basis has been demonstrated in several research breeds of birds:
• White Carneau pigeon: autosomal recessive trait. This pigeon breed has constitutive differences in the arterial wall compared to racing pigeons.
• Susceptible-to-experimental-atherosclerosis (SEA): Japanese quail. • Restricted ovulator (RO) chicken: mutation in the VLDL receptor gene.

INCIDENCE/PREVALENCE
• Only the postmortem prevalence has been determined. Antemortem prevalence of atherosclerotic diseases or subclinical atherosclerosis is not well known.
• Psittaciformes: common, raw prevalence around 7% but depends on many risk factors (see below), prevalence of atherothrombotic disease is low, approximately 0.1% (2% of advanced atherosclerotic lesions). In susceptible psittacine species, a 50% prevalence of advanced lesions is observed at 30–40 years. • Reported in most avian groups and common in a variety of species of falconiformes, accipitriformes, coraciiformes, piciformes, galliformes, columbiformes, and sphenisciformes. Turkeys have high prevalence of atherosclerosis in the wild.

GEOGRAPHIC DISTRIBUTION
N/A

SIGNALMENT
• **Species:** Most species are susceptible but the following have been found to be particularly predisposed: cockatiels, quaker parrots, Amazon parrots, African grey parrots, kite species, turkeys, pigeons, and falcons (gyrfalcons and insectivorous falcons).
• **Mean age and range:** Atherosclerosis has been reported in all ages but is more common in older birds, particularly in susceptible psittacine birds older than 20 years.
• **Predominant Sex:** Female psittaciformes have a higher risk than males, the prevalence of which lags behind that of females by an average of 4 years. In poultry, males seem more commonly affected than females.

SIGNS

General Comments:
• Atherosclerotic lesions are generally silent until either flow-limiting stenosis ensues (common and depends on the degree of narrowing) or lesion complications (rare) such as plaque rupture and thrombosis occur. The impact of subclinical atherosclerosis is unknown in birds and is likely a comorbid lesion to a variety of metabolic and endocrine disorders. • Clinical signs partly depend on the affected arteries and the types of atherosclerotic lesions such as large fibroatheromatous, thrombotic plaques or aneurysms.

Historical Findings:
• Sudden death. • Nonspecific signs: lethargy, anorexia, weight loss, decreased activity.
• Muscular signs: permanent or intermittent leg weakness (intermittent claudication), falling off perches, ataxia, gangrene of the legs. • Neurologic signs: obtundation, altered consciousness, seizures, ataxia, central neurological signs. • Cardiac signs: abdominal distention, exercise intolerance, dyspnea, syncopes. • Respiratory signs: dyspnea.

Physical Examination Findings:
• Generally no specific signs are detected on the physical examination in most instances but this depends on the type of atherosclerotic disease and the location of the lesions. • The bird may be overconditioned or thin. • Signs consistent with impaired cardiovascular function may be noticed such as abdominal distention, cardiac murmur and arrhythmias upon cardiac auscultation, cyanosis and dyspnea upon restraint. • Abnormalities during a neurological examination may be noted and are usually related to central lesions such as cranial nerves deficits, altered consciousness, and postural abnormalities.
• Intermittent claudication-like signs have been seen in Amazon parrots and other birds.

CAUSES
Risk factors associated with captive lifestyle may promote the occurrence and severity of atherosclerotic lesions.
See pathophysiology and risk factors section for further details.

RISK FACTORS
• Few studies have adequately investigated the epidemiology and risk factors of avian atherosclerosis and most information is found in psittaciformes. Findings in psittacine birds may not translate well to other avian groups, especially those with markedly different biology such as carnivorous birds. Likewise, the large body of human scientific information may not always be relevant to avian atherosclerosis, but some risk factors are anticipated to be similar when considering similarities in lesion progression, structure, and pathogenesis. In addition, risk factors are usually related to clinical atherosclerotic disease and not necessarily to subclinical

ATHEROSCLEROSIS (CONTINUED)

atherosclerosis. Since atherosclerotic lesions are challenging to diagnose, atherosclerotic disease is a usually rule out diagnosis suggested by supportive epidemiology and risk factors. • Epidemiologic risk factors ○ Age. ○ Psittaciformes: Female gender. This may be associated with vitellogenesis, high plasma lipid levels, and cholesterol transportation to the egg in reproductively active females. ○ Species: Quaker parrots, cockatiels, African grey parrots, Amazon parrots, gyrfalcons, lesser kestrel, black kites, Brahminy kites, pigeons, turkeys. Species predisposition may only reflect inadequate captive diet. In psittaciformes, susceptible species were found to have higher cholesterol than less susceptible species. Atherosclerosis has been reported in virtually all avian groups. • Lifestyle risk factors ○ A captive lifestyle characterized with *ad libitum* provision of food, unnatural diet, restricted exercise, increased stress may promote atherosclerosis. In some species, wild birds show a lower prevalence than their captive counterparts. Second-hand smoking in a household may be a risk factor in companion birds. ○ Diet ○ Psittaciformes: A high fat and nutrient-deficient diet such as all-seed diet may promote atherogenesis. Saturated fat, in particular from animal-based diet, may also be deleterious as most psittacine birds have evolved to eat seeds, fruits, and flower products. A lack of polyunsaturated fatty acids (omega 3 and 6) has also been implicated. ○ Birds of prey: The feeding of day-old chicks may promote atherosclerosis in susceptible species. Feeding unnatural prey items such as day-old chicks or mice to insectivorous species or limited fish to fish-eating species may also be atherogenic. • Physiopathological risk factors ○ Elevated blood pressure: Indirect blood pressure measurement is not reliable in most small to medium sized parrots. Turkeys have extremely high blood pressure. ○ Obesity. ○ Reproductive activity and diseases in female birds. ○ *Chlamydia psittaci* infection: Controversial and likely not a major risk factor. ○ Marek's disease (gallid herpesvirus 2 and 3) in chickens. • Dyslipidemic risk factors ○ In humans, dyslipidemic risk factors are the most important by far and include raised total, LDL, and non-LDL cholesterol, and low-HDL cholesterol. Other dyslipidemic risk factors of lesser importance include raised triglycerides, lipoprotein(a), apolipoprotein B, "small, dense" LDL, increased LDL particle number, and postprandial lipemia. ○ In birds, dyslipidemic risk factors and their importance have not been well characterized. Total cholesterol level has been found to be a risk factor in multiple species and predisposed species tend to have higher blood cholesterol levels. Lipoprotein risk factors are unknown and are suspected to differ from humans owing to different lipoprotein metabolism in birds. ○ Quaker parrots seem to be extremely susceptible to dyslipidemia in captivity and have higher cholesterol levels than most other psittacine species. ○ Female birds may exhibit physiological increase in total cholesterol, tryglycerides, VLDL, total calcium, total proteins during reproductive activity and vittelogenesis. While physiological, maladaptive and unregulated reproductive activity is common in captive female psittacine birds, which may be atherogenic.

DIAGNOSIS

DIFFERENTIAL DIAGNOSIS
Any cardiovascular or neurological disease may be a differential diagnosis. Since the diagnosis of atherosclerotic diseases is rarely achieved antemortem, all differential diagnosis must be ruled out first. • Cardiac disease: Dilated cardiomyopathy, valvular insufficiency, myocarditis, arrhythmias, other causes of congestive heart failure. • Hind limb paresis: Trauma, spinal disease, muscular disease, egg binding, viral neuritis.

CBC/BIOCHEMISTRY/URINALYSIS
• Hematologic and biochemistry tests are only intended to diagnose and assess the associated risk factors or consequences of atherosclerotic lesions but have limited value for atherosclerosis diagnosis itself. Only demonstration of the lesions is diagnostic. • Dyslipidemia: Elevated total cholesterol, elevated triglycerides, variable lipoprotein abnormalities. Needs to be differentiated from physiological increase in blood lipids in female birds. • Liver enzymes and bile acids: May be elevated in case of concurrent hepatic lipidosis, bile acids may be mildly elevated in hepatic congestion consecutive to congestive heart failure. • Creatinine kinase: May be elevated in repeated falling, ischemic myositis, or seizures.

OTHER LABORATORY TESTS
Lipoprotein profiles may be useful for diagnosis or management of the associated dyslipidemia, if any. In standard laboratory analyzer, LDL reagent is typically not available and must be calculated using the Friedewald formula: LDL=Total cholesterol – HDL – Triglycerides/x with x = 5 for American units and x = 2.18 for SI units. This formula is not reliable in case of lipemia. Inflammatory (acute phase proteins) and cardiac (troponins) markers have not been investigated in relation to avian atherosclerosis but may be useful.

IMAGING
• Diagnostic imaging is presently the only mean to diagnose atherosclerotic lesions, all laboratory tests focusing on dyslipidemic and inflammatory risk factors. Antemortem confirmed diagnosis is rarely achieved. • Radiographic findings: ○ Calcification of the large arteries at the base of the heart is very specific for advanced atherosclerotic lesions, susceptible to cause clinical signs. However, it is seldom detected on radiographs. ○ Enlargement and increased radiodensity of the large vessels at the base of the heart may be noticed but is highly subjective. The pulmonary veins, which are not susceptible to atherosclerosis, should not be confused with other vessels on the lateral view. ○ Abnormalities associated with congestive heart failure: pulmonary edema (uncommon), ascites, hepatomegaly, reduced air sac volume. • Angiographic findings: ○ Can be performed using radiography, fluoroscopy, and CT (a CT angiography protocol has been published in Amazon parrots). ○ Arterial luminal narrowing. ○ Aneurysmal dilation. ○ Cardiomegaly. • CT findings: ○ Calcification of the large arteries, more sensitive than radiographs at detecting calcification. ○ Cardiomegaly and other signs of congestive heart failure. • MRI findings: ○ Indicated in case of brain lesion, especially in the diagnosis of ischemic and hemorrhagic strokes. • Echocardiographic findings: ○ Hyperechogenicity at the base of the ascending aorta may indicate mineralization. Note that there are normal cartilages that can be mineralized at the base of the aorta and pulmonary arteries in birds. ○ Evidence of cardiac dysfunction and congestive heart failure—valvular insufficiency, decreased ventricular fractional shortening, chamber enlargement, pericardial effusion.

DIAGNOSTIC PROCEDURES
• Rigid endoscopy: Atherosclerotic lesions may be visualized on various arteries during a coelioscopic examination. The interclavicular approach allows the assessment of the brachiocephalic arteries and ascending aorta, but fat accumulation may be present in overweight birds at this location. • Electrocardiography: Arrhythmias consecutive to chronic ischemia and secondary cardiac effects of atherosclerotic lesions may be detected but have not been well characterized.

PATHOLOGIC FINDINGS
• Gross lesions: Thickened and hardened arterial walls, which may have a yellowish discoloration. This may not be obvious in small birds. Lesions occur most commonly on the brachiocephalic arteries, ascending aorta, carotid arteries but are also frequently found on the pulmonary arteries and intramyocardial coronary arteries. The thoracic and abdominal aorta, and other peripheral arteries may also be affected. Vascular aneurysm may occur and rupture leaving blood in the coelomic cavity. • Histopathology: ○ A classification system has been proposed for psittacine atherosclerotic lesions with seven lesion types: ○ Type I and II: minimal changes

characterized by isolated foam cells or multiple foam cell layers. ○ Type III: preatheromatous lesion with isolated pools of extracellular lipid forming confluent areas with minimal disruption of the arterial architecture. ○ Type IV: atheromatous lesion with the presence of a large lipid core with significant disruption of the arterial architecture, narrowing of the lumen, and calcification. ○ Type V: fibroatheroma characterized by the formation of a fibrous cap over the lipid core. ○ Type VI: complicated lesion by hematoma, fissure, or thrombosis. ○ Type VII: calcific lesion, osseous metaplasia, large calcium plaques but with minimal lipid deposition. • Nonvascular lesions have also been found in association with atherosclerosis: ○ Myocardial fibrosis, myocardial hypertrophy, myocardial infarction. ○ Hepatic diseases. ○ Pulmonary lesions.

TREATMENT

APPROPRIATE HEALTH CARE
• When clinical signs of atherosclerotic diseases are evident, the prognosis is usually guarded and treatment of limited efficacy. • Overall treatment is divided into the reduction of risk factors and the treatment of cardiovascular and ischemic consequences. • Drastic reduction of risk factors is advised and has shown to reduce vascular atherosclerotic burden in mammalian species.

NURSING CARE
• Nursing care depends on the clinical presentation and the reader is referred to corresponding sections on cardiac and neurological diseases. • Supportive care should be provided such as fluid replacement therapy, nutritional support, heat support, and oxygen therapy if dyspnea is present. Ascitic fluid should be drained to improve breathing.

ACTIVITY
Increased physical activity may be beneficial in prevention and treatment of dyslipidemia and atherosclerosis.

DIET
• The provision of a balanced diet supplemented with fruits and vegetables is probably the most important aspect of the treatment in captive parrots. • Decreasing the overall amount of food may be beneficial, especially in overconditioned birds. • Discontinue the feeding of animal products to psittacine birds. Nonanimal items do not contain any cholesterol. • Switch to a pelletized diet if the bird is on an all-seed diet. Low-fat pelletized diet (e.g., Roudybush low fat) may be beneficial in some cases. • Omega-3 fatty acids may help. They can be provided, among other items, with flaxseeds. • In birds of prey, day-old chicks should probably be limited or avoided in highly susceptible species such as gyrfalcons, insectivorous species, and kites.

CLIENT EDUCATION
• Atherosclerosis is a chronic disease with a long latency period that develops over decades. Offering a healthy captive lifestyle to pet birds and other captive birds may significantly reduce the likelihood of developing atherosclerotic diseases later on in life. • Owners should promote physical exercises by limiting time spent in a cage environment and providing increased physical space and complexity (e.g., free flying in the house, foraging tree, playground areas). As all birds are flying animals and their heart rate increase by more than four times during flight, flight activities may be beneficial. • The diet should be balanced and adapted to the species. Captive psittacine birds should preferentially be fed a pelletized diet supplemented with fruits and vegetables and omega-3 fatty acids. Some species are adapted to consume higher fat diet than others (e.g., macaws, palm cockatoos). • Decreasing reproductive stimuli in pet birds. • Owners should be made aware of major epidemiologic factors such species and gender susceptibility and increase frequency with age. It is wise to remind owners that older parrots may have severe subclinical lesions that may significantly impact medical management and anesthesia.

SURGICAL CONSIDERATIONS
N/A

MEDICATIONS

DRUG(S) OF CHOICE
Only limited information is available for most drugs listed below. Most is empirical and pharmacological information has rarely been obtained in birds. For treatment of heart failure, please see corresponding section.
• Statins: Lipid lowering agents, which are competitive inhibitors of the HMG-CoA reductase enzyme in hepatic cholesterol synthesis. Pharmacokinetic studies on rosuvastatin in Amazon parrots failed to reach therapeutic concentrations at 10–25 mg/kg. Other statins of interest include atorvastatin, simvastatin, lovastatin, pravastatin, and fluvastatin. Only rosuvastatin and atorvastatin have long half-lives in mammals and the suggested dose is 10–30 mg/kg pending further pharmacologic studies. Should not be taken while feeding grapefruits and concurrently with azole antifungals. Statins also have nonlipid effects such as beneficial properties on endothelial function, vascular inflammation, immune modulation, oxidative stress, thrombosis, and plaque stabilization.
• Fibrates: Hypolipidemic drugs, which are agonists of the PPARα that stimulates β-oxidation of fatty acids. Fibrates mainly cause a decrease in blood triglycerides.
• Other lipid lowering agents include cholesterol absorption inhibitors (ezetimibe but granivores and frugivores do not have a dietary source of cholesterol) and nicotinic acid. • Blood pressure medications: In case systemic hypertension can be documented, ACE-inhibitors (enalapril 1.25–5 mg/kg q12h PO or benazepril) or beta-blockers may be used. Beta-blockers have also been shown to reduce the progression of atherosclerosis in poultry. • Antithrombotic medications: Due to the low prevalence of atherothrombosis and emboli, preventive antithrombotic treatment is probably not warranted in birds. Aspirin, cilostazol, and clopidogrel may be employed.
• Peripheral vasodilators: For the treatment of signs attributable to peripheral arterial disease (e.g., intermittent claudication). Cilostazol is the drug of choice in humans and also inhibits platelet aggregation. Pentoxyfylline 15 mg/kg q12h PO and isoxsuprine may also be used. • GnRH agonists (leuprolide acetate, subcutaneous deslorelin implant) may be given to decrease reproductive hyperlipidemia and as part of the treatment of reproductive conditions.

CONTRAINDICATIONS
N/A

PRECAUTIONS
Statins and fibrates may cause rhabdomyolysis and myalgia.

POSSIBLE INTERACTIONS
• The combination of statins with fibrates may increase the risk of rhabdomyolysis.
• Some statins (e.g. atorvastatin) interact with drugs metabolized through the cytochrome P450 pathway such as azole antifungals.
• Grapefruit should not be fed to birds receiving statins, which may increase plasma levels.

ALTERNATIVE DRUGS
N/A

FOLLOW-UP

PATIENT MONITORING
• The lipid panel should be monitored at regular interval in case of dyslipidemia. Peak effects of statins may not be evident for several weeks. • The bird's weight should be monitored.

PREVENTION /AVOIDANCE
Reduction of risk factors: Balanced and healthy diet, physical activity, reduction of stress in captivity, reduction of reproductive stimuli. See sections on risk factors and client education.

POSSIBLE COMPLICATIONS
N/A

ATHEROSCLEROSIS (CONTINUED)

EXPECTED COURSE AND PROGNOSIS
- The long-term prognosis is guarded for birds displaying signs of atherosclerotic or cardiac diseases and most of them die in the next few weeks/months.
- The prognosis for subclinical atherosclerosis is variable and depends on the modifications of risk factors. Birds may live decades with clinically silent atherosclerotic lesions.

MISCELLANEOUS

ASSOCIATED CONDITIONS
- Reproductive disease.
- Hepatic disease.
- Cardiac disease.

AGE-RELATED FACTORS
Increased incidence with increasing age.

ZOONOTIC POTENTIAL
N/A

FERTILITY/BREEDING
N/A

SYNONYMS
Arteriosclerosis, coronary arterial disease, peripheral arterial disease

SEE ALSO
Cardiac disease
Dyslipidemia
Lameness
Neurologic conditions
Seizure
Lymphoid neoplasia

ABBREVIATIONS
N/A

INTERNET RESOURCES
http://www.athero.org/ (International Atherosclerosis Society)
http://my.americanheart.org/professional/StatementsGuidelines/ByTopic/Topics A-C/ACCAHA-Joint-Guidelines_UCM_321694_Article.jsp (American Heart Association Guidelines)

Suggested Reading
Beaufrere, H. (2013). Atherosclerosis: Comparative pathogenesis, lipoprotein metabolism and avian and exotic companion mammal models. *Journal of Exotic Pet Medicine*, **22**(4):320–335.

Beaufrère, H., Ammersbach, M., Reavill, D.R., et al. (2013). Prevalence of and risk factors associated with atherosclerosis in psittacine birds. *Journal of the American Veterinary Medical Association*, **242**(12):1696–1704. doi:10.2460/javma.242.12.1696.

Beaufrère, H. (2013). Avian atherosclerosis: parrots and beyond. *Journal of Exotic Pet Medicine*, **22**(4):336–347.

Pilny, A.A., Quesenberry, K.E., Bartick-Sedrish, T.E., et al. (2012). Evaluation of Chlamydophila psittaci infection and other risk factors for atherosclerosis in pet psittacine birds. *Journal of the American Veterinary Medical Association*, **240**(12):1474–1480. doi:10.2460/javma.240.12.1474.

St Leger, J. (2007). Avian atherosclerosis. In: Fowler, M.E., Miller, R.E. (eds), *Zoo and Wild Animal Medicine Current Therapy*, 6th edn. St Louis, MO: Saunders; pp. 200–205.

Author Hugues Beaufrère, Dr. Med. Vet, PhD, ECZM (Avian), ABVP (Avian), ACZM

 Client Education Handout available online

Avian Influenza

BASICS

DEFINITION
Avian influenza viruses (AIVs) are enveloped, segmented, single-stranded, negative sense RNA viruses of the family *Orthomyxoviridae*, genus *Influenzavirus A*. They are grouped into subtypes based on the surface glycoproteins hemagglutinin (HA) and neuraminidase (NA). There are 18 known HA subtypes (H1–H18) and 11 NA subtypes (N1–N11). AIV have one HA and one NA, expressed in any combination. Subtyping is denoted by listing HA and NA antigens together (e.g., H5N1). AIV are classified as low pathogenic (LPAI) or highly pathogenic (HPAI) based on genetic structure and ability to cause disease in chickens (*Gallus gallus*). HPAI viruses have historically been restricted to the H5 or H7 subtypes; LPAI may be caused by any subtype (including non-HPAI H5 and H7).

PATHOPHYSIOLOGY
• Host cell entry gained by viral HA attachment to receptors on host cell membrane, followed by receptor-mediated endocytosis, release of viral ribonucleoprotein, and transport thereof to the nucleus where replication occurs. Virions are released by budding. • LPAI virions bud from the host cell in a noninfectious state and are activated afterwards by extracellular host proteases (e.g., trypsin-like enzymes). This limits proliferation to areas where these proteases are present (e.g., digestive and respiratory tracts). • HPAI infectivity is activated by intracellular proteases prior to budding, allowing release of fully infectious virions and generalized proliferation. • Novel AIV result from two processes: ○ Antigenic drift—random mutations in the HA or NA genes. ○ Antigenic shift—reassortment of genetic material with other influenza A viruses, direct transfer of whole virus from one host to another, or re-emergence of a subtype previously found in the species but not currently in circulation. • Virus is shed in feces, saliva, and respiratory secretions. Shedding begins 1–2 days postinfection, may persist up to 36 days in chickens or 72 days in turkeys (*Meleagris gallopavo*). Transmission is by fecal-oral and aerosol routes. Fomites may enhance transmission. Incubation is approximately one to seven days but may vary by host species.

SYSTEMS AFFECTED
• Varies with host species.
• LPAI—gastrointestinal, respiratory, ophthalmic, renal/urologic, reproductive.
• HPAI—cardiovascular, endocrine/metabolic, gastrointestinal, hemic/lymphatic/immune, hepatobiliary, musculoskeletal, neuromuscular, nervous, ophthalmic, reproductive, respiratory, skin/exocrine.

GENETICS
N/A

INCIDENCE/PREVALENCE
• Varies greatly by geographic region, species, age, time of year, and environment.
• Common in anseriformes (ducks, geese, swans) and charadriiformes (gulls, terns, and shorebirds). Many of these species are reservoirs of LPAI. • Common in poultry in developing countries, rare in developed countries. • Rare in columbiformes (pigeons and doves), psittacines, raptors, and ratites.

GEOGRAPHIC DISTRIBUTION
• LPAI: Worldwide. • HPAI: Eradicated in most developing nations. Epidemic ongoing in parts of Asia, Africa, the Middle East, and Pacific.

SIGNALMENT
• **Species:** Any avian species may be infected. Anseriformes and Charadriiformes are reservoirs of AI. • **Mean age and range:** All ages susceptible, juveniles may show more severe disease. • **Predominant sex:** Both are susceptible.

SIGNS

General Comments
History, physical examination findings may vary markedly by host species.

Historical Findings
• **LPAI** ○ Combination of GI and respiratory signs generally noted. ○ Columbiformes, anseriformes, and charadriiformes generally asymptomatic. Domestic ducks or geese may show respiratory signs. ○ Poultry show high morbidity (>50%) and low mortality (<20%). General malaise, decreased egg production, malformed eggs, diarrhea, and/or respiratory signs may be reported. ○ Psittacine birds may show general malaise and GI signs. ○ Ratites primarily show respiratory signs with green diarrhea. • **HPAI** ○ GI, respiratory, and/or neurologic signs in combination with high mortality. ○ Columbiformes, passerines: Often asymptomatic, may show sporadic mortality. ○ Domestic ducks and geese: Respiratory and neurologic signs. ○ Poultry: Severe increase in mortality. Chickens and turkeys may be more severely affected than other Galliformes. Respiratory signs less prominent than in LPAI. May manifest in two phases: ▪ Peracute: Marked drop in egg production, malaise, diarrhea, birds die before clinical signs noted. ▪ Acute: As in peracute phase, but clinical signs noted prior to death. Neurologic signs common. 3–7 day average survival.

Physical Examination Findings
• **LPAI** ○ Columbiformes and wild waterfowl usually asymptomatic. ○ Domestic ducks and geese: Often asymptomatic, malaise, oculonasal discharge, coughing, sneezing, rales. ○ Poultry: Malaise, underweight, oculonasal discharge, sinusitis (esp. infraorbital sinuses), conjunctivitis, coughing, sneezing, rales, dyspnea, subcutaneous emphysema following air sac rupture. ○ Psittacines: Rarely seen. A case in a juvenile red-lored Amazon (*Amazona autumnalis autumnalis*) reported malaise, dehydration, melena, crop stasis, subcutaneous edema, and inability to stand. ○ Ratites: Oculonasal discharge, conjunctivitis, coughing, sneezing, rales, dyspnea. • **HPAI** ○ Columbiformes often asymptomatic ○ Domestic ducks and geese: Possibly asymptomatic, ataxia, excitation, nystagmus, head tilt, circling, opisthotonus, tremors, convulsions, paresis, paralysis, coughing, sneezing, rales, dyspnea, pulmonary crackles, death. ○ Passerines: Malaise, conjunctivitis, neurologic signs, death. ○ Poultry: Death, ataxia, excitation, cyanotic to necrotic wattles/combs/snoods, nystagmus, head tilt, circling, opisthotonus, tremors, convulsions, paresis, paralysis, petechiae/ecchymoses, multifocal subcutaneous edema, rales, coughing, sneezing. ○ Psittacines: Rarely seen. A case in budgerigars (*Melopsittacus ungulatus*) reported anorexia, malaise, neurologic signs, death. ○ Raptors: A report of AI in American kestrels (*Falco sparverius*) described anorexia, acute weight loss, fluffed feathers, head tilt, ataxia, tremors. ○ Ratites: Malaise, subcutaneous edema of head and neck, ataxia, head tilt, tremors of head and neck, paralysis of wings, sneezing, open-mouth breathing. ○ Wild waterfowl: May be asymptomatic, emaciation, diarrhea, sinusitis, death, wood ducks (*Aix sponsa*): Cloudy eyes, rhythmic myosis and mydriasis, unkempt feathers, ataxia, tremors, seizures, death. ○ Swans (mute [*Cygnus olor*] and whooper [*C. cygnus*]): Neurologic signs, death.

CAUSES
Influenzavirus A.

RISK FACTORS
• Exposure to the outdoors, wild birds, live bird markets, illegally-acquired birds, drinking water used by wild birds.
• Turkeys—exposure to swine. • Residence along wild waterfowl migration routes.
• Seasons with high waterfowl migratory activity.

DIAGNOSIS

DIFFERENTIAL DIAGNOSIS
• Newcastle disease (clinically indistinguishable from HPAI). • Avian metapneumovirus. • Infectious laryngotracheitis. • Infectious bronchitis. • *Chlamydia*. • *Mycoplasma*. • Acute respiratory infections.

AVIAN INFLUENZA (CONTINUED)

CBC/BIOCHEMISTRY/URINALYSIS
Results dependent on systems affected and severity of infection.

OTHER LABORATORY TESTS
- **Serology** ○ Indications: evaluate for flock-level exposure to AIV (often for trade purposes to certify as AI-free). ○ Contraindications: Determine individual bird immune status. ○ Positive results using one modality should be confirmed using another. ○ AGID ▪ Pros: Gold standard for exposure detection; early post-exposure Ab detection, high specificity. ▪ Cons: Moderate sensitivity, inconsistent results in some non-poultry species, testing period 24–48 hours. ○ ELISA ▪ Pros: Rapid results, quantitative. ▪ Cons: Moderate sensitivity and specificity. ○ HI and NI (for subtype identification of AIV Ab-positive sample) ▪ Pros: High specificity, rapid results, gold standard for subtyping ▪ Cons: Moderate sensitivity, specificity dependent on quality of reference sera • **Virus detection** ○ Virus isolation ▪ Indications: Rule-out AI in actively infected birds. ▪ Pros: Gold standard for diagnosis of AI, very high sensitivity. ▪ Cons: Moderate specificity – requires confirmation with other tests, requires BSL-3 laboratory, expensive, lengthy test period (1–2 weeks). ○ Antigen detection immunoassays (aka antigen capture) ▪ Indications: Rule-in AI in sick or recently dead birds (low viral load of asymptomatic birds may hinder detection). ▪ Pros: High specificity, rapid results, small and self-contained ("pen-side"). ▪ Cons: Moderate sensitivity, false positives caused by bacterial/blood contamination. ○ AGID ▪ Indications: Rule-in AI in suspect birds. ▪ Pros: High specificity, reasonable cost. ▪ Cons: Moderate sensitivity, testing period 24–48 h. ○ Molecular diagnostics (e.g., RT-PCR) ▪ Indications: Rule-out AI in suspect birds. ▪ Pros: High sensitivity and specificity, reasonable cost, rapid results. ▪ Cons: Does not differentiate between live and inactivated virus (contamination may reduce specificity for active disease), false negatives possible due to viral genetic mutation. ○ HI and NI ▪ Indications: Subtype identification of influenza A-positive sample. ▪ Pros: High specificity, rapid results. ▪ Cons: Specificity dependent on quality of reference antisera. • **Further virus characterization** ○ Sequence analysis: Characterization of HA gene cleavage site. ○ *In vivo* pathotyping: Inoculation of chicks to characterize an isolate as LPAI or HPAI. ○ Usually only performed for H5 and H7 isolates due to expense, facility requirements (BSL-3).

IMAGING
Findings dependent on systems affected and severity of infection.

DIAGNOSTIC PROCEDURES
N/A

PATHOLOGIC FINDINGS
Vary markedly by species and systems affected.

TREATMENT

APPROPRIATE HEALTH CARE
• Flock-level: Cull exposed poultry, quarantine of other exposed birds. • Individual-level: Strict isolation, therapy directed by affected systems. • Reduce stocking density, improve ventilation, and remove manure frequently. • Environment: ○ AIV survives best in moist environments at low temperatures. May survive indefinitely when frozen. ○ Remove litter, manure, and carcasses—burial, incineration, or composting for at least 10 days are acceptable methods. ○ Disinfection of equipment and affected premises: Oxidizing agents (e.g., bleach), povidone iodine, quaternary ammonium compounds, 70% ethanol, phenols, and lipid solvents. Buildings may be heated to 90–100°F (32–38°C) for one week to inactivate virus. Ionizing radiation or heating to 133°F (56°C) for ≥60 minutes will inactivate AIV.

NURSING CARE
Symptomatic care indicated by patient, situation. Fluid therapy, oxygen therapy, hand-feeding, supplemental heat may all be indicated.

ACTIVITY
N/A

DIET
N/A

CLIENT EDUCATION
Biosecurity measures appropriate to geographic location should be instituted. Educate owners on risk factors for AI exposure and zoonotic potential.

SURGICAL CONSIDERATIONS
N/A

MEDICATIONS

DRUG(S) OF CHOICE
Secondary infections are common, especially in the respiratory tract. Select antimicrobials based on pathogen, sensitivity, location of infection, and patient species.

CONTRAINDICATIONS
Use of anti-influenza drugs (e.g., amantadine, oseltamivir) in nonhumans is highly discouraged to prevent emergence of resistance.

PRECAUTIONS
Make drug selection for food animals in compliance with regulatory statutes.

POSSIBLE INTERACTIONS
N/A

ALTERNATIVE DRUGS
N/A

FOLLOW-UP

PATIENT MONITORING
Consult regulatory statutes for quarantine and follow-up testing requirements. HPAI and LPAI of H5 or H7 subtypes are considered notifiable diseases by the World Organization for Animal Health.

PREVENTION/AVOIDANCE
• Vaccination against HPAI has been employed in poultry and domestic waterfowl in countries where H5N1 HPAI is enzootic (i.e., China, Vietnam, Egypt, and Indonesia). • Vaccination of wild birds with poultry AI vaccines has been performed—effectiveness unknown. • Avoid exposure to most common potential virus sources: Manure, respiratory secretions, carcasses, unwashed eggs and associated packing material, and people/equipment contaminated by the above. • Proper biosecurity is essential.

POSSIBLE COMPLICATIONS
Secondary infections (esp. respiratory).

EXPECTED COURSE AND PROGNOSIS
Varies by species and pathogenicity of virus.

MISCELLANEOUS

ASSOCIATED CONDITIONS
• Influenza in nonavian species. • Acute respiratory infections.

AGE-RELATED FACTORS
Juveniles may be more susceptible to disease.

ZOONOTIC POTENTIAL
• HPAI infection in humans has been extensively documented since the outbreak of Asian lineage H5N1 HPAI. Severe respiratory disease results. Mortality rate is high, approximately 60% at the time of this writing. • Infection is commonly a result of direct/close contact with sick/dead infected poultry. Person-to-person transmission is rare, usually from prolonged contact with infected people. Transmission from wild birds to humans has not been documented. • Poultry products: pasteurization or cooking to an internal temperature of 165°F (74°C) will inactivate AIV

FERTILITY/BREEDING
May cause substantial drops in egg production.

SYNONYMS
Fowl plague, bird flu, avian flu, fowl pest

Avian Influenza

(Continued)

SEE ALSO
Infertility
Neurologic conditions
Paramyxoviruses
Pneumonia
Respiratory distress
Viral disease

ABBREVIATIONS
AGID—agar gel immunodiffusion
AI—avian influenza
AIV—avian influenza virus(es)
BSL—biosecurity level
GI—gastrointestinal
HA—hemagglutinin
HI—hemagglutinin inhibition
HPAI—high pathogenicity avian influenza
LPAI—low pathogenicity avian influenza
NA—neuraminidase
NI—neuraminidase inhibition
RT-PCR—reverse transcriptase PCR

INTERNET RESOURCES
- USDA - Biosecurity for Birds: http://www.aphis.usda.gov/animal_health/birdbiosecurity/
- OIE/FAO Network of Expertise on Animal Influenza: http://www.offlu.net/

Suggested Reading
Capua, I., Alexander, D.J. (eds) (2009). *Avian Influenza and Newcastle Disease: A Field and Laboratory Manual*. Milan, Italy: Springer-Verlag Italia.
Swayne, D.E. (2008). *Avian Influenza*. Ames, IA: Blackwell Publishing.
Swayne, D.E, Suarez, D.L., Sims, L.D. (2013). Avian influenza. In: Swayne, D.E. (ed.), *Diseases of Poultry*, 13th edn. Ames, IA: John Wiley & Sons, Inc., pp. 181–218.
Author Andrew D. Bean, DVM, MPH, CPH

Beak Injuries

BASICS

DEFINITION
Traumatic injuries to the beak are common clinical presentations. These injuries often are most often seen following bite wounds from other birds or predators, or household/aviary traumatic accidents.

PATHOPHYSIOLOGY
- Injuries to the beak, including damage to the keratin, its underlying dermis, and bone.
- Primary and secondary involvement of the prokinetic muscles and their innervation, and the multiple kinetic joints of prokinesis can also occur, resulting in abnormal form and function of the beak.
- With altered structure and force vector delivery, continuing keratin growth can lead to further deformities and functional deficits.

SYSTEMS AFFECTED
- Skin/Exocrine—the beak is covered by integument, including keratin and dermis.
- Musculoskeletal—the integumentary structures are supported by bone, and dynamic function of the upper and lower beak requires that all work in concert in order to maintain balanced force vector delivery.
- Behavioral—beak injuries may lead to the need for recurrent trimming procedures, which may result in undesired learned fear and impairment to quality of life as a result.

GENETICS
N/A

INCIDENCE/PREVALENCE
Incidence is not known. However, varying injuries to beaks are not uncommonly encountered in private practice and in zoological settings.

GEOGRAPHIC DISTRIBUTION
No specific distribution is recognized.

SIGNALMENT
- **Species:** All avian species. More prevalent in companion bird species. **Mean age and range:** N/A **Predominant sex:** N/A

SIGNS
- Visible injuries to the structures or function of the beak.
- Altered behaviors: Hyporexia, altered food selection (softer foods), avoidance of hard food items.
- Acute injuries to the distal rhinothecal tomium may present with blood in the oral cavity.
- Chronic injuries may be visibly apparent with healed injuries to the structures or function of the beak, or may manifest with abnormal keratin growth due to altered force vector delivery and uneven wear.

CAUSES
- Traumatic (primary): Crushing injuries—bite wounds (predator attacks, injuries from cage or house mates), blunt trauma (household or aviary accidents) or handling/restraint associated injuries to the upper or lower bill.
- Secondary potential contributing causes: ∘ Degenerative/Metabolic: Metabolic bone disease, abnormal keratin overgrowth secondary to hepatic functional deficits. ∘ Anatomic (congenital): N/A ∘ Nutritional: Chronic malnutrition, with secondary metabolic bone disease, or keratin malformation. ∘ Neoplastic: Squamous cell carcinoma, keratoacanthoma, melanoma, xanthoma, osteoma, fibrosarcoma. ∘ Immune mediated: N/A ∘ Infectious: Keratitis or osteomyelitis—bacterial (multiple organisms) or mycotic (Aspergillus sp, Cryptococcus sp), or parasitic keratitis (Knemidocoptes); sinusitis with or without direct extension to the structures of the beak, viral infections of beak and associated tissues: Psittacine Beak and Feather Disease, Pox. ∘ Toxic: N/A

RISK FACTORS
- Environmental factors: Husbandry—free-ranging activities in the home, cage or aviary hazards for entrapment.
- Behavioral factors: Aggressive behaviors resulting in inadvertent injury as a result of the bird's behavior.
- Medical conditions: Metabolic bone disease, chronic malnutrition, preexisting pathology of the structures of the beak.

DIAGNOSIS

DIFFERENTIAL DIAGNOSIS
There are comparatively few differential diagnoses for an acutely evident beak injury. The presence of soft or abnormally flexible bone of the beak(s), abnormal keratin or preexisting deformities should open the possibility of a beak injury that is secondary to a primary disease process.

CBC/BIOCHEMISTRY/URINALYSIS
- There are no hematologic or serum biochemistry abnormalities that are helpful in diagnosis of an acute traumatic injury to the beak. Anemia can occur secondary to blood loss.
- Serum bile acids: May be elevated in some species with beak keratin overgrowth associated with hepatic dysfunction

OTHER LABORATORY TESTS
Where there is reason to suspect underlying Psittacine Beak and Feather disease as a contributing factor, whole blood PCR is indicated.

IMAGING
Radiographic findings (plain radiography or computed tomography): Traumatic amputation of maxillary and/or nasal bones or mandibular rami, penetration to the nasal diverticulum of the frontal sinus or the nasofrontal hinge, fractures of the jugal, palatine and/or pterygoids, decreased bone density of the boney skull.

DIAGNOSTIC PROCEDURES
Where preexisting pathology may be suspected as a primary contributing factor in a specific case, biopsy and histopathology may be indicated.

PATHOLOGIC FINDINGS
Gross: Visible traumatic amputation or other injuries to the upper and lower bills, avascular necrosis of remaining distal beak, secondary opportunistic infections.

TREATMENT

APPROPRIATE HEALTH CARE
Where it is uncertain if the bird is capable of feeding normally due to the injury, inpatient management until it demonstrates ability for homeostasis is important. Comparatively minor and stable injuries may be treated on an outpatient basis.

NURSING CARE
- Outpatient: Should not require nursing or supportive care.
- Inpatient: ∘ Fluid therapy: Lactated Ringers or Normal Saline, 25–50 ml/kg maintenance 1–2 times daily SC. ∘ IV or IO (Lactated Ringers, Normal Saline or colloid fluids) if patient has sustained considerable blood loss or is in shock. ∘ Force feeding/Gavage feeding: As needed to assist in daily caloric intake.

ACTIVITY
- Where the beak is deemed to be painful, feeding activities should be modified to minimize re-injury.
- Psittacine species may benefit from restriction from climbing activities that require the use of the upper bill.

DIET
Where the beak is deemed to be painful, the diet may need to be altered to softer items to minimize the amount of force required for prehension and chewing behaviors

CLIENT EDUCATION
- Keys to treatment success are in large part dependent on the severity of the injuries sustained, and the ultimate probability of the bird regaining ability to feed on its own.
- Most birds, but not all, with injuries to the beak are ultimately capable of learning to feed on their own.
- Assurance that the bird can eat and drink on its own are important prior to outpatient management.
- Secondary opportunistic infections are comparatively less common, but possible.
- Comparatively minor injuries to the upper mandible typically heal well with conservative treatments.
- Surgical attempts at repairs of large fractures, re-attachments and prostheses carry an overall poor prognosis in most cases.

SURGICAL CONSIDERATIONS
- Most unilateral or bilateral compression fractures of the maxillary beak can be treated as open wounds, although some may benefit

Beak Injuries

(CONTINUED)

from the application of an acrylic cap.
• Damage to the vascular integrity of the beak(s) may not be initially apparent at the time of presentation, and there is some risk of avascular necrosis. • Where bone has been traumatically amputated, regrowth is unlikely.
• Normal force vector delivery is still possible in maxillary beak amputations, as long as the keratin occlusal ledge remains intact, enabling normal occlusion with the rostral gnathothecal tomium. • In acute injuries with partial fractures through the maxillary beak, surgical stabilization may have success. Lower beak symphyseal fractures carry a poor prognosis for successful repair. ○ Stabilization often requires the use of acrylic caps/prostheses, to best stabilize the entire upper beak. • Fractures of the mandibular rami may be successfully repaired in some cases.
○ KE fixation with cross pins through both mandibular rami provides best stability in most cases.

MEDICATIONS

DRUG(S) OF CHOICE
• Bimodal pain management:
○ Psittaciformes: Meloxicam 1 mg/kg IM q12–24h or 1.6 mg/kg PO q12–24h, Butorphanol 1–4 mg/kg IM q4–6h, Tramadol 30 mg/kg PO q8–12h. ○ Galliformes: Carprofen 1 mg/kg IM q8–12h, Ketoprofen 12 mg/kg IM q8–12h, Tramadol 7.5 mg/kg PO q12–24h. ○ Accipiteriformes: Tramadol 5 mg/kg PO q12h (bald eagles), 11 mg/kg PO q4h (Red tailed hawks). • Amtimicrobial management: Antibacterial choices are based on culture and sensitivity data, antifungals may include Itraconazole 5–10 mg/kg q12–24h, Voriconazole 12–18 mg/kg q12h.

FOLLOW-UP

PATIENT MONITORING
• Critical assessment of effectiveness of analgesia on a daily basis for inpatients.
• Physical assessment of outpatient form and function, and degree of healing on loosely a weekly basis.

PREVENTION/AVOIDANCE
• Reduce or eliminate household traumatic risks and hazards. • Prevent exposure to predators, aggressive birds. • Controlled, supervised observation in the home while out of cage.

POSSIBLE COMPLICATIONS
• In the acute injury: Anorexia without proper support leading to starvation and death.
• In the more chronic injuries: ○ Secondary infection, exposure of sinuses. ○ Exposure of tongue (galliformes, anseriformes) and desiccation following amputation of upper bill. ○ Avascular necrosis and loss of distal anatomy to the site of injury. ○ Abnormal beak alignment leading to keratin overgrowth problems in the future. ○ Permanent loss of the bird's ability to eat on its' own.

EXPECTED COURSE AND PROGNOSIS
• Compressive or penetrating injuries to the maxillary beak: Typically healed within weeks; prognosis is good for normal return of appearance and function. • Amputation of maxillary beak distal to its occlusal ledge (psittacines): Good for ability to feed on own and a lack of necessary follow up trimming of the rostral gnathothecal tomium overgrowth.
• Amputation of maxillary beak proximal to its occlusal ledge (psittacines): Fair for ability to feed on own, but will require regular trimming of the rostral gnathothecal tomium overgrowth. • Amputation of the lower beak: Guarded or poor for ability to feed on own; probable necessary trimming of rhinothecal overgrowth required. • Lower bill mandibular ramus symphyseal split: Heal quickly, but typically fail to reunite. Prognosis for ability to prehend and eat is good, but there will be probable be necessary trimming of rhinothecal and gnathothecal overgrowth.

MISCELLANEOUS

ASSOCIATED CONDITIONS
None.

AGE-RELATED FACTORS
None.

ZOONOTIC POTENTIAL
None.

FERTILITY/BREEDING
N/A

SYNONYMS
None.

SEE ALSO
Beak malocclusion
Bite wounds
Cere and skin, color changes
Hemorrhage
Liver disease
Overgrown beak and nails
Problem behavior
Trauma

ABBREVIATIONS
None.

INTERNET RESOURCES
None.

Suggested Reading
Speer, B.L. (2014). Beak deformities: form, function and treatment methods. *Proceedings of the Association of Avian Veterinarians Annual Conference*, New Orleans, LA, August 2–6, pp. 213–219.
Speer, B.L., Echols, M.S. (2013). Surgical procedures of the Psittacine skull. *Proceedings of the Association of Avian Veterinarians Annual Conference*, Jacksonville, FL, August 4–7, pp. 99–109.
Author Brian Speer, DVM, DABVP (Avian), DECZM (Avian)

Beak Malocclusion (Scissors Beak)

BASICS

DEFINITION
Scissors beak deformities are characterized by a bending of the upper beak and/or maxillary bone to one side to varying degrees, with bending of the lower beak to the opposing side. These deformities often are self-augmenting, due to abnormal force delivery and unchecked and imbalanced keratin growth of the rhinothecal and gnathothecal tomia.

PATHOPHYSIOLOGY
- Not clearly understood. • It is hypothesized by many that bruising of the rictal phalanges of the upper beak on one side or the other leads to a transient uneven growth in young hand feeding parrots, resulting in the ultimate development of a scissoring deformity.
- Possible causal factors that can lead to the onset of lateral asymmetrical growth of the upper beak include: ○ Hand feeding technique flaws resulting in bruising of the rictal phalanges. ○ Incubation or hatching problems. ○ Genetic predisposition. ○ Subclinical malnutrition. ○ Infectious sinusitis. ○ Trauma.

SYSTEMS AFFECTED
- Skin/Exocrine—the beak is covered by integument, including keratin and dermis.
- Musculoskeletal—the integumentary structures are supported by bone, and dynamic function of the upper and lower beak requires that all work in concert in order to maintain balanced force vector delivery.
- Behavioral—scissors beak deformities will lead to the need for recurrent trimming procedures, which may be accompanied by undesired learned fear and impairment to quality of life as a result.

GENETICS
There is no evidence of any genetic predisposition for this condition in young parrot species.

INCIDENCE/PREVALENCE
The true incidence or prevalence of scissors beak deformities is not known, however this specific acquired beak developmental deformity is probably the most commonly encountered in young hand reared psittacine birds.

GEOGRAPHIC DISTRIBUTION
No specific distribution is recognized. The problem is most commonly encountered in settings where young parrots are being hand reared, as opposed to parent reared.

SIGNALMENT
- **Species:** All psittacine bird species, but more commonly seen in the large macaws (*Ara* sp) and Hyacinth macaws (*Anodorhynchus*), and less commonly in other species. • **Mean age and range:** Onset of the problem is in subadults, typically prior to fledging; however, it can be an acquired problem as an adult, and adults with chronic deformities that have been present for years are not uncommonly encountered. • **Predominant sex:** N/A

SIGNS
- Deviation of the rostrum maxillare away from the midline, and away from a perpendicular angle with the nasofrontal hinge joint. • Secondary overgrowth of rhinothecal and gnathothecal tomia on opposing sides of the beaks, due to lack of opposing wear from normal occlusion. • Secondary overgrowth or abnormal wear of occlusal ledge keratin of the rhinotheca, where the gnathothecal rostral tomium is normally intended to contact. • Tertiary abnormal boney angles and growth of the upper and lower bills.

CAUSES
- Any event that causes anatomically uneven growth of the rhinothecal structures, resulting in abnormal force vector delivery and keratin overgrowth of the rhinotheca and gnathothecal tomia: ○ Hand-feeding-associated bruising of the rictal phalanges. ○ Unilateral traumatic injuries to the maxillary beak. ○ Incubation flaws with dehydration and secondary injury to beak tissues. • Secondary potential contributing causes: ○ Nutritional: Chronic malnutrition with secondary metabolic bone disease or keratin malformation. ○ Infectious: Unilateral bacterial (multiple organisms), mycotic (*Aspergillus* sp, *Cryptococcus* sp), parasitic (Knemidocoptes) or viral (Psittacine Beak and Feather disease virus) – all which can result in abnormal and uneven keratin growth, originating from the keratin, dermis, bone or sinuses. ○ Degenerative/Metabolic: Metabolic bone disease, asymmetric and abnormal keratin overgrowth secondary to hepatic functional deficits. ○ Neoplastic: Squamous cell carcinoma, keratoacanthoma, melanoma, xanthoma, osteoma, fibrosarcoma—all of which can result in abnormal and uneven keratin growth. ○ Immune mediated: N/A ○ Toxic: N/A

RISK FACTORS
- Behavioral factors: Aggressive feeding behaviors, inattentive hand feeding technique resulting in inadvertent injury to the rictal phalanges. • Environmental factors: N/A
- Medical conditions: Inadequate nutritional support, resulting in hungry chicks with stronger than normal feeding reflexes and a greater predisposition to traumatic injury of the rictus.

DIAGNOSIS

DIFFERENTIAL DIAGNOSIS
Primary deformities of the lower beak that result in abnormal force vector delivery and secondary deformity of the upper beak.

CBC/BIOCHEMISTRY/URINALYSIS
There are no hematologic or serum biochemistry abnormalities that are helpful in diagnosis of a scissors beak deformity.

OTHER LABORATORY TESTS
N/A

IMAGING
Radiographic findings (plain radiography or computed tomography): Asymmetrical rhinothecal overgrowth, lateral deviation of the maxillary bone.

DIAGNOSTIC PROCEDURES
Physical assessment and rule out of contributing underlying pathology.

PATHOLOGIC FINDINGS
Gross: Visible deviation of the maxillary beak away from its normal perpendicular articulation at the nasofrontal hinge joint. Lateral exposure of the rostral gnathothecal tomium outside of its normal point of contact at the occlusal ledge of the maxillary beak.

TREATMENT

APPROPRIATE HEALTH CARE
Where the deformity is noted in young hand feeding chicks, lateral pressure that is applied during hand feeding to direct the maxillary beak towards the midline may be effective for correction or slowing progress of minor scissors deformities. The goal with supportive health care is to enable normal growth until a more definitive corrective procedure can be done shortly before or after weaning.

Corrective procedures in young birds are designed to alter the forces that direct the rostral growth of the rhinotheca.

NURSING CARE
- Outpatient: Most birds should not require nursing or supportive care, although those birds with marked keratin overgrowth may benefit from corrective trimming. • Inpatient: N/A

ACTIVITY
There are no indicated alterations in activity, pending planned surgical corrective maneuvers for correction of a scissors beak deformity.

DIET
There are no indicated dietary alterations.

BEAK MALOCCLUSION (SCISSORS BEAK)

CLIENT EDUCATION
It is important to try to correct scissors beak deformities younger in life where at all possible, to reduce if not eliminate the need for corrective trimming or grinding procedures for the remainder of the bird's life.

SURGICAL CONSIDERATIONS
• In young birds that are at or close to weaning age, there are two procedures that can be considered for correction of this deformity: trans-sinus pin and tension band, or the use of a lower mandibular ramp prosthesis. • The trans-sinus pin and tension band procedure is the preferred method for correction, offering less time for the surgical procedure and a shorter duration of time until correction can be achieved. ○ The patient is induced under general anesthesia with isoflurane, desflurane or sevoflurane, maintained via endotracheal tube in dorsal recumbency. Pluck feathers and aseptically prepare the skin rostral to the eye and up to the junction of skin and rhinotheca. Use a positive or negative profile threaded pin, 0.035–0.045 mm to drill through the frontal bone just caudal and parallel to the naso-frontal hinge and parallel to it, caudal-ventral to the nares. Insert the pin from the side of the head that the maxillary beak is deviated away from, towards the side that it is deviated to. The pin is placed with careful attention to the angle of entry, as an incorrect angle of entry can result in inadvertent penetration of the naso-frontal kinetic joint or the eye. A hand-held pin driver may facilitate passage of the pin. Pass the pin evenly through the frontal sinus, and out the opposite side of the skull and skin at an identical exit site as the entrance location. On the side of the beak which the upper beak is deviated towards, cut the point from the threaded pin, and back it up to where it is seated in the frontal bone, but beneath the skin. On the opposite side, grasp the pin where it exits the skin with a firm clamp to prevent applying force to the bone itself, and bend the pin to a 90° angle, parallel with the rostral-caudal axis of the upper beak. At approximately the same length as the upper mandible, cut the pin off with wire cutters. This produces an "L"-shaped pin, with the short portion placed through the frontal sinuses, and the longer portion angling down the lateral length of the upper beak. Curl the end of this pin, at least 360°, producing a circle, which will enable fixation of your tension band. Variable sized rubber bands can be used to apply tension from the K wire to the tip of the beak. An important key to keep in mind is that gentle tension is usually all that is needed with young birds. More aggressively applied tension may actually have a higher failure rate with these youngsters. In some birds, a dremmel may be used to score a groove in the rhinotheca where the tension band is to rest. Adjust the tension by bending the angle of the pin or tightening the rubber band as needed to lightly pull the deviated upper beak towards midline. Enable normal feeding and prokinetic function during the period when there is a tension band in place. Most young birds will be able to be straightened in a matter of a few days to approximately two weeks. • A lower mandibular ramp prosthesis is also used for the surgical correction of scissors beak deformities in some large parrots, specifically where it is deemed that the trans-sinus pin technique is not a feasible option and that the lower mandible should be capable of supporting the apparatus. ○ The principle on which this method rests is by forming a mechanism that opposing force can be applied against the scissors deformity, from an apparatus affixed to the lower mandible. This procedure is also typically used in young birds. A method using dental composites has been described, which does not utilize anesthesia, and this method may offer advantages in some select cases. A second option involves the patient under general anesthesia with isoflurane, desflurane or sevoflurane, maintained via endotracheal tube in dorsal recumbency. The first component of this procedure is one of corrective grinding of the pressure bearing keratin of the upper and lower mandibles. The rostral gnathothecal tomium is ground and trimmed in an angle that is opposite the direction of overgrowth seen in a scissors deformity, leaving the highest point on the same side of the scissors deformity. Concurrently, any overgrowth of the occlusal ledge of rhinotheca is ground, in order to enable a more normal force vector delivery when the two beaks are closed together. A wire mesh foundation is then cut and shaped, specifically designed to fit over the entire outer aspect of the lower beak and extending upward in a ramp on the same side that the upper beak is deviated. This wire base is attached to the lower mandible using fine cerclage wires, which are applied by lacing through 22 gauge needles that are inserted through the wire and entire lower mandible and into the oral cavity, and then back out a second hole. These wires are lightly twisted tight enough to hold the wire mesh in place. Layers of light sensitive acrylic are then placed and hardened in order to produce a functional cap that encompasses the entire outer and inner aspects of the lower beak, and extends up the vertical ramp. As the ramp is cured, it is positioned against the upper beak carefully, to apply enough force to the upper so that the two beaks are showing normal occlusion when opened and closed. The height of the ramp is designed to be great enough to prevent the upper mandible from extending up and over it. Potential complications from the use of this ramp prosthesis include the development of abnormal lower beak angulation, fractures of the delicate mandibular ramus in young birds, or mechanical damage to the prosthesis caused by the bird's own behaviors. Functional correction of scissors deformities with this method may require 1–3 weeks on average. When the prosthesis is removed, a dremel is used to cut the acrylics, and suture removal scissors can be used to cut the wire mesh.

MEDICATIONS
DRUG(S) OF CHOICE
• Bimodal pain management (pre and postprocedurally): ○ Psittaciformes: Meloxicam 1 mg/kg IM q12–24h or 1.6 mg/g PO q12–24h; Butorphanol 1–4 mg/kg IM q6h; Tramadol 30 mg/kg PO q8–12h.

CONTRAINDICATIONS
N/A

PRECAUTIONS
N/A

POSSIBLE INTERATIONS
N/A

ALTERNATIVE DRUGS
N/A

FOLLOW-UP
PATIENT MONITORING
• Critical assessment of effectiveness of analgesia on a daily basis for inpatients. • Regular physical assessment of outpatient form and function on a weekly basis or more frequently as needed. There may be the need to add more acrylic to the medial side of the ramp to maintain contact with the upper beak as it begins to move towards the midline.

PREVENTION/AVOIDANCE
• Avoid or minimize risk of traumatic bruising of the rictal phalanges during hand feeding. • When upper beak lateral deviations are first detected, apply digital pressure to counter continued growth during hand feeding and hopefully avoid development of more significant deviations in time. • Utilize a surgical procedure to correct alignment once the gnathothecal rostral tomium continually rests laterally, outside of its normal contact point with the rhinothecal occlusal ledge.

POSSIBLE COMPLICATIONS
• Loosening of the trans-sinus pin in the frontal bone and inability to apply adequate lateral tension as a result. • Abnormal angulation of the lower beak towards the side bearing the ramp, and potential for worsened malocclusion as a result.

EXPECTED COURSE AND PROGNOSIS
• Trans-sinus pin tension band placement should be expected to have successful correction in a few days to two weeks in most cases, with an excellent return to normal for

BEAK MALOCCLUSION (SCISSORS BEAK) (CONTINUED)

and function. • Ramp prosthesis should require 1–3 weeks for correction, with a good prognosis for a permanent return of normal form and function.

MISCELLANEOUS

ASSOCIATED CONDITIONS
None.

AGE-RELATED FACTORS
None.

ZOONOTIC POTENTIAL
None.

FERTILITY/BREEDING
N/A

SYNONYMS
None.

SEE ALSO
Beak injuries
Bite wounds
Cere and skin, color changes
Overgrown beak and nails
Problem behavior
Trauma

ABBREVIATIONS
None.

INTERNET RESOURCES
None.

Suggested Reading
Speer, B.L. (2014). Beak deformities: form, function and treatment methods. *Proceedings of the Association of Avian Veterinarians Annual Conference*, New Orleans, LA, August 2–6, pp. 213–219.
Speer, B.L., Echols, M.S. (2013). Surgical procedures of the Psittacine skull. *Proceedings of the Association of Avian Veterinarians Annual Conference*, Jacksonville, FL, August 4–7, pp. 99–109.

Author Brian Speer, DVM, DABVP (Avian), DECZM (Avian)

Bite Wounds

BASICS

DEFINITION
A bite wound may be open (skin disruption from puncture, laceration, tearing) or closed (skin crushed). All bite wounds are considered contaminated with a microbial population representative of the biter's oral flora, the victim's skin and the environment. Damage to the underlying tissues is commonly more extensive than the often minor-appearing skin lesions.

PATHOPHYSIOLOGY
Devitalized tissue, dead space, compromised blood supply, and body fluid accumulation create a prime environment for bacterial growth. Wound infection is frequent if treatment is delayed. Mixed aerobic and anaerobic infections are most common. Cat bites are more likely to become infected than dog bites. Common aerobic isolates from dog and cat bite wounds include *Pasteurella* sp, *Streptococcus* sp, *Staphylococcus* sp, *Neisseria* sp, *Corynebacterium* sp and *Moraxella* sp. Common anaerobic isolates include *Fusobacterium* sp, *Bacteroides* sp, *Porphyromonas* sp, *Prevotella* sp, *Propionibacterium* sp and *Peptostreptococcus* sp. B-lactamase production is a common feature among anaerobes isolated from infected bites. Multiple or severe bite wound injuries may lead to systemic inflammatory response syndrome (SIRS).

SYSTEMS AFFECTED
- Skin—wound, bruising, infection, cellulitis, abscess, subcutaneous emphysema.
- Musculoskeletal—crushing, tearing, laceration, avulsion, necrosis of muscles, tendons and ligaments, luxation/fracture.
- Renal/Urologic—polyuria (stress, gastrointestinal infection, sepsis, SIRS).
- Cardiovascular—hemorrhage, hypotension, shock, vascular compromise.
- Nervous—nerve damage, head trauma.
- Respiratory—hemorrhage, edema, pulmonary contusion, perforated air sac/trachea/sinus, tachypnea, dyspnea (sepsis, SIRS). • Ophthalmic—traumatic ocular lesions. • Gastrointestinal—perforated esophagus/crop, vomiting, diarrhea, delayed crop emptying (bacterial infection from preening feathers soiled by mammalian saliva, sepsis).

GENETICS
N/A

INCIDENCE/PREVALENCE
Common

GEOGRAPHIC DISTRIBUTION
N/A

SIGNALMENT
No specific species, age or sex predilection.

SIGNS

General Comments
• If there is significant injury or long delay between the time of the injury and presentation to the veterinarian, the patient may require extensive stabilization therapy before complete physical examination can be performed. • Sedation or anesthesia may be indicated for birds that do not tolerate restraint or have very painful injuries.

Historical Findings
• Direct contact with a predator species or another bird. • Biting event witnessed by the owner.

Physical Examination Findings
• Skin—single or multiple wounds. After hemoglobin is broken down, biliverdin pigment accumulates giving bird bruises a greenish discoloration within 2–3 days post injury. Puncture wounds caused by feline predators are often very fine and difficult to see with the only outward sign of attack being moist feathers. Discolored skin (black or blanched white), absence of skin bleeding, no capillary refill time, cold to the touch and absence of bleeding from a cut toe nail are consistent with vascular compromise.
• Muscle, ligament, tendon—crushing, laceration, tearing; palpable fracture/luxation; lameness, inability to fly. • Polyuria.
• Lethargy, depression, pale mucous membranes, prolonged capillary refill time, tachycardia, hypotension, hemorrhage.
• Nerve damage, peripheral or central neurologic deficit, cranial nerve deficits.
• Tachypnea, dyspnea. • Vomiting, diarrhea, delayed crop emptying. • Palpebral, conjunctival, and corneal lesions; uveitis.

CAUSES
Traumatic.

RISK FACTORS
Environmental factors—cohabitation with another bird or predator species.

DIAGNOSIS

DIFFERENTIAL DIAGNOSIS
• Other traumatic event. • Nontraumatic infections resembling a bite.

CBC/BIOCHEMISTRY/URINALYSIS
• CBC—to characterize the inflammation/infection. • Biochemistry—to evaluate organ function.

OTHER LABORATORY TESTS
Culture—aerobic and anaerobic; indicated for all bite wounds to identify organisms present and their antimicrobial susceptibility; do not culture fresh bite wounds since infectious agents will most likely not be recovered during this time; growth and identification of anaerobic organisms are frequently a difficult task and should not be misinterpreted as absence of these organisms in infected wounds.

IMAGING
Whole body radiographs—to evaluate for internal or orthopedic injury.

DIAGNOSTIC PROCEDURES
• Orthopedic and neurological examination when the patient is stable. Sedation and analgesia may be considered for orthopedic evaluation only. • In-depth wound assessment when the patient is stable to determine the extent of the injury and remove any debris or devitalized tissue. Wound communication with body cavities must be determined. Sterile instruments and aseptic technique should be used. Consider sedation and analgesia.

PATHOLOGIC FINDINGS
Skin puncture, laceration, tearing with associated injury (bruising, crushing, contusion, edema, laceration, necrosis) to adjacent tissues; abscess; pulmonary contusion; generalized changes compatible with sepsis.

TREATMENT

APPROPRIATE HEALTH CARE
• Outpatient medical management—stable patient with minor wounds that can be closed primarily or do not require closure.
• Inpatient medical management—patient with acute wounds for which delayed primary closure is required. • Emergency inpatient intensive care management—patient with severe bite wounds requiring immediate assessment of the "ABCs" (airway, breathing, and circulation); patient with hemorrhagic wounds requiring application of a pressure bandage immediately to stop blood loss and prevent hypotensive shock. • Surgical management once patient is stable—patient with wounds requiring primary closure, delayed primary closure or secondary closure.

NURSING CARE
• Wound management—all wound management needs to be done as part of the overall patient assessment. • Feather cutting/removal—create a 2–3-cm circumferential featherless zone around the wound. Plucking may cause tearing of the skin. Consider applying sterile water-soluble gel within the wound to keep down feathers from adhering to the wound. • Wound cleansing—all bite wounds should be flushed copiously to mechanically dislodge nonviable tissue and bacteria from the wound surface. Topical antiseptic may be used initially. Sterile saline may be used once bacterial balance has been restored using topical antiseptic. Appropriate pressure can be obtained using a 20-mL syringe fitted with an 18-gauge needle. Consider sedation and analgesia. • Staged wound debridement—indicated if debris or

BITE WOUNDS (CONTINUED)

necrotic tissues persist within the wound after wound cleansing; facilitates identification of nonviable tissue while preserving potentially viable tissue; continued until all nonviable tissue has been removed from the wound. Consider local anesthesia by spraying a topical anesthetic on the wound. Consider sedation or anesthesia to minimize stress and to have optimal control of hemostasis and tissue handling. • Wound dressings—must be adapted to the phase of wound healing; numerous products available to enhance various phases of wound healing. Adherent bandages are used during the initial inflammatory stage of healing. Non-adherent bandages are applied for the proliferative and remodeling stages of healing. Dressings can be occlusive or nonocclusive. • Hydrotherapy—consider if mechanical nonselective debridement is indicated and to stimulate formation of granulation tissue, increase blood flow to the wound and decrease inflammation. • Bandage—must be adapted to the exact nature of the wound, its location, the patient and its functional needs. Temporary bandaging indicated if patient requires medical stabilization until more thorough wound treatment can be performed. Consider a distracting device consisting of a tape tag over the bandage that the bird can reach and damage while preserving the actual bandage until the patient gets use to the bandage. • Fluid therapy—may be provided SQ in birds exhibiting mild dehydration (5–7%) or IO/IV in birds with moderate to severe dehydration (>7%) or critically ill birds exhibiting signs consistent with shock. Avian daily maintenance fluid requirements are 50 ml/kg. Typically, maintenance plus half the fluid deficit is administered during the first 12 hours, with the remainder of the deficit replaced over the following 24–48 hours. In most patients, Lactated Ringers Solution or Normosol crystalloid fluids are appropriate. For birds with hypotensive shock, a colloidal solution may be administered concurrently with crystalloids. Blood transfusion or hemoglobin replacement products should be considered with severe blood loss. • Oxygen therapy—indicated if respiratory distress or anemia is present; beneficial for any sick avian patient. All methods of oxygen supplementation require humidification if used for more than a few hours. • Warmth (85–90°F/27–30°C)—to minimize energy spent to maintain body temperature. • Elizabethan collar—should be reserved for birds that insist on self-traumatizing their wounds; may suggest pain and the cause should be investigated.

ACTIVITY
Patient's activity should be altered according to its injuries.

DIET
Birds suffering severe injury may require assisted feeding. Enteral nutritional preparations may be administered initially at 20–30 ml/kg directly in the crop and the volume administered may be progressively increased according to patient tolerance.

CLIENT EDUCATION
Any bite wound is considered an emergency and requires immediate veterinary care.

SURGICAL CONSIDERATIONS
• Emergency management may be indicated before surgical debridement under anesthesia • Surgical debridement—if performed within 24 hours of a bird sustaining a severe biting injury, it might be placing the patient at unnecessary risk, especially if a more aggressive procedure is required later. Level of tissue damage and surgical risk can usually be assessed adequately 24 hours postinjury. Initial wound cleaning, minor debridement, and temporary wound dressing may be preferred over any aggressive surgical wound debridement to allow time for better assessment of tissue damage and patient stabilization. • Severed tendons, ligaments, or major nerves—attempt aggressive surgical debridement of the wound and primary re-anastomosis of these structures as soon as possible since tissue contraction may result in the impossibility to affix the ends without tension. • Penetration of the coelomic cavity—may require surgical exploration to identify and treat concurrent injury. • Primary closure—indicated for birds presented within 6–8 hours of injury with minimal contamination and tissue trauma. • Delayed primary closure—most acute wounds will be healthy enough for closure after 3–5 days of wound management with hydrophilic dressings. • Secondary closure, second-intention healing—extensive bite wounds may require wound management for longer periods. These wounds are closed once a healthy bed of granulation tissue is present, or left to heal by second intention. • Primary wound-healing time is 10–14 days for cutaneous healing in most avian patients.

MEDICATIONS

DRUG(S) OF CHOICE

Topical medications
• Medications used to enhance wound healing function in a variety of different ways. There is not one medication that is essential or best for every wound. • Topical antiseptic—chlorhexidine (0.05%), dilute povidone iodine (1% or less), hydrogen peroxide (3%). Chlorhexidine solution has minimal deleterious effects on wound healing and sustained residual activity. Povidone iodine at a concentration as low as 1–5% and hydrogen peroxide are toxic to fibroblasts and should be used as a single lavage. • Topical antimicrobial—use water-soluble topicals such as silver sulfadiazine cream (Silvadene, MarionLaboratories; Flamazine, Smith & Nephew). Silver sulfadiazine promotes epithelialization, penetrates necrotic tissues, but may impede wound contraction. Oil-based ointments should be avoided because when preened into the feathers, they inhibit normal thermoregulatory function.

Systemic medications
• Antibiotic therapy—must be efficacious against aerobic and anaerobic bacteria. Factors such as the ability of the antimicrobial to reach the wound at appropriate concentrations, bacterial resistance patterns, patient status and the existence of published pharmacokinetics in birds may dictate antibiotic choices. Examples of antibiotic therapy include: amoxicillin-clavulanic acid 125 mg/kg PO q12h for stable patients, piperacillin 100 mg/kg IM/IV q8–12h for unstable patients. • Analgesic therapy (butorphanol 1–4 mg/kg q4–12h) is indicated in most cases of bite wounds. Adverse effects in light of the patient's overall condition need to be considered. • Nonsteroidal anti-inflammatory medication (meloxicam 0.5 mg/kg PO/IM q12h) is indicated in most cases of bite wounds. Adverse effects in light of the patient's overall condition need to be considered.

CONTRAINDICATIONS
Do not flush wounds that could connect to the air sacs or lungs.

PRECAUTIONS
Fluid deficits should be addressed prior to using nonsteroidal anti-inflammatory medication.

POSSIBLE INTERACTIONS
None.

ALTERNATIVE DRUGS
N/A

FOLLOW-UP

PATIENT MONITORING
Varies with the severity of bite wound injuries and the patient's status.

PREVENTION/AVOIDANCE
• Birds should never be left unsupervised in the presence of a predator species. • Only indirect contact with predators should be permitted. • Only indirect contact between birds not tolerating each other should be permitted. • Do not allow a bird to land on another one's cage.

POSSIBLE COMPLICATIONS
• Abscess, septicemia—often secondary to closure of severely contaminated, infected, or traumatized wounds; also occurs following healing of puncture wounds.
• Mortality—generally due to either infection or concurrent trauma. • Disruption of the

wound, dressing or bandage materials—some birds may focus their attention on the wound site/dressing/bandage materials.

EXPECTED COURSE AND PROGNOSIS
• Varies with the severity of bite wound injuries. • Nerve and vascular damage—poor prognosis for healing and return to function, especially if the extremities are involved.

 MISCELLANEOUS

ASSOCIATED CONDITIONS
None.

AGE-RELATED FACTORS
N/A

ZOONOTIC POTENTIAL
N/A

FERTILITY/BREEDING
Avoid teratogenic antibiotics in laying hen.

SYNONYMS
None.

SEE ALSO
Crops stasis/ileus
Dehydration
Diarrhea
Fracture/luxation
Lameness
Polyuria
Regurgitation/vomiting
Trauma

ABBREVIATIONS
SIRS—systemic inflammatory response syndrome
SQ—subcutaneously
IO—intraosseously
IV—intravenously

INTERNET RESOURCES
http://www.merckmanuals.com/pethealth/special_subjects/emergencies/wound_management.html

Suggested Reading
Abrahamiam, F.M., Goldstein, E.J.C. (2011). Microbiology of animal bite wound infections. *Clinical Microbiology Reviews*, **24**(2):231–246.
Pavletic, M.M., Trout, N.J. (2006). Bullet, bite, and burn wounds in dogs and cats. *The Veterinary Clinics of North America. Exotic Animal Practice*, **36**(4):873–893.
Ritzman, T.K. (2004). Wound healing and management in psittacine birds. *The Veterinary Clinics of North America. Exotic Animal Practice*, **7**(1):87–104.

Author Isabelle Langlois, DMV, DABVP (Avian)

BORDETELLOSIS

BASICS

DEFINITION
Bordetellosis is a contagious disease caused by infection of birds with *Bordetella avium*. This organism is similar to other *Bordetella* species, as it tends to infect the tissues of the upper respiratory tract. In cockatiels and occasionally other psittacines, it presents as a syndrome known as "lockjaw", wherein the young birds are unable to open their mouths due to extension of sinusitis into the various surrounding tissues, often resulting in death unless aggressively treated. A dermonecrotic toxin produced by *B. avium* may play an important role in producing rhinitis, sinusitis, and "temporomandibular osteomyelitis" in these birds. It must be noted, however, that any serious sinus infection can result in this syndrome and that it is not always caused by *B. avium*. In young turkeys, it is the cause of coryza, a disease of high morbidity and low mortality, manifested mostly by tracheitis and sinusitis with excess mucous production. Bordetellosis has also been reported in ostrich chicks and quail and can be an opportunistic infection in chickens infected with bronchitis virus or exposed to poor conditions.

SYSTEMS AFFECTED
• Upper respiratory tract—rhinitis, frontal sinusitis (can be severe and necrotizing), tracheitis. • Lower respiratory tract—pneumonia and airsacculitis in turkeys can be seen with secondary colibacillosis or other coinfections. • Ophthalmic—conjunctivitis. • Musculoskeletal—in lockjaw syndrome – inflammation of all tissues in contact with the frontal sinus – muscles, nerves and bones and possibly joints.

GENETICS
N/A

INCIDENCE/PREVALENCE
Bordetellosis is reported sporadically in cockatiels, usually associated with a "lockjaw" type of presentation; it can manifest as an outbreak in some aviaries; more rarely it is reported in other psittacines. It is a significant problem in young turkeys and can be associated with outbreaks in turkey flocks.

GEOGRAPHIC DISTRIBUTION
Worldwide—in turkeys it appears in focal outbreaks; most common where turkeys are extensively reared.

SIGNALMENT
• **Species:** Most commonly seen in cockatiels as lockjaw syndrome; occasionally other psittacines; turkeys (coryza), quail are susceptible; opportunistic in chickens; has been reported in ostriches • **Mean age and range:** ○ Turkeys—generally less than six weeks of age ○ Cockatiels and other psttacines—usually 2–4 weeks.

• **Predominant sex:** Both sexes are susceptible.

SIGNS
Historical Findings
• In young turkeys, sudden onset after 7–10 day incubation period—up to 40% affected; low mortality. • In cockatiels with lockjaw—may initially have nasal or sinus signs, followed by gradual starvation due to inability to open mouth. May be seen in one or more birds, untreated birds die.

Physical Examination Findings
• Turkeys: ○ Swelling of sinuses. ○ Sneezing and nasal discharge. ○ Foamy eyes/ocular discharge. ○ Dyspnea, cough or "snick", loss or change of voice, excessive/abnormal respiratory sounds. ○ Decreased appetite and poor weight gain, thin, unthrifty. • In young cockatiels and other psittacine birds with lockjaw: ○ Weight loss/thin. ○ Inability to open the mouth (trismus) with resultant accumulation of material in the mouth—food and exudate. ○ Vocalizing due to hunger. ○ Signs of sinusitis—facial swellings, nasal discharge, abnormal respiratory sounds, dyspnea, sneezing.

CAUSES
Infection with *B. avium* in susceptible (young) birds.

RISK FACTORS
Unknown, but the age and general health status of each patient plays a role in whether it will become infected and, if infected, how sick the patient will become. Coinfections could play a role, as could environmental and genetic factors.

DIAGNOSIS

DIFFERENTIAL DIAGNOSIS
• Turkeys with respiratory disease: ○ Mycoplasmosis ○ Fowl cholera (Pasteurellosis) ○ Aspergillosis (brooder pneumonia) ○ Avian influenza ○ Newcastle disease ○ Chlamydiosis • Psittacines with lockjaw: ○ There are unlikely to be too many differential diagnoses to be considered when a young psittacine, especially a cockatiel, is presented with the typical clinical signs and findings. However, many different bacteria have been isolated in these cases. *Enterococcus sp.* infections may be overrepresented. ○ In theory, fungal infection, foreign bodies and introduction of irritants (food, etc.) into the nasal cavity could cause similar pathology.

CBC/BIOCHEMISTRY/URINALYSIS
Most of these patients would not be subjected to blood or urine testing, but if they were, there would likely be nonspecific findings. An inflammatory leukogram would be expected in patients with significant inflammation; in cases of lockjaw, there could be elevations in muscle enzymes

OTHER LABORATORY TESTS
• Turkeys: ○ Culture and sensitivity of affected tissues may reveal *B. avium*, but it can be difficult to isolate if there are other faster-growing bacteria present. ○ Serologic tests are available for turkeys. • Lockjaw cases: ○ Cytology of choana and or nasal/sinus flush material. ○ C&S (culture and sensitivity) of choana or sinus flush. ○ *B. avium* PCR testing is also available.

IMAGING
Lockjaw cases: Radiographs and skull CT might be helpful but are rarely necessary and generally cost prohibitive.

DIAGNOSTIC PROCEDURES
• Turkeys: Sampling of infected tissues—swab of or flush from nasal/sinus/trachea in order to procure samples for C&S, cytology, PCR, plus necropsy and histopathology of tissues of dead birds. • Lockjaw cases: Sampling of infected tissues—swab of or flush from nasal/sinus/trachea in order to procure samples for C&S, cytology, PCR.

PATHOLOGIC FINDINGS
• Turkeys: ○ Watery eyes, extensive mucous in sinuses and trachea, softening and flattening of the trachea; lesions rarely extend past the tracheal bifurcation. Pneumonia and airsacculitis in secondary infections (esp. colibacillosis). • Lockjaw cases: ○ Moderate subacute necrotizing to mucopurulent to fibrino-suppurative rhinitis and sinusitis and osteomyelitis. ○ Mild-to-moderate subacute necrotizing and fibrosing myositis, tendonitis, and perineuritis. ○ Some birds have inflammation of the quadrate-pterygoid apparatus and jugal arch, characterized by periosteal fibrosis and irregular lysis and mixed inflammatory infiltrates. ○ Rarely, peracute hepatic necrosis, chronic hepatic fibrosis, lymphocytic depletion, and subacute pneumonia with air sacculitis are present.

TREATMENT

APPROPRIATE HEALTH CARE
• Turkeys with coryza will generally get better unless coinfected with other viruses or bacteria. • Psittacines with sinusitis and secondary lockjaw syndrome need aggressive therapy—antibiotics PO or parenterally, nasal/sinus flushing, supportive care such as fluid therapy and gavage feeding, pain management and anti-inflammatory medications, as well as physical therapy of the complex masticatory apparatus.

NURSING CARE
• Turkeys with coryza need little nursing care. Individual cases may need supportive care, depending on whether they have secondary or coinfections. • Psittacines with lockjaw need

BORDETELLOSIS (CONTINUED)

extensive nursing care as mentioned above—they are generally young and need frequent feeding and medicating as well as vigilant monitoring of clinical signs and body weight. Physical therapy can be done several times a day by massaging the muscles of mastication and by opening the beak. This can be done digitally or with the use of hemostats or, in the case of larger birds a thumb-screw small mammal oral speculum, so that the fibrotic muscles can be gradually stretched. This is difficult to do in an awake patient; it can be done carefully in an anesthetized patient, but care must be taken to avoid aspiration.

ACTIVITY
N/A

DIET
• Turkeys: NA • Lockjaw cases: Generally gavaged with hand rearing formula—can be difficult due to difficulty passing the tube; alternatively an esophagostomy tube can be placed.

CLIENT EDUCATION
• Turkeys: General conversation about the illness and course to be expected; do not overtreat with antibiotics. Bring in recently deceased patients for necropsy to further evaluate the health of the flock. • Lockjaw cases: Depends on the situation—if owner is going to be treating the patient, they need a lot of guidance and should be given a poor prognosis in general. In aviary situations, the breeder needs to be educated about general aviary health management—cleanliness, disinfection, biosecurity, traffic flow, etc.

SURGICAL CONSIDERATIONS
N/A

MEDICATIONS

DRUG(S) OF CHOICE
• In turkeys: Antibiotics have generally not been very effective and disease will run its course. • In lockjaw cases: Selection of antibiotic should be based upon results of antimicrobial culture and sensitivity (C&S). However, this organism can be difficult to grow in the laboratory and, despite the selection of a proper antibiotic based on C&S, there may be minimal response to treatment due to various factors such as difficulty in achieving proper antibiotic concentration in the areas that are infected, presence of excess mucous, etc. • Various medications can be added to saline to flush the infraorbital sinus at least twice a day. • Anti-inflammatory medications such as meloxicam are essential—up to 1 mg/kg BID PO or by gavage. • Butorphanol can be given at up to 4 mg/kg as often as every four hours as needed for pain.

CONTRAINDICATIONS
N/A

PRECAUTIONS
N/A

POSSIBLE INTERACTIONS
N/A

ALTERNATIVE DRUGS
N/A

FOLLOW-UP

• Turkeys with coryza: N/A • Lockjaw cases: Quite a bit of follow up is needed, depending on whether the patient is treated in-house or by the client. Several months of treatment is likely to be needed. If the patient improves, and the bird is able to eat, the prognosis is much better and there is less need for continual follow-up.

PATIENT MONITORING
N/A

PREVENTION/AVOIDANCE
N/A

EXPECTED COURSE AND PROGNOSIS
In turkeys, the disease is usually self-limiting and will resolve with time, unless there are secondary infections such as *E. coli*, in which case there could be airsacculitis and pneumonaia and higher mortality. In psittacines with lockjaw, the disease is usually fatal unless aggressively treated. Since treatment is labor intensive, the risks and benefits of attempted treatment must be considered. With aggressive treatment, some birds can be saved, especially larger psittacines, but in general the prognosis is poor since most patients are cockatiels – thus the amount of pathology is great when compared to patient size and there are more difficulties in trying to perform physical therapy in these patients.

MISCELLANEOUS

ASSOCIATED CONDITIONS
N/A

AGE-RELATED FACTORS
In turkeys and cockatiels, this is generally seen in younger birds, but it is possible that older and/or recovered birds may be carriers.

ZOONOTIC POTENTIAL
B. avium may be a rare opportunistic infection in humans.

FERTILITY/BREEDING
N/A

SYNONYMS
In cockatiels—lockjaw, or lockjaw syndrome
In turkeys—infectious coryza or rhinotracheitis

SEE ALSO
Anorexia
Aspergillosis
Aspiration
Beak malocclusion
Chlamydiosis
Mycoplasmosis
Pasteurellosis
Rhinitis and sinusitis
Viral disease

ABBREVIATIONS
N/A

INTERNET RESOURCES
There are multiple good poultry disease websites, including the on-line Merck Veterinary Manual.

Suggested Reading
Clubb, S.L., Homer, B.L., Pisani, J., Head, C. (1994). Outbreaks of bordetellosis in psittacines and ostriches. *Proceedings of the Association of Avian Veterinarians Annual Conference*, pp. 63–68.
Fitzgerald, S.D., Hanika, C., Reed, W.M. (2001). Lockjaw syndrome in cockatiels associated with sinusitis. *Avian Pathology*, **30**:49–53.
Messenger, G. (2009). Successful management of lockjaw in a white-bellied caique (*Pionites leucogaster*). *Proceedings of the Association of Avian Veterinarians Annual Conference*.
Skeeles, J.H., Arp, L.H. (1977). Bordetellosis (turkey coryza). In: Calnek, B.W. (ed.), *Diseases of Poultry*, 10th edn. Ames, IA: Iowa State Press, pp. 275–288.
Schmidt, R.E., Reavill, D.R., Phalen, D.N. (2003). *Respiratory system. Pathology of Pet and Aviary Birds*. Ames, IA: Iowa State Press, pp. 17–40.
Author George Messenger, DVM, DABVP (Avian)

BOTULISM

BASICS

DEFINITION
- Botulism is a disease caused by the neurotoxin produced by *Clostridium botulinum*. It is an anaerobic, Gram-positive spore-forming, motile rod that commonly prospers in decaying organic matter and decomposing tissue. The disease is in essence a food poisoning caused by the ingestion of neurotoxin-laden food. In some cases the disease can be caused by enterotoxicosis caused by bacteria that colonized the gastrointestinal tract (GIT) of an individual bird and produced the neurotoxin. This form of the disease usually occurs in birds that can normally carry the bacteria in their GIT, such as chickens and ostriches. It is usually the result of preceding compromise to the normal gut flora.
- There are seven serotypes of *C. botulinum* but the majority of botulism outbreaks are caused by type C toxin with sporadic die-offs caused by types E and A toxins. *C. botulinum* spores are found in soil and wetland worldwide while type C spores are primarily found in freshwater and marine environment.
- The carcass-maggot cycle of botulism is also recognized; the disease occurs secondary to the ingestion of fly larvae from decomposing carrion that is loaded with the neurotoxin. This route of poisoning is identified in some outbreaks in waterfowl but it is the prominent route in raptors that are otherwise considered less susceptible to the disease. Vultures are considered resistant.
- Botulism is the most significant population disease in migratory birds, especially waterfowl and shorebirds and has important ecological effect. Yearly massive die-offs occur in localized outbreaks in wetlands all around the world. Die-offs of 50,000 birds or more are common in a single outbreak.
- Its epizoolotiological patterns are very complex and diverse depending on numerous factors such as high environmental temperature and alkaline water pH, fly density, carcass density, and foraging patterns. Botulism causes flaccid paralysis followed by death within hours or days.

PATHOPHYSIOLOGY
- The *Clostridium botulinum* exotoxin is a neurotoxin that causes flaccid paralysis of the neck and limbs, pharyngeal and respiratory muscles and subsequent death. Botulism toxins are the most toxic in nature. A protein complex against proteolysis in the GIT protects the neurotoxin and once in the systemic circulation it targets the peripheral cholinergic nervous system. It does not cross the blood-brain barrier. The extreme toxicity level is a result of its high affinity to the presynaptic membrane and specific and prolonged inhibition of neurotransmitter release from the nerve ending. The neurotoxin irreversibly binds to the neuronal membrane. It is then internalized into the cell by endocytosis with the final step in the intoxication being the prevention of acetylcholine release resulting in flaccid neuromuscular paralysis.
- GIT toxicosis is less common but occurs in chickens and may also occur occasionally in wild birds such as raptors, waterfowl and songbirds. Correlation with lead poisoning, vitamin A deficiency or other factors debilitating the host GIT function and flora was made. This route of exposure is important in individual birds in captive situations but is insignificant in the massive wild bird die-offs.
- Another pathogenic mechanism of *C. botulinum* type C is caused by C2 toxin. This is not a neurotoxin but rather a lethal hemorrhagic toxin. Its role in avian botulism is unclear although the toxin is prevalent in wetlands.

SYSTEMS AFFECTED
- Neuromuscular.
- Respiratory—paralysis of the respiratory muscles is usually the cause of death.
- Cardiovascular.
- Endocrine—the neurotoxin affects the adrenal glands.
- Gastrointestinal—severe ileus, megacolon, megacloaca.
- Renal/Urologic—dehydration, dysuria and acute renal failure are common.
- Hemic—recumbent birds are rapidly infected with leeches and other parasites and suffer severe anemia.

GENETICS
N/A

INCIDENCE/PREVALENCE
Outbreaks in wild birds involve thousands to millions of birds in a single event. Most outbreaks are seasonal and occur in the summer-fall when the temperatures are higher than 26°C.

GEOGRAPHIC DISTRIBUTION
Worldwide except the Antarctic. Most of the largest die-offs have occurred in North America and the frequency of outbreaks in higher in the USA and Canada compared to other countries. Avian Botulism type C is frequently associated with wetland sediments in the USA and Canada.

SIGNALMENT
- **Species:** All migratory waterfowl and shorebirds are highly sensitive. Botulism occurs less frequently in raptors and passerines and vultures are considered resistant. Outbreaks were described in farmed chickens and ostriches and in zoos.
- **Mean age and range:** N/A
- **Predominant sex:** N/A

SIGNS

General Comments
Neuromuscular paralysis is the main mechanism, secondarily all other organ systems are affected.

Historical Findings
- Massive acute die-offs with no earlier warnings. Within the flocks individual birds can show flaccid paralysis, inability to walk or fly, diarrhea, and shallow and slow breathing, all of which would be evident to field agents detecting and monitoring the outbreaks.

Physical Examination Findings
- Progressive paresis: flaccid neck and limbs.
- Ataxia.
- Dysphagia.
- Dyspnea.
- Dehydration.
- Green diarrhea.
- Acute death.

CAUSES
Clostridium botulinum neurotoxin type C (A and E in some cases) that binds irreversibly to the peripheral cholinergic nervous system.

RISK FACTORS
- Complex association among environmental factors and conditions is presumed to correlate with avian botulism outbreaks. However contradicting evidence continuously makes this association rather unclear.
- High environmental temperature, hence the summer and fall are correlated with the massive die-offs. But some outbreaks were reported in the winter and spring.
- Stagnant or brackish water in wetlands are optimal for proliferation of the bacteria in decaying matter; again few outbreaks occurred in areas of cold, well oxygenated running water.
- Bird crowdedness during migration and insect abundance are important risk factors as well.

DIAGNOSIS

DIFFERENTIAL DIAGNOSIS
In the individual bird, any systemic neurological disease can be considered as a differential diagnosis.

Other toxicities such as those caused by carbamates, organophosphorous, heavy metals or algae should be considered.

CBC/BIOCHEMISTRY/URINALYSIS
Hematological and biochemical changes are often absent due to the acute nature of the disease. Heterophilia, elevated liver enzymes and hyperuricemia may be present.

OTHER LABORATORY TESTS
- Presumptive diagnosis can be made based on the typical history and clinical signs alone.
- Definitive diagnosis is made by tissue toxin analysis, usually from blood, kidney or liver tissue.
- Mouse inoculation neutralization assay is still considered the most sensitive assay.
- Culture of the organism.
- PCR and rtPCR.
- Enzyme linked immunoassay (ELISA) can be used to detect the neurotoxin.

IMAGING
N/A

DIAGNOSTIC PROCEDURES
N/A

PATHOLOGIC FINDINGS
- Gross pathology of dead birds is often consistent with no obvious lesions.
- Marked

BOTULISM (CONTINUED)

pulmonary congestion and edema have been described. Maggots and leeches can be found in the GIT.

TREATMENT

APPROPRIATE HEALTH CARE
Treatment can be rewarding and most birds if caught early recover well with supportive care. Emphasis should be placed on protecting the patient from the elements, rehydration and nutritional support.

NURSING CARE
Aggressive fluid therapy is imperative. Recumbent birds should be provided with appropriate cushiony support.

ACTIVITY
Birds are usually recumbent or weak and have limited ability to walk or fly.

DIET
Providing easy access to food or gavage feeding in paralyzed birds is necessary.

CLIENT EDUCATION
N/A

SURGICAL CONSIDERATIONS
N/A

MEDICATIONS

DRUG(S) OF CHOICE
• Type A and type C antitoxin injections are available and are very effective in facilitating recovery and preventing the progression of clinical signs in affected birds. • Antibiotics against anaerobes such as metronidazole, penicillin and amoxicillin are useful in eliminating the bacteria from the GIT.

CONTRAINDICATIONS
N/A

PRECAUTIONS
N/A

POSSIBLE INTERACTIONS
N/A

ALTERNATIVE DRUGS
A vaccine against type C botulism during an outbreak was shown to aid in the healing process.

FOLLOW-UP

Botulism is a disease of wild birds and much more infrequently of smaller captive populations such as chickens and zoological collections. Once recovered, follow up is usually not necessary.

PATIENT MONITORING
N/A

PREVENTION/AVOIDANCE
• Vaccine derived from neurotoxin C was developed and can be useful in captive birds kept in zoological collections. • Other measures that can aid in prevention include carcass removal and drainage of prone wetlands prior to the warm seasons.

POSSIBLE COMPLICATIONS
Treatment of large number of birds is cost prohibitive.

EXPECTED COURSE AND PROGNOSIS
In the individual birds that received intensive supportive care the survival rate ranged between 75-90%. Full recovery can take several weeks.

MISCELLANEOUS

ASSOCIATED CONDITIONS
N/A

AGE-RELATED FACTORS
N/A

ZOONOTIC POTENTIAL
N/A

FERTILITY/BREEDING
N/A

SYNONYMS
Limberneck, Western duck disease, Alkali poisoning

SEE ALSO
Clostridiosis
Crop stasis/ileus
Dehydration
Diarrhea
Emaciation
Heavy metal toxicity
Neurologic conditions
Urate/fecal discoloration

ABBREVIATIONS
GIT—gastrointestinal tract

INTERNET RESOURCES
http://www.aav.org
http://www.merckmanuals.com
www.ncbi.nlm.nih.gov
http://www.cdc.gov/nczved/divisions/dfbmd/diseases/botulism/

Suggested Reading
Rocke, T.E., Bollinger, T.K. (2007). Avian botulism. In: Thomas, N.J., Hunter, B., Atkinson, C.T. (eds). *Infectious Diseases of Wild Birds*. Oxford: Blackwell, pp. 377–416.

Author Shachar Malka, DVM, Dipl ABVP (Avian)

Campylobacteriosis

BASICS

DEFINITION
Campylobacteriosis is caused by infection with bacteria from the genus *Campylobacter*, with *Campylobacter jejuni* most commonly associated with infection.

PATHOPHYSIOLOGY
Most birds infected with *Campylobacter* are carriers, which are infected, asymptomatic shedders. Clinical disease, such as subacute or chronic hepatitis, can occur, but often, factors such as parasitic or other infections likely predispose the bird to clinical disease.

SYSTEMS AFFECTED
- Gastrointestinal—hemorrhagic enteritis; diarrhea, sometimes with yellow staining.
- Hepatobiliary—subacute to chronic hepatitis.
- Non-specific—lethargy, anorexia, emaciation.

GENETICS
All bird species can serve as competent reservoirs for this thermophilic organism.

INCIDENCE/PREVALENCE
Prevalence varies with species but is generally high in poultry. One study found that 25.2% of galliformes, 12.9% of anseriformes, 8.3% of columbiformes, and only one of 179 psittaciformes examined on necropsy were positive for *Campylobacter jejuni* (Yohasundram et al., 1989).

GEOGRAPHIC DISTRIBUTION
Campylobacter has worldwide distribution.

SIGNALMENT
Species: All avian species.

SIGNS
General Comments
Most avian species are asymptomatic reservoirs of *Campylobacter*.

Historical Findings
- Possible exposure to feces of poultry or wild birds.
- Weight loss, diarrhea, decreased appetite.

Physical Examination Findings
- Poor body condition.
- Diarrhea, which might include blood or changes in color of urates to yellow or green.

CAUSES
Campylobacteriosis is caused by infection with the *Campylobacter* sp., most frequently the bacterium *Campylobacter jejuni*.

RISK FACTORS
Although many birds are infected with *Campylobacter*, few develop clinical disease. Birds that develop clinical disease, such as enteritis or hepatitis, often will have other predisposing factors such as infection with other infectious agents.

DIAGNOSIS

DIFFERENTIAL DIAGNOSIS
Other causes of enteritis, diarrhea, or hepatitis such as *Salmonella*, *E. coli*, *Yersinia* spp.

CBC/BIOCHEMISTRY/URINALYSIS
Asymptomatic patients are unlikely to have any changes. Animals with enteritis diarrhea may have indications of dehydration such as an increased total protein or packed cell volume. Animals with hepatitis may have elevated aspartate aminotransferase and bile acids or decreased albumin and protein.

OTHER LABORATORY TESTS
A sample of feces or tissue should be submitted for culture to isolate the organism. The diagnostic laboratory should be notified of suspected *Campylobacter* as an etiologic agent because special procedures are needed for growth. Pulsed field gel electrophoresis is used in investigations to determine the source of an outbreak.

IMAGING
Radiography may show an enlarged liver if hepatitis is present, but a sample must be collected for culture to determine the etiology of infection.

DIAGNOSTIC PROCEDURES
Collection of feces or a tissue sample for culture will be needed to definitively diagnose campylobacteriosis. A fine needle aspirate may be needed to collect a sample from the liver. Samples from affected tissues can be collected at necropsy.

PATHOLOGIC FINDINGS
The liver may be enlarged, pale, or greenish in color and often congested. Hemorrhage may be present. Enteritis can also be present. Diffuse inflammation may be present.

TREATMENT

APPROPRIATE HEALTH CARE
Most infected animals are asymptomatic and require no treatment. Treatment for those with clinical disease is described below.

NURSING CARE
Nutritional support, including fluids, will help correct dehydration associated with diarrhea and weight loss.

ACTIVITY
Movement of the animal should be limited to reduce fecal contamination, which could lead to exposure of other animals or humans to *Campylobacter*.

DIET
Although appetite may be decreased, oral intake of normal food and fluids is acceptable.

CLIENT EDUCATION
Campylobacter is zoonotic, and owners must take precautions to eliminate exposure to themselves, other humans, and other animals. Hand washing and cleaning and disinfection of contaminated areas and equipment are essential. Most human infections are associated with consumption of poultry meat.

SURGICAL CONSIDERATIONS
N/A

MEDICATIONS

DRUG(S) OF CHOICE
Treatment should be based on culture and sensitivity of the organism; asymptomatic patients are unlikely to be identified and treated. Oral erythromycin may be used to treat poultry at a dose of 10–30 mg/kg for four days or psittacines at 30–40 mg/kg. Azithromycin is currently used to treat infections in humans.

CONTRAINDICATIONS
Antibiotics that have not been shown to be appropriate based on culture and sensitivity.

PRECAUTIONS
Campylobacter can develop antibiotic resistance. Fluoroquinolones should not be used in poultry.

POSSIBLE INTERACTIONS
N/A

ALTERNATIVE DRUGS
N/A

FOLLOW-UP

PATIENT MONITORING
Fecal output and consistency should be monitored and will return to normal typically within a week. Attitude should also improve with fluid administration. Plasma biochemistry can be repeated to determine if indicators of hepatitis are improving.

PREVENTION/AVOIDANCE
Avoid exposure to *Campylobacter* by keeping environments clean and disinfected and purchasing food from reputable sources; wildlife should be excluded from contact with pets or livestock.

POSSIBLE COMPLICATIONS
Humans can become sick with campylobacteriosis.

EXPECTED COURSE AND PROGNOSIS
Most animals will not experience clinical disease from infection; those that develop hepatitis have a guarded prognosis and can suffer significant mortality in flock situations.

CAMPYLOBACTERIOSIS (CONTINUED)

MISCELLANEOUS

ASSOCIATED CONDITIONS
N/A

AGE-RELATED FACTORS
Young animals may be more likely to develop severe clinical disease.

ZOONOTIC POTENTIAL
There are approximately 1.3 million cases of human campylobacteriosis in the United States annually. Most of these cases are foodborne, particularly involving consumption of poultry meat and unpasteurized milk.

FERTILITY/BREEDING
N/A

SYNONYMS
N/A

SEE ALSO
Colibacillosis
Diarrhea
Emaciation
Liver disease
Pasteurellosis
Salmonellosis

ABBREVIATIONS
N/A

INTERNET RESOURCES
Centers for Disease Control and Prevention (CDC): http://www.cdc.gov/nczved/divisions/dfbmd/diseases/campylobacter/
Centers for Disease Control and Prevention (CDC) Healthy Pets: http://www.cdc.gov/healthypets/
U.S. Department of Agriculture Healthy Birds: http://healthybirds.aphis.usda.gov/
Merck Manuals: http://www.merckmanuals.com/vet/poultry/avian_campylobacter_infection/overview_of_avian_campylobacter_infection.html; http://www.merckmanuals.com/vet/digestive_system/enteric_campylobacteriosis/overview_of_enteric_campylobacteriosis.html

Suggested Reading
Keller, J.I., Shriver, W.G., Waldenstrom, J. et al. (2011). Prevalence of *Campylobacter* in wild birds of the mid-Atlantic region, USA. *Journal of Wildlife Diseases*, **47**:750–754.
Lin, J. (2009). Novel approaches for *Campylobacter* control in poultry. *Foodborne Pathogens and Disease*, **6**:755–765.
Yogasundram, K., Shane, S.M., Harrington, K.S. (1989). Prevalence of *Campylobacter jejuni* in selected domestic and wild birds in Louisiana. *Avian Diseases*, **33**:664–667.

Author Marcy J. Souza, DVM, MPH, DABVP (Avian), DACVPM

CANDIDIASIS

BASICS

DEFINITION
Candidiasis, "thrush" or "sour crop" is a common yeast infection reported in many species of birds. It is caused by *Candida spp* with *C. albicans* being most commonly reported. Other candida species have recently been reported as advanced molecular methods for species identification have become more readily available.

Candidiasis is commonly diagnosed in psittacine birds, pigeons and doves, songbirds, back-yard poultry, raptors, and waterfowl, but can cause disease in any bird species. Juvenile birds are most commonly affected. Adult birds can be affected at any age as well. Stress, malnutrition and immunosuppression are key factors. Rarely, infection can be seen even in adult birds without any apparent predisposing factors. Candidiasis is an opportunistic infection and candida spores can be part of normal gastrointestinal flora in apparently healthy birds. Clinical disease occurs with overgrowth, vitamin A deficiency, spoiled feed source, or secondary to primary damage to the mucosal membrane integrity. Infection is usually restricted to the upper GIT and the oropharynx and the esophagus are primarily affected, however infection can settle in all parts of the digestive system. Infection of the choana and the upper respiratory system is frequently reported. Rare involvement of other organ systems occurs.

PATHOPHYSIOLOGY
Overgrowth of budding candida yeasts on the surface of the gastrointestinal mucosal membrane results in thickening of the mucosal membrane. Invasion into the deeper cell layer results in ulcerative disease and dysfunction of the affected organ. Crop stasis, anorexia and regurgitation are the most common signs.

SYSTEM AFFECTED
- Gastrointestinal—oropharynx, esophagus/crop are most commonly affected. Lower gastrointestinal tract (GIT) infection in more severe chronic cases.
- Respiratory—choana, infraorbital sinus are frequently involved. Parabronchial pneumonia in severe disseminated cases.
- Skin—rare, usually in immunocompromised birds.
- Cardiovascular—rare, myocarditis in severe disseminated cases. • Nervous—rare, in severe disseminated cases. • Ophthalmic—rare.
- Reproductive—rare.

GENETICS
N/A

INCIDENCE/PREVELENCE
- All *Candida spp* have worldwide distribution and are considered part of normal gastrointestinal flora in birds both captive and free range. However, *C. albicans* is the predominant isolate worldwide. Other common *Candida* isolates include *C. parapsilosis*, *C. tropicalis*, *C. hemicola*, *C. famata* and *C. glabrata* is a common isolate in the USA and Europe. *C. kusei* has been reported in Australia. • In one study in healthy cockatiels *Candida spp.* isolates were found in 65% and *C. albicans* represented 32% of the isolates followed by *C. tropicalis* (20%) and *C. famata* (10%).

GEOGRAPHIC DISTRIBUTION
Worldwide.

SIGNALMENT
- **Species:** In psittacine birds, cockatiels, lovebirds and budgerigars are over represented. Also common in passerines, columbiformes, raptors, and back-yard poultry. Less common in waterfowl. • **Mean age and range:** All ages are susceptible but juvenile birds are highly susceptible.
- **Predominant sex:** No sex predilections.

SIGNS

General Comments
Signs are indicative of the organ affected. It can be just one clinical sign in the case of an isolated lesion or severe systemic illness in the case of disseminated gastrointestinal, respiratory or other organ system infection.

Historical Findings
- Anorexia • Regurgitation • Weight loss
- Exercise intolerance • Coughing and sneezing

Physical Examination Findings
- Nasal discharge • Coughing and sneezing
- Crop stasis • Halitosis • Tachypnea and labored breathing • Stomatitis (oral white plaques) • Melena/hematochezia • Partially digested stool

CAUSES
Candida spp

RISK FACTORS
- Hand feeding (psittacine birds). • Young age. • Stress: Overcrowding, raptors used for falconry and back-yard poultry are commonly affected due to inevitable increased stress level. • Malnutrition: Rancid feed, vitamin A deficiency. • Immunosuppression. • Poor environmental hygiene. • Concurrent disease (trichomoniasis, poxvirus, other). • Prolonged antibiotic therapy. • Feather destructive behavior. • Chronic skin infection.
- Cockatiels can harbor mixed population of *Candida spp.* in their GIT and can disseminate the organisms into the environment.

DIAGNOSIS

DIFFERENTIAL DIAGNOSIS
- Vitamin A deficiency (common in psittacines). • Capillariasis (common in nonpsittacines). • Mycobacteriosis.
- Chlamydiosis. • Trichomoniasis.
- Macrorhabdiosis. • Pox virus infection in the relevant species.
- Stomatitis/Ingluvitis/Esophagitis/Gastroenteritis of other etiology.

CBC/BIOCHEMISTRY/URINALYSIS
- Usually unremarkable. • Fibrinogen can be elevated.

OTHER LABORATORY TESTS
- Cytology from an affected lesion will show numerous budding yeasts and possibly pseudohyphae alongside inflammatory cells and bacteria. • Crop cytology is especially useful. • Gram stains are both effective in demonstrating infection on fecal or crop samples. • Culture and Sensitivity (C&S).
- PCR may be used to identify specific Candida species in cases of refractory infection or resistance. • Histopathology.

IMAGING
- Whole body radiographs may demonstrate signs that are compatible with ileus such as distended crop, gas filled loops of bowel.
- Delayed GIT emptying time can be demonstrated with contrast barium radiographs or fluoroscopy. • Skull radiographs are generally not useful in demonstrating sinusitis.

DIAGNOSTIC PROCEDURES
- Sinus flush and ingluvial lavage or swabbing can aid in collecting samples for cytology and C&S. • Endoscopy can be used to evaluate the thickened and inflamed esophagus/crop.
- "Turkish Towel" is a term used to describe the appearance of the esophagitis/ingluvitis caused by candidiasis.

PATHOLOGICAL FINDINGS
Thickened and ulcerative mucosal membrane with abundant yeasts, budding or forming pseudohyphae accompanied by necrosis and granulomatous inflammation.

TREATMENT

APPROPRIATE HEALTH CARE
In the individual patients, health care should be aimed at a specific affected system but also at alleviating underlying risk factors that contribute to the pathogenicity of candidiasis, such as: stress, overcrowding, and poor nutritional hygiene.

CANDIDIASIS (CONTINUED)

NURSING CARE
Outpatient in most cases. Fluid therapy and nutritional support may be needed in debilitated patients.

ACTIVITY
N/A

DIET
Nutritional support may be necessary with patients that suffer anorexia or weight loss. Vitamin A supplementation may promote healing of the mucosal membranes and will correct possible underlying deficiency.

CLIENT EDUCATION
Improve hygiene measures and reduce stress.

SURGICAL CONSIDERATIONS
N/A

MEDICATIONS

DRUG(S) OF CHOICE
• Nystatin is the first drug of choice. It acts on the mucosal membrane surface and does not get absorbed systemically, therefore it is a safe drug to use compared to other antifungals. 100,000–300,000 IU/kg PO q8-12h for 7–14 days is recommended. Lower dose preferred. • Amphotericin B: 100 mg/kg PO q12–24h is another recommended drug as it is not absorbed systemically from the GIT.
• Topical treatment ○ Topical antifungal (Miconazole, Amphotericin B, Terbinafine, Enilconazole are few that are available) gels or ointments can be applied directly on oral lesions to aid with systemic therapy. They can be used nasally or as part of a naso-choanal flush, q12-24h for 7–10 days. • Silver Sulfadiazine can be used topically. Emulsion combined with ciprofloxacin is available and well tolerated and easy to use in most birds.

PRECAUTIONS
• All drugs require direct contact with the lesion in order to be effective and are of low risk if there is no systemic absorption.
• Amphotericin B can be nephrotoxic and should be used to treat persistent infection in well hydrated patients, and when diffuse systemic ulcerative disease is not suspected, in order to avoid systemic absorption.

POSSIBLE INTERACTIONS
N/A

ALTERNATIVE DRUGS
• Azoles ○ Systemic use should be spared for severe infection where systemic infection is suspected or as a treatment against resistant strains. ○ Azoles should not be used when severe liver disease is suspected. ○ Azoles may cause anorexia and elevated liver enzymes. ○ Fluconazole 5–10 mg/kg PO q12h, 10 mg/kg PO q24h (cockatiels), higher dose of 20 mg/kg PO q48h was also reported in psittacines. ○ Itraconazole 5–10 mg/kg PO q24h, lower dose recommended, dosing can be divided q12h. ○ Voriconazole 10–20 mg/kg PO q24h. • Medicated water: Medicated water is a practice that may be used in a flock situation when it is impractical to treat the individual bird. ○ Fluconazole 100 mg/L drinking water for eight days.

FOLLOW-UP

PATIENT MONITORING
N/A

PREVENTION/AVOIDANCE
N/A

POSSIBLE COMPLICATIONS
Failure to thrive in cases of infected juvenile birds.

EXPECTED COURSE AND PROGNOSIS
• Follow up examination and crop/oral cytology preparations and fecal cytology evaluation is recommended after 10–14 days.
• Prognosis is good.

MISCELLANEOUS

ASSOCIATED CONDITIONS
N/A

AGE-RELATED FACTORS
N/A

ZOONOTIC POTENTIAL
Candida spp. is abundant in the environment. However, cockatiels were reported to be environmental asymptomatic shedders and can pose potential risk to humans, especially if immunocompromised. Close genetic relation was shown between *C. albicans* isolated from humans and chickens.

FERTILITY/BREEDING
N/A

SYNONYMS
Sour crop
Thrush
Turkish towel

SEE ALSO
Aspergillosis
Crop stasis/ileus
Diarrhea
Flagellate enteritis
Ingluvial hypomotility/ileus
Macrorhabdus ornithogaster
Nutritional deficiencies
Oral plaques
Poxvirus
Regurgitation/vomiting
Sick-bird syndrome
Trichomoniasis
Undigested food in droppings
Urate/fecal discoloration
Viral disease

ABBREVIATIONS
GIT—gastrointestinal tract

INTERNET RESOOURCES
http://www.merckmanuals.com
www.ncbi.nlm.nih.gov
http://www.aav.org

Suggested Reading
Ratzlaff, K., Papich, M.G., Flammer, K. (2011). Plasma concentrations of fluconazole after a single oral dose and administration in drinking water in cockatiels (*Nymphicus hollandicus*). *Journal of Avian Medicine and Surgery*, **25**(1):23–31.
Sidrim, J.J., Maia DC, Brilhante RS, *et al.* (2010). *Candida* species isolated from the gastrointestinal tract of cockatiels (*Nymphicus hollandicus*): *In vitro* antifungal susceptibility profile and phospholipase activity. *Veterinary Microbiology*, **145**(3–4):324–328.
Shrubsole-Cockwil, A.N., Millins, C., Jardine, C., *et al.* (2010). Avian pox infection with secondary *Candida albicans* encephalitis in a juvenile golden eagle (*Aquila chrysaetos*). *Journal of Avian Medicine and Surgery*, **24**(1):64–71.
Author Shachar Malka, DVM, DABVP (Avian)

CARDIAC DISEASE

BASICS

DEFINITION
Cardiac diseases include any disease state of the heart. Cardiac diseases are common in pet birds. As the population of birds ages, it is becoming an increasing proportion of health problems. The diagnosis and treatment of cardiac diseases can be complicated by technical limitations brought on by the small size and unique anatomy of birds. Even the extremely rapid heart rate can make some diagnostics challenging. Heart disease may include disorders of the myocardium, the heart valves, the conduction system, or the pericardium. Congestive heart failure may be the end result of many of the heart diseases.

PATHOPHYSIOLOGY
There are a variety of causes for heart disease. Diseases of the heart muscle can be brought on by a number of infectious agents. These would include *Sarcocystis*, bornavirus, avian serositis virus, *Chlamydia*, and other bacteria. Vitamin E and selenium deficiency may contribute to cardiac muscle degeneration, especially in emu and cockatiels. Hemochromatosis may affect the heart muscle in addition to the liver and other organs. Drugs such as furazolidone can be toxic to the myocardium. Avocado toxicosis also affects the heart muscle. Advancing age and inactivity may combine to result in myocardial degeneration. Atherosclerosis is common and although it generally does not directly affect the heart, it can occasionally result in infarcts. Bacteremia may result in vegetative endocarditis, resulting in valvular insufficiency. Degeneration of the valves may also occur idiopathically. Congenital defects have been rarely reported in birds. With most of these disorders, the end result may be heart failure. Heart failure may be left sided, right sided, or both.

SYSTEMS AFFECTED
Because the heart is responsible for supplying oxygen and nutrients, and for carrying away carbon dioxide and waste products, nearly all organs can be affected by heart disease to some degree. Some of these will be more readily evident than others, however.
• Cardiovascular output may be reduced. The perfusion of tissues may be poor. Blood pressure can be increased or decreased. Ascites or anasarca may occur. • Nervous signs may include syncope or seizures. • Neuromuscular weakness is common with heart disease.
• Respiratory signs may occur with left sided heart failure. • Hepatobiliary system is affected primarily by congestion during congestive heart failure.
• Hemic/Lymphatic/Immune. The spleen may become congested during congestive heart failure. Polycythemia may occur with chronic hypoxia. • Behavioral changes are often subtle, but include reduced activity.

GENETICS
Because of the rarity of congenital heart problems, the genetics are not known.

INCIDENCE/PREVALENCE
Heart disease appears to have 10–15% prevalence in pet birds.

GEOGRAPHIC DISTRIBUTION
There is no specific geographic distribution.

SIGNALMENT
• **Species:** No specific species are predisposed overall. Specific disease states may have predilections. • **Mean age and range:** Any age may be affected, but some disorders increase with age. • **Predominant sex:** There does not appear to be a sex predilection.

SIGNS

General Comments
Many of the signs of cardiac diseases occur when congestive heart failure occurs.

Historical Findings
Signs are often nonspecific including weakness, exercise intolerance, inactivity or lethargy. Occasionally syncope will be reported, or misreported as seizure activity.

Physical Examination Findings
Occasionally heart murmurs or arrhythmia may be noted during auscultation. There may be subcutaneous edema or ascites. Jugular or other veins may be excessively prominent. Conversely, there may be poor perfusion. The skin may be dark red and congested.

CAUSES
• Infectious causes of cardiomyopathy include *Sarcocystis*, avian bornavirus, avian serositis virus, *Chlamydia* and other bacteria.
• Infectious causes of valvular disease include bacterial endocarditis. This often results from chronic bacterial showers. • Toxic causes of cardiomyopathy include drugs such as furazolidone and natural toxins such as avocado. • Metabolic or nutritional causes of cardiomyopathy include hemochromatosis and vitamin E or selenium deficiency.
• Vascular disease may rarely result in myocardial infarcts. • Idiopathic changes occur to the heart muscle and valves, often associated with advanced age. • Congenital anomalies occur rarely.

RISK FACTORS
Risk factors for various heart disorders are different. Exposure to infectious agents is the obvious factor for these diseases. The opossum is the source of *Sarcocystis*. Pododermatitis or other chronic low-grade infections may be a risk factor for bacterial endocarditis. High iron diets, lack of dietary tannins, and a highly efficient iron extraction system of predisposed species, such as mynahs, toucans, birds of paradise, tanagers, and others, are the risk factors for hemochromatosis.

DIAGNOSIS

DIFFERENTIAL DIAGNOSIS
Differential diagnoses depend on the specific condition. For right-sided heart failure, differentials must include all causes of ascites or coelomic effusion; these may include reproductive disease, hepatic disease, hypoproteinemia, and sepsis. In addition causes for edema may include trauma or dermatopathies. Left-sided heart failure may mimic respiratory disease such as pneumonia. If syncope occurs, neurologic disorders and atherosclerosis should be ruled out.

CBC/BIOCHEMISTRY/URINALYSIS
• The hematology may show polycythemia if there has been chronic hypoxia. Hematocrit values of 60% or greater should be suspicious for chronic heart or respiratory diseases when obvious dehydration is not present. Other hematologic changes may be seen with infectious processes, usually reflecting inflammation or infection. • Biochemical changes may occur due to concurrent changes in the liver (elevated AST or bile acids) or kidneys (elevated uric acid) when systemic diseases such as chlamydia or hemochromatosis are present.

OTHER LABORATORY TESTS
N/A

IMAGING
• Radiographic findings may include cardiomegaly, pulmonary congestion or edema, coelomic effusion, hepatomegaly, or splenomegaly. • Ultrasound may demonstrate pericardial effusion, thinning cardiac muscle with dilated cardiomyopathy or thickened cardiac muscle with hypertrophic cardiomyopathy. Contractility can be subjectively assessed. Using a standard transcoelomic approach, only two views are available, a four chamber view, and an aortic outflow view. Cross-sectional views are not possible with this approach and cardiac measurements cannot be accurately made.

DIAGNOSTIC PROCEDURES
Electrocardiography may be helpful in identification of arrhythmia or other electrical activity of the heart.

PATHOLOGIC FINDINGS
Gross pathologic findings in heart disease may include pericardial effusion, hypertrophy or dilation of the ventricles, or valvular endocarditis. Histopathology may include inflammation of the myocardium, fibrosis, or atherosclerotic plaques.

CARDIAC DISEASE

(CONTINUED)

TREATMENT

APPROPRIATE HEALTH CARE
Treatment of cardiac diseases involves treatment of the underlying cause, if known, and treatment of the congestive heart failure that often results from these diseases.

NURSING CARE
Stress should be minimized in avian heart patients. Social arrangements should be adjusted to avoid conspecific aggression, and other sources of stress should be removed.

ACTIVITY
Although inactivity may contribute to development of heart disease, exercise may not be well tolerated by cardiac patients. Generally, birds will regulate their activity voluntarily. They should not be forced to exercise, however.

DIET
A nutritionally balanced, calorically appropriate diet should be provided. If hemochromatosis is suspected, iron should be restricted. Addition of tannins to the diet may reduce iron absorption.

CLIENT EDUCATION
General treatment and prognosis information should be discussed.

SURGICAL CONSIDERATIONS
Pericardiocentesis may be used if there is significant pericardial effusion. Endoscopic creation of a pericardial window may be useful for chronic pericarditis. Otherwise, surgical treatment is not practical for the treatment of heart disease in birds.

MEDICATIONS

DRUG(S) OF CHOICE
Treatment of congestive heart failure involves reducing the after pressure against which the heart must work, reduction of edema and ascites, and improving the contractility of the heart. The use of angiotensin converting enzyme (ACE) inhibitors can help reduce hypertension by inhibiting the conversion of angiotensin I to angiotensin II and aldosterone release. Diuretics, such as furosemide may be used to reduce edema and ascites. Positive inotropes may be warranted in dilated cardiomyopathy. Pimobendan works in this capacity and is also a vasodilator that will further reduce blood pressure. Digoxin can be used as a positive inotrope and can be used to suppress supraventricular arrhythmia. It has a low therapeutic index, and side effects are common.
- Enalapril: 1–5 mg/kg q 8–24h.
- Furosemide: 0.1–2 mg/kg q 6–24h.
- Pimobendan: 0.25 mg/kg q12h.
- Digoxin: 0.01–0.05 mg/kg q12–24h.

CONTRAINDICATIONS
ACE-inhibitors have renal disease listed as a contraindication. However, this may be based on the theoretical reduction in glomerular filtration rate that in mammals causes urea to rise. This may not be an issue in birds, and ACE-inhibitors may actually be of benefit in some renal conditions.

PRECAUTIONS
N/A

POSSIBLE INTERACTIONS
None.

ALTERNATIVE DRUGS
None.

FOLLOW-UP

PATIENT MONITORING
Heart patients should be regularly evaluated for attitude, exercise tolerance, vascular prominence, respiratory character, and edema. Auscultation should be done carefully to rate any arrhythmia or murmur. Body weight should be tracked. Electrocardiography and imaging may track changes in cardiac function. Blood counts and chemistries should be used to monitor infectious processes and for any adverse effects of drug therapy.

PREVENTION/AVOIDANCE
Prevention of reoccurrence depends somewhat on the cause of the heart disease. Control of infectious diseases is important for some etiologies. Iron restriction and iron binding agents may help reduce the reoccurrence of hemochromatosis.

POSSIBLE COMPLICATIONS
The primary complication is inadequate response to therapy.

EXPECTED COURSE AND PROGNOSIS
Generally by the time overt disease is present, the prognosis is poor to grave. Survival times are usually measured in weeks to months. Early detection and treatment may result in more prolonged survival.

MISCELLANEOUS

ASSOCIATED CONDITIONS
Atherosclerosis is a very common disease in birds and may be present in many birds with heart disease.

AGE-RELATED FACTORS
Although heart diseases may increase in frequency with age, there are no adjustments to the diagnosis and treatment associated with age.

ZOONOTIC POTENTIAL
There is no zoonotic potential for heart disease. If caused by *Chlamydia*, however, there would be a risk.

FERTILITY/BREEDING
Heart disease may affect a bird's ability to breed. However, there is no specific contraindication.

SYNONYMS
Heart failure

SEE ALSO
Atherosclerosis
Avocado/plant toxins
Chlamydiosis
Dyslipidemia
Obesity
Pododermatitis
Proventricular dilatation disease
Respiratory distress
Sarcocystis
Sick-bird syndrome

ABBREVIATIONS
None.

INTERNET RESOURCES
None.

Suggested Reading
Hanley, C.S., Murray, H.G., Torrey., S., et al. (1997). Establishing cardiac measurement standards in three avian species. *Journal of Avian Medicine and Surgery*, **11**:15–19.
Lumeij, J.T., Ritchie, B.W. (1994). Cardiology. In: Ritchie, B.W., Harrison, G.J., Harrison, L.R. (eds), *Avian Medicine: Principles and Application*. Lake Worth, FL: Wingers, pp. 695–722.
Oglesbee, B.L., Oglesbee, M.J. (1998). Results of postmortem examination of psittacine birds with cardiac disease: 26 cases (1991–1995). *Journal of the American Veterinary Medical Association*, **212**:1737–1742.
Pees, M., Krautwald-Junghanns, M.-E. (2005). Avian echocardiography. *Seminars in Avian and Exotic Pet Medicine*, **4**:14–21.
Pees, M., Krautwald-Junghanns, M.-E. (2009). Cardiovascular physiology and diseases of pet birds. *The Veterinary Clinics of North America. Exotic Animal Practice*, **12**:81–97.
Author Kenneth R. Welle, DVM, DABVP (Avian)

Client Education Handout available online

Cere and Skin, Color Changes

BASICS

DEFINITION
Color changes of the skin and cere are variations of normal pigmentation due to a disease process.

PATHOPHYSIOLOGY
The pathophysiology of color changes of the skin and cere varies depending upon the underlying etiology and includes pigment changes due to hormonal influences and the results of different disease processes including viral diseases and thermal damage.

SYSTEMS AFFECTED
Skin.

GENETICS
N/A

INCIDENCE/PREVALENCE
Varies with the underlying cause.

GEOGRAPHIC DISTRIBUTION
N/A

SIGNALMENT
• **Species:** Brown hypertrophy of the cere is most common in budgerigars. • **Mean age and range:** Brown hypertrophy of the cere is most common in older budgerigar hens. Crop burns are generally found in young birds being hand fed. • **Predominant sex:** N/A

SIGNS

Historical Findings
Owner reports changes in the color of the skin.

Physical Examination Findings
• Cere: The cere shows dark brown hypertrophy, hyperkeratosis and hyperplasia, sometimes with occlusion of the nares, rather than the normal pink or tan in a female budgerigar and blue in a male budgerigar. • Crop: The skin covering the crop shows variable color changes ranging from erythematous to dark or black. An ulcer may form in the crop allowing the contents to spill onto the surface of the feathers covering the crop. • Toes: The skin on the extremities, usually toes, may vary from pale and cold, to erythematous and edematous, to dark or black. In severe cases, skin or toes may slough. • Eyelid margins: The skin at the eyelid margins and conjunctiva show discrete erythematous macules that then become brown to black papules, then vesicles that erode and crust. In some species the only findings are hyperpigmented macules. • Abdomen: The featherless tracts of skin over the abdomen show petechia or ecchymoses. • Oral mucosa: White plaques are seen in the oral cavity due to hyperkeratosis of the oral mucosa, and there may also be concurrent blunted choanal papillae, upper respiratory and alimentary tract disease.

CAUSES
• Cere: Brown hypertrophy of the cere in budgerigar hens is due to increased estrogen levels and may be related to high fat diets. In male budgerigars, the change is related to a testicular (sertoli cell) tumor producing estrogen. • Crop: Crop burns are caused by thermal damage from hand feeding food that is too hot. • Toes: Frostbite, thermal injury caused by exposure to low temperatures, is often seen in birds housed outside during winter months. • Eyelid margins: Avian cutaneous poxvirus (see Poxvirus). • Abdomen: Avian Polyomavirus in young birds (see Polyomavirus). • Oral mucosa: Hypovitaminosis A (see Oral Plaques, Nutritional deficiencies).

RISK FACTORS
Hand feeding is a risk factor for crop burns, and housing birds outside in cold climates is a risk factor for frostbite. See appropriate chapters for risk factors for other causes.

DIAGNOSIS

DIFFERENTIAL DIAGNOSIS
• Brown hypertrophy of the cere—*Knemidocoptes* spp. mites. • Crop burn—foreign body penetration. • Frostbite—ulcerative pododermatitis, photosensitization, contact dermatitis, constricted toe. • See appropriate chapters for other causes.

CBC/BIOCHEMISTRY/URINALYSIS
N/A

OTHER LABORATORY TESTS
N/A

IMAGING
N/A

DIAGNOSTIC PROCEDURES
All of these conditions are often diagnosed based only on history and physical examination. See appropriate chapters for additional information.

PATHOLOGIC FINDINGS
• Brown hypertrophy of the cere—hyperkeratosis and hyperplasia of the cere. • Frostbite—histopathology shows coagulative necrosis with a peripheral inflammatory infiltrate of heterophils and macrophages and a well demarcated inflammatory margin between unaffected and necrotic tissue. • See appropriate chapters for other causes.

TREATMENT

APPROPRIATE HEALTH CARE
• Brown hypertrophy of the cere—in budgerigar hens no treatment is necessary apart from manual removal of any hyperkeratosis occluding the nares. In male budgerigars the gonadal tumor is generally considered inoperable. • Crop burns—full thickness crop burns are life threatening and aggressive therapy is needed. Stabilize the bird and initiate fluid therapy and appropriate antibiotic therapy for secondary bacterial infections. Monitor the patient for several days as the full extent of the burn might not be evident upon initial examination. Once the extent of the burn is identified, anesthetize the bird for surgical debridement of the necrotic tissue. Close the crop and overlying skin separately. Bandage the surgical site during initial healing stages and use an E-collar or similar protective device if the bird damages the surgical site. • Frostbite—stabilize the bird and initiate fluid therapy and appropriate antibiotic therapy for secondary bacterial infections. Monitor the patient for several days as the full extent of the frostbite might not be evident upon initial examination. Once the extent of the frostbite is identified, anesthetize the bird for surgical debridement of the necrotic tissue. Bandage the surgical site during initial healing stages and use an E-collar or similar protective device if the bird damages the surgical site. • See appropriate chapters for other causes.

NURSING CARE
Nursing care varies depending on the presentation. Nursing care may be particularly helpful in cases of burns and frostbite.

ACTIVITY
Activity should be limited if frostbite is suspected.

DIET
Appropriate formulated diet for the species.

CLIENT EDUCATION
• Crop burns—educate hand feeder in proper hand feeding formula heating and temperature testing protocols including not using a microwave which may create hot spots in the food. • Frostbite—educate clients on proper housing requirements during cold weather. • See appropriate chapters for other causes.

SURGICAL CONSIDERATIONS
N/A

MEDICATIONS

DRUG(S) OF CHOICE
• Antibiotics may be indicated in cases of crop burns and frostbite. • Peripheral vasodilators such as pentoxifylline or isoxsuprine can be considered to manage edema in cases of frostbite.

CONTRAINDICATIONS
N/A

CERE AND SKIN, COLOR CHANGES (CONTINUED)

PRECAUTIONS
N/A

POSSIBLE INTERACTIONS
N/A

ALTERNATIVE DRUGS
N/A

FOLLOW-UP

PATIENT MONITORING
N/A

PREVENTION/AVOIDANCE
• Crop burns—avoid hand feeding baby birds with food that is too hot. • Frostbite—house birds in outside enclosures to avoid thermal damage due to cold weather. • See appropriate chapters for other causes.

POSSIBLE COMPLICATIONS
Secondary infection is possible in the case of crop burns and frostbite.

EXPECTED COURSE AND PROGNOSIS
• Brown hypertrophy of the cere—good prognosis for budgerigar hens, guarded to poor long term prognosis for male budgerigars. • Crop burns and frostbite—severe cases have a poor prognosis if substantial tissue damage is involved. Crop burns detected early have a good to guarded prognosis if the patient is stabilized early, secondary infections are managed and adequate nutrition is provided. • See appropriate chapters for other causes.

MISCELLANEOUS

ASSOCIATED CONDITIONS
N/A

AGE-RELATED FACTORS
N/A

ZOONOTIC POTENTIAL
N/A

FERTILITY/BREEDING
N/A

SYNONYMS
N/A

SEE ALSO
Ectoparasites
Feather damaging behaviors/self-injurious behavior
Feather disorders
Nutritional deficiencies
Oral Plaques
Polyomavirus
Poxvirus
Rhinitis and sinusitis

ABBREVIATIONS
N/A

INTERNET RESOURCES
N/A

Suggested Reading
Girling, S. (2006). Dermatology of birds. In: Patterson, S. (ed.), *Skin Diseases of Exotic Pets*. Oxford: Blackwell Science Ltd, pp. 3–14.
Schmidt, R.E., Reavill, D.R., Phalen, D.N. (eds) (2003). *Pathology of Pet and Aviary Birds*. Ames, IA: Iowa State Press, pp. 177–196.

Authors Thomas M. Edling, DVM, MSpVM, MPH
Heide M. Newton, DVM, DACVD

Chlamydiosis

 ## BASICS

DEFINITION
• *Chlamydia psittaci* (formerly *Chlamydophila psittaci*) is an obligate intracellular Gram-negative bacterium. • At least eight serovars and nine genotypes described. • While the organism is not stable in the environment, it can remain infectious for at least a month if present within organic matter (dirty cages, fecal material). • Asymptomatic carriers are common, and the organism is only shed intermittently. • Stress may increase shedding of *C. psittaci*—this includes physiologic stress (egg-laying) and environmental stress.

PATHOPHYSIOLOGY
• *C. psittaci* is shed in nasal discharge and feces, and transmitted by inhalation of aerosolized secretions or fecal material. • The elementary body is the pathogen's infectious stage. Elementary bodies attach to and enter the host cells, where they develop into noninfectious reticulate bodies within the phagosomes of the host cell. These reticulate bodies multiply and enlarge, and after several intermediate forms will transform back to elementary bodies, which then rupture out of the cell. • Respiratory epithelial cells are affected first. The organism then spreads hematogenously to the reticuloendothelial system.

SYSTEMS AFFECTED
• Respiratory—upper respiratory signs, air sacculitis, pneumonia.
• Ophthalmic—secondary to upper respiratory disease.
• Gastrointestinal—enteritis.
• Hemic/Lymphatic/Immune—splenomegaly as a result of reticuloendothelial cell infection.
• Hepatobiliary—hepatitis.

GENETICS
N/A

INCIDENCE/PREVALENCE
Variable, depending on species.

GEOGRAPHIC DISTRIBUTION
Worldwide distribution.

SIGNALMENT
• *C. psittaci* has been isolated from over 460 species of birds. • It is most often diagnosed in psittacines, with Amazon parrots, macaws, cockatiels and budgerigars the most frequently infected. • Pigeons and doves have a higher rate of infection than other pet nonpsittacine birds. • Younger birds appear to be more susceptible to clinical disease.

SIGNS
General Comments
• Incubation period is three days to several weeks. • Some birds may be asymptomatic carriers.

Historical Findings
• Lethargy and depression • Fluffed feathers • Anorexia • Abnormal urates (green) • Diarrhea • Sneezing • Death.

Physical Examination Findings
• Ocular discharge • Conjunctivitis • Nasal discharge • Dyspnea • Biliverdinuria • Depression • Diarrhea • Emaciation • Dehydration.

CAUSES
Chlamydia psittaci.

RISK FACTORS
• Immunosuppression is a risk factor for development of clinical signs in an infected patient. This can result from stresses such as shipping, crowding, poor sanitation and husbandry, breeding, or other illnesses. • Disease transmission is facilitated by close spacing of cages (including stacked cages) and poor sanitation practices. • Birds that travel to settings where other birds will be in close quarters are at increased risk of exposure. This includes bird shows, fairs, bird fanciers' meetings, boarding facilities, veterinary facilities, and bird stores

 ## DIAGNOSIS

DIFFERENTIAL DIAGNOSIS
• Given the nonspecific clinical signs, numerous diseases can mimic chlamydiosis. Specific testing is required to reach a diagnosis. • Common rule-outs include other bacterial causes of rhinitis and keratoconjunctivitis should be considered, as well as ocular trauma, herpesviral infection, paramyxyoviral infection, and avian influenza.

CBC/BIOCHEMISTRY/URINALYSIS
• Leukocytosis, with heterophilia and monocytosis, is commonly seen. The leukocytosis is often severe, with a WBC count of 2–3 times normal or more possible. • A nonregenerative mild anemia may be seen in chronic cases. • Increases in AST, bile acids, and other hepatic markers may be seen with hepatic involvement.

OTHER LABORATORY TESTS
• Culture: This method is considered the gold standard, but specialized shipping and laboratory methods are required. Culture is not always the most practical option. Preferred specimens include combined swabs of the conjunctiva, choana, and cloaca, or liver biopsy samples in live patients; as well as samples of liver and spleen from necropsy specimens. • Serology: This is useful in detecting exposure to the organism, but a single titer may not be sufficient for definitive diagnosis, even if significantly elevated (it is considered a probable case if there is a single elevated titer with appropriate clinical signs or other laboratory findings). Paired samples demonstrating a four-fold titer increase over a two-week period are preferred for definitive diagnosis. False negatives may be seen with acute infection or antibiotic therapy initiated before sample collection. Methods include IFA and CF. • Antigen detection: May be performed by ELISA, IFA, or staining. False negatives are possible due to intermittent shedding or insufficient numbers of organism in the sample. Results should be interpreted in light of other clinical and laboratory findings. Identification of *C. psittaci* within a tissue sample by IFA is considered diagnostic, whereas ELISA can only detect probable cases. Many laboratories no longer offer EBA. The organism may also be identified in impression smears stained with Gimenez or Machhiavello stains, which is considered diagnostic. • PCR testing of combined conjunctival, choanal, and cloacal swabs or blood samples can be used to detect the DNA of *C. psittaci*. There can be differences between results from different laboratories, as PCR primers and amplification techniques will vary. The test cannot determine if the detected organisms are viable or not and thus can only identify probable cases.

IMAGING
• Radiography: Hepatomegaly and splenomegaly are commonly present.
• Coelomic ultrasound: Hepatomegaly may also be appreciable by this method. The spleen is not ordinarily readily identified on ultrasound, but it may be if it is enlarged.

DIAGNOSTIC PROCEDURES
Biopsy of the liver or spleen may demonstrate the presence of the organism by either IFA or staining (Gimenez, Macchiavello).

PATHOLOGIC FINDINGS
• Gross necropsy findings may include keratoconjunctivitis, rhinitis and nasal discharge, serositis with yellow exudate, pneumonia, air sacculitis, hepatomegaly, and splenomegaly. However, there are no pathognomonic findings that lead to confirmation of chlamydiosis. Pericarditis and tracheitis have been reported in turkeys and waterfowl. Enteritis may be noted in psittacines. • Histologically, infection is confirmed by detection of the organism within tissues by chromatic staining (Gimenez, Macchiavello) or IFA. Basophilic inclusion bodies may be noted, as well as basophilic and lymphocytic infiltrates within affected organs, and necrosis.

 ## TREATMENT

APPROPRIATE HEALTH CARE
• Confirmed or probable cases should be immediately isolated from other birds and treated. • Humans providing care to affected birds should use appropriate personal

CHLAMYDIOSIS (CONTINUED)

protective equipment (gloves, masks of N95 rating or higher).

NURSING CARE
• Appropriate nursing care depends on the severity of the bird's clinical signs. Fluid and nutritional support should be provided as needed. • Birds with respiratory signs or dyspnea may benefit from oxygen therapy.

ACTIVITY
• Activity should be limited as the patient's condition dictates (for example, a weak bird should only be given low perches if at all). • Contact with humans should be limited to the smallest number possible, and those people should wear appropriate personal protective equipment.

DIET
Avoid calcium-rich foods, as the gastrointestinal absorption of tetracyclines (e.g., doxycycline) is inhibited by calcium. Likewise, items such as cuttlebones, mineral blocks, and other calcium sources should be removed.

CLIENT EDUCATION
• It is very important that the client be made aware of the zoonotic risks of chlamydiosis, and instructed on proper hygiene and quarantine procedures to be used. • The client should be advised to consult their physician immediately regarding any indicated testing or treatment for themselves, especially if they develop symptoms consistent with influenza or respiratory disease. • In some localities, this disease is reportable in birds. • Clients must continue the full course of antibiotic treatment, even if clinical signs have resolved.

SURGICAL CONSIDERATIONS
N/A

MEDICATIONS

DRUG(S) OF CHOICE
• Orally-administered doxycycline (25–50 mg/kg PO every 24 hours for 45 days) is the only approved treatment for *C. psittaci* in the United States. The dose may vary by species. • A long-acting injectable doxycycline (Vibravenos®; 75 mg/kg IM every seven days for seven doses) may also be used. This drug is not commercially available in the United States but may be imported in small quantities. • Shorter courses of treatment (21 days) have been effective in treating experimentally-induced chlamydiosis, but have not been approved for routine treatment.

CONTRAINDICATIONS
N/A

PRECAUTIONS
If administered prior to sample collection, even a single dose of doxycycline and enrofloxacin can both cause the bird to stop shedding *C. psittaci*, leading to a false negative result by PCR. If possible, samples should be collected prior to treatment if *C. psittaci* is considered a differential diagnosis.

POSSIBLE INTERACTIONS
N/A

ALTERNATIVE DRUGS
• Azithromycin (40 mg/kg PO q48h for 21 days) has been shown to be effective in eliminating infection. • Enrofloxacin has also been suggested as an effective treatment.

FOLLOW-UP

PATIENT MONITORING
• Body weight and appetite should be monitored throughout the course of treatment. • If abnormalities were detected on the CBC or the biochemical profile, these should be rechecked while the bird is under treatment; the timing will depend on the severity of the initial changes and the bird's clinical status.

PREVENTION /AVOIDANCE
• Avoid acquiring birds that appear to be ill. • Testing and quarantine of all new birds, and quarantine those that are ill or have left and returned to the home or aviary (e.g., bird shows, fairs, and other settings where they may in in close proximity to other birds). • Isolate sick birds from other birds in the home, and follow quarantine and hygiene measures for any sick birds—handle them after healthy birds, use personal protective equipment, disinfect cages daily, avoid drafts or air currents in the room, moisten bedding before cleaning and mop the floor frequently. • Testing of all birds prior to boarding. • Husbandry measures—adequate distance between cages to present transfer of organic materials, avoid stacking cages, solid barriers between adjoining cages, use wire-bottomed cages and substrates that do not easily form dusts, daily cleaning of cages and bowls, disinfect cages between occupants.

POSSIBLE COMPLICATIONS
• Human zoonosis is possible. • In some locations, avian chlamydiosis is a reportable disease—consult your state veterinarian's office. • Re-infection is possible, necessitating a thorough cleaning and disinfection of the bird's environment. Contact with untested or untreated birds should be avoided.

EXPECTED COURSE AND PROGNOSIS
• The prognosis is generally good, provided that timely treatment is provided. • Individuals in whom disease is more severe or for whom diagnosis and treatment were delayed may experience worse clinical outcomes such as chronic air sacculitis and fibrosis, chronic hepatitis or fibrosis, or death.

MISCELLANEOUS

ASSOCIATED CONDITIONS
There is a possible association with *Chlamydia* infection and atherosclerosis in parrots.

AGE-RELATED FACTORS
N/A

ZOONOTIC POTENTIAL
• Humans may become infected with *C. psittaci*, resulting in influenza-like symptoms or other respiratory ailments. Those at highest risk include young children, elderly persons, and immunocompromised individuals. • The client should be advised to consult their physician immediately should they develop these symptoms, and alert their physician that they have been exposed to a confirmed or suspected case of avian chlamydiosis. Psittacosis is a nationally notifiable disease in humans. • If a necropsy is performed, personal protective equipment should be worn. If possible, the necropsy should be performed within a fume hood; alternatively, the bird's body may be wetted with a solution of detergent and water to avoid aerosolization of infectious agents.

FERTILITY/BREEDING
• As immunosuppression and stress can contribute to an individual developing chlamydiosis, breeding birds are susceptible to disease. Birds with chlamydiosis should not be bred until the infection has cleared. It is recommended that breeding pairs both be tested for *C. psittaci* prior to breeding. • Breeding stress may also lead to increased shedding, putting the hatchlings at risk of infection.

SYNONYMS
Psittacosis, ornithosis (humans), parrot fever (humans), chlamydophila

SEE ALSO
Atherosclerosis
Avian influenza
Diarrhea
Hepatitis
Herpesviruses
Infertility
Liver disease
Ocular lesions
Paramyxoviruses
Pneumonia
Respiratory distress
Rhinitis / sinusitis
Sick-bird syndrome
Urate / fecal discoloration

ABBREVIATIONS
CF—complement fixation
IFA—indirect fluorescent antibody

Chlamydiosis (Continued)

INTERNET RESOURCES

http://www.lafebervet.com/avian-medicine-2/general-avian-medicine/psittacosis-in-avian-patients/

http://www.nasphv.org/Documents/Psittacosis.pdf

Suggested Reading

Gerlach, H. (1999). Chlamydia. In: Ritchie, B.W., Harrison, G.J., Harrison, L.R., (eds), *Avian Medicine: Principles and Application*. Brentwood, TN: HBD International, Inc., pp. 984–996.

Guzman, D.S.M., Diaz-Figueroa, O., Tully, T. Jr., et al. (2010). Evaluating 21-day doxycycline and azithromycin treatments for experimental *Chlamydophila psittaci* infection in cockatiels (*Nymphicus hollandicus*). *Journal of Avian Medicine and Surgery*, **24**(1):35–45.

Pilny, A.A., Quesenberry, K.E., Bartick-Sedrish, T.E., et al. (2012). Evaluation of *Chlamydophila psittaci* infection and other risk factors for atherosclerosis in pet psittacine birds. *Journal of the American Veterinary Medical Association*, **240**(12):1474–1480.

Smith, K.A., Campbell, C.T., Murphy, J., et al. (2011). Compendium of measures to control *Chlamydophila psittaci* (formerly *Chlamydia psittaci*) infection among humans (psittacosis) and pet birds (avian chlamydiosis). *Journal of Exotic Pet Medicine*, **20**(1):32–45.

Author Kristin M. Sinclair, DVM, DABVP (Avian)

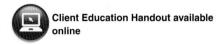 **Client Education Handout available online**

Chronic Egg Laying

BASICS

DEFINITION
Chronic and excessive egg laying without regard to clutch number, size, and nesting behaviors. This continuous laying occurs without the presence of a bonded mate or outside of a breeding season.

PATHOPHYSIOLOGY
- Chronic egg laying is a commonly seen disorder in clinical avian practice. • Most birds continue to lay regardless of clutch size, nest sites, or presence of a mate. • Egg production is a normal process in hens; however, calcium depletion frequently occurs.
- Oviductal and uterine muscle atony can develop and lead to complications such as egg binding and egg malformation. • Lack of enough vitamin D3 and depletion of vitamin E and selenium can also occur.

SYSTEMS AFFECTED
- Reproductive—fatigue and exhaustion, risk of egg retention leading to egg binding.
- Endocrine/Metabolic—electrolyte imbalances can develop leading to life-threatening conditions. Elevations of lipids can lead to strokes. Presence of an egg can lead to transient elevation of uric acid from renal parenchymal compression.
- Behavioral—hens will act broody and try to nest, may hide, and can show aggression while protecting eggs. • Gastrointestinal—hens typically retain feces while nesting.
- Integument—feather destructive behavior may result from chronic unaltered hormonal state and stress.

GENETICS
Some species may have a genetic predisposition for chronic and uncontrolled egg laying.

INCIDENCE/PREVALENCE
- Chronic egg laying is a common problem in single female bird households. • More commonly seen in smaller avian species, and most prevalent in cockatiels. • Females housed with other females or male birds can also be chronic egg layers.

GEOGRAPHIC DISTRIBUTION
N/A

SIGNALMENT
- **Species:** Most commonly seen in cockatiels, finches, canaries, and lovebirds. • **Mean age and range:** Sexually mature hens (age of maturity varies widely amongst species), some species may not start laying until later in life.
- **Predominant sex:** Occurs only in females of egg-laying age.

SIGNS

General Comments
Observation of excessive egg production without pause between clutches or disregard of number of eggs already laid.

Historical Findings
- Constant production of eggs. • Mate relationship with member of the household or another bird. • Broody behaviors may include hiding, paper shredding, nesting in spaces, and being cage aggressive. • Hens may sit on or nest inanimate objects. • Hens may spend a lot of time at the bottom of their cage.
- Some eggs may be misshapen or thin shelled. • Sudden and unexpected death is possible, especially in the smaller species.

Physical Examination Findings
- Normal to no abnormal findings are possible. • Abnormal coelomic palpation is consistent with reproductive tract enlargement. • Weight gain. • Sometimes a palpable egg is present in the caudal coelomic cavity. • Vent enlargement, engorgement, or a flaccid appearance. • Abdominal wall herniation may occur. • In some cases pathologic fractures can occur secondary to calcium depletion. • Hens may be obese, underweight, or normal body condition.
- Sudden death.

CAUSES

Behavioral anomaly
- Lack of ability to rear chicks to delay next egg laying cycle. • People or toys and mirrors may simulate a mate stimulus to lay eggs.
- Presence of nest site, box, or materials can contribute to chronic egg laying.

Disease of captivity
- Lack of normal stimuli for breeding and egg laying, such as length of daylight, food availability, temperature. • Lack of true nest sites, building, and entire process of rearing through weaning.

RISK FACTORS
- Poor nutritional state or incomplete diets lacking enough vitamins and calcium may lead to long term complications. • Absence of additional and enough calcium supplementation may lead to complications.
- Not seeking veterinary care or advice on methods to decrease or stop egg laying.

DIAGNOSIS

DIFFERENTIAL DIAGNOSIS
- Pathologic fractures may be indirectly related and secondary to osteoporosis and calcium depletion. • Coelomic distention with mass effect may be from an egg, organomegaly, ascites, neoplasia, or hernia.

CBC/BIOCHEMISTRY/URINALYSIS
- Hypercalcemia is a normal finding in an egg-laying hen. Calcium concentrations in the normal reference range indicate a calcium deficiency in an actively laying bird.
- Hypercholesterolemia is a common finding during egg production. • Hyperglobulinemia may be seen as a physiologic response while the hen is producing eggs. • Hyperuricemia can be seen with egg compression and space occupation. • Hypertriglyceridemia is seen during times of egg production.

OTHER LABORATORY TESTS
- Protein electrophoresis (EPH) may show hyperglobulinemia in egg laying birds characterized as a marked monoclonal increase in the beta-globulin fraction.
- Ionized calcium levels may provide a more exact measure of calcium status.

IMAGING
- Radiology is useful to confirm presence of a shelled egg. • Polyostotic hyperostosis or osteomyelosclerosis of the long bones due to medullary ossification is common.
- Pathologic fracture/metabolic bone disease may be evident as decreased overall bone opacity with obvious fractures, especially of the long bones. • Increased soft tissue opacity in the mid- to caudal coelomic cavity representing reproductive tract enlargement.
- Ultrasonography may be useful to fully evaluate the coelomic cavity.

DIAGNOSTIC PROCEDURES
N/A

PATHOLOGIC FINDINGS
- Only if there is disease associated with chronic egg laying such as ascites, ectopic egg production, or yolk-associated coelomitis.
- Metritis may occur but is often relative to chronic laying.

TREATMENT

APPROPRIATE HEALTH CARE
- Only needed if there are secondary complications to the chronic egg laying.
- Health care varies based on clinical findings.

NURSING CARE
Only needed if complications develop such as egg binding.

ACTIVITY
No change is required and exercise and flight should be encouraged.

DIET
- Appropriate calcium supplementation is mandatory, sometimes in several forms.
- Transition to a diet that includes pelleted foods and calcium rich supplements.
- Vitamin supplements that include omega fatty acids and Vitamins D3 and Vitamin E.

CHRONIC EGG LAYING (CONTINUED)

CLIENT EDUCATION
• Clients should be educated that egg formation and laying is a normal process in birds, but chronic or excessive laying is not and that steps need to be taken to reduce egg production. • Alteration of environmental stimuli, decreased photoperiod, alteration of nest sites, toys, and other inanimate objects that may be sexually stimulating and decreased physical stimulation from humans/mates. Encourage a flock relationship rather than with one person. Relocation of cage, redecoration, distraction techniques including boarding or taking the bird on a trip may be useful. Avoid any stimulatory petting or feeding. Any eggs should be left in an attempt to have the hen sit for periods of time where egg laying will temporarily cease. Remove male/other birds if present and related to increased egg laying behavior. • Hormonal therapy with leuprolide acetate and deslorelin implants should be discussed with clients as a means to decrease or stop egg production. • Periodic laying by healthy birds throughout the year can be acceptable and tolerated.

SURGICAL CONSIDERATIONS
• Salpinghysterectomy may be required in patients with chronic egg laying or undesirable behavioral changes secondary to chronic hormonal states. This is a complicated procedure that should only be performed by skilled avian veterinarians. • Surgical repair of fractures may be warranted.

MEDICATIONS

DRUG(S) OF CHOICE
• Leuprolide acetate, a long acting GnRH analog used to prevent ovulation and egg production. Doses vary:700–800 mcg/kg IM for birds <300 g, 500 mcg/kg IM for birds >300 g q 21–30d. • Calcium gluconate (10%) injectable—50–100 mg/kg SC, IM (diluted); for hypocalcemia; dilute 1:1 with sterile water for injections. • Calcium glubionate—25–100 mg/kg PO q 24h; daily oral calcium supplementation for hypocalcemia. • Deslorelin implants—may be surgically placed for prolonged control or arrest of egg laying.

CONTRAINDICATIONS
Use of medroxyprogesterone-acetate, levonorgestrel,hCG, testosterone, and tamoxifen are associated with risk of severe side effects.

PRECAUTIONS
N/A

POSSIBLE INTERACTIONS
N/A

ALTERNATIVE DRUGS
N/A

FOLLOW-UP

PATIENT MONITORING
• Extensive client communication, education, and review of effects of behavioral modification. • Periodic blood tests including biochemistry profile to monitor plasma calcium levels.

PREVENTION/AVOIDANCE
• Discourage further egg laying for the season by reducing the photoperiod to no more than eight hours per day. • Cover the cage at night to ensure quiet rest. • Keep the hen separate from any mate or perceived mates. • Reduce foods high in fat and overall quantity of food. • Removal of nests and nesting material may help. • Leuprolide acetate can be administered monthly/seasonally or when the hen is exhibiting reproductive behavior to prevent ovulation and egg laying. • A deslorelin implant can also be administered and may last for several months. • Avoidance could only occur by not having female pet birds or in hens that have undergone successful salpingohysterectomy.

POSSIBLE COMPLICATIONS
• Egg binding with inability to pass an egg; this can be life threatening. • Egg yolk coelomitis may develop from ectopic ovulation leading to ascites, peritonitis and sepsis. • Chronic hypocalcemia, osteoporosis, and pathologic fractures can develop without proper nutrition, calcium supplements, and appropriate care.

EXPECTED COURSE AND PROGNOSIS
Cessation of egg laying or decreased number of eggs laid for extended periods of time is highly recommended.
• Prognosis is good if egg laying can be prevented or slowed. • Prognosis is guarded in any hen that continues to lay excessively.

MISCELLANEOUS

ASSOCIATED CONDITIONS
• Egg binding. • Oviductal prolapse. • Lameness, fractures. • Ectopic egg production.

AGE-RELATED FACTORS
Any sexually mature bird can lay eggs. Some species, such as sun conures (sun parakeets), may begin egg laying later in life and continue to lay long term.

ZOONOTIC POTENTIAL
None.

FERTILITY/BREEDING
In some cases, allowing normal breeding and rearing of chicks may be the only method to stop or delay chronic egg laying but is undesirable for most pet owners and should be discouraged.

SYNONYMS
N/A

SEE ALSO
Atherosclerosis
Cloacal disease
Coelomic distention
Cystic ovaries
Dyslipidemia
Dystocia / egg binding
Egg yolk coelomitis / reproductive coelomitis
Hypocalcemia / hypomagnesemia
Lameness
Metabolic bone disease
Nutritional deficiencies
Osteomyelosclerosis / polyostotic hyperostosis
Problem behaviors
Regurgitation / vomiting
Salpingitis / uterine disorders
Sick-bird syndrome

ABBREVIATIONS
EPH—protein electrophoresis
GnRH—gonadotropin-releasing hormone

INTERNET RESOURCES
http://lafeber.com/vet/chronic-egg-laying/

Suggested Reading
Bowles, H.L. (2002). Reproductive diseases of pet bird species. *The Veterinary Clinics of North America. Exotic Animal Practice*, **5**(3):489–506.
Bowles, H.L. (2006). Evaluating and treating the reproductive system. In: Harrison, G., Lightfoot, T. (eds), *Clinical Avian Medicine*, Vol. 2. Palm Beach, FL: Spix Publishing, 519–540.
Pollock, C.G., Orosz, S.E. (2002). Avian reproductive anatomy, physiology and endocrinology. *The Veterinary Clinics of North America. Exotic Animal Practice*, **5**(3):441–474.
Author Anthony A Pilny, DVM, DABVP (Avian)

 Client Education Handout available online

CIRCOVIRUSES

BASICS

DEFINITION
The members of the *Circovidae* infect many species of birds. This discussion will be confined to those infecting parrots (Psittacine Beak and Feather Disease Virus [PBFDV]), pigeons (Pigeon Circovirus [PiCV]), and chickens (Chicken Anemia Virus [CAV]).

PATHOPHYSIOLOGY
• Viruses infect the host destroying cells in the immune system (PBFDV, PiCV and CAV) resulting in immune suppression and an increased susceptibility to other infectious agents. • Viruses (PBFDV in African grey parrots, and CAV) infect bone marrow cells causing a pancytopenia • PBFDV also has a tropism for growing epithelial cells resulting in dysplastic changes in growing feathers, rhinotheca, gnathotheca and nails.

SYSTEMS AFFECTED
• Skin (PBFDV)—progressive, becoming generalized, feather dystrophy, delayed molt, beak elongation and fracturing.
• Hemic/Lymphatic/Immune (PBFDV, PiCV, CAV)—immune suppression resulting in one or more secondary infections (PDFDV [African grey parrots] and CAV). Cause severe anemia and in some instances pancytopenia.

GENETICS
PBFDV infection has different outcomes in different species of parrots.

INCIDENCE/PREVALENCE
• PBFDV infection is common in many species of captive and wild parrots, particularly those from Africa (African grey parrots and lovebirds) and the Indopacific species. • PiCV infections are found in high prevalence in captive and wild pigeons (rock doves). • CAV is a common infection of chickens.

GEOGRAPHIC DISTRIBUTION
These viruses are found worldwide.

SIGNALMENT
PBFDV
• **Species:** Most common in psittacine birds from the Indopacific region, lovebirds, and African grey parrots. Rare in Neotropical parrots. • **Mean age and range:** Disease can occur at any age, but is most common in fledglings and in birds less than three years.
• **Predominant sex:** There is no sex predilection.
PiCV
• **Species:** Domestic and feral pigeons (*Columbia livia*). • **Mean age and range:** Fledglings and birds under one year.
• **Predominant sex:** There is no sex predilection.
CAV
• **Species:** Domestic chicken. • **Mean age and range:** Two to three week old chickens.
• **Predominant sex:** There is no sex predilection.

SIGNS
PBFDV
Historical Findings
• Abnormal feather development. • Feather loss with no regrowth. • Delayed molt.
• Elongate and fractured beak. • Nestlings: Painful to touch, inappetent. • African grey parrots: Depression, yellow urates.

Physical Examination Findings
• Feathers have thickened sheaths, hemorrhage within their shafts, pinch off at their base and stop growing or may not develop at all. Affected feathers may fall out.
• Beak overgrowth (white cockatoos) with fracturing of the rhinotheca and gnathotheca and separation of the gnathotheca from the oral mucosa. • African grey parrots may be depressed, have yellow urates and pallor.
• Failure to molt (eclectus parrots and lovebirds). • Powder down feathers in cockatoos stop growing. These birds may exhibit a shiny instead of dull powder-covered beak.

PiCV
Historical Findings
High morbidity and mortality of young birds in a pigeon flock.

Physical Examination Findings
• PiCV is immunosuppressive and thus signs are caused by secondary infectious agents.
• Pharyngitis. • Rhinitis and conjunctivitis.
• Anorexia. • Diarrhea. • Weight loss.
• Depression. • Death.

CAV
Historical Findings
Morbidity and mortality in 2–3-week-old chickens.

Physical Examination Findings
• Decreased food consumption. • Decreased activity, fluffed feathers and heat seeking.
• Pallor. • Poor response to vaccinations for other infectious agents.

CAUSES
PBFDV
• Feather and beak lesions are caused by virus growth in the developing feather and the germinal cells of the rhamphotheca.
• PBFDV damages the Bursa of Fabricius and the thymus as well as circulating cells of the immune system, causing immune suppression and secondary infections. • PBFDV causes hepatic necrosis in African grey parrots.
• Replication in the bone marrow in African grey parrots results in a pancytopenia.

PiCV
• PiCV damages the Bursa of Fabricius and the thymus as well as circulating cells of the immune system, causing immune suppression and secondary infections (Columbid herpesvirus, ornithosis, salmonellosis, candidiasis, trichomoniasis, coccidosis, and nematode infestations).

CAV
• Replication in the bone marrow destroys the red blood cell precursors resulting in anemia.
• CAV damages the Bursa of Fabricius and the thymus as well as circulating cells of the immune system, resulting in immune suppression and secondary infections.

RISK FACTORS
PBFDV
• The biggest risk factor is not quarantining and testing new birds coming into a collection. • PiCV is common in domestic and wild pigeons and therefore is difficult or impossible to keep out of a loft. • CAV occurs in flocks that are not vaccinated.

DIAGNOSIS

DIFFERENTIAL DIAGNOSIS
PBFDV
The pattern and type of feather disease is important in differential diagnosis:
• Birds that are damaging their own feathers will typically have normal feathers where they cannot reach (on their head and neck) whereas birds infected with PBFDV will have a generalized feather disease. • Birds damaging their own feathers will pull out healthy feathers, chew on them, and may damage their skin with their beak. PBFDV infected birds will have dysplastic feathers.
• Avian polyomavirus infections in nestling budgerigars often will result in identical signs to that of PBFDV infection and can only be differentiated with PCR-based diagnostics or histopathology. • Nutritional diseases can result in delayed molts and poor quality feathering that can resemble PBFDV in some species of birds, these birds will not have dysplastic feathers. • Endocrine diseases including hypothyroidism in parrots is rare, but these birds may gradually lose feathers and not replace them. They do not, however, exhibit feather dysplasia. • Overgrowth of the beak can also be caused by chronic liver disease, misalignment of the beak, and in some instances can be behavioral. • Neoplasia and bacterial and fungal infections of the beak can also result in abnormal beak formation.

PiCV
• Disease in birds with PiCV infection and immune suppression is caused by the secondary infectious agents. A necropsy of a representative bird or PCR testing for PiCV is necessary to determine if the observed diseases are primary or secondary.

CAV
• Many infectious diseases can cause the nonspecific signs seen in birds with CAV. A complete necropsy is necessary to distinguish between CAV and other infectious diseases.

CBC/BIOCHEMISTRY/URINALYSIS
PBFDV

CIRCOVIRUSES (CONTINUED)

- These tests are not helpful in PBFDV infected parrots, with the exception of African grey parrots that will often be anemic or pancytopenic and may have elevated liver enzymes.

PiCV
- These tests are not helpful in the diagnosis of PiCV infection.

CAV
- Affected chickens are often severely anemic.

OTHER LABORATORY TESTS
PBFDV
- PCR assays are available for detecting PBFDV in the blood of infected birds. • In Australia, hemagglutination assays are used to detect PBFDV in the feathers of parrots. • In Australia, hemagglutination inhibition assays are used to detect antibodies to PBFDV. Typically, birds that seroconvert will survive infection whereas birds that do not will die or persistently shed PBFDV.

PiCV
- PiCV can be detected in postmortem swabs of the liver by a PCR assay.

CAV
- Virus DNA can be detected by PCR assays in the thymus and bone marrow of birds at postmortem.

IMAGING
N/A

DIAGNOSTIC PROCEDURES
PBFDV: Feather and skin biopsies have been used to verify PBFDV infections in parrots with feather dysplasia.

PATHOLOGIC FINDINGS
- PBFDV: See signs, intracytoplasmic botryoid inclusion bodies are present in growing feathers and the Bursa of Fabricius.
- PiCV: Inclusion bodies are found in the Bursa of Fabricius. • CAV: Bone marrow and lymphoid tissue atrophy with inclusion bodies.

TREATMENT
There are no drugs that change the impact of PBFDV, PiCV, or CAV infection. Reduction of stress, feeding a high quality diet, and treating concurrent infections will prolong the life of birds with PBFDV and PiCV infections.

APPROPRIATE HEALTH CARE
Treat secondary infections as required.

NURSING CARE
N/A

ACTIVITY
N/A

DIET
N/A

CLIENT EDUCATION
These birds are a source of infection to the other birds in the owner's home.

SURGICAL CONSIDERATIONS
N/A

MEDICATIONS

DRUG(S) OF CHOICE
There are no treatments for these viruses.

CONTRAINDICATIONS
N/A

PRECAUTIONS
N/A

POSSIBLE INTERACTIONS
N/A

ALTERNATIVE DRUGS
N/A

FOLLOW-UP

PATIENT MONITORING
N/A

PREVENTION/AVOIDANCE
N/A

POSSIBLE COMPLICATIONS
Increased susceptibility to other infectious diseases

EXPECTED COURSE AND PROGNOSIS
- PBFDV: Most birds with clinical disease will die, uncommonly some (particularly lorikeets and budgerigars) will recover.
- PiCV: Increased morbidity and mortality in fledglings. • CAV: Increased morbidity and mortality.

MISCELLANEOUS

ASSOCIATED CONDITIONS
See above.

AGE-RELATED FACTORS
N/A

ZOONOTIC POTENTIAL
N/A

FERTILITY/BREEDING
N/A

SYNONYMS
PiCV: Young pigeon disease

SEE ALSO
Anemia
Beak malocclusion
Dermatitis
Feather cyst
Feather damaging behavior
Feather disorders
Overgrown beak and nails
Polyomavirus
Thyroid disease
Viral disease

ABBREVIATIONS
PCR—polymerase chain reaction

INTERNET RESOURCES
N/A

Suggested Reading
Balamurugan, V., Kataria, J.M. (2006). Economically important non-oncogenic immunosuppressive viral diseases of chicken—current status. *Veterinary Research Communications*, **30**:541–566.
Duchatel, JP, Todd, D, Smyth, JA, et al. (2006). Observations on detection, excretion and transmission of pigeon circovirus in adult, young and embryonic pigeons. *Veterinary Journal*, **172**:40–51.
Lierz, M. (2005). Systemic infectious disease. In: Harcourt-Brown, N., Chitty, J. (eds), *BSAVA Manual of Psittacine Birds*, 2nd edn. Quedgeley, UK: British Small Animal Veterinary Association, pp. 155–169.
Raue, R., Schmidt, V., Freick, M., et al. (2005). A disease complex associated with pigeon circovirus infection, young pigeon disease. *Avian Pathology*, **34**:418–425.
Author David N. Phalen, DVM, PhD, DABVP (Avian)

Cloacal Diseases

BASICS

DEFINITION
Cloacal diseases include any disease that involves or affects the cloaca.

PATHOPHYSIOLOGY
The cloaca is the collection chamber for the final products from the gastrointestinal tract (feces), urologic tract (urine and urates) and reproductive tract (eggs, sperm). The cloaca is comprised of three compartments: (from cranial to caudal) the coprodeum (the largest compartment), which communicates with the distal colon; the urodeum (the smallest compartment), which communicates with the distal ureters and oviduct(s) or vas deferens; and the proctodeum, which opens to the outside of the body through the vent (Figure 1).

Cloacal Diseases
• **Cloacitis:** Inflammation of the cloaca can be associated with primary or secondary bacterial, viral, mycotic, or parasitic infections. Foreign body granulomas have been reported but are rare in birds. • **Cloacal Neoplasia:** Psittacid herpesvirus 1 (PsHV 1), the etiologic agent of Pacheco's disease, is also associated with cloacal mucosal adenomatous polyps ("papillomas") and adenocarcinomas, particularly in neotropical psittacine species such as Amazon parrots. Cloacal adenocarcinomas are locally invasive and have potential for vascular invasion and distant metastasis. Other reported neoplasms involving the avian cloaca include lymphoma, squamous cell carcinoma, hemangioma, hemangiosarcoma, fibrosarcoma, lipoma, liposarcoma, and oviductal leiomyosarcoma.
• **Obstructive Diseases of the Cloaca:** Obstruction of the cloaca can be caused by cloacal uric acid or fecal concretions (cloacoliths), granulomatous or neoplastic masses, or retained eggs. In most cases, the underlying pathogenesis of cloacolithiasis is unknown. • **Cloacal Prolapse:** Cloacal prolapse is most often associated with excessive straining related to undesired behaviors such as excessive sexual stimulation. However, cloacal prolapse can also occur with other diseases involving the cloaca. Prolapsed tissues can include the cloaca (e.g., coprodeum), oviduct, colon, small intestine, or phallus (in species that possess them), as well as granulomatous or neoplastic masses.

SYSTEMS AFFECTED
• Gastrointestinal—infection, inflammation, obstruction, neoplasia, prolapse, dilatation.
• Reproductive—obstruction, prolapse.
• Renal/Urologic—obstruction.

GENETICS
None.

INCIDENCE/PREVALENCE
Variable to unknown.

GEOGRAPHIC DISTRIBUTION
None.

SIGNALMENT
• **Species:** ○ Cloacitis and Cloacal Neoplasia—adenomatous polyps are more common in neotropical psittacine species such as Amazon parrots and macaws than in other psittacine species. ○ **Obstructive Diseases of the Cloaca**—cloacoliths are relatively uncommon in birds in general but have been most often described in macaws, African grey parrots, Amazon parrots, and raptors. ○ **Cloacal Prolapse**—prolapse of the cloaca associated with undesired behaviors may be more frequent in cockatoos than in other psittacine species. • **Mean age and range:** Variable. • **Predominant sex:** Variable.

SIGNS

Historical Findings
Historical findings associated with cloacal disease include decreased appetite, lethargy, changes in posture, tenesmus, hematochezia, diarrhea, flatulence, malodorous feces, decreased fecal production, picking or excessive grooming around the vent, presence of blood or urofeces on the feathers or skin surrounding the vent or on the beak, or intermittent or persistent prolapse of tissues through the vent opening.

Physical Examination Findings
Examination findings are usually consistent with historical findings. Cloacoliths can often be palpated through the cloacal wall. Irregular, cobblestone cloacal mucosal lesions that bleed easily are characteristic of adenomatous polyps. Prolapsed oviductal tissue may arise from the left side of the cloaca and will have longitudinal striations that can help differentiate it from other tissues, such as the colon. A prolapsed colon lacks longitudinal folds and is generally centrally located within the cloaca. Intestinal prolapses and intussusceptions are difficult to differentiate from colonic prolapses.

A lubricated, sterile cotton-tipped applicator or finger (if the bird is large enough) can be inserted into the cloaca for further examination. A speculum or otoscope can be used to visually inspect the cloacal lumen, but the field of view is often limited. Diluted acetic acid (vinegar) can be applied to the cloacal mucosa to further evaluate for papillomatous changes.

CAUSES
• **Cloacitis** ○ Bacterial ■ *E. coli* ■ *Clostridium* spp. (eg. *Clostridium tertium*) ■ *Mycobacterium* spp. ○ Viral ■ PsHV 1 ○ Mycotic ■ *Candida albicans* ■ *Trichosporon begielli* ■ *Aspergillus* spp. ○ Parasitic ■ *Giardia* spp. ■ Helminths ■ Coccidia ○ Foreign body granuloma ○ Other cloacal diseases • **Neoplasia** ○ Adenomatous polyp (PsHV 1) ○ Adenocarcinoma ○ Squamous cell carcinoma ○ Lymphoma ○ Lipoma, liposarcoma ○ Oviductal leiomyosarcoma • **Obstructive Diseases of the Cloaca** ○ Cloacolithiasis ○ Egg binding ○ Neoplasia ○ Granulomatous disease ○ Vent stricture • **Cloacal Prolapse** ○ Cloaca ■ Spinal trauma, osteomyelitis, or neoplasia ■ Pudendal nerve trauma or neuritis ■ Peripheral neuropathy ■ Excessive sexual behavior ○ Oviduct ■ Egg binding ■ Salpingitis ■ Oviductal neoplasia ■ Chronic egg laying ■ Excessive sexual behavior ○ Colon ■ Colitis ■ Cloacitis ■ Neoplasia ■ Intestinal impaction ■ Intussusception ■ Intestinal parasitism ■ Granulomatous disease (eg. mycobacteriosis) ■ Megacolon ■ Excessive sexual behavior ○ Intestine ■ Trauma ■ Intussusception ○ Phallus ■ Excessive sexual behavior ■ Balanitis

RISK FACTORS
• **Cloacitis:** Concurrent cloacal or intestinal diseases. • **Cloacal Neoplasia:** Infection with PsHV 1 for adenomatous polyps or cloacal adenocarcinoma. • **Obstructive Disorders of the Cloaca:** Concurrent viral infection (e.g., PsHV 1), other concurrent cloacal disease, dysbiosis, fecal retention, chronic dehydration, nutritional disorders such as hypovitaminosis A. • **Cloacal Prolapse:** Excessive sexual stimulation, spinal trauma, spinal osteomyelitis, spinal neoplasia, peripheral neuropathy, egg binding, salpingitis, oviductal neoplasia, chronic egg laying, colitis, cloacitis, cloacal neoplasia, intestinal impaction, intestinal parasitism, trauma.

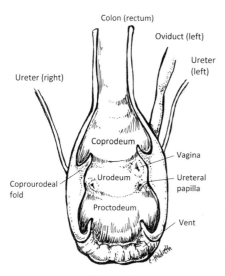

Figure 1.

Diagram of the avian cloaca and associated structures.

CLOACAL DISEASES (CONTINUED)

DIAGNOSIS

DIFFERENTIAL DIAGNOSIS
N/A

CBC/BIOCHEMISTRY/URINALYSIS
Hematologic and plasma biochemistry findings are usually within normal reference intervals for birds with cloacal disease. Inflammatory diseases may be associated with a leukocytosis and heterophilia. Excessive blood loss may result in anemia. Ureteral obstruction may result in hyperuricemia or azotemia.

OTHER LABORATORY TESTS
Cytology can be performed on cloacal swabs or feces to evaluate the cellular population of the cloaca. Gram stain and aerobic or anaerobic bacterial or fungal cultures can be performed on cloacal swabs or fecal samples to further characterize the cloacal microflora. Mineral analysis can be performed on extracted cloacoliths.

IMAGING
Survey and Contrast Radiography
Dilation of the distal intestinal tract can be associated with cloacal obstructive diseases. Cloacoliths can occasionally be identified radiographically. Contrast gastrointestinal radiography and contrast fluoroscopy can be useful to evaluate for filling defects, dilation, and changes in motility. Contrast media can also be instilled directly into the cloacal lumen.

Ultrasonography
Ultrasonography after instillation of warm saline or water into the cloacal can be helpful to assess intraluminal disease such as cloacal neoplasia and intussusception.

DIAGNOSTIC PROCEDURES
Endoscopic evaluation by infusion cloacoscopy can be helpful in visualizing the cloacal lumen and wall for pathology and for the collection of biopsy samples. Cloacoliths can be visualized endoscopically. Endoscopic fragmentation and removal of cloacoliths has been described in birds.

PATHOLOGIC FINDINGS
- **Cloacitis:** Affected tissues may appear red, ulcerated, thickened or edematous, and may bleed easily. Histologic findings include ulceration, inflammation, fibrosis, and variable presence of etiologic agents such as bacteria, yeasts, and fungi. • **Cloacal Neoplasia:** Adenomatous polyps are often raised or ulcerated, irregular, and thickened, and may have a cobblestone appearance on gross exam. Other neoplasms may be evident as visible or palpable masses within the cloaca or prolapsed through the vent opening. Histopathology is generally diagnostic for cloacal neoplasia. • **Obstructive Disorders of the Cloaca:** Grossly visible cloacal obstructions include cloacoliths, eggs, granulomas, or neoplasms. • **Cloacal Prolapse:** Prolapsed tissue can be identified by gross appearance. Histopathology on prolapsed tissue may be normal or demonstrate inflammatory or neoplastic changes.

TREATMENT

- **Cloacitis:** Bacterial cloacitis is treated with appropriate antibiotics, ideally based on microbiologic culture and sensitivity. Mycotic cloacitis is treated with appropriate antifungal agents. Parasitic infections are treated with appropriate antiparasitic agents. • **Cloacal Neoplasia:** Surgical excision, electrosurgery, silver nitrate cautery, cryosurgery, laser surgery, cloacoscopic diode laser ablation, and mucosal stripping have been described for the treatment of adenomatous polyps and other cloacal neoplasms in birds. A cloacotomy may be necessary for maximum visualization and surgical access. • **Obstructive Disorders of the Cloaca:** Cloacoliths can be manually, endoscopically, or surgically fragmented and extracted. • **Cloacal Prolapse:** Prolapsed tissues should be gently cleansed and lubricated prior to reduction, preferably under heavy sedation or anesthesia.
  ○ **Retention Sutures:** Once the prolapse has been reduced, two sutures can be placed transversely across the vent opening in an effort to prevent recurrence. The sutures must be placed far enough apart to allow the normal passage of urofeces. Purse-string sutures are not recommended in birds.
  ○ **Cloacopexy:** Rib and incisional cloacopexy techniques have been described. Intestinal entrapment has been reported with rib cloacopexy. ○ **Colopexy:** Colopexy has been described for the treatment of colonic prolapse in a sulfur-crested cockatoo (van Zeeland, 2014). Colopexy may be less likely to result in intestinal entrapment than cloacopexy. ○ **Ventplasty:** Ventplasty, which involves trimming and closure of one or more sections of the vent margin, permanently reduces the size of the vent opening.

APPROPRIATE HEALTH CARE
- Treatment must be specific to the underlying cause to be successful. • Mild disease can be treated as an outpatient, more moderate to severe disease generally requires hospitalization for inpatient nursing care.

NURSING CARE
Supportive care includes analgesic therapy, fluid and nutritional support, and appropriate antimicrobial therapy.

ACTIVITY
Activity should be restricted for birds with cloacal disease.

DIET
None specific to cloacal diseases.

CLIENT EDUCATION
If the cloacal disease is associated with excessive sexual stimulation, the client should be educated on appropriate husbandry and physical interaction with their pet bird.

SURGICAL CONSIDERATIONS
None.

MEDICATIONS

DRUG(S) OF CHOICE
Appropriate antibiotics should be used for primary and secondary bacterial infections based on culture and sensitivity. Metronidazole and clindamycin are often effective for clostridial cloacitis. Fluoroquinolones are often effective for infections with *E. coli*. Appropriate antifungal or antiparasitic agents should be used for mycotic and parasitic infections. Analgesics such as NSAIDs, opioids, tramadol, and local anesthetics should be considered as needed. Chemotherapy may be effective in some cases of cloacal neoplasia. Hormonal agents such as leuprolide acetate can be considered if excessive sexual stimulation is present.

CONTRAINDICATIONS
None specific to cloacal diseases.

PRECAUTIONS
None specific to cloacal diseases.

POSSIBLE INTERACTIONS
None specific to cloacal diseases.

ALTERNATIVE DRUGS
N/A

FOLLOW-UP

PATIENT MONITORING
Patients with cloacal disease should be closely monitored based on the nature of their disease. Patients should be examined at least every few months during and after treatment. Repeat testing such as fecal or cloacal swab Gram stain or culture and radiographs should be considered based on the nature of each case.

PREVENTION/AVOIDANCE
If the cloacal disease is associated with excessive sexual stimulation, the client should be educated on the proper care and husbandry and physical interaction for their pet bird.

POSSIBLE COMPLICATIONS
Potential complications of cloacopexy and ventplasty include dehiscence of the surgical sites(s), which may result in peritonitis.

CLOACAL DISEASES (CONTINUED)

EXPECTED COURSE AND PROGNOSIS
- **Cloacitis**: The prognosis is fair to excellent for most bacterial, mycotic, and parasitic causes of cloacitis.
- **Cloacal Neoplasia**: The prognosis for cloacal neoplasia is fair to guarded or poor depending upon the underlying cause, extent of disease, response to therapy, and presence of complicating diseases.
- **Obstructive Diseases of the Cloaca**: The prognosis for cloacolithiasis is fair to excellent if the cloacolith can be removed, although recurrence has been reported.
- **Cloacal Prolapse**: The prognosis for cloacal prolapse due to behavioral causes is fair to guarded, as often the underlying behavior is difficult to manage and control and recurrence and complications are common.

MISCELLANEOUS

ASSOCIATED CONDITIONS
Papillomatosis, egg binding, salpingitis, excessive sexual stimulation.

AGE-RELATED FACTORS
Variable.

ZOONOTIC POTENTIAL
None.

FERTILITY/BREEDING
Female birds with oviductal prolapse and associated disease and male birds with phallus prolapse should be rested or retired from breeding.

SYNONYMS
None.

SEE ALSO
Chronic egg laying
Clostridiosis
Colibacillosis
Diarrhea
Dystocia/egg binding
Enteritis/gastritis
Flagellate enteritis
Infertility
Intestinal helminthiasis
Neoplasia
Herpesviruses
Phallus prolapse
Salpingitis/uterine disorders
Tumors
Urate/fecal discoloration

ABBREVIATIONS
None.

INTERNET RESOURCES
None.

Suggested Reading

Gelis, S. (2006). Evaluating and treating the gastrointestinal system. In: Harrison, G.J., Lightfoot, T.L. (eds), *Clinical Avian Medicine*. Palm Beach, FL: Spix Publishing, Inc.

Graham, J.E., Tell, L.A., Lamm, M.G., *et al.* (2004). Megacloaca in a Moluccan Cockatoo (*Cacatua moluccensis*). *Journal of Avian Medicine and Surgery*, **18**(1): 41–49.

Lee, A., Lennox, A., Reavill, D. (2014). Diseases of the Cloaca: A Review of 712 Cloacal Biopsies. *Proceedings of the Association of Avian Veterinarians Annual Conference*, pp. 51–60.

Lumeij, J.T. (1994). Gastroenterology. In: Ritchie, B., Harrison, G.J., Harrison, L. (eds). *Avian Medicine: Principles and Application*. Lake Worth, FL: Wingers, pp. 509–512.

van Zeeland, Y.R.A., Schoemaker, N.J., van Sluijs, F.J. (2014). Incisional colopexy for treatment of chronic, recurrent colocloacal prolapse in a sulphur-crested cockatoo (*Cacatua galerita*). *Veterinary Surgery*, **9999**:1–6.

Author Lauren V Powers, DVM, DABVP (Avian / ECM)

 Client Education Handout available online

CLOSTRIDIOSIS

BASICS

DEFINITION
• Avian clostridiosis caused by *Clostridia spp* refers to enteric disease commonly with secondary cholangiohepatitis and is not to be confused with botulism that is caused by *C. botulinum*. *Clostridia* are anaerobic gram-positive spore-forming bacilli. They can produce a large number of toxins that are known to cause gas gangrene in affected tissues. However avian clostridiosis commonly refers to the necrotic enterocolitis form. • *C. perfringens* is the most commonly described clostridia in all avian species but other species such as *C. sordellii*, *C. tertium* and *C. colinum* have been described. *C. colinum* is considered the predominant pathogen in ulcerative enteritis in quails.
• Clostridiosis has been described in a wide range of captive or free-range avian species that are commonly presented for veterinary care including back-yard poultry, quails, psittacine birds, raptors and waterfowl, but is most researched in domestic chickens (broilers). • Clostridiosis can cause acute or peracute necrotizing enterocolitis that leads to a high mortality rate or chronic enterocolitis with lower mortality but with significant morbidity depending on the avian species that is affected. • It has long been thought that the wide range of endotoxins produced by *Clostridia* and used to classify different strains are responsible for their pathogenicity. Thus, *C. perfringens* type A and *C. difficile* type A were considered highly pathogenic and specific diagnostic tests were designed to detect the specific toxinogenic strains. Recent evidence establishes that a newly described toxin, B-like toxin (NetB), and various other factors are responsible for the necrotic enteritis form of the disease while alpha toxin was proved to mediate the gas gangrene disease. Other factors include preceding coccidial infection in chickens and nectar diet in Lories and lorikeets. • *Clostridia* are opportunistic pathogens and can be found in the GIT and in fecal samples of healthy raptors and ground dwelling birds such as poultry and waterfowl but are considered an abnormal finding in psittacine birds.

PATHOPHYSIOLOGY
• In infections caused by *C. perfringens*, NetB toxin, a member of a beta pore forming toxin family, causes cell rounding and lysis of the intestinal epithelial cells. The gene for its production is upregulated only when there is a high density population of the bacteria. Such high clostridial population is enabled by a high polysaccharide diet and by concurrent coccidial infection (*Eimeria spp*). The *Eimeria* parasites kill epithelial cells and induce leakage of proteins that are necessary for the enhanced colonization by the *Clostridia*. The mucin secreted in response to the *Eimeria* infection serves as a preferred growth substrate for the *Clostridia*. • The massive inflammatory response including heterophils, lymphocytes and plasma cells produces ulcerative necrotic enteritis and in many cases invades the biliary system resulting in severe cholangiohepatitis.

SYSTEMS AFFECTED
• Gastrointestinal—necrotic enteritis is the most common form. Colitis, megacolon and cloacitis are seen in psittacines (cockatoos, lories) and raptors.
• Hepatobiliary—cholangiohepatitis occurs in chronic or subclinical cases that allow colonization of the liver through the bile duct.
• Renal—in systemic end-stage disease, sepsis.

GENETICS
N/A

INCIDENCE/PREVALENCE
All *Clostridia spp* are distributed worldwide and found in soil. Subclinical infection with chronic diarrhea has been described in all affected species from broilers to cockatoos. This allows persistent flock infection, as many birds in a flock situation remain untreated.

GEOGRAPHIC DISTRIBUTION
Worldwide.

SIGNALMENT
• **Species:** Described in many species. Cockatoos and Lories are considered predisposed among psittacine birds, however clostridiosis was reported in many species of parrots. Raptors are considered unsusceptible and infection is rare. Disease is very common in commercial flocks, that is, broilers or quails. • **Breed predilections:** N/A • **Mean age and range:** Juvenile birds are more susceptible. In broiler chickens clostridiosis usually occurs at four weeks after hatching. In cockatoos the disease is commonly secondary to other factors and occurs in sexually mature birds. Hand-fed psittacine birds are predisposed. • **Predominant sex:** Higher predilection in phallus-containing birds, such as anseriformes and struthioniformes (ostrich, emu, etc.), should prolapse occur.

SIGNS

General Comments
Clinical signs can differ in accordance with the affected species or the form of the disease. For example, in cockatoos with chronic cloacitis, diarrhea or feather damaging behavior may be the only signs. In raptors, Lories, or chickens, peracute disease may result in death with no preceding clinical signs. Acute or chronic disease can manifest in severe diarrhea among other signs.

Historical Findings
• Poor digestion resulting in weight loss or poor weight gain in flocks. • Diarrhea, anorexia and weakness. • Malodorous feces is very common among psittacine birds.
• Feather picking. • Prolapsed cloaca.
• Bloody stool. • Regurgitation.

Physical Examination Findings
• Good body condition score in acute/peracute infection. • Lethargy.
• Weight loss in chronic infections.
• Prolapsed cloaca (psittacine birds, ostrich, waterfowl). • Hematochezia. • Malodorous feces. • Retained feces. • Soiled vent.
• Feather destructive behavior (psittacines).

CAUSES
Clostridia spp.

RISK FACTORS
• Nutritional: High polysaccharide diet is proposed as an important factor in promoting the proliferation of *C. perfringens*. Nectarivorous species such as Lories and lorikeets are highly susceptible.
  In falconry raptors, meat storage techniques were implicated to be a contributing factor in the proliferation of *C. perfringens*, as well as hand feeding practice used in rearing young psittacine birds.
• Stress and immunosuppression: Stress was shown to alter the intestinal flora and to elevate the risk for necrotizing enteritis.
• Cloacal diseases: Cloacal prolapse, enterolith, impaction. • Egg binding.
• Chronic salpingitis. • Intestinal/cloacal neoplasia. • Papillomatosis. • Behavioral: Owner-induced masturbatory behavior, delayed defecation behavior (psittacine species). • Concurrent coccidiosis (poultry).
• Cloacal/colonic atony. • Toxicosis (especially zinc or lead).

DIAGNOSIS

DIFFERENTIAL DIAGNOSIS
• Coccidiosis • Papillomatosis • Megacolon
• Other infectious enteritis (viral, bacterial, parasitic, fungal) • Toxicosis • Neoplastic or inflammatory disease.

CBC/BIOCHEMISTRY/URINALYSIS
• CBC may show leukocytosis with heterophilia and monocytosis. In acute infection leukopenia may occur due to the massive intraluminal leukocytosis. • Anemia may be present in subclinical or chronic infections. • Elevated liver enzymes may occur.

OTHER LABORATORY TESTS
• Fecal Gram's stained cytology is very typical of the large gram positive sporulated bacilli; however, sporulating bacilli are not always seen in cases of clostridial enteritis.
• Spore-forming bacteria can be seen in small numbers in the feces from healthy raptors or ground dwelling birds but should always be considered abnormal in psittacine species.
• Definitive diagnosis can be obtained by anaerobic culture/sensitivity. However, *Clostridia* organisms are fastidious and may result in false negative culture. • PCR is now available for specific strains or at the genus

CLOSTRIDIOSIS

(CONTINUED)

level. • Serum immunoassays targeting *C. perfringines* and *C. difficile* toxins can result in false negative results.

IMAGING
Coelomic radiographs of gas filled loops of bowel, megacolon and megacloaca are compatible with clostridiosis. Hepatomegaly may also be present.

DIAGNOSTIC PROCEDURES
N/A

PATHOLOGIC FINDINGS
Gross lesions of severe cases are usually restricted to the small intestines and/or the colon and cloaca, but can also be present in other organs such as liver and kidney. The intestines are usually thin walled and gas filled. The necrotic enteritis is characterized by extensive ulcerative necrosis covered with bile stained pseudomembrane while mild subclinical cases may have multiple smaller ulcers. Microscopic examination typically shows strong inflammatory reaction with heterophils and lymphocytic plasmacytic infiltration. This is different than gas gangrene caused by alpha toxin that usually lacks the microscopic inflammatory reaction. There is a clear demarcation line between viable and necrotic tissue. Large Gram-positive rods are associated with the area of necrosis.
The liver may be enlarged with signs of cholangiostasis and hepatitis.

TREATMENT

APPROPRIATE HEALTH CARE
• Treatment should be targeted to remove or alter any underlying condition that may have served as a predisposing factor in the development of clostridiosis. • Removing cloacal impactions, treating egg binding, correction of cloacal prolapse, etc. • Reducing stress. • Improve hygiene measures when feed is suspected as a source of infection (falconry, hand reared psittacines and lorikeets).

NURSING CARE
Fluid and nutritional support are necessary in debilitated patients.

ACTIVITY
N/A

DIET
Avoid high polysaccharide diets.

CLIENT EDUCATION
N/A

SURGICAL CONSIDERATIONS
In psittacine birds, especially cockatoo spp., with recurrent infections secondary to recurrent cloacal prolapse or chronically dystrophic cloaca, surgical modification of the cloacal lesion may be necessary.

MEDICATIONS

DRUG(S) OF CHOICE
• Pharmacological treatment is composed of antibiotics against anaerobic bacteria and GI protectants. GI promotility drugs are usually not necessary. • Metronidazole or penicillins are usually the drugs of choice, macrolides can be used as well. ○ Metronidazole 20–50 mg/kg PO q12h for 10–14 days. ○ Amoxicillin/clavulanate 125 mg/kg PO q12h for 10–14 days. ○ Azithromycin 10–30 mg/kg PO q24h for 10–14 days.

CONTRAINDICATIONS
N/A

PRECAUTIONS
N/A

POSSIBLE INTERACTIONS
N/A

ALTERNATIVE DRUGS
Other antibiotics that are effective against anaerobic bacteria can be used.

FOLLOW-UP

PATIENT MONITORING
Repeated fecal Gram's stain can be reviewed to confirm resolution of the infection once antibiotic therapy is completed.

PREVENTION /AVOIDANCE
• Avoid sexual stimulation in psittacine species. • Implement excellent hygienic feeding practice.

POSSIBLE COMPLICATIONS
Decreased meat and egg production in commercial flocks.

EXPECTED COURSE AND PROGNOSIS
• In mild and subclinical cases that are treated early, the prognosis is good. However subclinical infection in a flock situation usually becomes chronic. • Recurrent clostridiosis is common in psittacines that suffer from chronic cloacal dysfunction. • In acute clinical infections the mortality tends to be high in all avian species.

MISCELLANEOUS

ASSOCIATED CONDITIONS
• Coccidiosis • Cloacal prolapse • Cloacitis

AGE-RELATED FACTORS
Juvenile birds are more susceptible.

ZOONOTIC POTENTIAL
Cross-infection between humans and avian species with clostridiosis was not reported. However humans are susceptible to clostridiosis and good hygiene standards are preventive.

FERTILITY/BREEDING
N/A

SYNONYMS
Necrotic enteritis, quail disease

SEE ALSO
Botulism
Cloacal disease
Cloacal prolapse
Coccidiosis
Crop stasis/ileus
Dehydration
Dystocia/Egg binding
Emaciation
Heavy metal toxicity
Herpesviruses
Neurologic conditions
Phallus prolapse
Salpingitis/uterine disorders
Urate/fecal discoloration

ABBREVIATIONS
N/A

INTERNET RESOURCES
http://www.aav.org
http://www.merckmanuals.com
www.ncbi.nlm.nih.gov

Suggested Reading
Timbermont, L., Haesebrouck, F., Ducatelle, R., Van Immerseel, F. (2011). Necrotic enteritis in broilers: an updated review on the pathogenesis. *Avian Pathology*, **40**(4):341–347.
Van Immerseel, F., Rood, J.I., Moore, R.J., Titball, R.W. (2009). Rethinking our understanding of the pathogenesis of necrotic enteritis in chickens. *Trends in Microbiology*, **17**(1):32–36.
Author Shachar Malka, DVM, Dipl ABVP (Avian)

CNS Tumors, Brain Tumors, Pituitary Tumors

BASICS

DEFINITION
Primary tumors within the central nervous system are divided into those arising from neuroepithelial tissues, meninges, germ cell tumors, and the pituitary. Nonneuroepithelial neoplasms (craniopharyngioma) and cysts (epidermoid and dermoid cysts), as well as primary lymphomas of the central nervous system are also recognized; however, these will not be covered as they are rarely, if ever, recognized in birds.

The following are those CNS tumors described in birds:
- **Tumors of neuroepithelial tissue:**
  ○ *Gliomas* ■ Astrocytic tumors: derived from astrocytes. ▫ The high grade astrocytoma is a glioblastoma (Glioblastoma multiforme [GMF]). ■ Oligodendroglial tumor: derived from oligodendrocytes. ■ Other gliomas.
  ■ Mixed such as oligoastrocytoma that contains glial cells of different types.
  ■ Ependymal tumors: arise from ependymal tissue. ▫ Ependymoma. ○ *Choroid plexus tumors* ■ Choroid plexus papilloma. ○ *Pineal tumor* ■ Pineocytoma (benign).
  ■ Pineoblastoma (malignant). ○ *Embryonal tumors* ■ Medulloblastoma. ■ Primitive neuroectodermal tumors (PNETs)—a small cell tumor arising from a progenitor cell population capable of divergent differentiation along neuronal, ependymal, glial, and possibly mesenchymal cell lines. ▫ Neuroblastoma: an embryonal neoplasm with limited neuronal differentiation.
- **Tumors of the meninges:** Meningioma.
- **Germ cell tumors:** Teratoma.
- **Pituitary tumors:** Adenomas, adenocarcinomas.

PATHOPHYSIOLOGY
Predisposing factors have not been described with most avian CNS tumors. However, fowl glioma-inducing virus (FGV), which belongs to avian leukosis virus (ALV) subgroup A, induces "fowl glioma" in domestic chickens (*Gallus gallus domesticus*). This disease is characterized by multiple nodular gliomatous growths of astrocytes.

SYSTEMS AFFECTED
All tumors of the central nervous system, benign and malignant, can result in significant neurologic dysfunction simply as a mass effect within the skull. Some may also be functional tumors (pituitary and pineal tumors).
- CNS—changes in appetite, seizures, ataxia, paresis. • Endocrine—polydipsia and polyuria. • Ophthalmic—blindness, exophthalmia. • Reproductive—ovarian inactivity (functional pineal gland tumor).
- Respiratory—invasive growth into respiratory sinus. • Skin—feather color changes, altered molt.

GENETICS
No significant genetic predisposition has been identified.
Some tumors are congenital such as teratomas.

INCIDENCE/PREVALENCE
Tumors of the CNS are rare based on avian submissions to one pathology service retrospective study at less than 1% (6/557) of all tumors diagnosed (all avian species including domestic fowl). In an Australian survey of neoplasms in avian species excluding domestic fowl, 0.7% (3/383) were primary CNS tumors. A review of pet bird tumors both from the literature as well as submissions to a teaching hospital found CNS tumors at 0.9% (15/1539), 10 of which were pituitary tumors.

GEOGRAPHIC DISTRIBUTION
N/A

SIGNALMENT
No specific age or sex predilection with one exception. Pituitary tumors are more common in young to middle aged budgerigars.

SIGNS

Gliomas
- Astrocytoma: Transitory torticollis, retropulsion (walking backwards), and ataxia.
- Glioblastoma: Progressive weakness, weight loss, exophthalmos, ataxia, seizures, and a resting tremor. • Oligodendroglioma: Not reported. • Oligoastrocytoma: Unexpected death. • Ependymoma: Not reported.

Choroid plexus tumors
- Choroid plexus papilloma: Not reported.

Pineal tumor
- Pineocytoma: Ovarian inactivity.
- Pineoblastoma: Intermittent diarrhea, weight loss, polydipsia, head tilt, weakness.

Embryonal tumors
- Medulloblastoma: Ataxia, intention tremors. • Primitive neuroectodermal tumors (PNETs)
- Neuroblastoma: Mass in the left orbit of the eye.

Tumors of the meninges
- Meningioma: Not reported.

Germ cell tumors
- Teratoma: Head tilt, circling, and facial nerve paralysis.

Pituitary tumors
- Exophthalmia, ocular chemosis, blindness, localized pruritus, abnormal feather molt, polyuria, polydipsia, weight loss, ataxia, and depression.

CAUSES
Underlying causes or genetic alterations are unknown for most avian CNS tumors. The fowl glioma-inducing virus (FGV), which belongs to avian leukosis virus (ALV) subgroup A, induces "fowl glioma" in domestic chickens (*Gallus gallus domesticus*). This disease is characterized by multiple nodular gliomatous growths of astrocytes.

RISK FACTORS
Risk factors are unknown.

DIAGNOSIS
Antemortem definitive diagnosis has not been described.

DIFFERENTIAL DIAGNOSIS
Any disease process that involves the CNS.

Infectious

Virus
- Bornavirus: Cachexia, ingluvial stasis, gastric and intestinal dilatation, regurgitation, ataxia, ataxia, paresis, proprioceptive deficits, head tremors, and death. • Eastern Equine Encephalitis: High mortality, depression, anorexia, hemorrhagic gastroenteritis regurgitating food, sternal recumbency.
- Western Equine Encephalitis: Anorexia, weight loss, weakness, depression, drowsiness, stupor, ataxia, and incoordination.
- Paramyxovirus type 1: Polydipsia, ataxia, poor balance, torticollis, head tremors, inability to fly, and diarrhea. • West Nile Virus: Ataxia, tremors, weakness, seizures, and abnormal head postures and movement prior to death.

Bacteria
- *Listeria monocytogenes*, *Enterococcus* species, and *Salmonella* species.

Fungus
- *Aspergillus* species, *Dactylaria gallopava*.

Protozoa
- *Tetratrichomona gallinarum*: Ataxia, blindness, intermittent seizures.
- Sarcocystosis: Severe weakness, dyspnea, fresh blood in oral cavity, head twitching and torticollis. • Atoxoplasma: Respiratory distress, emaciation, dyspnea, tail-bobbing, and hypothermia. • Amebic meningoencephalitis: Anorexia, dyspnea, depression, head shaking, nasal discharge.
- *Toxoplasma gondii*: Anorexia, prostration, weight loss, diarrhea, and dyspnea.

Metazoa
- Schistosomiasis encephalitis: Circling, flapping wings, head tilt. • *Baylisascaris procyonis*: Seizures, torticollis, opisthotonus, head-tilt, circling, ataxia, paralysis, and visual deficits.

Noninfectious

Nutritional
- Encephalomalacia (hypovitaminosis E): Tremor, incoordination, and recumbency.

Toxin
- Lead toxicity: Dehydration, depression, polyuria, passive regurgitation, neurologic deficits (flipping backwards), ataxia, paralysis of wings or legs, seizures, head tilt, lime-green feces. • Sodium toxicity: Lethargic, stiff and posteriorly extended legs, dyspnea, anorexia.

CNS Tumors, Brain Tumors, Pituitary Tumors (Continued)

Miscellaneous
- Lafora bodies in a cockatiel. • Cerebellar degeneration and/or hypoplasia. • Ischemic stroke.

CBC/BIOCHEMISTRY/URINALYSIS
These tests would be used to identify other disease conditions and/or to monitor the effects of attempted therapy (chemotherapy or radiation).

OTHER LABORATORY TESTS
Bacterial and/or fungal cultures from the respiratory or digestive tracts to monitor for secondary disease conditions.

IMAGING
- Although antemortem diagnosis of a CNS tumor is not reported, both computed tomography (CT) and magnetic resonance imaging (MRI), with or without contrast would be potential diagnostic imaging techniques. Patient size may be a limiting factor; however, these techniques have been used to diagnose other CNS lesions in avian species. • Standard radiographic study of the skull may provide identification of any associated skeletal abnormalities. Two views (lateral and ventrodorsal) as well as angled views may help localize lesions.

DIAGNOSTIC PROCEDURES
Neurologic examination: Need to localize the lesions: brain, spinal cord, or neuromuscular system.

PATHOLOGIC FINDINGS

Gliomas
Astrocytoma: Astrocytomas are usually fairly uniform with hyperchromatic nuclei and abundant cytoplasm. Cell process can be seen with silver stains. Few mitotic figures are present.
Glioblastoma: It is cellular and has a pleomorphic histologic appearance. Some cells resemble differentiated astrocytes and others are elongated and fusiform with numerous mitotic figures. Multiple giant cells or multi-nucleated cells may be present. Glioblastoma typically occur in the cerebrum and occasionally the cerebellum.
Oligodendroglioma: The characteristic features of oligodendrogliomas include sheets of uniform cells with the "honeycomb" cell pattern which is described as a feature of delayed fixation of the tumor tissues. Microvascular proliferation is another characteristic feature of this tumor. Multifocal microcystic areas and accumulation of mucinous-like material are also frequent findings in oligodendrogliomas. Mitoses are infrequent. Immunohistochemistry is frequently used to further characterize these tumors. In poultry this is more common in adult birds and originates in the periventricular areas of the third ventricle.
Oligoastrocytoma: Oligoastrocytomas are mixed gliomas with neoplastic oligodendrocytes and astrocytes that are either intermingled or separated into distinct clusters.
Ependymoma: A mass arising from the lateral ventricle in a domestic fowl, having some characteristics of an ependymoma. The growth was expansive, non-invasive and sub-divided into major and minor lobules composed of neuroglial cells.

Choroid plexus tumors
Choroid plexus papilloma: These tumors present as well defined papillary structures within ventricles. They are gray-white to red. Histologically the tumors have a vascular connective tissue stroma covered by epithelial cells that are cuboidal or columnar and resemble normal choroid plexus.

Pineal tumor
Pineocytoma: Pineocytoma is a circumscribed encapsulated cellular mass generally embedded between folia of the rostral cerebellar vermis. These benign tumors present as lobulated nodular growths of the gland. The lobules are composed of columnar epithelium and surrounding parafollicular cells. Pineocytomas are rarely seen and best described in poultry.
Pineoblastoma: A friable suprathalamic grey mass located on midline that compresses adjacent structures. Composed of sheets and cords of cells having a round to oval nuclei and small to moderate amounts of vacuolated basophilic cytoplasm. Occasional rosettes.

Embryonal tumors
Medulloblastoma: Medulloblastoma is considered an embryonic neoplasm that grossly is well defined and histologically is composed of anaplastic cells that may form rosettes.

Primitive neuroectodermal tumors (PNETs)
Neuroblastoma: Greyish white soft nodular mass found in the rostral part of the left cerebrum.

Tumors of the meninges
Meningioma: These tumors are usually solitary within the meninges and are of variable shape. They are firm and gray-white to yellow. Microscopically tumor cells usually have abundant cytoplasm and are fusiform. Whorls and bundles of neoplastic cells are the most common pattern.

Germ cell tumors
Teratoma: A teratoma may be defined as a tumor composed of multiple cell types and tissues arising from the germ cells of two (diadermic) or three (tridermic) embryonic layers (ectoderm, endoderm and mesoderm). Extragonadal sites of development include the pineal gland, cerebellum, and spinal cord.

Pituitary tumors
These tumors are usually red-brown and may extend outside of the sella tursica if large. Histologically adenomas are comprised of large epithelial cells that are usually devoid of granules. These cells form nests and lobules with minimal stroma. Cells making up carcinomas are more anaplastic and there may be mitotic figures seen. Tumor lobules and cords are infiltrative into surrounding tissue. For many of the tumors in budgerigars, the IHC is commonly positive for growth hormone; consistent with somatotroph tumors. Distant metastasis (liver, midbrain and airsacs) have been recorded for the budgerigar somatotroph pituitary tumors.

TREATMENT
Internet searches via VIN and Pubmed (August 2014) have not identified any reports of therapeutic attempts for CNS tumors. Consultation with a veterinary oncologist is recommended in suspected cases.

APPROPRIATE HEALTH CARE
General supportive care measures warranted based on degree of signs.

NURSING CARE
Fluid therapy, nutritional support, seizure management.

ACTIVITY
Cage rest in the case of severe neurologic impairment.

DIET
Gavage feeding may be needed if bird is not eating enough to maintain weight.

CLIENT EDUCATION
Humane euthanasia may be indicated if quality of life is severely impaired. Adapting cage environment may be helpful (low perches, incubator set up, food and water bowls in easy reach).

SURGICAL CONSIDERATIONS
Surgical treatment has not been reported.

MEDICATIONS

DRUG(S) OF CHOICE
N/A

CONTRAINDICATIONS
N/A

PRECAUTIONS
N/A

POSSIBLE INTERACTIONS
N/A

ALTERNATIVE DRUGS
N/A

FOLLOW-UP

PATIENT MONITORING
N/A

CNS Tumors, Brain Tumors, Pituitary Tumors (Continued)

PREVENTION/AVOIDANCE
N/A

POSSIBLE COMPLICATIONS
N/A

EXPECTED COURSE AND PROGNOSIS
All the CNS tumors warrant a guarded to terminal prognosis. Distant metastasis (liver, midbrain and airsacs) have been recorded for the budgerigar somatotroph pituitary tumors.

MISCELLANEOUS

ASSOCIATED CONDITIONS
N/A

AGE-RELATED FACTORS
Pituitary tumors are more common in young to middle aged budgerigars.

ZOONOTIC POTENTIAL
None.

FERTILITY/BREEDING
N/A

SYNONYMS
N/A

SEE ALSO
Diarrhea
Emaciation
Heavy metal toxicity
Neurologic conditions
Ocular lesions
Polyuria
Polydipsia
Proventricular Dilatation Disease (PDD)
Seizure
Sick-bird syndrome
Tumors

ABBREVIATIONS
CNS—central nervous system

INTERNET RESOURCES
N/A

Suggested Reading
Garner, M.M. (2006). Overview of tumors: section II. In: Harrison, G.J., Lightfoot, T.L. (eds), *Clinical Avian Medicine*. Palm Beach, FL: Spix Publishing, pp. 566–571.
Langohr, I.M., Garner, M.M., Kiupel, M. (2012). Somatotroph pituitary tumors in budgerigars (*Melopsittocus undulatus*). *Veterinary Pathology*, **49**(3):503–507.
Leach, M.W. (1992). A survey of neoplasia in pet birds. *Seminars in Avian and Exotic Pet Medicine*, **1**(2):52–64.
Schmidt, R.E., Reavill, D.R., Phalen, D. (2003). *Pathology of Pet and Aviary Birds*. Ames, IA: Iowa State Press.
Suchy, A., Weissenböck, H., Schmidt, P. (1999). Intracranial tumours in budgerigars. *Avian Pathology*, **28**(2):125–130.

Author Drury R. Reavill, DVM, DABVP (Avian / Reptile & Amphibian), DACVP

COAGULOPATHIES AND COAGULATION

BASICS

DEFINITION
Coagulation, a physiologic defense mechanism which prevents hemorrhage, is highly conserved across species. It involves a cellular component, thrombocytes, and a biochemical component including the amplification (contact or intrinsic) pathway, the tissue activation (extrinsic) pathway, and the common pathway. These are activated by protein signals induced by vascular damage and inflammation. Fibrinolysis is as important as the initial coagulation event and results in return to basal homeostasis.

PATHOPHYSIOLOGY
- Coagulopathy may result from thrombocytopenia (cellular deficiency) or coagulation factor deficiency. • Coagulation factor deficiency may be inherited (rarely reported in birds) or acquired. • The most commonly reported acquired coagulopathies are toxin induced (anticoagulant rodenticides [AR]), disseminated intravascular coagulation (secondary to neoplasia or infectious disease), and hepatic failure (due to viral or toxic causes).

SYSTEMS AFFECTED
Experimental brodifacoum intoxication resulted in the following systems affected, though other differentials should be considered:
- Hemic/Lymphatic/Immune.
- Skin/Exocrine—subdermal hemorrhage in limbs as well as bleeding at new feather follicles. • Neurologic—subdural hemorrhage of the cranium, ataxia, obtundation.
- Reproductive—bloody vitellogenic follicles upon necropsy or endoscopy.
- Ophthalmic—hyphema, hemorrhage in sclera and orbital cavities. • Pulmonary or GI hemorrhage may result in epistaxis or blood in the mouth. • Gastrointestinal—found in brodifacoum intoxication as well as melena with thrombocytes <10,000/μl.

GENETICS
Inherited bleeding disorders are uncommonly reported in birds. However, birds do possess von Willebrand's factor and factor based deficiencies should be considered with hematoma formation.

INCIDENCE/PREVALENCE
- Coagulopathy in general is under reported due to lack of available diagnostic testing. • In raptors, one study documented 86% of 161 birds with detectable AR residues in liver tissue. In 6% of these necropsies, mortality was directly caused by AR. • Owls appear to be most severely impacted but all raptors may become secondarily intoxicated when prey contains AR with a half-life >300 days in the liver. Granivorous birds are primarily infected when they ingest bait.

GEOGRAPHIC DISTRIBUTION
Global distribution.

SIGNALMENT
Conure Bleeding Syndrome has been reported typically in young birds but some older birds affected without gender predilection. Calcium, Vitamin K, and Vitamin D deficiency have been implicated by researchers.

SIGNS

General Comments
In general, small hemorrhage such as petechiation is due to thrombocyte (cellular) deficiency while larger hemorrhage such as hematoma formation is more likely to be due to acquired factor deficiency.

Historical Findings
- Active hemorrhage. • Pallor. • Lethargy.
- Signs of infectious disease of intoxication.
- Collapse after jugular venipuncture. • No clinical signs.

Physical Examination Findings
- Ataxia. • Lethargy. • Inability to stand
- Petechiation, ecchymosis, hematoma formation—oral cavity, dermis, sclera, hyphema. • Epistaxis or oral bleeding—may originate in either respiratory or GI tract due to choanal communication. • Hematochezia or melena—consider genitourinary tract as well as gastrointestinal tract. • Hypovolemic shock.

CAUSES
- Vitamin K deficient diet—first coagulopathy described in chicks in 1935.
- Sulfa drugs—disrupt flora and decrease absorption of Vit K. Documented in poultry that are fed antibiotics for increased production, may occur in any species.

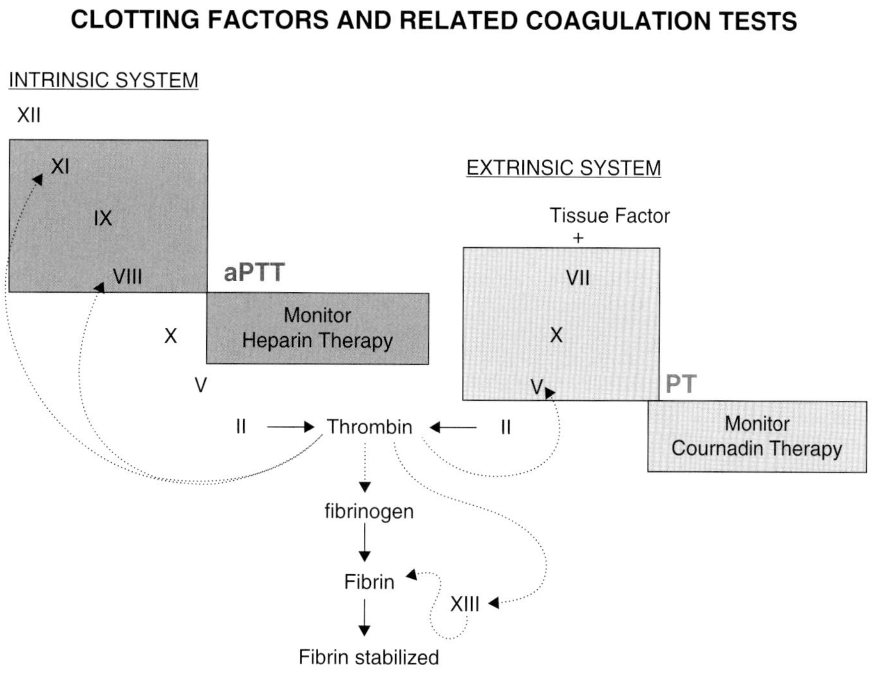

CLOTTING FACTORS AND RELATED COAGULATION TESTS

Figure 1.

Clotting factors and related coagulation tests.

COAGULOPATHIES AND COAGULATION (CONTINUED)

- Anticoagulant rodenticides. • Hepatic failure ○ Hepatic lipidosis ○ Aflatoxin ○ T-2 Toxin ○ Ochratoxin ○ Rapeseed ○ Hemachromatosis ○ Chlamydiosis ○ Mycobacteriosis ○ Polyomavirus ○ Adenovirus ○ Biliary adenocarcinoma associated with viral pathogen.
- Disseminated Intravascular Coagulation (DIC)—acquired syndrome characterized by intravascular activation of coagulation with loss of localization resulting from difference causes. ○ Diagnosis in birds may include thrombocytopenia, hypofibrinogenemia, hypoproteinemia, prolonged PT, ACT, or other abnormal hemostatic testing. ○ Circovirus ○ Reovirus ○ Herpesvirus ○ Polyomavirus ○ Adenovirus ○ Avian influenza ○ Infectious bursal disease ○ *Borrelia anserina* ○ *Salmonella pullorum* ○ *Escherichia coli* ○ *Erysipelothrix rhusiopathiae* ○ Snake envenomation ○ Neoplasia, for example, lymphocytic leukemia.

RISK FACTORS
- Overheparinization in smaller animals in the veterinary clinic may induce coagulopathy. • NSAID administered for pain may decrease thrombocyte function as well as result in GI bleeding. • Conure.

DIAGNOSIS

DIFFERENTIAL DIAGNOSIS
See above.

CBC/BIOCHEMISTRY/URINALYSIS
- Anemia • Thrombocytopenia
- Hypoproteinemia • Hypofibrinogenemia
- Evidence of hepatic damage or failure
○ Increased AST, GLDH, GGT, ALP activity
○ Increased bile acids concentration.
- Hematuria

OTHER LABORATORY TESTS

Activated Clotting Time (ACT)
- Point of care test performed in clinic using a tube with appropriate ratio of diatomaceous earth at 37°C. • Screen contact activation (intrinsic) pathway and common pathway. ○ Factors VIII, IX, V, X, II (thrombin), I (fibrinogen). • Not as sensitive as PT.
- Benefits: ○ No transportation to laboratory. ○ Quick turn-around time. ○ Inexpensive. ○ If sample clots quickly, measurement and interpretation is still possible. • Normal results in birds are typically <120 seconds with some specific variation.

Prothrombin Time (PT)
- Thromboplastin/calcium reagent. • Screens tissue activation (extrinsic) and common pathways. ○ Factors VII, V, X, II (thrombin), I (fibrinogen). • Requires avian specific reagent for diagnostic analysis. • When avian specific reagent is used, reference intervals are commonly 9–15 seconds in the adult bird.
- Younger birds commonly have extended PT.

Russell Viper Venom Time (RVV)
- Russell viper venom directly activates Factor X, and thereby the common pathway: V, II (thrombin), I (fibrinogen). • Can be found in human diagnostic labs as it is used in diagnosis of Lupus. • Requires 0.5–1 cc serum and volume may preclude use in smaller birds. • Should be validated on a species specific basis. • Normal times in pigeons are <200 seconds.

Activate Partial Thromboplastin Time (APTT)
- Typically not used in birds due to lack of reagent.

Fibrinogen
- Heat precipitation technique using two hematocrit tubes, one unheated and one heated to 56°C for three minutes. • Two measured TS values are subtracted providing fibrinogen estimated. • Estimate of fibrinogen useful in diagnosing DIC.

Thromboelastography
Thromboelastography (TEG) is a method of testing the efficiency of blood coagulation. More common tests of blood coagulation include prothrombin time (PT) and partial thromboplastin time (aPTT), which measure coagulation factor function, but TEG also can assess platelet function, clot strength, and fibrinolysis which these other tests cannot. TEG normals have been described in chickens and Hispaniolan Amazon parrots. Because few facilities have the ability to perform TEG this modality is not yet readily available to practitioners.

IMAGING
Radiographs may be warranted to rule out differentials such as trauma but imaging does not typically aid in diagnosis of the etiology of coagulopathy.

DIAGNOSTIC PROCEDURES
Surgical or endoscopic intervention for diagnosis or therapy should be approached cautiously due to risk of overt hemorrhage.

PATHOLOGIC FINDINGS
- Hemorrhage into cavities, musculature, etc.
- Hematoma formation. • Pulmonary thromboembolism. • Subdermal hemorrhage of the cranium and limbs as well as bleeding at new feather follicles. • Bloody vitellogenic follicles. • Hemorrhage in sclera and orbital cavities. • Pulmonary or GI hemorrhage may result in epistaxis or blood in the mouth.

TREATMENT

APPROPRIATE HEALTH CARE
- Treatment must be specific to the underlying cause to be successful.
- Hospitalization for inpatient nursing care is generally required.

NURSING CARE
- Gentle handling with soft blanket and pads is warranted to prevent hematoma formation.
- Optimal treatment of severe hemorrhage from anticoagulant rodenticide toxicosis is transfusion of whole blood to replace both red blood cells and clotting factors. ○ Though heterologous and homologous blood transfusions have been described, homologous are recommended due to short survival time of heterologous RBCs and possible acute immune reaction.

ACTIVITY
The patient's activity should be restricted in a padded cage and possibly in the dark or with hoods to decrease self-trauma.

DIET
- Vitamin K1 (phylloquinone) is regularly administered orally (0.13 mg/kg) to poultry as part of basic nutrition with documented decreased PT in comparison to controls.
○ This may be considered in conure bleeding syndrome or unexplained coagulopathy.

CLIENT EDUCATION
Typically with either hepatic disease or DIC there are underlying conditions that must be treated in order to resolve the condition.

SURGICAL CONSIDERATIONS
To be avoided if possible.

MEDICATIONS

DRUG(S) OF CHOICE
- Injectable or oral Vitamin K1 (phytonadione) is recommended for coagulopathies due to hypovitaminosis K, induced by rodenticide, nutritional deficiency, or antibiotic administration. • If marked liver disease is suspected, pretreatment with vitamin K for 24–48 hours prior to biopsy is recommended. • Reported dosage ranges are from 0.2-2.2 mg/kg SC q 4–8h though higher doses (2.5 mg/kg SC) q12h until hemostasis has been achieved have also been reported (Murray, 2011). Doses are similar but delivered IM according to Carpenter's formulary. • Oral treatment with Vitamin K1 (not K3) from 1 to 5 mg/kg is possible in birds but anticoagulant treatment dosages have not been validated with pharmacokinetic trials. • The K vitamins are essential cofactors for microsomal vitamin K-dependent carboxylases which catalyze posttranslational conversion of glutamyl residues to gamma carboxyglutamyl residues in mature proteins. Vitamin K hydroquinone is the active cofactor form required by factors II, VII, IX, and X for clot formation. Vitamin K hydroquinone is recycled by vitamin K reductase. This is the enzyme that is inhibited by the coumarin anticoagulants. • A 5 mg/kg Vitamin K administration at the time of 10 mg/kg warfarin administration resulted in no change

COAGULOPATHIES AND COAGULATION (CONTINUED)

in PT in dosed animals. This indicates that Vitamin K is an effective therapy. • Vitamin K3 (menadione) has not been effective against high doses of dicoumarol-type anticoagulants.

CONTRAINDICATIONS
Heparin, NSAIDs, Vitamin K3.

PRECAUTIONS
Anaphylactic reactions to injectable Vitamin K have been reported in mammals and are possible in birds though rare.

POSSIBLE INTERACTIONS
N/A

ALTERNATIVE DRUGS
N/A

FOLLOW-UP

PATIENT MONITORING
• Vitamin K therapy is typically instituted for extended periods in birds with rodenticide intoxication as PT assays may not be available. • Consider that half-life of brodifacoum in the liver is >300 days. • This may be discontinued when the PT is normal.

PREVENTION/AVOIDANCE
• Supervise birds at all times when out of their enclosure. • Birds with previously noted bleeding issues (i.e. conure bleeding syndrome) should have precautions taken to minimize risk during future venipunctures (i.e. use medial metatarsal veins only, so that a pressure wrap can be applied). If jugular venipuncture is required, consider premedication with vitamin K for 2–3 days prior to event.

POSSIBLE COMPLICATIONS
Conure bleeding syndrome–obtaining blood samples may result in life threatening situations.

EXPECTED COURSE AND PROGNOSIS
• DIC–Prognosis is guarded to poor and based on resolution of the underlying condition. • AR–See Anticoagulant Rodenticide. • Due to the extended half-life of many AR, the patient may clot normally during therapy and proceed to hemorrhage when Vitamin K therapy is discontinued. Owners should monitor this closely.

MISCELLANEOUS

ASSOCIATED CONDITIONS
N/A

AGE-RELATED FACTORS
N/A

ZOONOTIC POTENTIAL
Dependent upon underlying disease.

FERTILITY/BREEDING
Coagulopathy may result in abnormal oviposition and decreased chick survivability.

SYNONYMS
Hemostasis
Clotting or bleeding disorder
Bleeding diathesis

SEE ALSO
Anemia
Anticoagulant rodenticide
Heavy metal toxicity
Hemorrhage
Liver disease
Trauma
Sick-bird syndrome

ABBREVIATIONS
AR–anticoagulant rodenticide
PT–prothrombin time
ACT–activated clotting time
RVV–Russell viper venom time
DIC–disseminated intravascular coagulation
AST–aspartate aminotransferase
GLDH–glutamate dehydrogenase
GGT–gamma glutamyl transferase
ALP–alkaline phosphatase
TEG–thromboelastography

INTERNET RESOURCES
www.abaxisuniversity.com

Suggested Reading
Dam, H. (1935). The antihaemorrhagic vitamin of the chick. *Biochemistry Journal*, **29**:1273–1285.
Gentry, P.A. (2004). The comparative aspects of blood coagulation. *The Veterinary Journal*, **168**:238–251.
Harr, K.E. (2010). Avian coagulation. In: Weiss, D., Wardrop, J. (eds). *Schalm's Veterinary Hematology*. Hoboken, NJ: John Wiley & Sons, Inc.
Harr, K.E. (2003). Clinical chemistry of companion avian species: a review. *Veterinary Clinical Pathology*, **31**(3):140–151.
Murray, M. (2011). Anticoagulant rodenticide exposure and toxicosis in four species of birds of prey presented to a wildlife clinic in Massachusetts, 2006–2010. *Journal of Zoo and Wildlife Medicine*, **42**(1):88–97.

Author Kendal E. Harr, DVM, MS, DACVP
Acknowledgement Thanks to Dr. Anneliese Strunk for her thorough review of this consult.

Coccidiosis, Intestinal

BASICS

DEFINITION
• Coccidiosis is a clinical intestinal disease caused by nonmotile, protozoal parasites, coccidians, most commonly, *Eimeria* and *Isospora* species. • Coccidiasis is the presence of coccidians in the intestines without clinical disease. • Coccidia in avian species tend to be host specific and many coccidian species do not cause clinical disease. • *Eimeria* species tend to be the primary intestinal pathogens of avian species, and are most problematic in farmed birds (e.g., chickens, turkeys, ducks) raised in confinement, as well as zoo and pet bird species. • Subclinical coccidiosis in the poultry industry is associated with major economic losses. Seven species of *Eimeria* (*E. acervulina*, *E. brunetti*, *E. maxima*, *E. mitis*, *E. necatrix*, *E. praecox* and *E. tenella*) are recognized as infecting chickens. • Coccidian infections, while common, do not generally cause disease in free-ranging birds.

PATHOPHYSIOLOGY
• Most coccidia have a direct life cycle, in which infected birds shed noninfective oocysts (sporont) in feces. Oocysts sporulate within 48 hours and become infective (sporocysts containing infective sporozoites). • Susceptible birds ingest infective oocysts during feeding or drinking, the sporocysts walls break in the ventriculus, and the sporozoites invade intestinal epithelial cells. Within the intestines, oocysts may undergo several stages of development before becoming sexually mature male and female parasites. Mature female coccidia release noninfective oocysts into the environment through the feces and the direct life cycle continues. • The life cycle is complete in 4–8 days in most psittacine and poultry species, and 7–14 days in some free-ranging bird species. • In most cases, birds infected with coccidia will develop immunity and recover, although reinfection is common.
• The presence of disease depends on number of host intestinal epithelial cells invaded and the host immune status. • Severely infected birds may die acutely. Intestinal tissue damage can result in anorexia, diminished nutrient absorption, melena or passing of frank blood, anemia, dehydration, and immune compromise of the intestines leading to secondary bacterial and/or fungal infections.

SYSTEMS AFFECTED
Small and large intestines are affected; damage to intestinal epithelial cells disrupts normal function.

GENETICS
There is no known genetic basis for intestinal coccidiosis.

INCIDENCE/PREVALENCE
• The exact incidence and prevalence is currently unknown. • The presence of coccidia in the gastrointestinal tracts of many avian species is common, but mere presence is not always associated with disease. • Most free-ranging birds are likely infected, and farmed avian species are commonly infected and are more likely to exhibit clinical signs.

GEOGRAPHIC DISTRIBUTION
• Coccidia are found worldwide and any avian species are susceptible. • Crowding will exacerbate the spread of the parasite, likely leading to disease. This is typical of poultry monoculture situations.

SIGNALMENT
• **Species:** All avian species are potentially susceptible to infection. • **Mean age and range:** All ages are susceptible; in poultry, younger birds are considered to be at greater risk. • **Predominant sex:** Both sexes are susceptible.

SIGNS
• Frequently birds are asymptomatic.
• Clinical signs may range from nothing to one or more of the following: lethargy, fluffed feathers, anorexia, weight loss, diarrhea (watery, hemorrhagic, tarry), and death.
• Frank blood in droppings is a common finding in poultry. • Decreased egg production can be observed in breeding or laying birds.

CAUSES
• Direct exposure to oocysts in the environment, and ingestion.

RISK FACTORS
• Exposure to coccidian parasites.
• Overcrowding, especially in avian species used for food production or maintained in pet stores and aviculture. Heavily raced pigeons are also predisposed. • Stress may exacerbate infection. In free-ranging birds, stress can predispose to epizootics; typically associated with staging areas during spring migration.

DIAGNOSIS

DIFFERENTIAL DIAGNOSIS
• Coccidiosis should be included as a differential diagnosis for any bird with diarrhea and/or hematochezia. • Bacterial or fungal enteritis. • Mycobacterial enteritis.
• Gastrointestinal nematodiasis.
• Cryptosporidiosis.

CBC/BIOCHEMISTRY/URINALYSIS
• Most commonly, there will be no hematologic abnormalities with intestinal coccidiosis, with the exception of anemia in severe cases. • Nonspecific changes associated with dehydration and emaciation, which might include increased hematocrit and decreased electrolytes.

OTHER LABORATORY TESTS
N/A

IMAGING
N/A

DIAGNOSTIC PROCEDURES
• Finding oocysts in a fecal float and direct smear are the diagnostic methods of choice.
• However, routine fecal exams in avian species, without clinical signs, may be positive for coccidia. In these cases, some veterinarians choose to treat based on quantification of coccidia on the fecal smear of float, and taking in consideration of context (zoo exhibit, chicken flock, canary aviary, racing pigeons, etc.). This empirical approach is left up to the individual veterinarian. • In cases of death prior to antemortem diagnosis, a necropsy will be required with identification of coccidia on an intestinal scraping and/or histopathologic evidence of intestinal epithelial invasion by coccidia. • Endoscopy and biopsy: In larger bird species, in which an endoscope can be passed into the small intestine, a biopsy of the intestinal mucosa may be useful.

PATHOLOGIC FINDINGS
• Gross Lesions: Most commonly, hemorrhagic enteritis of the small intestine is observed, but location of lesions can vary with species of coccidia, avian species, and severity of infection. • Histopathologic Lesions: Depending on the coccidian species, parasite load, and gastrointestinal organ site of infection, coccidiosis can result in a range of pathologic conditions, including: mild enteritis; mild-to-moderate inflammation of the intestinal wall with pinpoint hemorrhages and mucosal sloughing; complete intestinal villar destruction resulting in extensive hemorrhage/necrosis; and death. Typically, schizonts are observed in the intestinal epithelial cells along with merozoites, infiltrating neutrophils and eosinophils.

TREATMENT

APPROPRIATE HEALTH CARE
Most birds can be treated as outpatients; critically ill birds may need hospitalization and more intensive supportive care.

NURSING CARE
• Supportive care may be necessary due to fluid loss associated with diarrhea and/or hematochezia and fecal occult blood. • Fluid support: Crystalloid fluid (e.g., Lactated Ringer's, 0.9% saline, Normosol®) replacement is standard of care; administered either subcutaneously (40–60 mL/kg depending on level of dehydration) or intravenously via IV catheter in emergency situations. • Warmth: Any individual bird should be maintained in a warm environment, which may include and incubator or heat lamp.

COCCIDIOSIS, INTESTINAL (CONTINUED)

ACTIVITY
N/A

DIET
N/A

CLIENT EDUCATION
Since treatment generally entails oral administration of medication, either directly or indirectly through medicated feed, the client must be instructed about methods of administration.

SURGICAL CONSIDERATIONS
N/A

MEDICATIONS

DRUG(S) OF CHOICE

General Comment
Drug resistance to anticoccidial medications is a constant concern, particularly in flock medicine.

Medications for individual patients
• **Amprolium** ◦ Chickens: 25 mg/kg PO q 24h for 5 days. ◦ Pigeons: 25 mg/kg PO q 24h for 5 days. ◦ Raptors: 30 mg/kg PO q 24h × 5 days. • **Sulfadimethoxine** ◦ Most species: 25 mg/kg PO q12h for 5–7 days. ◦ Raptors, psittacines: 50 mg/kg PO q24h for 5 days, then off 3 days, then on 5 days. • **Sulfaquinoxaline** ◦ Lories, pigeons: 100 mg/kg PO q 24h for 3 days, then off 2 days, then on 3 days. • **Toltrazuril** ◦ Budgerigars, raptors: 7–10 mg/kg PO q 24h for 2–3 days. ◦ Pigeons: 20–35 mg/kg PO once. • **Trimethoprim/sulfamethoxazole** ◦ Most species: 30 mg/kg PO q 12–24h for 5–7 days.

Medications for monocultured flocks and aviculture (food- and water-based)
• **Amprolium** ◦ Most species, including chickens, passerines, budgerigars: 50–100 mg/L drinking water for 5-7 days. ◦ Pigeons: 200 mg/L drinking water for 5–7 days. ◦ Poultry: 575 mg/L drinking water (using 9.6% solution) for 5–7 days. ◦ Poultry, cranes: 115–232 mg/kg feed for 5–7 days. • **Decoquinate** ◦ Poultry: 20–40 mg/kg feed q 5–7d. • **Diclazuril** ◦ Poultry: 5–10 mg/L drinking water for 5–7 days. ◦ Poultry: 1 mg/kg feed for 5–7 days. • **Dintolamide** ◦ Poultry: 40–187 mg/kg feed. • **Monensin** ◦ Poultry gamebirds, cranes: 70–100 mg/kg feed. • **Sulfachlorpyridazine** ◦ Passerines, pigeons: 300 mg/L drinking water for 5 days, then off 3 days, then on 5 days and repeat multiple times. ◦ Cockatiels, budgerigars: 400 mg/L drinking water for 30 days. • **Sulfadimethoxine** ◦ Poultry, pigeons: 250–500 mg/L drinking water for 5 days.
• **Sulfadimethoxine/ormetoprim** ◦ Poultry, game birds: 10 mg/kg feed or 320–525 mg/L drinking water. • **Sulfaquinoxaline** ◦ Poultry, pigeons: 250–500 mg/L drinking water for 6 days, then off 2 days, then on 6 days. • **Toltrazuril** ◦ Psittacines: 2–5 mg/L drinking water for 2 days, repeat in 14–21 days. ◦ Waterfowl: 12.5 mg/L drinking water for 2 days. ◦ Poultry: 25 mg/kg drinking water for 2 days. ◦ Passerines: 75 mg/L drinking water for 2 days/week for 4 weeks. ◦ Pigeons: 75 mg/L drinking water for 5 days. • **Trimethoprim/sulfamethoxazole** ◦ Poultry, pigeons: 320–525 mg/L for 5–7 days drinking water.

CONTRAINDICATIONS
There are now a number of drug resistant strains of coccidia in the world, especially in avian monoculture facilities.

PRECAUTIONS
N/A

POSSIBLE INTERACTIONS
N/A

ALTERNATIVE DRUGS
N/A

FOLLOW-UP

PATIENT MONITORING
• Fecal float and fecal direct smears 1–2 weeks posttreatment. • Hematochezia and fecal occult blood should be controlled after treatment.

PREVENTION /AVOIDANCE
• Vaccines are used in poultry operations, but not in individual pet birds. • Live attenuated and nonattenuated vaccines are available, with the goal of producing subclinical coccidiosis. However, this approach can is associated with decreased performance and increased bacterial enteritis in poultry operations.

POSSIBLE COMPLICATIONS
The primary complication associated with treatment is that the bird(s) may become lifelong carriers of coccidia, and have the potential to continue to infect other birds.

EXPECTED COURSE AND PROGNOSIS
With appropriate treatment, the prognosis for eliminating clinical signs (e.g., diarrhea, hematochezia) is good. However, completely eliminating the organisms from the patient, and the environment, may contribute to constant reinfection, particularly in poultry operations and aviaries.

MISCELLANEOUS

ASSOCIATED CONDITIONS
N/A

AGE-RELATED FACTORS
N/A

ZOONOTIC POTENTIAL
N/A

FERTILITY/BREEDING
N/A

SYNONYMS
N/A

SEE ALSO
Anemia
Clostridiosis
Flagellate enteritis
(Systemic) coccidiosis
Colibacillosis
Cryptosporidiosis
Diarrhea
Enteritis/gastritis
Gastrointestinal helminthiasis
Mycobacteriosis
Sick-bird syndrome
Urate/fecal discoloration

ABBREVIATIONS
N/A

INTERNET RESOURCES
For detailed information related to coccidian in free-ranging birds:
• http://www.nwhc.usgs.gov/publications/field_manual/chapter_26.pdf
For information about poultry coccidiosis:
• http://www.merckmanuals.com/vet/poultry/coccidiosis/overview_of_coccidiosis_in_poultry.html

Suggested Reading
Alnassan, A.A., Kotsch, M., Shehata, A.A., et al. (2014). Necrotic enteritis in chickens: development of a straightforward disease model system. *The Veterinary Record*, **174**:555–561.
Doneley, R.J.T. (2009). Bacterial and parasitic diseases of parrots. *The Veterinary Clinics of North America. Exotic Animal Practice*, **12**:417–432.
Dorrestein, G.M. (2009). Bacterial and parasitic diseases of passerines. *The Veterinary Clinics of North America. Exotic Animal Practice*, **12**:433–451.
Harlin, R., Wade, L. (2009). Bacterial and parasitic diseases of Columbiformes. *The Veterinary Clinics of North America. Exotic Animal Practice*, **12**:453–473.
Shirley, M.W., Lillehoj, HS. (2012). The long view: a selective review of 40 years of coccidiosis research. *Avian Pathology*, **41**:111–121.
Author Kurt K. Sladky, MS, DVM, DACZM, DECZM (Herpetology)

COCCIDIOSIS, SYSTEMIC

BASICS

DEFINITION
Systemic coccidiosis is caused by an extraintestinal infection with specific species of coccidia, generally in the genuses *Eimeria*, *Isospora*, and/or *Atoxoplasma*. The infective coccidians are host, tissue, and cell-specific and complete their development in both the intestinal tract and extra-intestinally. Infections cause significant morbidity and mortality. In cranes, this disease is called disseminated visceral coccidiosis.

PATHOPHYSIOLOGY
- The life cycle is direct. The bird ingests the sporulated oocysts lying in its environment. Oocysts undergo schizogony (merogony; asexual phase) in the intestine. If the schizonts produced are deep in the intestinal mucosa, they may damage the mucosa when they divide, causing enteritis. After schizogony, they differentiate into micro- and macrogametocytes, which fuse to form the zygote (gametogony or sexual phase, which is nonpathogenic). An oocyst wall forms around the zygote as it develops and matures. The oocysts are released into the intestinal lumen and passed with the feces onto the ground, where they sporulate under favorable conditions of moisture and temperature. In cases of systemic coccidiosis, the asexual phases occur both intra- and extra-intestinally; however, the sexual phase takes place only in the intestines. • The basic host response in systemic coccidiosis to coccidian development in the tissues is granulomatous inflammation. This change is primarily because of the infiltration of phagocytic mononuclear cells laden with developing asexual stages of coccidia in the organs and tissues. The consequent rupture of the host cells initiates a necrotizing inflammation.

SYSTEMS AFFECTED
- Gastrointestinal—diarrhea, enteritis, edema of the duodenum and jejunum, wet tail feathers. • Musculoskeletal—emaciation, severe pectoral muscle atrophy, "going light"/weight loss. • Behavioral—severe depression, ruffling of feathers, grouped birds huddle together.
- Hemic/Lymphatic/Immune—invasion of mononuclear cells with zoites, splenitis. • Hepatobiliary—hepatitis, "thick liver disease". • Respiratory—bronchopneumonia. • Renal/urologic—nephritis.
- Cardiovascular—myocarditis.
- Nervous—generalized neurologic signs, weakness.
- Endocrine/metabolic—pancreatitis.

GENETICS
None.

INCIDENCE/PREVALENCE
Not uncommon. More prevalent in captive birds due to stress, husbandry conditions, and overcrowding. Reported in numerous families/orders of birds (including passeriformes, anseriformes, gruiformes, sphenisciformes, procellariiformes, pelacaniformes, and raptors [uncommon]).

GEOGRAPHIC DISTRIBUTION
This disease has been seen in both captive and wild birds all over the world.

SIGNALMENT
More commonly seen in chicks and/or immunocompromised birds. No sex predilection.

SIGNS
- Asymptomatic. • Diarrhea—sometimes greenish in color. Can also be mucoid or mucosanguinous. • Weight loss/emaciation. Seen commonly as severe pectoral muscle atrophy. • Lethargy. • Weakness.
- Inappetance. • Wet tail feathers. • Cloacal swelling. • Coelomic distention associated with hepatomegaly. • Ruffled feathers.
- Huddling with other birds. • Leg weakness and tendency to tip forward with renal coccidiosis. • Nonspecific neurologic signs.
- Acute death.

Historical Findings
- Significant stressor in recent history (such as transport, recent hatching, etc.) or chronic or concurrent disease process. Onset of clinical signs is usually acute.

Physical Examination Findings
- Diarrhea. • Emaciation, pectoral muscle atrophy. • Lethargy. • Weakness. • Wet tail feathers, vent. • Cloacal swelling. • Coelomic distention, hepatomegaly. • Ruffled feathers.
- Dyspnea associated with bronchopneumonia. • Leg weakness and/or tendency to tip forward with renal coccidiosis.
- Non-specific neurologic signs. • Small, white, raised nodules in the oral cavity. • May not show any abnormalities until stressed.
- May be asymptomatic.

CAUSES
- Infection with *Eimeria* sp., such as *E. reichenowi* and *E. gruis* in cranes, *E. tenella* in chickens, *E. adenoides* in turkeys, *E. gaviae* in the common loon, *E. boschadis* in the mallard duck, *E. christianseni* in the mute swan, and *E. somaterie* in the common eider. • Infection with *Isospora* sp., such as *I. michaelbakeri* in sparrows, *I. canaria (serini)* in canaries, *I. robini* in American robins, or *I. gryphoni* in American goldfinches. • Infection with *Atoxoplasma* (formerly *Lankesterella*) sp.; some have also referred to it as *Isospora serini*.
- Infection (uncommon) with *Eimeria* and *Caryospora* in raptors.

RISK FACTORS
Poor husbandry, overcrowding, neonatal status, immunosuppression or concurrent infection, and/or environmental stressor(s).

DIAGNOSIS

DIFFERENTIAL DIAGNOSIS
- Lead poisoning • Botulism
- Trichomoniasis • Candidiasis • Tuberculosis (mycobacteriosis) • Aspergillosis
- Paratuberculosis • Avian cholera • Duck viral enteritis • Other protozoal infections (*Leukocytozoon*, *Sarcocystis*, *Toxoplasma*)
- Erysipelas • Campylobacteriosis
- Salmonellosis • Mycotoxins • Phytotoxins
- Neoplasia

CBC/BIOCHEMISTRY/URINALYSIS
- Possible hemoconcentration secondary to dehydration. • Possible anemia due to blood loss in feces, if any is present. • Possible elevations in liver enzymes if liver is affected by disease. • Possible elevation in uric acid if kidneys are affected. • May not see any hematologic abnormalities. • Merozoites may be seen in the cytoplasm of mononuclear cells on blood smear.

OTHER LABORATORY TESTS
Fecal direct smear and/or flotation to look for the presence of coccidia. Feces are best collected from the affected bird(s) in the afternoon (2–6 pm) due to oocyst shedding rhythms. It is possible for birds to have systemic coccidiosis without shedding oocysts in the feces.

IMAGING
Radiography
- Consider the use of dental radiography or mammography in very small patients. • May demonstrate the presence of masses in the thoracic and/or coelomic cavities; rule out granulomas vs. neoplasia vs. other. • May demonstrate asymmetry of the hepatic silhouette; rule out hepatomegaly, splenomegaly, neoplasia, other. • May locate a metal foreign body-important in ruling out heavy metal toxicosis as a differential diagnosis. • Whole-body radiographs: to identify any abnormalities seen, which may provide information in cases that are showing non-specific clinical signs of illness.

Ultrasonography
- Results limited depending on the size of the bird. • May reveal organ enlargement and/or architectural changes. • May reveal discrete masses. • May reveal thickened bowel walls.

DIAGNOSTIC PROCEDURES
- Fine needle aspirate and cytology of masses located vs. biopsy and histopathology.
- Exploratory celiotomy. • Impression smears of major organs can show mononuclear cells with zoites in cytoplasm. • Necropsy, especially if there is a flock of potentially exposed birds.

COCCIDIOSIS, SYSTEMIC
(CONTINUED)

PATHOLOGIC FINDINGS
Gross
• Grayish-white nodules (granulomas) disseminated throughout many organs and tissues-reportedly seen on the surface and parenchyma of the liver, lung, spleen, kidneys, heart; adventitia of the blood vessels, including the carotid and femoral arteries; serosa and mucosa of the esophagus, proventriculus, ventriculus, intestines, and cloaca; the mesentery and parietal peritoneum; submucosa of the trachea and mainstem bronchi; epimysium and parenchyma of pectoral and cervical muscles; and subcutaneous tissues of the thoracic and cervical regions. • Mottling of the liver and hepatomegaly; may see yellow-orange color to liver. • Congestion and enlargement of the spleen. • Hyperemia of the duodenal mucosa. • Lung consolidation. • Frothy material in the trachea. • Distention of the bowels. • Duodenum and jejunum creamy white in color with walls 4–5 times thickened. • Kidneys grossly swollen with white pinpoint foci of urates and cellular debris.

Histopathology
• Granulomas composed of inflammatory cells surrounded by a thin fibrous capsule with cellular infiltrate including macrophages, lymphocytes, plasma cells, and few heterophils. Numerous pale basophilic, round to oval uninucleated and multinucleated bodies, 5–10 mm in diameter scattered throughout the granulomas. Developing meronts in the cytoplasm of macrophages. Some granulomas do not contain meronts. • Liver—lymphoid aggregates protruding into the lumen of portal and hepatic veins, sometimes nearly occluding the smaller hepatic veins. Multifocal areas of granulomatous inflammation. Extensive areas of necrosis. • Small intestines—moderate to severe multifocal infiltration of the lamina propria by macrophages, lymphocytes, and few heterophils with gametogonic stages of coccidia in epithelial lining cells. Developing coccidial stages also found within macrophages in granulomatous infiltrates and in lymphatics or capillaries in the lamina propria and tunica muscularis. Inflammation and mild focal necrosis of the intestinal villi associated with coccidian oocysts and gametocytes in the epithelium. Large numbers of gamonts, oocysts, and meronts, some of which can be large, with large merozoites, present among necrotic intestinal glands. • Lungs—multifocal areas of consolidation, and parabronchi and bronchioles contain a pleocellular exudate. Airways may contain coccidian oocysts similar to those seen in the gut. • Spleen—multifocal areas of granulomatous inflammation. Extensive areas of necrosis. • Heart—multifocal areas of granulomatous inflammation. • Kidneys—multifocal areas of granulomatous inflammation. • Bursa of Fabricius—multifocal areas of granulomatous inflammation.

TREATMENT
APPROPRIATE HEALTH CARE
Emergency inpatient intensive care management. Birds in a flock should be separated immediately from all birds showing illness.

NURSING CARE
• Subcutaneous fluids can be administered (25–50 mL/kg) q8–12h, or as needed. IV access may be difficult to obtain and maintain depending on the size of the patient; consider intraosseus (IO) catheterization if intravascular fluids are needed. In most patients, Lactated Ringers solution or Normosol-R crystalloid fluids are appropriate. Maintenance fluids are estimated at ≥50 mL/kg/day. • Consider the addition of dextrose at 2.5–5% to the subcutaneous or IV fluid administration, especially if the bird is inappetant or anorexic. • Supplemental oxygen therapy may be warranted depending on clinical signs. • Supplemental heat should be provided at all times to hospitalized animals; temperature should be determined by the species that is being treated. • Gavage feeding should be considered if the patient will not eat on its own. • Strict hygiene should be practiced to decrease the parasite load in the patient's environment and to prevent cross-contamination between patients. • Handling should be kept as minimal as possible to avoid stressing the patient.

ACTIVITY
Dependent on how clinically affected the patient is. The patient's activity should be limited enough to maintain IV or IO catheter placement and patency and to provide supplemental heat to the patient, if needed. Consideration should also be given to the ability to properly disinfect the patient's environment when deciding how much space a patient should have access to.

DIET
The patient should continue to eat their normal diet, if possible. If the patient is inappetent or anorexic, consider gavage feeding an appropriate formula twice daily. This may also help in medicating the patient.

CLIENT EDUCATION
Prevention of this disease should be strongly stressed over reliance on treatment, as true systemic infection with coccidiosis (intracellular stages of life cycle) is untreatable. Decreasing the concentration of coccidia in the soil or environment is the best way to prevent birds from becoming infected. If possible, birds should be housed on concrete, tile, or other easily cleaned and disinfected artificial flooring, which is cleaned regularly. Suspended wire flooring is another option to help with sanitation. Food and water dishes should be kept off of the ground. Alternatively, if birds are to be housed outdoors, yearly to every third year pen rotation will help to reduce coccidia concentrations in the soil. Stress reduction and prevention of crowding are also of high importance in prevention.

SURGICAL CONSIDERATIONS
There is no surgical treatment for systemic coccidiosis. Anesthetizing a patient with systemic coccidiosis should be performed only when absolutely necessary, as the likelihood of patient death under anesthesia is very high.

MEDICATIONS
DRUG(S) OF CHOICE
Important to note that medications will treat the intestinal stages (i.e., reduce oocyst shedding), but generally are unable to influence the intracellular or extra-intestinal stages; monensin and possibly toltrazuril, however, may also have an effect against extraintestinal parasites. • Anseriformes ∘ Clazuril, sulfonamides. • Cranes ∘ Amprolium, 0.006% in drinking water or 0.0125–0.025 mg/kg in feed; intended for prevention of DVC. ∘ Monensin, 99 ppm in feed; for treatment or prevention of DVC. • Passerines ∘ Sulfachlorpyridazine, 150 mg/L in drinking water for 5 days a week until after molting; intended for treatment of extracellular stages of Atoxoplasma sp. ∘ Sulfachlorpyridazine, 300 mg/L in drinking water, 5 days a week for 2–3 weeks; intended for treatment of Eimeria sp. and Isospora sp. ∘ Amprolium, 50–100 mg/L in drinking water for 5 days; intended for treatment of Eimeria sp. and Isospora sp. ∘ Toltrazuril, 12.5 mg/kg PO daily for 2 days, off 5 days, then repeat the 7-day cycle as needed (months); for treatment of atoxoplasmosis/Isospora sp. • Raptors ∘ Toltrazuril, 10 mg/kg PO daily for 2 doses, then repeat in 2 weeks. ∘ Sulfadimethoxine, 55 mg/kg PO day 1, then 25 mg PO daily for 10 days. ∘ Clazuril, 10 mg/kg PO daily for 3 days, repeat 1–2 times with 2 days off. ∘ Toltrazuril, 0.25 mL/kg PO daily for 2 doses.

CONTRAINDICATIONS
Avoid the use of sulfonamides in animals with severe renal or hepatic impairment. Sulfonamides may also be teratogenic.

PRECAUTIONS
None.

POSSIBLE INTERACTIONS
None.

ALTERNATIVE DRUGS
None.

COCCIDIOSIS, SYSTEMIC (CONTINUED)

FOLLOW-UP

PATIENT MONITORING
- Fecal flotation and direct smear should be repeated weekly for four weeks or until smear and flotations are consistently negative.
- Patient should be monitored for resolution of clinical signs and ability to maintain weight and health status without nursing care.

PREVENTION/AVOIDANCE
Decreasing the concentration of coccidia in the soil or environment is the best way to prevent birds from becoming infected. If possible, birds should be housed on concrete, tile, or other easily cleaned and disinfected artificial flooring, which is cleaned regularly. Suspended wire flooring is another option to help with sanitation. Food and water dishes should be kept off of the ground. Alternatively, if birds are to be housed outdoors, yearly or every third year pen rotation will help to reduce coccidia concentrations in the soil. Stress reduction and prevention of crowding are also of high importance in prevention.

POSSIBLE COMPLICATIONS
Acute death.

EXPECTED COURSE AND PROGNOSIS
Dependent upon which organ systems are affected and how severely infected the patient is. Prognosis for systemic coccidiosis with organ involvement and/or granuloma formation is grave, despite treatment and nursing care.

MISCELLANEOUS

ASSOCIATED CONDITIONS
Secondary infections from parasites, bacteria, viruses, and/or fungi are all commonly seen.

AGE-RELATED FACTORS
None.

ZOONOTIC POTENTIAL
None.

FERTILITY/BREEDING
Avoid the use of sulfonamides in reproductive animals as they may be teratogenic.

SYNONYMS
Disseminated visceral coccidiosis

SEE ALSO
Aspergillosis
Botulism
Campylobacteriosis
Candidiasis
Flagellate enteritis
Heavy metal toxicity
Hemoparasites
Mycobacteriosis
Mycotoxicosis
Pasteurellosis
Salmonellosis
Sarcocystosis
Toxoplasmosis
Tumors

ABBREVIATIONS
DVC (disseminated visceral coccidiosis) in cranes

INTERNET RESOURCES
N/A

Suggested Reading
Atkinson, C.T., Thomas, N.J., Hunter, D.B. (eds.) (2008). *Parasitic Diseases of Wild Birds*. Hoboken, NJ: John Wiley & Sons, Inc., pp. 108–119, 162–180, 181–194, 204–222.
Coles, B.H. (ed.) (2007). *Essentials of Avian Medicine and Surgery*, 3rd edn. Hoboken, NJ: John Wiley & Sons, Inc., pp. 314–318.
Novilla, M.N., Carpenter, J.W. (2004). Pathology and pathogenesis of disseminated visceral coccidiosis in cranes. *Avian Pathology*, **33**:275–280.
Schrenzel, M.D., Mallouf, G.A., Gaffney, P.M., et al. (2005). Molecular characterization of isosporoid coccidia (*Isospora* and *Atoxoplasma* spp.) in passerine birds. *American Society of Parasitologists*, **91**:635–647.
Tully, T.N., Dorrestein, G.M., Jones, A.K. (eds.) (2009). *Handbook of Avian Medicine*, 2nd edn. Philadelphia, PA: Saunders Elsevier, pp. 182–392.
Authors Christine T. Higbie, DVM
James W. Carpenter, MS, DVM, DACZM

Coelomic Distension

BASICS

DEFINITION
Swelling of the coelomic cavity.

PATHOPHYSIOLOGY
• Distension of the coelom may be due to a mass effect (e.g., organomegaly, neoplasm, abscess, distension of bowel loops, or presence of an egg), abdominal herniation, obesity, or ascites (e.g., egg yolk peritonitis, congestive heart failure, or hepatopathy), • Enlargement of the coelomic organs may be a result of an infectious process, congestion, abscessation, impaction/obstruction, or neoplasm,
• Increased hydrostatic pressure in the blood vasculature, decreased oncotic pressure, leakage of vessels, or the combination of these factors can lead to ascites within a bird's coelomic cavity, • Moderate distension of the coelom is a normal finding associated with egg production, • Neonatal psittacine and passerine chicks normally have full and distended coelomic cavities due to prominent gastrointestinal tracts,

SYSTEMS AFFECTED
• Respiratory—increased respiratory effort due to compression of the air sacs by space occupying mass effect or fluid.
• Gastrointestinal—partial or complete obstruction of intestinal bowel loops by intraluminal mass or compression by an extraluminal mass; intestinal strangulation may occur if bowel loops are caught within a hernia; accumulation of feces may occur if the distension displaces the opening of the vent.
• Integument—localized plucking of feathers in the coelomic region in response to discomfort from the distension.
• Neuromuscular—masses in the coelomic cavity may compress the ischiatic nerves and lead to paresis or paralysis of the pelvic limbs.
• Renal/Urologic—damage to the kidneys and elevations in uric acid levels can develop due to compression of renal parenchyma and ureters by a coelomic mass.

GENETICS
N/A

INCIDENCE/PREVALENCE
• Distension of the coelom is a normal finding in hens during periods of oviposition.
• Neonatal/juvenile psittacine and passerine chicks normally have full and distended coeloms. • Coelomic distension is a common clinical sign for various disease processes, including ascites, organomegaly, ileus, neoplasms, obesity, and body wall hernias.

GEOGRAPHIC DISTRIBUTION
N/A

SIGNALMENT
• **Species predilections** ○ Ovarian neoplasms—cockatiels, chickens. ○ Ovarian cysts—cockatiels, budgerigars, canaries.
○ Testicular neoplasms—budgerigars.
○ Obesity—Amazon parrots, Quaker parrots, budgerigars. • **Mean age and range and Predominant sex** ○ Altricial neonate chicks (psittacines, passerines) normally have a full distended coelom. ○ Distension from an enlarged reproductive tract (either physiologic or pathologic) occurs in sexually mature hens.

SIGNS

Historical Findings
• Distended abdomen. • Tail bobbing and/or open-mouth breathing. • Generalized lethargy and depression; may be on bottom of the cage. • Decreased appetite. • Exercise intolerance, trouble flying. • May have history of recent egg laying. • May have history of straining (either to defecate or to lay an egg).

Physical Examination Findings
• Swollen coelom—may palpate egg, mass, or fluid. • Increased respiratory effort, especially after handling. • Wide-based stance or gait.
• Fecal and/or urate accumulation over vent and tail feathers. • Exercise intolerance.
• Anorexia. • Lethargic and depressed.
• Wheezing or crackles auscultated over lungs and caudal air sacs (ascites). • Holosystolic heart murmur (cardiomyopathy). • Yellow or green urates (hepatopathy).

CAUSES
Obesity (both intracoelomic and extracoelomic fat may be observed).
• Intracoelomic mass effect ○ Enlarged reproductive tract—egg, ovarian cysts, salpingitis, metritis, impacted oviducts with inspissated yolk ○ Hepatomegaly—hepatic lipidosis, hepatitis. ○ Splenomegaly—chlamydiosis, hemorrhage or hematomas.
○ Obstructed or distended gastrointestinal bowel loops or cloaca—fecalith, cloacal mass, ileus. ○ Neoplasia—ovarian, testicular, renal, pancreatic, hepatic, gastrointestinal, lymphoma/lymphosarcoma, lipomas.
○ Intracoelomic abscesses. ○ Granulomas.
• Ascites ○ Cystic ovarian disease. ○ Egg yolk coelomitis. ○ Hepatic cirrhosis.
○ Protein-losing nephropathy. ○ Protein-losing enteropathy. ○ Congestive heart failure. ○ Iron storage disease (ISD)/hemochromatosis.
○ Polyomavirus. ○ Mycobacteriosis.
○ Aspergillosis (granulomatous pneumonia leading to pulmonary hypertension). ○ Body wall herniation (with or without intestinal bowel loops).

RISK FACTORS
• Chronic egg layers and obese birds are at greater risk for egg binding and body wall herniation. • Birds fed a high fat diet in conjunction with a sedentary lifestyle are prone to obesity and are at risk for hepatic lipidosis and lipomas. • Ingestion of moldy peanuts and grain may lead to aflatoxicosis which can result in hepatic cirrhosis and ascites. • Birds in the ramphastidae (toucan) and sturnidae (mynah) families are at greater risk for ISD.

DIAGNOSIS

DIFFERENTIAL DIAGNOSIS
• Dyspnea/ respiratory distress—primary respiratory condition such as pneumonia, sinusitis, or tracheal obstruction; pain; heat exhaustion; cardiac disease. Diagnostic imaging such as ultrasound or radiographs used to differentiate. • Wide based stance/lameness—weakness in the pelvic limbs from either neurologic dysfunction, thromboembolic disease; musculoskeletal injury; generalized weakness. Radiographs to evaluate for musculoskeletal changes.

CBC/BIOCHEMISTRY/URINALYSIS
• Hypercalcemia is usually noted in reproductively active hens. • Moderate to severe hypoalbuminemia and decreased A:G ratio if a protein-losing nephropathy or enteropathy, or advance hepatic disease is present; neonates generally also have lower total protein and plasma albumin levels.
• May have severe leukocytosis if infectious disease process such as mycobacteria or chlamydia is involved; a higher white blood cell count can be a normal finding in juvenile psittacine birds. • Elevated bile acids if hepatic disease present. • Hyperlipidemia may be noted in obese birds and birds with hepatic lipidosis.

OTHER LABORATORY TESTS
• Cytologic analysis of fine needle aspirate or aspirated fluid; clear transudates are suggestive of heart failure or hypoproteinemia; exudates are suggestive of an infectious and/or inflammatory process; yolk or lipid globules are noted with egg yolk peritonitis; neoplastic cells may be noted with neoplastic effusions.
• Bacterial culture and sensitivity is recommended if purulent material or exudate is aspirated. • Histopathology of biopsied masses is used to differentiate between hyperplasia, inflammatory processes, and different types of neoplasms.

IMAGING

Radiographic Findings
• Distension of the caudal coelom can be visualized on lateral views. • Widening of the hepatic silhouette often noted on ventro-dorsal views. • May have poor serosal detail of the distended coelomic cavity, especially if ascites is present. • Upper gastrointestinal contrast study may help characterize the coelomic distension and help delineate a mass effect.

Ultrasonographic Findings
• Visualization of the coelomic cavity is enhanced if the coelomic distension is due to ascites or a mass effect. • Ultrasound allows for better evaluation of coelomic masses, unshelled eggs, cystic ovaries, distended/thickened intestinal loops, and loops of bowel within hernias. • A large heart,

COELOMIC DISTENSION (CONTINUED)

large hypoechoic liver, +/– ascites are noted in cases of CHF. • Changes associated with hepatic cirrhosis can include small hyperechoic liver with nodular margins, +/– ascites due to portal hypertension, small heart due to decreased perfusion. • Hepatic changes associated with hepatic lipidosis include a large hyperechoic liver.

Computed Tomography
• Cross-sectional imaging allows for coelomic cavity to be imaged without superimposition of anatomic structures; can also better differentiate tissue type and some masses may contrast enhance.

DIAGNOSTIC PROCEDURES
• If ascites is present, coelomocentesis is both diagnostic and therapeutic. • Laparoscopy allows for visualization and biopsy of coelomic masses or organomegaly. Eggs located ectopically within the coelom may also be visualized.

PATHOLOGIC FINDINGS
• Varies with cause of coelomic distension. • Ascites may be noted within one or more of the peritoneal cavities. Fluids may be transudate or exudate. • Organomegaly or neoplastic masses of the kidney, liver, spleen, pancreas, and oviduct may be seen. • Adhesions or carcinomatosis may be noted within coelomic cavity. • Cardiomegaly with right-sided, left-sided, or bi-ventricular enlargement noted in birds with cardiac disease; atherosclerosis of coronary, pulmonary, and systemic vasculature may also be noted.

TREATMENT

APPROPRIATE HEALTH CARE
• Birds demonstrating significant respiratory compromise from a mass effect or ascites may need to be hospitalized for oxygen supplementation. • Hens that are distended due to egg binding may respond to conservative medical management. If there is no response to medical management or if tissue is prolapsed, surgical intervention may be required to remove the egg. • Patients with lipomas can be managed conservatively on an outpatient basis with dietary management. However, surgical excision of lipomas may be necessary if conservative medical therapy fails. • Control of infection and inflammation needs to be addressed when medically managing cases of egg yolk peritonitis; salpingohysterectomy and removal of inspissated egg yolk may be required once the patient is stabilized. • Cystic ovarian disease is rarely treated surgically due to the vascular supply to the ovary; medical management with hormone therapy may provide symptomatic treatment. • Centesis of ascites helps alleviate pressure on the air sacs; centesis can be repeated as necessary on an outpatient basis. • Surgical excision of some neoplasms may alleviate symptoms if metastasis has not already occurred. • Phlebotomy (1% of body weight q7d) can be performed in birds with ISD to reduce stored liver iron levels.

NURSING CARE
• Oxygen supplementation if respiratory compromise is noted. • Fluid therapy if dehydrated—PO, SC, IV, IO routes; consider colloids intravenously if patient is hypoproteinemic. • Nutritional support—gavage feed if patient is anorexic or has significant weight loss. • If egg present in coelom, hen should be placed in warm, humidified incubator to encourage oviposition. • Therapeutic celiocentesis and diuretics may be beneficial if ascites is noted.

ACTIVITY
Minimize stress and activity since the patient may have limited respiratory capacity.

DIET
• Gavage feed if patient is not eating. • If patient is obese, reduce calories by limiting foods high in fat and reduce overall quantity of food. However, make sure that patient is continuing to eat if hepatic lipidosis is present. • Calcium supplementation should be provided for hens that are egg-laying to prevent calcium deficiency. • Birds susceptible to ISD should be on a specially formulated low iron diet.

CLIENT EDUCATION
• If patient is obese: Foraging activities should be incorporated into the bird's daily routine to encourage exercise to help the bird lose weight. • If patient is laying eggs: Hen should be monitored for prolonged straining, as this can be a sign for egg binding. • If ascites is present: Celiocentesis may need to be repeated as needed to provide relief for dyspnea • If an infectious process is suspected: Bird should be quarantined from other birds. Precautionary measures should be discussed for birds with zoonotic diseases, especially if they are exposed to any immunocompromised people.

SURGICAL CONSIDERATIONS
• Patient may have compromised ability to breathe due to compression of air sacs. Intubation with positive pressure ventilation is strongly recommended. • If ascites present, extra care must be taken to prevent damaging the air sacs during the surgical approach to minimize risk of fluid aspiration through the air sacs.

MEDICATIONS

DRUG(S) OF CHOICE
• An antibiotic may be considered if infection is suspected and ideally is based on culture and sensitivity results. • Meloxicam (1–2 mg/kg PO, IM, IV q12h) can be administered if an inflammatory process is involved or it the patient appears painful.

Treatments will vary greatly depending on cause of the distension.

For ascites
• Furosemide (0.1–2 mg/kg PO, SC, IM, IV q6–24h); diuretic to remove excess fluid in cases of CHF or liver insufficiency.

For reproductive related disorders
• Gonadotropin-releasing hormone (GnRH) agonist leuprolide acetate (200–800 μg/kg IM q2–6 weeks) should be administered to hens that are egg bound. This temporarily prevents further production of eggs. Some reproductive masses may also respond to the administration of leuprolide acetate (1500–3000 μg/kg IM q 2–4 weeks). • Deslorelin (4.7 or 9.4 mg SC implant)—GnRH agonist; long-term implant that can be administered to birds with ovarian cysts or reproductive neoplasms.

For congestive heart failure
• Benazepril HCl (0.25–0.5 mg/kg PO q24h)—ACE inhibitor used in the management of CHF and hypertension.
• Pimobendan (0.25–0.35 mg/kg PO q12h)—inodilator (has inotropic and vasodilator effects) used in the management of CHF.

For iron storage disease
• Deferoxamine (100mg/kg SC/IM q24h); iron chelator for birds with ISD.

For hepatic cirrhosis/fibrosis
• Colchicine (0.05–0.1 mg/kg PO q12–24h); anti-inflammatory.

CONTRAINDICATIONS
• Furosemide is contraindicated in patients with anuria. • Deferoxamine should not be used in patients with severe renal failure.

PRECAUTIONS
Overdose of furosemide can cause dehydration and electrolyte imbalances.

POSSIBLE INTERACTIONS
Colchicine may cause additive myelosuppression when used in conjunction with medications with bone marrow depressant effects (i.e. anti-neoplastics, immunosuppressants, chloramphenicol, amphotericin B).

ALTERNATIVE DRUGS
• Hepatoprotective medication such as silymarin may be beneficial if a hepatopathy is suspected. • L-carnitine has been used to help reduce obesity and the size of lipomas in budgerigars.

FOLLOW-UP

PATIENT MONITORING
• Repeat radiographs can be used to evaluate if the coelomic distension is decreasing or increasing in size. • If coelomic distension is due to obesity, weight loss through diet

COELOMIC DISTENSION

(CONTINUED)

restriction and exercise should be gradual and closely monitored with frequent weight checks. • Hens with a history of egg binding should be closely monitored during egg laying season. Administration of a GnRH agonist can be given seasonally whenever the hen begins to display reproductive behaviors to deter egg laying. • An echocardiogram should be repeated on patients with CHF to evaluate the response to heart medications. The patient's uric acid levels should also be monitored every few weeks to months to assess their renal function. • Bile acid levels should be monitored every few weeks to months to evaluate hepatic function in patients with suspected hepatopathies. • If a salpingohysterectomy is performed for the treatment of reproductive disorders, the patient may be at risk for internal laying and future cases of egg yolk peritonitis. The hen may need preventative doses of a GnRH agonist to prevent further laying.

PREVENTION/AVOIDANCE
• Obesity can be prevented by encouraging foraging behavior and providing a healthy and balanced diet. Owners should not be offering food items high in fat on a regular basis. • A GnRH agonist may be administered seasonally to deter egg laying.

POSSIBLE COMPLICATIONS
• Impaired fecal emptying or gastrointestinal obstruction. • Cloacal/oviductal prolapse or coelomic wall herniation can occur with excessive straining associated egg binding. • Compression of the kidney can lead to renal compromise and/or lameness of the pelvic limbs. • Lateral puncture of the coelomic cavity when performing celiocentesis may result in leakage of fluid directly into the air sac and can result in drowning of the patient. • Gastrointestinal obstruction may occur when fecaliths or eggs are lodged in the pelvic canal or if neoplasms originate in the cloaca.

EXPECTED COURSE AND PROGNOSIS
• Prognosis varies with cause of distension
• Prognosis is good if the distension is due to an egg and the hen is healthy. However, if there is prolonged egg binding or there is egg yolk peritonitis, the prognosis is fair to guarded, depending on if treatment is instituted in a timely manner. • Birds with ascites usually have a guarded to poor long-term prognosis as most causes (i.e., CHF, hepatic cirrhosis, ovarian neoplasia) are usually progressive even with treatment.
• Benign neoplasms such as lipomas have a good prognosis if treatment and diet modification is instituted early; other neoplasms generally have a guarded to poor long-term prognosis. • Hernias have a good prognosis as long as intestinal bowel loops have not become entrapped and compromised within the hernia.

 MISCELLANEOUS

ASSOCIATED CONDITIONS
• Unilateral or bilateral hind limb paresis may be noted if a coelomic mass is compressing the ischiatic nerve. • Ileus or an obstructive process of the gastrointestinal tract may lead to scant fecal production or malodourous feces from bacterial overgrowth.

AGE-RELATED FACTORS
• Altricial neonate and juvenile chicks will normally have a full and distended appearance to their coelom. • Congestive heart failure secondary to atherosclerosis usually affects mature, older birds.

ZOONOTIC POTENTIAL
Though no cases of bird to human mycobacteriosis are reported, owners must be advised of the potential zoonotic risk, especially to immunocompromised individuals.

FERTILITY/BREEDING
If oviduct is damaged during the course of the egg binding or during treatment, the hen may be predisposed to subsequent cases of egg binding/dystocia.

SYNONYMS
Abdominal distension (often used to describe, though technically birds do not have an abdomen)

SEE ALSO
Aspergillosis
Atherosclerosis
Cardiac disease
Chlamydiosis
Ovarian cysts/neoplasia
Dystocia/egg binding
Egg yolk coelomitis/reproductive coelomitis
Hepatic lipidosis
Iron storage disease
Lipomas
Liver disease
Lymphoid neoplasia
Mycobacteriosis
Obesity
Tumors
Appendix 7, Algorithm 5. Coelomic distension

ABBREVIATIONS
CHF—congestive heart failure
ACE—angiotensin-converting enzyme
ISD—iron storage disease
GnRH—gonadotropin-releasing hormone

INTERNET RESOURCES
N/A

Suggested Reading
Sedacca, C.D., Campbell, T.W., Bright, J.M., et al. (2009). Chronic cor pulmonale secondary to pulmonary atherosclerosis in an African grey parrot. *Journal of the American Veterinary Medical Association*, **234**:1055–1059.
Keller, K.A., Beaufrère, H., Brandão, J., et al. (2014). Long-term management of ovarian neoplasia in two cockatiels (Nymphicus hollandicus). *Journal of Avian Medicine and Surgery*, **27**(1):44–52.
Juan-Sallés, C., Soto, S., Garner, M.M., et al. (2011). Congestive heart failure in 6 African grey parrots (*Psittacus erithacus*). *Veterinary Pathology*, **48**(3):691–697.
Starkey, S.R., Morrisey, J.K., Stewart, J.E., et al. (2008). Pituitary-dependent hyperadrenocorticism in a cockatoo. *Journal of the American Veterinary Medical Association*, **232**(3):394–398.
Author Sue Chen, DVM, Dipl. ABVP (Avian)

Colibacillosis

BASICS

DEFINITION
• *Escherichia coli* is a gram-negative member of the Enterobacteriaceae with worldwide distribution. • Considered a common intestinal inhabitant of the GI tract in many avian species (especially raptors, pigeons, passerines, and ratites) and has been isolated from the cloaca of clinically normal psittacine species (especially *Cacatua* sp). • Isolated *E. coli* may be a pathogen, potential pathogen, commensal or transient flora. • Colibacillosis is often an acute bacterial disease reported in chickens, turkeys and ducks; it is the most common infectious bacterial disease of poultry. • *E. coli* infections result in significant economic loss for the poultry industry. • May cause disease in all avian species.

PATHOPHYSIOLOGY
• Avian pathogenic *E. coli* (APEC) strains may cause local or systemic infections and are commonly of the O1, O2, and O78 serogroups. • Extraintestinal pathogenic *E. coli* (ExPEC) strains cause disease outside of the intestinal tract. • Most APEC are ExPEC. • APEC tend to be less toxigenic than mammalian or human *E. coli*; although pigeons can be a source of shigatoxin producing *E. coli* strains • The virulence of *E. coli* serotypes and their enterotoxins varies; most are nonpathogenic. • Systemic infectious occur when bacteria spread hematogenously from the respiratory system or intestinal tract. • Avian *E. coli* enterotoxins cause diarrhea by hypersecretion of fluids into intestinal lumen. • Serotypes of *E. coli* are classified according to the Kauffmann scheme. In most serologic typing schemes only the O and H antigens are determined.

SYSTEMS AFFECTED
• Multiple body systems are affected; especially with septicemia; the name of disease is often based upon the tissue(s) or organ(s) affected; may manifest as a localized or systemic infection.
• Gastrointestinal—ingluvitis, enteritis, coligranuloma, typhlitis.
• Cardiovascular—pericarditis, endocarditis, myocarditis.
• Hepatobiliary—coligranuloma, hepatitis, perihepatitis cholangitis, cholecystitis, hepatic serositis. • Hemic/Immune—thrombi and microthrombi in liver and lungs, splenitis.
• Musculoskeletal—arthritis, osteomyelitis, polyserositis, synovitis. • Neurologic—meningitis, meningoencephalitis. • Special senses—labyrinthitis, anterior uveitis, panophthalmitis, hypopyon. • Renal/Urologic—glomerulonephritis, polyuria.
• Reproductive—infertility, salpingitis, oophoritis, orchitis, epididymitis.
• Respiratory—air sacculitis, pneumonia, coligranuloma, sinusitis, tracheitis.
• Skin/Exocrine—dermatitis, pododermatitis and abscesses. • Omphalitis /Yolk Sac infections—unretracted yolk sac.

GENETICS
• Genetic resistance or predisposition among companion (psittacine) species is not known. • Resistance is variable among genetic lines of poultry (chickens and turkeys).

INCIDENCE/PREVALENCE
• More commonly reported in poultry; however, all birds can suffer from this disease. • Young birds more likely to be affected with increased severity of disease. • Sporadic reports of disease in species other than poultry. • *E. coli* can be isolated from clinically normal psittacines, including free-ranging parrot species; considered more common in cockatoos than other psittacines. • *E. coli* is not normally found in intestinal tract of passerine species. • *E. coli* can be found in the intestinal tract and feces of raptors and pigeons and other free-ranging species.

GEOGRAPHIC DISTRIBUTION
Worldwide.

SIGNALMENT
• **Species:** All avian species are susceptible.
• **Mean age and range**: All ages are susceptible ○ Poultry—young birds are more frequently affected and disease severity is greater in younger birds, including embryos. ○ Colibacillosis in older birds often manifests as an acute septicemia. • **Predominant sex**: None.

SIGNS
• Clinical signs can be inapparent or nonspecific and may vary with age, tissue(s) or organ(s) involved, and concurrent disease; lethargy, depression, dehydration, anorexia, weakness, ruffled feathers, diarrhea, biliverdinuria, respiratory disease, retarded growth, weight loss, lameness, vomiting, moribund, and sudden death. • Localized infections have fewer or milder clinical signs than systemic disease.

CAUSES
• *E. coli.* • *E. fergusonii* and *E. albertii* have been isolated from birds and may cause disease or are of public health concern.

RISK FACTORS
• Healthy birds are generally resistant to infection. • Disease most commonly associated with environmental and host factors. • Poor hygiene and fecal contamination of food and water sources, perches, flooring and environment in general. • Mechanical transmission of *E. coli* following fecal contamination of the egg shell or poor hatchery hygiene. • Overcrowding. • Poor ventilation. • Concurrent disease (viral, bacterial, fungal, or parasitic infections; toxin exposure). • Immunosuppression.
• Nutritional deficiencies, maldigestion or malabsorption. • Damage to skin or mucosal barriers. • Overwhelming exposure due to environmental contamination. • Stress.

DIAGNOSIS

DIFFERENTIAL DIAGNOSIS
• Any number of infectious or inflammatory diseases, or toxicoses can resemble colibacillosis. • Hepatic, renal or pancreatic disease. • Bacterial diseases: *Chlamydia psittaci*, *Salmonella* spp, *Klebsiella* spp, *Pseudomonas* spp, *Proteus* spp, *Pasteurella* spp, *Yersinia* spp, *Clostridium* spp, *Campylobacter* spp, *Aeromonas* spp, *Citrobacter* spp, *Erysipelothrix rhusiopathiae*, *Vibrio* spp, *Mycoplasma* spp, *Staphylococcus* spp, *Streptococcus* spp, *Enterococcus* spp, *Mycobacterium* spp, *Listeria* spp, *Actinobacillus* sp, *Actinomyces* spp. • Parasitic diseases: Coccidiosis (*Isospora* sp, *Eimeria* sp, *Sarcocystis* sp and *Toxoplasma* sp), *Trichomonas* spp, *Giardia* sp, *Hexamita* sp, *Histomonas meleagridis*, *Crytposporium* spp, Hemoparasites (*Hemoproteus* spp, *Plasmodium* spp), Helminths (*Capillaria* spp, *Ascaridia* spp, *Microtetrameres* spp, *Heterakis* spp, *Cardiofilaria* spp, *Chandlerella* sp, *Syngamus* sp, *Seratospiculum* spp), Cestodes. • Viral diseases: adenovirus, bornavirus, herpesvirus, circovirus, polyomavirus, paramyxovirus, reovirus, West Nile virus. • Fungal diseases: aspergillosis, *Macrorhabdus ornithogaster* (avian gastric yeast), *Candida* spp.
• Neoplasia. • Toxicoses: heavy metals (lead and zinc), organophosphates, inhalant.
• Sudden diet change, dietary indiscretion, or starvation. • Gastrointestinal foreign body and obstruction. • Salpingitis, orchitis.
• Stress. • Maldigestion or malabsorption.
• Septicemia. • Trauma. • Iatrogenic: anthelmintic, antibiotic therapy.

CBC/BIOCHEMISTRY/URINALYSIS
• Few abnormalities may be noted in birds due to rapid death. • Chemistry abnormalities may indicate which organ or tissues are affected.

OTHER LABORATORY TESTS
Blood culture.

IMAGING
N/A

DIAGNOSTIC PROCEDURES
• Diagnosis based upon bacterial culture and identification of *E. coli*. • Previous antimicrobial therapy may affect culture results. • Isolation of pure culture of *E. coli* from blood (antemortem) or necropsy tissue (bone marrow, heart blood, liver, spleen, visceral lesions) is indicative of primary or secondary colibacillosis. • *E. coli* isolates grown on Blood and MacConkey agar; avian isolates of *E. coli* are usually nonhemolytic on sheep (5%) blood agar. • Addition of Congo

Colibacillosis (Continued)

Red to growth media may indicate pathogenicity of strain. • Pathogenicity of isolates determined using multiplex PCR.

PATHOLOGIC FINDINGS
Numerous tissues or organs affected.

Localized Infection
• Ingluvitis. • Coligranuloma. • Typhlitis.
• Enteritis—mucosal inflammation in small intestine; common in humans and other mammals but not poultry. • Omphalitis/Yolk sac infection—inflammation of the umbilicus and yolk sacculitis. • Coliform cellulitis—sheets of serosanguineous to caseated, fibrinoheterophilic exudate in subcutaneous tissues; associated with skin trauma. • Swollen Head Syndrome (SHS)—acute to subacute cellulitis of periorbital and adjacent subcutaneous tissues; inflammatory exudate accumulates beneath the skin as a result of E. coli infections following upper respiratory viral infections in poultry; portal of entry is the conjunctiva or inflamed mucous membranes of the infraorbital sinuses or nasal cavities.
• Salpingitis. • Orchitis, epididymitis.

Systemic Infection
Colisepticemia—presence of *E. coli* in the blood; associated with many lesions.
• Polyserositis—inflammation of many serosal surfaces within the coelomic cavity.
• Granulomatous disease. • Pericarditis often associated with myocarditis—serous and serofibrinous exudates accumulate within the pericardial sac, as disease progresses exudate causes adhesion of pericardial sac to epicardium; results in constrictive pericarditis.
• Respiratory colisepticemia—infectious bronchitis virus (IBV), Newcastle disease virus (NDV), mycoplasmas are common predisposing agents in poultry; lesions in the respiratory tract are found in the trachea, lungs, air sacs (thickened with caseous exudate) and are typical of subacute polyserositis, initial lesions consist of edema and heterophilic inflammation; phagocytes, macrophages and giant cells appear later in course of disease. • Hemorrhagic septicemia—turkeys; characterized by circulatory pathology, discoloration of serosal fat, hemorrhage on serosal surfaces, pulmonary edema and hemorrhage, and enlargement of the spleen, liver and kidneys.
• Neonatal colisepticemia—24–48 hours posthatch; initial pathologic lesions consist of congested lungs, edematous serous membranes, and splenomegaly.
• Meningitis/Meningoencephalitis—uncommon. • Panophthalmitis—uncommon but can be severe; hypopyon and hyphema are seen initially; retinal detachment, retinal atrophy and lysis of lens may occur.
• Osteoarthritis/osteomyelitis/polyarthritis.

TREATMENT

APPROPRIATE HEALTH CARE
• Stabilize patient and provide nursing care as indicated. • Critical patients may not be able to tolerate an extensive clinical examination.
• Provide appropriate supportive care (nutritional support and fluid therapy) to resolve anorexia, dehydration. • Provide appropriate antibiotic therapy based upon culture and sensitivity. • Minimize handling of debilitated/critically ill patients.

NURSING CARE
• Specific care depends upon location and severity of disease. • Administer warmed crystalloid fluid (PO, SQ, IV or IO) 50–150 mL/kg per day; provide daily maintenance plus half of calculated dehydration deficit and on-going losses (diarrhea, dehydration, regurgitation, or vomiting) during first 24 hours; repeat during second 24-hour period; route may vary depending upon severity of dehydration. • Colloids such as hetastarch may be administered to patients that are hypovolemic and hypoproteinemic; 10–15 ml/kg q8h for 1–4 treatments if indicated; reduce rate of crystalloid administration by half when giving colloids. • Debilitated birds should receive fluid therapy to correct dehydration before feeding. • Supplemental oxygen as indicated. • Nutritional support essential for birds that are anorexic.

ACTIVITY
Restrict activity to cage rest during hospitalization.

DIET
Provide appropriate diet for the species affected.

CLIENT EDUCATION
• Since disease is commonly associated with environmental factors discuss biosecurity, environmental hygiene, overcrowding, stress, air flow, and ventilation with owner to reduce risk and lessen exposure. • Eliminate underlying cause in flock situations and control other diseases such as intestinal parasitism. • Avoid fecal contamination of eggs. • Quarantine all new birds entering the flock or home. • Avoid exposure to free-ranging bird such as passerine species.
• Zoonotic potential.

SURGICAL CONSIDERATIONS
N/A

MEDICATIONS

DRUGS(S) OF CHOICE
• Antimicrobial therapy should be based upon culture and susceptibility (MIC) testing of *E. coli* isolates due to increasing resistance of *E. coli* to many antibiotics. • Empirical therapy should be instituted pending results of culture and sensitivity. • Consult Food Animal Residue Avoidance Databank for antibiotic choices when managing backyard flocks.
• Trimethoprim/sulfadiazine—psittacines, 30 mg/kg PO q8–12h, 20 mg/kg SC, IM q12h; raptors, 12–60 mg/kg PO q12h; pigeons, 60 mg/kg PO q12h, 475–950 mg/L drinking water; galliformes, 107 mg/L drinking water.
• Trimethoprim/sulfatroxazole—passerines, 10–50 mg/kg PO q12h.
• Trimethoprim/sulfamethoxazole—psittacines, 40–50 mg/kg q12h, 20 mg/kg PO q12h; raptors, 48 mg/kg PO, IM q12h; pigeons, 60 mg/kg PO q24h; passerines, 10–50 mg/kg PO q24h; most species 100 mg/kg PO q12h, 360–400 mg/L drinking water; geese, 400 mg/kg feed.
• Enrofloxacin—psittacines, 15 mg/kg PO q12–24h, 30 mg/kg PO, IM q24h, 100–200 mg/L drinking water; raptors, 15 mg/kg PO, IM, IV q24h; pigeons, 20-30 mg/kg PO, SC, IM q12h, 100–200 mg/L drinking water; passerines, 10–20 mg/kg PO q24h, 200 mg/L drinking water; waterfowl, 10–15 mg/kg PO, IM q12h. • Ciprofloxacin—psittacines, 15–20 mg/kg PO, IM q12h; raptors, 10–20 mg/kg PO q12h, 50 mg/kg PO q12h; pigeons, 5–20 mg/kg PO q12h, 250 mg/L drinking water; passerines, 20–40 mg/kg PO, IV q12h. • Marbofloxacin—most species, 5 mg/kg PO q24h; psittacines (blue and gold macaws), 2.5–5.0 mg/kg PO q24h; raptors, 10–15 mg/kg PO, IM Q12–24h, 2–3 mg/kg IV, IO q24h. • Amikacin—psittacines, 10–15 mg/kg IM, IV q8-12h, 10–15 mg/kg IM q12h; raptors, 10–15 mg/kg IM q24h; passerines, 15–20 mg/kg SC, IM, IV q8–12h.
• Chloramphenicol palmitate—psittacines, 30–50 mg/kg PO q6–8h; raptors, 50 mg/kg PO q6–12h; pigeons, 25 mg/kg q8h, 250 mg/kg PO q6h; passerines, 50–100 mg/kg PO q6–12h, 100–200 mg/L drinking water.
• Chloramphenicol succinate—psittacines, 50 mg/kg IM, IV q6–12h; raptors, 30 mg/kg IM q8h, 50 mg/kg IM, IV q6–12h; pigeons, 60–100 mg/kg IM q8h; passerines, 50–80 mg/kg IM q12–24h, 50 mg/kg IM q8–12h; ducks 22 mg/kg IM, IV q3h. • Ceftazidime—most species, 50–100 mg/kg IM, IV q4–8h
• Cefotaxime—psittacines, 75–100 mg/kg IM q12h; raptors, 75–100 mg/kg IM q12h; pigeons, 100 mg/kg IM q8–12h.
• Amoxicillin/clavulanate (Clavamox)—most species, 125 mg/kg PO q6h; psittacines, 125 mg/kg PO q12h; raptors, 125 mg/kg PO q12h; pigeons, 125 mg/kg PO q12h.
• Piperacillin—most species, 100–200 mg/kg IM, IV q6–12h; psittacines, 100 mg/kg IM q8–12h, psittacines (Amazon parrot), 75–100 mg/kg IM q4–6h, raptors, 100 mg/kg IM, IV q8–12h, raptors (red-tailed hawks and great horned owls), 100 mg/kg IM q4–6h; pigeons, 100 mg/kg IM, IV q8–12h.

Colibacillosis (Continued)

CONTRAINDICATIONS
- Fluoroquinolones are banned for use in food animals in the United States and many other countries. Cephalosporin use is restricted in chickens and turkeys. • Avoid long-term indiscriminate use of antibiotics in flocks or pet birds to prevent development of resistant strains of *E. coli*.

PRECAUTIONS
N/A

POSSIBLE INTERACTIONS
N/A

ALTERNATIVE DRUGS
- Alternative methods for control of colibacillosis include prebiotics, probiotics, enzymes, digestive acidifiers, vitamins, immune enhancers, anti-inflammatory drugs and other antimicrobial products.
- *Lactobacillus* spp administration may inhibit colonization of *E. coli* in the intestine.
- Essential oils and bacteriophage administration may be effective in reducing mortality associated with *E. coli*. • Vitamin E supplementation has been shown to have both prophylactic and therapeutic benefits for *E. coli* infections in poultry.

FOLLOW-UP

PATIENT MONITORING
- Monitor patients response to therapy, general behavior. • Repeat fecal examination.

PREVENTION/AVOIDANCE
- Quarantine all new additions to flock or pet birds for a minimum of 30 days. • Ensure good health and nutritional status of poultry and pet birds. • Ensure daily cleanliness of the environment, housing and cage conditions, reduce dust. • Avoid indiscriminate use of antibiotics in flock and pet birds. • Owners should wash hand and take appropriate biosecurity measures before handling healthy birds. • Vaccines (live, inactivated, and recombinant) are available for poultry.

POSSIBLE COMPLICATIONS
N/A

EXPECTED COURSE AND PROGNOSIS
- Prognosis depends upon the severity of disease and tissue(s) or organs(s) affected.
- Life threatening infections in young birds; adult birds may have less severe morbidity and mortality. • Severe infections carry a poor prognosis.

MISCELLANEOUS

ASSOCIATED CONDITIONS
N/A

AGE-RELATED FACTORS
Young birds are at greatest risk of infection and septicemia.

ZOONOTIC POTENTIAL
- *E. coli* are present in the intestinal tract of most animals and shed in feces. Always wash hands after handling. • Human infection usually associated with direct animal contact or ingestion of contaminated poultry products. • Illness in humans often manifests as an acute gastroenteritis. • Children, immunocompromised individuals should not come in contact with patients with diarrhea or handle young poultry. • All eggs and poultry products should be refrigerated and prepared according to FDA safety recommendations.

FERTILITY/BREEDING
N/A

SYNONYMS
E. coli septicemia.
Colisepticemia.
Hemorrhagic septicemia.
Coligranuloma (Hjarrre's disease).
Air sac disease (chronic respiratory disease, CRD).
Swollen Head syndrome.
Venereal colibacillosis.
Coliform cellulitis (inflammatory or infectious process, IP).
Peritonitis, salpingitis, orchitis, osteomyelitis/synovitis (including turkey osteomyelitis complex), panophthalmitis, omphalitis/yolk sac infection, and enteritis.

SEE ALSO
Aspergillosis
Anorexia
Botulism
Campylobacteriosis
Candidiasis
Chlamydiosis
Cloacal disease
Clostridiosis
Crop stasis/ileus
Diarrhea
Egg yolk coelomitis
Enteritis/gastritis
Flagellate enteritis
Intestinal helminthiasis
Macrorhabdus ornithogaster
Pasteurellosis
Proventricular dilatation disease
Regurgitation/vomiting
Salmonellosis
Salpingitis/uterine disorders
Sick-bird syndrome

ABBREVIATIONS
APEC—avian pathogenic *E. coli*
ExPEC–extraintestinal pathogenic *E. coli*

INTERNET RESOURCES
http://www.merckmanuals.com/vet/poultry/colibacillosis/overview_of_colibacillosis_in_poultry.html

Suggested Reading
Doneley, R.J.T. (2009). Bacterial and parasitic diseases of parrots. *The Veterinary Clinics of North America. Exotic Animal Practice*, **12**(3):217–232.
Dorrestein, G.M. (1997). Bacteriology. In: Altman, R.B., Clubb, S.L., Dorrestein, G.M. (eds), *Avian Medicine and Surgery*. Philadelphia, PA: WB Saunders Co., pp. 255–280.
Hawkins, M.G., Barron, H.W., Speer, B.L., et al. (2013). Birds. In: Carpenter J.W. (ed.), *Exotic Animal Formulary*, 4th edn. St Louis, MI: Elsevier, pp. 184–438.
Nolan, L.K., Barnes, H.J., Vaillancourt, J. (2013). Colibacillosis. In: Swayne, D.E. (ed.), *Diseases of Poultry*, 13th edn. Hoboken, NJ: John Wiley & Sons, Inc., pp. 751–795.
Randal, C.J., Reece, R.L. (1996). *Color Atlas of Avian Histopathology*. Philadelphia, PA: Mosby-Wolfe.
Author Michael P. Jones, DVM, Dipl. ABVP (Avian)

Cryptosporidiosis

BASICS
DEFINITION
Cryptosporidium is a contagious small coccidian parasite that infects most species of animals, including birds and humans. It has been shown to produce a wide variety of disease conditions (cryptosporidiosis) in birds, although it is generally considered to be a pathogen seen mainly in young and/or otherwise immunocompromised birds. *Cryptosporidium* has been found in over 30 species of birds and appears to be most often associated with respiratory and intestinal disease in chickens, turkeys and quail, where it can cause significant morbidity and mortality. There are also several reports of *Cryptosporidium*-associated disease in finches, canaries, falcons, and psittacine birds. In psittacine species, it appears to most commonly infect the proventriculus of smaller species such as cockatiels, budgerigars, and lovebirds. In falcons, it has been seen uncommonly as a cause of significant respiratory disease in young birds. A number of *Cryptosporidium spp* have been identified through PCR and other methods, including *C. meleagridis*, *C. baileyi*, *C. galli*, *C. parvum* and avian genotypes I, II and III.

PATHOPHYSIOLOGY
Cryptosporidium has a direct life cycle. Oocysts are infectious without having to go through sporulation. This also allows for autoinfection. Oocysts are inhaled or ingested. They can invade any epithelial surface, including the gastrointestinal, respiratory, and urinary tracts, the conjunctival sac and the bursa. They establish themselves intracellularly but extracytoplasmically. The lesions and clinical signs that result are due to mononuclear mucosal inflammation and hyperplasia, mucous production, as well as dysplasia, metaplasia and even neoplasia of glandular elements.

SYSTEMS AFFECTED
• Respiratory—mostly upper but also some lower respiratory signs – sinuses, glottis, trachea, air sacs and lungs. • Ophthalmic—conjunctivitis. • Gastrointestinal—enteritis, mainly proventricular and occasionally ventricular disease in small psittacines.
• Urinary—kidneys and ureters.
• Hemolymphatic—bursa.

GENETICS
N/A

INCIDENCE/PREVALENCE
The incidence of cryptosporidiosis is unknown. It is not commonly reported but it is also not generally looked for in clinically ill birds; diagnosis can be difficult and therefore it is probably underreported; in a number of studies the prevalence of cryptosporidia in fecal samples of wild and captive populations is about 5–10%. Note that the presence of the organism does not equate with disease, as there might be a high percentage of asymptomatic shedders. Occasional outbreaks in quail and chickens are seen.

GEOGRAPHIC DISTRIBUTION
Cryptosporidium spp have been found in about 5–10% of birds in most studies all over the world.

SIGNALMENT
• **Species:** *Cryptosporidium spp* have been isolated from at least 30 avian species; most reports involve chickens, quail, turkeys and finches but it has been found in many other birds; in psittacines it is most often seen in cockatiels, lovebirds and budgerigars; it has been seen in these and other psittacines in association with circovirus infection. • **Mean age and range:** It is generally seen in young birds (especially in chickens, turkey, quail and falcons), but in one study of small psittacines the age range was 8 weeks to 13 years; it is worth mentioning that chronic vomiting in older lovebirds has been seen as a result of gastric neoplasia related to cryptosporidial infection. • **Predominant sex:** Both sexes are susceptible.

SIGNS
General Comments
Findings are generally related to respiratory or gastrointestinal disease but also occasionally renal disease.

Historical Findings
Generally seen as isolated cases, but there may be an outbreak of respiratory disease or gastrointestinal disease (quail, turkeys, chickens) or respiratory disease in young falcons.

Physical Examination Findings
Respiratory cases:
• Sneezing, nasal discharge • Swelling of infraorbital sinuses • Coughing
• Conjunctivitis • Dyspnea with open mouth breathing • Stridor • Swelling of the glottis
• Weight loss
Gastrointestinal cases:
• Depression • Weight loss • Vomiting • True diarrhea • Passage of bulky stools with undigested food/seeds • Dehydration
• Abdominal pain

CAUSES
Infection with *Cryptosporidium spp*; contamination of environment; in falcons, possible link with ingestion of infected quail; underlying immunosuppression in some birds—due to age (young), circoviral infection and other factors

RISK FACTORS
Risk factors for developing this disease depend on immune status and transmission. Cryptosporidia can be transmitted via ingestion or inhalation of infectious material. Flock transmission is by direct and indirect contact and can happen swiftly; therefore, there would be an increased risk of acquiring the disease in overpopulated flocks.

DIAGNOSIS
DIFFERENTIAL DIAGNOSIS
Respiratory Disease
• Mycoplasmosis • Fowl cholera (Pasteurellosis) • Aspergillosis (brooder pneumonia) • Infectious bronchitis
• Infectious laryngotracheitis • Avian influenza • Avian pox • Newcastle disease
• Chlamydiosis • Trichomoniasis

Gastrointestinal Disease
• Intestinal parasites—coccidiosis, giardiasis, hexamitiasis, histomoniasis, helminthiasis
• Macrorhabdosis • Bacterial infection—salmonellosis, colibacillosis, pasteurellosis, *Clostridium colinum* (quail), other clostridia, other bacteria • Fungal infection—candidiasis, aspergillosis • Chlamydiosis
• Mycobacteriosis • Viral disease—Newcastle disease, avian influenza, other

CBC/BIOCHEMISTRY/URINALYSIS
There are no reports of laboratory findings with this disease. Dehydration could cause an elevation in total protein and hematocrit; an inflammatory leukogram would likely result from significant inflammation in the GI or respiratory tract.

OTHER LABORATORY TESTS
• Fecal flotation can reveal the small 2–6 μm oocysts but they can be difficult to find.
• Fecal direct smears should be done, especially in small psittacines to check for co-infection with other protozoa and *Macrorhabdus sp*. • Acid fast staining of feces can reveal the acid-fast organisms. • PCR testing can be performed on fecal or cytological samples. • IFA or ELISA analysis of fecal samples is uncommonly performed.
• Wright's staining of cytological samples (conjunctiva, glottis) may be helpful in identifying the organism. • Viral testing (especially circovirus) to check for underlying immunosuppressive disease.
• Culture/sensitivity of feces or respiratory secretions can be performed.
• Histopathology of deceased patients.

IMAGING
Imaging is not typically done to diagnosis this disease, but birds with respiratory infection could show evidence of airsacculitis and bronchopneumonia; patients with enteritis could have gas in the GI tract and/or thickening of the proventriculus; lovebirds with chronic vomiting due to proventricular cryptosporidiosis may show a filling defect of the proventriculus in contrast radiographs.

DIAGNOSTIC PROCEDURES
Cytological and microbiological sampling of infected tissues, such as conjunctiva, sinus flushes.

CRYPTOSPORIDIOSIS (CONTINUED)

PATHOLOGIC FINDINGS

Respiratory Disease
Gross—conjunctivitis, nasal discharge, swollen sinuses, tracheitis, swollen glottis, pneumonia, airsacculitis; microscopically: excess mucous production, proliferation of mucosa, presence of cryptosporidia.

Gastrointestinal Disease
Gas and excess liquid in the intestine, thickened proventriculus with excess mucous and liquid, mass in the wall of the proventriculus (lovebirds). Microscopic: the presence of *Cryptosporidium* on the surface of the mucosal epithelial cells, the apical surfaces of the mucosal glands, as well as within the glandular lumina and along their primary and secondary ducts, mild to moderate mononuclear infiltration of the lamina, mild fibrosis, mucosal hyperplasia, proliferation and exfoliation, glandular and ductular hyperplasia, dilatation, disorganization or disruption, and necrosis; metaplasia, dysplasia and neoplasia of glandular elements. Birds with concomitant macrorhabdosis have much more severe lesions.

TREATMENT

APPROPRIATE HEALTH CARE
Birds that are critically ill or that need an extensive workup should be hospitalized. Patients that are mildly ill can be treated as outpatients while the results of laboratory tests are pending.

NURSING CARE
Fluid, nutritional and thermal support may be necessary in birds that are anorexic and dehydrated due to this disease.

ACTIVITY
N/A

DIET
N/A

CLIENT EDUCATION
Cryptosporidia are difficult to eliminate in the environment. Strong bleach or ammonia solution or steam cleaning might be helpful.

SURGICAL CONSIDERATIONS
N/A

MEDICATIONS

DRUG(S) OF CHOICE
There are few reports of successful treatment.
- Paramomycin alone or with azithromycin has been recommended historically but there are few reports of success. The drug is expensive and difficult to obtain. It is poorly absorbed from the GI tract and therefore not a good candidate for respiratory disease. Suggested dosage is 100 mg/kg BID PO or gavage. • Nitazoxanide has been used in humans with cryptosporidiosis, but there is little information about its use in birds.
- Ponazuril can be compounded and given at 20 mg/kg by mouth every 24 hours. There was dramatic response within 24 hours in cases of respiratory cryptosporidiosis in falcons.

CONTRAINDICATIONS
N/A

PRECAUTIONS
N/A

POSSIBLE INTERACTIONS
N/A

ALTERNATIVE DRUGS
Treatment for secondary bacterial, fungal or macrorhabdial disease might be necessary, depending on clinical judgment or the results of culture and sensitivity.

FOLLOW-UP

PATIENT MONITORING
All sick patients should be routinely monitored to assess response to therapy, which involves performing periodic exams and recording body weight; follow up fecal testing would be useful.

PREVENTION/AVOIDANCE
Feeding of infected quail to falcons should be avoided; birds should not be placed in close contact with infected birds; disinfection of the environment should be performed in order to decrease or eliminate the presence of the organism (difficult).

POSSIBLE COMPLICATIONS
N/A

EXPECTED COURSE AND PROGNOSIS
In theory, if an accurate diagnosis is made, and appropriate therapy is used, there could be recovery from disease. Historically most birds are diagnosed after they have died. Also, birds that recover could be come carriers.

MISCELLANEOUS

ASSOCIATED CONDITIONS
Some birds with cryptosporidiosis have other disease, such as circoviral infection; some patients develop secondary disease, that is, colibacillosis. Small psittacines with concurrent macrorhabdosis or intestinal flagellates will have more severe disease than birds with cryptosporidiosis alone.

AGE-RELATED FACTORS
N/A

ZOONOTIC POTENTIAL
Zoonotic potential has been questioned, since there are at least two species of *Cryptosporidium* that have been found in both humans in birds (*C. meleagridis* and *C. parvum*), but this appears to be mostly speculation. It follows that immunocompromised (HIV) humans should not be in close contact with infected birds.

FERTILITY/BREEDING
Breeding, fertility rates, and egg production greatly decrease with cryptosporidial infection.

SYNONYMS
None.

SEE ALSO
Aspergillosis
Bordetellosis
Candidiasis
Chlamydiosis
Circoviruses
Diarrhea
Enteritis/gastritis
Flagellate enteritis
Intestinal helminthiasis
Macrorhabdus ornithogaster
Mycobacteriosis
Mycoplasmosis
Paramyxoviruses
Pasteurellosis
Pneumonia
Respiratory distress
Rhinitis/sinusitis
Salmonellosis
Trichomoniasis
Undigested food in droppings
Urate/fecal discoloration
Viral disease

ABBREVIATIONS
N/A

INTERNET RESOURCES
There are many websites that discuss cryptosporidiosis in poultry and humans. A PubMed search and VIN search will also reveal numerous references. The CDC (www.CDC.gov) has numerous discussion about the disease in humans.

Suggested Reading
Messenger, G., Garner, M. (2010). Proventricular cryptosporidiosis in small psittacines. *Proceedings of the Association of Avian Veterinarians Annual Conference*, pp. 55–58.

Doneley, R.J.T. (2009). Bacterial and parasitic disease of parrots. *The Veterinary Clinics of North America. Exotic Animal Practice*, **12**(3):423–431.

Reavill, D.R., Schmidt, R.E., Phalen, D.N. (2003). Gastrointestinal system and pancreas. In: Reavill, D.R., Schmidt, R.E., Phalen, D.N. (eds), *Pathology of Pet and Aviary Birds*. Ames, IA: Iowa State Press, pp. 41–65.

Patton, S. (2000). Avian parasite testing. In: Fudge, A.M. (ed.), *Laboratory Medicine–Avian and Exotic Pets*. Philadelphia, PA: WB Saunders, pp. 147–156.

Author George A. Messenger, DVM, DABVP (Avian)

Dehydration

BASICS

DEFINITION
Dehydration is defined as a loss of body fluids exceeding fluid intake.

PATHOPHYSIOLOGY
Loss of body fluids leads to both increased blood solute and increased osmolality due to increased sodium levels. Normally, water molecules shift out of cells and into blood to restore the balance between intra and extracellular spaces. Increased water intake and increased renal water retention serve to restore fluid balance. Dehydration results when restorative mechanisms fail. Severe dehydration can progress to hypovolemic shock.

SYSTEMS AFFECTED
• Circulatory • Digestive • Urogenital

GENETICS
No genetic associations.

INCIDENCE/PREVALENCE
Specific incidence data is lacking but dehydration can be seen more commonly in hotter months for birds housed in outdoor enclosures or when only one water bowl or dish is available in a multibird habitat.

GEOGRAPHIC DISTRIBUTION
No geographic distributions exist.

SIGNALMENT
• Species: Any avian species can present dehydrated. • Mean age and range: Birds of any age can present dehydrated.
• Predominant sex: Birds of any sex can present dehydrated.

SIGNS

Historical Findings
• Lethargy • Depression • Anorexia • Firm or smaller than normal stools.

Physical Examination Findings
• Skin tenting • Dry oral mucosa • Dry eyes
• Slow refill time of veins once pressure has been released.

CAUSES
• Degenerative/developmental: neonatal crop atony. • Anomalous/autoimmune.
• Metabolic/mechanical: foreign body, impaction, compression of crop, nematodiasis. • Nutritional/neoplastic: neoplasia, liver disease, renal disease.
• Inflammatory/infectious: proventricular dilatation disease, gastritis, septicemia.
• Traumatic/toxic: gastric ulcer, tube feeding.

RISK FACTORS
N/A

DIAGNOSIS

DIFFERENTIAL DIAGNOSIS
N/A

CBC/BIOCHEMISTRY/URINALYSIS
Packed cell volume (PCV) and total solids (TS) are used to estimate patient hydration. In addition to a physical examination, PCV/TS measurements should be taken daily to monitor the patient's progress. The author recommends microhematocrit tubes because only the amount of blood collected in the hub of a needle will be required to measure PCV and TS.

OTHER LABORATORY TESTS
N/A

IMAGING
Radiographs may show renal hypoplasia as evidenced by increased air sac diverticulum separating the dorsal renal surface from the ventral synsacrum.

DIAGNOSTIC PROCEDURES
N/A

PATHOLOGIC FINDINGS
N/A

TREATMENT

APPROPRIATE HEALTH CARE
In addition to treatment of underlying causes of dehydration, both the deficit and maintenance fluids for appropriate hydration are to be addressed. Suggested dehydration estimates are:
• 5% dehydration—subtle skin tenting and loss of elasticity. • 10–12% dehydration—pronounced skin tenting, dry eyes and mucous membranes. • 12–15% dehydration—skin remains tented, sunken eyes; upper eyelid does not fall back into place once lifted.

NURSING CARE
• For mild dehydration, crystalloids can be administered subcutaneously into the inguinal fold. Severely dehydrated patients and those presenting in hypothermic shock cannot be treated adequately with subcutaneous fluids. For moderate to severe dehydration, intravascular (IV)/intraosseous (IO) fluid administration is indicated; oral fluids are not recommended for critically ill patients. The fluid deficit (mL) equals the degree of dehydration (%) multiplied by the patient's body weight in grams. The fluid deficit should be replaced over 12–24 hours in birds without cardiopulmonary compromise and over 24–36 hours if cardiopulmonary compromise exists. Up to 50 mL/kg can be administered as an IV bolus to a bird. • In small birds (budgies, parakeets and passerines) an IO catheter is more sturdy and practical than an IV catheter. The distal ulna is the preferred site for IO catheterization. A 25-gauge needle can be used for parakeets and budgies while a 27-gauge is recommended for passerines and canaries. Landmarks for proper IO catheter placement are the distal ulna prominence located on the dorsal surface of the ulna just proximal to the carpal joint. Using the non-dominant hand the ulna is held between the thumb and index finger. With the dominant hand the needle is guided through the bone cortex with a slight twisting motion using the thumb and index finger. The needle should be seated up to the hub. To check proper catheter placement a small amount of fluid (0.5–1.0 mL) should be gently flushed through the needle. This fluid will be visible and palpable in the basilic vein. If no fluid flows, a new needle of the same or larger gauge may need to be seated due to a bone plug in the needle used to pierce the boney cortex. In passerines the IO catheter can be secured with tape. In larger birds the catheter can be secured with suture through both the skin and butterfly tape. • Lactated Ringers Solution (LRS) and other crystalloid solutions are indicated to correct dehydration and the recommended maintenance dose is 50mL/kg/d. Each 100 mL of LRS contains sodium chloride (0.6 grams), sodium lactate (0.31 grams), potassium chloride (0.03 grams), calcium chloride dehydrate (0.02 grams) and water qs. Per liter of LRS there are 130 mEq of sodium, 4 mEq of potassium, 3 mEq of calcium, 110 mEq of chloride and 28 mEq of lactate. • Sodium is the major cation of the extracellular fluid (ECF) and functions in the control of water distribution, fluid balance and osmotic pressure. Along with chloride and bicarbonate, sodium regulates the acid-base balance. Potassium, the major cation of intracellular fluid (ICF) is involved in protein synthesis and carbohydrate utilization. Additionally, potassium is important in both muscle contraction and nerve conduction. Chloride, the major ECF anion also participates in acid-base balance. The cation calcium is involved in bone development, blood clotting, and cardiac/neuromuscular function. Sodium lactate can be oxidized to bicarbonate and/or converted to glycogen.

DEHYDRATION (CONTINUED)

ACTIVITY
Hospitalized patients should have restricted activity to not only prevent over-exertion but also catheter damage.

DIET
Fresh vegetables and fruits added to a pelleted diet will increase the patient's water intake.

CLIENT EDUCATION
Clients should be encouraged to provide multiple water bowls/bottles and also to refill water dishes/bottles daily.

SURGICAL CONSIDERATIONS
N/A

MEDICATIONS

DRUG(S) OF CHOICE
N/A

CONTRAINDICATIONS
Avoid the use of any potentially nephrotoxic drugs in the face of dehydration.

PRECAUTIONS
N/A

ALTERNATIVE DRUGS
N/A

FOLLOW-UP

PATIENT MONITORING
N/A

PREVENTION/AVOIDANCE
Owners should provide fresh water daily. For birds that soil water with pellets, produce, bedding or excrement, water bottles can be used. Multiple water bowls at multiple levels in the cage may also be utilized. Water or food dishes should never be placed directly under a perch.

POSSIBLE COMPLICATIONS
N/A

EXPECTED COURSE AND PROGNOSIS
Mild dehydration has a better prognosis for being corrected as moderate and severe dehydration have as a consequence cell shrinkage as fluid is pulled out of intracellular spaces and into extracellular spaces. This cell shrinkage can be accompanied with varying degrees of major organ dysfunction affecting the liver and kidneys most often.

MISCELLANEOUS

ASSOCIATED CONDITIONS
Any problem or disease affecting birds can lead to dehydration. It is important to note that arthritis can lead to dehydration because of the associated pain when moving about the cage to get to the water dish. Visceral and peripheral uric acid deposition may result in severe dehydration and renal dysfunction.

AGE-RELATED FACTORS
There are no age-related factors that would be relevant to the diagnosis and treatment.

ZOONOTIC POTENTIAL
Dehydration does not have zoonotic potential.

FERTILITY/BREEDING
Because dehydration can impact all systems and organs in the body, fertility and breeding can both be negatively impacted due to dehydration.

SYNONYMS
N/A

SEE ALSO
Anorexia
Arthritis
Diarrhea
Emaciation
Ingluvial hypomotility/ileus
Lameness
Polydipsea
Polyuria
Proventricular dilatation disease
Regurgitation/vomiting
Renal disease
Sick-bird syndrome
Urate/fecal discoloration

ABBREVIATIONS
N/A

Suggested Reading
de Matos, R. Morrisey, J.K. (2005). Emergency and critical care of small psittacines and passerines. *Seminars in Avian and Exotic Pet Medicine*, **14**(2):90–105.
Lichtenenberger, M. (2004). Principles of fluid therapy in special species. *Seminars in Avian and Exotic Pet Medicine*, **13**(3): 142–153.

Author Kemba L. Marshall, DVM, DABVP (Avian)

Acknowledgement The author wishes to acknowledge the teachers, professors and patients who have allowed her to learn and live out the dreams she had as the eight-year old version of herself.

DERMATITIS

BASICS

DEFINITION
Dermatitis is inflammation of the skin. Dermatitis is a nonspecific term that encompasses multiple possible clinical presentations and is not a diagnosis or disease. Dermatitis can include crusts, scale, ulceration, laceration, nodular or proliferative changes, edema, and pruritus.

PATHOPHYSIOLOGY
The pathophysiology of dermatitis varies depending upon the underlying etiology. Characterization of the signs of dermatitis as crusts, scale, ulceration, laceration, nodular or proliferative changes, edema, and/or pruritus is the first step to creating a list of differential diagnoses and determining appropriate diagnostics and treatments. For details on pathophysiology, consult the chapters referenced below.

SYSTEMS AFFECTED
Skin.

GENETICS
N/A

INCIDENCE/PREVALENCE
Varies with the underlying cause.

GEOGRAPHIC DISTRIBUTION
N/A

SIGNALMENT
- **Species:** Avian. • **Mean age and range:** N/A for most causes. Constricted toe is seen in psittacines prior to fledging.
- **Predominant sex:** N/A

SIGNS
Signs of dermatitis may include one or more of: • Crusts and/or scale often honeycombed with small holes on face, cere, feet, or legs.
• Ulceration (defect in or absence of skin) or laceration of the skin anywhere on body, including plantar feet, patagium, axillary region, keel, and uropygial gland region.
• Nodular or proliferative changes in the skin anywhere on body (including eyelid margins, feet, corner of beak, face, uropygial gland) or mucosa of the oral cavity or cloaca. • Swelling or edema of legs or toes.
○ Photosensitization—the skin of the legs and feet may be erythematous, edematous, and pruritic, and may progress to vesicle and bulla formation, serum exudation, ulceration, crust formation, and skin necrosis. ○ Constricted toe–a thin (<1 mm) band of constricting tissue encircles the circumference of one or more toes with edema of the digit distally.
• Pruritus, often seen concurrently with erythema, papules and vesicles that may erode and crust.

General Comments
The veterinarian should strive to describe the lesions and lesion distribution as specifically as possible. The term "dermatitis" should not be used to describe lesions or as a differential diagnosis.

Historical Findings
Any of the signs listed above may be reported by the owner.

Physical Examination Findings
One of more of the clinical signs listed above may be noted on physical examination depending upon the underlying cause.

CAUSES
Crust/scale:
• *Knemidocoptes* spp. mite infestation (see Ectoparasites). • Papillomavirus (see Papilloma). • Ulcerative pododermatitis (see Pododermatitis). • Avian poxvirus (see Poxvirus).

Ulceration:
• Ulcerative dermatitis resulting from trauma with secondary bacterial or fungal infection (see Feather damaging behavior/self-injurious behavior) • Ulcerative pododermatitis (see Pododermatitis)

Laceration—created by trauma:
• Split keel (see Trauma). • Bite wounds from other birds or other animals (see Bite Wounds). • Self-mutilation (see Feather damaging behavior/self-injurious behavior).

Nodular or proliferative changes:
• Avian poxvirus (see Poxvirus). • Cutaneous papillomas caused by papillomavirus (see Papilloma). • Mucosal papillomas associated with Psittacid Herpesvirus 1 infections (see Herpesviruses, Oral Plaques, Cloacal Disease). • Feather follicle cysts (see Feather Cysts). • Neoplasia—including squamous cell carcinoma, basal cell tumor, lipoma, fibrosarcoma, fibroma, lymphosarcoma, melanoma, xanthoma, and uropygial gland tumors (see Lipoma, Lymphoid neoplasia, Squamous Cell Carcinoma, Tumors, Uropygial Gland Disease, Xanthoma).

Edema:
• Photosensitization due to ingestion of St John's Wort (*Hypericum perforatum*) by waterfowl: ○ St John's Wort contains hypericin. Exposure of the skin to direct sunlight allows photons to react with hypericin to create unstable, high-energy molecules. The high-energy molecules create free radicals that increase the permeability of outer cell and lysosomal membranes. This cell membrane damage results in leakage of cellular potassium and cytoplasmic extrusion, and lysosomal membrane damage releases lytic enzymes into the cell. These processes incite inflammation and lead to edema, skin ulceration and necrosis. • Constricted toe—No confirmed etiology is known although low humidity has been suggested as a contributing factor. • Frostbite (see Cere and Skin, Color Changes).

Pruritus:
• Hypersensitivity.

RISK FACTORS
Vary with underlying cause. The risk factor for photosensitization is ingestion of St John's Wort and exposure to direct sunlight. See appropriate chapters for risk factors for other causes.

DIAGNOSIS

DIFFERENTIAL DIAGNOSIS
See Causes above for differential diagnoses for common clinical signs, and see appropriate chapters for additional differential diagnoses for each etiology.

CBC/BIOCHEMISTRY/URINALYSIS
N/A

OTHER LABORATORY TESTS
N/A

IMAGING
N/A

DIAGNOSTIC PROCEDURES
N/A

PATHOLOGIC FINDINGS
• Vary with underlying cause.
• Photosensitization: Histopathology of skin reveals hyperkeratosis, parakeratosis and edema of the epidermis with areas containing vesicles, crusting and infiltration of polymorphonuclear leukocytes.
• Hypersensitivity: Histopathology of skin biopsies is nondiagnostic, and reveals perivascular inflammatory cell foci in the dermis and subcutis consisting primarily of mononuclear cells and possibly granulocytes.
• See appropriate chapters for pathologic findings from other causes.

TREATMENT

APPROPRIATE HEALTH CARE
• Photosensitization: Treatment is supportive and starts with removing the bird from access to direct sunlight. • Constricted toe: Stabilize the bird if necessary and anesthetize for surgical treatment of the constricting band of tissue. The constriction of the toe can be relieved by creating several longitudinal incisions through the band. Excision of the constricting band may lead to excessive hemorrhage depending on the width of the band. When excision is performed, the healthy opposing skin edges are sutured together to cover the defect. Bandage the affected foot to prevent self-trauma and keep the foot clean. Bandages should be changed daily. It is suggested to apply topical antibiotic to surgical sites to prevent secondary infection. • Hypersensitivity: Anecdotal evidence supports therapies such as dietary manipulations and essential fatty acid supplementation. There is no direct evidence

DERMATITIS (CONTINUED)

to support immunotherapy for hypersensitivity in avian species but research is ongoing. • See appropriate chapters for treatment for other causes.

NURSING CARE
Depends on the cause; see appropriate chapters listed in "Causes".

ACTIVITY
Depends on the cause; see appropriate chapters listed in "Causes".

DIET
• Photosensitization: Restrict access to St John's Wort. • Hypersensitivity: Perform an elimination diet trial as diagnostic test, and cease feeding any foods that are identified as allergens. Essential fatty acid supplementation may be helpful. • See appropriate chapters for other causes.

CLIENT EDUCATION
• Photosensitization: Restrict access to St John's Wort. • Constricted toe: Carefully monitor the toes of nestling psittacine birds. • Owners should be made aware that hypersensitivity in birds is not well understood. Keeping a journal of how their bird reacts to diets, environmental conditions, supplements and treatments is important in managing these cases. • See appropriate chapters for client education for other causes.

SURGICAL CONSIDERATIONS
N/A other than treatment listed for constricted toe.

MEDICATIONS

DRUG(S) OF CHOICE
• Photosensitization: Antimicrobial therapy based on culture and sensitivity and analgesia as needed. • Constricted toe: Antimicrobial therapy based on culture and sensitivity infections and analgesia as needed. • See appropriate chapters for medication recommendations for other causes.

CONTRAINDICATIONS
N/A

PRECAUTIONS
N/A

POSSIBLE INTERACTIONS
N/A

ALTERNATIVE DRUGS
N/A

FOLLOW-UP

PATIENT MONITORING
• Photosensitization: Birds should be monitored closely until completely healed. • Constricted toe: Birds should be monitored closely until completely healed. • Hypersensitivity: Owners should keep a journal recording how their bird reacts to diets, environmental conditions, supplements and treatments. • See appropriate chapters for monitoring recommendations for other causes.

PREVENTION/AVOIDANCE
• Photosensitization: Restrict access to St John's Wort • Constricted toe: Carefully monitor the toes of nestling psittacine birds. Monitor and raise humidity in the nursery if cases occur. • See appropriate chapters for prevention/avoidance listed for other causes.

POSSIBLE COMPLICATIONS
• Photosensitization: Secondary microbial infections can occur in severe cases. • Constricted toe: In severe cases, secondary microbial infections and loss of the affected digit can occur. • Hypersensitivity: If uncontrolled the bird will remain pruritic and secondary microbial infection due to self-mutilation is common. • See appropriate chapters for possible complications listed for other causes.

EXPECTED COURSE AND PROGNOSIS
• Photosensitization: The prognosis is good to guarded depending on amount of St John's Wort ingested and time in direct sunlight. • Constricted toe: The prognosis is good when detected early and treated and managed correctly. • Hypersensitivity: Hypersensitivity to food has a good prognosis if eliminated from the diet. The prognosis for hypersensitivity to environmental allergens is good to guarded depending on whether the allergen can be identified and controlled.

MISCELLANEOUS

ASSOCIATED CONDITIONS
N/A

AGE-RELATED FACTORS
N/A

ZOONOTIC POTENTIAL
N/A

FERTILITY/BREEDING
N/A

SYNONYMS
N/A

SEE ALSO
Bite wounds
Cloacal disease
Ectoparasites
Feather cysts
Feather damaging and self-injurious behavior
Herpesviruses
Oral plaques
Papilloma
Poxvirus
Trauma
Tumors

ABBREVIATIONS
N/A

INTERNET RESOURCES
http://www.merckmanuals.com/vet/integumentary_system/photosensitization/overview_of_photosensitization.html

Suggested Reading
Girling, S. (2006). Dermatology of Birds. In: Patterson, S. (ed.), *Skin Diseases of Exotic Pets*. Oxford, UK: Blackwell Science Ltd, pp. 3–14.
Nett-Mettler, C. (2014). Allergies in birds. In: Noli, C., Foster, A., Rosenkrantz, W. (eds), *Veterinary Allergy*. Chichester, UK: John Wiley & Sons Ltd, pp. 422–427.
Schmidt, R.E., Reavill, D.R., Phalen, D.N. (eds) (2003). *Pathology of Pet and Aviary Birds*. Ames, IA: Iowa State Press, pp. 177–196.
Carpenter, J.W. (2005). *Exotic Animal Formulary*, 3rd edn. St. Louis, MI: Elsevier Saunders, pp. 135–315.
Authors Thomas M. Edling, DVM, MSpVM, MPH
Heide M. Newton, DVM, DACVD

Diabetes Insipidus

BASICS

DEFINITION
Diabetes insipidus (DI) is characterized by the incapacity of the kidney to concentrate the urine, extreme thirst, and the production of large quantities of dilute urine.

PATHOPHYSIOLOGY
• Central DI: Lack of production of arginine vasotocin (AVT), the avian antidiuretic hormone. • Nephrogenic DI: Inability for the kidney to respond to AVT.

SYSTEMS AFFECTED
• Endocrine • Renal • Central nervous system.

GENETICS
Hereditary condition in certain strains of White leghorn chickens.

INCIDENCE/PREVALENCE
Unknown but DI is uncommonly seen.

GEOGRAPHIC DISTRIBUTION
N/A

SIGNALMENT
• **Species:** ◦ Central DI has been reported in an African Grey parrot; it has been diagnosed in other Psittaciformes. The authors have diagnosed DI in two other African grey parrots suggesting it may be more common in this psittacine species. ◦ It is suspected that diabetes insipidus may occur with pituitary neoplasia in budgerigars. ◦ Nephrogenic DI has not been reported in Psittaciformes. It is documented as being a hereditary condition in certain strains of White leghorn chickens and has been experimentally induced in quails. • **Mean age and range:** Any age.
• **Predominant sex:** N/A

SIGNS
Historical Findings
• Severe polyuria-polydipsia (PU-PD). The degree of PU-PD is usually highly suggestive of diabetes insipidus. Water intake may be as high as 1 L/kg/day. • Lethargy, behavioral changes.

Physical Examination Findings
• Usually no specific clinical signs are seen.
• Dehydration. • Blindness, anisocoria, mydriasis in case of a pituitary mass.

CAUSES
Central DI (extrapolated from mammals)
• Decreased production of AVT by the supraoptic and paraventricular nuclei of the hypothalamus. • Decreased release of AVT from the posterior pituitary gland (neurohypophysis): ◦ Trauma. ◦ Neoplasia (e.g., meningioma, pituitary adenoma) PU-PD has been associated in several species (e.g., budgerigars) with an intracranial neoplasia (pituitary adenoma or adenocarcinoma, ependymal tumors); it is possible that these cases had central DI, but this has not been confirmed. ◦ Infection or inflammation. ◦ Congenital.

Nephrogenic DI
• Congenital/hereditary (e.g., White leghorn chickens). • Kidney disease (not reported in companion birds). • Hypercalcemia.

RISK FACTORS
• Central DI: Unknown, trauma.
• Nephrogenic DI: Predisposition of White leghorn chickens.

DIAGNOSIS

DIFFERENTIAL DIAGNOSIS
PU-PD: • Behavioral: primary polydispia = psychogenic polydipsia (reported in an African grey parrot, turkeys). • **Other endocrine disease:** ◦ Diabetes mellitus ◦ Hyperadrenocorticism. • **Diet related:** ◦ Diets too rich in salt ◦ Large amount of fruits in the diet (polyuria without polydipsia). • **Organ dysfunction:** ◦ Renal failure/medullary wash-out ◦ Gastro-enteritis/cloacitis ◦ Cloacal prolapse ◦ Severe hepatopathy. • **Infectious diseases:** ◦ Paramyxovirus infection in pigeons. ◦ Proventricular dilatation disease.
• **Toxicosis:** Heavy metal toxicosis, ethylene glycol, aminoglycosides, hypervitaminosis D3. • **Iatrogenic:** Steroids, hormonal therapy.
• Hypokalemia, hypercalcemia. • **Egg laying**.
• **Hypoadrenocorticism** (not reported in birds).

CBC/BIOCHEMISTRY/URINALYSIS
• Results compatible with dehydration: increased electrolytes (Na, K, Cl), increased hematocrit and increased total solids despite polydipsia. • Mild hyperglycemia may be noted likely due to catecholamine response to hypovolemia. • No other abnormalities have been reported in central DI or nephrogenic DI in birds, but it is reasonable to think that nephrogenic DI could be associated with elevated renal parameters.

OTHER LABORATORY TESTS
• The typically severe PU-PD combined with plasma osmolality measurement and a modified water deprivation test is usually diagnostic. • Plasma osmolality: Elevated despite polydipsia (higher than 320 mmol/L in African grey parrots). It is highly suggestive of diabetes insipidus since with psychogenic polydipsia, the plasma osmolality tends to be decreased. • Urine osmolality (ϕ): Much lower than plasma osmolality (e.g., 40–60 mmol/L). • Urine specific gravity (USG): Can be useful to approximate the urine osmolality (in Hispaniolan Amazon parrots, $\phi = (-57) + 25749(USG-1)$ with USG measured with a canine specific gravity scale on a veterinary refractometer).

IMAGING
• MRI: Gold standard to investigate central DI—may reveal an intracranial lesion (neoplasia, vascular anomaly). • CT scan is not as sensitive as MRI but could also be useful to try and identify pituitary lesions.
• Radiographs and coelomic ultrasound: May be useful in suspected cases of nephrogenic DI to evaluate the size and aspect of the kidneys but are not sensitive. Avian kidneys are usually not visible on ultrasound without concurrent coelomic distention.
• Coelioscopy: Gold standard in suspected cases of nephrogenic DI, in order to visualize the kidneys and perform a biopsy for histology. Histopathology may not confirm nephrogenic diabetes insipidus in the absence of renal lesions.

DIAGNOSTIC PROCEDURES
• A water deprivation test should be performed in hospital to diagnose diabetes insipidus: ◦ Due to the significant risks of rapid dehydration during a water deprivation test, it is suggested to closely monitor the patient and adapt the water deprivation test to the patient's condition. ◦ Plasma and urine osmolalities should be measured prior to the procedure. ◦ Water is withdrawn from the bird. ◦ Plasma and urine osmolalities are measured 1–2 h after water removal (these tests can be repeated more frequently in larger birds). ◦ The bird's weight should be periodically monitored during the procedure and the patient should not lose more than 10% body weight overall. ◦ In central and nephrogenic DI, the plasma osmolality rapidly increases and the urine osmolality stays very low. In psychogenic polydyspia, the urine osmolality increases and the plasma osmolality may slightly increase. • An antidiuretic hormone response test is then performed to differentiate central DI from nephrogenic DI: ◦ Arginine vasopressin (suggested dose: 10–20 ug/kg) or arginine vasotocin (suggested dose 5–10 ug/kg) is administered to the bird (conjunctival, intranasal, or intramuscular). ◦ Plasma and urine osmolality are measured 1–2 h after the treatment (these tests may be repeated more frequently in larger birds). ◦ In central DI, the plasma osmolality should stabilize itself and the urine osmolality should increase. ◦ In nephrogenic DI, the bird continues to lose weight, the plasma osmolality continues to increase and the urine osmolality stays very low.

PATHOLOGIC FINDINGS
• Pituitary masses may be seen. • Renal lesions may be observed.

Diabetes Insipidus (Continued)

TREATMENT

APPROPRIATE HEALTH CARE
- Birds should have access to water at all time.
- If dehydrated, fluid therapy may be indicated (subcutaneous, or intravenous/intraosseous if the dehydration is severe).

NURSING CARE
N/A

ACTIVITY
N/A

DIET
- Central DI: N/A
- Nephrogenic DI: If kidney disease is suspected, a prescription diet with good quality proteins could be recommended.
- A diet low in sodium may be useful.

CLIENT EDUCATION
Inform the client that the clinical management of the disease is complex, that the treatment is lifelong, that the bird should have unlimited access to water at all time.

SURGICAL CONSIDERATIONS
N/A

MEDICATIONS

DRUG(S) OF CHOICE

Central DI
- Arginine vasopressin (mammalian hormone): 20ug/kg BID. Best results intramuscular, but intraconjunctival administration has been reported as well. May also be given intranasal.
- Only seems to be partially effective in birds.
- Arginine vasotocin (avian hormone): 0.5–12 ug/kg BID intranasal or conjunctival. The drug is available from research companies (e.g., Sigma) and needs to be compounded. Has been associated with successful results in a case of central DI in an African grey parrot seen by the authors. Start with a lower dose, monitor the progression of the disease and increase the dose if necessary.
- AVT decreases the GFR, mostly in the reptilian-type nephrons, increases water reabsorption through aquaporins in the collective duct, and it may also stimulate water reabsorption in the colon.
- With severe and long standing polyuria, response to antidiuretic therapy may be delayed due to renal medullary washout.

Nephrogenic DI
- Treatment of the underlying cause if possible.
- Hydrochlorothiazide: this is a diuretic that has been proven to decrease urine output in human patients with nephrogenic diabetes insipidus. The response may be mild, however.

CONTRAINDICATIONS
N/A

PRECAUTIONS
N/A

POSSIBLE INTERACTIONS
N/A

ALTERNATIVE DRUGS
N/A

FOLLOW-UP

PATIENT MONITORING
- Water intake monitoring at home by the owner.
- Plasma osmolality to monitor treatment efficacy.
- Urine osmolality to monitor treatment efficacy.
- Daily water consumption log.

PREVENTION/AVOIDANCE
N/A

POSSIBLE COMPLICATIONS
Dehydration and death if no access to water even for a short period.

EXPECTED COURSE AND PROGNOSIS
Guarded.

MISCELLANEOUS

ASSOCIATED CONDITIONS
N/A

AGE-RELATED FACTORS
N/A

ZOONOTIC POTENTIAL
N/A

FERTILITY/BREEDING
N/A

SYNONYMS
N/A

SEE ALSO
Chronic egg laying
Cloacal disease
Dehydration
Diarrhea
Egg yolk coelomitis
Enteritis/gastritis
Heavy metal toxicity
Hyperglycemia/diabetes mellitus
Liver disease
Neurologic conditions
Nutritional deficiencies
Pancreatic diseases
Paramyxoviruses
Pituitary/brain tumors
Polydipsia
Polyuria
Proventricular dilatation disease
Renal disease
Vitamin D toxicosis

ABBREVIATIONS
N/A

INTERNET RESOURCES
N/A

Suggested Reading

Brummermann, M., Braun, E.J. (1995). Renal response of roosters with diabetes insipidus to infusions of arginine vasotocin. *American Journal of Physiology. Regulatory, Integrative and Comparative Physiology*, **269**(1):R57–63.

Lumeij, J.T., Westerhof, I. (1998). The use of the water deprivation test for the diagnosis of apparent psychogenic polydipsia in a socially deprived African grey parrot (Psittacus erithacus erithacus). *Avian Pathology*, **17**(4):875–878. doi: 10.1080/03079458808436509.

Starkey, S.R., Wood, C., de Matos, R., et al. (2010). Central diabetes insipidus in an African Grey parrot. *ournal of the American Veterinary Medical Association*, **237**(4): 415–419. doi:10.2460/javma.237.4.415.

Authors Delphine Laniesse, Dr. Med. Vet, IPSAV
Hugues Beaufrère, Dr.Med.Vet., PhD, Dipl ABVP (Avian), Dipl ECZM (Avian), Dipl ACZM

Diarrhea

BASICS
DEFINITION
Abnormally increased frequency, liquidity and volume of fecal discharge.
PATHOPHYSIOLOGY
• Imbalance of intestinal absorption, secretion, and/or motility. • Impaired absorption can result from inflammation (enteritis), maldigestion (PDD, pancreatitis), dietary malassimilation, osmotically active agents (fiber), toxins (heavy metals), and hemorrhage (neoplasia). • Increased secretion and motility frequently result from infectious agents (bacterial and viral infections) that stimulate intestinal epithelial cell hypersecretion.
SYSTEMS AFFECTED
• Gastrointestinal. • Cardiovascular—compromise to collapse with excessive fluid losses. • Endocrine/metabolic—dehydration, fluid, electrolyte, acid/base abnormalities.
GENETICS
No specific genetic/breed predisposition.
INCIDENCE/PREVALENCE
Diarrhea is a relatively common problem encountered in birds.
GEOGRAPHIC DISTRIBUTION
N/A
SIGNALMENT
No specific breed, age, or sex predilection.
SIGNS
Historical Findings
• Abnormal liquid or unformed stool. • Abnormal coloration of stool. • Undigested food in stool. • Soiled vent feathers. • Polyuria/polydipsia. • Vomiting. • Anorexia. • Depression.
Physical Examination Findings
• Dehydration. • Emaciation. • Weakness. • Soiled vent feathers.
CAUSES
• Bacterial infection—*Campylobacter* spp., *Chlamydia psittaci*, *Escherichia coli*, *Pseudomonas* spp., *Aeromonas* spp., *Salmonella* spp., *Citrobacter* spp. • Viral infection—adenovirus, avian polyomavirus, herpesvirus, circovirus, proventricular dilatation disease (PDD), paramyxovirus. • Parasite infection—Coccidia, *Cryptosporidia* spp., *Microsporidia* spp., *Giardia* spp., *Hexamita* spp. • Obstruction—foreign bodies, neoplasia. • Metabolic disorders—liver disease (atoxoplasmosis), renal disease. • Dietary—dietary changes/intolerance, foods with increased water/fiber content, contaminated food. • Maldigestion—pancreatitis, hepatobiliary disease, malabsorption syndrome, stress and anorexia. • Drugs and toxins—heavy metals (lead, zinc), pesticides, plant toxins.

RISK FACTORS
• Exposure to other birds carrying disease. • Abrupt dietary changes. • Inappropriate diet. • Inappropriate environment (heavy metal exposure from cage, toys, paint).

DIAGNOSIS
DIFFERENTIAL DIAGNOSIS
• Normal stools can be loose, but without increased frequency, discoloration, or undigested food elements. • Certain species (e.g., Lories) normally have more liquid stools from diet. • Polyuria (formed stool with a large volume of surrounding urine) can mimic diarrhea. • High volume secretory diarrhea suggests possible infectious etiology. • Concomitant neurologic signs suggests possible heavy metal toxicity or toxins.
CBC/BIOCHEMISTRY/URINALYSIS
• Leukocytosis with heterophilia may be seen with gastroenteritis resulting from bacterial infection. • Hypochromic, regenerative anemia can be seen with lead toxicity. • Amylase may be elevated with underlying pancreatitis. • Abnormal renal and/or hepatic lab values may indicate additional etiologic causes.
OTHER LABORATORY TESTS
• Gram's stains and cultures of fecal, cloacal, crop, and/or proventricular swabs. • Cytologic examination of fecal, cloacal, crop, and/or proventricular lavage fluid. • Fecal Gram's stain for assessment of intestinal flora. • Direct fecal examination for protozoal parasite evaluation. • Fecal floatation for intestinal parasite examination. • Heavy metal testing: Normal lead blood level in birds is <2.5 ppm. 0.25 ppm is significant when associated with clinical signs; >0.5 ppm is high regardless of findings. Normal zinc blood level in parrots is <2.5 ppm. 2.6–3.4 ppm is above normal; 3.5–4.4 ppm is high; >4.5 ppm is toxic.
IMAGING
• Survey whole-body radiographs may indicate obstruction, foreign body, mass, organomegaly. • Contrast radiography or ultrasonography may find mucosal irregularities, thickening of the bowel wall or mass effect. • Contrast radiography can be utilized to assess gastrointestinal motility.
DIAGNOSTIC PROCEDURES
• Endoscopy of the upper gastrointestinal tract or cloaca can be performed if evidence of masses. • Removal of foreign bodies, biopsy/impression smears of abnormal masses or other organs (e.g., kidney, liver) if indicated.
PATHOLOGIC FINDINGS
Evidence of enteritis (mucosal inflammation and damage/ulceration) should be present in mucosal biopsies from patients with infectious diarrhea.

TREATMENT
APPROPRIATE HEALTH CARE
• Treatment must be specific to the underlying cause to be successful. • Mild disease (dietary modifications, mild infections) can be treated as an outpatient, more moderate to severe disease (infection, chronic disease, toxins) generally require hospitalization for parenteral therapy and fluid management.
NURSING CARE
• Administer warmed crystalloid fluids SC, IV, IO (50–150 mL/kg/d maintenance plus dehydration deficit correction) at a rate of 10–25 mL/kg over a five-minute period or at a continuous rate of 100 mL/kg/q24h. • Increase environmental temperature to 85–90°F (29–32°C). • Provide a humidified environment by placing warm moist towels in the incubator. • Nutritional support is required in most cases.
ACTIVITY
Patients should be rested in a quiet environment to minimize additional stressors.
DIET
Dependent on the underlying cause and patient ability to consume food:
• Diets developed for recovery and nutritional support designed to provide easily digestible nutrients and high energy. • Gavage/tube feeding formula (25 mL/kg q6–12).
CLIENT EDUCATION
• Review of proper dietary requirements. • Review of sources of heavy metal toxicity, when indicated. • Care for additional birds in the home, especially for infectious diarrhea.
SURGICAL CONSIDERATIONS
• Exploratory endoscopy and other surgical procedures should only be pursued if there is evidence of obstruction/mass. • For heavy metal foreign bodies, patients should be stabilized and on chelation therapy before any surgical procedure is attempted. • Surgical removal of objects from the gastrointestinal tract can be done endoscopically (or with magnets for metals), but proventriculotomy or enterotomy procedures may be necessary if other attempts to remove foreign objects fail.

MEDICATIONS
DRUG(S) OF CHOICE
• Appropriate antimicrobial therapy based on culture and sensitivity and suitable diagnostic testing. • Appropriate antiparasitic medication based on culture and sensitivity. • Chelation therapy, if indicated.

DIARRHEA (CONTINUED)

CONTRAINDICATIONS
• D-Penicillamine is extremely unpalatable even when sweetened. Birds may have severe reaction (e.g., retching, convulsions) when given orally. • Cathartics may cause diarrhea and are contraindicated in birds that present with diarrhea, dehydration, and hypovolemia.

PRECAUTIONS
Use caution when initiating antibiotic therapy unless bacterial infection is suspected from initial diagnostic testing, as this may worsen diarrhea or cause anorexia.

POSSIBLE INTERACTIONS
D-Penicillamine should not be given if lead is present in the gastrointestinal tract (increases absorption of heavy metal particles).

ALTERNATIVE DRUGS
• Sucralfate (25 mg/kg PO q8h)—esophageal, crop, and gastrointestinal protectant.
• Kaolin/pectin (2 mL/kg PO q6–12h)—intestinal protectant, antidiarrheal.
• Bismuth subsalicylate (1–2 mL/kg PO q12h)—intestinal protectant, antidiarrheal.
• Cimetidine (5 mg/kg PO, IM q8–12h)—proventriculitis, gastric ulceration.

FOLLOW-UP

PATIENT MONITORING
• Monitor hydration status and weight on a daily basis while hospitalized. • Monitor eating habits and/or response to gavage/tube feeding. • Monitor stool volume, consistency, color, and odor as a response to treatment. • Repeat CBC and plasma chemistry panel to monitor treatment response and recovery approximately one week following initiation of treatment. • If parasites and/or microorganisms are the primary cause of diarrhea, retest for treatment response and status of infection following completion of treatment.

PREVENTION/AVOIDANCE
• Provide an appropriate fresh diet and clean (filtered) drinking water. • Reduce environmental stress and sources of toxin exposure. • Quarantine all new birds and those that have been exposed to other avian species for a minimum of 30 days.

POSSIBLE COMPLICATIONS
Diarrhea may worsen/persist following treatment due to severe infection, continued environmental exposure, additional stressors and exposures, and multifactorial disease.

EXPECTED COURSE AND PROGNOSIS
• Acute cases with a diagnosed treatable cause have a good prognosis toward complete resolution. • Chronic cases in sick patients have a more guarded prognosis. • For severe diarrhea in patients with untreatable conditions (e.g., circovirus, proventricular dilatation disease) the prognosis is eventually grave despite apparent recovery of the acute event.

MISCELLANEOUS

ASSOCIATED CONDITIONS
• Dehydration, hypovolemic cardiovascular and renal failure. • Septicemia. • Anorexia, cachexia. • Anemia, hypoalbuminemia.

AGE-RELATED FACTORS
N/A

ZOONOTIC POTENTIAL
• *Chlamydia psittaci*. • *Giardia*.
• *Salmonellosis* sp.

FERTILITY/BREEDING
N/A

SYNONYMS
Pasty vent

SEE ALSO
Cloacal disease
Enteritis/gastritis
Flagellate enteritis
Gastrointestinal foreign body
Gastrointestinal helminthiasis
Heavy metal toxicity
Liver disease
Polyuria
Proventricular dilatation disease (PDD)
Renal disease
Sick-bird syndrome
Undigested food in droppings
Urate/fecal discoloration
Appendix 7, Algorithm 2. Diarrhea

ABBREVIATIONS
PDD—proventricular dilatation disease

INTERNET RESOURCES
LafeberVet.com

Suggested Reading
Bauck, L. (2000). Abnormal droppings. In: Olsen, G.H., Orosz, S.E. (eds), *Manual of Avian Medicine*. St Louis, MI: Mosby Inc., pp. 62–70.
Gelis, S. (2006). Evaluating and treating the gastrointestinal system. In: Harrison, G.J., Lightfoot, T.L. (eds), *Clinical Avian Medicine*. Palm Beach, FL: Spix Publishing, pp. 411–440.
Jones, M.P., Pollock, C.G. (2000). Supportive care and shock. In: Olsen, G.H., Orosz, S.E. (eds), *Manual of Avian Medicine*. St Louis, MI: Mosby Inc., pp. 17–46.
Monks, D. (2005). Gastrointestinal disease. In: Harcourt-Brown, N., Chitty, J. (eds), *BSAVA Manual of Psittacine Birds*, 2nd edn. Quedgeley, UK: British Small Animal Veterinary Association, pp. 180–190.
Oglesbee, B.L. (1997). Differential diagnosis. In: Altman, R.B., Clubb, S.L., Dorrestein, G.M., Queensberry, K. (eds), *Avian Medicine and Surgery*. Philadelphia, PA: WB Saunders, pp. 225–226.
Author Julie DeCubellis, DVM, MS

Dyslipidemia/Hyperlipidemia

BASICS

DEFINITION
• Dyslipidemia are abnormalities in the plasma lipid profile, usually an increase in one or several lipid fractions caused by disturbances in lipid metabolism. • Most commonly recognized are elevations in plasma cholesterol, very low-density lipoprotein cholesterol (VLDL-C), low-density lipoprotein cholesterol (LDL-C), high-density lipoprotein cholesterol (HDL-C), and triglycerides. The benefits of high HDL-C in birds, as it is the case in mammals, is unknown and elevated HDL-C is often seen in dyslipidemic psittacine birds. • Other dyslipidemia characterized in humans but of unknown occurrence and significance in birds include, among others, decrease in HDL-C, raised lipoprotein(a), raised apolipoprotein B, increased ratio apoB/apoA1, increased ratio non-HDL/HDL, raised "small,dense" LDL, and increased LDL particle numbers. • LDL and HDL subclasses are also present but their characterization and diagnostic values have been poorly investigated in birds.

PATHOPHYSIOLOGY
• Dyslipidemia in birds are presumably caused by disturbances in normal lipid metabolism associated with inadequate nutrition, captive lifestyle, and possibly other species and individual factors. • Dyslipidemia are presumed risk factors or comorbid conditions to atherosclerosis, hepatic lipidosis, hepatic diseases, pancreatic diseases, xanthomatous lesions, obesity, and possibly others. However, the association has not been confirmed in most instances. Birds are also suspected to develop a similar metabolic syndrome as described in mammals but this has not been investigated. • Avian lipid metabolism has marked differences from mammals: larger ability to store fat, poorly developed lymphatic system, lipogenesis is mostly restricted to the liver (include adipose tissue in mammals), portomicrons instead of chylomicrons, HDL-C is the predominant cholesterol type in healthy birds (vs. LDL-C in humans), and different apolipoproteins are present.

SYSTEMS AFFECTED
• Hepatic. • Adipose.

GENETICS
• Restricted ovulator (RO) chicken—mutation in the VLDL receptor gene. • Experimentally inbred laboratory strains of quails and pigeons have higher cholesterol than other breeds. • Genetic factors may play a role in some individuals. In mammals, deficient LDL receptor, enzymes, or apolipoproteins are the causes of various familial dyslipidemias.

INCIDENCE/PREVALENCE
Unknown but suspected to be prevalent in Quaker parrots and older psittacine birds.

GEOGRAPHIC DISTRIBUTION
N/A

SIGNALMENT
• **Species:** ○ All birds are susceptible to dyslipidemia. Among psittaciformes, Quaker parrots seem particularly susceptible. Amazon parrots, African grey parrots, and cockatiels seem to have slightly higher cholesterol than other psittacine species. Psittacine birds adapted to high-fat diet (e.g., macaws, palm cockatoos) are expected to show a lower susceptibility to dyslipidemia. ○ Female birds undergoing vitellogenesis have lipidemic changes that may be characterized as dyslipidemic when chronic. • **Mean age and range:** Any age may be affected but older psittacine birds may be more susceptible. In mammals, plasma lipid fractions increase with age. • **Predominant sex:** Both but female birds are prone to dyslipidemia due to physiologic or abnormal vitellogenesis.

SIGNS
Signs are related to the causes or consequences of altered lipid metabolism or associated conditions (see corresponding sections) and may include: • Reproductive signs: ○ Ascites ○ Reproductive behavior ○ Chronic egg laying ○ Feather damaging behavior. • Cutaneous signs: Cutaneous xanthomatosis. • Neurological signs: Central neurological signs caused by lipid emboli or carotid atherosclerosis. • Cardiovascular signs due to atherosclerosis: ○ Congestive heart failure ○ Ataxia, hind limb weakness ○ Dyspnea. • Hepatic signs: ○ Abdominal distention ○ Gastrointestinal signs. • Nonspecific: ○ Lethargy, depression ○ Decreased appetite ○ Overweight, obesity.

CAUSES AND RISK FACTORS
• Vitellogenesis in female birds: This may be physiological but is often maladaptive in captive psittacine birds and associated with inappropriate chronic reproductive stimulation. Under estrogen stimulation, the liver produces two very-low density lipoproteins, vitellogenin and VLDLy. • Diet: An inappropriate diet may contribute to the development of dyslipidemia and other diseases ○ Excessive dietary intake. ○ Postprandial hyperlipidemia. ○ High fat and multideficient diet, typically all-seed diet in granivorous/frugivorous birds. ○ Animal products containing animal fat and cholesterol (egg, meat) in granivorous/frugivorous birds. ○ Overreliance on day-old chicks in certain susceptible raptors (controversial). Inappropriate prey items for certain birds of prey and possibly others (e.g. rats and day-old chicks for insectivorous or piscivorous birds). ○ Lack of lipotropic factors that promote lipid catabolism such as methionine, choline, and vitamin Bs. • Lack of physical activity, obesity. • Hepatobiliary disease may lead to changes in plasma lipid fractions as the liver is the source of cholesterol, lipid, and apolipoprotein synthesis. In particular, biliary obstruction often results in elevated plasma cholesterol. • Pancreatitis has been shown to cause dyslipidemia in mammals. • Endocrine diseases ○ Hypothyroidism is not well characterized in birds but is a prominent cause of hypercholesterolemia in companion mammals. ○ Diabetes mellitus is an important cause of dyslipidemia in human patients but this association is not well characterized in birds, especially since diabetes mellitus is not common in birds. • Certain drugs may alter some lipid fractions such as steroids, progestins, thiazides, β-blockers, and cyclosporine. • Second hand smoke in pet birds may also contribute to dyslipidemic disorders.

DIAGNOSIS

DIFFERENTIAL DIAGNOSIS
• Laboratory errors. • Physiological female reproductive activity. • See Causes.

CBC/BIOCHEMISTRY/URINALYSIS
Plasma lipemia may lead to an elevation in total solids and may interfere with some biochemical assays.

OTHER LABORATORY TESTS
• Various lipoprotein assays are used in mammals to measure the different lipoprotein fractions but most have not been properly validated in the different avian species. Some laboratory tests may not be accurate for measuring cholesterol fractions in birds. It is recommended that birds are fasted prior to measurement as some lipid fractions may greatly increase after a meal. Selected methods classically used for cholesterol fraction/lipoprotein determinations are: ○ Ultracentrifugation: Gold standard but laboratory intensive method and classically not commercially available. Plasma lipoproteins are separated according to their density by a series of centrifugal steps using density gradient (salt solutions) at relative centrifugal fields of 200–300,000 G. Then cholesterol is measured in each fraction. Density limits have not been published for parrot plasma but have been used in pigeons. Modified techniques using different compounds to generate density gradients have been described. ○ Laboratory analyzers: Enzymatic and precipitation colorimetric assays are classically used to measure the total cholesterol, the HDL cholesterol, and the triglycerides. In a standard laboratory analyzer, direct LDL reagent is typically not available and must be indirectly calculated using the Friedewald formula: LDL = Total cholesterol – HDL – Triglycerides/x, with

Dyslipidemia/Hyperlipidemia (Continued)

x = 5 for American units and x = 2.18 for SI units. This formula is not reliable in case of lipemia (especially with triglycerides >4.5 mmol/L). While it has not thoroughly been validated against ultracentrifugation techniques, this is the most convenient assay and appears to be of clinical value in birds. ○ Electrophoresis: Lipoproteins can be separated in a single operation but this has not been validated in birds. Electrophoretic techniques have been used in research in birds, including Quaker parrots. ○ Magnetic resonance spectroscopy: Has not been described in birds.

IMAGING
Dyslipidemia has classically been associated with atherosclerosis and hepatic lipidosis. As such, hepatomegaly may be observed on radiographs. Please consult the atherosclerosis chapter for specific diagnostic findings on vascular imaging.

DIAGNOSTIC PROCEDURES
N/A

PATHOLOGIC FINDINGS
• Histopathologic lesions that may be concurrent with dyslipidemia include hepatopathies, xanthomatous lesions, and atherosclerosis.

TREATMENT

APPROPRIATE HEALTH CARE
Decreasing reproductive stimuli in pet birds may be beneficial.

NURSING CARE
Supportive care.

ACTIVITY
Activity level should be as high as possible. Opportunities for increased activity may be scarce in captivity but caretakers should be encouraged to promote physical exercise with flying, playgrounds, and foraging setups.

DIET
• The provision of a balanced diet supplemented with fruits and vegetables is probably the most important aspect of the treatment in captive parrots. • Decreasing the overall amount of food may be beneficial, especially in overconditioned birds.
• Discontinue the feeding of animal products to psittacine birds. Nonanimal items do not contain any cholesterol. • Switch to a pelletized diet if the bird is on an all-seed diet. Low-fat pelletized diet (e.g., Roudybush low fat) may be beneficial in some cases. Some species are adapted to consume higher fat diet than others (e.g., macaws, palm cockatoos).
• Omega-3 fatty acids may help. They can be provided, among other items, with flaxseeds.

• In birds of prey, day-old chicks should probably be limited or avoided in certain species such as gyrfalcons, insectivorous species, and kites.

CLIENT EDUCATION
Inform the clients about proper diet for respective species, the need for adequate physical exercise, and how to reduce reproductive stimulation in females.

SURGICAL CONSIDERATIONS
N/A

MEDICATIONS

DRUG(S) OF CHOICE
• Only limited information is available for most drugs listed below. Most is empirical and pharmacological information has rarely been obtained in birds. For treatment of heart failure, please see corresponding section.
• Statins—lipid lowering agents, which are competitive inhibitors of the HMG-CoA reductase enzyme in hepatic cholesterol synthesis. A pharmacokinetic study on rosuvastatin in Amazon parrots failed to demonstrate therapeutic concentrations at 10–25 mg/kg. Other statins of interest include atorvastatin, simvastatin, lovastatin, pravastatin, and fluvastatin. Only rosuvastatin and atorvastatin have long half-lives in mammals and the suggested dose is 10–30 mg/kg pending further pharmacologic studies. Should not be taken while feeding grapefruits and concurrently with azole antifungals. Statins also have nonlipid effects such as beneficial properties on endothelial function, vascular inflammation, immune modulation, oxidative stress, thrombosis, and atherosclerotic plaque stabilization.
• Fibrates—hypolipidemic drugs, which are agonists of the PPARα that stimulates β-oxidation of fatty acids. Fibrates mainly cause a decrease in blood triglycerides.
• Other lipid lowering agents include cholesterol absorption inhibitors (ezetimibe but granivores and frugivores do not have a dietary source of cholesterol) and nicotinic acid. • GnRH agonists (leuprolide acetate, subcutaneous deslorelin implant) may be given to decrease reproductive hyperlipidemia and as part of the treatment of reproductive conditions.

CONTRAINDICATIONS
N/A

PRECAUTIONS
N/A

POSSIBLE INTERACTIONS
• The combination of statins with fibrates may increase the risk of rhabdomyolysis.

• Some statins (e.g., atorvastatin) interact with drugs metabolized through the cytochrome P450 pathway, such as azole antifungals. • Grapefruits should not be fed to birds receiving statins; this may increase plasma levels.

ALTERNATIVE DRUGS
N/A

FOLLOW-UP

PATIENT MONITORING
Blood cholesterol and lipoproteins should be monitored on a regular basis. Peak effects of statins may not be evident for several weeks.

PREVENTION/AVOIDANCE
• See diet and activity sections. • Reduction of captive reproductive stimuli for female birds.

POSSIBLE COMPLICATIONS
• Atherosclerotic diseases • Obesity
• Diabetes mellitus • Hepatic lipidosis
• Xanthomatosis, xanthomas.

EXPECTED COURSE AND PROGNOSIS
N/A

MISCELLANEOUS

ASSOCIATED CONDITIONS
• Atherosclerosis • Obesity • Diabetes mellitus • Hepatic lipidosis • Xanthomatosis, xanthomas.

AGE-RELATED FACTORS
Plasma lipid levels may increase with age.

ZOONOTIC POTENTIAL
N/A

FERTILITY/BREEDING
Plasma cholesterol and lipids increase with reproductive activity and is expected with physiological vitellogenesis in female breeding birds.

SYNONYMS
N/A

SEE ALSO
Atherosclerosis
Chronic egg laying
Hepatic lipidosis
Lipoma
Nutritional deficiencies
Obesity
Pancreatic diseases
Xanthoma

ABBREVIATIONS
N/A

INTERNET RESOURCES
N/A

Suggested Reading
Beaufrere, H. (2013). Avian atherosclerosis: parrots and beyond. *Journal of Exotic Pet Medicine*, **22**(4):336–347.

Beaufrere, H. (2013). Atherosclerosis: Comparative pathogenesis, lipoprotein metabolism and avian and exotic companion mammal models. *Journal of Exotic Pet Medicine*, **22**(4):320–335.

Facon, C., Beaufrere, H., Gaborit, C., *et al.* (2014). Cluster of atherosclerosis in a captive population of black kites (*Milvus migrans subsp.*) in France and effect of nutrition on the plasma lipidogram. *Avian Diseases*, **58**(1):176–182.

Petzinger, C., Heatley, J.J., Cornejo, J., *et al.* (2010). Dietary modification of omega-3 fatty acids for birds with atherosclerosis. *Journal of the American Veterinary Medical Association*, **36**(5):523–528. doi:10.2460/javma.236.5.523.

Petzinger, C., Bauer, J. (2013). Dietary considerations for atherosclerosis in avian species commonly presented in clinical practice. *Journal of Exotic Pet Medicine*, **22**(4);358–365.

Reiner, Z., Catapano, A., de Backer, G., *et al.* (2011). ESC/EAS Guidelines for the management of dyslipidaemias. *European Heart Journal*, **32**:1769–1818.

Author Hugues Beaufrère, Dr.Med.Vet., PhD, Dipl ABVP (Avian), Dipl ECZM (Avian), Dipl ACZM

DYSTOCIA AND EGG BINDING

BASICS
DEFINITION
Egg binding is the failure of the egg to pass through the reproductive tract within a normal period of time. Related to egg binding is dystocia, which is the mechanical obstruction of oviposition.

PATHOPHYSIOLOGY
• Egg binding and dystocia are among the most frequently seen obstetric disorders in clinical practice. Normal transit time through the reproductive tract varies between species and even among individuals, but is generally between 24 and 48 hours for the species most commonly kept as pets. Shell formation occurs in the uterus, also known as the shell gland. • Delayed transit of the egg through the oviduct can be due to issues with the reproductive tract or due to issues with the egg being oversized, malformed, or abnormally positioned. • Dysfunction of the oviduct can be a result of inflammation or infection, damage to the oviductal wall, inertia secondary to calcium deficiency, torsions, and the presence of a persistent cystic right oviduct. • Masses of the reproductive tract, cloaca, or other coelomic organs can compress the lumen of the oviduct, thus obstructing the passage of an egg. • Chronic egg laying results in physical exhaustion of the reproductive tract and depletion of body stores of calcium, vitamin E, and selenium. Subsequently, oviductal muscle atony develops and results in delayed oviposition. • The distal uterus, vagina, and vaginal-cloacal junction are the most common areas for dystocia. • Musculoskeletal deformities of the pelvic canal, either congenital or acquired after injury, can physically obstruct the passage of an egg.

SYSTEMS AFFECTED
• Reproductive/Urogenital—difficulty or inability to pass an egg. Prolonged straining can result in a prolapsed oviduct or the cloaca. Pressure necrosis or tear of the oviduct wall may occur. • Respiratory—increased respiratory effort can be seen due to compression of the air sacs by the space-occupying egg. • Gastrointestinal—hen may not be able to defecate if the egg is occluding the cloaca.
• Renal/Urologic—damage to kidney or elevations in uric acid levels can develop due to compression of renal parenchyma and ureters by the retained egg.
• Neuromuscular—unilateral or bilateral paresis or paralysis of the pelvic limbs can result from compression of the ischiatic nerves if the egg is lodged within the pelvic canal.
• Integument—localized plucking of feathers from the caudal coelom and/or over the back may occur in response to discomfort from the egg.

GENETICS
Some breed lines may have a genetic predisposition for egg binding/dystocia.

INCIDENCE/PREVALENCE
• Common, especially in smaller species and hens that are chronic egg layers. • Higher incidence noted when birds breed out of their natural season or are first time egg producers.

GEOGRAPHIC DISTRIBUTION
N/A

SIGNALMENT
• **Species predilections:** Cockatiels, budgerigars, lovebirds, finches, canaries, cockatoos, and Amazon parrots have increased risk for egg binding and dystocia.
• **Mean age and range:** Hens of egg-producing age (sexual maturity varies between species). • **Predominant sex:** Females of egg-laying age.

SIGNS
Historical Findings
• There may be a history of recent egg laying with a sudden cessation of egg production. In other cases, the birds are novice egg layers.
• Recently laid eggs may be abnormally sized, misshapen, thin-shelled, or unshelled. • The hen may exhibit broody behavior, such as shredding paper, seeking dark places, or being cage protective. • Hen may have a wide-based stance or appears to be actively straining to pass the egg. • Scant or no fecal production. Fecal matter may be adhered to tail. Blood may be noted in the droppings. • Increased respiratory effort may be noted, especially after activity (such as flying). Tail bobbing or open-mouth breathing may be noted. • A decreased appetite, lethargy, and weakness may be noted if the egg binding has been prolonged. • In some cases, the cloaca or oviduct may be prolapsed. An egg may be adhered to the prolapsed tissue. • Some birds may have trouble perching or are on the bottom of the cage. They may be favoring one or both limbs. • Sudden death may be noted, especially in the smaller species such as finches and canaries.

Physical Examination Findings
• An egg is typically palpable in caudal coelom, which may be distended; eggs that are soft-shelled, shell-less, or located cranially in the oviduct or coelom are often not palpable. • Dyspnea, especially after handling, is a common finding. • Straining or "winking" of the vent; vent may also be flaccid. • Lethargy and/or anorexia may be noted. • Wide-based stance • Unilateral or bilateral hind limb weakness or paresis; feet may feel cold; patient may present in sternal recumbency due to weakness in the legs.
• Fecal and/or urate stained vent and tail feathers. • Hen may be obese, thin, or of normal body condition. • Sudden death.

CAUSES
Reproductive
• Chronic egg laying. • Salpingitis or metritis. • Oviductal torsion. • Tears or damage to the oviduct. • Adhesions of the oviduct.
• Persistent cystic right oviduct. • Oversized, malformed, or abnormally positioned eggs.
Metabolic
• Calcium deficiency (metabolic or nutritional). • Vitamin E deficiency.
• Selenium deficiency.
Neoplasia
• Neoplasia of the reproductive tract, cloaca, or other coelomic organs.
Other
• Stress. • Musculoskeletal deformities of the pelvic canal (congenital or acquired).
• Genetic predisposition. • Lack of exercise.
• Obesity. • Malnutrition. • Breeding out of natural season. • Secondary to systemic disease.

RISK FACTORS
• Previous episodes of egg binding or dystocia may result in damage to the oviduct. • All seed diet, which predisposes the hen to calcium deficiency and obesity. • Chronic egg laying, often due to environmental cues such as prolonged exposure to daylight, presence of a mate or perceived mate (e.g., owner or toy), and access to nesting material or nesting site. Untreated chronic egg laying may lead to depletion of calcium stores, which can result in pathologic fractures and thin or unshelled eggs.

DIAGNOSIS
DIFFERENTIAL DIAGNOSIS
• Coelomic distension—ascites; egg-yolk coelomitis; body wall hernias; neoplasia, abscesses, or granulomas of any of the coelomic organs; diagnostic imaging and/or celiocentesis used to differentiate. • Straining to lay egg—can be confused with gastrointestinal signs such as gastroenteritis, fecaliths, or constipation; cloacal masses; uroliths. • Dyspnea—primary respiratory condition such as pneumonia, sinusitis, or tracheal obstruction; pain; heat exhaustion; cardiac disease; space-occupying masses or fluid in the coelomic cavity; radiographs may differentiate. • Hind limb lameness or paresis—traumatic injury to pelvic limbs or spine; pododermatitis; osteomyelitis; lead toxicity; intramedullary or compressive spinal neoplasia; damage to the ischiatic nerve due to renal infection or renal or ovarian neoplasms; claudication secondary to atherosclerosis.

CBC/BIOCHEMISTRY/URINALYSIS
• Hypercalcemia is a normal finding in an egg-laying bird and a calcium level in normal reference ranges may be indicative of a relative calcium deficiency. • A leukocytosis

DYSTOCIA AND EGG BINDING (CONTINUED)

characterized by a relative heterophilia is not uncommon, especially if an inflammatory or infectious process is involved. • Muscle tissue damage may lead to elevated aspartate aminotransferase and creatine phosphokinase levels. • Uric acid levels may be elevated if the retained egg is compressing the ureters or kidneys. • Hypercholesterolemia can be a normal finding in an ovulating hen.

OTHER LABORATORY TESTS
• Hyperglobulinemia may be noted on protein electrophoresis (EPH) in egg-laying birds. It is characterized by marked monoclonal increases in the beta-globulin fraction caused by selective increases in transferrin. • Aerobic culture or gram stain of the oviduct may be indicated to rule out an infection of the reproductive tract. Swabs of the cloaca more likely reflect bacterial populations of the proctodeum, urodeum, or coprodeum instead of the microflora of the oviduct.

IMAGING
Radiographic findings
• Whole body radiographs are useful in determining the size, location, position, and number of eggs present in the oviduct. Radiographs can also help evaluate if the egg is thin-shelled, shell-less, or malformed.
• Polyostotic hyperostosis (osteomyelosclerosis) of the long bones is due to medullary ossification. This provides a calcium reserve for the formation of eggshells and is a normal finding in an egg laying bird. Hens that are calcium depleted may have normal or osteopenic bone density. • The reproductively active ovary and oviduct are seen as increased soft-tissue opacity in the mid-to-cadual region of dorsal coelom on lateral radiographs and cause a widening of the hepatic silhouette on ventrodorsal radiographs.
Ultrasonographic findings
• Nonmineralized eggs (i.e., soft shelled, unshelled eggs) that are not visible on radiographs can be evaluated by ultrasonography. • The extraoviductal location of ectopic eggs can be confirmed by the egg's independent position from the oviduct.

DIAGNOSTIC PROCEDURES
Endoscopy may be utilized in larger birds to visualize the lumen of the oviduct to assess for torsion or masses. Additionally, eggs located ectopically within the coelom may be visualized by laparoscopy.

PATHOLOGIC FINDINGS
• Salpigitis and metritis, either as the cause or secondary to the egg binding, may be noted.
• An ectopic egg may result if an egg is retropulsed up the oviduct or through a tear in the oviduct wall into the coelomic cavity

TREATMENT
APPROPRIATE HEALTH CARE
• Hens that are demonstrating mild clinical signs for 24 hours or less often respond to supportive measures alone, but should be carefully monitored for worsening of clinical signs. Some cases may be treated on an outpatient basis if the patient is otherwise in stable clinical condition. • For hens straining for over 24–48 hours or demonstrating severe clinical symptoms, earlier surgical intervention may be necessary.
Surgical treatment
• If the egg is caudal enough in the oviduct, manual extraction through the cloaca can be attempted; to facilitate removal, ovocentesis and collapse of the egg may be necessary.
• Percutaneous ovocentesis can be used for eggs that cannot be visualized through the cloacal opening; however, it has a higher risk for damage to the oviduct with subsequent tears and adhesion formation.
• Salpingohysterectomy can be performed to remove the oviduct and egg. This is a technically challenging procedure and should be performed by veterinarians familiar with avian reproductive anatomy. • A salpingotomy can be performed to remove the egg if the egg laying potential of the hen needs to be preserved.

NURSING CARE
• Should be attempted first if patient is otherwise healthy and stable. • Hen should be kept in a humidified incubator warmed up to 85–90°F; ideally in a quiet and low-light setting to minimize stress. • Fluid supplementation—parental fluids (e.g., subcutaneous); oral fluids though gavage tube. • Pain management with analgesics and anti-inflammatory medications. • Calcium supplementation if calcium deficiency suspected. • Broad-spectrum antibiotics if an underlying infectious process or coelomitis is suspected.

ACTIVITY
Minimize stress and activity by keeping the hen in a quiet and low-light setting (i.e., incubator or cage that is partially covered with a towel).

DIET
• Provide calcium supplementation in diet (i.e., cuttlebone, oyster shell). • May need to be gavage fed a critical care formula if hen is not eating or not producing feces.

CLIENT EDUCATION
• Discourage further egg laying for the season to allow the reproductive tract to heal—reduce the photoperiod to no more than 8–10 hours per day by covering the cage at night, keep the hen separate from any mate or perceived mates, reduce foods high in fat and overall quantity of food, remove nests and nesting material, move cage to unfamiliar environment. • Leuprolide acetate can be administered seasonally if the hen starts to exhibit reproductive behaviors.

SURGICAL CONSIDERATIONS
• Compromised ability to breathe due to compression of air sacs by egg and enlarged reproductive tract. Intubation with positive pressure ventilation is strongly recommended.
• Placement of an intraosseous catheter may be difficult due to medullary bone formation in the ulna and tibiotarsal bones. • If the hen is stable and the surgery can be delayed, presurgical administration of leuprolide acetate 1–2 weeks prior can help reduce the size and vascularity of the reproductive tract. Administration of leuprolide acetate also will decrease the bone density of the long bones allowing for easier placement of an intraosseous catheter.

MEDICATIONS
DRUG(S) OF CHOICE
• Calcium gluconate (10%): 50-100 mg/kg SC, IM (diluted); for hypocalcemia; dilute 1:1 with saline or sterile water for injections.
• Calcium glubionate: 25–150 mg/kg PO q12–24h; calcium supplementation for hypocalcemia. • Meloxicam: 0.5–2 mg/kg PO, IM q12–24h, nonsteroidal anti-inflammatory medication. • Butorphanol tartrate: 1–5 mg/kg q2–3h PRN IM, IV; opioid angonist-antagonist for pain management. • Leuprolide acetate: 100–1200 µg/kg IM q2–4 weeks; long-acting synthetic GnRH analog used to prevent ovulation.
• Prostaglandin E_2 gel: 0.02–0.1 mg/kg topically to uterovaginal sphincter; causes relaxation of the uterovaginal sphincter and oviductal contractions; decreased systemic effects compared to $PGF_{2\alpha}$. • Antibiotics if an infection is suspected; ideally to be based on culture and sensitivity.

CONTRAINDICATIONS
Use of oxytocin and prostaglandin $F_{2\alpha}$ are contraindicated if uterovaginal sphincter is not dilated and if adhesions are present.

PRECAUTIONS
• Prostaglandin $F_{2\alpha}$ ($PGF_{2\alpha}$)—0.02–0.1 mg/kg IM, once; causes uterine muscle contractions; must check that uterovaginal sphincter is open, otherwise uterine rupture may occur; also contraindicated in cases of adhesions. • Oxytocin 0.5–10.0 U/kg IM—causes uterine muscle contractions; should be preceded by calcium administration; must check that uterovaginal

Dystocia and Egg Binding (Continued)

sphincter is open, otherwise rupture of shell gland or reverse peristalsis may occur; also contraindicated in cases of adhesions.

POSSIBLE INTERACTIONS
Pregnant woman and asthmatics should avoid handling PGE_2; gloves should be worn when administering this medication.

ALTERNATIVE DRUGS
• Misoprostil can be considered if PGE_2 unavailable. • Acupuncture and chiropractic adjustments have been described as adjunct therapies. • Homeopathic remedies include calcarea carbonica, kali carbonicum, and pulsatilla pratensis.

FOLLOW-UP

PATIENT MONITORING
Repeat radiographs to assess if egg has progressed down the oviduct with medical management. Repeat films can be taken several days later if patient is stable. However, if the hen is declining the films should be taken earlier to see if surgical intervention is required.

PREVENTION/AVOIDANCE
• Discourage further egg laying for the season—reduce the photoperiod to no more than 8–10 hours per day by covering the cage at night, keep the hen separate from any mate or perceived mates, reduce foods high in fat and overall quantity of food, and remove nests and nesting material. • Leuprolide acetate can be administered seasonally or when the hen is exhibiting reproductive behavior to prevent ovulation and egg formation. A deslorelin implant can also be administered and may have longer lasting effects.

POSSIBLE COMPLICATIONS
• Ectopic eggs may result if reverse peristalsis or rupture of the oviduct occurs. • Prolonged straining can result in prolapse of the oviduct, cloaca, and/or the rectum. • Multiple eggs may be present in the oviduct if the hen continues to ovulate despite the presence of an egg already in the oviduct. • Hens that have undergone salpingohysterectomy may be at risk for internal ovulation and subsequent egg yolk peritonitis. Measures to discourage reproductive behavior should be implemented if a hen is spayed.

EXPECTED COURSE AND PROGNOSIS
• Prognosis is good if the hen is otherwise in good health and supportive care is instituted early in the course of the disease. • Some cases of egg binding/dystocia can be prolonged (i.e., weeks to months) without any outward clinical signs of illness. Though not ill, these hens are unlikely to pass their egg without surgical intervention and are at risk for additional issues such as multiple eggs in the oviduct, ectopic eggs, and egg yolk peritonitis. • Prognosis is fair-to-guarded if the health status of the hen is compromised or if the egg binding is secondary to an ongoing disease process. • Eggs that are not palpable on physical examination and retained higher up in the oviduct require more invasive treatment, such as transcutaneous implosion of the egg or salpingohysterectomy, both of which are higher risk procedures. • Damage to the oviduct from the disease process itself or iatrogenically during removal of the egg can predispose the hen to future bouts of egg binding/dystocia.

MISCELLANEOUS

ASSOCIATED CONDITIONS
• Ectopic eggs • Oviductal or uterine prolapse • Lameness.

AGE-RELATED FACTORS
First-time layers and older birds may have increased risk.

ZOONOTIC POTENTIAL
N/A

FERTILITY/BREEDING
If oviduct is damaged during the course of the egg binding or during treatment, the hen may be predisposed to subsequent cases of egg binding and dystocia.

SYNONYMS
Egg retention.

SEE ALSO
Chronic egg laying
Cloacal disease
Coelomic distention
Egg yolk coelomitis / reproductive coelomitis
Lameness
Neurologic conditions
Nutritional deficiencies
Ovarian cysts / ovarian neoplasia
Osteomyelosclerosis / polyostotic hyperostosis
Salpingitis / uterine disorders
Appendix 7, Algorithm 6. Egg binding

ABBREVIATIONS
EPH—protein electrophoresis
GnRH—gonadotropin-releasing hormone
PGE_2—prostaglandin E_2
$PGF_{2\alpha}$—prostaglandin $F_{2\alpha}$

INTERNET RESOURCES
http://www.lafebervet.com/emergency-medicine/birds/presenting-problem-shelled-egg-palpable/

Suggested Reading
Bowles, H.L. (2006). Evaluating and treating the reproductive system. In: Harrison, G.J., Lightfoot, T.L. (eds), *Clinical Avian Medicine*, Vol. II. Palm Beach, FL: Spix Publishing, pp. 519–539.
Hadley, T.L. (2010). Management of common psittacine reproductive disorders in clinical practice. *The Veterinary Clinics of North America. Exotic Animal Practice*, **134**:429–438.
Crosta, L., Gerlach, H., Bürkle, M, Timossi, L. (2003). Physiology, diagnosis, and diseases of the avian reproductive tract. *Vet Clin Exot Anim The Veterinary Clinics of North America. Exotic Animal Practice*, **6**:57–83.
Mans, C., Sladky, K.K. (2013). Clinical management of an ectopic egg in a Timneh African grey parrot (Psittacus erithacus timneh). *Journal of the American Veterinary Medical Association*, **242**:963–968.
Rosen, L.B. (2012). Avian reproductive disorders. *Journal of Exotic Pet Medicine*, **12**:124–131.

Author Sue Chen, DVM, Dipl. ABVP (Avian)

ECTOPARASITES

BASICS

DEFINITION
Ectoparasites of birds are arthropods (insects and arachnids) that live on or in the skin and feathers. They are represented by acarids such as mites (sucking and biting), ticks (hard and soft) and insects (lice, mosquitos, flies and bedbugs). There are a large number and variety of external parasites that have been found in all types of birds. Most of these are not commonly seen in private practice, but it is worth discussing some of the most common and serious conditions that are associated with these pests.

PATHOPHYSIOLOGY
External parasites live on the skin and/or feathers, where they can cause direct irritation with resultant dermatitis and feather lesions. In severe cases, birds are irritable and can develop anemia. Some species can act as vectors for blood-borne diseases such as tularemia, encephalitis, spirochetosis (borreliosis), piroplasmosis, anaplasmosis, dirofilariasis, rickettsial diseases and leucocytozoonosis and West Nile virus.

SYSTEMS AFFECTED
- Integument—dermatitis, scabs, dandruff, hyperkeratosis, overgrowth and disfiguration of the beak. • Hemolymphatic—blood loss/anemia in certain species.
- Reproductive—reduced egg production.
- Nervous system—tick paralysis.

GENETICS
N/A

INCIDENCE/PREVALENCE
Incidence is dependent on a number of factors; some of these parasites cause common and serious problems in pigeons, poultry and nestlings of many species in aviary settings; *Knemidokoptes* in budgerigars is sporadic and does not appear to be highly contagious.

GEOGRAPHIC DISTRIBUTION
Worldwide; different species have different distribution.

SIGNALMENT
- **Species:** Ectoparasites are found in essentially all classes of birds; those of the most importance or the most recognized will be mentioned, along with the parasites that affect them. ○ Poultry: red mite (*Dermanyssus gallinae*), northern fowl mite (*Ornithonyssus sylvarium*), several other mite species, numerous louse species, bedbugs (*Cimex lectularius*), *Knemidokoptes mutans*, sticktight flea (*Echidnophaga gallinacea*), other fleas, mosquitoes, midges, blackflies, hippoboscid flies. ○ Budgerigars and sometimes canaries, finches: *Knemidokoptes pilae*. ○ Grey cheeked parakeets and rarely in Amazon parrots: sarcoptic mite *Myialges spp*. ○ Ostriches: the louse *Struthioliperirus struthionis*, also ticks (*Argus persicus*). ○ Canaries and finches: *Dermanyssus gallinae* and *Ornithonyssus sylvarium*, *Knemidokoptes pilae*, quill mites, epidermotic mites, lice. ○ Pigeons: lice—*Columbicola columbae* (wings), *Campanulotes bidentatus* (tail), feather mites (*Megninia columbae* and *Falculifer rostratus*), *Knemidocoptes mutans* (leg mange), ticks (*Argus reflexus*), bedbugs, hippoboscid flies.
○ Falcons: hippoboscid flies • **Mean age and range:** All ages are susceptible, but young birds, especially nestlings, are most susceptible to morbidity and mortality. • **Predominant sex:** Both sexes are susceptible.

SIGNS

General Comments
Clinical signs are related to the presence of parasites and their effect on skin and feathers, but also in severe cases, anemia and death can occur, especially in nestlings.

Historical Findings
In flocks, there may be a history of pruritus, restlessness, death of nestlings, reduction of egg production, identification of mites or other parasites or their droppings in or around birds or on eggs.

Physical Examination Findings
- *Knemidokoptes* infections result in honeycomb-like hyperkeratotic lesions on the beak, cere, cloaca, legs, feet. • Pruritus.
- Presence of parasites on skin, feathers.
- Dermatitis and associated lesions including flakiness, redness, sores and scabs. • Feather abnormalities including holes in feathers, lines in feathers, hemorrhagic area in quills.
- Depression. • Anemia.

CAUSES
Listed under "Signalment".

RISK FACTORS
Risk factors for developing ectoparasites are mostly related to the exposure to infested animals or materials. Mixing of domestic birds with wild populations can result in the spread of parasite infections. Overcrowding of birds increases the risk or the spread of external parasites. *Knemidokoptes* infestation might be more common in immunocompromised animals.

DIAGNOSIS

DIFFERENTIAL DIAGNOSIS
- Dermatitis due to other causes: ○ Pox ○ Dermatophytosis ○ Hyperkeratosis ○ Papillomavirus ○ Poxvirus. • Circovirus.
- Feather destructive behavior due to other causes.

CBC/BIOCHEMISTRY/URINALYSIS
Generally not done. Findings would be nonspecific, although anemia may be noted in some cases.

OTHER LABORATORY TESTS
• Skin scrapings. • Microscopic or macroscopic examination of feather preparations.

IMAGING
N/A

DIAGNOSTIC PROCEDURES
• Skin scrapings or acetate tape preparations. • Direct microscopic examination of parasites. • Fecal examination will occasionally show adults or eggs of mite or lice (due to ingestion by the host). • Sampling of feathers for preparations. • Examination of nest boxes and other area around caging to look for mites and excrement.

PATHOLOGIC FINDINGS
Honeycomb-like hyperkeratotic and acanthotic lesions of the beak, cere, legs and sometimes other areas, as well as beak abnormalities with *Knemidokoptes* infestations. Most other infestations cause skin inflammation, with erythema, dandruff, hyperkeratosis, scabbing, ulceration and bleeding. Presence of parasites may be seen. Feather lesions such as fraying or holes in the feathers may also result from damage caused by mites and lice.

TREATMENT

APPROPRIATE HEALTH CARE
Most patients with ectoparasites can be treated as outpatients.

NURSING CARE
Theoretically, would be necessary in cases of severe infestations.

ACTIVITY
N/A

DIET
N/A

CLIENT EDUCATION
Proper administration of medications as well as their side effect should be explained. In flock situations various husbandry and management practices should be outlined.

SURGICAL CONSIDERATIONS
N/A

MEDICATIONS

DRUG(S) OF CHOICE
• Ivermectin is generally used for many ectoparasites, especially *Knemidokoptes* infestations. Dosage is 0.2–0.4 mg/kg. It can be difficult to dose in small species. It is given PO, by gavage, topically or IM. It is often diluted in propylene glycol to achieve a uniform concentration; it should be properly mixed. It does not mix well with saline.

ECTOPARASITES (CONTINUED)

- Moxidectin can be used for the same at 0.2 mg/kg PO or by gavage. It can also be used topically in budgerigars, 1 mg/bird.
- Fipronil and selamectin can be used topically to treat ectoparasites in birds.
- Poultry: Permethrin can be used topically in poultry species for ectoparasites. Examples of products that contain permethrin that can be used in poultry include "Prozap Garden and Poultry Dust" and "Permectrin II"—these products have zero day meat or egg withdrawal times. If ivermectin is used in poultry, withdrawal times should be observed.
- Diatomaceous earth can be used in poultry houses and is helpful to manage internal and external parasites. In a recent study, hens with access to dust baths containing either diatomaceous earth, kaolin clay, or sulfur had a reduction in ectoparasites by 80–100% after one week when compared to hens not provided dust baths.

CONTRAINDICATIONS
Although carbaryl 5% powder (Sevin dust) or sprays have historically been used for treatment of ectoparasites in avian species, carbaryl is regulated by the EPA and has a zero tolerance policy for use in poultry/food animals.

PRECAUTIONS
Ivermectin needs to be carefully mixed and administered and should be kept away from exposure to light.

POSSIBLE INTERACTIONS
N/A

ALTERNATIVE DRUGS
Paraffin oil and mineral oil have been applied to *Knemidokoptes* lesions in an attempt to suffocate the mites; it also helps to soften some of the larger and more disfiguring lesions

 FOLLOW-UP

PATIENT MONITORING
Patients with ectoparasites should be examined at regular intervals for the presence of parasites and retreated if necessary.

PREVENTION/AVOIDANCE
Treatment of the premises with insecticidal sprays and/or powders should be performed as needed. Cleaning and disinfection of nest boxes and other materials should be done prior to breeding season.

POSSIBLE COMPLICATIONS
- Decreased egg production. • Loss of toes or beak tissue can occur in advanced *Knemidokoptes* cases in budgerigars and finches. • Secondary infections due to vector potential of some ectoparasites.

EXPECTED COURSE AND PROGNOSIS
Ectoparasites generally respond well to treatment; prognosis should be good except in cases where there is overcrowding or contaminated premises that have not been properly cleaned and/or treated; *Knemidokoptes* in finches does not always respond as well as in budgerigars, perhaps due to secondary bacterial infection.

 MISCELLANEOUS

ASSOCIATED CONDITIONS
Some species of ectoparasites can act as vectors for blood-borne diseases, such as tularemia, encephalitis, spirochetosis (borreliosis), piroplasmosis, anaplasmosis, dirofilariasis, rickettsial diseases, leucocytozoonosis, and West Nile virus.

AGE-RELATED FACTORS
N/A

ZOONOTIC POTENTIAL
There have been reports of humans becoming bitten by argassid ticks; the sticktight flea might act as a vector for human (murine) typhus; some pigeon mites and bedbugs associated with pigeons and chickens have been shown to bite humans.

FERTILITY/BREEDING
Breeding, fertility rates, and egg production could decrease in poultry that are heavily parasitized.

SYNONYMS
Knemidokoptes in budgies—scaly face and leg mite
Knemidokoptes in finches—tassle foot

SEE ALSO
Air sac mites
Cere and skin, color change
Circoviruses
Dermatitis
Feather disorders
Organophospate/carbamate toxicity
Overgrown beak and nails

ABBREVIATIONS
N/A

INTERNET RESOURCES
http://www.merckmanuals.com/vet/poultry/ectoparasites/mites_of_poultry.html
http://www.merckmanuals.com/vet/poultry/ectoparasites/lice_of_poultry.html
There are numerous other Internet websites that contain extensive information about poultry ectoparasites.

Suggested Reading
Reavill, D.R., Schmidt, R.E., Phalen, D.N. (2003). Integument. In: Reavill, D.R., Schmidt, R.E., Phalen, D.N. (eds) *Pathology of Pet and Aviary Birds*. Ames, IA: Iowa State Press, pp. 177–195.
Arends, J.J. (1997). External parasites and poultry pests. In: Calnek, B.W. (ed), *Diseases of Poultry*, 10th edn. Ames, IA: Iowa State University Press, pp. 785–814.
Doneley, R.J.T. (2009). Bacterial and parasitic disease of parrots. *The Veterinary Clinics of North America. Exotic Animal Practice*, **2**(3):423–431.
Clyde, C.I., Patton, S. (2000). Parasitism of caged birds. In: Olsen, G.H., Orosz, S.E. (eds), *Manual of Avian Medicine*. St. Louis, MO: Mosby, Inc., pp. 424–448.
Author George A. Messenger, DVM, DABVP (Avian)

Egg Yolk and Reproductive Coelomitis

BASICS

DEFINITION
The presence of egg yolk material within the coelomic cavity outside of the reproductive tract. It can encompass egg-related peritonitis, ectopic ovulation-associated coelomitis and septic coelomitis.

PATHOPHYSIOLOGY
• Egg yolk coelomitis is a commonly seen reproductive disorder in pet birds. • An ovulated follicle (yolk) fails to enter the reproductive tract, inducing an inflammatory response, peritonitis, and secondary bacterial infection. • Ascites, organ inflammation, and fibrinous peritonitis can result. • Yolk coelomitis causes ascites, which causes dyspnea from fluid space occupation. • Yolk coelomitis can be life threatening.

SYSTEMS AFFECTED
• Reproductive—ovulated yolks do not enter the reproductive tract and are free in the coelomic cavity. • Respiratory—dyspnea, increased respiratory effort, or distress occur from fluid accumulation or space occupation. • Endocrine/Metabolic—ovulating hens typically show hypercalcemia, hyperglobulinemia, and elevated lipids in the blood stream.

GENETICS
It is not definitively known although it is speculated this may be a hereditary condition and certain color mutations may be predisposed.

INCIDENCE/PREVALENCE
• There are no scientific data based on large scale studies; however, the condition could develop in any female bird. • Most commonly seen in smaller species, typically cockatiels.

GEOGRAPHIC DISTRIBUTION
N/A

SIGNALMENT
• **Species predilections:** Cockatiels, budgerigars, waterfowl. • **Mean age and range:** Sexually mature hens of egg producing age (varies greatly between species). • **Predominant sex:** Only affects females past egg-laying age.

SIGNS

General Comments
Clinical signs may be vague and nonspecific, such as lethargy and inappetence, or directly attributable to development of peritonitis, ascites, and coelomic distension. Birds may have varying degrees of yolk peritonitis and be unaffected and still show typical clinical signs and physical examination findings.

Historical Findings
• Bird may have a reproductive history that includes past or chronic egg laying. • Recent egg laying or recent cessation of egg laying are commonly seen, birds may still be exhibiting hormonal behaviors. • Owners may report loss of appetite, staying at the bottom of the cage, and increased sleeping. • Nonspecific signs, such as loss of appetite/anorexia, weakness, depression/lethargy, fluffed feathers, and decreased-to-absent vocalizations are reported. • Sudden death.

Physical Examination Findings
• Coelomic distension is the most common exam finding. • Ascites is typically identified as both the cause of distension and dyspnea. • Dyspnea/hyperpnea are observed due to space occupation with fluid or organ enlargement and air sac compression and may become worse with handling. • Lethargy and depression may be noted. • Wide-based stance is sometimes seen. • Enlarged, flaccid vent can be seen. • Hen may be obese, thin, or normal body condition.

CAUSES

Reproductive
• Aberrant reproductive behavior. • Chronic/excessive egg laying. • Abnormal ovulation. • Cystic ovarian disease. • Oviductal and ovarian granulomas. • Reproductive tract diseases, such as salpingitis, metritis, oviductal rupture, and neoplasia.

Other
• Genetic predisposition. • Obesity. • Lack of exercise. • Stress.

RISK FACTORS
• Only female birds can develop the condition. • Uncontrolled excessive egg laying resulting from prolonged daylight, lack of sleep, presence of real or perceived mate, and nest sites.

DIAGNOSIS

DIFFERENTIAL DIAGNOSIS
• Coelomic distension—ascites secondary to hepatic disease, cardiac disease, and neoplasia. • Abdominal wall herniation. • Coelomic bacterial infection, abscessation of the coelomic organs. • Dyspnea secondary to primary respiratory disease, tissue space occupation, pain/discomfort. Radiographs will aid in determination. • Neoplasia of the reproductive tract or other organs.

CBC/BIOCHEMISTRY/URINALYSIS
• Leukocytosis with heterophilia is not uncommonly seen on the complete blood count. • Hypercalcemia is a normal finding in an egg-laying hen; values within normal reference ranges indicate deficiency. • Elevated hepatic enzyme activity from reactivity or related to egg production. • Hyperuricemia may be associated with reproductive tract enlargement or previous egg laying. • Lipemic plasma is seen clinically with elevated lipids related to egg production.

OTHER LABORATORY TESTS
• Hyperglobulinemia may be noted on protein electrophoresis. • Celiocentesis for fluid analysis and cytologic examination to obtain a definitive diagnosis. Aspirated fluid can be stained and evaluated in-house for the presence of protein globules indicative of egg yolk. • Culture and sensitivity testing of coelomic fluid samples may aid in treatment.

IMAGING

Radiographic findings
• Polyostotic hyperostosis is typically seen as increased ossification of the long bones relating to calcium storage in the hormonally active hen. • Radiography is also useful to determine the presence of a shelled egg. • Contrast radiography is useful and may show a space occupying mass and organ displacement.

Abdominal/coelomic ultrasonography
• Ultrasound will show coelomic fluid, ovarian cysts/follicles, and masses. • Ultrasonography is useful for ruling out other disease, such as tumors, reproductive tract torsions, and any other concurrent disease conditions.

CT scan
• Will show organ enlargement or presence of organ disease, especially ovarian or uterine.

DIAGNOSTIC PROCEDURES
• Sample collection for cytologic analysis is diagnostic showing a septic exudate with yolk globules. • Endoscopy may be utilized in larger birds to visualize the oviduct to assess for torsion or masses. Additionally, eggs located ectopically within the coelom may be seen.

PATHOLOGIC FINDINGS
• Ascites secondary to peritonitis. • Fibrinous peritonitis. • Organomegaly. • Visible yolk material in the coelomic cavity. • Presence of an egg(s) in the reproductive tract.

TREATMENT

APPROPRIATE HEALTH CARE
• Treatment varies based on severity, as some patients will resorb the yolk without complication. • Outpatient treatment with home care instructions may be warranted in stable patients. • Hospitalization is needed in patients that are anorexic, severely dyspneic, or showing severe clinical signs. • Surgery consisting of a salpingohysterectomy may be indicated in patients with chronic reproductive disease.

NURSING CARE
• Hen should be kept in a humidified incubator warmed up to 90°F; ideally in a quiet area to minimize stress. • Fluid supplementation—parental fluids (e.g. subcutaneous); oral fluids though gavage tube

EGG YOLK AND REPRODUCTIVE COELOMITIS (CONTINUED)

as feedings. • Pain management with analgesics and anti-inflammatory medications. • Broad spectrum antibiotics when infection is suspected. • Supportive care should include celiocentesis as indicated. • Hormonal treatment with Leuprolide acetate may help some patients. • Some patients may benefit from short-term oxygen therapy.

ACTIVITY
Patients should be cage rested and monitored for appetite and levels of dyspnea.

DIET
• No changes should be made until the patient has recovered, at which time needed improvements can be addressed. • Hospitalized patients will require assisted feedings with clinical care or hand-rearing formula. • Transition to a diet that included formulated pellets and calcium supplementation after recovery.

CLIENT EDUCATION
• Clients should understand some patients have a guarded prognosis. • Attempts to prevent or decrease further egg laying are encouraged. Reduction of photoperiod, increase sleep time, removal of nests and nesting materials, and reduction in sexual stimulation by a perceived mate is recommended. • Hormonal therapy with Leuprolide acetate or deslorelin implants may be warranted.

SURGICAL CONSIDERATIONS
• Dehydrated and anorexic patients need appropriate pre-anesthetic care. • Compromised ability to breathe due to compression of air sacs by egg and enlarged reproductive tract. Intubation with positive pressure ventilation is strongly recommended. • Patients with ascites are at increased risk of anesthetic complications should they be placed under anesthesia.

MEDICATIONS

DRUG(S) OF CHOICE
• Antibiotic therapy should be based on culture and sensitivity results; however, antibiotics such as fluoroquinolones (enrofloxacin 15 mg/kg PO q12h) or potentiated penicillins (amoxicillin with clavulanic acid 125 mg/kg q8–12h), can be initiated. • Analgesics such as butorphanol (1–4 mg/kg IM q2–4h to PRN) are needed during initial treatment.

• Anti-inflammatories such as meloxicam (0.5–2 mg/kg PO, IM q12–24h) are useful and should be continued for at least five days after diagnosis. • GnRH agonists such as Leuprolide acetate (100–1200 μg/kg IM q3–4 weeks) are helpful in both short- and long-term management. Note that doses vary widely with this medication.

CONTRAINDICATIONS
N/A

PRECAUTIONS
N/A

POSSIBLE INTERACTIONS
N/A

ALTERNATIVE DRUGS
• Various antibiotics may be useful, and may be chosen based on culture and sensitivity testing. • Deslorelin implants can be used for long-term prevention of ovulation and egg laying. • Holistic therapies can be recommended by specialized veterinarians as adjunct therapy.

FOLLOW-UP

PATIENT MONITORING
• Physical examination to determine degree of ascites accumulation and recurrence pattern. • Celiocentesis may be required periodically until the condition resolves.

PREVENTION/AVOIDANCE
• Owners should take steps to discourage and/or minimize egg laying. • Treatment with Leuprolide acetate injections or a deslorelin implant to prevent ovulation and egg laying.

POSSIBLE COMPLICATIONS
• Sepsis can occur from bacterial translocation and fluid accumulation. • Adhesions may result from fibrinous peritonitis. • Hens that have undergone salpingohysterectomy may still be at risk for developing yolk associated coelomitis. • Sudden death.

EXPECTED COURSE AND PROGNOSIS
• Patients diagnosed quickly have a good prognosis with appropriate medical care and owner compliance. • Prognosis is guarded in all patients that present with severe dyspnea. • Patients who survive surgery have an excellent prognosis.

MISCELLANEOUS

ASSOCIATED CONDITIONS
• Ectopic eggs • Adhesion formation.

AGE-RELATED FACTORS
N/A

ZOONOTIC POTENTIAL
N/A

FERTILITY/BREEDING
Although this can be a random occurrence, birds who experience recurrences should not be used for breeding.

SYNONYMS
Egg yolk peritonitis

SEE ALSO
Atherosclerosis
Chronic egg laying
Cloacal disease
Coelomic distention
Cystic ovaries
Dyslipidemia
Dystocia / egg binding
Hypocalcemia / hypomagnesemia
Lameness
Metabolic bone disease
Nutritional deficiencies
Osteomyelosclerosis / polyostotic hyperostosis
Problem behaviors
Regurgitation / vomiting
Salpingitis / uterine disorders
Sick-bird syndrome

ABBREVIATIONS
EPH – protein electrophoresis

INTERNET RESOURCES
http://www.lafebervet.com/avian-medicine-2/avian-emergency-medicine/reproductive-emergencies/

Suggested Reading
Bowles, H.L. (2002). Reproductive diseases of pet bird species. *The Veterinary Clinics of North America. Exotic Animal Practice*, **5**(3):489–506.
Bowles, H.L. (2006). Evaluating and treating the reproductive system. In: *Clinical Avian Medicine*, Vol 2, Harrison, G., Lightfoot, T. (eds). Palm Beach, FL: Spix Publishing, 519–540.
Pollock, C.G., Orosz, S.E. (2002). Avian reproductive anatomy, physiology and endocrinology. *The Veterinary Clinics of North America. Exotic Animal Practice*, **5**(3):441–474.

Authors Anthony A Pilny, DVM, DABVP (Avian)

EMACIATION

BASICS

DEFINITION
Starvation is the prolonged absence of nutrition, the physiologic result of starvation is emaciation. Patients can become emaciated as a result of malnutrition. Malnutrition is defined as any disorder with unbalanced or inadequate nutrition associated with nutritional deficiencies or excess. Malnutrition leads to immunocompromise, altered drug metabolism and lower rates of tissue synthesis and repair.

PATHOPHYSIOLOGY
This condition is defined as the loss of body fat and muscle as a result of severe malnourishment or starvation. During starvation normal blood glucose is maintained by increasing hepatic glycogenolysis, increasing gluconeogenesis and decreasing glycogen stores. Animals decrease blood insulin concentrations and increase blood glucagon concentrations. Glucose is the initial, primary fuel source for the central nervous system, bone marrow and injured tissue. In the absence of nutrition intake, glycogen stores are rapidly depleted within 24 hours of a fast. New glucose is derived from gluconeogenesis via the breakdown of protein from organs like the liver and also muscle tissue. Once hepatic glycogen stores are depleted, amino acids become the energy source for gluconeogenesis. Muscle catabolism releases glucogenic amino acids, pyruvate and lactic acid for energy production in the liver. Amino acids are obtained from the breakdown of skeletal muscle and visceral proteins. Eventually the liver uses immunoglobulins and lymphokines as a protein source for energy, compromising immune function. Severe fatigue accompanies emaciation due to a severe depletion of muscle mass along with most vitamins, minerals and glucose required for homeostasis.

SYSTEMS AFFECTED
- Cardiovascular • Digestive
- Neuromuscular.

GENETICS
No genetic associations exist.

INCIDENCE/PREVALENCE
Specific incidence data are lacking but emaciation can be seen in nestlings of first-time parents who are not feeding appropriately. Emaciation can also be seen in adult, injured raptors that are unable to hunt and young-of-the-year raptors who are not skilled hunters.

GEOGRAPHIC DISTRIBUTION
No geographic distributions exist.

SIGNALMENT
- **Species predilections:** Birds of any species can present emaciated. • **Mean age and range:** Birds of any age can present emaciated.
- **Predominant sex:** Birds of any sex can present emaciated.

SIGNS
Ravenous or no appetite, scant fecal output.

General Comments
All seed diets where only hulls are left in the food bowl and foods that are allowed to become moldy or rancid in outdoor enclosures are frequently noted in cases of emaciation. Birds of prey may become emaciated as a result of decreased prey availability and/or extreme weather conditions. Raptors are tolerant of acute food deprivation for various amounts of time. Larger species are more tolerant than smaller species.

Historical Findings
Lethargy, depression, anorexia, decreased vocalization/energy, smaller than normal stools, undigested seeds in stools.

Physical Examination Findings
Loss of muscle around the keel, muscle wasting, extreme weight loss, decreased/absent grip, inability to fly and/or sit upright.

CAUSES
- Degenerative/developmental: Neonatal crop atony.
- Anomalous/autoimmune
- Metabolic/mechanical: Foreign body, impaction, compression of crop, nematodiasis or other internal parasites.
- Nutritional/neoplastic: Cancer cachexia, neoplastic conditions, liver disease, and renal disease.
- Inflammatory/infectious: Proventricular dilation syndrome, macrorhabdus, gastritis, septicemia, papilloma virus, rancid hand feeding formula/produce, coccidiosis, giardiasis.
- Traumatic/toxic: Lead poisoning, gastric ulcer, aggressive gavage feeding, thermal trauma due to overheated (microwaved) hand feeding formula, trauma (i.e., vehicular/gunshot) that renders birds flightless or unable to hunt.

RISK FACTORS
Fledgling birds being hand fed, birds housed in outdoor aviaries, young raptors that have not learned to hunt prey successfully.

DIAGNOSIS

A thorough patient history and a minimum database (physical examination, comprehensive fecal examination, CBC and chemistry) are essential to diagnose causes for emaciation. Normal gastrointestinal resident microflora maintain an acidic environment which inhibits the proliferation of Gram-negative rods and yeast.

DIFFERENTIAL DIAGNOSIS
Critical illness where birds are in a hypermetabolic state and unable to take in adequate nutrition, eventually becoming hypoglycemic and acidotic.

CBC/BIOCHEMISTRY/URINALYSIS
Hypophosphatemia, hypokalemia, hypomagnesemia, hypoglycemia, hypoproteinemia, acidosis.

OTHER LABORATORY TESTS
Anemia may be present.

IMAGING
N/A

DIAGNOSTIC PROCEDURES
N/A

PATHOLOGIC FINDINGS
Atrophied pectoral musculature with minimal adipose stores (lack of coelomic and cardiac fat).

TREATMENT

APPROPRIATE HEALTH CARE
Emaciation can often occur with dehydration so appropriate care addresses both fluid and nutritional requirements. Emaciated patients are often dehydrated with concurrent slowed digestion times so fluid imbalances should be addressed first.

NURSING CARE
- Fluid deficits should be corrected before supplemental, enteral nutrition is offered. The author uses Lafeber's emerald (Lafeber Company, Ordell, IL) for parental nutritional support of critically ill psittacine patients. When critically ill patients are being treated, gavage/tube feeding should always be done last and the patient placed immediately back in its cage. Regurgitation and resultant aspiration pneumonia are always risks associated with gavage feeding. It is always better to feed small amounts frequently (2–4 times daily) as opposed to overfilling the crop.
- In order to determine daily nutritional support required, start by calculating the basal metabolic rate: Nonpasserine: kcal/day = $73.5 \times kg^{0.734}$; passerine: kcal/day = $114.8 \times kg^{0.726}$. Next calculate maintenance requirements as $1.5 \times BMR$; adjust maintenance requirements for stress. The adjustments to nutritional maintenance for stress are represented by the following multiples of maintenance energy requirements:
  ○ Starvation: 0.5–0.7, Elective surgery: 1.0–1.2, Mild trauma: 1.0–1.2, Severe trauma: 1.1–2.0, Growth: 1.5–3.0, Sepsis: 1.2–7.5, Burns: 1.2–2.0, Head injuries: 1.0–2.0.

EMACIATION (CONTINUED)

Kcal/day divided by Kcal/mL of formula equals the amount of formula in mL required per day.

- Once the enteral diet has been prepared appropriately for the species in question, the bird should be restrained upright by an assistant. The assistant will need to gently straighten the neck so that the esophagus will also be straightened and tube passage facilitated. The feeder should palpate the crop (when present) at the thoracic inlet to be sure it is empty prior to introducing the tube. Starting from the bird's left, direct the tube toward the right and down into the crop. When properly placed the tube goes easily and should never be forced. Ball tipped tubes will displace feathers and can be seen in place through the skin of the neck. Tube feeding should be done in a slow and steady pace, if food is seen swelling at the back of the throat the procedure should be halted immediately and the bird put down right away. Though tube feeding is a common practice in avian and exotic practices, it should never be done hurriedly or haphazardly. The patient's safety must take top priority at all times.
- For birds of prey determine minimum daily nutritional support required by calculating the basal metabolic rate: (BMR) = $K(W^{0.75})$ is one formula that can be used for this. K is the amount of kilocalories required for 24 hours and is 78 for birds of prey. W represents the weight of the bird in kilograms. Fresh water should be available to convalescing patients at all times. Intravenous fluid therapy is recommended for patients unable to sit upright in cages or drink normally from a dish. As small amounts of prey are reintroduced to the diet, 1 g of prey fed can be roughly approximated to equal 1 kcal. Allometric scaling, known also as the determination of the daily minimum energy cost (MEC) is similar to the basal metabolic rate in determining the daily caloric patient needs.
- For emaciated birds of prey, appropriate supplemental diets include, among others, Emeraid Carnivore (Lafeber Company, Ordell, IL). Enteral diets should be diluted with an isotonic electrolyte solution and warmed. Never use a microwave to warm enteral diets because of the risk of varying "pockets" of warmth in the food which can lead to esophageal and crop thermal injury.
- Initially, 3% of the bird's body weight in grams should be offered in milliliters of enteral diet.
- If tolerated, the volume of the enteral diet can be increased; do not exceed 5% of the patient's body weight.
- If the enteral diet is being well tolerated and the patient is gaining weight, whole prey items void of casting material (fur/feathers) can be offered in small, frequent amounts so that $\frac{1}{4}$ of the MEC is fed in 3–4 feedings and held down.
- Feedings of MEC should be increased by $\frac{1}{4}$ of the MEC daily until the bird is eating $2\times$ MEC to promote weight gain.
- If the bird is passing normal feces and gaining weight, casting material can be reintroduced at this point.

ACTIVITY
Hospitalized patients should have exercise restrictions not only to allow patients time to regain their strength but also to protect indwelling catheters.

DIET
As compared to parenteral nutrition, human research has concluded that enteral nutrition is associated with reduced costs, better nutritional outcomes, improved wound healing and less mucosal permeability.

CLIENT EDUCATION
Clients should be encouraged to provide food daily and completely discard old food as opposed to adding new food on top of old food.

SURGICAL CONSIDERATIONS
N/A

MEDICATIONS
DRUG(S) OF CHOICE
N/A
CONTRAINDICATIONS
N/A
PRECAUTIONS
N/A
ALTERNATIVE DRUGS
N/A

FOLLOW-UP
PATIENT MONITORING
Both gram and baby scales are inexpensive and can easily be fitted with a perch to obtain patient weights daily. During initial recovery patients are expected to stop losing weight and then weight should steadily increase. The patient's attitude and fecal output should be monitored daily as well. Fecal exams should be done to rule out intestinal parasitism.

PREVENTION/AVOIDANCE
For psittacine birds at home, owners should be encouraged to monitor food intake by completely changing out foodstuffs daily.

POSSIBLE COMPLICATIONS
N/A

EXPECTED COURSE AND PROGNOSIS
If an underlying cause is identified and treatable the prognosis is good-to-excellent. Neoplasias typically do not have a favorable prognosis.

MISCELLANEOUS
ASSOCIATED CONDITIONS
Refeeding Syndrome: In human medicine, refeeding syndrome affects malnourished patients receiving enteral or parenteral nutrition. During starvation, people use protein and fat as opposed to carbohydrates for their main energy sources. Refeeding syndrome is characterized by fluid and electrolyte shifts that occur following increased insulin and decreased glucagon secretion in response to suddenly available glucose. During this time the body switches from a catabolic to anabolic state. Overfeeding, excessive carbohydrate feeding and excessive fat administration have all been associated with negative outcomes in human patients.

Birds are shown to have lower insulin and higher glucagon levels than mammals. Glucagon appears to be the main hormone regulating avian glucose metabolism. Carnivorous birds have lower insulin levels than chickens. For these reasons direct correlations cannot be made between nutritional support requirements of humans and birds of prey. Pertinent similarities between human patients and birds of prey are however apparent. Chronically malnourished patient support must be done methodically. Supplemental heat and fluid therapy are the first order of business in the avian patients' supportive care needs. Once fluid deficits have been addressed over 12–24 hours, small volumes of easily digestible food should be offered over multiple (3–4) feedings per day. Thiamine (1–2mg/kg) containing B vitamin complexes may be a beneficial supplement to improve patient metabolism.

AGE-RELATED FACTORS
Young raptors that have not learned to hunt prey successfully may be more susceptible to emaciation as well as those facing harsh winter conditions.

ZOONOTIC POTENTIAL
Emaciation does not have zoonotic potential.

FERTILITY/BREEDING
Because emaciation can impact all systems and organs in the body, fertility and breeding can be negatively impacted.

SYNONYMS
Starvation, Malnutrition

SEE ALSO
Anemia
Anorexia
Dehydration
Ingluvial hypomotility/ileus
Nutritional deficiencies

ABBREVIATIONS
N/A

(CONTINUED)

INTERNET RESOURCES
http://emeraid.com/critical-care-nutrition
http://emeraid.com/downloads/Emeraid_Feeding_Guidelines.pdf

Suggested Reading
Murray, M. (2014). Raptor gastroenterology. *The Veterinary Clinics of North America. Exotic Animal Practice*, **17**:211–234.
Orosz, S. (2013). Critical care nutrition for exotic animals, *Journal of Exotic Pet Medicine*, **22**:163–177.
Tully, T.N. (2000). Psittacine therapeutics, *The Veterinary Clinics of North America. Exotic Animal Practice*, **3**:59–90.

Author Kemba L Marshall, DVM, DABVP (Avian)

Acknowledgement The author wishes to acknowledge the teachers, professors and patients who have allowed her to learn and live out the dreams she had as the eight-year-old version of herself.

ENTERITIS AND GASTRITIS

BASICS

DEFINITION
Gastritis and enteritis are common conditions in pet birds. Sometimes they are part of a more systemic condition, and other times they are isolated problems. Gastritis is any inflammatory condition of the stomach. In birds, there are two sections of the stomach: the proventriculus or glandular stomach, and the ventriculus or gizzard, which is the muscular stomach. Enteritis is any inflammatory condition of the intestines. Typhlitis is inflammation of the ceca. Many pet birds lack ceca, but typhlitis may be evident in species where ceca are present.

PATHOPHYSIOLOGY
Gastritis and enteritis are caused by a wide array of infectious agents. Their pathophysiology varies somewhat, but generally involves some degree of inflammatory cell infiltrate and various degrees of tissue damage to the mucosa, villi, or crypt cells. In some cases, inflammation can occur without infectious agents being present. Proventricular dilatation disease is a common clinical condition involving the stomach. However, strictly speaking, the disease involves neuritis rather than the stomach itself; the splanchnic ganglia and nerves are usually involved, causing poor gastric tone. In some cases, secondary infections are caused by alterations in motility caused by foreign bodies, papillomas, or other mechanical obstructions.

SYSTEMS AFFECTED
- Behavioral changes involving the appetite, such as anorexia, hyporexia, or polyphagia, can occur with gastritis or enteritis.
- Cardiovascular changes may include hypovolemia in severe cases.
- Hepatobiliary infection may occur secondary to enteritis, either via the biliary tree, portal circulation, or septicemia.
- Musculoskeletal signs may include weight loss and weakness.
- Feathers may be matted around the vent of the bird and there may be a generally unkempt appearance of the feathers.

GENETICS
There does not appear to be a genetic component to enteritis and gastritis.

INCIDENCE/PREVALENCE
Since they are broad categories of disease, enteritis and gastritis are common. Specific diseases vary in their prevalence.

GEOGRAPHIC DISTRIBUTION
Avian enteritis and gastritis are worldwide conditions. Individual diseases have more variation in their distribution.

SIGNALMENT
- **Species:** All birds of all ages are susceptible to some forms of gastritis or enteritis. Individual diseases have more specific predilections. For example, Macrorhabdus infections are common in budgerigars. Candidiasis is most common in young birds, while mycobacteriosis is most common in older birds.

SIGNS
Diarrhea is the most common sign associated with enteritis. It should be distinguished from polyuria, which is common with many other conditions. Melena may occur with enteritis or gastritis. This suggests a severe and erosive condition. Hematochezia may occur with enteritis of the more distal intestines. It can also occur with cloacal disorders. Fetid fecal odor may occur with some types of enteritis or gastritis. This is especially common with clostridial enteritis. Mucus-laden stools may occur, especially with lower intestinal inflammation. Matting of the vent area with feces or urates will be commonly seen with these conditions. Coelomic pain or discomfort may occur. This often presents as very vague signs of malaise and anorexia. Undigested grains or seeds may be passed in cases of gastritis. Generally grains are not released from the ventriculus until they are ground to a fine particle size. Passage of whole grains is a sign of dysfunction of the ventriculus. Although not consistently seen, vomiting may occur with gastritis. If ingesta backs up into the crop, regurgitation may occur. Dehydration and weight loss can occur due to the loss of fluids and nutrients.

Historical Findings
Many of the signs of enteritis and gastritis are readily evident to the owner. Regurgitation, vomiting, diarrhea, and fetid fecal odor are all relatively obvious signs. Weight loss can be easily missed by owners. Anorexia or polyphagia may be noted by astute owners.

Physical Examination Findings
The physical examination should include inspection of the droppings. This may reveal loose stools, melena, hematochezia, strong odor to the feces, undigested grains or other changes to the fecal portion of the droppings. The crop may be thickened or fluctuant on palpation. Fecal or urate matting of the pericloacal feathers may be found. Loss of muscle mass, poor feather condition, or other signs of poor nutrition may be noted.

CAUSES
Proventricular dilatation disease caused by avian bornavirus (ABV) should be mentioned here. However, strictly speaking this is neuritis rather than a true gastritis. Mycobacteriosis can occur anywhere in the gastrointestinal tract and systemically. In birds, the gastrointestinal tract is the most common entry site for the organism. Gram-negative bacteria including *Salmonella spp.*, *E. coli*, *Yersinia pseudotuberculosis*, and others can be either primary or opportunistic infections of the stomach or intestines. *Clostridium spp.* commonly infects the intestinal tract. Depending on the strain and the host, the disease can be mild (loose, fetid feces), to extremely severe (hemorrhagic enteritis). *Campylobacter spp.* can cause enteritis and occurs most commonly in passerine species. *Macrorhabdus ornithogaster* or avian gastric yeast (AGY) is one of the most common causes of gastritis in small species of birds. Candidiasis is a common gastric and intestinal infection. Although it usually causes little inflammation, it is still considered gastritis or enteritis, depending on the site of infection. Coccidiosis is common in some nonpsittacine species, but is uncommon in most pet species. It is a concern for backyard poultry, and some zoo or aviary birds. Giardiasis is somewhat common in budgerigars, but rare in other pet species. *Giardia psittaci* is the causative organism. Hexamitiasis (spironucleosis) is common in cockatiels and a few nonpsittacine birds. Ascarids, *Capillaria spp.*, *Heterakis gallinarum* (cecal worm), and other helminth parasites can affect the stomach and intestines. Heavy metal ingestion can cause erosive changes in the stomach that often result in secondary inflammation.

RISK FACTORS
Poor sanitation increases transmission of parasites and primary pathogens, as well as increasing exposure to opportunistic pathogens. In particular, food and water sanitation is critical. The use of moist, high carbohydrate diets, such as lory nectar diets or hand-feeding formula, promotes bacterial growth. Immunocompromise resulting from underlying viral diseases, such as psittacine beak and feather disease (PBFD), stress, and poor nutrition, allows opportunists to cause disease. Conditions promoting fecal retention, such as egg laying, cloacal prolapse, or cloacal papilloma often result in secondary enteritis. High population density increases contact with primary infectious agents.

DIAGNOSIS

DIFFERENTIAL DIAGNOSIS
Since both enteritis and gastritis are categories of disease, most differentials include etiologies of these conditions. However, there are a few disorders that have similar signs without inflammation of the gastrointestinal tract. Such conditions may include hepatic disease, neoplasia, or dietary indiscretion.

CBC/BIOCHEMISTRY/URINALYSIS
Hematology can vary from normal to having severe leukocytosis, depending on the etiology of the condition. Mycobacteriosis often results in dramatic leukocytosis. Dehydration may result in slight elevations in the hematocrit. Gastrointestinal bleeding or chronic inflammation can result in anemia. Chemistry changes also vary substantially. If

there is significant muscle wasting, creatine kinase (CK), aspartate transaminase (AST), and uric acid may have mild-to-moderate elevations. Protein can be elevated if there is dehydration, or can be reduced if there is protein loss in the gastrointestinal tract. Other chemistries are not likely to be substantially altered, unless there is involvement of other organs in the disease process.

OTHER LABORATORY TESTS
A variety of other diagnostic tests may aid in the diagnosis of gastritis and enteritis. Addition of iodine or Sudan stain to the feces can reveal undigested starches (amylorrhea) or fats (steatorrhea) respectively, suggesting maldigestion. Simple tests may reveal cellular changes or organisms that may shed light on the etiology. Direct saline preparations of feces or other gastrointestinal (GI) samples may help identify flagellates, *Macrorhabdus* yeasts, or other distinctive organisms. Gram-stained preparations of the same samples help to delineate the flora further. Culture can help further identify specific bacteria and determine their antimicrobial susceptibility Other tests that may be performed on feces or other GI samples include special stains (e.g., acid fast), and polymerase chain reaction (PCR) tests for a variety of viral, bacterial, fungal, or parasitic agents.

IMAGING
Since palpation is often very limited in birds, imaging is critical to developing a full evaluation of the digestive tract. Radiography is helpful in identifying anatomic changes in the gastrointestinal tract. Survey radiographs are often adequate for evaluation, but contrast studies can be used when needed to improve identification of the GI tract. Fluoroscopy can give real-time information on the motility of the GI tract when used with contrast. Ultrasound is often limited by artifacts caused by the air sac system of birds. However, if there is coelomic effusion or swelling, better views may be obtained. Advanced imaging techniques such as plain or contrast computed tomography (CT) can give very detailed information about anatomical changes which can suggest particular etiologies. For example, thickening of the proventricular or enteric walls can suggest infiltrative diseases such as mycobacteriosis.

DIAGNOSTIC PROCEDURES
Although a diagnosis can frequently be gained by simpler diagnostic samples, collection of gastric washes, either with or without the use of an endoscope, may be indicated in some cases. Histopathology of the intestine is infrequently performed anetmortem because of the high morbidity and mortality of such procedures. Crop biopsies are often used to diagnose proventricular dilatation disease (PDD).

PATHOLOGIC FINDINGS
Since this is a spectrum of diseases, the pathologic findings can be highly variable. However, inflammation of the intestine, proventriculus, or ventriculus is how these diseases are defined. Grossly, this may be represented by thickened walls, edema, hyperemia, distension, hemorrhage, erosions, or ulcers.

TREATMENT
Gastritis and enteritis encompass a wide variety of diseases and the treatments will depend somewhat on the etiology. However, there are some common threads when designing treatment plans for this type of health problem.

APPROPRIATE HEALTH CARE
Severely affected patients may require hospitalization for stabilization of their condition. Indications for hospitalization include anorexia, vomiting or regurgitation, severe fluid loss, secondary aspiration pneumonia, or other life-threatening conditions. If there are advanced treatments required that the owner is unable to perform, inpatient care may be advisable until the owner can be instructed to do them or they are no longer required. Outpatient care is appropriate for more mildly affected patients with treatments that are within the comfort level of the caregiver.

NURSING CARE
In all but the mildest cases, fluid therapy is appropriate. The degree of dehydration should be estimated and that deficit should be added to the maintenance levels. Crystalloid fluids will be sufficient in mild-to-moderate cases. The route of administration depends on perfusion and the tolerance of the gastrointestinal tract. Routes of administering fluids include intravenous (IV), intraosseous (IO), subcutaneous (SC), and crop gavage. Severely affected patients, and those that have low protein levels, may benefit from colloids, which must be given IV or IO. Nutritional support can prove challenging in these cases, as the alimentary tract may not tolerate regular food. Although parenteral nutrition can be used for a short time, alimentation should start early. The use of antiemetic drugs can be used to facilitate this. Small, frequent feedings are tolerated best. While balanced nutrition is important, the diet may have to be simplified when the GI tract is distressed. In the short term, calories and protein can be provided by a simple rice or oatmeal baby cereal diet. As the GI tract heals, more complex dietary components can be added.

ACTIVITY
While there is no reason to specifically restrict activity, some birds may benefit from added warmth, so a smaller hospital cage with supplemental heat is often appropriate. Additionally, having a small enclosure where food and water are readily accessible will benefit many sick birds.

DIET
Prolonged fasting is not practical or advisable for most birds. Brief withholding of oral fluids and food may be indicated for patients that are vomiting or regurgitating. When returning to oral fluids and food consumption, it should be done gradually. For less severe conditions, dietary modifications may require extensive modification or none at all, dependent on the etiology of the condition and the current diet of the patient. The long-term goal is to have a nutritionally balanced diet that is tolerated by the GI tract of the bird, and that promotes a healthy and balanced microbial flora. The use of soft, moist, high-carbohydrate foods tends to promote extensive bacterial and yeast growth that occasionally result in GI disease. If these dietary components are deemed essential to the diet of the bird, they should be offered for only brief periods of time. Mycobacterial infections and other conditions that result in weight loss or malabsorption may necessitate an increase in caloric intake.

CLIENT EDUCATION
Clients should be educated about any potentially zoonotic diseases, such as mycobacteriosis, salmonellosis, and others. Sanitation is often a factor in the development of these diseases, and clients should be taught appropriate sanitation procedures.

SURGICAL CONSIDERATIONS
Gastritis and enteritis are generally treated medically. However, there are certain surgical procedures that may facilitate diagnosis or treatment. Endoscope-assisted gastric lavage (EAGL) may be performed, often in conjunction with a crop incision and biopsy, to gain a diagnosis and aid in the treatment of gastric conditions. The placement of feeding tubes can facilitate oral treatments and alimentation in selected cases. Generally, esophagostomy tubes are most applicable for this purpose. However, if temporarily bypassing the proventriculus and ventriculus would be beneficial for the patient, a duodenostomy tube can placed.

MEDICATIONS
DRUG(S) OF CHOICE
Because of the wide variety of diseases included in this category, there are potentially hundreds of drugs that could be appropriate for treatments. In addition, dosages vary between species of birds. Individual treatment recommendations for all the diseases in this category would be beyond the scope of this chapter. However some generalizations can be

ENTERITIS AND GASTRITIS (CONTINUED)

made. • Eliminate pathogens and restore the microflora balance in the GI tract. ◦ Antimicrobial or antiparasitic drugs should be used to treat birds with gastrointestinal bacterial, fungal, protozoal, or helminth infections. Mycobacterial infections require multiple concurrent antibiotics. Many birds have mixed infections and may require combination therapies. It is important to consider anaerobes such as clostridium that may not be recognized on standard cultures as potential pathogens. In some cases, probiotics can help restore appropriate microbial flora. • Protect the GI mucosa from further damage. ◦ Histamine-2 (H2) blockers such as ranitidine can be used to reduce acid damage to the stomach and small intestine. ◦ Adsorbents such as bismuth subsalicylate, kaolin, or similar products coat the mucosa and absorb bacterial toxins. Cholestyramine may absorb bacterial toxins more effectively. Sucralfate adheres to exposed submucosal tissues and provide a protective layer. • Control inflammation. ◦ Control of inflammation should proceed carefully as many systemic anti-inflammatory drugs can predispose to gastric erosions and ulcers. Bismuth subsalicylate can topically treat the intestine. Nonsteroidal anti-inflammatory drugs (NSAIDs) may be used carefully, especially when there is nonspecific inflammation without GI bleeding. Corticosteroids should generally be avoided.

CONTRAINDICATIONS
There are no specific contraindications for this category of disease.

PRECAUTIONS
As described, NSAIDs should be used cautiously and only after the patient has had hydration and perfusion restored.

POSSIBLE INTERACTIONS
As many drugs have GI signs as potential side effects, it can be difficult to monitor for adverse effects.

ALTERNATIVE DRUGS
As there are numerous drugs in each category of therapy, this should not prove to be an issue.

FOLLOW-UP
PATIENT MONITORING
During hospitalization, weight, appetite, regurgitation or vomiting, and fecal quality should all be monitored. Once stable and sustaining weight with voluntary intake, the patient can be discharged. Follow-up timing depends upon the chronicity of the condition. Often rechecks should be scheduled at 1–4 week intervals. Parameters to be monitored include physical examination, weight, stool quality and appetite. Any noninvasive laboratory findings that contributed to the diagnosis should be followed until they return to normal. Repeating more invasive diagnostics may be indicated in some cases, but should be weighed against the risks involved.

PREVENTION/AVOIDANCE
As a diverse collection of disorders, only general steps can be recommended. Exposure to infectious agents should be minimized by quarantine, sanitation, food safety measures, and other general management procedures. Maintenance of a healthy immune system should be promoted by proper nutrition, stress reduction, and provision of an appropriate habitat.

POSSIBLE COMPLICATIONS
Dehydration and weight loss are common sequelae of gastritis and enteritis. Ascending infections to the liver can occur through the biliary tree. Bacteremia and septicemia can occur when the GI mucosa is compromised. The liver and kidneys can become infected first because of the hepatic and renal portal systems respectively. Protein-losing enteropathies can lead to hypoproteinemia. Gastrointestinal hemorrhage can result in hypovolemic shock or anemia. Anemia can be either regenerative or nonregenerative, depending on the severity and chronicity.

EXPECTED COURSE AND PROGNOSIS
There is a wide range in the ways that these diseases may progress. Some will be transient and self-limiting. Others may be acute and fatal. Still others can be chronic, some causing minimal to mild clinical signs, while others insidiously cause severe tissue damage, ultimately leading to death. Likewise, the prognosis varies from excellent to grave.

MISCELLANEOUS
ASSOCIATED CONDITIONS
N/A

AGE-RELATED FACTORS
N/A

ZOONOTIC POTENTIAL
Some diseases within this category are zoonotic. These include mycobacteriosis, chlamydiosis, salmonellosis, and some strains of other bacteria.

FERTILITY/BREEDING
N/A

SYNONYMS
N/A

SEE ALSO
Anorexia
Campylobacteriosis
Candidiasis
Cloacal disease
Clostridiosis
Dehydration
Diarrhea
Emaciation
Flagellate enteritis
Intestinal helminthiasis
Macrorhabdus ornithogaster
Mycobacteriosis
Polyuria
Proventricular dilatation disease
Regurgitation/vomiting
Trichomoniasis
Undigested food in droppings
Urate/fecal discoloration

ABBREVIATIONS
N/A

INTERNET RESOURCES
N/A

Suggested Reading
Denbow, D.M. (2000). Gastrointestinal anatomy and physiology In: Whittow, G.C. (ed.), *Sturkie's Avian Physiology*, 5th edn. San Diego, CA: Academic Press, pp. 299–325.
Fudge, A.M. (2000). Avian liver and gastrointestinal testing In: Fudge, A.M. (ed.) *Laroratory Medicine: Avian and Exotic Pets*. Philadelphia, PA: W.B. Saunders, pp. 47–55.
Girling, S. (2004). Diseases of the digestive tract of psittacine birds. *In Practice*, **26**:146–153.
Hadley, T.L. (2005). Disorders of the psittacine gastrointestinal tract. *Veterinary Clinics of North America. Exotic Animal Practice*, **8**:329–349.
Author Kenneth R. Welle DVM, Diplomate ABVP (Avian)

Feather Cyst

BASICS

DEFINITION
A smooth "lump" or "swelling" filled with white, caseous material, which results from the inability of a feather to erupt from the feather follicle (comparable to an ingrown hair in humans).

PATHOPHYSIOLOGY
Feather cysts often develop as a result of the inability of a feather to break through the skin surface. While the feather continues to grow, it curls up within the follicle and degenerates to a caseous mass. Cysts may continue to grow slowly until they eventually rupture. Secondary infections or hemorrhage may occur.

SYSTEMS AFFECTED
• Skin/Exocrine—usually affects the feather follicles of the wing, pectoral or scapular (back) feather tracts; secondary infections may occur. • Behavioral—cysts may irritate, thereby resulting in feather damaging behavior or automutilation. • Musculoskeletal—large cysts that are present on the wings may result in drooping of the wing and tripping over the wing. These cysts or large cysts on the body may also interfere with flying. • Hemic—trauma to the cyst may result in severe bleeding and subsequent anemia.

GENETICS
In canaries, the condition is believed to be hereditary, as it is typically seen in "soft-feathered" or "type" breeds. The exact mode of inheritance is unknown.

INCIDENCE/PREVALENCE
A true incidence is unknown but the condition is commonly seen in canaries and less so in other bird species.

GEOGRAPHIC DISTRIBUTION
N/A

SIGNALMENT
• **Species:** Commonly seen in smaller passerines and psittacines, such as canaries and budgerigars. Particularly soft-feathered canary breeds, e.g. Norwich and Gloucester, are prone. • **Mean age and predominant sex:** No age or sex predilection reported, although condition is generally not seen before the first molt.

SIGNS

Historical Findings
• A slowly growing, white-to-yellowish mass or lump on the body or wings • May develop following a normal molt. • Other signs reported by the owner may include inability to fly, drooping of the wing and/or feather damaging behavior or self-mutilation in the region of the affected feather, resulting in skin damage and/or hemorrhage.

Physical Examination Findings
• Feather cysts present as an oval or elongated swelling of the feather follicle with accumulation of dry, concentrically layered yellow-white material (keratin). • Although cysts may occur on any part of the body they are often found on the wing, back or pectoral region. • One or multiple cysts may be present at a time. • Secondary hemorrhage and/or infection may be present. In addition, feather damaging or automutilation of the skin with secondary ulceration and crust formation may be visible. • Birds may present with a wing droop or inability to fly if the mass is large.

CAUSES
Multiple etiologic factors may be involved including one or more of the following:
• Hereditary factors—particularly in soft-feathered canary breeds.
• Trauma—damage to the feather follicle resulting in abnormal growth, either iatrogenic or by the bird itself (e.g., due to feather damaging behavior). • Nutritional deficiencies, particularly hypovitaminosis A or amino acid deficiencies. • Viral (e.g., PBFD, polyomavirus), bacterial and parasitic infections may potentially also influence the occurrence of these cysts • Any other condition that interferes with the normal growth and development of a feather.

RISK FACTORS
N/A

DIAGNOSIS

DIFFERENTIAL DIAGNOSIS
• Any disorder resulting in the formation of a "mass" or "lump" may be considered in the differential diagnosis of feather cysts. • In particular, abscesses or tumorous growths such as xanthomas and lipomas may resemble feather cysts in outer appearance, although the latter often have a more firm consistency and a layered appearance (when incised).
• Fine needle aspirates or excisional biopsies may help to differentiate between feather cysts and other causes.

CBC/BIOCHEMISTRY/URINALYSIS
N/A

OTHER LABORATORY TESTS
N/A

IMAGING
N/A

DIAGNOSTIC PROCEDURES
• Diagnosis may usually be made based on the gross morphological appearance of the lump. • If necessary, fine needle aspiration or excisional biopsies may be performed to obtain a definite diagnosis. • Cytologic findings may vary, dependent on chronicity of the lesion: Early stages may reveal marked numbers of erythrocytes and erythrophagocytosis, whereas chronic stages are characterized by presence of mixed-cell inflammation with a marked amount of debris, feather fragments and occasional multinucleated giant cells. • In case (secondary) infection is suspected, material from the cyst may be submitted for a culture and sensitivity testing.

PATHOLOGIC FINDINGS
• Grossly, feather cysts present as a round or elongated white-yellow mass in the skin. Upon incising the mass, the concentrically layered structure of keratin material becomes apparent. • Histologically, the mass consists of an outer layer of proliferated stratified squamous epithelium that lines the keratinaceous material of the deformed feather. In case of secondary infection, inflammatory cells and micro-organisms may be present as well.

TREATMENT

APPROPRIATE HEALTH CARE
• Treatment of choice consists of removal of the affected feather follicle by excision. Other options include lancing and subsequent curettage and marsupialization or fulguration using radiosurgery or laser. These techniques, however, frequently result in recurrence or damage to the adjacent feather follicles. • A tourniquet may be used to aid in hemostasis when removing feather cysts on the wing. Cysts are excised, with wounds usually left open to heal by secondary intent. Take care not to damage any adjacent follicles or their blood supply during the procedure, as this may result in formation of new cysts.
• Hemorrhage is preferably controlled using ligatures rather than radiocautery as the latter may damage adjacent follicles or their blood supply. • Feather cysts of the tail may require amputation of the pygostyle. • Feather cysts on the body are generally easy to remove using an elliptical or fusiform excision, followed by primary closure of the skin. In case multiple cysts are present, removal of the whole feather tract may be considered. • In older birds or birds that do not seem to experience any discomfort of the cyst, treatment may not be necessary.

NURSING CARE
• Stabilize any patient that is critical prior to initiating the surgery. • Provide appropriate fluid therapy (SC, IV, IO) prior to, during and after the procedure. • After removal of the cyst, bandages (e.g., figure-of-eight bandage and/or body wrap) may be placed to prevent movement at the surgery site, provide extra hemostasis and protect the wound from contamination or mutilation by the bird.
• The use of e-collars may be indicated in birds that pick at the surgical site. • The

Feather Cyst (Continued)

wound is allowed to heal by secondary intention. Once adjacent feathers begin to regrow debris may be removed on a daily basis using cotton swabs or warm sterile saline flushes.

ACTIVITY
Birds in which a feather cyst has been removed from the wing may benefit from flight restraint for the first days postsurgery.

DIET
In the event that malnourishment is suspected to play a role in the development of a feather cyst, correction of the diet to a well-balanced diet is advised

CLIENT EDUCATION
Owners need to be informed about the heritable nature of this condition in canaries and should be aware that recurrence is common.

SURGICAL CONSIDERATIONS
• Provide fluids and appropriate nutritional support to any patient with substantial blood loss and stabilize first prior to performing the surgery. • Also provide adequate fluid support (SC, IV, IO) throughout the procedure and maintain proper hemostasis, particularly in smaller birds.

MEDICATIONS

DRUG(S) OF CHOICE
Treatment is surgical. Additional pain relief (e.g., NSAIDs) and/or appropriate antibiotics (preferably based on results from a culture and sensitivity testing) are indicated in case of concurrent inflammation and secondary infections.

CONTRAINDICATIONS
N/A

PRECAUTIONS
N/A

POSSIBLE INTERACTIONS
N/A

ALTERNATIVE DRUGS
N/A

FOLLOW-UP

PATIENT MONITORING
Recommend regular check-ups of the bird (e.g., once every 6–12 months) to detect formation of new cysts in an early stage.

PREVENTION/AVOIDANCE
N/A

POSSIBLE COMPLICATIONS
• Excessive hemorrhage (e.g., from trauma, self-mutilation or during surgical excision).
• Secondary infections of the cysts, particularly when the skin surface is damaged.
• Automutilation and/or poor wound healing following surgical excision.

EXPECTED COURSE AND PROGNOSIS
Prognosis is usually good following complete surgical excision. Recurrence, however, is common, particularly in birds in which a genetic background is suspected.

MISCELLANEOUS

ASSOCIATED CONDITIONS
Feather damaging birds may develop one or multiple cysts as a result from chronic trauma to the feather follicles.

AGE-RELATED FACTORS
N/A

ZOONOTIC POTENTIAL
N/A

FERTILITY/BREEDING
Although it is advisable not to breed with affected canaries, show standards accept these types of breeds and award prizes to birds prone to developing feather cysts, thereby resulting in persistence of this condition in the population.

SYNONYMS
Feather follicle cyst, Feather folliculoma, Plumafolliculoma

SEE ALSO
Cere and skin, color changes
Dermatitis
Feather damaging behavior/self-injurious behavior
Feather disorders
Tumors
Xanthoma

ABBREVIATIONS
PBFD—psittacine beak and feather disease
NSAID—nonsteroidal anti-inflammatory drug

INTERNET RESOURCES
N/A

Suggested Reading
Bauck, L. (1987). Radical surgery for the treatment of feather cysts in the canary. *AAV Today*, **1**(5):200–201.
Bennett, R.A., Harrison, G.J. (1994). Soft tissue surgery. In: Ritchie, B.W., Harrison, G.J., Harrison, L.R. (eds), *Avian Medicine: Principles and Application*. Lake Worth, FL: Wingers Publishing, pp. 1096–1196.
Fraser, M. (2006). Skin diseases and treatment of caged birds. In: Patterson S. (ed.) *Skin Diseases of Exotic Pets*, Ames, IA: Blackwell Science Ltd, pp. 22–47.
Wheeldon, E.B., Culbertson, M.R. (1982). Feather Folliculoma in the Canary (*Serinus canarius*). *Veterinary Pathology*, **19**(2):204–206.

Authors Yvonne van Zeeland, DVM, MVR, PhD, DECZM (Avian, Small Mammal)
Nico Schoemaker, DVM, PhD, DECZM (Small mammal, Avian), DABVP-Avian

Feather Damaging Behavior and Self-Injurious Behavior

BASICS

DEFINITION
Feather damaging behavior is a condition in which the parrot mutilates its feathers with the beak and may involve chewing, biting, plucking and/or fraying. Typically, the feathers of the head and crest remain unchanged as these are inaccessible to the bird's beak. In case of self-injurious behavior or automutilation, skin damage is also present.

PATHOPHYSIOLOGY
In general, three underlying pathophysiological mechanisms should be taken into consideration for this condition:
1. *Maladaptive behavior*, resulting from attempts of the animal to cope with an abnormal or inadequate environment, which is lacking the appropriate stimuli needed and/or in which stressors are present.
2. *Malfunctional behavior*, resulting from an abnormal psychology, brain development, or neurochemistry, which may have developed as a result of the bird's living conditions, particularly in early life.
3. Abnormal behavior resulting from *underlying medical (physical) problems*. Any disease causing pain, discomfort, irritation and/or pruritus may result in the bird damaging its feathers.

SYSTEMS AFFECTED
- Behavioral—birds may spend more time on preening and/or preen more intensively, which results in damaged feathers and/or skin; other abnormal repetitive behaviors (such as stereotypic behavior) may also be seen. • Skin/Exocrine—feathers may be pulled and/or frayed resulting in generalized or patchy alopecia in areas that are accessible to the bird's beak. Covert and/or down feathers are the main target, although remiges and/or rectrices may be targeted as well. Skin damage and/or (secondary) infections may also be present.
- Endocrine/Metabolic—metabolic needs may be increased due to lack of insulation and decreased thermoregulatory abilities.
- Hemic/Lymphatic/Immune—blood loss may occur in birds with self-injurious behavior.

GENETICS
Genetic factors are thought to be involved because of species predilections. Results of a study in Amazon parrots that demonstrated high heritability estimates support the hypothesis that a genetic basis indeed exists.

INCIDENCE/PREVALENCE
Feather damaging behavior (including self-mutilation) is estimated to occur in 10–15% of captive parrots.

GEOGRAPHIC DISTRIBUTION
N/A

SIGNALMENT
- **Species predilections:** ○ Although the condition may occur in any parrot species, Grey parrots (*Psittacus spp.*), Eclectus parrots (*Eclectus roratus*) and cockatoos (*Cacatua spp.*) appear particularly prone. ○ The condition is less common in budgerigars (*Melopsittacus spp.*). ○ Aside from parrots, the condition may also be seen in other bird species, including birds of prey, with the Harris hawk (*Parabuteo unicinctus*) identified as a highly susceptible species. ○ Automutilation appears most common in cockatoos, particularly umbrella (*Cacatua alba*) and Moluccan cockatoos (*Cacatua moluccensis*). • **Mean age and range:** Although feather damaging behavior may occur at any age, it has been suggested that the age of onset lies around the time when parrots become sexually mature.
- **Predominant sex:** Feather damaging may occur both in male and female parrots, with a suggested predilection for the female gender.

SIGNS

General Comments
- The behavior is usually self-inflicted but in some cases it can be directed to cage mates or nestlings. • Severity of the disease may vary from mild or localized feather damage or alopecia to severe forms with generalized feather damage, alopecia and/or self-mutilation.

Historical Findings
- The most noticeable signs in birds with feather damaging or self-injurious behavior are the presence of featherless areas and/or skin damage. • Owners may note the bird plucking, biting or pulling its feathers or damaging its skin, but the behavior may be difficult to distinguish from normal preening and also occur when the owner is not present. • Extensive history taking is needed to identify potential underlying medical and/or behavioral causes.

Physical Examination Findings
- Presence of featherless areas and/or damage to the feathers (fraying, chewing) and/or skin. • Feathers are mainly plucked in the easy accessible regions of the neck, chest, flank, inner thigh and ventral wing surface; feathers of the head and crest are typically unaffected as these are inaccessible to the bird's beak.
- Contour and down feathers are usually affected, but some birds may also damage the tail and flight feathers. • A thorough physical examination (including a dermatologic and fecal examination) is needed to determine the extent of the feather and/or skin damage and identify potential underlying medical causes (e.g., ecto- or endoparasites). • In some birds with skin damage, secondary infections and/or hemorrhage may be present.

CAUSES
- Numerous causes for feather damaging and/or self-injurious behavior have been reported. Generally, however, definitive proof of causal relationships is lacking for most conditions. • The following *medical causes* have been associated with feather damaging behavior: ○ Ecto- and endoparasites (e.g., *Knemidocoptes*, feather or quill mites, lice, *Giardia*/protozoal infection [particularly in cockatiels]). ○ Bacterial or fungal dermatitis and/or folliculitis (including *Staphylococcus, Aspergillus, Candida, Malassezia spp.*). ○ Viral infections such as polyomavirus, PBFD (circovirus) and poxvirus. ○ Infectious skin and/or feather disease (bacterial, fungal, viral). ○ Skin neoplasia (e.g., xanthoma, lipoma, squamous cell carcinoma). ○ Nutritional deficiencies (e.g., hypovitaminosis A) and/or dietary imbalances. ○ Low humidity levels, lack of bathing opportunities. ○ Airborne, topical and/or ingested toxins, including cigarette smoke, scented candles, air fresheners, hand lotions and creams, heavy metal ingestion (e.g., lead, zinc). ○ Hypersensitivity, skin allergy. ○ Systemic diseases such as Chlamydiosis, Proventricular Dilatation Disease. ○ Internal disease involving the respiratory tract (e.g., air sacculitis, pneumonia), liver (e.g., hepatitis), gastrointestinal (e.g., colic, gastroenteritis) or urogenital tract (e.g., renomegaly, cystic ovaries, egg binding). ○ Endocrine and/or metabolic conditions (e.g., hypothyroidism, diabetes mellitus, hypocalcemia). ○ Orthopedic disorders (e.g., osteosarcoma, fracture, osteomyelitis). ○ Improper wing trim or (iatrogenic) trauma. • *Behavioral (psychogenic) causes* leading to the development of feather damaging behavior may include: ○ Social isolation or overcrowding. ○ Inability to perform species-specific behaviors (e.g., foraging) resulting in redirected feather damaging behavior. ○ Small cage or poor cage design, which provides the parrot with little space to move around. ○ Sudden changes to the environment, lack of predictability and/or controllability of the environment. ○ Sleep deprivation, abnormal photoperiod. ○ Boredom. ○ (Sexual) frustration, hormonal imbalance (often cyclic or seasonal changes occurring). ○ Stress (feather damaging behavior serves as a coping mechanism, resulting in dearousal). ○ Anxiety, phobias. ○ Attention seeking behavior; reinforced by responses of the owner. ○ Abnormal repetitive behavior resulting from neurotransmitter deficiencies and/or excesses (e.g., serotonin, dopamine, endorphins), similar to obsessive-compulsive or impulsive disorders in humans.

RISK FACTORS
- Feather damaging behavior and self-injurious behavior are generally regarded

FEATHER DAMAGING BEHAVIOR AND SELF-INJURIOUS BEHAVIOR

as multifactorial disorders that may be influenced by a number of medical, genetic, neurobiologic and/or socioenvironmental factors. • Any type of disease causing pain, discomfort, irritation or pruritus may result in the bird displaying abnormal behaviors such feather damaging behavior. • Early living conditions, particularly hand rearing, inadequate socialization and deprivement, may predispose the bird to develop feather damaging behavior. • Lack of an appropriate living environment (particularly lack of foraging opportunities) and/or presence of stressors, both in early life and present living conditions, may influence the onset of abnormal behavior. • Abnormal behavior may unintentionally be reinforced by the owner, as his/her responses may be rewarding the bird. • Neurotransmitter (e.g., serotonin, dopamine, endorphin) deficiencies and/or excesses may play a role in the onset and/or maintenance of the abnormal behavior, although little is currently known about the neuropathophysiologic mechanism underlying the behavior.

DIAGNOSIS

DIFFERENTIAL DIAGNOSIS
• Feather damaging behavior should be distinguished from other causes of alopecia or feather loss including normal molt and/or apteria (not recognized by an inexperienced owner), cage mate plucking (excessive allopreening), (iatrogenic) trauma, feather loss due to parasitism or bacterial, mycotic or viral infections, and plucking related to normal brooding behavior (bird preparing a brood patch). In addition, causes for lack of feather growth (e.g., hypothyroidism, malnutrition, PBFD) should be ruled out as well. • A thorough history and physical examination (including a thorough dermatologic examination of skin and feathers) are needed to identify potential underlying causes for feather damaging or self-injurious behavior.
• Before a (definite) diagnosis of feather damaging as a primary behavioral disorder can be made, the presence of underlying medical conditions disease should be ruled out (see Causes). • In case a psychologic or behavioral origin of the disorder is likely, effort should be made to identify the potential underlying triggers (antecedents) and reinforcing factors (consequences) that may have contributed to the onset and maintenance of FDB.

CBC/BIOCHEMISTRY/URINALYSIS
• Leukocytosis, heterophilia, and/or monocytosis may be seen in case of (secondary) infections. • In case of blood loss, hematocrit values may be decreased. • Plasma creatine kinase values may be elevated.
• Dependent on the underlying cause, other abnormalities may be noted.

OTHER LABORATORY TESTS
Other laboratory tests that may be performed to diagnose or rule out any medical underlying conditions may include one or more of:
• Fecal examination (including wet mount and flotation), for example, for diagnosis of Giardiasis or other endoparasites. • PCR testing on full blood (PBFD, circovirus) or cloacal swab (Polyomavirus, avian bornavirus, *Chlamydia*). • Serology testing for avian bornavirus, *Chlamydia*. • Heavy metal screening for lead or zinc toxicosis. • TSH stimulation tests for hypothyroidism (only if feather growth is absent).

IMAGING
• Whole body radiographs may be useful to identify various underlying causes including heavy metal intoxication, reproductive disorders (e.g., egg binding), hepato-, spleno- or renomegaly, proventricular dilatation disease, pneumonia, air sacculitis, neoplastic conditions, musculoskeletal disease (e.g., osteoarthritis, osteomyelitis, fractures, osteosarcoma). • Ultrasound may be indicated to rule out or diagnose hepatomegaly, reproductive disorders (e.g., egg peritonitis, cystic ovary), neoplastic conditions, cardiac disease, ascites.

DIAGNOSTIC PROCEDURES
Additional diagnostic tests that may be useful to diagnose underlying medical conditions include:
• Cytologic examination of skin lesions (scrapings, tape strip, impression smear, swab, fine needle aspirate) or feather pulp (feather digest) to diagnose bacterial or fungal folliculitis or dermatitis, pox virus, ectoparasites (feather or quill mites, *Knemidocoptes*), neoplasia. • Culture and sensitivity of skin lesions and/or feather pulp to diagnose bacterial or fungal dermatitis or folliculitis. • Skin and/or feather follicle biopsy (histopathology) to diagnose a variety of infectious, inflammatory and/or neoplastic skin diseases, for example, PBFD, polyomavirus, bacterial and fungal folliculitis, quill mite infestation, xanthomatosis, squamous cell carcinoma, feather follicle cysts. • Intradermal skin testing to diagnose hypersensitivity reactions, allergic skin disease. Thus far these tests are, however, not found to be reliable due to the bird's diminished reaction to histamine. • Endoscopy, for example, to diagnose air sacculitis, hepato- or nephropathy, splenomegaly, pancreatic disorders, reproductive disease.

PATHOLOGIC FINDINGS
• Gross pathological finding: Patchy or generalized alopecia or feather damage with or without skin damage and/or secondary infections, with the head typically remaining unaffected. • Histopathology may help to distinguish between inflammatory and traumatic (self-inflicted) skin disease: • Inflammatory skin disease is characterized by presence of perivascular inflammation in the superficial or deep dermis of clinically affected and unaffected sites (i.e., outside of the reach of the bird's beak). • Traumatic skin disease is characterized by superficial dermal scarring with or without inflammation in the affected sites and an absence of inflammation in the unaffected sites. • In case inflammatory cells are present, these typically include lymphocytes and occasionally plasma cells, histiocytes, and granulocytes.

TREATMENT

APPROPRIATE HEALTH CARE
• Correct the diet and/or modify the bird's housing and living conditions to address any environmental factors that may be involved.
• Promote a more stimulating environment by means of increasing the size of the enclosure, taking the bird outdoors, providing social contact, (chewing) toys, puzzle feeders and other forms of environmental enrichment. In particular, foraging enrichment has been shown to effectively reduce feather damaging behavior. • Training and behavior modification techniques, such as desensitization, counter-conditioning and differential reinforcement of alternate (desired) behaviors, may be employed to alter the behavior of the bird and provide the bird a mentally stimulating challenge or task.
• Address medical conditions appropriately, if present. • In case of severe automutilation, mechanical barriers such as Elizabethan collars, neck braces, "ponchos", "jackets" and "vests", can be used. These, however, merely provide symptomatic treatment and may cause additional stress. • Local application of foul-tasting substances is controversial as these may sometimes result in deterioration. • In case of (secondary) skin infections, the use of appropriate antibiosis (preferably based on results of a culture and sensitivity testing) and/or NSAIDs may be considered. • Drug therapy may be warranted in cases that are refractory to treatment with enrichment and/or behavior modification (see Medication).

NURSING CARE
• In case of severe skin damage, wound management followed by placement of a protective wound dressing and/or "jacket" to prevent the bird from further mutilating itself is indicated. • In case of (severe) blood loss, fluid therapy and/or blood transfusions may be considered.

ACTIVITY
Promote the bird's species-specific behaviors and provide the bird with sufficient exercise and mental challenge to satiate its needs.

Feather Damaging Behavior and Self-Injurious Behavior

DIET
• Nutrition of the bird should be optimized to correct for any potential deficiencies that may exacerbate the behavior. If fed a seed mixture, owners should be advised to convert the bird to a pelleted diet. • Provide the bird with mental and/or physical challenge by offering the food in a puzzle feeder, hiding and/or mixing it with inedible items. Larger-sized food particles and/or treats (e.g., walnuts, pine cones) may promote the bird's foraging activities as well.

CLIENT EDUCATION
• Provide the bird with adequate nutrition and a low stress, stimulating environment in which it is able to display its species-specific behaviors. • To create a stimulating environment, various types of enrichment that stimulate the bird's senses and provide a physical and mental challenge may be offered to the bird. Particularly puzzle-feeders and other types of foraging enrichment promote the bird's natural behavior and can help to satiate the behavioral need to forage. • Create awareness with the owner that he/she may (unintentionally) reinforce the behavior by paying attention to the bird when it is displaying the abnormal behavior in an attempt to distract the bird or getting it to stop plucking or biting its skin and/or feathers. • Create awareness that, once a bird is plucking or damaging its feathers and/or skin, the condition may be challenging to treat, with relapses occurring frequently.

SURGICAL CONSIDERATIONS
N/A

MEDICATIONS

DRUG(S) OF CHOICE
• In case of refractory cases, pharmacologic intervention using one of the following psychotropic agents may be considered:
○ Dopamine antagonists, for example, haloperidol. ○ Opiate receptor antagonists, for example, naloxone, naltrexone. ○ Tricyclic antidepressants, for example, amitriptylline, clomipramine, doxepin. ○ Serotonergic reuptake inhibitors, for example, fluoxetine, paroxetine. ○ Benzodiazepines, for example, alprazolam, clonazepam, diazepam, lorazepam. ○ Anxiolytic drugs, for example, buspirone. ○ Hormone therapy using GnRH agonists such as leuprolide acetate and deslorelin implants may be indicated when feather damaging behavior is suspected to be sexual or hormonally related. ○ Of the aforementioned drugs, clomipramine thus far appears the best investigated drugs. For most – if not all – others, placebo-controlled, double-blind, randomized studies concerning dosages, pharmacokinetics, toxicity and efficacy are lacking, thereby limiting the ability to make recommendations at this stage. • In case of underlying medical conditions and/or secondary infections, other types of medication may need to be given.

CONTRAINDICATIONS
N/A

PRECAUTIONS
Since limited information is available on the use of psychotropic drugs these should be used with caution and titrated carefully to effect while also monitoring closely for any adverse effects. Many of these drugs may take several weeks to take effect and should be gradually weaned off to prevent withdrawal symptoms from occurring.

POSSIBLE INTERACTIONS
Interactions between the various psychotropic drugs are common and may potentially be hazardous. Simultaneous use should, therefore, be avoided. When switching from one drug to another, the bird should first be fully weaned off of one drug prior to starting the new drug to reduce the risk of undesired side effects.

ALTERNATIVE DRUGS
N/A

FOLLOW-UP

PATIENT MONITORING
• Regular rechecks are recommended. As it may take at least 3–4 weeks for feathers to regrow once they are pulled, monthly rechecks seem appropriate, unless the condition is severe and/or worsens (in which case more frequent rechecks may be scheduled). • During rechecks a detailed inspection of the plumage and skin condition should be made, with photographs and/or feather scoring systems aiding in monitoring any changes in the behavior as the latter are usually difficult to evaluate directly. • Owners are advised to keep a log and document when the feather damaging behavior occurs, which may help identify any underlying causes. • In case of (secondary) infections or underlying medical conditions, a CBC and/or biochemistry may be performed to monitor changes in the bird's health status. • In case psychotropic drugs are used, owners are instructed to monitor their bird carefully for potential side effects.

PREVENTION/AVOIDANCE
• Provide the bird with a stimulating, low-stress living environment and adequate nutrition. Particularly, the provision of (novel) enrichment, training and exercise (e.g., taking the bird outside for walks) may help to create and maintain a stimulating and controllable environment that allows the bird to perform its species-specific behaviors and make its own decisions, while simultaneously preventing the development of abnormal behaviors. • Hand-rearing of young birds should be avoided as this may increase the risks of the bird developing abnormal behavior later on in life.

POSSIBLE COMPLICATIONS
• Severe feather loss may lead to a compromise of thermoregulatory abilities, and affect the bird's metabolic needs and immune system, thereby resulting in increased susceptibility to disease. • Hemorrhage. • (Secondary) infections (particularly if skin damage is present).

EXPECTED COURSE AND PROGNOSIS
Feather damaging and self-injurious behavior are challenging conditions to treat, with recurrence and relapses commonly occurring, especially if one is unable to identify and eliminate the underlying cause. Prognosis is considered guarded, with chances of a successful outcome decreasing once the condition becomes more chronic and ritualized.

MISCELLANEOUS

ASSOCIATED CONDITIONS
Birds with feather damaging behavior may also display other forms of abnormal behavior, such as stereotypic behaviors (e.g., circling, tumbling, tongue playing, head bobbing or twirling).

AGE-RELATED FACTORS
N/A

ZOONOTIC POTENTIAL
N/A

FERTILITY/BREEDING
Since genetic factors are thought to be involved, breeding with feather damaging birds may best be avoided.

SYNONYMS
Feather picking, feather plucking, feather pulling, feather destructive behavior, pterotillomania

SEE ALSO
Appendix 7, Algorithm 12. Feather damaging behavior
Cere and skin, color changes
Circoviruses
Dermatitis
Ectoparasites
Feather disorders
Flagellate enteritis
Nutritional deficiencies
Problem behaviors

ABBREVIATIONS
PBFD—psittacine beak and feather disease

INTERNET RESOURCES
http://www.parrots.org/index.php/referencelibrary/behaviourandenviroenrich/
http://www.behaviorworks.org/
http://www.parrotenrichment.com/

FEATHER DAMAGING BEHAVIOR AND SELF-INJURIOUS BEHAVIOR

Suggested Reading
Nett, C.S., Tully, T.N. (2003). Anatomy, clinical presentation and diagnostic approach to feather-picking pet birds. *Compendium on Continuing Education for the Practicing Veterinarian*, **25**(3):206–219.

Orosz, S.E. (2006). Diagnostic work-up of suspected behavioural disorders. In: Luescher, A.U. (ed.) *Manual of Parrot Behavior*. Oxford, UK: Blackwell Publishing, pp. 195–210.

Rubinstein, J., Lightfoot, T.L. (2012). Feather loss and feather destructive behavior in pet birds. *Journal of Exotic Pet Medicine*, **21**(3):219–234.

Seibert, L.M. (2006). Feather-picking disorder in pet birds. In: Luescher, A.U. (ed.) *Manual of Parrot Behavior*. Oxford, UK: Blackwell Publishing, pp. 255–265.

van Zeeland, Y.R.A., Spruit, B.M., Rodenburg, T.D., et al. (2009). Feather damaging behaviour in parrots: a review with consideration of comparative aspects. *Applied Animal Behaviour Science*, **121**(2):75–95.

Authors Yvonne van Zeeland, DVM, MVR, PhD, Dip. ECZM (Avian, Small Mammal)
Nico Schoemaker, DVM, PhD, Dip. ECZM (Small mammal, Avian), Dipl. ABVP-Avian

Client Education Handout available online

Feather Disorders

BASICS

DEFINITION
• Feather dystrophy is characterized by the formation of abnormally shaped feathers, which may be present as a localized or generalized disorder. • Feather discoloration is a change of the normal coloration of a feather, which may involve only part of the feather (e.g., spots) or the feather as a whole, and can be limited to a single or a few feathers, or more extensive, involving multiple feathers or the entire plumage.

PATHOPHYSIOLOGY
• Abnormal growth of dystrophic feathers is the result of (in)direct damage to the feather follicle, follicular collar or developing feathers. • Feathers derive their coloration from pigments (melanins, porphyrins and carotenoids) and/or structural conditions of the feather that modify or separate the components of white light (e.g., Tyndall effect, iridescence). Lack of certain nutrients that serve as a precursor for the aforementioned pigments, or a condition resulting in a change of the feather structure will result in a change of color of the feather.

SYSTEMS AFFECTED
• Skin/Exocrine—one or multiple abnormally formed or shaped feathers and/or follicles (feather dystrophy) or abnormally colored feathers. • Feather dystrophy and/or discoloration does not affect any organ system itself. See polyomavirus and circovirus for their effect on the Hepatobiliary, Hemic/Lymphatic/Immune and/or Renal/Urologic system.

GENETICS
• Feather duster disease is a feather dystrophic condition which occurs in English show budgerigars and is caused by a recessive gene. • Genetic mutations may result in change of color of the feathers. Generalized color mutations such as lutinos are bred specifically for the show.

INCIDENCE/PREVALENCE
• Feather duster disease is a rare condition. • Due to proper management in breeding facilities, the incidence of circovirus and polyomavirus infections have decreased considerably throughout the past decade. • With proper nutrition, the incidence of feather discoloration should be minimal.

GEOGRAPHIC DISTRIBUTION
N/A

SIGNALMENT
• Feather duster disease is seen in English show budgerigars during the development of the first feathers. • Feather dystrophy due to polyomavirus infections is commonly seen in smaller psittacines (particularly budgerigars, lovebirds) during the development of the first feathers, from the age of 10 days onwards. • Feather dystrophy due to circovirus infections are seen in all psittacine breeds up to the age of three years. Australasian species are most prone followed by African and then South American species. • Feather discoloration can occur in all birds, regardless of the species, age or gender. Species that are known to require carotenoids in their diet to obtain a normal plumage color include flamingos and specific color-bred canaries. If white birds (such as ducks) are fed carotenoids, their plumage color may change.

SIGNS
• In feather duster disease the feathers continue to grow, resulting in extremely long, curly feathers covering the entire body of the budgerigars. • See "Polyomavirus" for the feather dystrophic signs seen in birds infected with this virus. • See "Circoviruses" for the progressive feather dystrophic signs seen in birds infected with this virus. • Feather discoloration: any discoloration from the normal plumage color (e.g., red feathers in the normal grey plumage of Grey parrot, yellow discoloration of feathers in Amazon parrot, overall/generalized dull plumage due to lack of carotenoid diet in canaries).

CAUSES
• Multiple etiologic factors may be involved in feather dystrophy, including one or more of: ◦ Hereditary—in budgerigars with feather duster disease. ◦ Trauma—damage to the feather follicle resulting in abnormal growth, either iatrogenic or by the bird itself (e.g., due to plucking). ◦ Viral (e.g., PBFD, polyomavirus). ◦ Any other condition that interferes with the normal growth and development of a feather (e.g. nutritional deficiency, liver disease). • Multiple etiologic factors may be involved in feather discoloration, including one or more of: ◦ Nutritional deficiencies—that is, carotenoids, nonheme iron, tyrosine, copper, choline, lysine, methionine and/or riboflavin. ◦ Hereditary—genetic mutations. ◦ Inflammation—early circovirus infection. ◦ Metabolic disease—liver disease, hypothyroidism, neoplasia (e.g., pituitary tumor). ◦ Intoxication—chronic lead toxicosis, drug administration (e.g., thyroxine, fenbendazole). ◦ Localized color changes involving a single or few feathers are most likely the result of a localized inflammatory process or trauma that affected the growing feather or its follicle during its development.

RISK FACTORS
N/A

DIAGNOSIS

DIFFERENTIAL DIAGNOSIS
N/A

CBC/BIOCHEMISTRY/URINALYSIS
N/A

OTHER LABORATORY TESTS
N/A

IMAGING
N/A

DIAGNOSTIC PROCEDURES
• Histological evaluation of a feather follicle biopsy will provide the most accurate diagnosis for the cause of feather dystrophy. • Diagnostic procedures for specific diseases which may cause feather dystrophy and/or discoloration are described under the specific conditions

PATHOLOGIC FINDINGS
The histologic findings of feather follicle biopsies from birds infected with circovirus and polyomavirus infections can be found under the specific topics.

TREATMENT

APPROPRIATE HEALTH CARE
• Treatment should always be aimed at eliminating the underlying cause. • Treatment options for the specific diseases which may lead to feather dystrophy and/or feather discoloration can be found under the specific topics.

NURSING CARE
N/A

ACTIVITY
N/A

DIET
When malnutrition is suspected as a cause for the feather discoloration, conversion to a complete, pelleted diet is advised, which may result in a change of color back to normal after the subsequent molt. In some birds, a second molt may be necessary for the full plumage to return to normal.

CLIENT EDUCATION
Owners need to be informed about: • Heritability of feather duster disease and the need for supportive care. • The infectious nature of circovirus and polyomavirus infections (i.e., avoid contact with other birds, especially neonates and juveniles). • The necessity of feeding a complete, pelleted diet.

SURGICAL CONSIDERATIONS
N/A

MEDICATIONS

DRUG(S) OF CHOICE
N/A

CONTRAINDICATIONS
N/A

FEATHER DISORDERS (CONTINUED)

PRECAUTIONS
N/A

POSSIBLE INTERACTIONS
N/A

ALTERNATIVE DRUGS
N/A

FOLLOW-UP

PATIENT MONITORING
• Budgerigars with feather duster disease are not able to fly and may have difficulty eating. Euthanasia should be advised, although some owners prefer providing supportive care. • For patient monitoring of birds infected with a circovirus or polyomavirus infection, the reader is referred to these specific topics. • To monitor the effect of nutritional treatment of feather discoloration, a recheck after at least one molt is needed.

PREVENTION/AVOIDANCE
• To prevent dissemination of feather duster disease, do not breed with the parents or siblings of affected birds. • Use a closed aviary concept to prevent birds from contracting infection with circovirus and/or polyomavirus infections. • Feed birds a well-balanced (pelleted) diet to prevent feather discoloration due to malnutrition.

POSSIBLE COMPLICATIONS
• Birds with feather duster disease may have difficulty eating and are, therefore, at risk of starvation. • Birds with circovirus infections are prone to develop secondary infections due to the associated immune suppression.

EXPECTED COURSE AND PROGNOSIS
• The prognosis for birds with feather duster disease and circovirus infections is unfavorable. • Feather regrowth may occur in small psittacines infected with polyomavirus. • After correction of the diet, feather discoloration will generally return to normal after one (or two) molts.

MISCELLANEOUS

ASSOCIATED CONDITIONS
Conditions associated with circovirus and polyomavirus infections and nutritional deficiencies can be found under the specific topics.

AGE-RELATED FACTORS
Feather duster disease, circovirus and polyomavirus infections are diagnosed at an early age.

ZOONOTIC POTENTIAL
N/A

FERTILITY/BREEDING
• It is advised not to breed with the parents or siblings of budgerigars with feather duster disease. • It is advised to (temporarily) stop breeding during an outbreak of polyomavirus infections in small psittacines.

SYNONYMS
Budgerigars with feather dystrophy due to circovirus or polyomavirus infections: French moult; runners; hoppers, creepers; BFD (budgerigar fledgling disease)
Feather duster disease: Chrysanthemum disease

SEE ALSO
Cere and skin, color changes
Circoviruses
Dermatitis
Ectoparasites
Feather cyst
Feather damaging behavior
Nutritional deficiencies
Polyomavirus

ABBREVIATIONS
PBFD—psittacine beak and feather disease
BFD—budgerigar fledgling disease

INTERNET RESOURCES
http://www.birds-online.de/gesundheit/gesgefieder/featherduster_en.htm

Suggested Reading
Cooper, J.E., Harrison, G.J. (1994). Dermatology. In: Ritchie, B.W., Harrison, G.J., Harrison, L.R. (eds), *Avian Medicine: Principles and Application* (eds). Lake Worth, FL: Wingers Publishing, pp. 607–639.
Harrison, G.J., McDonald, D. (2006). Nutritional disorders. In: Harrison, G.J., Lightfoot, T.L. (eds), *Clinical Avian Medicine*. Palm Beach, FL: Spix Publishing, pp. 108–140.
Pass, D.A. (1995). Normal anatomy of the avian skin and feathers. *Seminars in Avian Exotic Pet Medicine*, **4**:52–160.
Roudybush, T. (1986). Growth, signs of deficiency, and weaning in cockatiels fed deficient diets. *Proceedings of the Annual Meeting of the Association of Avian Veterinarians*, Miami, FL, pp. 333–340.
van Zeeland, Y.R.A., Schoemaker, N.J. (2014). Plumage disorders in psittacine birds – part 1: Feather abnormalities. *European Journal of Companion Animal Practice*, **24**(1):34–47.

Authors Nico Schoemaker, DVM, PhD, Dip. ECZM (Small mammal, Avian), Dipl. ABVP-Avian
Yvonne van Zeeland, DVM, MVR, PhD, Dip. ECZM (Avian, Small Mammal)

FLAGELLATE ENTERITIS

BASICS

DEFINITION
Enteritis is defined as inflammation of the small intestine. There are several flagellated protozoa that infect birds. Those that cause intestinal disease are limited to *Giardia spp.*, *Spironucleus (Hexamita) spp.* and *Cochlosoma spp.* While *Trichomonas spp* and other trichomonads as well as *Histomonas meleagridis* cause significant problems in a variety of birds, they do not generally cause enteritis and will be left out of this discussion.

PATHOPHYSIOLOGY
There is generally a direct fecal oral life cycle for all of these parasites. Infectious organisms are ingested and establish themselves in the intestines, where they multiply by binary fission. They cause enteritis with resultant clinical signs and possible dermatologic manifestations; there may be malabsorption of nutrients and fat-soluble vitamins, which can cause complications. Secondary bacterial and/or yeast infections can occur also.

SYSTEMS AFFECTED
• Gastrointestinal—small intestinal lesions and dysfunction. • Integument— pruritis and feather/skin lesions secondary to some flagellate infections.

GENETICS
N/A

INCIDENCE/PREVALENCE
Not well documented. Many studies are done to determine the prevalence of these organisms in wild populations, but the finding of the organism does not equate with evidence of disease. Giardiasis is most common in budgerigars, with many asymptomatic carriers; nestlings die at 10–28 days; cochlosomiasis is reported sporadically and appears to be more common in cockatiels in Australia, *Spironucleus spp.* infections in turkeys are not as common as they were several decades ago; in pigeons they are a fairly common problem.

GEOGRAPHIC DISTRIBUTION
These parasites are all likely to be found worldwide.

SIGNALMENT
• **Species:** ○ *Giardia spp.*: budgerigars, cockatiels, lovebirds, rarely other parrots, finches, great blue herons, raptors, waterfowl, poultry and toucans. ○ *Spironucleus spp.*: pigeons (*S. columbae*), turkeys (*S. meleagridis*), budgerigars, cockatiels, Australian King Parrots, Splendid Grass parakeets, cranes, pheasant, quail, chukar partridge, peafowl. ○ *Cochlosoma spp*: finches, esp. Gouldians, cockatiels. • **Mean age and range:** All ages are susceptible, but young birds, especially nestlings, are most susceptible to morbidity and mortality. Adult carriers are extremely common with most of these parasites. Cochlosomiasis is seen mostly in finches 10 days to 6 weeks of age. • **Predominant sex:** Both sexes are susceptible.

SIGNS
General Comments
Clinical signs are related to the enteritis caused by these organisms; disease is generally seen in young animals.

Historical Findings
• Flock problem usually in aviaries. • Disease outbreaks in nestlings mostly. • *S. columbae* in pigeons seen in young squabs in spring and summer. • Cochlosomiasis in finches most commonly seen as a result of fostering under Bengalese finches. • 10–100% mortality in budgies. • Occasional outbreak occurs after introduction of new birds, but often will occur in closed flock after stressor such as moulting, breeding.

Physical Examination Findings
Giardiasis:
• Chronic-to-intermittent watery whitish diarrhea. • Malodorous mucoid stools. • Nestlings caked with feces. • Listlessness. • Weak. • Depression. • Poor feathering. • Ruffled feathers. • Anorexia. • Distended crop, rarely vomiting and regurgitation. • Weight loss and emaciation. • Neonatal mortality. • Parents may look normal. • Feather picking can be seen in cockatiels—ferociously pluck from wings, flanks, legs—often screaming while doing so.

Spironucleus spp.:
• Chronic, often intractable diarrhea—copious, foamy, malodorous, yellow or intensely green. • Dehydration. • Weight loss and emaciation. • Young birds may die; older birds do not. • Poults/squabs at first nervous and active, but later listless and huddled, then convulsions and death.

Cochlosomiasis:
• Diarrhea—moist bulky droppings. • Weight loss and emaciation. • Shivering due to dehydration and hypothermia. • Debilitation. • Difficulties with moulting. • Death. • Feather picking in cockatiels.

CAUSES
• Giardiasis: In budgerigars the organism appears to be *G. psittaci*, but it is unclear as to which species of *Giardia* inhabit and/or cause disease in which species of bird. • *Cochlosoma spp*, *Spironucleus (Hexamita) spp.*

RISK FACTORS
• Raising Australian finches under society finches puts them at risk for cochlosomiasis. • Overcrowding and stress is a risk factor for all of these infections. • Mixing species (peafowl with turkeys) increases the risk of spironucleosis. • Adding new stock to a closed population is a risk.

DIAGNOSIS

DIFFERENTIAL DIAGNOSIS
Any causes of enteritis, depending on the species: • Bacterial, such as *E. Coli*, *Salmonella spp.* clostridial infection, *Campylobacter sp.*, yersiniosis, mycobacteriosis, other gram negative organisms. • Other protozoal diseases such as coccidiosis and cryptosporidiosis. • Macrorhabdosis. • Viral diseases— many—in pigeons, adenovirus, paramyxovirus.

CBC/BIOCHEMISTRY/URINALYSIS
Usually not performed, but eosinophilia has been found in cockatiels and budgerigars with giardiasis; hypoproteinemia can result from malabsorption.

OTHER LABORATORY TESTS
• General: diagnosis is by identification of the organism in fecal samples or scrapings from the intestines of a fresh carcass. • Saline wet mounts of fresh feces are best—within 10 minutes of voiding if possible; adding iodine might help in identifying *Giardia;* warming the saline might be helpful for *Spironucleus spp.* Shedding is often intermittent and it might be necessary to perform multiple fecal smears. *Giardia* and *Spironucleus spp.* trophozoites can be identified by their size and shape and characteristic movement; sometimes *Giardia* cysts can be seen on direct smears but are hard to identify. • Fecal trichrome staining—several stools are collected over a three-day period; these are stored in polyvinyl alcohol or 10% formalin and stained with a fecal trichrome stain. • Zinc sulfate centrifugation technique— *Giardia* cysts can often be seen. • Fecal cytology—Wright's/Giemsa/Dif-Quik staining may reveal *Spironucleus spp.* or other organisms. • Gram's staining—may demonstrate *Giardia* trophozoites. • PCR analysis may be useful for *Giardia* diagnosis, but it is unclear at this time. • Fecal ELISA *Giardia* testing is available for mammalian fecal samples, but its use in birds is controversial.

IMAGING
If radiographs were performed, it is unlikely to help in the diagnosis, but could help to rule out other problems. Gas-filled intestines might be seen.

DIAGNOSTIC PROCEDURES
Collection of feces, cloacal contents or intestinal scrapings.

PATHOLOGIC FINDINGS
Giardiasis
• Sometimes no lesions. • Distended small intestine with excessive fluid and mucous and/or yellowish creamy material.

- Histologically, a mononuclear inflammatory infiltrate; organisms found in mucous between the villi, villar atrophy.

Spironucleosis
- Severe catarrhal enteritis in squabs. • Watery intestinal contents with large numbers of organisms seen on microscopic examination of material from the crypts. • Atony and distension, especially in the upper small intestine. • Small ulcerative lesions in ileum and rectum have been noted. • Secondary bacterial infections are commonly seen.

Cochlosomiasis
- Lesions are not described, but likely similar to the other flagellates.

TREATMENT

APPROPRIATE HEALTH CARE
Patients that are critical need to be hospitalized and treated accordingly; patients that are not critical can be treated as outpatients.

NURSING CARE
Nursing care can be done in the sicker patients, including providing thermal and nutritional support and fluid therapy. In flock situations, this might be impractical.

ACTIVITY
N/A

DIET
N/A

CLIENT EDUCATION
Proper administration of medications should be explained. Hygiene and husbandry and management practices need to be outlined in the case of flock and aviary situations.

SURGICAL CONSIDERATIONS
N/A

MEDICATIONS

DRUG(S) OF CHOICE
Several drugs of the nitroimidazole class are suggested for these diseases; dosages and response will vary. Many of these drugs are available through internet pigeon supply companies.
- Metronidazole in water for 5 days—or individual treatment. 25 mg/kg PO BID or 50 mg/kg QD for 5–10 days; birds can be hospitalized and given metronidazole IM at 5–10 mg/kg QD-BID.
- Ronidazole—2.5–20 mg/kg PO.
- Carnidazole—20 mg/kg PO.
- Fenbendazole—50 mg/kg PO QD for three days. • Ipronidazole—doses range from 125–250 mg/L of water. • For cochlosomiasis in finches: Ronidazole at 400 mg/kg egg food and 400 mg/L drinking water for five days.

After a pause of two days, the regimen is repeated.

CONTRAINDICATIONS
N/A

PRECAUTIONS
- Dimetridazole—if exceeding 100 mg/l for five days, torticollis can be a sign of toxicosis
- Metronidazole—can cause toxicosis in finches. • Fenbendazole—can be toxic to cockatiels, pigeons and doves, lories and other birds. Use with caution or at lower dosages.

POSSIBLE INTERACTIONS
N/A

ALTERNATIVE DRUGS
Treat secondary bacterial infections with trimethoprim-sulfa or enrofloxacin or as based upon results of culture and sensitivity.

FOLLOW-UP

PATIENT MONITORING
Patients with intestinal flagellate infections should be evaluated at regular intervals. Physical examination of affected flocks and individuals as well as repeated fecal examinations should be performed as needed.

PREVENTION/AVOIDANCE
- Environmental hygiene, such as thorough cage cleaning to remove organic debris, disinfecting with quaternary ammonium compounds or 10% bleach, keeping aviaries dry to reduce the number of infectious cysts; insect control, prevent overcrowding (low stocking density), elevate food and water dishes, use floor grates. Isolate and quarantine all new birds as well as any infected birds.
- For spironucleosis, consider removing carrier birds, separate older stock from poults, exclude other avian host species from poultry flock. • For cochlosomiasis, avoid fostering Australian finches under Bengalese (society) finches.

POSSIBLE COMPLICATIONS
Secondary bacterial infections should be suspected in cases where antiprotozoal treatment seems to be inadequate. Re-infection is common also.

EXPECTED COURSE AND PROGNOSIS
Untreated nestlings will often die, with a variable degree of mortality; recovered birds can remain as carriers. Giardiasis appears to be more treatable than cochlosomiasis and spironucleosis.

MISCELLANEOUS

ASSOCIATED CONDITIONS
There can be secondary bacterial and/or yeast infections associated with any of these diseases; cockatiels can have a secondary pruritus and associated feather picking disorder; it has been speculated that there is a malabsorption of fat soluble vitamins, with potential vitamin E deficiency and muscle weakness in cockatiels

AGE-RELATED FACTORS
The very young are most affected in all of these protozoal diseases; parents/adults are often asymptomatic carriers.

ZOONOTIC POTENTIAL
It is unclear as to whether there is spread of *Giardia* from birds to humans. It is poorly documented, and it remains to be determined.

FERTILITY/BREEDING
N/A

SYNONYMS
N/A

SEE ALSO
Cere and skin, color changes
Diarrhea
Feather damaging behavior/self-injurious behavior
Gastritis/enteritis
Gastrointestinal helminthiasis
Nutritional deficiencies
Problem behaviors
Trichomoniasis
Undigested food in droppings
Urate/fecal discoloration

ABBREVIATIONS
N/A

INTERNET RESOURCES
For more information, searches on VIN, PubMed and other veterinary medical search engines will reveal much useful information.

Suggested Reading
Brandão, J., Beaufrère, H. (2013). Clinical update and treatment of selected infectious gastrointestinal diseases in avian species. *Journal of Exotic Pet Medicine*, **22**:101–117.
Clyde, C.I., Patton, S. (2000). Parasitism of Caged Birds. In: Olsen, G.H., Orosz, S.E. (eds), *Manual of Avian Medicine*. St Louis, MO: Mosby, Inc., pp. 424–448.
Doneley, R.J.T. (2009). Bacterial and parasitic disease of parrots. *Veterinary Clinics of North America. Exotic Animal Practice*, **12**(3):423–431.
Patton, S. (2000). Avian parasite testing. In: Fudge, A.M. (ed.), *Laboratory Medicine–Avian and Exotic Pets*. Philadelphia, PA: WB Saunders, pp. 147–156.
Reavill, D.R., Schmidt, R.E., Phalen, D.N. (2003). Gastrointestinal system and pancreas. In: Reavill, D.R., Schmidt, R.E., Phalen, D.N. (eds), *Pathology of Pet and Aviary Birds*. Ames, IA: Iowa State Press, pp. 41–65.

Author George A. Messenger DVM, DABVP-Avian

FRACTURES AND LUXATIONS

BASICS

DEFINITION
• Fracture—complete or partial disruption of the integrity of bone. Can be further classified as a closed fracture (absence of a wound communicating with deeper tissue and bone) or an open fracture (presence of a wound communicating with deeper tissue and bone). Can be described based on fracture characteristics (e.g., transverse, oblique, comminuted, etc.) and fracture location (e.g., proximal, distal, articular, etc.).
• Luxation—complete dislocation of the bony elements comprising a joint. Partial dislocation is termed a *subluxation*. Can be described based on the direction of the dislocation (e.g., medial, craniodorsal, etc.).

PATHOPHYSIOLOGY
• Fracture—the majority of fractures in birds, particularly free-living birds, are of traumatic etiology. Pathologic conditions that weaken bone, such as nutritional secondary hyperparathyroidism and neoplasia, can be causal factors but are not the focus of this chapter. • Luxation—the majority of luxations in birds, particularly free-living birds, are of traumatic etiology. Tearing or stretching of ligaments and damage to surrounding muscle results in instability of a joint. Congenital, developmental, or degenerative abnormalities can contribute to joint instability but are not well described in birds.

SYSTEMS AFFECTED
• Musculoskeletal. • Skin—if fracture/luxation is open.
• Neuromuscular—injury to nerves can accompany fractures and luxations.
• Cardiovascular—generalized trauma can result in signs of shock.

GENETICS
None.

INCIDENCE/PREVALENCE
Unknown; however, traumatic injuries and resulting fractures are seen commonly in free-living birds of prey admitted to wildlife clinics or rehabilitation centers.

GEOGRAPHIC DISTRIBUTION
None.

SIGNALMENT
• **Species:** Any. • **Mean age and range:** N/A
• **Predominant sex:** N/A

SIGNS

General Comments
The signs associated with traumatic fractures and luxations can be similar if the fracture is close to a joint. In these cases the two can only be distinguished by careful palpation of the anesthetized bird and properly positioned, orthogonal view radiographs. Signs of generalized trauma may be present in birds suffering fractures or luxations. The overall status of the bird must be addressed along with the fracture or luxation.

Historical Findings
Free-living birds are often found in or on the side of a road due to having collided with a motor vehicle. Captive or pet birds may have a witnessed event of flying into an object, such as a window, becoming entangled, being stepped on, or being attacked by another animal.

Physical Examination Findings
• Wing droop or lameness. • Inability to fly in the absence of a wing droop can occur.
• Swelling. • Bruising. • If complete fracture—palpable sharp bone fragments and discontinuity of bone cortices. • If luxation—abnormal motion of the elements of a joint. • Laceration of skin if fracture or luxation is open.

CAUSES
• Motor vehicle collision (free-living birds).
• Collision with an object such as a window.
• Gunshot injuries (free-living birds).
• Attack by other animals. • Crushing injuries. • Entanglement.

RISK FACTORS
In pet birds, allowing unsupervised and unrestricted access to the home environment increases the chances of accidental traumatic injury.

DIAGNOSIS

DIFFERENTIAL DIAGNOSIS
• Soft tissue injury. • Sprain or strain.

CBC/BIOCHEMISTRY/URINALYSIS
N/A

OTHER LABORATORY TESTS
N/A

IMAGING
Radiography is generally sufficient for diagnosing fractures and luxations. Although there may be cases in which advanced imaging such as computed tomography (CT) may be beneficial, given the cost associated with CT and the adequacy of standard radiographs in the majority of cases, CT is not routinely indicated for the diagnosis of fractures and luxations. Fractures are usually readily visualized on properly positioned, orthogonal view radiographs, although incomplete fractures may be less obvious. Luxations may appear as displacement of bones from their normal articular positions or as increased joint space, depending on the degree of dislocation of the involved bones. Soft tissue swelling is often present with both fractures and luxations. It is beneficial to obtain whole body radiographs in birds that have suffered trauma, rather than focusing on the area of injury, in order to rule out concurrent injuries. General anesthesia or sedation is usually necessary to obtain diagnostic orthogonal view radiographs; however, the bird's overall condition must first be considered and addressed.

DIAGNOSTIC PROCEDURES
N/A

PATHOLOGIC FINDINGS
N/A

TREATMENT

APPROPRIATE HEALTH CARE
• Inpatient medical management—appropriate therapy for concurrent traumatic injuries; temporary stabilization of fracture or luxation. • Surgical management once patient is stable—necessary for best prognosis for some fractures and luxations. • Outpatient medical management—once fracture or luxation is stabilized either surgically or through external coaptation.

NURSING CARE
• Fluid therapy—indicated in patients suffering from generalized trauma.
• Bandaging—temporary bandaging of the affected bone(s) or joint is indicated to prevent further soft tissue trauma, to prevent closed fractures from becoming open, and to increase patient comfort level. Injuries of the wing distal to the elbow can be immobilized with a figure-of-eight bandage. Injuries proximal to the elbow or involving the elements of the pectoral girdle (coracoid, clavicle, and scapula) can be immobilized with a figure-of-eight bandage along with a body wrap. Injuries of the leg distal to the hock can be immobilized by taping the tarsometatarsus to the tibiotarsus with the hock in flexion. Injuries proximal to the hock can be immobilized via this type of bandage with the leg also wrapped against the body. Some fractures and luxations can be successfully managed and carry a good prognosis for full return to function when treated with bandaging alone. Examples include midshaft fractures of the ulna if the radius is intact, fractures of the bones of the pectoral girdle, and tarsometatarsal fractures in birds weighing less than 100 grams. In birds weighing less than 50 grams, such as passerines, bandaging and/or cage rest may be the only realistic treatment options in some cases due to the small size of the patient.
• Wound care—indicated for open fractures or luxations.

ACTIVITY
Patient activity should be restricted throughout the duration of treatment and gradually increased after surgical fixation or external coaptation is removed.

Fractures and Luxations (Continued)

DIET
N/A

CLIENT EDUCATION
If return to flight or normal ambulation is imperative, owners should be counseled regarding treatment options that carry the best prognosis for restoration of full function. Owners should be made aware that certain injuries, such as articular fractures and luxations that are not easily reduced, carry a poor to grave prognosis for return to full function and may result in degenerative joint disease over time. Owners should be informed of the monitoring and follow-up care that will be required during the course of management. For birds in which mobility may be permanently impaired due to the injury, owners should be counseled regarding modification of housing to accommodate the bird's needs.

SURGICAL CONSIDERATIONS
• Successful management of orthopedic injuries in birds requires knowledge of avian anatomy and bone healing along with the characteristics of avian bone and of the avian patient as a whole. • Orthopedic repair in birds should not be undertaken without careful study of techniques and careful consideration of the needs of the individual bird. The references listed at the end of this chapter provide comprehensive information on avian orthopedics and will help guide clinicians' decision making regarding specific fracture/luxation management. • Best course of treatment is always determined on an individual basis. Factors to consider include patient size, the type and location of the fracture/luxation, the need for full return to function, the proximity of a fracture to a joint, the surgical experience level of the clinician, and the availability of orthopedic referral services. • Avian practitioners who refer patients for surgical repair should be prepared to provide guidance to the surgeon regarding the factors that differentiate orthopedic surgery in birds versus mammals if the surgeon has limited experience with birds. • Open fractures are common in birds due to the lack of soft tissue coverage over the distal limbs; however, open fractures can have a good prognosis if treated promptly and appropriately. • Some avian bones are pneumatic. Which bones are pneumatic varies among species. The humerus is commonly pneumatic, as well as the femur in many species. Irrigation of these bones during triage of open fractures or during surgical repair must be performed very cautiously to avoid introducing irrigation solutions into the connecting air sacs. • Avian cortices are very thin. This characteristic can affect the ability of the bone to hold hardware. • The high calcium content of avian bones makes them brittle and prone to fissures. • The ideal fixation device is lightweight and entirely removable. Due to the small size of many avian patients, heavy fixation devices will result in discomfort and morbidity. In free-living birds intended for release, all hardware should be completely removed in most cases. An external skeletal fixator with an intramedullary pin incorporated into the connecting bar (ESF-IM pin tie-in, or tie-in fixator) has been used successfully in many long bone fractures in multiple species of birds. • The small size of many birds (e.g., passerines) and the fact that the bones of the wing are nonweight bearing are considerations affecting the choice of appropriate stabilization or repair. • Bone healing in otherwise healthy birds receiving appropriate nutrition occurs rapidly. Clinical union can be complete in as little as three weeks and rarely requires longer than six weeks. • Treatment of luxations always entails reduction of the luxation along with stabilization. If luxations are easily reduced, stabilization with bandaging may be appropriate; otherwise surgical intervention is indicated.

MEDICATIONS

DRUG(S) OF CHOICE
• Analgesics—pain management is of the utmost importance in emergent patients and throughout the course of treatment. Opioids can be used in initial stages of treatment as well as peri- and postoperatively. Nonsteroidal anti-inflammatories (NSAIDs) can be used postoperatively and during the course of management. Opioids or tramadol can be combined with NSAIDs for multimodal analgesia. • Antibiotics—indicated peri-operatively, for open fractures, and generally for the duration of time an ESF-IM pin tie-in is in place. Enrofloxacin and amoxicillin-clavulanic acid are appropriate empiric choices. If infection is present or suspected, culture and sensitivity should guide antibiotic selection. • Antifungals—should be considered prophylactically for species at higher risk of respiratory aspergillosis secondary to antibiotic therapy and for species of free-living birds at higher risk due to the stress of being held in captivity for treatment and rehabilitation.

CONTRAINDICATIONS
Consult Food Animal Residue Avoidance Databank (FARAD.org) for antibiotic choices when managing backyard flocks.

PRECAUTIONS
NSAIDS—avoid or use with caution in hypovolemic patients and in patients with hepatic or renal impairment.

POSSIBLE INTERACTIONS
N/A

ALTERNATIVE DRUGS
N/A

FOLLOW-UP

PATIENT MONITORING
• Radiographic monitoring should be performed every 7–10 days during the course of treatment to monitor healing. For fractures managed with external coaptation, bandage changes should be performed every 7–10 days as well. • Appropriate therapy for and monitoring of wounds associated with open fractures must be performed. For fractures managed with external coaptation, initial wound care may necessitate more frequent bandage changes. • Clinical union by formation of fibrous callus, which may not be visible on radiographs, precedes bony callus formation. Palpation of a firm callus at the fracture site despite lack of visible callus on radiographs indicates appropriate healing. • Inspection and cleaning of pin sites in patients with external skeletal fixators should be performed daily by owners for pet birds, if possible. Free-living birds in a rehabilitation setting should be examined in hand at least weekly, but daily handling is best avoided, if possible, to decrease patient stress level. • In free-living birds, full return to function is a requirement for releasing the bird back to its natural habitat. Birds must be flight tested, observed for normal flight or locomotion/perching ability, and flight conditioned in a flight cage of appropriate size for the species.

PREVENTION/AVOIDANCE
Modification of the home environment may be necessary for pet birds that incurred an injury via an accident in the home.

POSSIBLE COMPLICATIONS
• Decreased range of motion in joints secondary to bandaging—reversible in most cases following removal of bandage. • In birds with leg fractures, the contralateral foot may develop bumblefoot if weight bearing is not restored to the injured leg during or following treatment. • Degenerative joint disease and loss of range of motion with articular fractures or luxations. • Malunion or permanent decreased range of motion resulting in inability to fly or chronic lameness. • Nonunion. • Osteomyeltitis or soft tissue infection.

EXPECTED COURSE AND PROGNOSIS
Highly variable and dependent on injury, selected course of treatment, and need for full function. If return to full function is not required, prognosis is for healing and acceptable outcome for most orthopedic injuries with appropriate treatment is good. For free-living birds requiring full function, prognosis for many orthopedic injuries can be also good if appropriate management and flight reconditioning is provided. Prognosis for healing is poor for chronic fractures.

Prognosis for full return to function is poor for articular fractures and luxations that are not easily reduced.

 MISCELLANEOUS

ASSOCIATED CONDITIONS
N/A

AGE-RELATED FACTORS
None.

ZOONOTIC POTENTIAL
None.

FERTILITY/BREEDING
None.

SYNONYMS
None.

SEE ALSO
Air sac rupture
Arthritis
Bite wounds
Lameness
Metabolic bone disease
Pododermatitis
Procedures: www.fiveminutevet.com/avian
Splay leg/slipped tendon
Trauma

ABBREVIATIONS
N/A

INTERNET RESOURCES
N/A

Suggested Reading
Beaufrère, H. (2009). A review of biomechanic and aerodynamic considerations of the avian thoracic limb. *Journal of Avian Medicine and Surgery*, **23**(3):173–185.

Harcourt-Brown, N.H. (2005). Orthopaedic and beak surgery. In: Harcourt-Brown, N., Chitty, J. (eds), *BSAVA Manual of Psittacine Birds*. Gloucester, UK: British Small Animal Veterinary Association, pp. 120–135.

Helmer, P., Redig, P. (2006). Surgical resolution of orthopedic disorders. In: Harrison, G.J., Lightfoot, T.L. (eds), *Clinical Avian Medicine*. Palm Beach, FL: Spix Publishing Inc., pp. 761–774.

Orosz, S.E., Ensley, P.K., Haynes, C.J. (1992). *Avian Surgical Anatomy: Thoracic and Pelvic Limbs*. Philadelphia, PA: WB Saunders.

Redig, R., Cruz, L. (2008). Fractures. In: Saymour, J. (ed.), *Avian Medicine*, 2nd edn. Edinburgh, UK: Mosby Elsevier, pp. 215–247.

Author Maureen Murray, DVM, DABVP (Avian)

GASTROINTESTINAL FOREIGN BODIES

 ## BASICS

DEFINITION
Gastrointestinal foreign bodies (GIFB) are relatively common in captive birds. Both indoor and outdoor birds are at risk of ingesting foreign material. Outdoor housed nonpsittacine birds are more likely to ingest sand, bedding, plant material, and large metallic objects such as screws, coins, and wires. Of the indoor pet psittacine birds, young cockatoos seem to be at a greater risk of foreign object ingestion when compared to other species. While any household item has the potential to become a foreign body, commonly ingested items include pieces of cage toys, cage hardware, grit, perches, and bedding. Flighted pet birds may chew ledges, woodwork, and other objects in the vicinity where they land and perch. Old paint may contain lead.

PATHOPHYSIOLOGY
Not all foreign ingested material results in disease. Clinical signs and disease state are based on size, amount, and type of material ingested, location within the gastrointestinal (GI) tract, and potential toxicity of the foreign matter. Small pieces of nontoxic material may pass through the GI tract without consequence. Ingested toxic material containing lead or zinc can lead to a systemic illness unrelated to particle size. Ingested wires, although uncommon, can lead to perforation and may additionally contain toxic metals.

SYSTEMS AFFECTED
- Gastrointestinal—irritation, ulceration, perforations, partial or full obstruction, decreased motility/crop stasis, peritonitis, pancreatitis. • Hemic—anemia (lead/zinc).
- Renal—hemoglobinuria (lead).
- Nervous—seizures, paresis (lead).

GENETICS
N/A

INCIDENCE/PREVALENCE
Ingestion of foreign material is relatively common in indoor pet birds. Of the outdoor birds, ratites on sand are at high risk for sand impaction. Waterfowl risks include lead sinkers and hooks in fishing areas. Back-yard poultry pets can pick up metal objects, small rocks, or overeat straw bedding.

GEOGRAPHIC DISTRIBUTION
There is no geographic distribution unique to the development of foreign bodies; however, lead-associated GIFB most likely seen in older urban dwellings where lead-based paint exists (lead was outlawed in paint in the early 1970s).

SIGNALMENT
- **Species:** Psittacines (cockatoos), galliformes, ratites, waterfowl. • **Mean age and range:** All ages susceptible but young cockatoos overrepresented in the literature.
- **Predominant sex:** Both sexes at risk.

SIGNS

General comments
Signs of illness are generally nonspecific "sick bird".

Historical Findings
Anorexia, lethargy, fluffed feathers, diarrhea, vomiting, decreased vocalizations, not perching, falling, seizures, behavior changes.

Physical Examination Findings
- Some birds have no physical examination abnormalities. • Weakness. • Dehydration.
- Weight loss. • Crop stasis. • Crop impaction. • Melena. • Hemoglobinuria (lead). • Anemia. • Paresis in limbs.

CAUSES
N/A

RISK FACTORS
Uncaged and unsupervised pet birds are at risk. Caged birds with inappropriate toys, sandpaper or cement perches, overexposure to grit, access to bedding like corncob or sand. Older urban environments are more likely to have lead-based paint.

 ## DIAGNOSIS

DIFFERENTIAL DIAGNOSIS
- Proventricular dilatation disease (PDD).
- Renal disease. • Liver disease. • Neoplasia (proventricular/esophageal). • Thyroid tumor (for crop stasis/regurgitation.) • Epilepsy (seizures).

CBC/CHEMISTRY/URINALYSIS
Depending on foreign material ingested potential findings include: Hyperuricemia, anemia, elevated liver enzymes, and hematuria.

OTHER LABORATORY TESTS
Metal toxicity best diagnosed by specific toxin blood testing: Lead (whole blood), zinc (plasma).

IMAGING
Whole body radiographs taken in lateral and ventro-dorsal projections enable visualization and location of foreign material. Metallic foreign bodies are usually still in the GI tract at time of presentation but can pass out in feces or be absorbed. Mineral-based gravel and grit are readily visible but radiolucent material such as bedding, fibers, string, wood or plastic may require contrast studies using barium sulfate or iohexol to visualize. Sick patients unable to tolerate sedated radiographs can have a "box shot" or standing film to rule out metal pieces. Fluoroscopy can be performed in the unsedated bird to evaluate motility of the GI tract. Dilation and low motility of the gastrointestinal tract is suggestive for, but not diagnostic of proventricular dilatation disease

DIAGNOSTIC PROCEDURES
Endoscopic examination for identification and removal of proventricular or ventricular foreign bodies.

PATHOLOGIC FINDINGS
Zinc toxicity can result in acute sloughing of the koilin layer of the ventriculus. Zinc also causes acute hemolytic anemia and necrotizing pancreatitis. Lead toxicity can be challenging to diagnose via necropsy so in suspect cases take radiographs of the carcass to evaluate for metal ingestion.

 ## TREATMENT

APPROPRIATE HEALTH CARE
Pet birds that are sick from GIFB's need to be treated as inpatients until stable. Some birds require surgery or endoscopy but will need to be stabilized before prolonged anesthetic procedures.

NURSING CARE
- Birds with foreign body disease are usually painful and dehydrated. Parenteral analgesics and fluid therapy are indicated. • Lactated Ringer's Solution (LRS) maintenance dose is 50 ml/kg divided q8–12h but needs to be increased if diarrhea, vomiting, or hemoglobinuria from lead toxicity. • If using injectable chelation therapy (e.g., calcium EDTA), do not give in same location as LRS as subcutaneous chelation may occur. • Do not use nonsteroidal anti-inflammatory drugs (NSAIDs) in cases with potential gastric erosions or ulcers. • Parenteral butorphanol at 1–5 mg/kg IM q4–8h can be used in most cases but can result in transient dose-dependent sedation. Indoor pet birds need warm, quiet, dark incubator without a perch if weak. • Nutritional support will vary depending on function of the GI tract.
- Gavage feeding is generally needed in sick birds that are not regurgitating. • Use antibiotics if compromise of the GI tract is suspected (ulceration, perforation, melena).

ACTIVITY
Sick birds should be confined with limited activity and no flying and no perch. Remove water bowls if ataxic or seizuring.

DIET
Bird should be NPO if obstructed, impacted, or perforated in the GI tract. Sick birds benefit from gavage feedings of an enteral diet until able to resume normal diet. Use an easy to absorb enteral formula such as Emeraid® (Lafeber).

CLIENT EDUCATION
It is beneficial to inspect the cage where the bird may have picked up foreign material. Have the owner remove sandpaper, grit,

Gastrointestinal Foreign Bodies

inappropriate bedding and toys as needed. Clients need to assess premises if lead exposure took place indoors.

SURGICAL CONSIDERATIONS
Small particles can often pass the GI tract unassisted or with the use of laxatives/lubricants. Foreign objects that are not amenable to endoscopic retrieval must be removed surgically. Foreign objects and impactions in the crop can be easily and safely removed with an ingluviotomy. Surgery of the proventriculus and ventriculus carries a much greater risk and deserves careful consideration. Complications include leakage at the incision site, peritonitis and dehiscence. For ventricular foreign bodies that require surgical removal, this author prefers the ventral approach as described by Altman (1997).

MEDICATIONS

DRUG(S) OF CHOICE
- Small particles in the lower GI tract can be treated with laxatives (cat hairball type, lactulose, or Epsom salts/magnesium sulfate) or lubricants (mineral oil/peanut butter). Magnesium sulfate is used when there is still lead in the GI tract to bind and decrease absorption (500–1000 mg/kg PO q12–24h × 1–3 days). • Patients with lead or zinc toxicity are best treated with parenteral chelation using calcium EDTA (calcium disodium versenate/edetate calcium disodium) at 30–35 mg/kg SQ q12h, in saline only (not LRS) to decrease pain at injection site. Direct intramuscular injections are done at home if hospitalization is not possible. For enteral therapy in stable birds use oral preparations of DMSA (2,3 dimercaptosuccinic acid or succimer) at 30 mg/kg PO q12h for 7–21 days. • Birds with gastric ulcers should be treated with sucralfate suspension (1 ml/kg PO q8h) and metoclopromide (0.5 mg/kg IM q8–12h).

CONTRAINDICATIONS
N/A

PRECAUTIONS
Do not use NSAIDs like meloxicam or carprofen in patients with gastric foreign bodies, melena, or hemoglobinuria.

POSSIBLE INTERACTIONS
Do not give calcium EDTA chelation therapy in fluid pocket with Lactated Ringer's solution (LRS) to avoid chelation in the subcutis and decreased bioavailability.

ALTERNATIVE DRUGS
For the treatment of heavy metal or zinc toxicity, penicillamine (Cuprimine, Merck) can be used in most species at 50 mg/kg PO q24h × 1–6 weeks once stable or after treatment with CaEDTA.

FOLLOW-UP

PATIENT MONITORING
Repeat radiographs to evaluate location or elimination of foreign material as needed. Droppings can also be checked for the presence of foreign material. In birds with toxic blood levels of lead or zinc, follow-up blood testing is generally recommended after the cessation of chelation therapy. Chelation can cause a transiently high metal level as the toxin is being eliminated from the bloodstream.

PREVENTION/AVOIDANCE
Pet birds should not be unsupervised in the household, and should be caged when owners are away. Environmental clean-up for outdoor housed birds.

POSSIBLE COMPLICATIONS
• Perforations from sharp metallic GIFB are possible. • Seizures from heavy metal toxicity. • Intestinal obstructions and death. • Renal failure from severe acute hemoglobinuria.

EXPECTED COURSE AND PROGNOSIS
Most patients with small pieces of foreign material pass them without complications. Lead and zinc toxic patients often survive and recover well if diagnosed early and appropriately treated. Birds that require surgical intervention of the lower GI tract carry a more guarded prognosis.

MISCELLANEOUS

ASSOCIATED CONDITIONS
N/A

AGE RELATED FACTORS
Young birds may be more prone to foreign body ingestion.

ZOONOTIC POTENTIAL
N/A

FERTILITY/BREEDING
N/A

SYNONYMS
N/A

SEE ALSO
Anemia
Aspiration
Coelomic distention
Enteritis/gastritis
Heavy metal toxicosis
Ingluvial hypomotility/ileus
Proventricular dilatation disease
Regurgitation/vomiting
Sick-bird syndrome

ABBREVIATIONS
GIFB—gastrointestinal foreign body
NPO—nil per os

INTERNET RESOURCES
Lafebervet.com: Heavy metal poisoning in birds and other related topics.

Suggested Reading
Altman, R.B. (1997). Soft tissue surgery. In: Altman, R.B., Clubb, S.L., Dorrestein, G.M., Quesenberry, K.E. (eds) *Avian Medicine and Surgery*. Philadelphia, PA: W.B. Saunders, pp. 704–732.
Hoefer, H.L. (2005). Management of gastrointestinal tract foreign bodies in pet birds. *Exotic DVM*, **7.3**:4–27.
Hoefer H.L., Levitan, D. (2013). Perforating foreign body in the ventriculus of an umbrella cockatoo (*Cacatua alba*). *Journal of Avian Medicine and Surgery*, **27**(2):128–135.
Author Heidi L. Hoefer, DVM, DABVP (Avian)

Gastrointestinal Helminthiasis

BASICS

DEFINITION
Gastrointestinal helminthiasis is any macroparasitic disease in birds in which any part of the gastrointestinal tract is infected with parasitic worms known as helminths. These parasites are further classified as nematodes (roundworm-like), cestodes (tapeworms) or trematodes, although acanthocephalans are also occasionally seen.

PATHOPHYSIOLOGY
Transmission is either direct by ingestion of embryonated eggs or indirect by ingestion of an intermediate host, depending on the parasite species. Cestodes require an intermediate host, such as pillbugs, cockroaches, beetles, flies, etc. Once inside the host, the life cycle begins, generally with the attachment of the larva to the mucosa. Damage occurs through direct irritation and/or blood loss; parasites can also cause obstruction of the gastrointestinal tract in heavy infestations.

SYSTEMS AFFECTED
• Gastrointestinal—helminth parasites can be found throughout the gastrointestinal tract, from the oral cavity to the intestines.
• Hepatobiliary—trematodes may infect the liver and bile and pancreatic ducts.

GENETICS
N/A

INCIDENCE/PREVALENCE
Approximately 100 worm species have been recognized in wild and domestic birds in the USA. It is estimated that 10% of helminths cause significant disease in birds. Nematodes (roundworms) are the most significant in number of species and in economic impact. Of species found in commercial poultry, the common roundworm (*Ascaridia galli*) is by far the most common. In surveys of poultry raised under nonconfinement conditions throughout the world, an incidence of infection >80% is not uncommon. In confinement operations, parasitism is quite uncommon. In waterfowl, there are an extremely large number of parasites that have been discovered: 264 species of tapeworms, 52 species of acanthocephalans, and 536 species of trematodes. Nematodes are particularly a problem in wild-caught birds or those housed in planted aviaries that favor the parasite's life cycle.

GEOGRAPHIC DISTRIBUTION
Helminth parasites are seen in birds all over the world. Some parasites are more common in certain parts of the world, which is most likely as a result of the presence of the intermediate hosts.

SIGNALMENT
• **Species:** All species of birds can develop gastrointestinal helminthiasis. Some examples are: ○ Chickens—over 100 species of parasites seen; ascarids are by far the most common. ○ Anseriformes—over 800 species of helminths seen. ○ Ostriches—*Libyostrongylus douglassi*. ○ Pigeons—ascarids most common, *Capillaria* spp., *Dispharynx spiralis* (proventriculus), *Ornithostrongylus* spp. (proventriculus), *Tetrameres* spp., tapeworms, trematodes. ○ Parrots—ascarids, *Capillaria* spp., tapeworms. ○ Insectivorous finches—cestodes. • **Mean age and range:** Highly variable. • **Predominant sex:** Both sexes are susceptible

SIGNS

General Comments
Signs and findings include those that may be seen in conjunction with this condition and those that occur concurrently as a result of the inciting problem.

Historical Findings
There are many possible historical findings in association with this condition depending on the specific disorder.

Physical Examination Findings
• Lethargy. • Poor condition. • General debilitation. • Diarrhea. • Undigested seed in droppings. • Vomiting. • Weight loss.
• Death in some species with some parasites.
• Anorexia, dysphagia, head flicking and anemia in capillariasis of the upper GI tract.
• In Anseriformes with *Sphaeridiotrema globulus*—bloody cloacal discharge, wing droop, death in 5–6 days, chronic enteritis possible, weight loss, lameness.

CAUSES
There are thousands of species of helminth parasites in avian hosts. Some of the more common or important parasites are:
• Ascarids—especially young raptors, budgies, cockatiels, princess parrots, Australian grass parakeets, chickens. • Capillarids—raptors, pigeons, pheasant, vulturine guinea fowl.
• Cestodes—insectivorous finches, parrots of wild stock, especially African greys, eclectus parrots, and cockatoos—*Raillietaenia*, *Choanataenia*, *Gastronemia*, *Idiogenes*, and *Amoebataenia*. • *Dispharynx nasuta*—wild game birds, pigeons, passerines.
• *Libyostrongylus douglassi* (ostrich)—proventricular worms.
• *Amidostomum* spp.—anseriformes.
• *Eustrongylides ignotus*—herons and egrets—proventricular worms. • *Tetrameres* spp.—ducks, chickens, pigeons, aquatic birds—proventricular worms.
• Trematodes—especially *Sphaeridiotrema globulus* in water birds—scaup, mallards.

RISK FACTORS
Metazoan parasites are more commonly seen in birds housed in cages with access to the ground and in wild-caught birds. Presence of intermediate hosts. Parasites will be seen more frequently in situations where there are poor hygienic conditions, such as inadequately maintained pigeon lofts and chicken coops. Planted aviary and zoo situations are harder to clean and, therefore, birds can become reinfected by ingesting ova and intermediate hosts.

DIAGNOSIS

DIFFERENTIAL DIAGNOSIS
• Bacterial infections of the GI tract—salmonellosis, Gram negative infections, *Enterococcus* spp., clostridial infection. • Flagellate enteritis. • Candidiasis.
• Cryptosporidiosis. • Chlamydiosis.
• Macrorhabdosis. • Mycobacteriosis. • Viral infections—adenovirus, paramyxovirus, reovirus, coronavirus, rotavirus, herpesvirus, and more. • Gastrointestinal neoplasia.
• Numerous other diseases that result in poor body condition and weight loss.

CBC/BIOCHEMISTRY/URINALYSIS
Findings are not commonly documented, but in theory eosinophilia might be present with parasitism; leukocytosis with heterophilia might be seen in cases where there is significant inflammation; certain organisms tend to cause anemia; hypoproteinemia could result from malabsorption. Liver involvement, as in the case of trematodiasis, could result in elevated bile acids and liver enzymes.

OTHER LABORATORY TESTS
• Fecal flotation. • Direct smears with or without saline. • Intestinal or proventricular scrapings on necropsy. • Other tests to check for concurrent bacterial or viral infection, such as fecal Gram's staining, fecal acid-fast stain and other stains, culture and sensitivity, PCR—*Cryptosporidium* spp., viral diseases.

IMAGING
Imaging is not generally carried out, but in cases of GI parasitism where there are no fecal sample results, or the results are negative, radiographs might be performed; findings would be nonspecific and would not necessarily be helpful except to rule out some other disorders.

DIAGNOSTIC PROCEDURES
N/A

PATHOLOGIC FINDINGS
• In general, pathological findings are confined to the gastrointestinal tract, from the mouth to the ceca. • There are varying degrees of inflammation, thickening of the intestinal walls, epithelial hyperplasia, ulceration, necrosis, sloughing, hemorrhage, or plaque-like lesions, and occasional perforation of the stomach or intestinal wall, depending upon the parasite and the location. In all cases, there are accumulations of the parasites, either in the lumen or imbedded in the wall

Gastrointestinal Helminthiasis

of the intestine. • Depending on the parasite and the location in the body, various lesions are found. Examples include: ○ *S. gobulus*—fibrinohemorrhagic ulcerative enteritis in the lower small intestine in scaup, canvasback long-tailed ducks, swans and mallards. ○ Roundworms—intestinal blockage can occur. ○ *Tetrameres* spp.—raspberry-like appearance of the stomach of pigeons. ○ *Dispharynx* and *Ornithostrongylus* in pigeons—severe proventricular hemorrhage. ○ *Capillaria* spp.—oral inflammatory masses, diphtheritic oral lesions, hemorrhagic inflammation around commissures of beak.

TREATMENT

APPROPRIATE HEALTH CARE
Patients that are critical need to be hospitalized for work-up and treatment. Patients that are unthrifty and are diagnosed with parasitism at the time of the office visit can be treated as outpatients.

NURSING CARE
Supportive care must be done in the sicker patients. This might include providing warmth, fluid therapy, nutritional support/gavage feeding and fluid therapy.

ACTIVITY
N/A

DIET
N/A

CLIENT EDUCATION
Proper administration of medications should be explained. For poultry, provide appropriate withdrawal times. Discuss control of intermediate hosts and reduction of risk factors for reinfection.

SURGICAL CONSIDERATIONS
N/A

MEDICATIONS

DRUG(S) OF CHOICE
It is difficult to draw conclusions about the proper use of anthelmintics for all birds, due to the wide variety of species and types of parasites involved. Medications that are commonly used in mammals are generally safe and effective in birds, including ivermectin, moxidectin, fenbendazole, pyrantel pamoate, levamisole and praziquantel. Examples of specific and general dosages that are suggested for some conditions are: • Anseriformes: For cestodes—praziquantel 10–20 mg/kg PO; repeat in 10–14 days. • Anseriformes: For roundworms, gizzard worms, stomach worms—ivermectin 200 ug/kg PO once. • Acanthocephalans in anseriformes: Thiabendazole. • Parrots: For tapeworms—praziquantel 8 mg/kg IM, PO and fenbendazole. • Pigeons: For ascarids, *Capillaria*, *Tetrameres*, strongyles—fenbendazole 10–12 mg/kg PO QD for three days. • Pigeons: For nematodes, including ascarids—pyrantel pamoate 25 mg/kg PO; repeat in 14 days. • Pigeons: Best for all nematodes, excellent for *Capillaria* spp. and *Tetrameres* spp., but less effective for ascarids—ivermectin 200 ug/kg PO, SQ, IM. • Pigeons: For cestodes—praziquantel 10–30 mg/kg PO; repeat in 10–14 days. • *Capillaria* (difficult to eradicate) ○ fenbendazole 100 mg/kg once or 25 mg/kg daily for five days – repeat in 14 days. ○ oxfendazole 10 mg/kg, levamisole 40 mg/kg (narrow safety margin); repeat in 14 days. ○ moxidectin 20 ug/kg, ivermectin 200 ug/kg. • Finches: For gizzard worms—80 mg levamisole or 50 mg fenbendazole/L of drinking water for three days. • Chickens and turkeys: For ascarids—piperazine is approved. Give as a single dose, 50 mg/bird (<6 wk of age), 100 mg/bird (≥6 weeks of age), in the feed at 0.2–0.4% or in the drinking water at 0.1–0.2%; it may be administered to turkeys at 100 mg/bird (<12 weeks old) or 200 mg/bird (≥12 weeks old). For severe cases, treatment can be repeated after 14 days. These medications must be withdrawn 14 days before slaughter. Piperazine is not approved in the USA for birds producing eggs for human consumption. • Chickens: For ascarids, cecal worms and capillarids—hygromycin B given at 8–12 g/ton in feed. A withdrawal time of three days is required; fenbendazole (not approved) but is effective against *Ascaris* when administered once at 10–50 mg/kg; if needed the treatment can be repeated after 10 days; at 10–50 mg/kg, fenbendazole when administered daily over five days is effective against *Capillaria*. Fenbendazole is also effective against other nematodes when administered at 10–50 mg/kg/day for 3–5 days or as a single dosage of 20–100 mg/kg, or added to the drinking water at 125 mg/L for five days or to the feed at 100 mg/kg. • Chickens: For capillarids—coumaphos, 0.004% in feed for 10–14 days for replacements, or 0.003% in feed for 14 days for layers • Turkeys: For ascarids—fenbendazole is approved for use in growing turkeys at the rate of 14.5 g/ton of feed (16 ppm), fed continuously as the sole ration for six consecutive days. No withdrawal time is required.

CONTRAINDICATIONS
Consult Food Animal Residue Avoidance Databank (FARAD.org) for drug choices when managing backyard flocks.

PRECAUTIONS
• Fenbendazole can cause toxicosis in a number of species, including pigeons and doves, vultures, storks, cockatiels and lories. This can manifest as bone marrow suppression and direct intestinal tract cell damage in affected species. Fenbendazole should not be administered to poultry during molt, because it may interfere with feather regrowth. • Ivermectin needs to be diluted properly. It is best to dilute in propylene glycol and is sensitive to light.

POSSIBLE INTERACTIONS
Hygromycin B is a cholinesterase inhibitor, and treated birds should not be exposed to other cholinesterase inhibitors (drugs, insecticides, pesticides, or chemicals) within three days before or after treatment.

ALTERNATIVE DRUGS
The use of diatomaceous earth supplemented at 2% in feed and fed continuously lowers numbers of *Heterakis* and *Capillaria* in chickens.

FOLLOW-UP

PATIENT MONITORING
All patients should be monitored for signs of toxicity during treatment. Follow-up examinations, including fecal examinations and/or further deworming should be performed as clinical experience dictates.

PREVENTION/AVOIDANCE
• Control is achieved by preventing access to fecal matter, and fecal-contaminated food and water. If this is not feasible, regular worming (every 2–3 months) may be necessary. • Deworm pigeons prior to the racing and breeding season. • Environmental control to prevent access to insects (or other intermediate hosts) and feces is essential. • Insect control is essential to prevent cestode reinfestations. • Removal of earthworms should ideally be done. • Pigeon lofts must be thoroughly cleaned regularly.

POSSIBLE COMPLICATIONS
Secondary bacterial infections can occur with some of these parasites.

EXPECTED COURSE AND PROGNOSIS
Many cases of parasitism in birds are seen in wildlife and zoo and aviary birds. Therefore, detection is not necessarily done until the bird has died. In cases of flock parasitism, once a diagnosis is made, most parasites can be quite effectively treated.

MISCELLANEOUS

ASSOCIATED CONDITIONS
There could be secondary bacterial infections with some of these parasites. In chickens, *Heterakis* is commonly a carrier of *Histomonas meleagridis* also. Neurologic impairment may be caused by aberrant migration of the raccoon roundworm, *Baylisascaris procyonis*, into the central nervous system of avian species which can serve as an intermediate host.

Gastrointestinal Helminthiasis

AGE-RELATED FACTORS
Young or immunocompromised birds would be more likely to suffer the effects of helminth parasitism.

ZOONOTIC POTENTIAL
There are no known zoonoses associated with gastrointestinal helminths in birds.

FERTILITY/BREEDING
Birds that are seriously ill have a decreased potential for reproduction. Heavy parasitism of several types can result in a drop in egg production in poultry.

SYNONYMS
N/A

SEE ALSO
Anemia
Diarrhea
Emaciation
Enteritis/gastritis
Liver disease
Oral plaques
Proventricular dilatation disease
Regurgitation/vomiting
Undigested food in droppings
Urate/fecal discoloration
Sick-bird syndrome

ABBREVIATIONS
N/A

INTERNET RESOURCES
http://www.merckmanuals.com/vet/poultry/helminthiasis/overview_of_helminthiasis_in_poultry.html
For a list of approved drugs for poultry: www.fda.gov

Suggested Reading
Brandão, J., Beaufrère, H. (2013). Clinical update and treatment of selected infectious gastrointestinal diseases in avian species. *Journal of Exotic Pet Medicine*, **22**:101–117.
Clyde, C.I., Patton, S. (2000). Parasitism of caged birds. In: Olsen, G.H., Orosz, S.E. (eds), *Manual of Avian Medicine*. St Louis, MO: Mosby, Inc, pp. 424–448.
Doneley R.J.T. (2009). Bacterial and parasitic disease of parrots. *The Veterinary Clinics of North America. Exotic Animal Practice*, **12**(3):423–431.
Patton, S. (2000). Avian parasite testing. In: Fudge A.M. (ed.), *Laboratory Medicine–Avian and Exotic Pets*. Philadelphia, PA: WB Saunders, pp. 147–156.
Reavill, D.R., Schmidt, R.E., Phalen, D.N. (2003). Gastrointestinal system and pancreas. In: Reavill, D.R., Schmidt, R.E., Phalen, D.N. (eds), *Pathology of Pet and Aviary Birds*. Ames, IA: Iowa State Press, pp. 41–65.

Author George A. Messenger, DVM, DABVP (Avian)

Heavy Metal Toxicity

BASICS

DEFINITION
Toxicities induced by the ingestion of certain forms and doses of metals. Lead and zinc poisonings frequently occur in pet and wild birds, while copper and mercury toxicities are mainly seen in the wild.

PATHOPHYSIOLOGY
• Heavy metals are absorbed mainly by the gastrointestinal tract (rarely by the lungs after inhalation). Due to the poor rate of absorption, lead shot embedded in muscles does not usually induce toxicity. • Lead is widely distributed in soft tissues, and bones serve as the long-term storage site for the metal. Lead is toxic to multiple enzymatic systems and interferes with the numerous cellular functions that require calcium. In particular, lead will affect the red blood cells (microcytic anemia and/or porphyrinuria), the renal tubules (Fanconi syndrome nephropathy), and the central and peripheral nervous system (encephalopathy and polyneuropathy by demyelination and neuronal necrosis). Both the immune and reproductive systems can also be impaired (infertility, embryonic mortality, and teratogenicity). • Though the exact mechanism of zinc toxicity is not completely understood, zinc has a high affinity for the pancreas, liver, and the kidneys. Through both its direct and indirect effects, zinc can cause anemia, pancreatitis, hepatic failure, and renal failure. Zinc salts have direct irritant and corrosive effects on tissue, interfere with the metabolism of other ions such as copper, calcium, and iron, and inhibit erythrocyte production and function.

SYSTEMS AFFECTED
• Gastrointestinal • Nervous • Neuromuscular • Hemic/Lymphatic/Immune • Renal/Urologic • Behavioral • Hepatobiliary • Reproductive

GENETICS
N/A

INCIDENCE/PREVALENCE
• Lead and zinc poisonings are both common in pet birds. • Lead toxicity is very common in wild birds, while mercury poisoning occurs less frequently.

GEOGRAPHIC DISTRIBUTION
Worldwide.

SIGNALMENT
• **Species:** All pet birds, wild and captive waterfowl, poultry, raptors. • **Mean age and range:** Lead will accumulate throughout life in wild birds, so adults are more at risk for chronic toxicity. • **Predominant sex:** No known sex predilection.

SIGNS
General Comments
• Subacute toxicosis can impair flight capacity and affect the behavior. Therefore, wild raptors with chronic lead exposure may present for trauma and not display the typical signs of acute toxicity. • The severity of clinical signs does not always correlate well with blood metal concentration. • Signs of lead toxicosis are generally more severe than those seen with zinc poisoning.

Historical Findings
• Nonspecific signs: Lethargy, anorexia, depression. • Gastrointestinal signs: Regurgitation, diarrhea. • Neurologic signs: Seizures, weakness, head down posture. • Urologic signs: Polyuria/polydipsia, red urine. • Sudden death.

Physical Examination Findings
• Abnormal mentation. • Ataxia, twitching, circling, paresis, paralysis, blindness. • Crop stasis. • Green staining around the cloaca. • Hemoglobinuria, porphyrinuria. • Weight loss (chronic).

CAUSES
Ingestion of particles of heavy metals, either directly (i.e., parts of toys, galvanized cages, lead-based paints, fishing weights, lead-coated seeds, pennies (>1982)) or indirectly (i.e., ingestion of a prey shot with lead pellets).

RISK FACTORS
Inappropriate supervision, exposure to old lead paints, drapery weights, jewelry, galvanized wires, and so on (pet birds). Presence of lead in the environment, hunting season (wild birds).

DIAGNOSIS

DIFFERENTIAL DIAGNOSIS
• Meningoencephalitis: Viruses (bornavirus, West Nile virus, paramyxovirus, reovirus), bacteria (*Chlamydia psittaci*, *Salmonella*), fungal (*Aspergillus* spp.), parasites (*Baylisascaris*, *Toxoplasma*, *Sarcocystis*). • Other toxicoses: Pesticides, cannabis, iatrogenic (itraconazole in African grey parrots), toxins (botulism). • Metabolic disorders: Hepatic and renal failure, diabetes mellitus, hypocalcemia. • Trauma. • Pancreatitis. • Foreign body ingestion. • Neoplasia: Central nervous system, digestive system, lymphoma. • Atherosclerosis.

CBC/BIOCHEMISTRY/URINALYSIS
• Regenerative microcytic hypochromic anemia. • Greater number of immature RBC, poikilocytosis, and nuclear abnormalities (fusiform, elongated, and irregular nuclei are seen with zinc toxicosis in birds). • Elevation of uric acid, AST, LDH and CK. • Hemoglobinuria and porphyrinuria (Amazon parrots).

OTHER LABORATORY TESTS
Reference ranges vary with species.

Lead
• Collect whole blood on EDTA or heparin. Do not submit serum or plasma. • Blood levels: As a general rule, intoxication is confirmed if the lead concentration in whole blood is >0.2 ppm (20 µg/dL). Subclinical toxicosis occurs at lower levels. Some species (pigeons, waterfowl) seem to be able to tolerate higher levels before showing clinical signs of acute toxicity. • Tissue levels >3–6 ppm in liver or kidneys. • Decreased ALAD activity.

Zinc
• Collect blood on serum tubes without rubber components. • Most pet birds have normal zinc levels below 3.5 ppm (0.35 mg/dL). • Zinc toxicity is suspected when liver concentrations >75 ppm (mg/kg) wet weight.

IMAGING
Metal particles can be visualized on radiographs. However, their absence does not rule out heavy metal toxicity.

DIAGNOSTIC PROCEDURES
None.

PATHOLOGIC FINDINGS
• Gross lesions: Nonspecific and not commonly reported (bile stasis, swollen kidneys, mottled pancreas, pectoral atrophy). • Microscopic findings (lead): Acute tubular necrosis (sometimes associated with characteristic acid-fast intranuclear inclusion bodies), myocardial and hepatocellular necrosis, brain edema, peripheral nerve degeneration, and necrosis of the ventriculus muscles. • Microscopic findings (zinc): Pancreatic necrosis and vacuolation/degranulation of acinar cells, hepatic biliary retention and hemosiderosis to multifocal necrotizing hepatitis, acute tubular necrosis, enteritis, erosive ventriculitis with koilin degeneration.

TREATMENT

APPROPRIATE HEALTH CARE
Patients with acute heavy metal poisoning should be hospitalized and require intensive medical care.

NURSING CARE
• Aggressive fluid therapy (renal failure), IV/IO if possible. • Nutritional support adapted to the species. • Keep the bird in an environment appropriate to its neurologic condition (padded cage, if having seizures; comfortable bedding, if sternally recumbent; etc.).

ACTIVITY
Birds with neurologic signs should not be allowed to fly.

HEAVY METAL TOXICITY (CONTINUED)

DIET
Fine or coarse grit, as well as cathartic emollients (peanut butter) may be added to the diet to hasten the passage of any metallic particles in the gastrointestinal tract.

CLIENT EDUCATION
If oral chelation therapy is administered at home, put an emphasis on the potential for drug toxicity.

SURGICAL CONSIDERATIONS
• Proventricular and ventricular saline lavages have been used to remove lead particles in birds. • In rare cases, endoscopic or surgical removal of the metallic particles may be warranted.

MEDICATIONS

DRUG(S) OF CHOICE
• Edetate Calcium Disodium ($CaNa_2EDTA$, injectable sterile solution) 35 mg/kg q12h IV or IM for five days, followed by a "rest" period of 3–5 days to allow a redistribution of tissue and fluid lead concentrations and to prevent excessive chelation of endogenous minerals. Assessment of blood concentrations after the rest period will indicate if the protocol needs to be repeated. • Drug of choice for lead, zinc, and mercury intoxications. • Midazolam 0.1–0.2 mg/kg IM can be used to control seizures. • Antibiotics (amoxicillin/clavulanic acid 125 mg/kg q12h PO, enrofloxacin 15 mg/kg q12h PO) might be indicated due to the immunosuppressive effects of lead and in cases with a severe enteritis.

CONTRAINDICATIONS
Previous history of acute renal failure (not associated with the current toxicity).

PRECAUTIONS
• Potential nephrotoxicity. • Chelation of other minerals such as zinc, magnesium, and copper with long-term use. • Might require additional dilutions for IV injections (if concentration is ≥ 150 mg/ml).

POSSIBLE INTERACTIONS
Use with caution with other nephrotoxic compounds.

ALTERNATIVE DRUGS
• Dimercaptosuccinic acid (succimer, DMSA, needs to be compounded) 20–35 mg/kg q12h PO for 5–7 days. Reassessment of metal concentrations is recommended. Not as effective for zinc chelation. Does not chelate other essential minerals. Not nephrotoxic. Narrow margin of safety (80 mg/kg can be lethal in cockatiels). • D-penicillamine 55 mg/kg PO q12h for 1–2 weeks, followed by a "rest" week before reassessing the blood concentrations. Not recommended if metallic particles still present in the GI tract as could increase their absorption. • Dimercaprol has a very narrow margin of safety and should not be used in birds.

FOLLOW-UP

PATIENT MONITORING
Heavy metal blood concentration should be reassessed a few days after chelation therapy is discontinued. If still high, the protocol is repeated until blood levels are considered within an acceptable range.

PREVENTION/AVOIDANCE
• Prevent access to any source of lead and zinc. • Birds should be under close supervision when free in the house.

POSSIBLE COMPLICATIONS
• Seizures can lead to severe trauma in birds. • Dehydration can be secondary to renal and gastrointestinal losses.

EXPECTED COURSE AND PROGNOSIS
• Prognosis is poor without chelation therapy and supportive care. • Birds respond generally very rapidly to chelation therapy with neurologic status being back to normal within 24–36h. • Medical intensive care and nutritional support in the hospital can be expected to be required for 3–5 days. • Outcome is usually positive if the bird is diagnosed and treated in a timely and appropriate manner.

MISCELLANEOUS

ASSOCIATED CONDITIONS
None.

AGE-RELATED FACTORS
None.

ZOONOTIC POTENTIAL
None.

FERTILITY/BREEDING
• Lead and mercury have been shown to reduce fertility and cause embryonic deaths. • Chelation therapy might be considered in breeding birds, even with subclinical blood lead concentrations.

SYNONYMS
Plumbism, saturnism (lead)
New wire disease (zinc)
Hydrargyria (mercury)

SEE ALSO
Anemia
Anorexia
Diarrhea
Enteritis/gastritis
Gastrointestinal foreign bodies
Ingluvial hypomotility/ileus
Liver disease
Neurologic conditions
Proventricular dilatation disease
Regurgitation/vomiting
Seizure
Sick bird syndrome
Urate/fecal discoloration

ABBREVIATIONS
N/A

INTERNET RESOURCES
N/A

Suggested Reading
Bauk, L., LaBonde, J. (1997). Toxic diseases. In: Altman, R.B., Clubb, S.L., Dorrestein, G.M., Quesenberry, K. (eds), *Avian Medicine and Surgery*. Philadelphia, PA: WB Saunders, pp. 604–613.
Friend, M., Franson, J.C. (1999). Chemical toxins. In: USGS, *Field Manual of Wildlife Diseases. General Field Procedures and Diseases of Birds*. Biological Resources Division, U.S. Geological Survey, Department of Interior, pp. 284–353.
Lightfoot, T.L., Yeager, J.M. (2008). Pet bird toxicity and related environmental concerns. *The Veterinary Clinics of North America. Exotic Animal Practice*, **11**(2):229–259.
Puschner, B., Poppenga, R.H. (2009). Lead and zinc intoxication in companion birds. *Compendium : Continuing Education for Veterinarians*, **31**(1):E1–E12.
Richardson, J.A. (2006). Implication of toxic substances in clinical disorders. In: Harrison, G.J., Lightfoot, T.L. (eds), *Clinical Avian Medicine*. Palm Beach, FL: Spix Publishing, Inc., pp. 711–719.

Author Marion Desmarchelier, DMV, MSc, DACZM
Acknowledgement Shannon Ferrell, DVM, DABVP (Avian), DACZM

 Client Education Handout available online

HEMOPARASITES

BASICS

DEFINITION
Most avian hemoparasites are protozoal and of little clinical significance; however, many types of avian hemoparasites can become pathologic under stressful conditions (e.g., captivity, breeding season, migration), when they infect a host species that is out of its natural ecosystem (e.g., captivity), or when vector species invade new geographic areas (e.g., due to climate change).

PATHOPHYSIOLOGY
• Lifecycle of *Plasmodium*: An infected vector (typically a *Culex* spp. mosquito) bites an uninfected bird; parasite sporozoites are passed into the bird's blood and via the bloodstream reach the liver; in the liver the sporozoites develop into pre-erythrocytic schizonts, which then become merozoites; merozoites enter erythrocytes and develop into macrogametocytes (female), microgametocytes (male), or segments (schizonts). Schizonts divide in erythrocytes (intraerythrocytytic merogony) indefinitely until the bird dies or the bird's immune system responds, therefore there is potential for persistence of infection with frequent relapses. Second-generation and subsequent generation exoerythrocytic schizonts can be seen in tissues other than the liver. Birds typically undergo an acute phase of infection where parasitemia increases steadily to a peak at 6–12 days after infection, then the host immune system begins to bring the infection under control; chronic infection then persists for the life of the bird, with recurrence of clinical disease possible. • Lifecycle of *Haemoproteus*: A vector (typically a midge or hippoboscid) ingests gametocytes in RBCs of an infected bird; inside the insect vector the parasites migrate from the insect's GI tract to the bloodstream, then to the salivary glands as sporozoites; sporozoites are injected into the bloodstream of a new bird when the insect feeds; sporozoites migrate from the bird's bloodstream into endothelial cells of various tissues (lung, liver, bone marrow, spleen) where they develop into schizonts; each schizont contains many merozoites that are released into the bloodstream when the endothelial cell dies; merozoites in the bloodstream enter RBCs to become gametocytes. Gametocytes in a bird's RBCs can become infective in as little as seven days after they enter the bird's RBCs; parasitemia in a host bird peaks at 10–21 days after infection and falls rapidly within seven days to a low intensity. • Lifecycle of *Leukocytozoon*: A vector (typically a black fly) ingests gametocyte-containing blood from an infected bird; gametocytes develop into sporozoites inside the fly; the fly injects sporozoites into the bloodstream of a new bird; sporozoites travel from the bloodstream of the new bird to invade endothelial and parenchymal cells of various tissues, such as liver, heart and kidney; sporozoites develop into schizonts, which then rupture and release merozoites that infect RBCs and leukocytes. Alternatively, released merozoites may be ingested by macrophages to become megaloschizonts in tissues such as the liver, lung and kidney, and from that point the megaloschizonts may release merozoites that develop into gametocytes.
• Possible clinical signs due to direct blood cell effects such as anemia ○ *Plasmodium* spp. ○ *Aegyptianella* spp. ○ *Leukocytozoon* spp. (not common).
• Possible clinical signs due to multiorgan and muscle tissue destruction as parasites progress through life cycle stages ○ *Plasmodium* spp. ○ *Leukocytozoon* spp. ○ *Atoxoplasma* spp. ○ *Haemoproteus* spp. (unusual).
• Transmitted via mosquitoes (*Culex* spp., *Mansonia crassipes*, *Aedeomyia squamipennis*) ○ *Plasmodium* spp. ○ Some *Trypanosoma* spp.
• Transmitted via hippoboscid flies ○ Some *Haemoproteus* spp. ○ Some *Trypanosoma* spp.
• Transmitted via biting midges (ceratopogonids, *Culicoides* spp.) ○ Most *Haemoproteus* spp. ○ *Leukocytozoon caulleryi*.
• Transmitted via black flies (simuliids) ○ Most *Leukocytozoon* spp. ○ Some *Trypanosoma* spp.
• Transmitted via mites, ticks, fleas or other arthropods ○ Some *Trypanosoma* spp. ○ *Hepatozoon* spp. ○ *Babesia* spp. ○ *Aegyptianella* spp. ○ *Borrelia anserina*.
• Transmitted via ingestion of sporulated oocysts (feces-contaminated water or food) ○ *Atoxoplasma* spp.

SYSTEMS AFFECTED
• Behavioral—lethargy and weakness are seen in symptomatic infections of most avian hemoparasites.
• Cardiovascular ○ Hemolytic anemia—*Plasmodium* spp., *Aegyptianella* spp., *Haemoproteus* spp. (unusual), *Leukocytozoon* spp. (unusual), *Borrelia anserina*. ○ Lymphocytosis, leukocytosis—*Plasmodium* spp., *Leukocytozoon* spp. (unusual).
• Hemic/Lymphatic/Immune ○ Spleen—*Atoxoplasma* spp., *Leukocytozoon* spp.
• Hepatobiliary ○ Liver—*Plasmodium* spp., *Atoxoplasma* spp., *Leukocytozoon* spp.
• Nervous ○ Central nervous system signs—*Plasmodium* spp., *Leukocytozoon* spp.
• Neuromuscular ○ Loss of balance, lameness or reluctance to move in galliformes—*Plasmodium* spp.
• Respiratory ○ Lungs—*Atoxoplasma* spp.

GENETICS
N/A

INCIDENCE/PREVALENCE
Prevalence of *Atoxoplasma* spp. can approach 100% in some passerine collections. *Haemoproteus* spp. are the most common blood parasite genus in birds. *Haemoproteus* spp., *Leukocytozoon* spp. and *Plasmodium* spp. are relatively common in wild birds. Seasonality of parasitemia generally coincides with vector prevalence.

GEOGRAPHIC DISTRIBUTION
• *Aegyptianella* spp. usually affects birds of tropical or subtropical climates.
• *Haemoproteus* spp. are distributed worldwide in temperate, tropical and subtropical climates.
• *Plasmodium* spp. and *Leukocytozoon* spp. are found in all zoogeographic regions except Antarctica (due to lack of mosquito vectors).

SIGNALMENT
• **Species** ○ *Atoxoplasma* spp. are especially pathogenic in small passerines, especially the families Fringillidae and Sturnidae. ○ *Aegyptianella pullorum* affects galliformes (chickens, turkeys) and anseriformes (ducks, geese). ○ *Trypanosoma* spp. usually affect passerines, galliformes, waterfowl and pigeons. ○ *Borrelia anserina* usually affects galliformes or waterfowl. ○ *Haemoproteus* spp. are found in many species, especially passerines, strigiformes and columbiformes. ○ *Plasmodium* spp. have been found in birds from nearly all avian orders (not yet reported in struthioniformes, coliiformes, or trogoniformes). *Plasmodium relictum* has been found in natural infections of birds of at least 70 avian families. *Plasmodium* spp. appears to be especially pathogenic in penguins, small passerines, galliformes (chickens, turkeys) and anseriformes (ducks, geese). ○ *Leukocytozoon* infections have been most often reported in passerines, galliformes and coraciiformes, but appear most pathogenic in anseriformes (ducks, geese, swans), galliformes (chickens), columbiformes, and less commonly in falconiformes. *Leukocytozoon simondi* is especially pathogenic in ducks and geese, and *L. caulleryi* is especially pathogenic in chickens in Asia. ○ *Aegyptianella* spp. have been reported in many species including galliformes, pigeons, crows, anseriformes, ratites, falcons, passerines and psittacines. *A. pullorum* is pathogenic in chickens. • **Mean age and range:** Atoxoplasmosis is usually a disease of young birds, particularly fledglings, and adults are usually asymptomatic.
• **Predominant sex:** N/A

SIGNS

General Comments
Most avian hemoparasites are of little clinical significance; however, many types of avian hemoparasites can become pathologic under stressful conditions (e.g., captivity, breeding season, migration), when they infect a host species that is out of its natural ecosystem (e.g., captivity), or when vector species invade new geographic areas (e.g., due to climate change).

HEMOPARASITES (CONTINUED)

Historical Findings
- Lethargy, listlessness—*Plasmodium* spp., *Leukocytozoon* spp., *Babesia shortti*, *Haemoproteus* spp. (unusual), *Leukocytozoon* spp. (unusual), *Borrelia anserina*.
- Labored breathing—*Leukocytozoon simondi* (unusual).
- Central nervous system signs (ataxia, convulsions)—*Plasmodium* spp., *Leukocytozoon* spp.
- Diarrhea—*Leukocytozoon simondi* (unusual), *Aegyptianella pullorum*.
- Erratic flight or other neurologic signs, vomiting—*Leukocytozoon toddi*.
- Loss of balance, lameness or reluctance to move in galliformes—*Plasmodium* spp.
- Acute death—*Leukocytozoon simondi* in juveniles.

Physical Examination Findings
- Weight loss—*Haemoproteus* spp., *Leukocytozoon* spp.
- Pale mucous membranes—*Plasmodium* spp., *Babesia shortti*, *Haemoproteus* spp. (unusual), *Leukocytozoon* spp. (unusual).
- Jaundice—*Aegyptianella* spp., *Babesia shortti*.
- Typically asymptomatic—*Trypanosoma* spp., *Hepatozoon* spp., most *Babesia* spp., *Haemoproteus* spp., *Leukocytozoon* spp., *Trypanosoma* spp., microfilaria of filarial nematodes.

CAUSES
See pathophysiology section above.

RISK FACTORS
Likelihood of clinical signs due to avian hemoparasites increases with seasonal changes in photoperiod, increased vector prevalence, increased reproductive activity, and exposure to predators. Likelihood of clinical signs is inversely correlated with host immunocompetence.

DIAGNOSIS

DIFFERENTIAL DIAGNOSIS
Most avian hemoparasites are differentiated using their appearance in blood smears. Multiple genera of hemoparasites may be present in the same patient. Numerous other nonparasitic etiologies exist for the nonspecific clinical signs of lethargy and weight loss.

CBC/BIOCHEMISTRY/URINALYSIS
- Parasites inside red blood cells
  ○ *Haemoproteus* spp.—elongate pigmented gametocyte, usually alongside or wrapping around rather than deforming the RBC nucleus. The degree of parasitemia can be used as a gauge of general immunocompetence of the host (inverse correlation). ○ *Plasmodium* spp.—usually a round pigmented gametocyte, trophozoite or schizont that may displace the RBC nucleus, but may be elongate and not displace the RBC nucleus. In contrast to *Haemoproteus* spp., *Plasmodium* can show schizogony in RBC's and endothelial cells of various organs, gametocytes can displace the RBC nucleus, and parasite stages can be seen within thrombocytes and leukocytes as well as in RBCs. ○ *Leukocytozoon* spp.—a gametocyte is sometimes round but is typically large, elongate, with wispy ends, and without pigmented granules; may distort the infected host cell so much that the cell's original identification is difficult. ○ *Aegyptianella* spp.—tiny nonpigmented vacuole appearance in RBCs. ○ *Babesia* spp.—nonpigmented white vacuole.
- Parasites inside white blood cells
  ○ *Hepatozoon* spp.—in monocytes or lymphocytes. ○ *Atoxoplasma* spp.—a single merozoite or a meront in monocytes or lymphocytes, causing an indentation in the cell's nucleus. ○ *Leukocytozoon* spp. ○ *Plasmodium* spp.
- Parasites inside thrombocytes ○ *Plasmodium* spp.
- Extracellular parasites—*Haemoproteus* spp. (if several hours elapsed between blood collection and smear preparation), *Trypanosoma* spp. (long, flagellated, with an undulating membrane), microfilaria of filarial nematodes, *Borrelia anserina* (spirochete with loose spirals).
- Hemoglobinuria—*Plasmodium* spp.
- Elevation of AST or ALT—*Leukocytozoon* spp. (unusual), *Atoxoplasma* spp., *Haemoproteus* spp., *Plasmodium* spp.
- Hypoalbuminemia—*Plasmodium* spp.
- Hypergammaglobulinemia, therefore hyperproteinemia—*Plasmodium* spp.
- Anemia—*Plasmodium* spp., *Aegyptianella* spp., *Haemoproteus* spp. (unusual), *Leukocytozoon* spp. (unusual), *Borrelia anserina*, *Aegyptianella* spp.
- Lymphocytosis, leukocytosis—*Plasmodium* spp., *Leukocytozoon* spp. (unusual), *Haemoproteus* spp.

OTHER LABORATORY TESTS
- Buffy coat smear, looking for parasites amongst white blood cells from a centrifuged hematocrit tube of whole blood—especially used for *Atoxoplasma* spp. and *Trypanosoma* spp.
- PCR of tissues or whole blood.
- Fecal direct smear and centrifugation/flotation with Sheather's sugar solution—sometimes used for *Atoxoplasma* spp. The oocysts of organism cannot be differentiated in this manner from typical enteric species of *Isospora* (two sporocysts containing four sporozoites each); however, the relative prevalence of *Atoxoplasma* versus enteric *Isospora* in some avian species is so disproportionate that the finding of oocysts in feces is diagnostic for *Atoxoplasma*. However, there is no correlation between presence of *Atoxoplasma* oocysts in feces and the presence of mononuclear merozoites in the same bird at the same time.
- Increased intensity of green color (biliverdin) in feces.
- Several serological tests (agar gel precipitation, counter-immunoelectrophoresis, immunofluorescence, enzyme-linked immunosorbent assay, immunoblot analysis, latex agglutination) have been developed for *L. caulleryi*, but not for other *Leukocytozoon* spp.

IMAGING
N/A

DIAGNOSTIC PROCEDURES
See blood analysis recommendations above.

PATHOLOGIC FINDINGS
- Splenomegaly: *Atoxoplasma* spp., *Haemoproteus* spp., *Plasmodium* spp., *Leukocytozoon* spp., *Aegyptianella* spp., *Borrelia anserina*.
- Hepatomegaly: *Atoxoplasma* spp., *Haemoproteus* spp., *Plasmodium* spp., *Leukocytozoon* spp., *Aegyptianella* spp., *Borrelia anserina*.
- Hepatic necrosis: *Atoxoplasma* spp., *Haemoproteus* spp., *Leukocytozoon* spp., *Aegyptianella* spp.
- Renal necrosis: *Aegyptianella* spp.
- Lung lesions: *Haemoproteus* spp.
- Muscle necrosis (white or hemorrhagic streaks): *Haemoproteus* spp.
- Atoxoplasmosis: also may see necrotic foci in spleen and/or heart; pancreatic edema and/or hemorrhage; fluid accumulation in intestines; ascites.
- Impression smears of spleen, liver or lung can be used to detect *Atoxoplasma* spp. sporozoites.
- *Atoxoplasma* can be confirmed in infected tissues via PCR.

TREATMENT

APPROPRIATE HEALTH CARE
See medications section below.

NURSING CARE
If a patient is symptomatic, general supportive care should be provided (nutrition, hydration, appropriate temperature, calm environment).

ACTIVITY
See "nursing care" section.

DIET
See "nursing care" section.

CLIENT EDUCATION
Chronic infection with some avian hemoparasites (e.g., *Haemoproteus*) may stimulate immunity to reinfection with homologous parasites of the same species; therefore, treatment may be elected against in cases of asymptomatic infection. However, immunosuppression due to stress or other factors may cause recrudescence of parasitemia and clinical signs.

SURGICAL CONSIDERATIONS
N/A

HEMOPARASITES (CONTINUED)

MEDICATIONS

DRUG(S) OF CHOICE
- *Atoxoplasma* spp.—reported treatment options include: ○ Sulfachlorpyrazine (ESB3^R) in drinking water at a dosage of 300 ppm (1 gram of 30% powder added to each 1 liter of drinking water). If sulfachlorpyrazine is used, a vitamin B_6 supplement should be given during the treatment. ○ Sulfachlorpyridazine (Vetisulid^R) in drinking water at a dosage of 300 ppm. If sulfachlorpyridazine is used, a vitamin B_{12} supplement should be given during the treatment. ○ Toltrazuril 12.5 mg/kg PO SID for 14 days. ○ Diclazuril. ○ Ponazuril should show some efficacy, although this remains anecdotal. • *Haemoproteus* spp.—reported treatment options include atebrine, plasmochin, chloroquine sulfate, quinacrine, primaquine, mefloquine, buparvaquone, pyrimethamine, pyrimethamine-sulfadoxine combinations, and tetracyclines.
- *Plasmodium* spp.—reported treatment options include chloroquine phosphate, primaquine phosphate, pyrimethamine-sulfadoxine combinations, mefloquine, sulfamonomethoxine, sulfachloropyrazine, doxycycline, halofuginone and atovaquone-proguanil combinations (Malarone™). • *Leukocytozoon* spp.—reported treatment options include pyrimethamine, pyrimethamine-sulfamonomethoxine in combination, clopidol, atebrine, trimethoprim-sulfamethoxazole combination, melarsomine, and primaquine • *Aegyptianella* spp.—doxycycline. • *Babesia shortti* in falconiformes—imidocarb dipropionate 5–13 mg/kg IM q7days for 2–3 weeks.

CONTRAINDICATIONS
N/A

PRECAUTIONS
N/A

POSSIBLE INTERACTIONS
N/A

ALTERNATIVE DRUGS
N/A

FOLLOW-UP

PATIENT MONITORING
See blood analysis section above.

PREVENTION/AVOIDANCE
- Two vaccines were developed for *Plasmodium relictum*; both provided protection for penguins and canaries against natural infection, but immunity was short-lived in canaries and immunity waned to that of unvaccinated control birds when challenged with mosquito vectors a year later.
- Two vaccines have been developed for protection of chickens against *L. caulleryi*.
- Frequent replacement of drinking water, bathing bowls and enclosure substrate to prevent ingestion of sporulated oocysts—*Atoxoplasma* spp. • Protection from flying insect vectors with screening—*Plasmodium* spp., some *Trypanosoma* spp., *Haemoproteus* spp., *Leukocytozoon* spp. • Protection from tick vectors with use of acaricides—some *Trypanosoma* spp., *Hepatozoon* spp., *Babesia* spp., *Aegyptianella* spp. • Absolute prevention of infection may be counterproductive for protection from some hemoparasite-induced diseases, as birds naïve to infection are much more likely to experience morbidity and mortality if subsequently infected.

POSSIBLE COMPLICATIONS
N/A

EXPECTED COURSE AND PROGNOSIS
Atoxoplasmosis is often a diagnosis made postmortem. Most other hemoparasites are either asymptomatic, or their numbers are able to be reduced in the host with appropriate medications.

MISCELLANEOUS

ASSOCIATED CONDITIONS
N/A

AGE-RELATED FACTORS
N/A

ZOONOTIC POTENTIAL
None

FERTILITY/BREEDING
N/A

SYNONYMS
- The term "malaria" is most accurately associated with *Plasmodium* spp., but some older literature uses the word "malaria" to refer to *Haemoproteus* spp. infections.
- Haemoproteosis, hematozoan disease, haemosporidian disease, blood parasite disease, Bangkok hemorrhagic disease (*L. caulleryi*), "going light" (*Atoxoplasma* spp.) in passerines, "black spot disease" (*Atoxoplasma* spp.—due to liver visible through body wall).
- Note, some older reports of *Lankesterella* infection may have been actually due to *Atoxoplasma* spp.

SEE ALSO
Anemia
Coagulopathies
Coccidiosis (systemic)
Ectoparasites
Emaciation
Hemorrhage
Liver disease
Sick-bird syndrome

ABBREVIATIONS
- RBC—red blood cell • AST—aspartate aminotransferase concentration in serum or plasma • ALT—alanine aminotransferase concentration in serum or plasma

INTERNET RESOURCES
Norton, T.M., Greiner, E., Latimer, K. Little, S.E. (2004). Medical protocols recommended by the U.S. Bali Mynah SSP. Accessed on July 20, 2014. Available at: http://www.aazv.org/?547; last accessed October 5, 2015.
VanWettere, A.J. (2013). Merck Veterinary Manual: Overview of blood borne organisms in poultry. Last accessed on September 23, 2015 at: http://www.merckmanuals.com/vet/poultry/bloodborne_organisms/overview_of_bloodborne_organisms_in_poultry.html

Suggested Reading
Atkinson, C.T., Thomas, N.J., Hunter, D.B. (eds). (2008 [print] or 2009 [e-book]), *Parasitic Diseases of Wild Birds*. Hoboken, NJ: John Wiley & Sons, Inc. Relevant chapters include: 2: *Haemoproteus*, by Carter T. Atkinson. 3: Avian Malaria, by Carter T. Atkinson. 4: Leukocytozoonosis, by Donald J. Forrester and Ellis C. Greiner. 5: *Isospora, Atoxoplasma* and *Sarcocystis,* by Ellis C. Greiner.
Campbell, T.W. (2012). Hematology of birds. In: Thrall, M.A., Weiser, G., Allison, R., Campbell, T.W. (eds), *Veterinary Hematology and Clinical Chemistry*, 2nd edn. Hoboken, NJ: John Wiley & Sons, Inc., pp. 262–266.
Peirce, M.A. (2000). Hematozoa. In: Samour, J. (ed.), *Avian Medicine*. London: Mosby pp. 245–252.
Remple, J.D. (2004). Intracellular hematozoa of raptors: a review and update. *Journal of Avian Medicine and Surgery*, **28**(2):75–88.
Valkiunas, G. (2004). *Avian Malaria Parasites and Other Haemosporidia*. Boca Raton, FL: CRC Press.

Author Lisa Harrenstien, DVM, DACZM

Hemorrhage

BASICS

DEFINITION
• Any condition that results in the loss of blood from the vascular space into surrounding tissues or from body surfaces. • The LD_{50} of blood loss in ducks (*Anas platyrhynchos*) is 60% of their total blood volume.

PATHOPHYSIOLOGY
• Hemorrhage is often acute in nature. • Clinical signs of hemorrhage result from one of two mechanisms: Either blood loss from damaged or diseased vessels or bleeding diatheses (defects of normal hemostatic processes). • Blood vessel defects arise after trauma or recent surgery. They can also occur secondary to infectious, inflammatory or neoplastic processes that cause vessel erosion and infiltration. • Bleeding diathesis are due to failure of platelet plug formation and coagulopathies. • The exact cause of bleeding diathesis of conures is not known but believed to occur when normal bone marrow is replaced by immature RBCs. The exact cause is unknown; but may be due to a Vitamin K, calcium or other nutritional deficiency. • Microvascular hemorrhage can result from hypertension, anemia or hyperviscosity. • Polyomavirus of neonatal budgerigars are especially susceptible to gastrointestinal hemorrhage.

SYSTEMS AFFECTED
• Cardiovascular • Hemic/Lymphatic/Immune • Respiratory • Hepatobiliary • Gastrointestinal • Skin/Exocrine

GENETICS
N/A

INCIDENCE/PREVALENCE
N/A

GEOGRAPHIC DISTRIBUTION
Polyomavirus and PMV-3 are found worldwide. PMV-3 has rare outbreaks in the US due to illegal transport of poultry from Mexico.

SIGNALMENT
• **Species**: All avian species can be affected. ○ Blood loss: None. ○ Bleeding diathesis of conures is more commonly found in blue-crowned, peach-fronted, orange-fronted and Patagonian species. ○ Polyomavirus is more common in neonatal budgerigars less than 15 days old. ○ PMV-3 is common in poultry, African grey parrots, finches. ○ EEE/WEE in Lady Gouldian finches.

SIGNS
General Comments
Vascular injury is the most common cause of hemorrhage.

Historical Findings
• Nonspecific signs such as weakness, lethargy, anorexia and death. • Recent history of trauma or surgery. • Respiratory changes. • Multiple members of a flock are affected.

Physical Examination Findings
• Visible hemorrhage • Tachycardia (compensatory) • Hypertension (compensatory) • Depression • Pale mucus membranes • Cold limbs • Coelomic distention • Tachypnea • Hematuria (secondary to renal neoplasia or toxicities) • Melena • Acute gastric hemorrhage • Solid masses • Bruising • Petechiation • Retinal Hemorrhage • Epistaxis • Bradycardia in late stages of shock • Hypotension in mid to late stages • Collapse • Shock • Multiorgan failure • Death

CAUSES
• **Trauma** ○ Vehicular trauma ○ Bite wounds or inter-mate aggression ○ Cannibalism due to overcrowding ○ Self mutilation due to pain, behavior abnormalities ○ Electrical shock ○ Chemical exposure ○ Recent fall or night fright ○ Neuromusculoskeletal disease that causes weakness ○ Broken blood feather • **Toxicity** ○ Heavy metal ○ Anticoagulant rodenticides ○ Sulfa-containing drugs in gallinaceous birds ○ Estrogen ○ Medications (NSAIDs, clopidogrel, sulfonamindes, heparin, warfarin, plasma expanders, estrogens, cytotoxic drugs) ○ Recent surgery and failure to use proper hemostatic techniques ○ Neoplasia causing hemorrhage in the affected organ • **Coagulopathy** (primary) ○ Anticoagulant rodenticides ○ Color mutation cockatiels-factor deficiency ○ Liver disease • **Metabolic** ○ Gastrointestinal Disease ▪ Ulcers ▪ GI foreign bodies ▪ Pancreatitis ○ Genitourinary Disease ▪ Cloacal papillomatosis ▪ Cloacitis ▪ Egg laying ▪ Cloacal/uterine prolapse • **Viral** ○ Polyomavirus ○ EEE/WEE ○ PMV-3 • **Nutritional deficiencies** ○ Diet deficient in Vitamin K ○ Malnutrition leading to squamous metaplasia ○ Bleeding diathesis of conures ○ Starvation

DIAGNOSIS

DIFFERENTIAL DIAGNOSIS
• Respiratory disease. • Other causes of shock.

CBC/BIOCHEMISTRY/URINALYSIS
Hemogram
• Anemia secondary to loss and thrombocytopenia due to platelet consumption. In general anemia is defined as a PCV <35%; however, the normal range can vary with different species. • Thrombocytopenia is also seen in nonbudgerigar parrots with polyomavirus. • Polychromasia is noted in mallard ducks 12 hours posthemorrhagic event. • Biochemistry panel—liver enzyme abnormalities (elevated bile acids, AST, and ALT) may be seen. Other nonspecific laboratory abnormalities will be noted in multiorgan failure. • Note: Caution must be taken in birds with excessive hemorrhage. Only 1% of the bird's weight in grams is recommended for blood work (i.e., 1 mL of blood for every 100g of body weight). Therefore, all blood loss from hemorrhage and bruising must be taken into account.

Biochemistry profile
• Elevation in AST and/or bile acids may be seen with hepatic trauma.

OTHER LABORATORY TESTS
• Blood pressure measurement to access cardiovascular status and detect hypotension. • PT measurement; however, this has limited availability. • Fecal blood testing will be positive in gastrointestinal hemorrhage. • Although not widely available, thromboelastography has been described in some avian species including poultry and Hispaniolan Amazon parrots.

IMAGING
• Radiographs: May be useful if initial place of hemorrhage is not readily identifiable. • Endoscopy: To evaluate abdominal masses, the upper gastrointestinal tract or cloaca.

PATHOLOGIC FINDINGS
• Bleeding diathesis of conures ○ Histologic examination shows acute and chronic hemorrhages within various tissues. ○ Erythrocyte proliferation in the bone marrow, hepatic sinusoids, and splenic red pulp. • Trauma ○ Superficial hematomas and hemarthrosis or massive hemorrhage into body cavities.

TREATMENT

APPROPRIATE HEALH CARE
• Outpatient medical management is possible if the hemorrhage is minimal, the laceration/abrasion is small. • Inpatient medical management: ○ Stop active hemorrhage with compression, pressure wrap application (except around keel), silver nitrate application on a bleeding nail, the removal of the broken blood feather, hemostatic matrix or emergency surgery. ○ Note: Although pulling a blood feather is commonly recommended, this can result in damage to the feather follicle. ○ Minimize stress by placing in a quiet, dark cage. ○ Stop access to feathers in self/mutilation cases with an e-collar designed for birds or bandaging. Next work up the patient for self mutilation (See Feather Damaging Behavior). ○ Place in warm, humid environment (85°F and 70% humidity), unless there is evidence of head trauma.

HEMORRHAGE

(CONTINUED)

NURSING CARE
• Endoscopic or surgical removal of foreign bodies, bleeding masses or uncontrolled internal hemorrhage. • Identify and correct underlying metabolic disease and treat accordingly. • Nutritional support via gavage feeding. • Dyspneic birds place in incubator with 78–85% oxygen supplementation at 5 L/min. • Supplemental heat at 85–90°F • Fluid therapy: ○ SQ, IO or IV depending on level of dehydration and vascular access. ○ 50–150 mL/day maintenance depending on species. ○ Plus any additional fluids to correct for dehydration and ongoing losses.

ACTIVITY
• Restrict activity until the PCV normalizes or recovery from surgery. • Place in a smaller cage and restrict flight.

DIET
• No long term diet modification needed in most diseases. • Liver disease and bleeding diathesis of conures being the exceptions. Bleeding conures should be placed on a nutritionally complete diet. • All anorectic patients should have assisted feedings.

CLIENT EDUCATION
• Heavy metal toxicosis: Identify sources of heavy metals. • Debilitated birds have a poor prognosis despite treatment.

SURGICAL CONSIDERATIONS
• Necessary if bleeding cannot be stopped with pressure. • Patients with hemorrhage/hypovolemic shock should be stabilized first. • Hypotension should be corrected before surgery. • If patient is clinical for its anemia or excessive blood loss, stabilization should occur before surgery. • Electrocautery can be used to minimize blood loss.

MEDICATIONS

DRUG(S) OF CHOICE
• **Fluid therapy**. ○ Subcutaneous fluids can be given for mild levels of dehydration (<4–5%). ○ IO/IV fluids should be given for moderate levels of dehydration (approximately 4–8% dehydration). 80% of fluid deficit should be replaced over 6–8 hours in acute losses and over 12–24 hours in chronic losses. • **Management of hypovolemic shock** (>10% dehydration): First determine that phase by assessing the BP, HR, mm and CRT. ○ For treatment of all phases, warm the fluids to 100–103°F. ○ Compensatory Phase—the first stage of shock ▪ Typical examination findings—tachycardia and hypertension. ▪ This phase is typically seen with blood loss of less than 20% of total blood volume. ▪ Treatment includes volume replacement of the deficit with crystalloids IV/IO over a 12-hour period. Where maintenance equals 50–150 mL/day (depending on the species) and deficit equals % dehydration × BW_{kg} × 1000. ○ Early Decompensatory Phase—second stage of shock in which there has been decreased blood flow to the kidneys, GI tract, skin and muscles. ▪ Typical examination findings include tachycardia, +/– hypothermia and normal to decreased blood pressure, pale mm, prolonged CRT, cool limbs and depression. ▪ This phase typically occurs with blood loss of greater than 25–30% of total blood volume. ▪ Treatment: Crystalloid bolus (10ml/kg) and HES bolus (3–5 mL/kg) or Oxyglobin (5 mL/kg) repeated until the blood pressure is greater than 90mm/Hg (approximately 3–4 total) IV/IO. ○ Decompensatory Phase—final stage of shock ▪ The LD_{50} of ducks (Anas platyrhynchas) was 60% of their total blood volume. ▪ Typical exam finding include bradycardia, hypothermia, hypotension, pale mm and absent CRT. ▪ Treat in the following order: □ HSS (3–5 mL/kg) over 10 minutes +/– HES (3 mL/kg) IV/IO. □ Warm patient with supplemental heat such as a Bair hugger. □ Give crystalloids (10ml/kg) and HES (3–5 mL/kg) bolus IV/IO. Repeat until blood pressure is greater than 90 mm/Hg (3–4 boluses total). • **Once stable**, place on crystalloids at maintenance rate + deficits and ongoing losses. • **If there** is no response to above treatments, check the BG, PCV, TP and ECG. • **If hypoglycemic** give 50% dextrose 50–100 mg/kg IV slow to effect. Dilute 1:1 with 0.9% saline. • **If abnormal** cardiac contractility give nitroglycerin—place a 1/8-inch/2.5-kg strip on the skin. Wear gloves. • **If PCV** <20%, consider a blood transfusion or use of a blood alternative. Note: blood transfusions are rarely used in emergency situations due to lack of availability. Multiple transfusions may carry increased risk of a fatal reaction. ○ Calculate the patient's total blood volume (8% of body weight) and replace 10% of the blood volume IV/IO. ○ Alternatively give Oxyglobin (not currently available) at 5 mL/kg boluses IV slowly over one minute, every 15 minutes. • **Pain management** ○ Butorphanol: 0.05–6 mg/kg IM IV q1–4h (dosing varies on species). ○ Meloxicam: 0.1–2.0 mg/kg PO q 12–24 hours (may be contraindicated if actively hemorrhaging). • **Nutritional** support via gavage feeding of hand feeding formula. • Antibiotics: ○ Chose broad spectrum antibiotics for lacerations and abrasions. ○ For bites chose: piperacillin or a fluoroquinolone. • **Medications** for bleeding diathesis of conures. ○ Vitamin K 2.5 mg/kg IM q24h for several days. ○ Vitamin D_3 3300U/kg IM q7 days prn (caution when giving long term due to toxicity). ○ Calcium gluconate 5–10 mg/kg SQ, IM q12. ○ Antibiotics-broad spectrum antibiotics. ○ Transition to a well balanced diet. ○ There is no determined end point for medications. Consider continuing until on a better plain of nutrition.

CONTRAINDICATIONS
• Avoid drugs with anticoagulant or antiplatelet effects (NSAIDs, clopidogrel, sulfonamides, heparin, warfarin, plasma expanders, estrogens and cytotoxic drugs). • Adequan (polysulfated glycosaminoglycan) (PSGAG) has been associated with fatal hemorrhage/bleeding diathesis in multiple avian species.

PRECAUTIONS
Indirect blood pressure measurement does not always correlate with direct arterial measurement; but it can provide information on blood pressure trends.

FOLLOW-UP

PATIENT MONITORING
• Monitor for cessation of active bleeding and petechiation formation. • Stabilization/normalization of hematocrit/PCV and total solids. Typically occurs within six days. • Correction of thrombocytopenia or prolonged clotting times. • Hemorrhagic/hypovolemic shock: resolution of tachycardia, bradycardia, hyper or hypotension. • Normalization of PCV/TS within 3–6 days.

PREVENTION/AVOIDANCE
• Remove access to heavy metals. • Perform routine wing trimmings if needed (may not apply to all cases). • Discuss with owner the risk of unsupervised interaction with another animal or activity.

EXPECTED COURSE AND PROGNOSIS
• Depends on the following: ○ Initial stabilization and correction of hemorrhagic shock. ○ Ability to identify and control active hemorrhage. ○ Minimization of trauma.

MISCELLANEOUS

ASSOCIATED CONDITIONS
N/A

AGE-RELATED FACTORS
N/A

ZOONOTIC POTENTIAL
N/A

FERTILITY/BREEDING
N/A

SYNONYMS
Bleeding diathesis
Conure bleeding syndrome

SEE ALSO
Anemia
Anticoagulant rodenticide
Bite wounds

Hemorrhage (Continued)

Cloacal disease
Cloacitis
Coagulopathies
Ectoparasites
Feather damaging behavior/self-injurious behavior
Fracture/luxation
Gastritis/enteritis
Heavy metal toxicity
Hemoparasites
Intestinal helminthiasis
Liver disease
Nutritional deficiencies
Oil exposure
Pancreatic diseases
Paramyxoviruses
Polyomavirus
Sick-bird syndrome
Trauma
Tumors

ABBREVIATIONS

PCV—packed cell volume
HCT—hematocrit
HES—hydroxyethyl starch
HSS—hypertonic saline solution
EEE/WEE—Eastern/Western equine encephalomyelitis
LD_{50}—amount of blood loss required to cause mortality in 50% of the patients
PVM-3—Newcastle disease
PT—prothrombin time
RBC—red blood cells
ALT—alanine transferase
AST—aspartate aminotransferase

INTERNET RESOURCES

N/A

Suggested Reading
Carpenter, J.W. (2013). *Exotic Animal Formulary*, 4th edn. St Louis, MO: Elsevier.
Harrison, G.J., Lightfoot, T.L. (2006). *Clinical Avian Medicine*. Palm Beach, FL: Spix Publishing.
Lennox, A. (2013). *Avian Critical Care*. Proceedings BSAVC (http://www.vin.com/doc/?id=5742316).
Lichtenberger, M. (2004). Response to fluid resuscitation after acute blood loss in Peking ducks. *Proceedings of the Association of Avian Veterinarians Annual Conference*.
Lichtenberger, M. (2005). Avian shock: Recognition and treatment. Presented at the 11th International Veterinary Emergency and Critical Care Symposium (IVECCS 2005), September 7-11, Atlanta, GA.

Author Erika Cervasio, DVM, DABVP (Avian)
Acknowledgements Marjory B. Brooks, DVM, DACVIM
Stephen Dyer, DVM, DABVP (Avian)
Angela Lennox, DVM, DABVP (Avian)
Marla Lichtenberger, DVM, DACVECC
Drury Reavill, DVM, DABVP (Avian), DACVP

HEPATIC LIPIDOSIS

BASICS

DEFINITION
Excessive accumulation of triglycerides in the liver resulting in hepatic dysfunction.

PATHOPHYSIOLOGY
• Metabolic imbalance leads to excessive hepatic lipid accumulation/storage. • Altered lipid metabolism and accumulation can result from increased intake (high fat diets, obesity), increased lipogenesis (hormonal, stress), and/or decreased lipid metabolism (nutritional deficiencies). • Intrahepatic lipid accumulation causes toxic changes in hepatocytes leading to impaired function, including cholestasis.

SYSTEMS AFFECTED
• Hepatobiliary. • Gastrointestinal. • Skin/Exocrine. • Hemic/Lymphatic/Immune (coagulopathy in advanced disease). • Nervous (advanced disease).

GENETICS
N/A

INCIDENCE/PREVALENCE
Relatively common problem encountered in caged birds.

GEOGRAPHIC DISTRIBUTION
N/A

SIGNALMENT
• More common in Amazon parrots, cockatoos (especially galah cockatoos), budgerigars, and lorikeets. • All ages and sexes equally affected.

SIGNS
Historical Findings
• Nonspecific sickness. • Anorexia and/or regurgitation/vomiting. • Polyuria/polydipsia. • Dyspnea. • Green stools.

Physical Examination Findings
• Slightly overweight to obese. • Lethargy and weakness, encephalopathy (rare, end-stage). • Dehydration. • Poor feather condition (pigment changes, stress bars, feather picking). • Abnormal and discolored nails, overgrown rhinotheca with degenerative keratin changes (especially budgerigars). • Dyspnea (from hepatic enlargement and/or intracoelomic fat). • Abdominal distension (hepatic enlargement and/or ascites). • Palpable hepatic enlargement. • Polyuria, diarrhea, regurgitation, and/or vomiting. • Increased biliverdin green pigment in urine and stool (from cholestasis). • Melena or bloody droppings (end-stage coagulopathy).

CAUSES
• Can be multifactorial, avian lipid metabolism is complex and incompletely understood. • Imbalance of lipid intake, endogenous lipid production, and lipid metabolism. • Increased lipid intake—high fat, low protein diets (seed-based), overfeeding, excessive high-energy intake in neonates (cockatoos, macaws). • Increased lipid production—hormonal lipogenesis (estrogens during egg laying, hormone-sensitive lipase in diabetes mellitus), drug-induced lipogenesis (corticosteroids, pesticides), and peripheral lipolysis (catecholamines in stress, thyroxine with thyroid dysfunction, but not seen with rapid weight loss). • Impaired lipid metabolism—multinutrient deficient diets, including essential fatty acids (linoleic acid), sulfur amino acids (choline, methionine, cysteine), lipotrophic factors (L-carnitine), and vitamins (biotin, vitamins B1, B2, B6, and B12, E, and folic acid).

RISK FACTORS
• Inappropriate diet and overfeeding. • Restricted exercise, sedentary lifestyle. • Obesity. • Chronic stress. • Thyroid disease. • Genetic predisposition.

DIAGNOSIS

DIFFERENTIAL DIAGNOSIS
Hepatic lipidosis must be distinguished from other causes of hepatomegaly.
• Vascular congestion (passive congestion, portal hypertension in cardiovascular disease). • Toxin-induced (mycotoxins such as aflatoxins, plants, pesticides, heavy metals, environmental toxins). • Drug-induced (e.g., antifungals, volatile anesthetics, some antibiotics, corticosteroids, vitamin A). • Hepatic masses (vascular anomalies, neoplasia or metastases). • Storage or breakdown product accumulation (amyloidosis—rare in psittacines; iron storage disease—especially in lorikeets, Sturnidae, and Ramphastidae). • Infectious hepatitis—bacterial (*Chlamydia psittaci*, *Mycobacteria* spp., Gram-negative hepatitis), viral (polyomavirus, herpesvirus, adenovirus, reovirus), and parasitic (trematodes, protozoa).

CBC/BIOCHEMISTRY/URINALYSIS
• **Mild CBC changes** (mild nonregenerative anemia, leukocytosis or leukopenia) vs. inflammatory hepatopathies. • **Evidence of hepatocellular damage** ○ Aspartate aminotransferase (AST) elevation—Not consistently elevated in lipidosis, may be normal in advanced disease (high sensitivity). ▪ Should always be interpreted with creatine kinase (CK). ○ Glutamate dehydrogenase (GLDH)—Mitochondrial enzyme increased with severe damage (high specificity). • **Evidence of impaired hepatocellular function** ○ Bile acids should be moderate to highly elevated (high sensitivity and specificity). ○ Decreased synthetic function results in low total protein, albumin, coagulation factors, uric acid. • **Evidence of altered lipid metabolism** ○ Lipemic serum (can interfere with biochemical testing). ○ Hyperlipidemia with increased triglycerides and cholesterol. • **Other findings** ○ Hypoglycemia (impaired gluconeogenesis, starvation, diabetes, chronic disease). ○ Hypokalemia (vomiting/regurgitation, polyuria).

OTHER LABORATORY TESTS
• Infectious hepatitis studies should be negative. • Coagulation studies will be abnormal in advanced disease.

IMAGING
• Radiographs demonstrate hepatic enlargement with compression of coelomic air sacs and over inflation of more cranial air sacs. Ascites may be present but without evidence of cardiomegaly. • Ultrasonography should document an enlarged liver with smooth contours and diffuse hyperechoic parenchymal alteration. Ascites can be documented, but passive congestive changes should not be present. Increased visceral fat deposits and/or atherosclerosis may also be present.

DIAGNOSTIC PROCEDURES
• Endoscopic visualization of the liver by lateral (from caudal thoracic air sac into peritoneal cavity) or direct ventral approach will find an enlarged, rounded liver with pale to mottled yellow-tan parenchyma. • Ultrasound-guided fine needle aspiration of the liver in a study in Amazon parrots showed low hepatocyte yield due to hemodilution; however, samples were adequate for determining vacuolation as seen in hepatic lipidosis.

PATHOLOGIC FINDINGS
• Definitive diagnosis of hepatic lipidosis requires a liver biopsy in a stable patient (severe lipidosis/metabolic crisis is a contraindication). Ultrasound guided fine needle aspiration of the liver may show evidence of vacuolation if hepatic lipidosis is present. • Histopathology documents vacuolization and degenerative changes of hepatocytes, with areas of parenchymal destruction and inflammation. • Overall prognosis correlated to severity of histopathology.

TREATMENT

APPROPRIATE HEALTH CARE
• Stabilize and improve patient's physical condition and nutritional status. • Treat secondary conditions that may be causing hepatic lipidosis. • Formulate a plan for gradual weight loss (including increased activity) and improved nutrition. • Limit stressors.

HEPATIC LIPIDOSIS (CONTINUED)

NURSING CARE
- Supportive care with administration of warmed fluids (avoid lactate and high glucose/dextrose infusions) (50–150 mL/kg/d maintenance plus dehydration deficit correction) at a continuous rate of 100 mL/kg/q24h. • Provide a warm incubator, including oxygen if dyspneic or depressed. • Nutritional support and dietary changes (see below) are required.

ACTIVITY
Sick patients should be rested in a quiet environment to minimize additional stressors. Once stabilized, activity should be increased as part of a weight loss program.

DIET
- Nutritional management/supplementation is critical for treatment. • Well-balanced diet—formulated diets with correct quantities of fresh fruits and vegetables facilitate management, supplemental vitamins can be added (see below). • Increased protein content to reduce hepatic lipid accumulation (unless concern for hepatic encephalopathy) using a recovery or neonatal psittacine formula (e.g., Recovery Formula, Harrison Bird Food, crude protein 35%; Exact, Kaytee, crude protein 22%; A21, Nutribird, Versele-laga, crude protein 21%; Neonate formula, Harrison Bird Food, crude protein 26%). • Match intake with resting energy needs (to inhibit peripheral lipolysis) but avoid excessive caloric intake. • Gavage feeding may be necessary to restore nutritional balance in anorectic birds.

CLIENT EDUCATION
- Hepatic lipidosis is most often the result of chronic nutritional problems. • Provide a dietary and nutritional supplementation plan. • Discuss ways to increase activity (enrichment, flight, etc.).

SURGICAL CONSIDERATIONS
- Liver biopsy should only be performed in a stable patient without evidence of metabolic crisis, coagulopathy, or hepatic vascular congestion. • Ascitic fluid should not be removed (protein reservoir) unless there is dyspnea or it is needed for diagnostic purposes.

MEDICATIONS

DRUG(S) OF CHOICE
- Nutritional supplementation—vitamins (especially B complex, E, K1, and biotin), essential amino acids, and lipotrophic factors (choline and methionine, 40–50 mg/kg q24h). • Nausea/vomiting should be treated with antiemetic drugs (metoclopramide, 0.5 mg/kg q6h; metopimazine, 0.5 mg/kg q24h). • Ascities can be managed with furosemide (1–2mg/kg, up to twice daily as needed). • Severe hypoproteinemia may require supplementation to increase colloid oncotic pressure (hetastarch, 10–15 ml/kg/d IV or 5 ml/kg bolus). • If present, hepatic encephalopathy can be managed with lactulose (150–650 mg/kg q12h). • Estrogen-induced lipidosis from chronic egg laying should be suppressed with leuprolide acetate (100–1250 ug/kg IM q14 days) or hCG (500–1000 IU q3–5 weeks). Failure to suppress risks rupture of hepatic vessels during egg laying (fatty liver hemorrhagic syndrome).

CONTRAINDICATIONS
Avoid protein supplementation if there is concern for hepatic encephalopathy.

PRECAUTIONS
- Anabolic steroids inhibit bile flow and may increase the risk of hepatic lipidosis (shown in cats). Anabolic steroids and glucocorticoids also have an inhibitory influence on beta-oxidation. • Tetracyclines (e.g., doxycycline) have lipogenic effects on hepatocytes in mammals and should be used with caution in birds with hepatic lipidosis.

POSSIBLE INTERACTIONS
Furosemide and other loop diuretics should be used with caution in hypokalemic birds.

ALTERNATIVE DRUGS
- L-Carnitine (100–250 mg/kg q 24h) is a component of the mitochondrial membrane and necessary for beta-oxidation. Although regularly used in feline hepatic lipidosis, its efficacy is debated. • Antioxidant agents may be helpful in preventing hepatocellular oxidative stress and stabilizing membranes. N-acetylcysteine (NAC) and S-adenosyl-methionine (SAMe) (15–20 mg/kd q24h) are glutathione precursors, sources of sulfur-containing amino acids, and may assist in hepatocyte lipid metabolism. SAMe is also a precursor of L-carnitine. • Ursodeoxycholic acid (15 mg/kg q24h) has cytoprotective, anti-inflammatory, antioxidant (via glutathione), and anti-fibrotic effects on hepatocytes. It is used to treat human and feline cholestatic disorders, but it is not useful for ameliorating triglyceride accumulation.

FOLLOW-UP

PATIENT MONITORING
- Assess progression of dietary changes and gradual weight loss at regular rechecks until stable. • Reassess hepatic function/damage by hepatic biochemistry monitoring at rechecks. • A recheck biopsy may be indicated after completion of therapy.

PREVENTION/AVOIDANCE
- Initiate a nutritionally-balanced pelletized diet, avoiding excessive energy intake. • Avoid excessive seed intake. • Encourage exercise and play. • Minimize stressors. • Suppress chronic egg laying. • Monitor for signs of hepatic disturbance.

POSSIBLE COMPLICATIONS
Failure to correct diet and weight management can lead to worsening hepatic lipidosis with risk of death from liver failure and/or encephalopathy.

EXPECTED COURSE AND PROGNOSIS
- Hepatic lipidosis is a chronic disease and, following initiation of treatment and modifications, takes time to reverse. • After resolution of any acute crisis with a guarded prognosis, outcome is dependent on the severity of lipidosis at presentation and the ability to maintain dietary and other changes to correct the underlying predisposing factors.

MISCELLANEOUS

ASSOCIATED CONDITIONS
- Obesity • Atherosclerosis • Neurologic signs • Diabetes mellitus • Dyspnea • Weakness.

AGE-RELATED FACTORS
N/A

ZOONOTIC POTENTIAL
N/A

FERTILITY/BREEDING
Chronic egg laying and estrogenic influence can cause hepatic lipidosis.

SYNONYMS
Fatty liver syndrome, hepatic steatosis, fatty infiltration of the liver

SEE ALSO
Ascites
Atherosclerosis
Chronic egg laying
Coagulopathies
Coelomic distention
Dyslipidemia
Egg yolk coelomitis/reproductive coelomitis
Hemorrhage
Iron storage disease
Lipoma
Liver disease
Nutritional deficiencies
Obesity
Overgrown beak and nails
Pancreatic diseases
Seizure
Sick bird syndrome
Thyroid diseases
Undigested food in droppings

ABBREVIATIONS
N/A

INTERNET RESOURCES
N/A

Suggested Reading
Fudge, A.M. (2000). Avian liver and gastrointestinal testing. In: Fudge, A.M. (ed.), *Laboratory Medicine: Avian and Exotic*

Pets. Philadelphia, PA: WB Saunders, pp. 47–55.

Jaensch, S. (2000). Diagnosis of avian hepatic disease. *Seminars in Avian and Exotic Pet Medicine*, **9**(3):126–135.

Lumeij, J. (1994). Hepatology. In: Ritchie, B.W., Harrison, G.J., Harrison, L.R. (eds), *Avian Medicine: Principles and Application*. Lake Worth, FL: Wingers Publishing, pp. 522–537.

Redrobe, S. (2000). Treatment of avian liver disease. *Seminars in Avian and Exotic Pet Medicine*, **9**(3):136–145.

Scherk, M.A, Center SA (2005). Toxic, metabolic, infectious, and neoplastic liver diseases. In Ettinger, S.J., *et al.* (eds), *Textbook of Veterinary Internal Medicine*, Vol. 2, edn 6. St Louis, MO: Elsevier/Saunders, pp. 1464–1477.

Author Julie DeCubellis, DVM, MS

HERPESVIRUSES

BASICS

DEFINITION
• Herpesviruses are responsible for a variety of diseases in birds. Individuals that recover from their first infection will usually develop latent infections for prolonged periods of time. *Psittacid-herpesvirus 1* (PsHV-1) causes Pacheco's disease (upon first exposure) and internal papillomatosis (after chronic latent infection). *Gallid herpesvirus 1* (GHV-1) is the etiologic agent for infectious laryngotreacheitis (ILT) in chicken and pheasants. A possible mutation of the GHV-1 has been involved in Amazon tracheitis disease. • Other avian herpesviruses (See Viral Appendix) include *Psittacid-herpesvirus 2* (which induces papillomas in African grey parrots), parakeet herpesvirus, finch cytomegalovirus, *Columbid herpesvirus 1* in pigeons and raptors, and *Anatid herpesvirus 1* (duck plague).

PATHOPHYSIOLOGY
• *Pacheco's disease*: An acute viremia causes necrotizing lesions in multiple organs, including the liver, the gastrointestinal (GI) tract, and the spleen. Peracute death is common with a high mortality rate in naïve flocks, but this can vary with the species affected and the virus genotypes and strains. • *Papillomatosis*: Birds that survived an outbreak of Pacheco's disease can develop papillomas in their GI tract many years later, possibly following various stress factors. Depending on their size and location, papillomas can interfere with food ingestion, normal GI transit and absorption, and reproduction. Pancreatic, bile duct, and GIT carcinomas have also been reported with PsHV-1 latent infections. • *ILT/Amazon tracheitis*: Accumulation of necrotic debris in the trachea can result in dyspnea, occlusion, and possibly asphyxiation.

SYSTEMS AFFECTED
Pacheco's disease and internal papillomatosis
• Gastrointestinal • Hepatobiliary
• Reproductive.

ILT/Amazon tracheitis
• Respiratory.

GENETICS
Unknown.

INCIDENCE/PREVALENCE
Pacheco's disease outbreaks were frequent in the 1970s and 1980s until importation of wild-caught birds started to be more closely regulated. Birds imported 20–30 years ago are now more commonly seen with internal papillomatosis.
  Infectious laryngotracheitis is commonly observed in poultry, while Amazon tracheitis remains rare in psittacines.

GEOGRAPHIC DISTRIBUTION
Pacheco's disease was first described in Brazil and is most commonly associated with South American parrot species.

SIGNALMENT
• **Species predilections:** Varies with the genotypes and strains of the PsHV-1. In general, macaws, Amazon parrots, cockatoos, and African Grey parrots are highly susceptible. Conures are healthy carriers of some strains, but can be susceptible to others. Old World psittacines appear to be more resistant. Hybrid and fancier chickens, Indian peafowl, guineafowl, canaries, peafowls, and pheasants are the most susceptible to GHV-1. Amazon parrots and Bourke's parakeets are susceptible to a virus related to the GHV-1.
• **Mean age and range:** Birds of all ages can be affected. However, internal papillomatosis generally occurs in older birds.
• **Predominant sex:** Unknown.

SIGNS
Historical Findings
Pacheco's disease:
• Sudden death • Lethargy • Anorexia
• Regurgitation • Green to yellow diarrhea
• Neurologic signs.
Internal papillomatosis:
• Chronic weight loss • Regurgitation
• Infertility • Papilloma in the oral cavity or in the cloaca. These can wax and wane in size.
• Tenesmus • Blood in the droppings
• Passing whole seeds • Dysphagia
• Wheezing.
ILT/Amazon tracheitis:
• Coughing • Gasping • Rales • Anorexia
• Lethargy • Nasal discharge • Ocular discharge.

Physical Examination Findings
Pacheco's disease:
• Severe depression. • Green staining of the cloacal feathers.
Internal papillomatosis:
• Poor body condition. • Some papillomas can be observed directly or visualized by cloacoscopy.
ILT/Amazon tracheitis:
• Severe dyspnea. • Mouth and beak can be bloodstained. • Weight loss. • Conjunctivitis.

CAUSES
• *Pacheco's disease:* Direct or indirect contact with infected birds that are shedding PsHV-1 in their feces, their growing feather quills, and/or their pharyngeal secretions. Can occur in a closed collection after the virus is reactivated by stress in a latent carrier.
• *Internal papillomatosis:* Reactivation of PsHV-1 months or years after the primary infection. Stress and other causes of immunosuppression are suspected to play a role in this condition. • *ILT/Amazon tracheitis:* Direct contact with the respiratory secretions of infected birds (sick or latent carriers). Indirect infection also occurs (contaminated crates, litter spread in pastures, humans, etc.).

RISK FACTORS
• *Pacheco's disease*: Mixing of birds of different origins, especially wild-caught specimens and/or from South America. Large quarantine facilities. Contact with a latent carrier of PsHV-1. Indoor collections. High stocking density. • *Internal papillomatosis*: Previous non-lethal exposure to the PsHV-1. Stress.
• *ILT/Amazon tracheitis:* Recent introduction of new birds. Inadequate biosecurity.

DIAGNOSIS

DIFFERENTIAL DIAGNOSIS
Pacheco's disease:
• Causes of sudden death potentially associated with outbreaks: ◦ Bacterial infections causing potentially fatal hepatitis and systemic disease/septicemia: chlamydiosis, salmonellosis. ◦ Viral infections causing mortality: polyomavirus and circovirus (if young parrots are affected), adenovirus, paramyxovirus, bornavirus.
◦ Acute renal failure (gout): high mortality, if failure in the water system. ◦ Toxicity.
• Causes of sudden death in one individual (not associated with outbreaks): ◦ Acute hepatic failure, heart failure, atherosclerosis, stroke, or aneurysm rupture. ◦ Trauma.
◦ Obstructive foreign body. ◦ GI perforation.
◦ Salpinx rupture/coelomitis.
Internal papillomatosis:
• For the presence of papilloma(s):
◦ Poxvirus. ◦ Papillomavirus (African grey parrots) ◦ Cloacal prolapse (breeding behavior, enteritis, cloacitis). ◦ Dystocia.
◦ Fungal, parasitic (*Trichomonas*) or bacterial granuloma. ◦ Abscess. ◦ Neoplasia. • For the chronic weight loss: ◦ Viral infections: bornavirus, circovirus. ◦ Mycobacterial infections. ◦ Bacterial infections: *Chlamydia psittaci*. ◦ Parasitic and fungal enteritis (*Macrorhabdus, Candida*). ◦ Chronic hepatitis +/− pancreatitis. ◦ Nonobstructive foreign body. ◦ Neoplasia. ◦ Chronic renal insufficiency. ◦ Hepatic insufficiency.
◦ Atherosclerosis. ◦ Inappropriate diet quality and quantity/competition for food.
ILT/Amazon tracheitis:
• Aspergillosis (*Aspergillus spp.*). • Chronic respiratory disease (*Mycoplasma gallinarum* and *Escherichia coli*). • Infectious coryza (*Hemophilus paragallinarum*). • Infectious bronchitis (coronavirus). • Avian influenza (orthomyxovirus). • Newcastle disease (paramyxovirus). • Swollen head syndrome (pneumovirus). • Avian adenovirus.
• Fowlpox (diphtheritic form). • Parasitic tracheitis (*Syngamus trachea*). • Tracheal foreign body. • Trichomoniasis.
• Hypovitaminosis A.

CBC/BIOCHEMISTRY/URINALYSIS
Pacheco's disease:
• It is rare that a bird with Pacheco's disease survives long enough to have blood collected.

HERPESVIRUSES (CONTINUED)

- In experimental infections, birds develop leukopenia and a markedly increased AST.

Internal papillomatosis:
- No specific findings known. • Birds with concurrent bile duct carcinomas can show elevated GGT.

ILT/Amazon tracheitis:
- Rarely performed.

OTHER LABORATORY TESTS
PsHV-1:
- Dead birds: swab of liver or spleen for PCR +/- culture. • Live birds: cloacal/oral swabs and blood for PCR.

ILT:
- Tracheal and conjunctival swabs for PCR +/- culture.

IMAGING
Pacheco's disease:
- It is rare that a bird with Pacheco's disease survives long enough to have radiographs or ultrasound taken.

Internal papillomatosis:
- Depending on the papilloma location, radiographs could show proventricular and/or ventricular dilation and gas in the intestines (enteritis). • Ultrasonography may show some abnormal findings in the liver in the case of a bile duct carcinoma.

ILT/Amazon tracheitis:
- Imaging is rarely useful for the diagnosis. • Contrast tracheal radiographs could be used to differentiate ILT from a tracheal foreign body. • Whole body radiographs could be used to differentiate ILT from other diseases.

DIAGNOSTIC PROCEDURES
- "Vinegar test"—papillomatous tissue will turn white when acetic acid (5%) is directly applied. • Biopsy of the papilloma. • Biopsies for pancreatic and bile duct carcinoma.

PATHOLOGIC FINDINGS
Pacheco's disease:
- May or may not have gross lesions.
- Generally in good body condition with ingesta in the digestive tract. • Enlarged, congested liver, with areas of hemorrhage and abnormal coloration (from green to brown).
- Enlarged discolored spleen. • Multifocal congestion and hemorrhages within the kidneys, intestines, and brain. • Necrotizing lesions in the liver, spleen, and GI tract, but also many other organs, including the respiratory tract. • Minimal inflammatory response. • Intranuclear eosinophilic inclusion bodies (Cowdry type A).

Internal papillomatosis:
- Mucosal papillomas are typically raised, pink, and have a cauliflower-like surface.
- Papillomas can be unique or multiple in the GI tract. • They may ulcerate and bleed.
- Microscopically, papillomas are formed of multiple fimbriae, each composed of a vascular core surrounded by a pseudostratified or stratified, cuboidal to columnar, epithelium. • Bile duct carcinomas are multifocal and coalescing. They often replace the majority of the liver before the bird dies. They generally do not metastasize.
- Pancreatic duct carcinomas are grey, nodular, and coalescing.

ILT/Amazon tracheitis:
- Presence of blood, mucus, yellow caseous exudates, or a hollow caseous cast in the trachea. • Microscopically, a desquamative, necrotizing tracheitis is characteristic of acute disease.

TREATMENT

APPROPRIATE HEALTH CARE
- Birds presented alive with Pacheco's disease, ILT, or Amazon tracheitis may require emergency inpatient intensive care. • Birds with papillomatosis are generally outpatients.

NURSING CARE
Birds presented alive with Pacheco's disease, ILT, or Amazon tracheitis might be in shock and/or dyspnea and will require aggressive supportive care, including IO/IV fluid therapy and placement in an incubator with oxygen, if needed.

ACTIVITY
N/A

DIET
- Assisted feeding might be required in some patients. • Highly digestible pellets and formulas might be used in birds with internal papillomatosis.

CLIENT EDUCATION
No treatment can cure a herpesvirus infection. The bird will remain a carrier if it survives the initial infection.

SURGICAL CONSIDERATIONS
If they cause a significant problem to the bird, papillomas can be removed by surgery, cryosurgery, or staged cauterization with silver nitrate sticks (rinse with copious amounts of fluids during the procedure). However, no technique prevents recurrence.

MEDICATIONS

DRUG(S) OF CHOICE
For Pacheco's disease:
- Acyclovir: ◦ 80–100 mg/kg PO q8h for 10–14 days. ◦ 40 mg/kg IV/SC/IM q8h, if cannot be given orally. ◦ 1 mg/mL in drinking water + 240 mg of medication/kg of food for seven days (use after a single IM injection).
- Use of acyclovir might help reduce the morbidity/mortality, but will not cure the infection.

For ILT/Amazon tracheitis:
- ILT: Eye-drop live-attenuated vaccines or SC, in ovo, or wing-web viral vector recombinant vaccines. • ILT vaccines may help reduce mortality in Amazon parrots, but more studies are needed.

CONTRAINDICATIONS
For Pacheco's disease: any drug with potential hepatoxicity should be avoided.

PRECAUTIONS
Acyclovir:
- Should be used with caution in patients with pre-existing renal conditions. • IM injections are very irritating. • Potential side effects include GI signs, leukopenia, anemia, and renal failure.

POSSIBLE INTERACTIONS
Acyclovir use is contraindicated with nephrotoxic drugs and with zidovudine (antiretroviral drug).

ALTERNATIVE DRUGS
Though other antiherpesviral drugs could potentially be used, none of them have been studied in birds.

FOLLOW-UP

PATIENT MONITORING
Physical examinations as needed.

PREVENTION/AVOIDANCE
Infection with PsHV-1:
- Avoid mixing birds from unknown sources, especially conures and other New World psittacines. • PCR testing of all birds (but latent carriers can be missed). • Strict hygiene and sanitation in quarantine. • Vaccine is no longer available.

Infection with GHV-1 and related viruses:
- Strict hygiene. • Quarantine all new birds.
- Buy birds from breeding facilities with ILT-negative status. • Vaccination.

POSSIBLE COMPLICATIONS
- Potentially life-threatening. • Cloacal strictures following the surgical removal of papillomas.

EXPECTED COURSE AND PROGNOSIS
- Pacheco's disease is associated with a very high mortality. • Papillomas vary from being insignificant to causing death after chronic wasting disease. • ILT and Amazon tracheitis are associated with mild to high mortality depending on the viral strains and other risk factors. • Survivors of all these infections become latent carriers and pose a risk to other birds.

MISCELLANEOUS

ASSOCIATED CONDITIONS
Bile duct, pancreatic, and GI tract carcinomas have been linked to the PsHV-1.

AGE-RELATED FACTORS
None.

HERPESVIRUSES (CONTINUED)

ZOONOTIC POTENTIAL
None.

FERTILITY/BREEDING
Cloacal papillomas can interfere with breeding.

SYNONYMS
None.

SEE ALSO
Adenoviruses
Anemia
Appendix 4: Avian Viral Infections
Aspergillosis
Atherosclerosis
Avian influenza
Circoviruses
Cloacal disease
Cloacitis
Coagulopathies
Gastrointestinal foreign bodies
Heavy metal toxicity
Hemorrhage
Hepatic lipidosis
Infertility
Iron storage disease
Liver disease
Marek's disease
Nutritional deficiencies
Overgrown beak and nails
Papilloma
Paramyxoviruses
Polyomavirus
Poxvirus
Renal disease
Respiratory distress
Trichomoniasis
Tumors
Undigested food in droppings
Urate/fecal discoloration

ABBREVIATIONS
ILT—infectious laryngotracheitis
PsHV-1—psittacine herpesvirus type 1
GGT—gamma-glutamyl transferase
GHV-1—gallid herpesvirus type 1

INTERNET RESOURCES
N/A

Suggested Reading
Gerlach, H. (1994). Viruses. In: Ritchie, B.W., Harrison, G., Harrison, L.R. (eds), *Avian Medicine: Principles and Applications*. Lake Worth, FL: Wingers Publishing, pp. 874–885.
Kaleta, E.F, Docherty, D.E. (2007). Avian herpesviruses. In: Thomas, N.J., Hunter, D.B., Atkinson, C.T. (ed), *Infectious Diseases of Wild Birds*. Oxford, UK: Blackwell Publishing, pp. 63–86.
Ou, S.C., Giambrone, J.J. (2012). Infectious laryngotracheitis in chickens. *World Journal of Virology*, **1**(5):142–149.
Phalen, D., Woods, R. (2009). Psittacid Herpesviruses and mucosal papillomas of psittacine birds in Australia: Fact Sheet. Australian Wildlife Health Network. https://www.wildlifehealthaustralia.com.au/Portals/0/Documents/FactSheets/Herpesviruses%20(Psittacine)%20Aug%202009%20(2.2).pdf. Last accessed on September 23, 2015.
Tomaszewski, E.K., Wigle, W., Phalen, D. (2006). Tissue distribution of psittacid herpesviruses in latently infected parrots, repeated sampling of latently infected parrots and prevalence of latency in parrots submitted for necropsy. *Journal of Veterinary Diagnostic Investigation*, **18**(6):536–544.

Author Marion Desmarchelier, DMV, MSc, DACZM
Acknowledgement Shannon Ferrell, DVM, DABVP (Avian), DACZM

Hyperglycemia and Diabetes Mellitus

BASICS

DEFINITION
• Hyperglycemia is defined as an abnormal elevation of serum blood glucose. Though reference ranges may vary between species, hyperglycemia is generally suspected in birds when glucose levels exceed 20 mmol/L (350 mg/dL) and confirmed when over 28 mmol/L (500 mg/dL). • Diabetes mellitus is a disease caused by (i) an inherited or an acquired deficiency in insulin production by the pancreas, (ii) by the ineffectiveness of the insulin produced on target tissues, or (iii) by glucagon excess, which can all result in hyperglycemia and associated clinical signs.

PATHOPHYSIOLOGY
• Hyperglycemia either results from:
◦ Diabetes mellitus caused by either defects in insulin secretion (destruction/absence of the Langerhans pancreatic islets), insulin action (insulin resistance), or both of these mechanisms. ◦ Action of stress hormones or drugs on insulin and glucose metabolism.
• The relative importance of glucagon and insulin in avian metabolism and the pathogenesis of diabetes mellitus have been extensively discussed in several avian species. Most studies suggest that avian diabetes mellitus may be caused by an excess of glucagon and not a deficiency of insulin. However, recent cases using validated insulin assays and immunohistochemistry on tissue showed the existence of type 1 diabetes mellitus in some individual birds. • When hyperglycemia persists over time, the glucose reabsorption threshold is reached within the kidneys, and glucose will be excreted in the urine. This leads to an increase in urine osmotic pressure, which inhibits the renal reabsorption of water, resulting in polyuria, secondary dehydration, and compensating polydipsia. • The lack of insulin or insulin resistance will promote the release of free fatty acids from lipid storage sites as a source of energy, which is converted by the liver into ketone bodies. A ketoacidosis will occur once the body reaches its buffering capacity.

SYSTEMS AFFECTED
• Endocrine/Metabolic • Renal/Urologic
• Behavioral • Gastrointestinal • Hemic/Lymphatic/Immune • Neuromuscular
• Reproductive.

GENETICS
Unknown.

INCIDENCE/PREVALENCE
• Stress hyperglycemia is very common.
• Diabetes mellitus is rare.

GEOGRAPHIC DISTRIBUTION
None.

SIGNALMENT
• **Species:** Macaws, toucans, cockatiels, and budgerigars seem to be more susceptible.
• **Mean age and range:** More common in adults. • **Predominant sex:** None.

SIGNS
Historical Findings
• Polyuria • Polydipsia • Polyphagia • Weight loss • Lethargy • Abnormal behavior.
Physical Examination Findings
Poor body condition.

CAUSES
For moderate hyperglycemia:
• Stress and pain. • Corticosteroids, medroxyprogesterone. • Female reproductive activity/disease.
For severe hyperglycemia/diabetes mellitus:
• Obesity • Inappropriate diet • Pancreatitis
◦ Bacterial (*Chlamydia,* Gram negative)
◦ Viral (paramyxovirus 3, herpesvirus, polyomavirus, adenovirus, poxvirus)
◦ Inflammatory ◦ Hemosiderosis
◦ Amyloidosis ◦ Hypervitaminosis A
◦ Secondary to egg-yolk coelomitis
• Genetics • Neoplasia • Pancreatectomy.

RISK FACTORS
• Female in reproductive phase.
• Inappropriate diet (e.g., dog food diet for toucans, high refined sugar diet for parrots).

DIAGNOSIS

DIFFERENTIAL DIAGNOSIS
PU/PD associated with systemic signs:
• Renal disease ◦ Heavy metal poisoning
◦ Nephritis ◦ Nephrotoxicity ◦ Gout ◦ Urolith
◦ Neoplasia ◦ Hypercalcemia • Reproductive disease ◦ Egg binding • Sepsis • Nutritional
◦ Hypovitaminosis A ◦ Hypervitaminosis D
◦ Excess dietary sodium • Iatrogenic
◦ Aminoglycosides ◦ Allopurinol
◦ Sulfonamides ◦ Tetracyclines
◦ Cephalosporins • Cloacolith.
PU/PD associated without systemic signs:
• Can be physiologic during breeding season.
• Stress associated with handling, transport, and the environment. • High water content of food (fruits). • Psychogenic polydipsia.

CBC/BIOCHEMISTRY/URINALYSIS
• Hematology: varies with the underlying cause. • Biochemistry: ◦ Increased blood glucose, triglycerides, cholesterol and hepatic enzymes. ◦ Possibly elevated lipase and amylase with pancreatitis. ◦ Blood gas analysis for ketoacidosis. • Handheld glucometers have been shown to be unreliable in birds. Use of laboratory analyzer should therefore be favored as much as possible.

OTHER LABORATORY TESTS
• Fructosamines result from an irreversible, nonenzymatic, insulin-independent binding of glucose to serum proteins. This glycosylation reaction occurs throughout the life span of albumin and is proportional to the glucose concentration over that period of time (1–3 weeks in cats and dogs). Because the half-life of avian serum proteins is notably lower than in domestic mammals, fructosamines may reflect the blood glucose state of diabetic birds over a shorter period. Fructosamine blood levels in clinically normal psittacine birds were between 113 and 238 µmol/L, and diabetic birds usually have values >300 µmol/L. • Urinalysis: Usually not helpful as samples are often contaminated with feces. However, the massive polyuria observed in diabetic birds allows for better separation of the urine from the feces to reduce contamination. Monitoring of glucose and ketones in the urine is not invasive and may be useful in small species like budgerigars. However, interpretation remains questionable. • Choanal and cloacal swabs, blood, and feathers for PCR for both *Chlamydia psittaci* and avian polyomavirus, which are noted as potential causes of pancreatitis in birds. • Measurement of serum insulin concentration by radioimmunoassay.

IMAGING
Radiography and ultrasonography can be useful to rule out other causes of PU/PD.

DIAGNOSTIC PROCEDURES
• Pancreatic biopsy. • Liver biopsy in toucans (hemosiderosis).

PATHOLOGIC FINDINGS
• No gross lesions. • Hypoplasia, atrophy, and/or vacuolization of islet cells.
• Inflammation is rare with diabetes mellitus.

TREATMENT

APPROPRIATE HEALTH CARE
Diabetic patients generally require emergency inpatient care management as they often present with ketoacidosis.

NURSING CARE
• Aggressive fluid therapy, including correction of any acid-base or electrolyte imbalance (potassium, bicarbonate).
• Placement in a warm incubator in a low stress environment. • Assisted feeding as needed. • Control of concurrent inflammatory and/or infectious diseases.

ACTIVITY
N/A

DIET
Low carbohydrate and low fat diet is generally recommended. However, drastic changes in the diet should be made only when the bird is stable.

CLIENT EDUCATION
Treatment of diabetes mellitus is difficult and requires multiple adjustments over time.

HYPERGLYCEMIA AND DIABETES MELLITUS (CONTINUED)

Insulin therapy might require teaching proper IM injection techniques to the clients.

SURGICAL CONSIDERATIONS
Birds with diabetes mellitus should ideally be stabilized before undergoing a surgery.

MEDICATIONS

DRUG(S) OF CHOICE
- Insulin: ○ 0.3–1.3 IU/kg q12h IM. ○ To be adjusted according to the clinical signs and glucose curves. ○ Insulin preparations available to the practitioner are in a constant state of change. Many varieties of insulin that have been used in avian cases in the past are no longer available. • Glipizide: ○ 0.3–1 mg/kg q12h PO. ○ Highly variable results.

CONTRAINDICATIONS
- Insulin is contraindicated in case of hypoglycemia and severe allergy to bovine, porcine or human insulin proteins.
- Glipizide is contraindicated in case of severe burns, severe trauma, severe infection, diabetic coma or other hypoglycemic conditions, major surgery, ketosis, ketoacidosis, or other significant acidotic conditions.

PRECAUTIONS
- Insulin should not be injected at the same site every time or lipodystrophic reactions could occur. • Insulin overdose can induce hypoglycemia. • Glipizide should only be used with extreme caution with untreated endocrine dysfunctions; renal or hepatic function impairment; prolonged vomiting; emaciation or debilitated condition.

POSSIBLE INTERACTIONS
- Insulin has shown interaction in mammals with many drugs including beta blockers, enalapril, diltiazem, MAO inhibitors, fluoxetine, sulfonamides, and corticosteroids.
- Glipizide has shown interaction in mammals with many drugs including antifungal azoles, beta blockers, chloramphenicol, corticosteroids, MAO inhibitors, and sulfonamides.

ALTERNATIVE DRUGS
Synthetic somatostatins have been used successfully in one avian case.

FOLLOW-UP

PATIENT MONITORING
- Daily monitoring of clinical signs and body weight. • Blood glucose curves. No standard protocol exists for birds. Therefore, blood glucose curves can be done using the clinician judgment, as it is not always possible to draw blood from birds every hour for a prolonged period of time. • Fructosamine levels.

PREVENTION/AVOIDANCE
- Feed an appropriate diet for the species.
- Avoid contact with other birds to diminish the risks of an infectious pancreatitis.

POSSIBLE COMPLICATIONS
Diabetic ketoacidosis and insulin overdose are life-threatening conditions.

EXPECTED COURSE AND PROGNOSIS
- Prognosis greatly varies with the primary cause, which might never been known.
- Birds are very unpredictable in their response to insulin therapy. Adjustments of the therapy may take weeks.

MISCELLANEOUS

ASSOCIATED CONDITIONS
Hemosiderosis in toucans.

AGE-RELATED FACTORS
None.

ZOONOTIC POTENTIAL
None.

FERTILITY/BREEDING
Birds with diabetes mellitus might not be breeding actively if not stable.

SYNONYMS
None.

SEE ALSO
Adenovirus
Ascites
Chlamydiosis
Chronic egg laying
Coelomic distention
Dehydration
Diabetes insipidus
Diarrhea
Egg yolk coelomitis/reproductive coelomitis
Hepatic lipidosis
Herpesviruses
Iron storage disease
Lipoma
Liver disease
Nutritional deficiencies
Obesity
Ocular lesions
Pancreatic diseases
Paramyxoviruses
Polydipsia
Polyuria
Poxvirus
Renal disease
Salpingitis/uterine disorders
Sick-bird syndrome
Urate/fecal discoloration

ABBREVIATIONS
None.

INTERNET RESOURCES
N/A

Suggested Reading
Desmarchelier, M., Langlois, I. (2008). Diabetes mellitus in a Nanday conure (*Nandayus nenday*). *Journal of Avian Medicine and Surgery*, **22**(3):246–254.
Hazelwood, R.L. (2000). Pancreas. In: Whittow, G.C. (ed.), *Sturkie's Avian Physiology*, 5th edn. Academic Press, pp. 539–555.
Lumeij, J.T. (1994). Endocrinology. In: Ritchie, B.W., Harrison, G.J., Harrison, L.R. (eds), *Avian Medicine: Principles and Application*. Lake Worth, FL: Wingers Publishing, pp. 582–606.
Oglesbee, B. (1997). Diseases of the endocrine system. In: Altman, R.B., Clubb S.L., Dorrestein G.M., Quesenberry K. (eds), *Avian Medicine and Surgery*. Philadelphia, PA: WB Saunders, pp. 482–488.
Rae, M. (2000). Avian endocrine disorders. In: Fudge, A.M. (ed.), *Laboratory Medicine: Avian and Exotic Pets*. Philadelphia, PA: WB Saunders, pp. 76–89.

Author Marion Desmarchelier, DMV, MSc; DACZM
Acknowledgement Shannon Ferrell, DVM, DABVP (Avian), DACZM

HYPERURICEMIA

BASICS

DEFINITION
• Birds are uricotelic. ○ End products of nitrogen metabolism excreted in bird urine include urates, ammonia, urea, creatinine, amino acids, and others. ○ Uric acid (UA) is the main end product of nitrogen metabolism, representing 80% or more of the nitrogen excreted. • Hyperuricemia is defined as any plasma UA concentration higher than the calculated limit of solubility of sodium urate in plasma. ○ In birds, this theoretical limit of solubility of sodium urate is estimated to be 600 μmol/L. ○ In chickens, UA renal tubule transport system does not appear to become saturated until plasma UA levels exceed 60 mg/dL.

PATHOPHYSIOLOGY
• As the final breakdown product of dietary or endogenous purines, UA is generated by xanthine dehydrogenase (xanthine oxidase). ○ The liver is the primary source of UA. ○ Uric acid is also synthesized in the kidneys, intestines, and pancreas. • Uric acid is excreted by the cortical or reptilian nephrons. ○ Kidneys regulate UA by reabsorption, secretion, and postsecretory absorption in the proximal convoluted tubules. ○ Uric acid excretion is largely independent of glomerular filtration, water resorption, and urine flow rate. • Uric acid levels may elevate when glomerular filtration decreases by more than 70% to 80% (severe dehydration) or when large numbers of renal tubules have been compromised

SYSTEMS AFFECTED
• Musculoskeletal (articular gout). • Renal.

GENETICS
N/A

INCIDENCE/PREVALENCE
N/A

GEOGRAPHIC DISTRIBUTION
N/A

SIGNALMENT
• **Species:** Carnivorous and piscivorous birds, such as raptors and penguins, show a marked postprandial hyperuricemia in the first hours after ingestion of prey. • **Mean age and range:** A significant decreasing trend was noted with increasing age in UA. The higher levels found in nestlings were theorized to be the result of increased protein synthesis. • **Predominant sex:** In the collared scope owls (*Otus lettia*), UA levels were significantly higher among males than females.

SIGNS
General Comments
Affected birds are often asymptomatic.

Historical Findings
Is the bird molting? Uric acid levels tend to be lower when there is higher physiologic demand for protein such as the time of molting in anseriformes.

Physical Examination Findings
• Evidence of dehydration. • Signs of articular gout: pain and swelling in the synovial joints, nodules may appear white because of tophi formation.

CAUSES
The causes of hyperuricemia can be multifactorial but fall into two main categories:
1. Reduced renal tubular secretion of UA.
• Prolonged dehydration causes slight elevations in UA. • Hypothermia. • Renal disease. ○ Damage of the proximal convoluted tubules, nephrosis. ○ Any potential cause of nephritis such as nephrotrophic strains of infectious bronchitis virus, influenza virus, avian nephritis virus, chicken astrovirus, cryptosporidium, bacteria. ○ Nephrotoxin exposure causing renal tubular damage: nonsteroidal anti-inflammatory drugs (e.g., diclofenac, ketoprofen), mycotoxins (e.g., oosporein, ochratoxin A), heavy metals (e.g., lead and zinc toxicity), allopurinol. ○ There are a number of nutritional imbalances that can damage the kidneys, including hypovitaminosis A, hypervitaminosis A, hypervitaminosis D, excess dietary calcium, high dietary fat, dietary sodium and potassium imbalances, dietary magnesium and phosphorus deficiency as well as excess of trace elements, for example, zinc, and sodium bicarbonate supplementation. • Obstructive post-renal disease.
2. Excessive production of UA. • Physiologic hyperuricemia: Carnivorous birds show a normal, marked postprandial hyperuricemia within the first hours after ingestion of prey. ○ Raptors demonstrate a decline of plasma UA concentration between 8 and 23 hours after feeding. ○ This change has also been demonstrated in broiler chickens.
• Pathologic hyperuricemia: An important cause of elevated UA in humans but less important in birds can be seen with long-term, excessive high dietary protein levels.

RISK FACTORS
• **Medical conditions** ○ Duration of hyperuricemia: prolonged elevations increase the risk of gout developing. ○ Glomerular filtration rate, conditions that reduce: ■ Blood loss ■ Severe dehydration, water deprivation ■ Shock. ○ Mycotoxicosis. ○ Renal tubular disease. • **Medications** ○ Nephrotoxins: NSAIDs, intravenous contrast media. • **Environmental factors** ○ Nutritional imbalances: ■ Excessive dietary protein. ■ High dietary protein + high dietary calcium. ■ Vitamin A deficiency.

DIAGNOSIS

DIFFERENTIAL DIAGNOSES
• Lipemia can cause an artifactual elevation in UA measured by the photometric uricase method. • Blood samples taken via nail trim of a foot soiled with droppings can also reveal falsely elevated UA levels.

CBC/BIOCHEMISTRY/URINALYSIS
• **CBC**: Look for signs of infection or inflammation. • **Biochemistry panel** ○ Uric acid: ■ Slight or mild elevations are observed in dehydrated birds because of an increase in solute concentration: □ In a study of dehydrated chickens, UA levels increased after 24–48 h of water restriction but only in birds allowed free access to food. □ Recheck UA values after the patient has been clinically stabilized and rehydrated. ■ Moderate to severe elevations can be caused by renal disease or a physiologic postprandial increase in carnivorous birds: □ UA increases and peaks approximately 2 hours after ingesting a natural meal. □ Levels remain elevated for at least 12 hours after feeding. ■ Isoflurane anesthesia resulted in decreased UA values when compared with controls in American kestrels (*Falco sparverius*). ○ Urea ■ Increases in blood urea nitrogen (BUN) levels have been linked to dehydration in the bird. ■ Some studies have also reported elevations in BUN post-feeding. ○ Urea:UA ratio ■ In vulture species studied, this ratio has been described as a useful tool for evaluating both prerenal azotemia and renal damage. ○ Creatinine ■ Increases in creatinine have been linked to dehydration in pigeons. ○ Electrolytes ■ In birds with gout induced by sodium bicarbonate intoxication, hypernatremia, hyperuricemia, hypokalemia, and hypochloremia were common findings among exposed birds. ■ Moderate metabolic acidosis can be linked to hyperuricemia and can be exacerbated by damage to the proximal convoluted tubules. • **Urinalysis** ○ The presence of renal casts can indicate renal pathology. ○ Persistent hematuria has been associated with renal neoplasia, avian polyomavirus, bacterial and viral nephritis, and some forms of toxic nephropathy. ○ Myoglobinuria can cause false positive hematuria. ○ Porphyrinuria, as seen in lead-poisoned Amazon parrots, can result in urine that mimics hemoglobinuria, hematuria. ○ Although proteinuria is the hallmark sign of glomerulonephritis in mammals, voided urine samples are "normally" positive for protein due to fecal contamination in the cloaca.

OTHER LABORATORY TESTS
• Murexide test:
1. Aspirate a suspect articular gout lesion, and place the sample on a microscope slide.

HYPERURICEMIA (CONTINUED)

2. Mix with nitric acid and allow to dry over flame.
3. Add a drop of ammonia.
4. Mauve color indicates UA crystals are present.

IMAGING
Increased opacity may occur as a result of dehydration or renal gout.

DIAGNOSTIC PROCEDURES
Laparoscopic examination of the kidneys is often recommended in cases in which there is persistent hyperuricemia.

PATHOLOGIC FINDINGS
There may be no pathologic changes however profound, pathologic hyperuricemia can lead to: • Swollen kidneys with prominent lobules due to marked accumulation of urates in the tubules. • Milky white kidneys. • White crystalline deposits on the serosal surfaces of the pericardium, or on serosal surfaces of viscera like the liver and spleen, and skeletal muscles. • White crystals within synovial joints.

TREATMENT

APPROPRIATE HEALTH CARE
• Treat the underlying cause of renal disease whenever possible. • Patients suffering from a hyperuricemic crisis should be managed as inpatients.

NURSING CARE
• Fluid therapy: ○ If not treated promptly and aggressively, dehydration can rapidly exacerbate renal disease. ○ Fluids can be given PO, SC, IV, or IO routes depending on clinical circumstances. • Provide nutritional support as needed.

ACTIVITY
N/A

DIET
Offer a balanced diet appropriate for your species of interest.

CLIENT EDUCATION
• Monitor appetite, body weight, droppings. • Stress the importance of keeping the patient hydrated.

SURGICAL CONSIDERATIONS
N/A

MEDICATIONS

DRUG(S) OF CHOICE
• **Reduce UA levels** ○ Allopurinol: there are many anecdotal reports on the use ■ Appears to be relatively safe in galliformes, psittaciformes, columbiformes. ■ Use with caution in carnivorous birds, has been shown to increase hyperuricemia and cause visceral gout. ○ Colchicine: 0.01–0.2 mg/kg PO q12–24h; may potentiate gout formation. ○ Urate oxidase: 100–200 U/kg IM; considered a safer and more effective alternative to allopurinol, based on studies in pigeons and red-tailed hawks. • **Parenteral** vitamin A for oliguric and anuric renal patients with hyperuricemia.

CONTRAINDICATIONS
Stop or avoid all nephrotoxic drugs that could cause or aggravate renal disease such as aminoglycosides or sulfonamides.

PRECAUTIONS
N/A

POSSIBLE INTERACTIONS
Dehydration and renal disease may result in adverse responses to NSAIDs, especially COX inhibitors like meloxicam.

ALTERNATIVE DRUGS
Nonsteroidal anti-inflammatory agents to alleviate the pain of articular gout.

FOLLOW-UP

PATIENT MONITORING
• Appetite, water intake. • Droppings. • Hydration status. • CBC/chemistry panel/urinalysis.

PREVENTION/AVOIDANCE
Ensure adequate water intake.

POSSIBLE COMPLICATIONS
• Moderate to severe hyperuricemia can result in articular or visceral gout; however, the formation of UA crystals is a complex process. It is not clear why under normal circumstances urate deposits do not occur in carnivorous birds that show postprandial hyperuricemia after ingesting a natural meal. • In human patients, hyperuricemia is also an independent risk factor for hyperlipidemia, obesity, atherosclerosis, hypertension, coronary heart disease, and diabetes mellitus.

EXPECTED COURSE AND PROGNOSIS
Prolonged hyperuricemia increases the risk for development of gout.

MISCELLANEOUS

ASSOCIATED CONDITIONS
• Gout, articular: UA crystals accumulate within synovial capsules and the tendon sheaths of joints. • Gout, visceral. • Renal failure. • Urolithiasis.

AGE-RELATED FACTORS
A significant decreasing trend was noted with increasing age in UA. The higher levels found in nestlings were theorized to be the result of increased protein synthesis.

ZOONOTIC POTENTIAL
N/A

FERTILITY/BREEDING
N/A

SYNONYMS
Uricacidemia

SEE ALSO
Anorexia
Avian influenza
Coccidiosis
Cryptosporidiosis
Dehydration
Diabetes insipidus
Diarrhea
Emaciation
Heavy metal toxicity
Hyperglycemia
Hyperuricemia
Lameness
Mycotoxicosis
Nutritional deficiencies
Paramyxoviruses
Polydipsia
Polyuria
Renal disease
Sick-bird syndrome
Urate/fecal discoloration
Viral disease
Vitamin D toxicosis
Appendix 7, Algorithm 14. Hyperuricemia.

ABBREVIATION
UA—uric acid

INTERNET RESOURCES
N/A

Suggested Reading

Calabuig, C.P., Ferrer, M., Muriel, R. (2010). Blood chemistry of wild Brazilian Coscoroba swans during molt. *Journal of Wildlife Diseases*, **46**(2):591–595.

Chan, F.T., Lin, P.I., Chang, G.R., et al. (2012). Hematocrit and plasma chemistry values in adult collared scops owls (*Otus lettia*) and Crested Serpent Eagles (*Spilornis cheela hoya*). *Journal of Veterinary Medical Science*, **74**(7): 893–898.

Hernandez, M., Margalida, A. (2010). Hematology and blood chemistry reference values and age-related changes in wild bearded vultures. *Journal of Wildlife Diseases*, **46**(2):390–400, 2010.

Sinclair, K.M., Church, M.E., Farver, T.B., et al. (2012). Effects of meloxicam on hematologic and plasma biochemical analysis variables and results of histologic examination of tissue specimens of Japanese quail (*Coturnix japonica*). *American Journal of Veterinary Research*, **73**(11):1720–1727.

Author Christal Pollock, DVM, Dipl. ABVP (Avian)

Hypocalcemia (and Hypomagnesemia)

BASICS

DEFINITION
Low blood calcium levels are responsible for a variety of clinical presentations, including skeletal malformations in developing birds, a well-described neurologic/seizure disorder in African greys (*Psittacus erithacus*), and reproductive issues in laying hens. Low blood magnesium levels are rare, but when present cause severe concurrent hypocalcemia.

PATHOPHYSIOLOGY
• Calcium homeostasis is a complex, well-regulated process in birds utilizing multiple target organs, hormones and receptors. Overall, the main cause of hypocalcemia in birds is nutritional secondary hyperparathyroidism. • Bone contains 99% of calcium as hydroxyapatite. Extracellular calcium is known as total calcium and is a combination of ionized calcium and calcium bound to protein (primarily albumin) and anions. Ionized calcium is the physiologically active form and levels are very tightly regulated through hormonal mechanisms. Total calcium levels fluctuate widely with changes in protein levels and acid/base balance. • The parathyroid gland secretes parathyroid hormone (PTH) in response to low circulating calcium levels, leading to decreased renal loss of calcium, increased calcium resorption from bones, and increased calcitriol production leading to increased intestinal absorption of dietary calcium.
• PTH works in concert with Vitamin D. Pro-vitamin D_3 is present in the skin and uropygial gland secretions. This is converted to Vitamin D_3 (25-hydroxycholecalciferol, referred to as cholecalciferol) secondary to exposure to ultraviolet B wavelengths (290–315 nm). Cholecalciferol is primarily converted into calcitriol (1,25-dihydroxycholecalciferol) in the kidneys in response to low calcium levels. Calcitriol is the most active vitamin D3 metabolite acting on the intestine to increase calcium absorption, and osteoclasts to promote bone resorption and release of calcium and phosphorous into the blood stream. Calcitriol also stimulates bone formation and inhibits PTH production during periods of eucalcemia. • Ultimobranchial glands produce calcitonin in response to high blood levels of calcium. Calcitonin appears to inhibit osteoclast activity in bones and increase urinary calcium excretion. • Egg laying birds acquire 30–40% of required calcium from medullary bone, which is released through calcitriol and estrogen activity. 10% of calcium reserves may be required for egg production within a 24-hour period. • Vitamin D_3 is able to be absorbed through the GI tract with 60–70% efficiency, making oral supplementation important (especially for birds not exposed to UV light).
• Hypocalcemia leads to enlargement of the parathyroid glands and elevated PTH. Bones become weak from calcium resorption to maintain adequate calcium levels (leads to folding fractures seen in African grey nestlings). In adults, low calcium levels can lead to neurologic signs and seizures. In chronic egg layers, osteomalacia may lead to spontaneous fractures and decreased muscle function may lead to egg binding. • Excessive dietary vitamin A and E may impact absorption of vitamin D, but this is uncommon. • Magnesium is an essential dietary element. Magnesium is an intracellular cation and is integral as an activator or catalyst in over 300 enzymatic processes, most importantly in those associated with ATP. Hypomagnesemia has been induced in chickens fed on a deficient diet. Signs include hyperexcitablity, twitching, weakness, lethargy, tremors and death. Hypomagesemic birds are concurrently hypocalcemic, even if fed normal amounts of calcium. Their bone density (and calcium content) is increased, circulating PTH levels are high and calcitriol levels are not elevated. The exact mechanism of how magnesium inhibits circulating calcium levels is not known (although it may be associated with inhibition of calcitriol production, osteocytic dysfunction or resistance to PTH at its organs of action). Magnesium is present in adequate levels in seeds, nuts and formulated diets, making deficiency rare. However, hypomagnesemia has been reported in one neurologic hypocalcemic African grey that did not respond to either calcium or vitamin D3 supplementation alone, but did recover after a single dose of parenteral magnesium.

SYSTEMS AFFECTED
• Endocrine/Metabolic system is the main site of the disease process, leading to effects in other systems. • Musculoskeletal—system affected in juveniles and severe chronic hypocalcemia seen in egg laying females.
• Nervous—system most affected in adult birds with hypocalcemia. • Reproductive—system affected in egg laying females.

GENETICS
There is no genetic predisposition.

INCIDENCE/PREVALENCE
Common in birds fed low quality diets that consist primarily of seeds.

GEOGRAPHIC DISTRIBUTION
May affect birds anywhere.

SIGNALMENT
• **Species:** Any avian species are susceptible, but African grey parrots appear to be highly susceptible (the cause is not clear, but is postulated to be related to feeding seed diets (low in calcium and vitamin D3) and/or a higher need for UVB light). • **Mean age and range** ○ Young birds are susceptible to osteodystrophy relating to transient hypocalcamia in the hen. ○ Hypocalcemic induced seizures in African Greys have been reported in birds 2–15 years old. However, any adult bird on an inadequate diet (and little to no UVB exposure) may develop hypocalcemia and demonstrate the classic signs of weakness, ataxia and seizures. Any laying hen on an inadequate diet (and little to no UVB exposure) may develop osteomalacia and egg binding. • **Predominant sex:** The juvenile osteodystrophy and adult neurologic signs may occur in any sex.

SIGNS

Historical Findings
Juvenile birds may come out of the nest box unable to perch or stand normally due to folding fractures. Adult birds may show (and be presented for) general sick bird signs before the neurologic signs manifest (weakness, decreased appetite, decreased vocalization, exercise intolerance, and sitting fluffed on the bottom of the cage). Neurologic signs that owners may notice include ataxia, tremors, falling off the perch and seizures. In addition to exhibiting the above signs, affected hens may present for straining to lay, dyspnea (due to coelomic distention), cloacal prolapse, decreased stool production, bloody eliminations and lameness.

Physical Examination Findings
• Juveniles—Folding fractures of the long bones or curvature of the spine. • Adults ○ Weakness ○ Ataxia ○ Tremors ○ Seizures.
• Females ○ Weakness ○ Egg present on coelomic palpation ○ Cloacal prolapse ○ Metabolic fractures.

CAUSES
• Low dietary calcium. • Low dietary vitamin D_3 (or decreased vitamin D absorption due to over-supplementation of vitamins A and E).
• High dietary phosphorous. • Chronic egg laying. • Little to no UVB radiation (from specialized lights or direct sunlight).
• Gastrointestinal disease that leads to maldigestion or malabsorption.

RISK FACTORS
N/A

DIAGNOSIS

DIFFERENTIAL DIAGNOSIS
Hypocalcemia is generally straightforward to rule in with total and ionized calcium levels in symptomatic birds.

Juveniles
• Trauma—differentiate with history, physical examination and radiographs.

Adults
• Metabolic (hypoglycemia, hepatic encephalopathy, renal failure)—differentiate with chemistry panel. • Toxicosis (lead,

HYPOCALCEMIA (AND HYPOMAGNESEMIA) (CONTINUED)

mercury, organophosphates/carbamates)—differentiate with history of exposure, chemistry panel and ionized calcium, and heavy metal testing. • Infection (meningitis, meningoencephalitis): bacterial (including *Chlamydia psittaci*), parasitic, viral (including *Bornavirus* (Proventricular Dilatation Disease (PDD)), fungal—differentiate with history, physical examination, CBC, chemistry panel, radiographs, barium study and specialized testing for specific etiologies. • Neoplasia—differentiate through physical examination (which may demonstrate persistent neurologic deficits) and through the exclusion of all other etiologies and possibly through advanced imaging (computed tomography [CT]). • Idiopathic epilepsy—differentiate through exclusion of all other etiologies. • Trauma—differentiate with history and physical examination (use alcohol to moisten feathers on head and over spine to visualize any bruising).

CBC/BIOCHEMISTRY/URINALYSIS
Low plasma total calcium may or may not be present.

OTHER LABORATORY TESTS
• Ionized calcium is the most important value to determine true hypocalcemia. It should ideally be analyzed immediately in a point of care analyzer (such as the i-Stat (Abbott Laboratories, Abbott Park, IL)). However, if the sample is frozen immediately after being spun down and sent to the laboratory on ice, useful results can be obtained (contact your laboratory for special instructions). The normal range in captive African grey parrots is 0.96–1.22 mmol/L. • Vitamin D_3 levels (25-hydroxycholecalciferol and 1,25 dihydroxycholecalciferol) are becoming more commonly measured, but require very specialized handling (drawn into heparin and frozen until test performed). They are considered to be useful but levels vary widely based on dietary Vitamin D and UVB exposure. Reference ranges for poultry (in nmol/L): ○ 25-hydroxycholecalciferol (14.5–20.0). ○ 1,25 dihydroxycholecalciferol (100.0–332.8). • PTH levels are not currently clinically feasible to obtain but may be useful in the future. • Plasma magnesium levels are available through most large laboratories. Reference ranges have been determined in several psittacine species (in mmol/L): ○ African Grey (nonreproductive) (0.82–1.07). ○ Hispanolian Amazon (nonreproductive) (0.82–1.07). ○ Hyacinth Macaw (1.2 +/– 0.5). ○ Blue and Yellow Macaw (1.0+/– 0.2). ○ Green Winged Macaw (1.3+/– 0.5).

IMAGING
Whole body radiographs should be performed to evaluate bone malformation in young birds, look for possible alternative diagnoses in neurologic birds, evaluate bone density (decreased or hyperostosis), and look for presence of eggs in reproductive females.

DIAGNOSTIC PROCEDURES
N/A

PATHOLOGIC FINDINGS
• Parathyroids: Classic appearance of nutritional secondary hyperparathyroidism (NSHP)—enlarged parathyroids with enlarged chief cells that become foamy to clear in severe disease. The chromatin becomes less condensed and the nucleus enlarges. A trabecular pattern may also be seen. • Bony changes in patients with osteodystrophy: Classic appearance of rickets—elongated, disorganized zone of proliferation and wide seams of unmineralized osteoid at the trabecular periphery. Large numbers of osteoblasts are present along the trabecular periphery. Fibrous tissue proliferation is present in diseased bone.

TREATMENT

APPROPRIATE HEALTH CARE
• Juvenile birds are generally treated on an outpatient basis. Most birds are clinically normal and require diet counseling and cage modification advice. Severely affected cases may require surgery to restore better function to pelvic limbs. • Adults with hypocalcemic seizures require inpatient care until the seizures resolve and the bird is able to eat and maintain hydration on its own. Follow up with dietary counseling and supplement recommendations on an outpatient basis. • Hens with egg binding—see treatment in "dystocia/egg binding"

NURSING CARE
Base supportive care on physical examination and the clinical picture. Patients may need fluid therapy based on hydration status and severity of hypovolemia (subcutaneous for mild cases, intravenous (IV)/interosseous (IO) for severely affected birds; having vascular access also allows for more rapid administration of anticonvulsants). Supplemental gavage feedings may be required in birds that are unable to eat on their own.

ACTIVITY
Strict cage rest until the seizures resolve and weakness improves. Making cage modifications may help prevent further injury (wrapping the perches for better grip, padding the bottom of the cage to minimize trauma if the patient falls).

DIET
• Focus on encouraging a balanced diet based on formulated pellets and a wide variety of vegetables, fruits, nuts and grains. • Calcium should be around 0.5% of the diet (higher in laying hens). The calcium to phosphorous ratio should ideally be 1.5:1. Calcium recommendations range from 0.3–1% in laying psittacines, 1.88–3.25% in laying chickens (high end for those that lay daily). • Recommended dietary magnesium levels should be over 0.06%. This should be very easy to achieve in any diet. Sunflower seeds contain over 0.3% Mg, millet contains over 0.11% Mg and peanuts contain over 0.16% Mg. Several pellet companies were contacted during the writing of this chapter and Mg levels in pellets range 0.12–0.15% (personal communications with Dr. Greg Harrison of Harrison Bird Foods and Cyndi Wheeler of Lafeber Company).

CLIENT EDUCATION
• Diet modification should be the primary focus (see above). • Encouraging the addition of direct sunlight exposure to the daily routine provides the bird with the best source of UVB. Alternatively, adding a UVB lamp is beneficial when direct sunlight exposure is not feasible (make sure all shielding is removed from the fixture). Discuss proximity of the lamp to birds; the bulb should be no closer than six inches and no greater than 18 inches. Long fluorescent tube bulbs are the most consistent, and should be changed every six months. Other bulbs are likely to be effective, but may not be as safe (high UVB levels may cause irritation to the cornea and/or skin around the eyes). • For birds with clinical hypocalcemia with weakness and intermittent seizure activities, caging may need to be modified to provide padding in case of falls, more secure gripping surfaces for perches, and easier access to food and water. • For birds that have survived a transient hypocalcemia that led to fractures, cage modifications should focus on allowing the best mobility, such as wrapping perches and placing platforms in the cage. Additionally, it is important to monitor the plantar surfaces regularly for development of pododermatitis if the bird does not ambulate normally.

SURGICAL CONSIDERATIONS
Calcium levels should be restored before significant anesthetic procedures. Very short periods of anesthesia should be used with caution if necessary for sample collection. Consider injectable or intranasal sedation instead of an inhalant.

MEDICATIONS

DRUG(S) OF CHOICE
• Calcium gluconate—dilute 1:1 with saline or sterile water for IV, SQ, IM injection; 5–10 mg/kg slow IV for tetany; 10–100 mg/kg SC, IM, slow IV. • Vitamin D_3 injection (typically given in combination with Vitamins A and E [Vital E+A&D])—3300–6600 U/kg IM once. • Calcium glubionate—150 mg/kg PO q12h or 750 mg/L drinking water, change

HYPOCALCEMIA (AND HYPOMAGNESEMIA) (CONTINUED)

daily. • Magnesium sulfate injection—20 mg/kg IM once; should be considered if the plasma magnesium levels are low, or if calcium and Vitamin D_3 supplementation do not resolve clinical signs or cause an appropriate increase in calcium. • For acute seizure control, benzodiazepines are useful:
◦ Midazolam—0.1–2 mg/kg IM or IV; better IM absorption than diazepam.
◦ Diazepam—0.05–0.5 mg/kg IV.

CONTRAINDICATIONS
N/A

PRECAUTIONS
N/A

POSSIBLE INTERACTIONS
Calcium may alter absorption of some medications, including doxycycline and enrofloxacin.

ALTERNATIVE DRUGS
If cost is an issue, work towards oral supplementation as soon as possible. There is not an alternative to injectable calcium for treatment of hypocalcemic seizures.

FOLLOW-UP

PATIENT MONITORING
• Seizures and neurologic signs should resolve very quickly (within hours to days) after starting treatment. • Total calcium and/or ionized calcium levels should be monitored often initially to make sure the levels are rising (potentially every 1–3 days). Once the levels are noted to be rising, check every 1–2 weeks until levels are normal then monitor yearly.
• Vitamin D levels—if they are measured and noted to be low, the values should be monitored every 1–2 months to evaluate return to normal levels. • Magnesium may be measured if the patient did not respond to the calcium and vitamin D3 therapy as expected. If it is low, the magnesium level should be checked within a week. Low magnesium levels are very uncommon so if they are noted in a bird, the values should be evaluated yearly (along with routine blood testing).

PREVENTION/AVOIDANCE
Providing a formulated diet with calcium to phosphorous ration of 1.5:1 and magnesium level of greater than 0.6% (or higher, if this is determined for nonpoultry species in the future). Do not oversupplement with calcium or phosphorous, which may compete with Mg, leading to hypomagnesemia. Do not oversupplement with vitamins A or E, which may affect absorption of vitamin D.

POSSIBLE COMPLICATIONS
• Seizure activity that is persistent/uncontrollable may lead to death. • Development of a seizure focus leading to reoccurring seizures and the need for anticonvulsant therapy. • Bony changes in juvenile birds may cause them to ambulate on different parts of their feet, leading to possible development of wounds or sores (pododermatitis/bumble foot) or arthritis.

EXPECTED COURSE AND PROGNOSIS
If calcium levels return to normal and clinical signs resolve, the prognosis is good. If neurologic signs persisted for a prolonged period, they may continue even after resolution of calcium deficiency.

MISCELLANEOUS

ASSOCIATED CONDITIONS
N/A

AGE-RELATED FACTORS
Juveniles with bony changes do not often have continued hypocalcemia (it was likely related to transient hypocalcemia in the hen).

ZOONOTIC POTENTIAL
None.

FERTILITY/BREEDING
Hypocalcemia directly affects a hen's ability to produce and lay eggs. It must be controlled to allow for normal reproductive function.

SYNONYMS
Calcium tetany, osteoporosis, cage layer fatigue (poultry).

SEE ALSO
Chronic egg laying
Cloacal disease
Dystocia/egg binding
Egg yolk coelomitis
Fracture/luxation
Metabolic bone disease
Neurologic conditions
Nutritional deficiencies
Seizure
Sick-bird syndrome

ABBREVIATIONS
N/A

INTERNET RESOURCES
USDA National Nutrient Database for Standard Reference: http://ndb.nal.usda.gov

Suggested Reading
de Carvalho, F.M., Gaunt, S.D., Kearney, M.T., et al. (2009). Reference intervals of plasma calcium, phosphorus, and magnesium for African grey parrots (*Psittacus erithacus*) and Hispaniolan parrots (*Amazona ventralis*). *Journal of Zoo and Wildlife Medicine*, **40**(4):675–679.
de Matos, R. (2008). Calcium metabolism in birds. *The Veterinary Clinics of North America. Exotic Animal Practice*, **11**(1):59–82.
Kirchgessner, M.S., Tully, T.N., Nevarez, J., et al. (2012). Magnesium therapy in a hypocalcemic African grey parrot (*Psittacus erithacus*). *Journal of Avian Medicine and Surgery*, **26**(1):17–21.
Klasing, K.C. (1999). *Comparative Avian Nutrition*. Oxfordshire, UK: CAB International.
Schmidt, R.E., Reavill, D.R., Phalen, D.N. (2003). *Pathology of Pet and Aviary Birds*. Ames, IA: Iowa State Press.
Author Anneliese Strunk, DVM, DABVP (Avian)

 Client Education Handout available online

INFERTILITY

BASICS

DEFINITION
- Infertility in birds can be defined as the failure to produce fertile eggs. • In a wider perspective, the term infertility is erroneously used to define avian pairs that do not produce eggs, or do not seem to have interest for breeding. • Finally, the widest definition may include birds that do produce fertile eggs, but there is a high percentage of embryonic deaths.

PATHOPHYSIOLOGY
- Multifactorial • All disorders that involve courtship, mating, copulation and directly egg laying can lead to infertility.

SYSTEMS AFFECTED
- Reproductive. • Musculoskeletal. • Gastrointestinal. • Endocrine/Metabolic. • Behavioral. • Management (owner/curator).

GENETICS
Possible, but unlikely. Eventually limited to overbred species with possible inbreeding problems (canaries, budgerigars, chicken).

INCIDENCE/PREVALENCE
- Important in breeding flocks and rare species, especially when involved in recovery programs. • Not relevant in pet birds.

GEOGRAPHIC DISTRIBUTION
Worldwide.

SIGNALMENT
- **Species:** All avian species. • **Species predilections:** Canary, budgerigar, all psittacine birds, cranes, vultures. Less frequently diurnal birds of prey.

SIGNS
- No eggs are laid. • All (or most of the) eggs that are laid are infertile. • Alternatively there is no mating. • By extension: high (or total) per cent of embryonic deaths.

Historical Findings
- Inadequate husbandry for the species. • Inadequate nest. • Inappropriate diet (type and food rotation).

Physical Examination Findings
- Often no abnormalities are noticed on physical examination. • If some findings are present, they are usually related to the musculoskeletal system, the cloaca, or the eyes.

CAUSES
- Reproductive apparatus—orchitis, oophoritits, oviductitis, hermaphroditism (true hermaphroditism is extremely rare, but wrong development of deferens is not). Lesion of the phallus (limited to ratites and waterfowl). • Musculoskeletal apparatus—arthritis, arthrosis, limb deviation, bumblefoot, missing limbs, kyphosis, scoliosis. • Gastrointestinal system—cloacitis, cloacal paralysis (e.g., neoplasia), cloacal papillomas.
- Endocrine/Metabolic causes—low calcium blood level, pituitary gland deficiency.
- Behavioral causes—wrong rearing (alone, no social contacts), wrong imprinting (esp. birds of prey, waterfowls). • Husbandry (owner/curator)—wrong cage/aviary, wrong nest (site, size and shape), inadequate neighbors, wrong lighting in indoor facilities, wrong facilities according to geographical location (too cold/warm; too wet/dry).
- Nutritional causes—Ca/P unbalance, low dietary calcium, low protein in diet.
- Toxic—aflatoxins and other mycotoxins, some drugs, maybe anecdotal (metronidazole, tetracyclines, ivermectin). • Generalized/chronic diseases—chlamydiosis (may cause orchitis), hepatitis (limited production of yolk proteins), many chronic diseases may limit egg production/laying, as well as inhibit mating behavior.

RISK FACTORS
- Nutritional deficiencies. • Bad pair bonding. • Wrong husbandry.
- Species-specific needs are not fulfilled.

DIAGNOSIS

DIFFERENTIAL DIAGNOSIS
The first important step is to determine if there is a disease, physical impairment to breeding or fertile egg production, or a problem related to husbandry. This may also require a differentiation between eggs that do not hatch when incubated artificially, but do develop normally when incubated by the parents.

Once this has been clarified, a differential diagnostic list and plan can be designed.
- *Normal eggs are laid, but they are all infertile:* ○ Viruses: Any virus able to alter the shape, pH and flora of the cloaca (herpesvirus/papillomatosis) may impede sperm vitality and fertilization of the egg. ○ Orchitis: Bacteria, *C. psittaci*, neoplasm (especially cloaca and preen gland). ○ No pair bonding: Healthy birds, but without copulation. ○ Musculoskeletal problems: Arthritis, impaired limbs, inhibition to copulation due to pain and/or mechanical problems. ○ Homosexual couple: Two males or two females housed together. DNA sexing can fail in some species, especially if poor collection technique. ○ Abnormal development of gonads in the male bird: Testicles and deferens may be altered in function and anatomy. Often seen at endoscopy. • *Normal eggs are laid, but there is a high number of infertile eggs:* ○ Inadequate husbandry: Wrong perches, overly long peri-cloacal feathers (limited to some breeds/mutations), and any other factor related to management, that may impede a normal copulation. ○ Competitions/disturbances of neighboring pairs, this also may disturb birds during copulation.
○ Musculoskeletal problems: Minor forms of arthritis/arthrosis, bumblefoot, limb deviation, and lipomas may render copulation difficult, but not impossible. ○ Viruses: Sperm vitality and fertilization of the egg may be lowered, but not completely stopped.
- *Abnormal/misshapen eggs are laid:* ○ Metritis: By definition, metritis is an inflammatory process of the uterine portion of the oviduct (egg shell chamber). Can be bacterial, fungal, mycoplasmal, or viral. Metritis can cause abnormal shell formation. ○ Low dietary calcium/low blood calcium: Quality of the eggshell is directly related to bioavailability of calcium. ○ Parathyroid problems: Also involving the calcium metabolism.
- *Embryonic deaths:* ○ Viruses: Species related, but adenovirus, reticuloendotheliosis virus, orthomyxovirus, paramyxovirus, polyomavirus, circovirus and herpesvirus, have all been suspected, or proven, to cause embryonic death, at different embryonic development stages. ○ Bacteria: Can cause cloacitis, oviductitis, and infect the egg content. Sometimes, bacteria can enter the eggshell directly from the environment, but in this case, a crack, or a minor fissure in the shell should be suspected. ○ Mycoplasmas: Can cause metritis and embryonic death. ○ Fungi: Can occasionally be cultured from dead embryos, in this case suspect that the eggshell is not intact. ○ Wrong incubation parameters: Limited to artificial incubation and egg storing before incubation starts. Too low or too high temperature and relative humidity can lead to embryonic death. • *No eggs are laid:* ○ Oophoritis: Can be infectious (viral, bacterial, or fungal), or neoplastic in origin. Whatever the cause, usually egg formation is inhibited and no eggs will be laid. ○ Salpingitis: An inflammatory process of the proximal portion of the oviduct. Being localized in a narrow tube, like the salpinx, it usually results in the sealing of the oviduct lumen. ○ Chronic egg binding: Uncommon, but if an egg is located in the proximal coelom, not causing problems with passing feces and urine, a female bird may become egg bound for months, or years. This will stop the subsequent formation and/or passage of new eggs. ○ Homosexual pair: Two males will not produce any egg. Juvenile gonads may be wrongly interpreted at endoscopy by nonexperienced avian endoscopists. ○ Cystic ovary: Supposedly due to endocrine unbalances. Outcome similar to oophoritis. ○ Psittacosis: Most often the male gonads are affected, but also Chlamydial oophoritis has been described. ○ Thyroid problems: Cause obesity and sexual inactivity (rare). ○ Pituitary gland diseases: May affect the hypothalamic–hypophyseal–ovarian axis and cause the ovary to stop its activity. ○ No pair bonding: When two birds have been arbitrarily selected as "a pair", mating may not occur. ○ Very poor diet: Nutrients adequate for survival but

INFERTILITY (CONTINUED)

insufficient to form and lay eggs. ° Wrong husbandry: Birds may be missing most/all the environmental and social stimuli that would lead to breeding. ° Competition/disturbances of neighboring pairs: Birds (especially males) may be so deeply involved in the competition with birds in other aviaries that they ignore their cage mate.

CBC/BIOCHEMISTRY/URINALYSIS
• CBC may be altered in case of infectious diseases affecting several organs. Rarely it will be altered in the case of poor husbandry.
• Biochemistry will be altered only when a specific system/organ is affected. Apart from calcium, ionized calcium and cholesterol, very little diagnostic value in avian infertility.
• Urinalysis: N/A.

IMAGING
• Radiology: Useful in cases of musculoskeletal problems, or when previous, chronic egg binding is suspected.
• Endoscopy: To evaluate possible homosexual pairs, or pathologies of the gonads. • Ultrasound: For cystic ovaries and coelomitis.

DIAGNOSTIC PROCEDURES
• *Normal eggs are laid, but they are all infertile:*
° Egg necropsy to verify whether the eggs are infertile, or there is a very early stage embryonic death. ° Virology with samples collected from egg and/or parents.
° Bacteriology with samples collected from egg and/or parents. ° Testicular biopsy to rule out orchitis or testicular tumor. ° Collection and evaluation of semen via electro ejaculation, and/or manual massage.
° Analysis of behavior, eventually using a video recording system. ° Radiology to diagnose musculoskeletal problems,
° Endoscopy to rule out homosexual pairs or when an alteration of the male gonads is suspected. • *Normal eggs are laid, but there is a high number of infertile eggs:* ° Careful inspection of the aviary and facilities to evaluate all the risk factors that may limit a normal copulation. ° Analysis of behavior for competitions/disturbances of neighboring pairs. ° Radiology of birds to diagnose musculoskeletal problems. ° Viral screening.
• *Abnormal/misshapen eggs are laid:*
° Bacteriology to rule out metritis. Additional testing including fungal, mycoplasmal, or viral screening as warranted. ° CBC to rule out infections. ° Blood chemistry for evaluation of calcium and ionized calcium levels • *Embryonic deaths:* ° Viral screening of egg/embryo or parents. ° Microbiology, incl. *C. psittaci* and mycoplasma. ° Histopathology.
° In case of artificial incubation evaluate if the incubators are working well and monitor incubation parameters. • *No eggs are laid:*
° CBC/protein electrophoresis and microbiology when oophoritis and salpingitis are suspected. ° Radiology to rule out ovostasis. ° Ultrasound to rule out cystic ovary and/or celomitis (in the latter, aspiration of the coelomic fluid is an option).
° Endoscopy to rule out homosexual pair, cystic ovary, or other pathology. ° Psittacosis testing. ° Thyroid function testing. ° Analysis of behavior, eventually using a video recording system.

PATHOLOGIC FINDINGS
Gross and histopathology findings will differ depending upon the main cause.

TREATMENT
APPROPRIATE HEALTH CARE
Hospitalization and nursing of avian patients with infertility is unlikely, unless the primary cause is a generalized, or severe disease (for example, chlamydiosis, tumor of the gonads, chronic egg binding).

NURSING CARE
Refer to specific chapter about generalized diseases.

ACTIVITY
N/A

DIET
• Diet is a significant consideration when infertility has to be addressed in a bird breeding flock. • Diet will be adjusted/modified/monitored according to species, season, and geographic location.

CLIENT EDUCATION
• If the main cause of the problem is inadequate husbandry, the client must be instructed about the best way to manage the breeding flock. This may vary dramatically with different avian species. • Furthermore, when viral, bacterial and fungal diseases are considered the major cause of the problem, a careful education plan must be set up for the client.

SURGICAL CONSIDERATIONS
• Refer to specific publications on surgery.
• There is not a specific procedure for avian infertility; however, the main surgical procedures that might be considered are:
° Coelioscopy for evaluation of gonadal abnormalities and biopsies. ° Hysterectomy to treat chronic egg binding (see specific chapter). ° Surgery for cloacal papillomas.

MEDICATIONS
DRUG(S) OF CHOICE
• Antibiotic therapy—ideally based on culture and sensitivity results if bacterial disease.
• Antifungal therapy—therapy warranted if fungal disease diagnosed. • Analgesics—may be warranted if painful condition prohibits normal mating behavior. • Other therapies based on etiology, varies based on cause.

CONTRAINDICATIONS
NSAIDs should not be used if gastrointestinal ulceration is known or suspected to be present.

PRECAUTIONS
NSAIDs should not be administered if renal disease is known or suspected to be present or discontinued immediately if clinical signs suggestive of renal dysfunction (such as polyuria/polydipsia) are observed.

POSSIBLE INTERACTIONS
N/A

ALTERNATIVE DRUGS
N/A

MISCELLANEOUS
ASSOCIATED CONDITIONS
Many conditions are associated with secondary infertility. All disorders that involve courtship, mating, copulation and egg laying can be associated with infertility.

AGE-RELATED FACTORS
Mostly young birds (inexperience) and old birds (generalized aging), are affected.

ZOONOTIC POTENTIAL
Only chlamydiosis is a serious threat.

FERTILITY/BREEDING
As listed.

SYNONYMS
N/A

SEE ALSO
Arthritis
Ascites
Chlamydiosis
Chronic egg laying
Circoviruses
Cloacal disease
Coelomic distention
Colibacillosis
Cystic ovaries
Dystocia/egg binding
Egg yolk coelomitis/reproductive coelomitis
Herpesviruses
Hypocalcemia/hypomagnesemia
Lameness
Metabolic bone disease
Mycoplasmosis
Nutritional deficiencies
Obesity
Phallus prolapse
Pododermatitis
Polyomavirus
Poxvirus
Salpingitis/uterine disorders
Thyroid diseases
Viral disease
Vitamin D toxicosis
Xanthoma

ABBREVIATIONS
N/A

INFERTILITY (CONTINUED)

INTERNET RESOURCES
N/A

Suggested Reading
Clubb, S., Phillips, A. (1992). Psittacine embryonic mortality. In: Schubot, R.M., Clubb, K.J., Clubb, S.L. (eds), *Psittacine Aviculture, Perspectives, Techniques and Research*. Loxahatchee, FL: Aviculture Breeding and Research Center, pp. 318–325.

Crosta, L., Gerlach, H, Bürkle., M., Timossi, L. (2003). Physiology, diagnosis and diseases of the avian reproductive tract. *The Veterinary Clinics of North America. Exotic Animal Practice*, **6**:57–83.

Fischer, D., et al. (2013). Assisted reproduction in two rare psittacine species – the Spix's Macaw and the St. Vincent amazon. *Proceedings of the International Conference on Avian, Herpetological and Exotic Mammal Medicine (ICARE 2013)*, Wiesbaden, Germany, April 20–26, pp. 295–296

Stelzer, G., Crosta, L., Bürkle, M., Krautwald-Junghanns, M.E. (2005). Attempted semen collection using the massage technique and semen analysis in various psittacine species. *Journal of Avian Medicine and Surgery*, **19**(1):7–13.

Whittow, G.C. (ed.) (2000). *Sturkie's Avian Physiology*, 5th edn. Elsevier.

Authors Lorenzo Crosta, DVM, PhD
Petra Schnitzer, DVM

Ingluvial Hypomotility, Crop Stasis and Ileus

BASICS

DEFINITION
• Ingluvial hypomotility is a relatively common clinical concern observed in a large number of avian species. In the majority of clinical presentations, ingluvial hypomotility is secondary to additional systemic pathology, but occasionally, primary pathology may be present. • Common noninfectious causes of ingluvial hypomotility include fistulas, trauma, toxicoses and foreign bodies.
• Common infectious causes include bacterial infections, mycotic infections, parasitic, and viral infections. • There are many etiologies which may result in secondary or tertiary ingluvial stasis which must be clinically distinguished from primary causes. • The ingluvies (or crop) is a dilation of the cervical esophagus and is used primarily to store food. It is well developed in many grainivorous birds but minimally developed in birds of prey and not present in piscivorous and many passerine species.

PATHOPHYSIOLOGY
• The pathophysiology of ingluvial stasis is complicated as the majority of clinical presentations are due to primary disease outside of the ingluvies itself. Stasis can be associated with both primary noninfectious and infectious pathology. Primary ingluvial pathology is more common in juvenile and neonatal birds and less common in adults.
• Primary peristalsis controls movement of food from the esophagus to the stomach and in fasted birds the esophagoingluvial fissura is closed which prevents food entering the crop. Lacking a true sphincter, crop motility is commonly affected by aborad pathology. The destination of food is controlled by the contractile state of the ventriculus, mediated by the extrinsic nervous system, and rate of passage is correlated to particle size. Any disease process that affects these systems will result in secondary ingluvial stasis.

SYSTEMS AFFECTED
• Behavioral—nonspecific but commonly includes voluntary regurgitation. Other behavior changes that may be observed in patients with ingluvial stasis include lethargy, anorexia, hyporexia or polyphagia.
• Gastrointestinal—distension of the ingluvies with fluid or ingesta, regurgitation, altered output of droppings with typically decreased feces, decreased or less commonly increased appetite. • Nervous—segmental and peripheral neuropathies. • Neuromuscular—wasting; weakness and weight loss, typically associated with inadequate calorie absorption.
• Renal/Urologic—renal tubular constipation secondary to chronic dehydration.
• Respiratory—open-mouth breathing and increased respiratory effort due to ingluvial or proventricular distension, secondary aspiration pneumonia or tracheitis.
• Immune—proventricular dilatation disease (PDD).

GENETICS
No proven genetic component.

INCIDENCE/PREVALENCE
• Handfeeding birds are more likely to develop ingluvial trauma, atonic ingluvies, and/or fistulas. • Birds of prey commonly develop ingluvial bacterial dysbiosis (sour crop).

GEOGRAPHIC DISTRIBUTION
None.

SIGNALMENT
• **Species:** All avian species could present for concerns of ingluvial hypomotility.
◦ Galliformes (chickens, turkeys).
◦ Psittaciformes (cockatiels, budgerigars, lovebirds, grey parrots, etc.). ◦ Picivorous species (distension of the cervical esophagus). ◦ Anseriformes (ducks, geese, swans). ◦ Accipitiformes (birds of prey). • **Mean age and range:** Neonatal and juvenile birds most commonly. • **Predominant sex:** No known gender predilection.

SIGNS

General Comments
The clinical presentation of ingluvial stasis is most often secondary to a primary etiology, which may have other clinical signs as well. It is best to not regard ingluvial stasis or hypomotility as a primary diagnosis, although it is an important clinical sign to manage in balance with the overall patient.

Historical Findings
• Weight loss, ingluvial distension, decreased fecal volume, increased respiratory effort and altered appetite. • Hand-fed (gavage) neonates may show a discolored swelling in the neck in cases of cervical esophageal perforation. • Sudden death (secondary to aspiration, sepsis). • Erythematous/discolored ingluvies. • Cervical masses.

Physical Examination Findings
• Mild to severely distended crop. • Adequate to severely decreased muscle mass. • Labored (open-mouth) breathing/Dyspnea/Tachypnea in some cases. • Regurgitation. • Adequate to severely decreased appetite. • Swollen and/or discolored cervical region. • Weak/Moribund.
• Ingluvial foreign material. • Decreased fecal component of the droppings. • Polyuria on occasion.

CAUSES
• Infectious: *Candida albicans*, bacterial (large number of potential pathogens, most of which are opportunist secondary infectants), viral (herpesvirus, poxvirus, bornavirus, avian polyomavirus), and parasitic (*Trichomonas gallinae*, *Capillaria* spp, *Echinura uncinata*, *Gongylonema ingluvicola*, *Dispharynx nasuata*, *Onciola canis*). • Anomalous: Atonic ingluvies (overstretched), esophageal strictures, iatrogenic (feeding liquid formula to granivorous birds), pendulous crop in turkeys.
• Trauma: Feeding tube trauma and perforation, bite wounds, and crop burns.
• Foreign bodies: Weeds, ingluvioliths, grit, seeds, nuts, metal (nontoxic), feathers, toys, etc. • Metabolic: Hypovitaminosis A, inadequate maintenance of body temperature in neonatal birds, cold food, starvation/generalized weakness. • Toxicosis: Lead, zinc, omeprazole, chlorphacinone (anticoagulant).
• Neoplasia: Papillomatous disease, squamous cell carcinoma, basal cell carcinoma, adenocarcinoma, leiomyosarcoma, fibrosarcoma.

RISK FACTORS
• All birds being tube fed are at potential risk of ingluvial trauma. • Starving birds allowed free access to food. • Access to lush grass, sprouted grains, dried oatmeal and soybeans (poultry species).

DIAGNOSIS

DIFFERENTIAL DIAGNOSIS
• The most common challenge is differentiating primary, secondary and tertiary factors involved in causality of ingluvial stasis.
• Cervical airsac rupture/trauma: Differentiated by physical examination and passage of a feeding tube to help define the esophagus and ingluvies. • Cervical hematoma: Presents as discoloration of the cervical skin, typically following phlebotomy of the jugular vein. • Peri-ingluvial fat/soft tissue mass at the thoracic inlet: Can be differentiated using physical examination, radiographs, CT, MRI, and/or contrast imaging. • Thyroid hypertrophy/neoplasia.
• Aerophagia.

CBC/BIOCHEMISTRY/URINALYSIS
• Typically should not be expected to be diagnostic for a cause of the problem. • AST and/or CK may be elevated due to muscle damage associated with trauma.

OTHER LABORATORY TESTS
• Crop cytology and culture: To identify potential primary or secondary pathogens.
• Blood lead and zinc levels.

IMAGING

Radiographic findings
• Increased ingluvial size with ingesta or fluid present. • Ingluvial or lower GIT foreign material/metal may be visible, but radiolucent material may not be easily identified without further imaging. • Contrast imaging may used to help define the presence and location of radiolucent foreign material and the nature of peristalsis.

Ultrasonography
- Fluid and some intramural or intraluminal soft tissue masses may be visible within the crop.

Advanced imaging
- Computed tomography (CT). • Magnetic resonance imaging (MRI). • Both of these imaging modalities, in some specific clinical settings, may help identify contributing primary causes of ingluvial hypomotility.

DIAGNOSTIC PROCEDURES
- Ingluvial flush: To obtain diagnostic samples (parasitology, microbiology, cytology) and potentially remove ingluvial foreign material. • Ingluviotomy and biopsy: To remove foreign material, and obtain diagnostic samples for histopathology, microbiology, parasitology.

PATHOLOGIC FINDINGS
- Bacterial: Colonization of mucosa but typically little deeper inflammation or lesions. • Virus: Identification of viral inclusions in appropriate locations. Many involved viruses may not have identifiable lesions in the ingluvies if it is secondarily involved in the process. • Proventricular dilatation disease (PDD): Lymphoplasmacytic infiltration of the central and peripheral nervous systems. • Foreign bodies and trauma: Typically coagulative necrosis of the crop wall. • Vitamin A deficiency: Squamous metaplasia of crop mucosa.

TREATMENT

APPROPRIATE HEALTH CARE
Patients that are self-sustaining and ambulatory can be treated as outpatients, but hospitalization and inpatient management is often appropriate when patient is not stable. Debilitated patients may require extended periods of hospitalization for supportive care until stable prior to some surgical procedures.

NURSING CARE
Supportive care may include fluid therapy (subcutaneous fluids can be administered at 5–10% of bodyweight), supplemental heat, analgesia, and nutritional support (gavage feeding of small amounts).

ACTIVITY
Activity should be restricted if there is any increase in respiratory effort, or risk of mechanically induced regurgitation due to pressure on a distended crop.

DIET
- Diet may need to be modified to a liquid, complex carbohydrate, medium fiber content diet to allow easier digestion while ingluvial stasis is being resolved. • Carnivores and piscivores may require food to be deboned and cleaned or temporary use of liquid carnivore diets

CLIENT EDUCATION
- It is important to review with the clients that ingluvial hypomotility is most commonly secondary to other problems, and typically the clinical sign of ingluvial stasis/hypomotility returns if additional pathology is present and has not been identified. Regardless, secondary factors that are identified still must be addressed in the process of management of clinical cases. • Ingestion of foreign bodies is often repeated, unless future patient access to foreign material is controlled

SURGICAL CONSIDERATIONS
- Ingluviotomy/endoscopy: Surgical evaluation to identify ingluvial and proventricular contents and potentially biopsy ingluvial wall where indicated. • Ingluvial repair: Should be performed to resolve crop burn/fistula only after the extent of the necrotic tissue has been defined.
- Esophagostomy or ingluviostomy tube placement may facilitate more regular feeding in some patients.

MEDICATIONS

DRUG(S) OF CHOICE
- Antibiotic therapy—systemic and/or parenteral; ideally based on culture and sensitivity results from known infectants or on a prophylactic basis. • Antifungal therapy – Nystatin 200,000-333,000 IU/kg PO q8–12h, fluconazole 2–5 mg/kg PO q12h • Analgesics ○ Psittacines: Butorphanol 1–4 mg/kg IM q1–4h (primarily kappa antagonist); Tramadol 20–30 mg/kg PO q8–12h (synthetic mu-receptor opiate-like agonist that also inhibits reuptake of serotonin and nor-epinephrine). ○ Raptors: Tramadol 5–11mg/kg PO 4–6h (extrapolated from red-tailed hawks, American bald eagle, and Indian peafowl pharmacokinetics study data); Butorphanol 2 mg/kg IM q2h (extrapolated from red-tailed hawk and Great Horned owl pharmacokinetic study data but had no supported pharmacodynamic effect in American kestrels). • Nonsteroid antiinflammatory drugs (NSAIDs) – Inhibit prostaglandin synthesis by the COX enzyme: Meloxicam 1.6 mg/kg PO q12–24h (Hispaniolan Amazon pharmacokinetic study), carprofen 1–5 mg/kg PO q12–24h (extrapolated from Hispaniolan Amazon and chicken pharmacokinetic studies), ketoprofen 1 mg/kg IM q12h. • CaEDTA 35 mg/kg IM q12h for lead and zinc toxicosis. • Dopamine D2-receptor antagonist used as antiemetic and gastroprokinetic: Metoclopramide 0.5 mg/kg PO, IM, IV q8–12h.

CONTRAINDICATIONS
NSAIDs should not be used if gastrointestinal ulceration is known or suspected to be present.

PRECAUTIONS
- NSAIDs should not be administered if renal disease is known or suspected to be present or discontinued immediately if clinical signs suggestive of renal dysfunction (such as polyuria/polydipsia) are observed.
- Metoclopramide-induced improvement in prokinetic function has not been shown to be consistent in studied avian species, and abnormal central nervous signs have been observed in multiple species treated with metoclopramide. If this drug is being used and CNS signs are noted, it should be discontinued. • Sedation has been anecdotally reported in a number of species associated with the administration of tramadol.

POSSIBLE INTERACTIONS
Do not administer CaEDTA that has been diluted in lactated ringers to avoid chelation into the subcutis.

ALTERNATIVE DRUGS
Crop suspension bandages (bra) may need to be placed if there is atonic ingluvies in some specific species (mostly psittaciformes).

FOLLOW-UP

PATIENT MONITORING
Repeat physical examination should be performed to confirm patient progress, and to monitor for reason to consider further investigations toward identification of primary etiologies where indicated.

PREVENTION/AVOIDANCE
Recommendations to prevent further events will be based on the identified or hypothesized primary etiology.

POSSIBLE COMPLICATIONS
Atonic ingluvies may not return to normal function, or may require considerable time and nursing care.

Repeated clinical events are strongly suggestive that delayed ingluvial emptying is secondary to an underlying primary condition.

EXPECTED COURSE AND PROGNOSIS
Most presentations of primary delayed ingluvial emptying will resolve with balanced medical management.

MISCELLANEOUS

ASSOCIATED CONDITIONS
Many conditions are associated with secondary ingluvial hypomotility, and these can maintain the outward clinical appearance of the problem at times in and of their own.

AGE-RELATED FACTORS
Primary delayed ingluvial emptying is more common in neonatal birds.

INGLUVIAL HYPOMOTILITY, CROP STASIS AND ILEUS
(CONTINUED)

ZOONOTIC POTENTIAL
None.

FERTILITY/BREEDING
N/A

SYNONYMS
Sour crop

SEE ALSO
Air sac rupture
Anorexia
Aspiration
Coelomic distention
Emaciation
Enteritis/proventriculitis
Gastrointestinal foreign bodies
Heavy metal toxicity
Macrorhabdus ornithogaster
Obesity
Regurgitation/vomiting
Sick bird syndrome
Thyroid diseases

ABBREVIATIONS
N/A

INTERNET RESOURCES
N/A

Suggested Reading
Gelis, S. (2006). Evaluating and treating the gastrointestinal system. In: Harrison, G.J., Lightfoot, T.L. (eds), *Clinical Avian Medicine*. Palm Beach, FL: Spix Publishing, Inc, pp. 426–430.
Morrisey J. (2010). Gastrointestinal diseases of birds. American Board of Veterinary Practitioners Symposium 2010, Online Proceedings. (http://www.vin.com/doc/?id=4465431; last accessed October 5, 2015).
Reavill D. (2009). Differential diagnosis of common clinical presentations in pet birds. Western Veterinary Conference 2009, Notes online. (http://www.vin.com/doc/?id=3985296; last accessed October 5, 2015).

Author Geoffrey P Olsen, DVM, Dip ABVP (Avian)

Iron Storage Disease (ISD)

BASICS

DEFINITION
Iron storage disease (ISD) is used to describe the spectrum of physiologic and pathologic changes that occur with excessive iron accumulation.

PATHOPHYSIOLOGY
• Body iron content is balanced by control of dietary iron at the level of the enterocyte; there are no effective physiological methods for the removal of excess iron. • Excess iron absorbed from the diet is typically stored within the hepatocytes, but can also be found in phagocytic cells in the liver and spleen, as well as the cardiomyocytes. Increased iron accumulation without evidence of associated disease is called hemosiderosis. • As iron levels increase, oxidative damage to membranes and proteins can occur within the cells, resulting in cellular death and replacement by fibrosis. Disease occurs and clinical signs develop when accumulated iron affects organ function (liver, heart, spleen, pancreas). In veterinary medicine the term hemochromatosis may be used to describe iron overload with accompanying disease. In human medicine, however, hemochromatosis refers specifically to a group of hereditary primary iron storage disease. • Clinical disease is often related to liver dysfunction and failure. Iron storage disease is also suspected to promote other disease processes in birds such as diabetes mellitus (documented in toucans and parrots, iron induces oxidative stress to the B cells), and heart failure (mynahs). Other conditions reported in humans may also be relevant in birds such as neoplasia (lymphoproliferative disorders and hepatocellular carcinoma) and myopathies.

SYSTEMS AFFECTED
• Liver • Endocrine • Cardiac.

GENETICS
N/A

INCIDENCE/PREVALENCE
Unknown but seen commonly in susceptible species.

GEOGRAPHIC DISTRIBUTION
N/A

SIGNALMENT
• **Species:** ○ Iron storage disease is most common and most severe in species of frugivorous birds. ○ Birds most susceptible to ISD are Ramphastidae (toucans, in particular Toco toucans) and some other Piciformes, Thraupidae (tanagers), Bucerotidae (hornbills, only the more frugivorous species), Coraciidae (rollers), Momotidae (motmots), Trogonidae (trogans, quetzals), Paradisaeidae (birds of paradise), Sturnidae (mynahs, starlings), Pipridae (manakins), and some other members of the Passeriformes. ○ ISD, although not frequently reported in Psittaciformes, has also been described in the Loriidae and New World Psittacidae (e.g., Amazon parrots, Hawk-headed parrot, macaws). • **Mean age and range:** Any age but iron accumulation increases over time. • **Predominant sex:** N/A

SIGNS
Historical Findings
Usually nonspecific.

Physical Examination Findings
Clinical signs are generally secondary to hepatic, pancreatic, and cardiac dysfunction and include:
• Dyspnea • Abdominal distension • Ascites • Emaciation • Depression • Sudden death without premonitory signs (most frequently in toucans) • Reduced hemostasis • Hepatomegaly (liver enlargement may be seen transcutaneously, directly or with the use of abdominal transillumination).

CAUSES
• Diets containing high levels of available iron compared to the natural levels of iron intake for the species. • Presumptive lack of ability of ISD-susceptible species to adequately downregulate the absorption of dietary iron. Most frugivorous species have evolved eating a diet low in available iron. This may result from low absolute iron levels, or the presence of natural iron chelating compounds. • Iron accumulation can also occur in association with systemic conditions that affect iron uptake and sequestration mechanisms, for example infectious diseases and some neoplasia.

RISK FACTORS
• Frugivorous birds and, to a lesser extent, insectivorous birds: Passeriformes (in particular Sturnidae, Paradisaeidae, Thraupidae, Pipridae), Trogoniformes, Psittaciformes (in particular Loriidae and South American parrots), Coraciiformes (in particular Coraciidae, Momotidae), and Bucerotiformes (in particular Bucerotidae). • Diets high in iron. • Diets containing heme-based sources of iron (meat) provided to noncarnivorous species. • Diets containing high levels of Vitamin C (e.g., citrus fruits), which promote intestinal iron absorption. • Diets containing high levels of fructose or sucrose.

DIAGNOSIS

DIFFERENTIAL DIAGNOSIS
• Hepatic disease: Hepatic lipidosis, hepatic fibrosis of other causes, hepatic neoplasia, hepatic congestion (congestive heart failure), and amyloidosis, chronic hepatitis and hepatic necrosis. • Ascites: Reproductive disease, coelomic neoplasia, congestive heart failure, and infectious coelomitis including viral diseases. • Diabetes mellitus: Idiopathic destruction of B cells, corticosteroid use, and pancreatitis.

CBC/BIOCHEMISTRY/URINALYSIS
• Hematologic and biochemical changes are not consistent and may only be present in advanced disease. • Anemia of chronic disease. • Hypoproteinemia/hypoalbuminemia (protein electrophoresis is the only reliable measure of avian albumin). • Hypoglycemia (liver failure) or hyperglycemia (diabetes mellitus). • Decreased uric acid in terminal stages of liver failure. • Elevated bile acids. • Elevated liver enzymes (AST, GGT, GLDH, LDH). • Elevated muscle enzymes (CK, AST, LDH). • Increased coagulation time (in cases of hepatic failure).

OTHER LABORATORY TESTS
• Serum ferritin (SF) is considered the best noninvasive measurement of body iron status in mammals; however, an avian specific ferritin assay has not been developed or assessed for use in birds. • Other plasma or serum analytes used in mammals, including total iron binding capacity (TIBC), iron saturation, transferrin, and plasma iron, have been used to limited degrees in birds to monitor treatment effectiveness but they appear to be of low value in initial diagnosis. • Hepcidin assays have not been developed for birds and their usefulness is unknown.

IMAGING
• Radiography: Hepatomegaly, coelomic fluid, decreased air sac space, enlarged cardiac silhouette. • Coelomic ultrasound: Enlarged and hyperechoic liver, coelomic fluid. • Echocardiography: Enlarged cardiac chambers, systolic dysfunction, pericardial effusion. • MRI: Has been used to estimate hepatic iron concentration in hornbills and then monitor the effects of treatment. Of limited usefulness in general clinical practice.

DIAGNOSTIC PROCEDURES
• Biopsy samples of liver are collected directly via coeliotomy (particularly if the liver is enlarged) or through coelioscopy using a rigid endoscope. The liver can be approached using the standard left lateral, right lateral, or midline approaches. The midline approach through the ventral hepatoperitoneal cavities provides the best view of both liver lobes (after incising the ventral mesenteric membrane). The liver may appear enlarged with rounded borders and may have yellowish or greenish patches that can be associated with iron accumulation. • Evaluation by atomic absorption spectrophotometry (AAS), inductively coupled plasma mass spectrometry (ICP-MS) or another analytic technique is the gold standard for assessing hepatic iron levels.

PATHOLOGIC FINDINGS
Histopathology can be used to visually estimate the amount of iron present in

Iron Storage Disease (ISD)

(CONTINUED)

hepatocytes and Kuppfer cells. Staining with Perls' Prussian blue increases the sensitivity of detection. Digital image analysis and morphologic grading scales have been used to categorize the amount of iron present. Histopathology allows the identification of hepatic pathology associated with iron overload; including necrosis, inflammation, fibrosis, and neoplasia.

TREATMENT

APPROPRIATE HEALTH CARE
• To reduce systemic iron levels and reduce subsequent accumulation: ○ Reduce dietary iron (see below). ○ Phlebotomy (1% of body weight weekly). The PCV should be monitored to ensure iatrogenic anemia is not induced, though lowering of the PCV is rarely encountered in birds. Regular phlebotomy of highly susceptible species not yet suffering from iron storage disease has been suggested for prevention. ○ Iron chelation (see below). • Other ○ Treatment of concurrent disease processes; for example, liver or cardiac failure and diabetes mellitus. ○ Abdominocentesis may be needed to relieve dyspnea in cases with air sac compression by ascitic fluid.

NURSING CARE
N/A

ACTIVITY
N/A

DIET
• Diets low in available iron (30–65 mg/kg dry matter). Most avian food companies offer a low-iron pelletized diet. These diets may not consistently prevent the disease in highly susceptible species such as toucans and mynahs. • Avoid heme-based sources of iron for noncarnivorous species. • Limit the amount of vitamin C in the diet (recommended dose is 50–150mg/kg dry matter), and ideally try and give the vitamin C source separate from the iron source (vitamin C only enhances iron absorption when given at the same time). • Addition of tannins to the diet can help but this should be done with caution as they can also chelate other essential trace minerals and may change palatability. The addition of decaffeinated Ceylon black tea leaves to the diet has been shown to limit iron absorption in starlings. • Iron testing of the diet: As some bags of food may have a higher iron content than advertised, it may be advisable to measure the iron level in every new batch of food for highly susceptible species.

CLIENT EDUCATION
Inform the client about species susceptibility and dietary causes.

SURGICAL CONSIDERATIONS
N/A

MEDICATIONS

DRUG(S) OF CHOICE
• Iron chelators: ○ Deferiprone: 75mg/kg PO q24h for 90 days (studied in hornbills, chickens, and pigeons). ○ Deferoxamine: 100mg/kg SC q24h for 16 weeks (studied in the European starling). These treatments should always be associated with diet modifications. Urine may have a rusty color during treatment. • Liver support medications such as antioxidants may be beneficial. • Treatment for diabetes mellitus with insulin therapy, dietary recommendations, or glipizide may be indicated. Requirement for drugs may decrease as iron overload improves.

CONTRAINDICATIONS
N/A

PRECAUTIONS
N/A

POSSIBLE INTERACTIONS
N/A

ALTERNATIVE DRUGS
N/A

FOLLOW-UP

PATIENT MONITORING
• Serial liver biopsies (for iron level measurement or histopathology) may be performed once or twice a year. • Further developments in the use of MRI to assess hepatic iron levels may provide a noninvasive way of monitoring birds with ISD.

PREVENTION/AVOIDANCE
See "Diet".

POSSIBLE COMPLICATIONS
• Liver failure. • Cardiac failure. • Diabetes mellitus. • Possible decrease in fertility in Rhamphastidae. • ISD has been associated with higher risk of liver neoplasia in humans and in bats, but this association has not yet been identified in birds.

EXPECTED COURSE AND PROGNOSIS
• If treated early, the prognosis is very good. • If damage to the liver/heart has already occurred, the prognosis is poor. • Sudden death without premonitory clinical signs has been reported in Ramphastidae.

MISCELLANEOUS

ASSOCIATED CONDITIONS
• Liver disease • Diabetes mellitus • Cardiac disease

AGE-RELATED FACTORS
N/A

ZOONOTIC POTENTIAL
N/A

FERTILITY/BREEDING
N/A

SYNONYMS
N/A

SEE ALSO
Anemia
Ascites
Cardiac disease
Coelomic distention
Hepatic lipidosis
Hyerglycemia/diabetes mellitus
Liver disease
Polydipsia
Polyuria
Sick bird syndrome
Tumors
Viral disease

ABBREVIATIONS
N/A

INTERNET RESOURCES
N/A

Suggested Reading
Cork, S.C. (2000). Iron storage diseases in birds. *Avian Pathology*, **29**(1):7–12. doi: 10.1080/03079450094216.
Mete, A., Hendriks, H.G., Klaren, P.H.M., et al. (2003). Iron metabolism in mynah birds (*Gracula religiosa*) resembles human hereditary haemochromatosis. *Avian Pathology*, **32**(6):625–632. doi: 10.1080/03079450310001610659.
Olsen, G.P., Russell, K.E., Dierenfeld, E., Phalen, D.N. A (2006). Comparison of four regimens for treatment of iron storage disease using the European starling (*Sturnus vulgaris*) as a model. *Journal of Avian Medicine and Surgery*, **20**(2):74–79. doi: 10.1647/2004-033.1.
Seibels, B., Lamberski, N., Gregory, C.R., et al. (2003). Effective use of tea to limit dietary iron available to starlings (*Sturnus vulgaris*). *Journal of Zoo and Wildlife Medicine*, **34**(3):314–316. doi: 10.1638/02-088.
Sheppard, C., Dierenfeld, E. (2002). Iron storage disease in birds: Speculation on etiology and implications for captive husbandry. *Journal of Avian Medicine and Surgery*, **16**(3):192–197. doi: 10.1647/1082-6742(2002)016 [0192:ISDIBS]2.0.CO;2.
Authors Delphine Laniesse, Dr.Med.Vet, IPSAV
Hugues Beaufrère, Dr.Med.Vet., PhD, Dipl ABVP (Avian), Dipl ECZM (Avian), Dipl ACZM
Dale A. Smith, DVM, DVSc.

LAMENESS

 ## BASICS

DEFINITION
Impaired ability in locomotion, typically in response to pain, physical defect, or dysfunction of the musculoskeletal or nervous system

PATHOPHYSIOLOGY
• Traumatic injuries to the musculoskeletal system from predators, cage mate aggression, and flight injuries are commonly seen in clinical practice. The thin cortical bone and large medullary canal combined with their high calcium content makes avian bones prone to shattering upon impact. • Severe and acute injuries of the pelvic limbs present as nonweight bearing of the affected limb. Mild pain may be more difficult to discern if the bird is putting less weight on the affected limb, has a weaker grip, or is reluctant to walk. Injuries to the thoracic limbs present as a wing droop, reluctance to extend the wing, and/or the inability to fly. • The paucity of soft tissue surrounding the appendicular skeleton makes it prone to vascular compromise and peripheral neuropathies after traumatic injuries. • Inflammation, demyelination, or compression of a nerve can also result in paraesthesia of the limb and changes in locomotion. • Pain from pathologic fractures (osteopenia), lytic bony lesions (osteomyelitis, neoplasms of the bone), and articular gout also lead to clinical signs of lameness.

SYSTEMS AFFECTED
• Musculoskeletal—fractures, soft tissue swelling or disuse atrophy of the muscle can develop. • Nervous system—proprioceptive deficits; paresis; paralysis. • Skin—bruising; abrasions; ulceration.

GENETICS
Congenital deformities that lead to abnormal bone development have been reported.

INCIDENCE/PREVALENCE
Impaired use of the legs and/or the wings are common presentations in clinical practice.

GEOGRAPHIC DISTRIBUTION
N/A

SIGNALMENT
• **Species predilections:** ○ Renal neoplasms – budgerigars ○ Articular gout – cockatiels, budgerigars ○ Ovarian neoplasia – cockatiels, chickens ○ Foot necrosis syndrome – Amazon parrots ○ Pododermatitis – Amazon parrots, waterfowl, raptors. • **Mean age and range** ○ Splay leg and dyschrondroplasia affect neonatal and juvenile birds ○ Arthritis and neoplasm more commonly affects geriatric birds • **Predominate sex:** N/A

SIGNS
General Comments
Clinical signs will vary depending on the cause and location of the lameness.

Historical Findings
• Clarify with owner the onset, duration, and progression of the lameness; also determine which limb(s) may be affected. • May have history of recent trauma. • May have decreased appetite and activity. • Difficulty balancing and perching; may be falling off perch or spending more time on the bottom of the cage. • Trouble flying. • May be feather picking or self-mutilating the affected limb.

Physical Examination Findings
• Abnormal gait or posture. If forelimb is involved, the bird is often unable to fly and may have a drooped wing. If the hind limb is involved, the bird may have a weak grip, be non-weight bearing, or unable to stand. • Crepitus or laxity may be noted on palpation of the bones and joints. • Bruising, swelling, bleeding, missing feathers, and open wounds may be noted on the affected limb. • Paresis or paralysis of limb. • Extremities may be cold to the touch. • Decreased range of motion or contracture of limb may be noted in chronic cases. • Pressure sores may be noted on the contralateral leg as a result of increased usage of the limb. • Full coelomic cavity may be noted on palpation if a mass effect (renal, ovarian, testicular neoplasm) is present.

CAUSES
• Degenerative—arthritis; degenerative joint disease; claudication secondary to atherosclerosis; cerebrovascular ischemia. • Anatomic—splay leg; dyschondroplasia. • Metabolic—articular gout; anemia; thromboembolic disease. • Nutritional—rickets (hypovitaminosis D, calcium-phosphorus imbalance); hypocalcemia; selenium-Vitamin E deficiency; perosis (manganese deficiency); anemia from iron deficiency. • Neoplasia—primary or metastatic bone neoplasia; ovarian, testicular, or renal neoplasia may compress ischiatic nerve; spinal neoplasia. • Immune-mediated—Amazon foot necrosis. • Infectious—bacterial or fungal osteomyelitis; PDD; West Nile Virus; mycobacteria; paramyxovirus; Marek's disease virus. • Trauma—fractures; sprains/strain; joint luxations; pododermatitis; injuries to the patagium; nerve plexus avulsions. • Toxic—lead toxicity.

RISK FACTORS
• Pododermatitis ("bumblefoot")—obesity in heavy bodied birds (e.g. Amazon parrots, waterfowl); birds kept on concrete perches or hard surfaces; malnutrition. • Osteopenia—chronic egg laying, diet deficient in calcium. • Lead toxicity—pet birds allowed unsupervised free-roam in a home; ingestion of lead shot (raptors) or lead weights in ponds (waterfowl).

 ## DIAGNOSIS

DIFFERENTIAL DIAGNOSIS
• Weakness from systemic disease or a neurologic condition may result in ataxia that could be confused as a lameness. • Wing trimming and pinioning are procedures performed to render the bird flightless. • Some birds will normally hold one foot up while at rest.

CBC/BIOCHEMISTRY/URINALYSIS
• Elevated CK and AST, especially in cases of traumatic injury. • Leukocytosis: Inflammatory leukogram; severe elevations with certain infectious diseases. • Hypocalcemia (osteopenia). • Elevated uric acid (articular gout).

OTHER LABORATORY TESTS
• Blood lead levels. • PCR tests for bornavirus, West Nile Virus, and mycobacteria.

IMAGING
Radiographic findings
• Used to evaluate for fractures, dislocations, osteomyelitis, and other bony lesions. • Osteopenia noted in birds that are hypocalcemic. • In acute injuries, soft tissue swelling may be noted around the affected limb; alternatively, in subacute or chronic disease, disuse muscle atrophy may be noted. • Soft tissue densities may be noted in the gonadorenal region if renal, ovarian, or testicular neoplasia is present; a gastrointestinal barium contrast series or computed tomography may help differentiate the soft tissue densities. • Metal densities in the gastrointestinal tract are suggestive of lead toxicity; however other metals can have the same appearance; likewise, the absence of metal densities does not rule out lead toxicity.

Ultrasonographic findings
• If soft-tissue densities are noted in the coelomic cavity, ultrasonographic examination may help differentiate between different disease processes. • Echocardiography should be performed if a cardiomyopathy is suspected.

DIAGNOSTIC PROCEDURES
• Fine needle aspirate or biopsy of any bony lesions for cytologic or histopathologic examination respectively; samples should also be collected for aerobic bacterial culture and sensitivity and anaerobic bacterial and fungal culture. • Electromyography and nerve conduction studies to measure muscle and nerve function. • Laparoscopy allows for visualization and biopsy of any intracoelomic masses (i.e., renal, ovarian, testicular) that may be compressing the ischiatic nerve.

LAMENESS
(CONTINUED)

PATHOLOGIC FINDINGS
Varies depending on cause.

TREATMENT

APPROPRIATE HEALTH CARE
Fracture stabilization
- Sedation or general anesthesia may be required if a patient is especially fractious to prevent further damage to the injured limb and/or manage pain during manipulation.
- External coaptation, internal fixation, external fixation or a combination of these methods are used to stabilize fractures.
- Modified Schroeder–Thomas splint or tape splints (in smaller birds) can be applied for stabilization of tibiotarsal or metatarsal fractures.
- Figure-of-eight bandages are used for radial or ulnar fractures; figure-of-eight bandages are used in conjunction with body bandage for humeral and coracoid fractures.
- May need distractors (tape tabs) or Elizabethan collars to prevent the bird from chewing the bandage off.
- Sedation may be necessary when applying and at bandage changes for fractious and/or painful birds.
- Surgical fixation with intramedullary pins, external fixators, and/or plates may be necessary.

Pododermatitis
- Cover perches with bandage material or artificial turf; move bird off hard surfaces (i.e., concrete).
- Gradually reduce the bird's body weight; correct dietary deficiencies.
- Ball or other type of bandages are used to prevent and treat pododermatitis.

Systemic disease
- Treat underlying disease.
- Assess the bird for shock and take appropriate measures as necessary.

NURSING CARE
- Parental fluid supplementation (SC, IV, IO) if blood loss is noted in conjunction with fractures and open wounds.
- Aggressive fluid diuresis recommend for patients with articular gout to help clear hyperuricemia.
- Low-level laser therapy (cold laser) may help alleviate pain and decrease inflammation.

ACTIVITY
Strict cage/incubator rest for birds with fractures or having difficulty moving around.

DIET
- If modifications have been made to the cage (i.e., removal of perches) to prevent the bird from falling, bowls should be placed so food and water are easily accessible.
- Some birds do not eat well when bandaged or placed in an Elizabethan collar; have owners weigh daily.

CLIENT EDUCATION
- Bandages and splints must be kept dry and clean and should be assessed regularly to minimize complications.
- If the bird has difficulty balancing, enclosures will need to be modified to keep the bird comfortable and prevent it from falling; all perches should be lowered or removed; some birds may need to be kept in a smooth-sided enclosure where they are not able to climb up the side to decrease movement and facilitate healing.

SURGICAL CONSIDERATIONS
- Osteopenia should be addressed before attempting surgical fixation.
- The humerus and femur (in some species) are pneumatic and are interconnected with the respiratory system; take care when irrigating to prevent washing infected material further into the bone or air sac.
- Vascular compromise and peripheral neuropathy are possible complications due to the paucity of soft tissue surrounding the limbs.

MEDICATIONS

DRUG(S) OF CHOICE
Treatment for pain and arthritis
- Meloxicam—0.5–2 mg/kg PO, IM q12–24h, nonsteroidal anti-inflammatory medication.
- Butorphanol tartrate—1-5 mg/kg q2–3h PRN IM, IV; opioid agonist–antagonist for pain management.
- Tramadol—5–30 mg/kg PO q6–12h; synthetic opioid.

Treatment for articular gout
- Colchicine—0.01–0.2 mg/kg PO q12–24h; anti-inflammatory.

Treatment for osteomyelitis
- Antibiotic based on culture and sensitivity.

Treatment for lead toxicity
- Calcium EDTA—30–35 mg/kg IM, SC q12h; parental chelator for lead toxicity, maintain hydration.
- DMSA—40 mg/kg PO q12h; oral chelator for lead toxicosis; can be used alone or in conjunction with CaEDTA.

Treatment for hypocalcemia
- Calcium glubionate—25–150 mg/kg PO q12–24h; calcium supplementation.
- Calcium gluconate—dilute 1:1 with saline or sterile water for IV, SQ, IM injection; 5–10 mg/kg slow IV for tetany; 10–100 mg/kg SC, IM, slow IV.

Treatment for testicular or ovarian neoplasms
- Leuprolide acetate—100 to 1200 μg/kg IM q 2–4 weeks; long-acting synthetic GnRH analog that may decrease size of tumor.

CONTRAINDICATIONS
N/A

PRECAUTIONS
Adequan (polysulfated glycosaminoglycan) (PSGAG) has been used, but fatal hemorrhage has been reported.

POSSIBLE INTERACTIONS
N/A

ALTERNATIVE DRUGS
- Oral glucosamine and chondroitin has been used anecdotally for the treatment of arthritis.
- Acupuncture may be useful for cases with chronic pain.

FOLLOW-UP

PATIENT MONITORING
- Weekly assessment of bandages and splints is recommend to ensure that the limb is healing appropriately and or for any necessary adjustments.
- Repeat radiographs can be performed in 3–5 weeks to evaluate fracture healing. Endosteal callus bone formation may be difficult to visualize radiographically.

PREVENTION/AVOIDANCE
N/A

POSSIBLE COMPLICATLONS
- Malunion or nonunion can occur with inadequate immobilization of the fracture.
- External coaptation may lead to joint stiffness, muscle atrophy, tendon or skin contracture, and fracture misalignment.
- Open fractures are at higher risk for bacterial contamination and osteomyelitis.
- Misalignment and synostosis of radial and ulnar fractures as well as patagial contraction may prevent future ability to fly.
- Development of pressure sores on contralateral leg, especially in heavy-bodied birds such as waterfowl and raptors.
- Neurologic or vascular derangements can occur secondary to traumatic injuries.

EXPECTED COURSE AND PROGNOSIS
- Well aligned, stable fractures have sufficient endosteal callus to stabilize the fracture as early as three weeks; heal over 4–6 weeks.
- Lameness with renal gout and neoplasia is usually progressive over several weeks to months.

MISCELLANEOUS

ASSOCIATED CONDITIONS
N/A

AGE-RELATED FACTORS
- Splay leg and dyschrondroplasia affect neonatal and juvenile birds.
- Arthritis and neoplasm more commonly affects geriatric birds.

ZOONOTIC POTENTIAL
N/A

FERTILITY/BREEDING
- May interfere with the bird's ability to mate.
- Hypocalcemic hens are at risk for egg binding and dystocia.

SYNONYMS
N/A

LAMENESS (CONTINUED)

SEE ALSO
Arthritis
Atherosclerosis
Dystocia/egg binding
Fracture/luxation
Heavy metal toxicity
Hyperuricemia
Hypocalcemia/hypomagnesemia
Proventricular dilatation disease
Marek's disease/viral neoplasia
Mycobacteria
Neurologic conditions
Pododermatitis
Splay leg/slipped tendon
Trauma
Tumors
West Nile Virus
Appendix 7, Algorithm 13. Lameness

ABBREVIATIONS
PDD—proventricular dilatation disease
DMSA—meso-2,3-dimercaptosuccinic acid
GnRH—gonadotropin-releasing hormone

INTERNET RESOURCES
http://www.lafebervet.com/emergency-medicine/birds/external-coaptation/

Suggested Reading
Harcourt-Brown, N.H. (2002). Orthopedic conditions that affect the avian pelvic limb. *The Veterinary Clinics of North America. Exotic Animal Practice*, **5**(1):49–81.
Orosz, S.E. (2002). Clinical considerations of the thoracic limb. *The Veterinary Clinics of North America. Exotic Animal Practice*, **5**(1):31–48.

Author Sue Chen, DVM, Dipl. ABVP (Avian)

LIPOMAS

BASICS

DEFINITION
Lipomas are benign, soft, rubbery, encapsulated mesenchymal tumors of adipose tissue, composed of mature fat cells. They may appear to be large depot sites of normal fat. Lipomas are more common than similar-appearing aggressive neoplastic growths. Larger growths can impair normal activities/ functions and may develop complications of ulceration or central necrosis.

PATHOPHYSIOLOGY
• Lipomas typically develop during obesity and can be found externally. • Usually slow growing but can develop quickly. • Less commonly they can develop near internal organs. • Larger complicated masses may require surgical removal if medical management insufficient.

SYSTEMS AFFECTED
• Skin/Exocrine—common location of lipomas is subcutaneous. • Behavioral—in reaction to the weight and location of the mass as well as mobility changes, and reaction to any wounds present. • Respiratory—if mass is internal it may impair respiration. • Gastrointestinal—if mass is involving cloacal area defecation can be affected; if intracoelomic can cause extraluminal mass effects of the digestive tract. • Renal/Urologic—if mass is intracoelomic or involving cloaca it can impair urate elimination. • Reproductive—mechanical interference with mating or oviposition. • Nervous—if contained intracranially such as found in crested ducks.

GENETICS
Possibly in budgerigars.

INCIDENCE/PREVALENCE
Moderately common in captive birds.

GEOGRAPHIC DISTRIBUTION
Worldwide.

SIGNALMENT
• **Species:** Any but especially budgerigars, rose-breasted cockatoos, Amazon parrots, cockatiels, other cockatoos, crested ducks.
• **Mean age and range:** Tends to occur after maturity and seems to be more likely with age. • **Predominant sex:** Both sexes affected.

SIGNS

Historical Findings
• Nutritional history: Excessive caloric consumption. • Small cage and/or insufficient exercise. • A lump may or may not be mentioned by the owner. • Round masses around the abdomen, cloaca or tail may present as an "egg". • Bleeding and/or site mutilation if the mass ulcerates. • Difficulty defecating or malodorous droppings. • Dyspnea if located in the thoracoabdominal cavity. • Nervous system signs. • Infertility or dystocia.

Physical Examination Findings
• Obesity. • Soft yellow mass, usually in the sternal or abdominal subcutis but can be anywhere. • Mass may have thickened yellow skin over it, or ulceration. • If mass involves cloacal area there may be abnormal defecation behavior or soiled pericloacal feathers.
• Lipomas can also develop in the coelom, and intracranially (reported in crested ducks).
• May see dyspnea, digestive tract-related, or neurologic signs related to masses in body cavities.

CAUSES
• Typically develops due to obesity and an enlarging fatty deposit developing into a lipoma. • Obesity caused by abnormal nutrition (excessive caloric intake) combined with insufficient exercise. • Rarely endocrine; hypothyroidism has been suggested as cause but little evidence available.

RISK FACTORS
Inadequate diet, close confinement, high-risk species, higher age.

DIAGNOSIS

DIFFERENTIAL DIAGNOSIS
• Obesity with large fat deposits.
• Xanthomatous skin. • Masses of any type: abscesses, feather cysts, myelolipomas, osteolipomas, hemangiolipomas, liposarcomas, fibrosarcomas, leiomyosarcomas, others. • Abdominal herniation.

CBC/BIOCHEMISTRY/URINALYSIS
Changes associated with obesity such as biochemical evidence of hepatic lipidosis, hyperlipidemia/dyslipidema; otherwise none specific.

OTHER LABORATORY TESTS
Thyroid testing, but thyroid conditions are rare; see "Thyroid diseases" for further discussion. Document lipidemia with a lipoprotein panel.

IMAGING
• Radiography can confirm presence and extent of mass. • Ultrasound can confirm mass in abdominal area. • May need contrast series to determine whether ventral abdominal swelling is lipoma vs. herniation with intestinal displacement into mass.
• Concurrent assessment of hepatic and cardiac silhouettes can provide more information on other obesity related organ changes.

DIAGNOSTIC PROCEDURES
• Aspirate cytology: smear has "greasy" look on slide. • "Quick" stain adequate but some fat is lost with alcohol. • Optimal: water-based stains such as new methylene blue; fat stains (ex: Sudan IV). • Smear contains background fat droplets, variably sized fat cells. • Surgical biopsy and histopathology confirmatory. • Internal lipomas can be biopsied via laparoscopy.

PATHOLOGIC FINDINGS
• Grossly excised lipomas are pale yellow, soft, round or with multiple lobules: fatty, thinly encapsulated. • Cytologically lipocytes have multiple vacuoles and pyknotic nuclei pushed to edge of cell. • Mass histopathology: encapsulated lobules of well-differentiated lipocytes; there can be necrosis within the mass. • May be difficult to differentiate between lipoma vs. necrotic/inflammatory change leading to fatty area swelling.
• Lipoma cells can occur with other cells present in the mass and create masses such as myelolipomas, osteolipomas, and hemangiolipomas.

TREATMENT

APPROPRIATE HEALTH CARE
• Chronic issue unless acute change (such as ulceration with hemorrhage). • Small or noncomplicated lipoma managed at home with husbandry adjustments. • Masses may take additional time to reduce after other body fat depositions are reduced.
• Hemostasis for hemorrhage and control of secondary infections may be required.
• Complicated masses may require surgical resection or reduction. • Open wounds may require antimicrobials, analgesics.

NURSING CARE
• Chronic mild cases: Husbandry management consultation. • Hemorrhage: Fluid support (crystalloids, colloids, or transfusion) for replacement of blood loss as necessary. • Ulcerated skin: Wound management. • May require barrier (bandaging, Elizabethan Collar) to prevent self-mutilation.

ACTIVITY
• Small lipomas best managed by slowly increasing activity and exercise. • Increase exertion rates and times slowly to avoid sudden cardiovascular stress and possible cardiac events.

DIET
• Optimize diet and caloric intake for activity level to provide slow weight loss. • Objective: Slow weight loss (for example, 5% of body weight loss per month) until body condition score is ideal for species.

CLIENT EDUCATION
• Nutrition and husbandry adjustments.
• Clients to monitor for growth, ulceration, change in character.

LIPOMAS (CONTINUED)

SURGICAL CONSIDERATIONS
Masses may impair respirations by being within the thoracoabdominal cavity or adding weight and added pressure to chest excursions during anesthesia. Comorbidities such as hepatic lipidosis/arteriosclerosis can complicate anesthesia as well. Hemostasis is an important surgical consideration. Recent surgical technologies (CUSA [Cavitron ultrasonic surgical aspirator]) can simplify removal.

MEDICATIONS

DRUG(S) OF CHOICE
Pharmaceuticals seldom indicated.

CONTRAINDICATIONS
None.

PRECAUTIONS
Some references state levothyroxin can be used to help reduce lipomas; without laboratory documentation of a hypothyroid state, levothyroxin is not indicated.

POSSIBLE INTERACTIONS
None.

ALTERNATIVE DRUGS
N/A

FOLLOW-UP

PATIENT MONITORING
Masses can be evaluated visually unless within a body cavity. Body condition and weight should be monitored weekly to verify adequate but not excessive weight loss.

PREVENTION/AVOIDANCE
• Appropriate diet and activity. • Periodic monitoring of body weight with a scale. • Ongoing monitoring for fatty depositions.

POSSIBLE COMPLICATIONS
• Concurrent obesity complications can impair recovery, such as hepatic lipidosis. • Surface ulcerations. • Central vascular compromise and necrosis. • Impaired: Mobility, egg laying, internal organ function. • Transition to more aggressive neoplasia such as liposarcoma. • Difficulty of full surgical excision with extensive masses.

EXPECTED COURSE AND PROGNOSIS
• Mild to moderate cases: Good prognosis with husbandry management. • Cases requiring surgery: Good prognosis unless mass is large, or concurrent disease present. • Continued growth, ulceration or necrosis if no management is chosen.

MISCELLANEOUS

ASSOCIATED CONDITIONS
Obesity
Other nutritional diseases
Hepatic lipidosis
Arteriosclerosis and other cardiovascular disease
Xanthomatosis
Hyperlipidemia
Liposarcoma
Hypothyroidism
Myelolipomas
Hemangiolipomas
Osteolipomas

AGE-RELATED FACTORS
Aging increases rate of occurrence as well as surgery complication risk.

ZOONOTIC POTENTIAL
None.

FERTILITY/BREEDING
Lipomas that involve the cloacal area can mechanically interfere with successful mating and oviposition.

SYNONYMS
Fatty tumor

SEE ALSO
Atherosclerosis
Coelomic distention
Hepatic lipidosis
Dyslipidemia
Nutritional deficiencies
Obesity
Thyroid diseases
Tumors
Xanthoma

ABBREVIATIONS
N/A

INTERNET RESOURCES
N/A

Suggested Reading
Altman, R., Clubb, S., Dorrestein, G., Quesenberry, K. (1997). *Avian Medicine and Surgery*. Philadelphia, PA: WB Saunders.
Harrison, G., Lightfoot, T. (2006). *Clinical Avian Medicine*, Volume II. Palm Beach, FL: Spix Publishing.
Ritchie, B., Harrison, G., Harrison, L. (1994). *Avian Medicine: Principles and Application*. Lake Worth, FL: Wingers Publishing.
Rosskopf, W., Woerpel, R. (1996). *Diseases of Cage and Aviary Birds*, 3rd edn. Baltimore, MD: Williams and Wilkins.

Author Vanessa Rolfe, DVM, ABVP (Avian)

Liver Disease

BASICS

DEFINITION
• The liver is involved with numerous systems in the body and as such is part of a wide variety of disease processes. A large number of diseases have been reported in the avian liver, and most systemic diseases will have hepatic components. • Liver disease is very common in birds, both specifically and as part of systemic disease. Although generally thought of as a digestive organ, the liver plays a part in many organ systems, resulting in very wide and divergent clinical effects of liver disease.
• There are many types of liver diseases in birds but most can be placed into five categories: infectious, metabolic, toxic, neoplastic, and degenerative. • Many liver diseases can be caught when subclinical by routine clinical evaluation. In these cases, the prognosis is much better than when treating overt clinical disease.

PATHOPHYSIOLOGY
• Because of the numerous liver diseases, the pathophysiology is widely variable.
• Exposure of the liver to infectious agents may occur from sepsis, by an ascending infection from the gastrointestinal (GI) tract, or through the portal system. • Toxins may affect the liver. • Metabolic processes may cause accumulation of substances within the hepatic cells, resulting in damage to the cells.
• Hepatic fibrosis can occur secondary to any chronic liver condition.

SYSTEMS AFFECTED
The liver has numerous functions and the signs can give apparent changes in many systems.
• Behavioral changes such as lethargy or dullness are common. • Gastrointestinal problems such as diarrhea, steatorrhea, or anorexia are often noted in liver disease.
• Hemic/Lymphatic/Immune changes include hypoproteinemia which may result in ascites or edema. • Musculoskeletal changes often include a loss of muscle mass • Nervous problems are theoretically possible with hepatic encephalopathy. This appears to be rare in birds. • Renal/Urologic problems include biliverdinuria. This is one of the more specific signs of liver disease. • Respiratory effort can be increased in cases where the liver is enlarged or there is ascites. • Skin/Exocrine changes such as poor feather quality and beak overgrowth or degradation are common.

GENETICS
Genetically related liver diseases are not commonly reported.

INCIDENCE/PREVALENCE
Liver disease is extremely common in birds.

GEOGRAPHIC DISTRIBUTION
There appears to be no geographic distribution of liver disease overall, although specific diseases have higher prevalence in some areas compared to others.

SIGNALMENT
• **Species:** All birds are susceptible, although some etiologies are more common in specific bird species. • **Mean age and range:** Varies with disease. • **Predominant sex:** Most have no sex predilection.

SIGNS

General Comments
Depending on the severity of the disease physical findings may be subtle or severe.

Historical Findings
General malaise, often referred to as "sick bird syndrome" is common with liver disease. Because of the digestive function of the liver, diarrhea, steatorrhea, and weight loss can occur with liver disorders. Biliverdinuria is one of the most specific of the clinical signs. When the liver fails to conjugate and excrete the biliverdin in the bile, it accumulates and is excreted in the urine.

Physical Examination Findings
Many birds with liver disease have a "pot-bellied" appearance due to either ascites or severe enlargement of the liver. If there has been weight loss, this appearance will be accentuated. Coagulopathies are rare but can occur with liver disease. Overgrowth and dystrophy of the beak keratin is commonly seen in chronic, low grade liver diseases, although the reason is unclear.

CAUSES
• **Infections** account for a large proportion of the liver disease found in birds. ◦ Viruses ▪ Pacheco's disease is caused by a herpesvirus and results in death without premonitory clinical signs. ▪ Polyomavirus is a systemic disease, but can affect the liver. ▪ Adenovirus, reovirus and many other viruses have been implicated in hepatitides of birds. ◦ Bacteria ▪ Chlamydiosis caused by *Chlamydia psittaci* is a systemic infection, but hepatic signs are often prominent in the presentation.
▪ Mycobacteriosis resulting from organisms in *M. avium intracellulare* complex also is a systemic disease with significant hepatic involvement. ▪ Opportunistic infections can occur by direct extension from the intestine via the biliary tree, hematogenously during sepsis, or through the portal system. In many cases, the organism is not recovered, but the inflammatory changes within the liver suggest a previous infection. ◦ Parasites
▪ Atoxoplasmosis is a common and severe disease of some passerine birds. This is a systemic disease, but the liver is often involved and is one of the better sites to identify the parasites. • **Metabolic conditions** ◦ Hepatic lipidosis is extremely common in pet birds. It often results in very chronic and low-grade disease. Excessive calorie consumption, inactivity, and nutritional imbalances appear to be responsible. ◦ Hemochromatosis is the most common disease process in many species, including mynahs, birds-of-paradise, and many toucan species. It can affect many organs, but often the liver is the most severely affected. Although uncommon in psittacidae, it can occur. Lories are the most commonly affected psittacidae. • **Toxins** ◦ Aflatoxicosis is one of the most common naturally occurring hepatotoxins. It can cause centrilobular necrosis of the liver. ◦ Iatrogenic hepatotoxins such as itraconazole will often cause acute anorexia and elevation of hepatic enzymes. • **Neoplasia** ◦ Biliary adenocarcinoma is a common hepatic neoplasm. It has a strong association with the herpesvirus responsible for cloacal papillomas. ◦ Lymphoma is another common neoplasm of the liver. It often is multisystemic.
• **Degenerative** ◦ Chronic hepatic fibrosis is commonly found. It is a common sequela of any chronic disease state of the liver and generally remains after the primary disease has resolved.

RISK FACTORS
• Exposure to the infectious agents is the most important factor for many hepatitides.
• Unbalanced, high calorie diets combined with inactivity predispose birds to hepatic lipidosis. • High iron diets, lack of dietary tannins, and species predilection predispose to hemochromatosis. • Moldy foods predispose to aflatoxicosis. • Herpesvirus infections predispose to biliary adenocarcinoma.

DIAGNOSIS

There are two phases of diagnosis. First, it must be determined if liver disease is present, then it must be determined what is causing the liver disease. The cause of liver disease may sometimes be found with tests for specific pathogens, or can be presumptive and based on risk factors such as species (hemochromatosis) or obesity (lipidosis).

DIFFERENTIAL DIAGNOSIS
There are many clinical signs of liver disease and each has its own set of differentials. Differentials for diarrhea, weight loss, and malaise include systemic disease, GI disease, or pancreatic disease. Biliverdinuria may occur with hemolysis. Enlarged coelom can occur with heart disease, reproductive disease and others. Overgrown beak may be due to knemidokoptic mange, fungal infections, or lack of wear.

CBC/BIOCHEMISTRY/URINALYSIS
The presence and severity of liver disease can often be determined by evaluating biochemistries such as enzymes (AST, GGT, etc.), bile acids, protein, and glucose:

LIVER DISEASE (CONTINUED)

- AST elevation (hepatocellular damage).
- GGT elevation (biliary cell damage).
- GLDH elevation (hepatocellular necrosis).
- Bile acid elevation (functional impairment).
- Hypoproteinemia (functional impairment).
- Hypoglycemia (functional impairment).

Hematology may support an inflammatory or infectious etiology although it will not be specific.

OTHER LABORATORY TESTS
Screening for infectious diseases may be helpful in identifying the etiology of hepatic diseases. *Chlamydia psittaci* can be detected via serology or PCR. *Mycobacterium spp.* may be detected with acid fast stains or with PCR. Herpesvirus may also be identified using PCR.

IMAGING
Radiographs may show alterations in size of the liver. Ultrasound will give more information regarding the internal texture of the liver. CT is less commonly used, but can give a great deal of anatomical information about the liver and other internal structures. The size of the liver may be either increased or decreased in liver disease. Hepatomegaly is much more common. Even some chronic fibrotic livers are quite enlarged.

DIAGNOSTIC PROCEDURES
Liver biopsy is frequently required to definitively identify the cause of liver disease. Fine needle aspiration can give some information, but a definitive diagnosis often requires histopathology. Endoscopic sampling of the liver is most common in birds, but open surgical biopsy or ultrasound guided sampling are also possible.

PATHOLOGIC FINDINGS
Gross changes range from enlargement to atrophy. There can be a variety of color changes in the liver, depending on the condition. Histopathology also varies with the specific disease process. Vacuolar changes occur with hepatic lipidosis. Inflammatory infiltrates occur with various infectious processes. Hemosiderin accumulates in the macrophages with hemochromatosis. Centrilobular necrosis occurs with aflatoxicosis.

TREATMENT

APPROPRIATE HEALTH CARE
The treatment of liver disease depends on treatment of the underlying etiology of the disease, as well as symptomatic management of the patient. Early cases may require very little, beyond some husbandry alterations. Others may require hospitalization and intensive care.

NURSING CARE
Fluid therapy to achieve appropriate hydration and perfusion is appropriate. Many patients with hepatic disease will be anorectic and nutritional support is critical. Most often alimentation in the form of gavage feeding is used. Small frequent feedings may be needed to avoid regurgitation in some cases. Despite the need for nutritional improvement, the time when a patient is acutely ill may not be the appropriate opportunity to do this.

ACTIVITY
Unless fluid lines or other supportive care prevents it, normal activity can be maintained for most birds. If the liver is extremely large, the bird should be carefully protected from falls. The liver can fracture easily, resulting in fatal hemorrhage in these cases. If there is ascites, the fluid could rupture into the air sacs and drown the bird. Increased activity is recommended for obese birds provided the previous situations are not present.

DIET
The diet should be gradually converted to a balanced, calorically appropriate, and toxin-free diet. Diet conversion may take several weeks and close monitoring is warranted. Certain nutritional supplements may be appropriate for specific conditions of the liver. L-carnitine can improve fat metabolism in cases of hepatic lipidosis. The addition of tannins to the diet may reduce iron absorption in cases of hemochromatosis.

CLIENT EDUCATION
A few of the liver diseases are zoonotic and this should be conveyed to the owner. There is a great deal of education involved in the dietary conversion.

SURGICAL CONSIDERATIONS
Liver biopsy is often required for diagnosis, but is not generally used for treatment of liver disease. The only exception may be if there is an isolated tumor in an accessible part of the liver.

MEDICATIONS

DRUG(S) OF CHOICE
- Chlamydia: Doxycycline; dosing schedules vary with preparation and species.
- Mycobacteria: Combined therapy with ethambutol, rifabutin, clarithromycin, or others. Long-term therapy is recommended but likely does not eliminate disease; therefore, treatment is controversial.
- Gram-negative bacteria: Enrofloxacin or others based on culture and sensitivity.
- Anti-oxidants such as silymarin (milk thistle) may help reduce oxidative damage and further degeneration of the hepatocytes.
- S-adenosyl methionine (SAMe) is often recommended for liver disease but the mechanism is unclear. • L-carnitine can improve fat metabolism in cases of hepatic lipidosis. • Ursodiol can improve bile flow.
- Lactulose is often recommended, but its effect is questionable. It is generally thought to reduce gastrointestinal production of ammonia, which does not appear to be a concern for most pet birds.

CONTRAINDICATIONS
None.

PRECAUTIONS
Hepatotoxic drugs such as itraconazole should be used cautiously in birds with hepatic disease.

POSSIBLE INTERACTIONS
None.

ALTERNATIVE DRUGS
N/A

FOLLOW-UP

PATIENT MONITORING
The patient's general appearance, appetite, and weight should be monitored until stable. The hepatic enzymes and bile acids should be followed to determine if therapy is effective. Frequency depends on the severity and chronicity of the condition. Acute or severe changes should be monitored more frequently (e.g., weekly) while milder or chronic disease can be monitored over a longer time frame (e.g., monthly or quarterly).

PREVENTION/AVOIDANCE
Many liver diseases can be prevented with sound nutrition and avoidance of toxins and infectious agents.

POSSIBLE COMPLICATIONS
The most common complications are failure to respond to treatment, and secondary fibrosis of the liver.

EXPECTED COURSE AND PROGNOSIS
Subclinical liver disease often has a good prognosis. Acute hepatic diseases caused by readily treatable conditions, such as chlamydiosis, have a fair to good prognosis. However, overt signs of liver disease occur when a large proportion of the liver is damaged or nonfunctional. Because of this chronic liver disease has a guarded to poor prognosis when the patient has started to show clinical signs.

MISCELLANEOUS

ASSOCIATED CONDITIONS
The liver is involved in many systemic and gastrointestinal disorders, primarily infectious diseases.

LIVER DISEASE

(CONTINUED)

AGE-RELATED FACTORS
Older patients are more likely to have chronic disorders of the liver.

ZOONOTIC POTENTIAL
Chlamydia psittaci and *Mycobacterium spp.* are zoonotic.

FERTILITY/BREEDING
No specific effects.

SYNONYMS
Hepatic disease; hepatopathy

SEE ALSO
Adenoviruses
Anticoagulant rodenticide
Ascites
Chlamydiosis
Coagulopathies
Coccidiosis (intestinal and systemic)
Coelomic distention
Hepatic lipidosis
Herpesviruses
Iron storage disease
Lymphoid neoplasia
Mycobateriosis
Nutritional deficiencies
Obesity
Overgrown beak and nails
Polyomavirus
Regurgitation/vomiting
Sick bird syndrome
Tumors
Viral disease
Urate/fecal discoloration
Appendix 7. Algorithm 15. Hepatopathy

ABBREVIATIONS
N/A

INTERNET RESOURCES
N/A

Suggested Reading
Davies, R.R. (2000). Avian liver disease: Etiology and pathogenesis. *Seminars in Avian and Exotic Pet Medicine*, **9**:115–125.
Grunkemeyer, V.L. (2010). Advanced diagnostic approaches and current management of avian hepatic disorders. *The Veterinary Clinics of North America. Exotic Animal Practice*, **13**:413–427.
Jaensch, S. (2000). Diagnosis of avian hepatic disease. *Seminars in Avian and Exotic Pet Medicine*, **9**:126–135.
Nordberg, C., O'Brien, R.T., Paul-Murphy, J., et al. (2000). Ultrasound examination and guided fine-needle aspiration of the liver in Amazon parrots (Amazona species). *Journal of Avian Medicine and Surgery*, **14**:180–184.
Author Kenneth R. Welle, DVM, Diplomate ABVP (Avian)

 **Client Education Handout available online**

LYMPHOID NEOPLASIA

BASICS

DEFINITION
• Lymphoid neoplasia is a type of cancer that originates from lymphoreticular cells that usually arise in primary or secondary lymphoid tissues but may arise from any tissue in the body. • Leukemia is used for lymphoid neoplasms presenting with widespread involvement of the bone marrow, usually accompanied by the presence of large numbers of tumor cells in the peripheral blood. • Lymphoma, on the other hand, is used to describe proliferations arising as discrete tissue masses.

PATHOPHYSIOLOGY
Lymphoid neoplasms arise from lymphocytes at different stages of differentiation. Gene mutations promote changes in these cells changing the gene expression and promoting a neoplastic formation. The causes of these gene mutations are multifactorial and include genetic as well as environmental factors.

SYSTEMS AFFECTED
• Hemic/Lymphatic/Immune—this type of cancer originates in lymphoreticular cells and can involve the bone marrow. The most commonly affected organs in birds are the spleen and liver. • Hepatobiliary—the most commonly affected organs in birds are the spleen and liver. • Skin/Exocrine—lymphoma, specifically T cell in origin, has been documented in the skin of birds.
• Lymphoid neoplasms have also been documented in the gastrointestinal tract, kidney, thyroid gland, oviduct, lungs, air sacs, sinuses, thymus, testes, brain, palate/choana, trachea, heart, fat, periorbital muscles and pancreas in birds.

GENETICS
N/A

INCIDENCE/PREVALENCE
Prevalence is unknown, but lymphoid neoplasia is the most common form of hematopoietic neoplasia in avian species.

GEOGRAPHIC DISTRIBUTION
N/A

SIGNALMENT
• **Species:** Reported in chickens, turkeys, ducks, geese, quail, pheasants, emu, Great horned owls, partridges, pigeons, owls, ostriches, egrets, canaries, budgerigars, cockatiels, African grey parrots, mynah birds, cockatoos, pionus, lovebirds, corellas, parakeets, rosellas, lorikeets, macaws, caiques, doves, conures, parrotlets and Amazon parrots. • **Species predilections:** Among avian species, canaries have the highest reported incidence of lymphoreticular neoplasms with an increased reported prevalence in males. • **Mean age and range:** N/A • **Predominant sex:** N/A

SIGNS

General Comments
Lymphoid neoplasms can affect any tissue in the body; therefore, historical and physical examination findings are variable and depend on the location and extent of disease.

Historical Findings
Reported historical findings include sinus or cutaneous swellings, depression, anorexia, weight loss, paresis, coelomic swelling, diarrhea, regurgitation, exercise intolerance, straining to defecate, melena, hematochezia and blindness.

Physical Examination Findings
Multicentric disease with hepatic and splenic involvement is the most common presentation of this disease—examination findings are nonspecific but include pectoral muscle wasting, depression and weakness. Depending on the organ system affected findings can include:
• Skin: Nodules, plaques and dermal ulcers more commonly reported on the head and neck but can be diffuse presentation in cockatoos. • Bone marrow: Pale mucous membranes, petechiation, ecchymosis, epistaxis. • Liver: Ascites, biliverdinuria, coelomic mass. • Spleen: Coelomic mass.
• Respiratory: Dyspnea, sneezing, nasal discharge, epistaxis; choanal masses reported in Amazon parrots. • Gastrointestinal: Regurgitation, diarrhea, melena, hematochezia, straining to pass droppings.
• Urogenital: Coelomic effusion, hematuria, coelomic mass. • Ocular: Exophthalmos, ocular discharge, blindness.

CAUSES
• Viral diseases in poultry that are linked to lymphoproliferative disease: ○ DNA oncogenic herpesvirus, Marek's disease virus (MDV) serotype 1. ○ Avian Lymphoid Leukosis viruses of the Retroviridae family. ○ Retroviruses of the reticuloendotheliosis virus (REV) group. • Viral implications in nonpoultry avian species: ○ Multicentric lymphoma was diagnosed in a European starling presenting for emaciation, dyspnea and abdominal distension. PCR amplification of the tumor indicated the presence of retroviral sequences. • Hepatic lymphosarcoma in association with hemochromatosis has been documented in a mynah bird.

RISK FACTORS
N/A

DIAGNOSIS

DIFFERENTIAL DIAGNOSIS
• Other soft tissue tumors (both benign and malignant). • Localized abscess/granuloma (fungal or bacterial). • Multisystemic infection such as *Mycobacteriosis spp.* or *Chlamydia psittaci* that can cause hepato-splenomegaly.

CBC/BIOCHEMISTRY/URINALYSIS

CBC
• Nonregenerative anemia secondary to chronic inflammation/disease is most commonly seen and is a nonspecific finding.
• If the bone marrow is involved, moderate-to-marked lymphocytosis consisting of either small, mature lymphocytes or medium-large sized lymphocytes and lymphoblasts can be seen. This is called leukemia. • Pancytopenia with nonregenerative anemia, heteropenia and thrombocytopenia was reported in a Pekin duck with lymphocytic leukemia.

BIOCHEMISTRY
Depending on the degree of organ involvement, changes may also be seen on the serum biochemistry. Note that the absence of elevations in specific enzymes does not preclude severe involvement of that organ system. • Hepatic involvement (most common organ involved in birds): Elevations in serum aspartate transaminase (AST), gamma glutamyl transferase (GGT) and bile acids. • Renal involvement: Elevations in uric acid. • Nervous system involvement: Elevations in AST and creatinine phosphokinase (CPK). • Generalized nonspecific abnormalities: Elevations in AST and CPK in if the bird is weak and/or anorexic and has muscle break down.
• Hypercalcemia and hyperglobulinemia have been reported occasionally in birds with malignant lymphoma.

OTHER LABORATORY TESTS
In poultry species, serology or polymerase chain reaction analysis for Marek's Disease virus, Avian Leukosis virus and Reticuloendotheliosis virus can be performed on tissue samples collected via biopsy or during necropsy.

IMAGING
As with the biochemistry results, coelomic radiographic and ultrasonographic changes depend on the organ affected but can include:
• Hepatosplenomegaly with or without ascites; on ultrasound the liver may be heterogeneously mottled or hypoechoic; distinct nodules/masses may be seen.
• Splenomegaly; on ultrasound the spleen may be heterogeneously mottled or hypoechoic; distinct nodules/masses may be seen. • Renomegaly; unilateral or bilateral.
• CNS; although not currently reported in the literature, advanced imaging such as MRI or CT can be considered for suspected CNS lesions.

DIAGNOSTIC PROCEDURES
• If organomegaly or dermal lesions are detected, aspiration and/or biopsies of the affected organ are indicated. See below for histopathologic findings. • If ascites is

LYMPHOID NEOPLASIA

(CONTINUED)

present, diagnostic coelomocentesis should be performed. Rarely neoplastic lymphocytes can be found on cytology of the fluid. • If there is a peripheral lymphocytosis, bone marrow aspiration should be performed looking for neoplastic cells. Decreased/absence of other precursors can also be seen in advanced cases.

PATHOLOGIC FINDINGS

Application of the World Health Organization(WHO) classification system of hematopoietic and lymphoid neoplasms is established in several veterinary species. This system integrates information about tumor topography (affected anatomical site), cell morphology, immunophenotype, genetic features, and clinical presentation and course. Unfortunately, this has not been established in avian patients, but a similar approach can be taken when examining avian patients to include interpretation of tumor topography, cell morphology and immunophenotype.

Gross Pathololgic Findings
• Distinct soft, pale-tan to white nodules noted in organs; reported sites in birds include: liver, kidney, spleen, skin, gastrointestinal tract, kidney, thyroid gland, oviduct, lungs, air sacs, sinuses, thymus, testes, brain, palate/choana, trachea, heart, fat, periorbital muscles and pancreas.
• Generalized organomegaly and pallor of affected organs, most commonly the liver, spleen and kidneys. • Coelomic effusion and secondary thickening of serosal surfaces are often noted. • Dermal lesions present as gray to yellow, multifocal to coalescing thickened regions, plaques and/or ulcers. • The head and neck are more commonly affected.
• Lesions such as muscle and fat atrophy and bacterial or fungal infections are seen secondary to lymphoma.

Histopathologic Findings
• The affected tissue will have an infiltrative population of neoplastic round cells that may efface the normal parenchyma. • Depending on the lymphoma subtype, the neoplastic lymphocytes will be described based on size (small, intermediate or large), nuclear features (including, but not limited to centroblastic, immunoblastic, or anaplastic), and grade (indolent, low, mid, or high). Grading is based on the number of mitotic figures seen in a single high power field (400×). Secondary necrosis, ulcers or infection may also be present. • Immunohistochemistry has been used in poultry medicine to distinguish between B- and T-cell lymphomas and has been used in several cases of lymphoma in psittacine birds. Antibodies directed against CD3 and BLA-36 have been used successfully in psittacine birds. • In the future, hopefully this information can be used to direct treatment and assess prognosis, as it can in humans, canines and felines with lymphoma.

TREATMENT

APPROPRIATE HEALTH CARE
In animals, the goal of lymphoid neoplasia treatment is to induce a complete and durable remission. "Remission" indicates that the animal's cancer is controlled for greater than six months. If the animal is diagnosed with multicentric lymphoma, combination systemic chemotherapy is the treatment of choice (see below). For localized lymphoma, both radiation and/or surgical therapy can be utilized.

Surgical Therapy
Lymphoid neoplasia is commonly a multicentric disease that is not amenable to surgical therapy. For rare instances of solitary lymphoid tumors that are able to be surgically removed, this is an appropriate treatment option.

Radiation Therapy
Round cells are typically very sensitive to radiation therapy and can be used in concert with surgical therapy or chemotherapy for large tumors. There are two reported cases of the use of radiation therapy for lymphoid neoplasia in birds • Orthovoltage teletherapy: 40 Gy given in 10 fractions of 4 Gy/fraction ○ Case: African gray parrot that had a malignant lymphoreticular neoplasia in the right periorbital region. ○ Outcome: Feather loss in the area of the radiation field as well as ocular changes of the right eye and right-sided nasal discharge; decreased significantly but then returned to approximately 50% its original size in two months. • Megavoltage teletherapy: 32 Gy given in 4 fractions of 8 Gy/fraction ○ Case: Bursal lymphoma diagnosed in a three-year-old intact female Congo African grey parrot. ○ Outcome: Tumor did not respond.

NURSING CARE
• If the patient is dehydrated or hypovolemic secondary to systemic illness or anorexic/regurgitating, fluid therapy should be initiated. Subcutaneous, intravenous, intraosseous and oral routes can be utilized.
• Oxygen support is necessary if the patient is dyspneic secondary to air sac compression from organomegaly or ascites or primary respiratory involvement. • If ascites is present and causing respiratory distress, therapeutic coelomocentesis can be performed. • If the neoplasia is causing pain or discomfort appropriate analgesia should be administered.
• If there is a secondary fungal or bacterial infection, appropriate antimicrobials should be administered based on culture and sensitivity results.

ACTIVITY
If the patient is weak or systemically ill, activity should be limited. A cage can be set up with low/no perches to conserve energy and the patient should not be permitted to fly.

DIET
The patient should be encouraged to eat. Supportive feeding may be necessary in the form of gavage or syringe feeding in anorexic patients or those that are losing weight.

CLIENT EDUCATION
It is important to discuss the owners' goals in treatment of the neoplasia—whether it be curative or palliative.

SURGICAL CONSIDERATIONS
N/A

MEDICATIONS

DRUG(S) OF CHOICE
The following chemotherapeutic agents that have been used in birds with lymphoid neoplasms: • Prednisone (25 mg/m^2 PO daily), vincristine (0.75 mg/m^2 IO weekly), cyclophosphamide (200 mg/m^2 IO weekly), doxorubicin (30 mg/m^2 IO every 3 weeks), L-asparaginase (400 IU/kg IM weekly) and alpha interferon (15,000 units/m^2 SC q 48h × 3 treatments). ○ Case: Systemic lymphosarcoma with leukemic profile in Moluccan cockatoo. ○ Outcome: Gastrointestinal toxicity from the cyclophosphamide and L-asparaginase; complete resolution of solid tumors, stable leukemia. • Vincristine sulfate (0.5–0.75 mg/m^2 × 4 weeks), prednisone (0.45 mg/kg PO q12h) and chlorambucil (1 mg PO twice weekly). ○ Case: Lymphocytic leukemia and malignant lymphoma in a Pekin duck. ○ Outcome: Initially successful then euthanized due to poor quality of life. • Vincristine (0.1 mg/kg IV once weekly/biweekly) and chlorambucil (2 mg/kg PO twice weekly) × 17 weeks ○ Case: Nonepitheliotropic cutaneous B-cell lymphoma with leukemic profile in an umbrella cockatoo. ○ Outcome: Successful treatment—complete remission at week 29, no known relapse eight years later.
• Chlorambucil (1 mg/kg PO twice weekly × 6 weeks), prednisone (1 mg/kg PO q24h × 29 weeks), cyclophosphamide (5 mg/kg 4d/week × 29 weeks). ○ Case: T-cell chronic lymphocytic leukemia in a green-winged macaw. ○ Outcome: Thrombocytopenia with chlorambucil, lymphocytosis was stable for 29 weeks. • Predinsolone (0.45 mg/kg PO q12h) × 32 weeks. ○ Case: Multicentric lymphoma and leukemia in an Amazon parrot. ○ Outcome: Favorable initial response.
• Chlorambucil (20 mg/m^2 PO q14 days × 3 treatments). ○ Case: Cutaneous pseudolymphoma in a juvenile blue and gold macaw. ○ Outcome: Successful treatment × 3 years, initial elevation in ALT.

LYMPHOID NEOPLASIA (CONTINUED)

• Chlorambucil (2 mg/kg PO twice weekly × 5.5 weeks). ○ Case: T-cell chronic lymphocytic leukemia in a double yellow-headed Amazon parrot. ○ Outcome: No response after 40 days, patient euthanized.

CONTRAINDICATIONS
N/A

ALTERNATIVE DRUGS
Prednisone/Prednisolone: Must be used judiciously in avian patients as significant side effects such as immunosuppression, diabetes mellitus, hyperadrenocorticism and hepatic damage have been reported following administration.

FOLLOW-UP

PATIENT MONITORING
• In cases of solitary tumors that are not surgically removed, the size of the tumor should be monitored for response to treatment. Frequency depends on the type of treatment being used. • In cases of leukemia, the white blood cell count should be monitored. This should be performed based on the nadir for the chemotherapeutic agent. • In cases of organomegaly, follow up should include measurements based on radiographs and/or ultrasound. Frequency depends on the type of treatment being used.

PREVENTION/AVOIDANCE
N/A

POSSIBLE COMPLICATIONS
• Recurrence of the neoplasia can occur.
• Side effects of chemotherapeutic agents can be severe and can include toxicity as well as infections secondary to immunosuppression.

EXPECTED COURSE AND PROGNOSIS
The prognosis for domestic species diagnosed with lymphoma is variable. Overall, animals that already have clinical signs of systemic disease at the time of diagnosis tend to have a shorter survival time than those without systemic clinical signs.

To date, prognosis in avian patients with lymphoid neoplasia is difficult to determine due to the low number of published cases although successful treatment of avian lymphoid neoplasia has been achieved with chemotherapy, surgery and radiation therapy.

MISCELLANEOUS

ASSOCIATED CONDITIONS
Neoplasia of the lymphoreticular system can lead to lymphocyte dysfunction with increased incidence of secondary bacterial and fungal infections. This is most commonly reported in wildlife.

AGE-RELATED FACTORS
As with other neoplastic processes, incidence increases with age.

ZOONOTIC POTENTIAL
N/A

FERTILITY/BREEDING
If the reproductive tract (ie oviduct) is involved this can decrease fertility.

SYNONYMS
Lymphoma, malignant lymphoma, lymphoproliferative neoplasia, leukemia

SEE ALSO
Anemia
Ascites
Cere and skin, color changes
Coelomic distention
Infertility
Liver disease
Neurologic conditions
Sick bird syndrome
Tumors
Viral disease
Viral neoplasms

ABBREVIATIONS
N/A

INTERNET RESOURCES
N/A

Suggested Reading
Hammond, E.E., Sanchez-Migallon Guzman, D., Garner, M.M., et al. (2010). Long-term treatment of chronic lymphocytic leukemia in a Green-winged macaw (*Ara chloroptera*). *Journal of Avian Medicine and Surgery*, **24**(4):330–338.
Osofsky, A., Hawkins, M.G., Foreman, O., et al. (2011). T-Cell chronic lymphocytic leukemia in a double yellow-headed Amazon parrot (*Amazona ochrocephala oratrix*). *Journal of Avian Medicine and Surgery*, **25**(4),286–294.
Rivera, S., McClearen, J.R., Reavill, D.R. (2009). Treatment of nonepitheliotropic cutaneous B-cell lymphoma in an umbrella cockatoo (*Cacatua alba*). *Journal of Avian Medicine and Surgery*, **23**(4):294–302.
Valli, V.E., Kass, P.H., et al. (2013). Canine lymphomas – Association of classification type, disease stage, tumor subtype, mitotic rate, and treatment with survival. *Veterinary Pathology*, ;**50**(5):738–748.
Vail, D.M., Pinkerton, M.E., Young, K.M. (2013). Hematopoietic tumors. In: Withrow, S.J. MacEwen, E.G. (eds), *Small Animal Clinical Oncology*, 5th edn. Philadelphia, PA: WB Saunders Company, pp. 608–678.

Author Nicole R. Wyre, DVM, DABVP (Avian)

MACRORHABDUS ORNITHOGASTER

BASICS

DEFINITION
Macrorhabdus ornithogaster is a cause of proventriculitis in birds. Although this organism was originally termed Megabacterium due to its large, rod-like appearance, the organism has since been classified as an anamorphic ascomycetous yeast. Clinical signs may be variable and include sudden death or chronic wasting. Diarrhea or enteritis has also been reported in birds colonized by this organism; however, these birds can have concurrent parasites, bacterial infections, or other diseases that could cause diarrhea.

PATHOPHYSIOLOGY
• *Macrorhabdus ornithogaster* is a large anamorphic ascomycetous yeast, a long thin rod with rounded ends that is 3–4 µm × 20–80 µm in size. Branching is rarely seen. • Fecal–oral transmission. • Organism colonizes the isthmus between the proventriculus and ventriculus, causing ulceration, increased mucus secretion, and shortening gastrointestinal transit time. This results in maldigestion and malabsorption, regurgitation, and diarrhea. • *M. ornithogaster* has also been identified in clinically healthy birds. There is some debate whether it is a true pathogen or a commensal organism.

SYSTEMS AFFECTED
Gastrointestinal.

GENETICS
N/A

INCIDENCE/PREVALENCE
In infected budgerigar aviaries, prevalence may range from 27–65%.

GEOGRAPHIC DISTRIBUTION
Worldwide distribution.

SIGNALMENT
• **Species:** Wide species distribution, but most often reported in budgerigars, parrotlets, lovebirds, cockatiels, canaries, and finches. It is sporadically reported in larger parrots, chickens, turkeys, geese, ducks, ostriches, and ibis. • **Mean age and range:** Clinical signs are usually seen in middle-aged birds, but infection appears to occur much earlier in life. • **Predominant sex:** None.

SIGNS

Historical Findings
• Regurgitation. • Undigested seeds in the feces. • Diarrhea. • Weight loss despite apparent good appetite. • Lethargy and depression. • Melena as an end-stage finding. • Acute anorexia, regurgitation, and death are uncommon, but reported in budgerigars and parrotlets. • There may be no clinical signs reported.

Physical Examination Findings
• Emaciation. • Regurgitated material and saliva on the dorsal head and face. • Diarrheic feces. • Dehydration. • Depression. • Melena. • The bird may also be asymptomatic.

RISK FACTORS
Infected breeding birds may pass *Macrorhabdus* to their offspring.

DIAGNOSIS

DIFFERENTIAL DIAGNOSIS
• Trichomoniasis • Candidiasis • Bacterial or fungal ingluvitis • Proventriculitis • Proventricular dilatation disease is an important differential for undigested seed in the feces; identification of *Macrorhabdus* on fecal sampling can help to rule this diagnosis out. • Gastric neoplasia (see Associated Conditions) • Enteritis • Foreign body ingestion • Heavy metal toxicosis • Renal disease • Hepatic disease.

CBC/BIOCHEMISTRY/URINALYSIS
• Typically nonspecific changes • Anemia • Hyponatremia • Hypochloremia • Hypophosphatemia • Hypoglycemia • Hypocholesterolemia • Decreased AST • Low total protein if gastric ulceration present.

OTHER LABORATORY TESTS
• Fecal wet mount—most effective in clinically affected birds, as they are likely to be shedding larger numbers of the organism. Examine the sample at 100× and 400×, narrowing the stage diaphragm for best viewing. Multiple samples may be required for diagnosis. • Gram's stain (feces, proventricular scraping) or Romanowsky stains—*Macrorhabdus* is Gram-positive, but does not consistently stain well with either Romanowsky or Gram's stains. The organism may also wash off the slide during preparation; heat fixing can help prevent this. • The organism is intermittently shed, so it may be missed if only a single sample is examined. • Fecal PCR is also available.

IMAGING
• Plain radiography and ultrasonography—generally of limited use, aside from ruling out other causes of regurgitation, abnormal feces, and weight loss. • Contrast radiology—may see a "sandglass-like" appearance at proventricular-ventricular isthmus, dilation of proventriculus, increased GI transit time.

DIAGNOSTIC PROCEDURES
Necropsy—individual and flock diagnosis.

PATHOLOGIC FINDINGS
• Dilation of the isthmus and thinning of its wall. • Proventriculitis with ulceration +/– hemorrhage. • Ventriculitis with ulceration.

• Increased gastric mucus secretion. • Emaciation. • Histologically, *Macrorhabdus* is found on the surface of the isthmus glands and koilin layer, and may be located deeper within the glands and koilin layer in severe infections. The glands may atrophy or necrose, and disruption of the koilin layer can occur. Inflammatory cells may be minimal. The organism will be eosinophilic when stained with hematoxylin and eosin stain. • Morphologic and statistical data support a positive correlation between macrorhabdiosis and proventricular adenocarcinoma in budgerigars.

TREATMENT

APPROPRIATE HEALTH CARE
• Inpatient medical management may be required; fluid and nutritional support are indicated if the patient is debilitated, and treatment is ideally administered via gavage. • Stable birds may be managed on an outpatient basis with a water-based treatment.

NURSING CARE
• Fluid support (SC, IV, IO) may be indicated if the bird is debilitated. • Nutritional support via gavage may be indicated if anorexic or passing undigested feed.

ACTIVITY
No restrictions indicated.

DIET
If the patient is passing undigested seed, swapping to a pelleted diet may improve digestion and absorption of nutrients.

CLIENT EDUCATION
If multiple birds are present in the home, all should be screened for Macrorhabdus infection.

SURGICAL CONSIDERATIONS
N/A

MEDICATIONS

DRUG(S) OF CHOICE
• Amphotericin B—100 mg/kg q12h × 30 days by gavage; a shorter course of 14 days by direct oral administration has been suggested, as well as a lower dose (25 mg/kg). A drinking water treatment (0.9 mg/ml) has been used to reduce shedding in healthy birds, but results have been inconsistent in sick birds. • Nystatin reported to be effective in goldfinches (5,000 U/bird PO q12h × 10 days). Resistance may occur.

CONTRAINDICATIONS
Medications that may be associated with gastrointestinal ulceration (e.g., NSAIDs) should be used with caution.

MACRORHABDUS ORNITHOGASTER (CONTINUED)

PRECAUTIONS
Sodium benzoate (5 ml per 1 L drinking water)—this drug has a very bitter taste and must be introduced gradually. Birds must be monitored closely during treatment to ensure adequate water intake. Overdosage may cause neurologic signs and death. Breeding birds may drink more water than normal, and doses should be reduced by $\frac{1}{4}-\frac{1}{2}$. Shedding and clinical signs do not consistently resolve with this treatment.

POSSIBLE INTERACTIONS
None reported.

ALTERNATIVE DRUGS
• Altering gastric acidity has been proposed as a method for treating *Macrorhabdus*, by way of adding various vinegars or other organic acids to the water supply. However, this has not been shown to be efficacious.
• *Macrorhabdus* has responds poorly to other systemic antifungals. While fluconazole has shown promise in chickens, even low doses have resulted in toxicosis in budgerigars.

FOLLOW-UP

PATIENT MONITORING
Serial fecal samples should be monitored for the presence of the organism. However, as *Macrorhabdus* may be intermittently shed, it is difficult to prove that it has been eradicated.

PREVENTION /AVOIDANCE
• Eggs may be pulled from the parent birds, cleaned, and hand-raised to avoid transmission of *Macrorhabdus* from parent to chick. • Selectively breeding birds that do show clinical signs may be of value. • Flock treatment can be challenging, as amphotericin B is ideally administered via gavage, and due to the risks of adverse effects with sodium benzoate.

POSSIBLE COMPLICATIONS
There has been an association with *Macrorhabdus* infection and gastric adenocarcinoma.

EXPECTED COURSE AND PROGNOSIS
• Birds that are treated earlier in the course of disease tend to have a better prognosis.
• More debilitated birds, or those with acute onset of signs, may continue to deteriorate and die.

MISCELLANEOUS

ASSOCIATED CONDITIONS
Gastric adenocarcinoma has frequently been found in association with *Macrorhabdus* infection during necropsy of budgerigars.

AGE-RELATED FACTORS
N/A

ZOONOTIC POTENTIAL
None known.

FERTILITY/BREEDING
Infected parents may pass the organism to their offspring. Removing and cleaning eggs, followed by hand-raising the chicks, prevents transmission to the chicks.

SYNONYMS
Avian gastric yeast, megabacteriosis, virgamycosis.

SEE ALSO
Ingluvial hypomotility
Diarrhea
Emaciation
Enteritis/gastritis
Proventricular dilatation disease
Regurgitation/vomiting
Sick bird syndrome
Undigested food in droppings

INTERNET RESOURCES
http://vetbook.org/wiki/bird/index.php/.

Suggested Reading
Hoppes, S. (2011). Treatment of *Macrorhabdus ornithogaster* with sodium benzoate in budgerigars *(Melopsittacus undulatus)*. Proceedings of the Association of Avian Veterinarians Annual Conference, p. 67,
Phalen, D.N. (2006). Implications of Macrorhabdus in clinical disorders. In: Harrison, G.J., Lightfoot, T.L. (eds), *Clinical Avian Medicine*. Palm Beach, FL: Spix Publishing, Inc., pp. 705–709.
Phalen, D.N. (2014). Update on the diagnosis and management of Macrorhabdus ornithogaster (formerly Megabacteria) in avian patients. *Veterinary Clinics of North America: Exotic Animal Practice*, **17**(2):203–210.
Author Kristin M. Sinclair, DVM, DABVP (Avian)

 Client Education Handout available online

Metabolic Bone Disease

BASICS

DEFINITION
Metabolic bone disease is a pathology affecting the skeleton due to a diet deficient in vitamin D, calcium, phosphorous or due to an incorrect calcium to phosphorous ratio. This condition may be exacerbated by limited exposure to sunlight (or another source of UV light in the range of 290–315 nm) which is required for synthesis of vitamin D. During growth, there is high bone turnover and demands for calcium, vitamin D, and phosphorous are much greater than needs for adult maintenance; therefore, this is a particularly sensitive period.

PATHOPHYSIOLOGY
Nutritional deficiency in vitamin D, calcium, or phosphorus could manifest as various skeletal abnormalities including thin bone cortices, abnormal bone curvature causing limb deformity, stunted growth, or increased risk of bone fractures. The active form of vitamin D is important for absorption of calcium and phosphorus from the gastrointestinal tract and reabsorption of calcium in the kidneys. Circulating calcium is then used by osteoblasts and concentrates in the bone to provide physical stability while the level of calcium in the circulation is maintained in homeostasis by interaction of several hormones including estrogen and parathyroid hormone (with calcitonin likely having a minor role). Phosphorus is maintained at homeostasis by the action of the parathyroid hormone and active vitamin D; both increase phosphorus absorption from the gastrointestinal tract. Parathyroid hormone also releases phosphorus from the bone and acts to decrease its reabsorption from the urinary filtrate. A balance between dietary calcium and phosphorus is imperative to maintain the appropriate metabolism of these minerals as dietary increase in one may decrease the net absorption of the other. With a deficiency in one or more of the aforementioned nutrients, hormonal balance affecting bone turn-over is altered, thereby causing abnormal bone mineralization and balance between osteoclast and osteoblast activity. This can potentially lead to decreased bone strength and bone deformity.

SYSTEMS AFFECTED
- Musculoskeletal ∘ Bone deformity. ∘ Decreased mineral density. ∘ Multiple fractures. ∘ May cause troubles in ambulation, general weakness, tremors. ∘ Beaks may be soft and bent. • Neuromuscular ∘ Neuromuscular changes may accompany angular limb deformity. ∘ Hypocalcaemia may cause muscle tetanus in juvenile birds. • Skin/Exocrine ∘ Poor feather quality. ∘ Poor molting.

GENETICS
N/A

INCIDENCE/PREVALENCE
Incidence of this condition is lower than it was in the past thanks to increased awareness for the need to provide appropriate nutrition. However, this is still a common condition in avian practice.

GEOGRAPHIC DISTRIBUTION
N/A

SIGNALMENT
- **Species:** All avian species may be affected.
- **Mean age and range:** Much more common in juvenile and sub-adult birds.
- **Predominant sex:** N/A

SIGNS
Historical Findings
Failure to thrive, difficulty in ambulation, deformed limbs.

Physical Examination Findings
Angular limb deformity, poor body condition, generalized weakness, poor plumage, hypocalcemic tetany.

CAUSES
• Nutritional causes for metabolic bone disease include calcium, vitamin D, and phosphorus dietary intake deficiency. Most commonly this is iatrogenic due to inadequate nutritional intake of these nutrients. • The form of vitamin D also plays an important role as vitamin D2 is much less bioavailable than vitamin D3. • Unbalanced diets comprised mostly of seeds have been found to be deficient in calcium, vitamin D, and phosphorus. In addition, the calcium to phosphorus ratio is low, thereby limiting calcium absorption. Similarly, reared carnivorous birds fed a nonsupplemented all-meat diet are at risk for calcium deficiency. Vitamin D deficiency may be made worse by lack of UV light exposure.

RISK FACTORS
• High fat diets are considered as an increased risk for metabolic bone disease. • All seed diets or unbalanced home-prepared diets should be considered as risk factors. • An unsupplemented all-meat diet in a carnivorous bird or an all seed diet in other orders are risk factors for developing this condition.

DIAGNOSIS

DIFFERENTIAL DIAGNOSIS
Trauma, developmental skeletal deformities.

CBC/BIOCHEMISTRY/URINALYSIS
• *Decreased ionized calcium*—ionized calcium provides an assessment of body calcium status whereas total calcium may be affected by other physiological factors such as plasma proteins. • *Hyperphosphatemia*—phosphorous may be increased as a result of increased parathyroid hormone which is secondary to calcium depletion or in cases of a diet with highly skewed calcium to phosphorous ratio. • *Hypophosphatemia*—may be seen as a result of body phosphorus depletion.

OTHER LABORATORY TESTS
Measurements of parathyroid hormone levels and 25-hydroxycholecalciferol have been used to varying levels of success in different avian species. Due to the intricacy of these tests and the possibility of false positive results, they should not be used as screening tools, rather interpreted as part of the full clinical picture and the combination of clinical findings.

IMAGING
• Radiography is the most useful imaging modality to evaluate bone mineralization, bone deformity, and fractures. • Computed tomography may also be used, particularly in larger birds, and may be a useful modality to evaluate changes in bones of the skull and the spine.

DIAGNOSTIC PROCEDURES
N/A

PATHOLOGIC FINDINGS
Bones may be deformed or fractured. Osteomalacia may cause soft, under-mineralized long bones. The cortices may be diminished with large numbers of osteoclasts. The metaphyses may have unmineralized osteoid especially in the periphery, and the trabeculae may be lined with osteoblasts and abnormally large amounts of fibrous tissue and retained cartilage.

TREATMENT

APPROPRIATE HEALTH CARE
• Animals with severe bone deformity may have a very guarded prognosis. • External coaptation may be used to support limbs and to help correct deformities. • Fractures need to be stabilized either with external cooptation, bandaging, or surgery; however poor bone quality may decrease chances of success of a surgical procedure. • Analgesia may be indicated especially if fractures are present.

NURSING CARE
• Bandaging and analgesia may be indicated to treat fractures due to metabolic bone disease. In severe cases, poor bone quality warrants limiting the bird's activity and cage padding to avoid additional trauma. • Weak or debilitated birds may not consume sufficient food and water voluntarily. Therefore, assisted gavage feeding and fluid support may be needed.

Metabolic Bone Disease (Continued)

ACTIVITY
Limited activity in a padded cage is indicated if a young bird has severe bone lesions or fractures. Once the bird is in a better nutritional plane and bone quality appears radiographically improved, moderate activity and gentle physiotherapy may be beneficial in establishing bone and muscle strength.

DIET
A complete diet assessment will provide an indication of this condition and could help in devising a plan for appropriate therapy and nutritional modifications. Diets for carnivorous birds based on meat alone are unlikely to provide sufficient calcium, whereas a diet for companion pet birds comprised of seeds alone is likely to be deficient in calcium, vitamin D and phosphorous among other deficiencies. Formulated diets of high quality are expected to provide sufficient nutrients. Diets formulated to support the requirements of growing birds are nutrient dense and could be a good choice for a bird with apparent deficiencies. These provide a good alternative to all-seed diets or for unbalanced diets based on home-prepared foods. The nutrient requirements of carnivorous birds can be met by consumption of whole vertebrate prey. Mature prey is generally preferred to ensure calcium content and appropriate intake of fat. The recommended calcium:phosphorous ratio is generally between 1.4:1 and 2:1 with 4:1 still well tolerated. Egg laying hens require calcium supplementation.

CLIENT EDUCATION
Client education is highly important to manage this condition as it is usually the result of lack of client awareness of the importance of proper nutrition. The significance of a complete diet needs to be explained to the owner as most cases of nutritional diseases, including metabolic bone disease are not only treated by proper nutrition, but are also completely preventable.

SURGICAL CONSIDERATIONS
Some fractures that result from metabolic bone disease may need to be managed surgically. However, very poor bone quality may have a negative impact on surgical success and may affect treatment choice.

MEDICATIONS
DRUG(S) OF CHOICE
Treatment of nutritional deficiencies should be done primarily through improvement of the diet. Additional supplementation may be considered to better address specific deficiencies for the short term; however, it should not replace improved nutrition, and care must be taken to avoid over supplementation.
• Calcium glubionate: 25–150 mg/kg PO up to twice daily. • Calcium gluconate (10%): 25–100 mg SQ or IM (diluted), injection may be painful. • Vitamin D3: 3300 IU/kg IM every seven days as needed; beware of excessive use leading to hypervitaminosis D. • Sunlight exposure: 11–30 minutes of sunlight per day; a safer alternative to vitamin D injections. Keep the bird in appropriate temperature and avoid overheating.

CONTRAINDICATIONS
N/A

PRECAUTIONS
N/A

POSSIBLE INTERACTIONS
N/A

ALTERNATIVE DRUGS
N/A

FOLLOW-UP
PATIENT MONITORING
Monitoring of dietary changes is important to ensure owner compliance. Follow up radiographs may be indicated every two weeks to monitor improvement in bone quality or even at higher frequency if there are fractures present. Monitoring should also include ionized calcium (if decreased), body weight and body condition, plumage quality, and ambulation.

PREVENTION/AVOIDANCE
Feeding an appropriate commercially prepared balanced diet to pet birds as the main portion of the daily intake appears to be the safest way to prevent nutritional deficiencies. Wildlife rehabilitators should provide wild birds with proper nutrition according to the species. Carnivorous birds should not be fed an all-meat diet without appropriate supplementation.

POSSIBLE COMPLICATIONS
Bone deformities may be permanent and may have severe consequences on birds including possible decrease in ambulation, risk of concurrent diseases such as pododermatitis, and risks of dystocia. Complications may vary according to the body part affected.

EXPECTED COURSE AND PROGNOSIS
Prognosis is highly correlated with severity of the disease. Mild cases of decreased bone density but without bone deformity or fractures have a good prognosis and will usually respond well to dietary improvement. Moderate cases with one or more fractures, poor bone mineralization or bone deformities may need extensive supportive care but have good chances of recovery albeit that bone deformities may be permanent. Severe cases with severe bone deformity or multiple fractures or complicated fractures may have a guarded prognosis, requiring much longer treatment. Euthanasia may be considered in severe cases where decreased ambulation has a significant impact on quality of life.

MISCELLANEOUS
ASSOCIATED CONDITIONS
In many cases, the bird may be affected by multiple nutritional deficiencies while some deficiencies may not be clinically apparent. For this reason it is important to improve the bird's overall diet rather than supplement a single nutrient in the diet.

AGE-RELATED FACTORS
Juvenile birds are most susceptible to this condition since bone metabolism and calcium, phosphorous and vitamin D requirements are highest during growth. Milder clinical signs may also be seen in adult birds following a long duration of nutrient depletion.

ZOONOTIC POTENTIAL
N/A

FERTILITY/BREEDING
Calcium is vital for shell mineralization and smooth muscle contractions during egg laying. Since the bone is the predominant calcium store in the body, poor bone mineralization may predispose reproductive females to decreased reproductive success or even dystocia. For this reason it is important to correct the calcium, phosphorous or vitamin D status of the bird before breeding.

SYNONYMS
Nutritional secondary hyperparathyroidism, osteomalacia, fibrous osteodystrophy, rickets, juvenile osteoporosis, bone atrophy.

SEE ALSO
Angel wing
Beak malocclusion
Chronic egg laying
Fracture/luxation
Lameness
Neurologic conditions
Nutritional deficiencies
Osteomyelosclerosis/polyostotic hyperostosis
Splay leg/slipped tendon
Trauma

ABBREVIATIONS
MBD—metabolic bone disease.

INTERNET RESOURCES
http://avianmedicine.net/content/uploads/2013/03/05_calcium.pdf

Suggested Reading
Adkesson, M.J., Langan, J.N. (2007). Metabolic bone disease in juvenile Humboldt penguins (Spheniscus humboldti): investigation of ionized calcium, parathyroid hormone, and vitamin D3 as diagnostic parameters. *Journal of Zoo and Wildlife Medicine*, **38**(1):85–92.

de Matos, R. (2008). Calcium metabolism in birds. *Veterinary Clinics of North America: Exotic Animal Practice*, **11**(1):59–82.

Koutsos, E.A., Matson, K.D, Klasing, K.C. (2001) Nutrition of birds in the order Psittaciformes: a review. *Journal of Avian Medicine and Surgery*, **15**(4):257–275.

Author Jonathan Stockman, DVM, DACVN

Mycobacteriosis

BASICS

DEFINITION
Mycobacteriosis is infection with one or more organism belonging to the genus Mycobacteriae. The great majority of infections in birds are atypical, or non-tuberculous species such as *Mycobacterium genavense*, and *Mycobacterium avium*.

PATHOPHYSIOLOGY
Source of infection is most likely environmental contamination, as atypical mycobacterial organisms are present in soil and water. In humans and other species, immunosuppression appears to be a prerequisite for infection; this has not been established in birds. Bird to bird transmission is unlikely. Rare cases of *M. tuberculosis* were thought to have been obtained via contact with infected humans.

SYSTEMS AFFECTED
- Hepatobiliary—hepatomegaly. Liver dysfunction. • Lymphopoeitic—splenomegaly. • Gastrointestinal—weight loss, abnormal stools. • Respiratory—air sacculitis. • Skin—cutaneous nodules.
- Ocular—conjunctival swelling or masses.

GENETICS
N/A

INCIDENCE/PREVALENCE
Incidence is unknown, as many infections may be subclinical.

GEOGRAPHIC DISTRIBUTION
Worldwide.

SIGNALMENT
- **Species:** Avian, including passerine, waterfowl and many wild/zoo species.
- **Species predilections:** A 2013 retrospective study of 123 cases indicate the most commonly affected psittacines were Amazon parrots (*Amazona sp*) and grey-cheeked parakeets (*Brotogeris pyrropherus*). • **Mean age and range:** All ages are susceptible but tends to be a disease of older birds, as organism is slow growing and disease is chronic. • **Predominant sex:** Both sexes are susceptible.

SIGNS

Historical Findings
Owners may note nonspecific chronic signs including weight loss and lethargy.

Physical Examination Findings
- Signs are often nonspecific and refer to the affected body system, but in all advanced forms include weight loss and lethargy.
- Birds with primary GI disease may present with weight loss, abnormal stool and other symptoms referred to the GI tract. • Birds with respiratory infections may present with abnormal respirations (rate/effort), audible respiratory sounds (wheeze/click). • The ocular or dermal forms may present as conjunctival/skin masses. • Birds with hepatic form present for vague illness culminating in signs attributable to hepatic enlargement (enlarged coelom, respiratory distress, coelomic fluid). • As disease is generally chronic, owners may report long-term vague symptoms. • Physical examination findings are again dependent on body system affected, and can include reduce pectoral muscle mass (weight loss), dyspnea (increased respiratory rate and effort), enlarged coelomic space, and evidence of a coelomic mass or fluid. Masses may appear on the conjunctiva or any portion of the skin. • Several anecdotal cases reported masses at the site of tattoo placement used to permanently identify gender after surgical sexing (ventral patagium).

CAUSES
In most cases, source of atypical mycobacterial organisms is unknown, but can be presumed to be water, food and soil. Multiple infections within groups of birds are rarely reported. The source of confirmed *Mycobacterium tuberculosis* cases is apparently infected human owners.

RISK FACTORS
Risk factors for birds have not been determined, but may be exposure to soil. However, surveys of tap water have been positive for atypical mycobacterial organisms as well. Infections of immunocompetent humans are extremely rare, and most infections are a consequence of severe immunocompromise. The role of immunosuppression in atypical mycobacterial infections in birds is uncertain.

DIAGNOSIS

DIFFERENTIAL DIAGNOSIS
Due to the wide range of body systems affected and chronic nature of the disease, the differential diagnosis list is extensive in most cases and includes numerous other infectious, neoplastic or metabolic diseases. In cases of dermal or conjunctival masses, differential diagnoses include inflammation, infection, cysts and neoplasia.

CBC/BIOCHEMISTRY/URINALYSIS
- A commonly reported finding in birds with systemic mycobacterial infections is marked elevation of the leukogram, with monocytosis. However, birds with local lesions, and not all birds with systemic disease, will present with CBC abnormalities. • Biochemistry alterations reflect disease of the body system affected and general debilitation, for example, elevated AST and bile acids in advanced hepatic forms, and hypoalbuminemia in chronic GI cases. • All chronic inflammatory and infectious processes tend to produce anemia in birds, in many cases, nonregenerative in nature.

OTHER LABORATORY TESTS
- Protein electrophoresis may support chronic inflammatory/infectious disease. • In birds with gastrointestinal infections, nonstaining bacterial rods may appear in fresh fecal cytologic preparations. Suspect samples can be submitted for acid-fast staining. • Culture and sensitivity of mycobacterial organisms is possible, but challenging, as most organisms are difficult to grow. Culture is expensive, and may take many weeks to months for results. For this reason, culture is seldom performed in veterinary medicine.

IMAGING
- Plain radiography may reflect abnormalities of the affected system, for example, enlarged liver, thickened bowel loops and pulmonary nodules. • Contrast radiography is extremely useful to distinguish between a widened proventricular and hepatic silhouette, and to give more information on GI transit time.
- Ultrasound may help confirm hepatomegaly.

DIAGNOSTIC PROCEDURES
Polymerase chain reaction (PCR) testing of specimens remains the most useful diagnostic test both for confirmation of infection, and speciation of the organism in question, which is a critical consideration when considering patient treatment. Samples include feces, mass and organ specimens (liver, lung, bowel). Many exotic diagnostic laboratories offer PCR for mycobacterial organisms.

PATHOLOGIC FINDINGS
Histopathologic changes consistent with mycobacteriosis include granulomatous inflammation with macrophages containing acid-fast microorganisms.

TREATMENT

APPROPRIATE HEALTH CARE
Prior to considering treatment, the organisms should be positively identified by species to rule out rare cases of *M. tuberculosis*, which is of severe zoonotic risk to humans. Most experts agree birds with *M. tuberculosis* should be euthanized. It should be kept in mind that treatment protocols require daily administration of medications for a year or more. It is likely that even birds treated with long-term therapy may never clear disease; as such, prolonged therapy with multidrug regimens has the potential to result in multidrug resistant mycobacterial organisms. Treatment should never be considered without consultation with a human physician. A 2014 study in ring neck doves showed poor organism clearance in birds treated for 180 days with a three-drug combination.

Mycobacteriosis

(CONTINUED)

NURSING CARE
Birds with chronic disease are often debilitated and require advanced supportive care. In severe cases, this may require hospitalization to correct fluid deficits and provide nutritional support.

ACTIVITY
For very ill birds, activity should be limited until the bird is well enough to avoid injury from falls.

DIET
All ill birds benefit from an appropriate diet based on current research. For most psittacine birds, this consists of a formulated diet, with limited produce and other table foods. Anorexic or hyporexic birds benefit from tube feeding with a highly-digestible, liquid diet designed for ill or recovering birds.

CLIENT EDUCATION
Clients must be aware of the zoonotic potential of this disease, and discuss this with their physician (see treatment and zoonoses below). Owners should also be aware that treatment is long term, as long as one year or more, and is unlikely to be curative.

SURGICAL CONSIDERATIONS
Local skin or conjunctival masses may be removed entirely. Removal of individual coelomic masses may be of benefit; however, disease is often systemic.

MEDICATIONS

DRUG(S) OF CHOICE
Reviews of outcomes of large numbers of pet birds treated for mycobacteriosis with specific drug combinations at specific dosages are unavailable. A 2014 study in ring neck doves showed poor organism clearance in birds treated for 180 days with a three-drug combination. Treatment is based upon drug combinations used in humans and other animals for similar species. Without benefit of culture and sensitivity, (which is available but difficult and expensive), the practitioner is advised to research medications currently used for similar organisms in human patients. Drugs reported used for mycobacteriosis in birds include enrofloxacin, rifampin, ethambutol, clarithromycin, and others. Drug dosages are entirely extrapolated from other species. It should be noted that some countries restrict the use of anti-tuberculous drugs for humans in other species due to the growing risk of antibiotic resistance.

CONTRAINDICATIONS
There is too little information on the use of antituberculous drugs in birds to determine contraindications.

PRECAUTIONS
Owner's physician should be consulted prior to instituting any treatment. Euthanasia should be considered in cases of definitive diagnosis of mycobacteriosis.

POSSIBLE INTERACTIONS
N/A

ALTERNATIVE DRUGS
N/A

FOLLOW-UP

PATIENT MONITORING
Treatment efficacy is sometimes difficult to judge, due to the widespread nature of some forms of the disease. Clinical abnormalities may appear to resolve. Birds that exhibit changes in the complete blood count, such as marked heterophilia and anemia, may show marked improvement with treatment. Birds with gastrointestinal forms of the disease may be screened for the presence of acid-fast bacteria; however, it should be kept in mind that shedding of mycobacterial organisms is sporadic. Birds with hepatic form diagnosed via biopsy and histopathology may benefit from repeat biopsy. Regardless, it is difficult to prove "cure" even after long-term therapy. Current literature suggests elimination of disease is unlikely.

PREVENTION/AVOIDANCE
As source of nontypical mycobacteriosis is often unknown, prevention of re-infection is difficult. Organisms have been commonly found in soil, food and water.

POSSIBLE COMPLICATIONS
Outcome likely depends on the chronicity of the disease and degree of body system damage. Bird with severe hepatic mycobacteriosis, for example, may succumb to liver failure despite elimination of the organisms. Poor treatment compliance in humans and other species is linked to risk of organism resistance and ultimately treatment failure. Therefore, treatment must be administered regularly, and for the recommended course (often more than one year). This may prove difficult for some owners. Recent data suggests even prolonged therapy in avian species is unlikely to clear disease.

EXPECTED COURSE AND PROGNOSIS
Long-term retrospective outcome studies are unavailable in birds. There are numerous single reports (both documented and anecdotal) of apparent treatment successes. However, other reports indicate the disease may be difficult to completely eradicate and may reappear after many years. Excision of single dermal or conjunctival masses may be curative. The prognosis for birds with systemic disease is uncertain.

MISCELLANEOUS

ASSOCIATED CONDITIONS
N/A

AGE-RELATED FACTORS
Due to chronicity, diagnosis is rarely obtained in young birds (under 2 years of age).

ZOONOTIC POTENTIAL
As mentioned above, *M. tuberculosis*, is of severe zoonotic risk to humans. Most experts agree birds with *M. tuberculosis* should be euthanized. Treatment of birds with atypical mycobacterial infections should be considered in light of a current publication in ring neck doves which states: "We do not recommend at this time the medical treatment of birds with mycobacteriosis in uncontrolled clinical settings" (Saggese, 2014).

FERTILITY/BREEDING
Birds with GI forms of mycobacteriosis should not raise chicks.

SYNONYMS
TB (tuberculosis), NTM (nontuburculous mycobacteria)

SEE ALSO
Anemia
Anorexia
Ascites
Coelomic distention
Diarrhea
Enteritis/gastritis
Infertility
Lameness
Liver disease
Sick bird syndrome

ABBREVIATIONS
TB—tuberculosis
NTM—nontuburculous mycobacteria

INTERNET RESOURCES
Most state department of agriculture websites have specific information on mycobacteriosis. In some states, mycobacteriosis is a reportable disease.

Suggested Reading
Lennox, A.M. (2007). Mycobacteriosis in companion psittacines birds: A review. *Journal of Avian Medicine and Surgery*, **21**(3):181–187.
Palmieri, C., Roy, P., Dhillion, A.S., Shivaprasad, H.L. (2013). Avian mycobacteriosis in psiattacines: A retrospective study of 123 cases. *Journal of Comparative Pathology*, **148**(203):126–138.
Saggese, M.D., Tizard, I., Gray, P., Phalen, D.N. (2014). Evaluation of multidrug therapy with azithromycin, rifampin, and ethambutol for the treatment of *Mycobacterium avium* subsp *avium* in ring-necked doves (*Streptopelia risoria*): an uncontrolled clinical study. *Journal of Avian Medicine and Surgery*, **28**(4):280–289.

Mycobacteriosis (Continued)

VanDerHeygen, N. (1997). Clinical manifestations of mycobacteriosis in pet birds. *Seminars in Avian and Exotic Pet Medicine*, **6**:18–24.

Washko, R.M., Hoefer, H., Kiehn, T.E., *et al.* (1998). Mycobacterium tuberculosis infection in a green-winged macaw (Ara chloroptera): report with public health implications. *Journal of Clinical Microbiology*, **36**(4):1101–1102.

Author Angela M. Lennox, DVM, Dipl. ABVP (Avian, Exotic Companion Mammal)

MYCOPLASMOSIS

BASICS

DEFINITION
Mycoplasmosis is disease caused by any one of a number of species of *Mycoplasma*. *Mycoplasma spp.* are bacteria without a cell wall that renders this group of organisms resistant to antibiotics that attack cell wall synthesis. *Mycoplasma spp.* are the smallest bacteria and have characteristics that require specialized transport and nutrient media for growth in the laboratory. For these reasons, mycoplasma infections can be difficult to definitively identify as being caused by mycoplasma. *Mycoplasma spp* infect people and many species of animals, including birds. One of the more significant species in terms of morbidity and mortality is *Mycoplasma gallisepticum*. This organism causes chronic respiratory disease (CRD) in chickens and infectious sinusitis in other galliformes, passerines, and other species. *Mycoplasma synoviae* is another serious cause of morbidity and mortality in gallformes leading to debilitating joint and limb disease and other respiratory complications. *Mycoplasma meleagridis* is seen in turkeys and can cause both respiratory and limb disease. Since *Mycoplasma spp.* are so difficult to culture, it is suspected that cases of unknown etiology in psittacines and passerines may be due to mycoplasma but remain undiagnosed.

PATHOPHYSIOLOGY
Mycoplasma attach to avian epithelial cells, some species have a proclivity for the respiratory tract others for joint tissues. This leads to cell damage and an eventual inflammatory response by the body. In turn, this causes swelling, discharge, and other inflammatory changes that prohibit normal organ function. Severe untreated cases can be fatal.

SYSTEMS AFFECTED
• Musculoskeletal—*M. synoviae* and *M. meleagridis* cause joint swelling, limb deformities, abnormal gait.
• Ophthalmic—*M. gallisepticum* (and other *Mycoplasma spp.*) cause conjunctivitis and ocular discharge. • Reproductive—decreased egg production. • Respiratory—both upper and lower respiratory signs.

GENETICS
N/A

INCIDENCE/PREVALENCE
Both *M. gallisepticum* and *M. synoviae* are worldwide in distribution as carried by wild birds and can infect domesticated flocks. Once established in a domestic flock, all infected birds remain carriers despite treatment. Therefore, in infected flocks, the incidence rate is very high.

GEOGRAPHIC DISTRIBUTION
Worldwide as carried by passerines, galliformes, raptors, anseriformes.

SIGNALMENT
• **Species:** Galliformes, house finch, columbiformes, psittacines, raptors, anseriformes. • **Mean age and range:** All ages are susceptible. • **Predominant sex:** Both sexes are susceptible.

SIGNS

Historical Findings
In flocks, a previous history of respiratory disease or birds with swollen joints is common if mycoplasma is suspected. Exposure to wild birds, especially those with signs of respiratory disease, is also significant.

Physical Examination Findings
Mycoplasma spp.
• Sneezing. • Dyspnea with open-mouth breathing. • Conjunctivitis with swelling of ocular tissues and crusted discharge around the eyes. Feather loss around the eyes. • Nasal discharge. • Tracheitis, especially in galliformes. • Air sacculitis. • Sinusitis. • Pneumonia. • Decreased egg production. • Weight loss. • Reduced growth rate.
Mycoplasma synoviae specifically:
• Swollen joints in legs. • Pain on palpation of joints. • Lameness. • Abnormal gait.
Mycoplasma meleagridis
• Skeletal disease including slipped tendons.

CAUSES
N/A

RISK FACTORS
Risk factors for developing this disease depend on immune status and transmission. *Mycoplasma spp.* can be transmitted in poultry eggs or via infectious aerosols in feed and water. Humans can also act as fomites. Flock transmission is by direct and indirect contact and can happen swiftly.

DIAGNOSIS

DIFFERENTIAL DIAGNOSIS
• Avian influenza. • Chlamydiosis.
• Newcastle disease. • Infectious bronchitis.
• Avian poxvirus. • Vitamin A deficiency.
• *Bordetella avium*. • Reovirus (Viral arthritis). • Staphylococcus. • Pasteurella.
• Gout. • Marek's Disease.

CBC/BIOCHEMISTRY/URINALYSIS
• The white blood cell count may show a heterophilia and in chronic disease, monocytosis may also be present. • Chronic cases may also show a nonregenerative anemia. • Biochemistry is usually normal.

OTHER LABORATORY TESTS
M. gallisepticum can be isolated by using swabs of infected tissue. Serology can be done using immunofluorescence, agglutination inhibition tests, ELISA testing, and PCR. Culture of mycoplasma needs to be done on specialized media including both transport and growth media.

IMAGING
Whole body radiographs may exhibit signs of air sacculitis, including air sac lines on both lateral and ventrodorsal projections. The appearance of the lungs on radiographs may show increased radio-opacity of the lung fields. Soft tissue swelling of joints may also be present. Skull radiographs may not have enough resolution to show sinusitis. Skull CT or MRI can show evidence of sinusitis.

DIAGNOSTIC PROCEDURES
• Tracheal or bronchial lavage for culture and serology. • Fine needle aspirate of joint swellings for cytology, serology, and culture.

PATHOLOGIC FINDINGS
Histologically, inflammation including lymphocyte infiltration and necrosis will be present in tissues infected with mycoplasma.

TREATMENT

APPROPRIATE HEALTH CARE
• In individual patients, health care is directed to the effected system. Medical management of respiratory signs includes oxygen therapy, sinus flushes, nasal cavity flushes, and nebulization. • It is recommended with flocks that patients are not treated as birds become carriers for life and can always infect other birds. It is recommended that infected birds and the entire flock be euthanized.

NURSING CARE
Fluid and nutritional support may be necessary in birds that are anorexic and dehydrated due to this disease.

ACTIVITY
If there are skeletal and joint signs, activity is restricted until the patient improves during treatment.

DIET
N/A

CLIENT EDUCATION
N/A

SURGICAL CONSIDERATIONS
N/A

MEDICATIONS

DRUG(S) OF CHOICE
Tetracylines and quinolones can be given PO, topically, and by injection but most birds are not treated as flock eradication is recommended. Quinolones cannot be used in food animal species (poultry/waterfowl).

MYCOPLASMOSIS (CONTINUED)

CONTRAINDICATIONS
N/A

PRECAUTIONS
N/A

POSSIBLE INTERACTIONS
N/A

ALTERNATIVE DRUGS
Eradication, not treatment, is recommended for flocks.

FOLLOW-UP

PATIENT MONITORING
N/A

PREVENTION/AVOIDANCE
Vaccines are available for some species of mycoplasma which can be used in flocks of chickens and turkeys. Vaccines have various levels of effectiveness and may not be legal to give depending on state regulations.

POSSIBLE COMPLICATIONS
• Decreased egg production. • Decreased meat production. • Long-term respiratory and skeletal disease.

EXPECTED COURSE AND PROGNOSIS
Treated birds are considered lifelong carriers.

MISCELLANEOUS

ASSOCIATED CONDITIONS
N/A

AGE-RELATED FACTORS
N/A

ZOONOTIC POTENTIAL
No zoonotic complications have been shown with avian species of mycoplasma. But humans can act as fomites bringing mycoplasma disease to unaffected flocks. Caution should be taken when going to different flocks when mycoplasma is suspected.

FERTILITY/BREEDING
Breeding, fertility rates, and egg production greatly decrease with mycoplasma infection.

SYNONYMS
Mycoplasma synoviae: Infectious Synovitis, Avian Mycoplasmosis, Infectious Sinusitis and Mycoplasma Arthritis

SEE ALSO
Arthritis
Infertility

Lameness
Rhinitis and sinusitis
Respiratory distress
Splay leg/slipped tendon

ABBREVIATIONS
N/A

INTERNET RESOURCES
Most state department of agriculture websites have specific information on mycoplasma. In some states, mycoplasma is a reportable disease.

Suggested Reading
Ley, D.H. (2013). Overview of Mycoplasmosis in Poultry. http://www.merckmanuals.com/vet/poultry/mycoplasmosis/overview_of_mycoplasmosis_in_poultry.html (last accessed September 28, 2015).

Lierz, M. (2010). Importance of Mycoplasmas in non-poultry birds. *Proceedings of the Association of Avian Veterinarians Annual Conference.*

Tully, T., Lawton, M., Dorrestein, G. (2000). *Avian Medicine*. Woburn, MA: Butterworth-Heinemann.

Author Karen Rosenthal, DVM, MS

Mycotoxicosis

BASICS

DEFINITION
Mycotoxins are secondary metabolites of toxigenic fungi on avian feed or other foodstuffs. These mycotoxins cause adverse biological effects, when consumed in adequate quantities. Many fungal genera and species in nature are known to grow and produce toxins, when environmental factors (temperature, moisture and aeration) are favorable. While there have been several hundred mycotoxins documented, only some of the most common affecting birds will be mentioned here.
• **Aflatoxins** from corn, cottonseed, ground nuts and tree nuts are produced by toxigenic strains of *Aspergillus flavus, Asperpgillus parasiticus,* and *Aspergillus nominus.*
• **Ergotism** (*Claviceps sp.*) from infected rye or triticale and other cereal grains, results in vasoconstriction and gangrene from ergopeptine alkaloids. • **Trichothecene mycotoxicosis** from *Fusarium* sp. Toxins produced include DON (deoxynivalenol, vomitoxin) T-2 toxin and DAS (diacetoxyscirpenol). Trichothecene toxins are relatively nontoxic in poultry, when compared to swine. • **Ochratoxins**, produced by *Penicillium viridicatum* and *Aspergillus ochraceus,* can be quite toxic to poultry.

PATHOPHYSIOLOGY
Best pathophysiology data from poultry; pet bird cases often based on histopathology/case reports and occasional toxin assays. Systemic changes vary with the toxin, the severity of exposure, the avian species/breed and the age.

SYSTEMS AFFECTED
• Behavioral—refusal to feed across several toxins. • Cardiovascular—vasoconstriction associated with some toxins.
• Endocrine/Metabolic—ergotism.
• Gastrointestinal—primarily liver; some toxins affect proventriculus, gizzard, and spleen. Oral mucosal necrosis with Trichothecene toxicosis.
• Hemic/Immune—bone marrow suppression, microcytic anemia, hemorrhage, immune suppression particularly turkeys.
• Hepatobiliary—necrosis, fatty change, hemorrhage, bile duct proliferation; fibrosis in chronic cases. • Nervous—ophisthotonus, seizures with some toxins. • Renal—enlarged, pale, necrotic kidneys.
• Reproductive—decrease in egg production, decreased hatchability, decreased male fertility, delayed hen sexual maturity. • Skin—necrosis and ulcers in accessory dermic structures, toes.
• Neoplastic—unlike mammals, mycotoxins are not suspected to be a major cause of neoplastic disease in poultry. The question of chronic intoxication in pet birds which results in neoplasms, particularly of the liver, has not been subjected to long-term studies.

GENETICS
Susceptibility to intoxication varies with the wide range of toxins and subject species; however, with the most common toxicant group—Aflatoxins—ducklings, turkeys and pheasants are susceptible. Chickens, bobwhite/Japanese quail, chukar partridge and guinea fowl relatively resistant. Acute and chronic mycotoxicoses cases in pet birds have been documented or suspected.

INCIDENCE/PREVALENCE
Natural disease occurs in domestic poultry and waterfowl; free-living species worldwide. Clusters of mortality have been documented in free-living birds, when environmental conditions led to the availability of contaminated feedstuffs. Acute mycotoxicosis is probably most recognized in production facilities, due to active surveillance and diagnostic methods. Chronic intoxication is more difficult to document and may be an important factor in pet birds fed primarily a seed-based diet.

Modern management methods are designed to minimize the incidence of mycotoxicosis; however, these methods have not precluded significant "outbreaks" of morbidity and mortality in poultry. In pet birds, incidence of mycotoxicosis is poorly documented; however, sublethal levels of some common mycotoxins are commonly detected in feed and grain samples.

GEOGRAPHIC DISTRIBUTION
Should be considered worldwide, with variability on climactic, seasonal and management methods of feedstuffs. Similar distribution is possible in free-living species in nature or exposure to man-made mass storage of foodstuffs.

SIGNALMENT
• **Species:** Poultry, waterfowl, pet psittacines and passerines, zoological/wildlife species.
• **Species predilections:** It is difficult to generalize across many avian orders, as mycotoxicoses are best studied in production poultry and some wildlife species. • **Mean age and range:** All ages can be affected; however, this depends on the species/breed, toxin and the age of the subjects. • **Predominant sex:** There is no predominant gender predisposition.

SIGNS
Aflatoxicosis
Hepatic, immunologic, digestive and hematopoietic dysfunction and or pathology. Signs vary from general sick bird signs to high morbidity and mortality. Aflatoxins are carcinogenic, not a problem in poultry but may be in pet psittacines due to long-term aflatoxin exposure.
Ergotism
Chicks—discolored toes due to vasoconstriction. In older poultry, vasoconstriction affects the comb, wattles, face and eyelids. In canaries, partial loss of digits.

Trichothecene mycotoxicosis
Feed refusal, caustic injury of the oral mucosa and skin contacted skin, egg drop, depression, recumbency.
Waterfowl—pseudomembranous inflammation of the upper GI tract. Gross pathology can include GI mucosal reddening, hepatic mottling, splenic atrophy and visceral hemorrhages.
Ochratoxicosis
Severe intoxication can result in inactivity, huddling, hypothermia, diarrhea, rapid weight loss and death. Moderate intoxication results in weight loss, pigmentation, decreased egg production, reduced fertility and decreased hatchability. The toxin affects the kidney primarily but also liver, immune system and bone marrow.

Historical Findings
A history of exposure to wet, moldy, or spoiled feed may increase suspicion of mycotoxicosis.

Physical Examination Findings
No examination findings are diagnostic for mycotoxicosis. Clinical signs vary greatly, depending, on the type of toxin, the dosage of the toxin, the avian species involved and the organ systems affected.

CAUSES
Opportunistic or natural production of mycotoxins can occur when favored by environmental conditions.

RISK FACTORS
Feeding wet, spoiled or moldy feed.

DIAGNOSIS

DIFFERENTIAL DIAGNOSIS
Many noninfectious and infectious disorders due to the wide variety of clinical signs and lesions.

CBC/BIOCHEMISTRY/URINALYSIS
Blood panel values typically will not be helpful in the diagnosis of mycotoxicosis.

OTHER LABORATORY TESTS
Toxin assays are discussed under specific toxins. Feed analysis is generally more useful than organ or blood analysis for toxic levels.

IMAGING
Not useful except to possibly rule out other syndromes or assessment of organ size.

DIAGNOSTIC PROCEDURES
• Complete necropsy, including histopathology and microbiology. Be sure to hold back frozen tissues for virology or toxicologic testing, in a manner recommended by your diagnostic laboratory.
• Feed analysis for mycotoxins is often available from state veterinary laboratories. Consultation with the toxicologist is helpful in order to select appropriate assays. In many cases histopathologic changes will not be

Mycotoxicosis (Continued)

diagnostic of mycotoxicosis but some are suggestive and some are going to show minimal lesions due to a mycotoxin.
• Generally feed analysis is most helpful, compared to organ analysis. A qualitative mycotoxin (feed) screen is offered by the California Animal Health and Food Safety Toxicology Laboratory and includes the most important US livestock mycotoxins: aflatoxin B1, ochratoxin, fumonisins, zearalenone, DAS, and T-2. Your state veterinary diagnostic laboratory may offer a similar screen. http://www.aavld.org/accredited-laboratories

Toxin-associated laboratory diagnostic findings
• *Aflatoxicosis:* ○ Feed samples typically show elevation in Aflatoxin B1 or related aflatoxins. Organ levels of aflatoxin may or may not be present. ○ Microcytic hypochromic anemia, leukocytosis with concurrent lymphopenia. Decreased total protein, cholesterol, triglycerides, uric acid, calcium, phosphorus, zinc and LDH. Theoretically, increases in hepatic-associated enzymes may occur. Bile acids may elevate in chronic cases of hepatic fibrosis. Histopathology of hepatocytes includes fatty degeneration, necrosis, and hemorrhage. Bile duct proliferation appears rapidly with fibrotic changes in chronic suspected pet bird cases. Kidney: membranous glomerulopathy and interstitial fibrosis occurs. • *Ergotism:* Necropsy will show dermal necrosis, vesicles and ulcers; visceral congestion. Finding significant numbers of sclerotia of Claviceps purpurea microscopically in feed samples is suggestive of intoxication. Feed samples may contain ergot alkaloids. • *Trichothecene mycotoxicosis*: Toxic levels of trichothecenes (T-2, DAS, DON, nivalenol) in assayed feed samples are suggestive of intoxication. Histopathology or oral lesions include mucosal necrosis/ulceration; superficial crusts of exudate, bacterial colonies and submucosal inflammation. Systemic histopathology can include lymphoid/hematopoietic necrosis/depletion, liver and gall bladder necrosis and hemorrhage with bile duct proliferation. GI lesions include necrosis of mucosal epithelium of proventriculus, gizzard, and intestine. • *Ochratoxicosis:* Toxic levels of ochratoxin in assayed feed samples may suggest intoxication. Nonspecific elevations in plasma CK, amylase and uric acid may be observed associated with underlying pathology. Anemia, leukopenia and prolonged clotting time have been reported. Histopathologic lesions include acute tubular nephrosis with interstitial nephritis. Hepatocyte vacuolar change and necrosis is common. Lymphocyte depletion occurs in lymphoid organs.

PATHOLOGIC FINDINGS
Pathology varies among the selected toxins as listed above.

TREATMENT

APPROPRIATE HEALTH CARE
N/A

NURSING CARE
Supportive care.

ACTIVITY
N/A

DIET
Discontinue suspected contaminated feedstuffs and replace with new lots.

CLIENT EDUCATION
Bird keepers should be made aware of safe feed storage, acquisition and quality control measures.

SURGICAL CONSIDERATIONS
N/A

MEDICATIONS

DRUG(S) OF CHOICE
There are no specific antidotes to any of the mycotoxins, nor specific drugs to use. Therapy is focused on supportive measures and removing the suspected toxic foodstuff.

CONTRAINDICATIONS
Some mycotoxins cause marked hepatic and/or renal pathology that might temper the clinician's choices of supportive drugs.

PRECAUTIONS
N/A

POSSIBLE INTERACTIONS
N/A

ALTERNATIVE DRUGS
None.

FOLLOW-UP

PATIENT MONITORING
Clinical assessment of mentation, neurological signs, core temperature, hydration, fecal and urine output is recommended. Upon presentation a blood panel is recommended in order to rule out other causes of presented clinical signs. Subsequent blood panels may aid in assessment of patient organ systems.

PREVENTION/AVOIDANCE
• Producers should avoid any feed that is obviously tainted with mold growth; however mycotoxins can be present in the absence of visual changes. Feed assays should be part of a producer's supply chain. Producers and hobbyists should store feedstuffs in a way to control moisture content. In practical terms this includes protection from rain and vermin.
• Research with poultry, using various binders as feed additives, has attempted to address preventive measures for mycotoxin exposure. Overall, these binders are not very effective.

POSSIBLE COMPLICATIONS
Patients who survive acute mycotoxin exposure may suffer permanent organ dysfunction, particularly the liver. However numerous systems can be affected, depending on the toxin and the target species.

EXPECTED COURSE AND PROGNOSIS
As laboratory confirmation of specific mycotoxin contamination in feed or organ residues may take several days, suspected cases must be monitored until the course of the morbidity/mortality subsides. Removal of the suspected source of the mycotoxin(s) will often provide the most rapid course of clinical resolution.

MISCELLANEOUS

ASSOCIATED CONDITIONS
None.

AGE-RELATED FACTORS
Immature avian patients may be more susceptible to given mycotoxins.

ZOONOTIC POTENTIAL
None. Public health concerns can arise when an avian species raised for meat, accumulates sufficient mycotoxins chronically, that it poses a health risk to the human subject. This is particularly true for aflatoxins.

FERTILITY/BREEDING
Several mycotoxins can reduce male fertility, reduce egg production and reduce hatchability.

SYNONYMS
N/A

SEE ALSO
Anemia
Anorexia
Cardiac disease
Coagulopathies
Hemorrhage
Hepatic lipidosis
Infertility
Liver disease
Neurologic conditions
Renal disease
Seizure
Tumors

ABBREVIATIONS
N/A

INTERNET RESOURCES
http://portal.nifa.usda.gov/web/crisprojectpages/0217858-mycotoxinsbiosecurity-and-food-safety-nc129.html
http://www.knowmycotoxins.com/poultry.htm

http://fsrio.nal.usda.gov/pathogen-biology/natural-toxins/aflatoxins
http://www.cahfs.ucdavis.edu/services/toxicology.cfm

Suggested Reading
Hoerr, F.J. (2013). Mycotoxicoses. In: Swayne, D.E., Glisson, J.R., McDougald (eds), *Diseases of Poultry*, 13th edn. John Wiley & Sons Ltd, pp. 1271–1286.

Hoerr, F.J. (2014). Overview of mycotoxicoses in poultry. http://www.merckmanuals.com/vet/poultry/mycotoxicoses/overview_of_mycotoxicoses_in_poultry.html (last accessed September 28, 2015).

Puschner, B. (2002). Mycotoxins. *The Veterinary Clinics of North America. Small Animal Practice*, **32**(2):409–419.

Puschner, B. (2008). Mycotoxins: Moldy feed and forages. *Proceedings of the Western Veterinary Conference Annual Conference*.

Author Alan M Fudge, DVM, DABVP (Avian)

Acknowledgements The author acknowledges and thanks the following for manuscript review:
Robert E. Schmidt, DVM PhD DACVP
Birgit Puschner, DVM PhD DABVT

Neurologic Conditions

 BASICS

DEFINITION
Any condition that causes damage to the nervous system, via trauma, toxin, infection/inflammation, or organ failure.

PATHOPHYSIOLOGY
• Trauma to the nervous system results in injuries to intracranial or spinal structures such as cerebral lacerations from skull fractures, direct neuronal-axonal damage, and intracranial hemorrhage. Secondary injuries to neuronal tissues occur due to ATP depletion and the failure of autoregulation systems that help maintain adequate blood flow to the brain. • Toxicities: Organophosphates act as anticholinesterases, allowing excessive aceteylcholine at the neuromuscular junctions resulting in tremoring, weakness or paralysis. Heavy metals such as lead or zinc will damage the nervous system by interfering with calcium metabolism, cause vascular damage resulting in cerebral edema, neuronal necrosis, and demyelination. • Infectious diseases damage to the nervous system, usually by causing cell death. In some cases this is due to direct infection (e.g., *Chlamydia psittaci*, paramyxovirus-3), or indirectly by damage from the host immune system (avian bornavirus). Marek's disease virus (a herpesvirus of chickens) causes lymphoma that is tropic to nerves. • Atherosclerosis causes hypoxic damage to the CNS and secondary seizures. • Neoplasia causes space occupying lesions (e.g., Marek's disease, pituitary adenomas, cerebellar neuroectodermal tumors). • Seizures may result from nutritional deficiencies in Vitamins B1, B2, B6, B12, D3, E, as well as selenium, calcium, and phosphorus.
• Dyslipidemia/lipid emboli secondary to hyperestrogenism of reproductively active hens, characterized by excessive lipid circulating, causing vascular sludging and lipid emboli. The resulting hypoxic cell damage can lead to neurologic signs such as head tilt.

SYSTEMS AFFECTED
• Nervous.
• Musculoskeletal—paresis/paralysis.
• Cardiovascular—vascular damage by lead, or due to atherosclerosis.
• Gastrointestinal—regurgitation, undigested seeds, weight loss.

GENETICS/INCIDENCE/PREVALENCE
• African Grey parrots are prone to hypocalcemia as a cause of seizures. • Silky chickens are more susceptible to Marek's disease. • Budgerigars are prone to the formation of pituitary adenomas that may results in damage to the optic nerve.
• Epilepsy of Red-lored Amazons may have a genetic origin. • Amazons and African grey parrots may be more prone to atherosclerosis and secondary seizures. • Muscovy ducks may be particularly sensitive to duck viral enteritis, mallard ducks are more resistant to the disease.

GEOGRAPHIC DISTRIBUTION
Locations of the country where older houses are prevalent may experience higher rates of lead toxicosis due to the persistence of lead-based paints (e.g., Northeast United States).

SIGNALMENT
• **Species:** All avian species can be affected.
• **Species predilections**: African grey parrots, silky chickens, budgerigars, Red-lored Amazons, Muscovy ducks. • **Mean age and range:** Any age can be affected, but older patients may be more susceptible to atherosclerotic or neoplastic causes of neurologic signs. • **Predominant sex:** Reproductively active females are susceptible to dyslipidemia/lipid emboli, and will be more likely to succumb to hypocalcemia. Other causes of neurologic disease show no sex predilection.

SIGNS

Historical Findings
Owners may note acute or chronic progression of neurologic signs.

Physical Examination Findings
• Seizures • Blindness • Ataxia
• Paresis/paralysis • Changes in mentation or equilibrium • Muscle tremors or fasciculation
• Acute death with no clinical signs

Cranial nerve (CN) exam may reveal:
• CN I (Olfactory) deficits may result in altered appetite and eating habits. • CN II (Optic) damage will result in blindness, decreased or absent menace. • CN III (Oculomotor) dysfunction may result in ventrolateral deviation of globe, or drooping of the upper eyelid. Iris constriction may also be affected, but less so due to the presence of striated muscle within the iris. • CN IV (Trochlear) dysfunction will lead to dorsolateral deviation of the globe. • CN V (Trigeminal) has three branches with sensory and motor functions. Dysfunction of the ophthalmic branch will show loss of sensation around the face and beak. Dysfunction of the maxillary branch will lead to loss of sensation of the face, beak, and/or mouth/palate. Dysfunction of the mandibular branch will lead to decreased bite strength and closing, and inability to close eye. • CN VI (Abducens) deficits will result in inability to move the nictitating membrane (third eyelid). • CN VII (Facial) deficits will result in lack of response to facial sensation and possible facial droop or asymmetry. Taste sensation is also innervated by CN IX, and may be difficult to evaluate in birds. • CN VIII (Vestibulocochlear) deficits will result in hearing loss, and/or loss of equilibrium (nystagmus, head tilt or circling toward the affected side). • CN IX (Glossopharyngeal) and CN X (Vagal) nerves anastomose in birds and are evaluated together. Deficits will result in lack of tongue sensation, dysphagia, regurgitation, voice changes or loss. • CN XI (Accessory) dysfunction may result in poor neck mobility. • CN XII (Hypoglossal) dysfunction will result in deviation of the tongue toward the affected side, decreased mobility or tone, or change in or absence of voice • Eye movement is the result of interactions with a variety of cranial nerves (III, IV, VI and VIII), the brainstem and the cerebellum. Problems with any of these nerves may lead to strabismus. • The palpebral reflex can be evaluated to assess both CN V and CN VII. • CN V, IX, X, XI, and XII control tongue movement, beak strength and swallowing.

Postural/gait examination findings:
• Vestibular or cerebellar lesions may result in postural changes like head tilts, or twisting of the body, or intention tremors. Central vestibular signs may also be associated with changes in mental status, postural reaction deficits, or vertical nystagmus. • Decreased or asymmetric wing retraction or grip strength.
• Crossed extensor reflexes are associated with upper motor neuron lesions and can be an indicator of severe spinal cord lesions.
• Knuckling over, or a loss of conscious proprioception, indicates that there is a lesion in the peripheral and/or central nervous system. • Dysfunction of the pudendal nerves results in decreased cloacal reflex. • Muscle atrophy.

CAUSES
• Toxicities such as organophosphates, heavy metals (see lead toxicity section).
• Infectious/inflammatory: ○ Viral: Proventricular dilatation disease (PDD, a bornavirus), Marek's disease (herpesvirus in poulty), Newcastle's disease (paramyxovirus), Duck Viral Enteritis (herpesvirus in waterfowl), and Duck Viral Hepatitis (picornavirus). ○ Bacterial: Botulism caused by *Clostridium botulinum* enterotoxin, chlamydiosis caused by *Chlamydia psittaci*, general septicemia. ○ Fungal, such as aspergillosis. ○ Parasitic (aberrant larval migrans with *Baylisascaris*, toxoplasmosis, sarcocystosis, etc.). ○ Yolk emboli secondary to egg yolk ceolomitis. • Neoplasia (e.g., pituitary adenomas, neuroectodermal tumor in the cerebellum). • Trauma (see trauma section). • Dietary or nutritional:
○ Hypocalcemia (especially in African grey parrots). ○ Specific vitamin, calcium, phosphorus, or selenium nutritional deficiencies in young psittacines. • Metabolic causes, such as hepatic encephalopathy, though controversy exists as to whether hepatic encephalopathy can occur in birds, as nitrogenous waster products are not the end product in birds as it is in mammals. Neurologic signs can occur with severe

Neurologic Conditions (Continued)

hepatic disease. • Atherosclerosis. • Lafora body neuropathy— reported in cockatiels. • Primary epilepsy.

RISK FACTORS
• Supervised or unsupervised time outside of a cage that might allow exposure to toxins or injuries. • Exposure to other birds or animals. • Nutritionally inadequate diet. • Reproductive activity.

DIAGNOSIS

DIFFERENTIAL DIAGNOSIS
• Any disease or trauma that leads to depression, ataxia or muscle weakness. • Myopathy/muscular dystrophy-like syndrome (rare).

CBC/BIOCHEMISTRY/URINALYSIS
CBC
• A marked leukocytosis, often with a monocytosis, is sometimes seen in chlamydiosis. • Anemia may be seen with lead toxicosis due to increased red cell fragility.

Biochemistry Profile
• Decreased total blood calcium levels may be seen with hypocalcemia. • Elevations in AST and/or bile acids may be elevated in hepatic encephalopathy. • Cholesterol/triglycerides may be elevated in atherosclerosis and dyslipidemia syndrome.

Urinalysis
• Hematuria may be seen with lead toxicosis, particularly in amazon parrots.

OTHER LABORATORY TESTS
• Ionized calcium levels may be useful in suspect cases when total calcium is normal but history or clinical signs warrant further investigation for hypocalcemia. • Avian blood lead levels: >10 mg/dL are suggestive of lead toxicosis in the presence of clinical signs. >20 mg/dL are considered diagnostic. • Avian serum zinc levels: Elevated levels suggest zinc toxicosis. • Avian bornavirus PCR testing may be helpful in birds showing clinical signs. • Chlamydiosis PCR testing for the presence of *Chlamydia psittaci*. • Electromyography (EMG) values can determine the health of the muscle cell and the integrity of the motor unit (lower motor neuron and associated muscle fibers) to differentiate peripheral nerve damage versus primary muscle disease.

IMAGING
Radiographs
• Metallic densities consistent with heavy metal toxicity. • Fractures of the spine or skull suggesting a traumatic cause to neurologic signs. • Dilated proventriculus/ventriculus in the case of PDD.

CT or MRI
• Useful in evaluating structures within the head and spine, space occupying lesions, areas of trauma or other abnormalities.

DIAGNOSTIC PROCEDURES
Histopathology (crop/GI biopsy or necropsy) remains the most definitive diagnostic tool in PDD.

PATHOLOGIC FINDINGS
PDD:
• Gross examination reveals dilation of the proventriculus, the presence of undigested food in the colon or cloaca, and poor body condition. Histologic examination reveals accumulation of lymphocytes and plasma cells in association with nerves of the gastrointestinal tract or central nervous system.

Marek's Disease, a herpesvirus:
• Gross examination reveals distended peripheral nerves that are edematous or grayish in appearance. Often the sciatic nerves are affected first. Grey nodules may also be present on the bursa, thymus, ovary, kidney, spleen, skin, liver, and skeletal muscle. The bursa may be atrophied. Histologic findings reveal neoplastic lymphocytes in the peripheral nerves.

Duck Viral Enteritis, a herpesvirus:
• Hemorrhage and erosions under the tongue, esophagus, intestinal mucosa and cloaca. Annular rings of hemorrhage may be present in the intestines. Intranuclear inclusion bodies may be seen in liver and areas of the gastrointestinal mucosa.

Newcastle's disease, a paramyxovirus:
• Hemorrhage throughout the GI tract and tracheitis. Pericardial effusion and cardiomegaly may be seen. Inflammation of the kidneys and brain is seen microscopically.

Lead toxicosis:
• Cerebral edema from vascular damage, neuronal necrosis and axonal degeneration and demyelination. Renal nephrosis may be present, and multifocal myocardial degeneration may be seen.

TREATMENT

APPROPRIATE HEALTH CARE
• Identify and treat the underlying condition. • Inpatient treatment is often required cases of ongoing seizure activity, anorexia, lethargy, recumbency, dysphagia, ataxia, or coma. • Hypoglycemia, even after seizure activity, is rare, but normoglycemia must be maintained. • Gastric lavage, endoscopic or surgical removal of lead foreign bodies.

NURSING CARE
• Chelation for heavy metal toxicosis. • Fluid therapy if unable to maintain hydration. This may be delivered via SC, IV or IO routes. • Nutritional support via gavage or feeding tube if the patient is unable to feed itself. • A padded cage may need to be provided if the bird is unable to stand, ataxic, or seizures repeatedly.

ACTIVITY
If the patient is weakened or has difficulty climbing, then perches, food and water sources may need to be placed lower in the cage to allow the patient access.

DIET
In cases where the patient is eating, it is advisable to offer the patient's regular diet, even if it is substandard to assure that the patient continues to eat. Mineral and vitamin deficiencies may be supplemented orally or parenterally. When the patient has recovered, transitioning to a better balanced diet may be performed. If the patient is unable to feed itself, nutritional support via gavage or feeding tube should be performed using a resuspended balanced hand feeding formula.

CLIENT EDUCATION
• Exposure to sources of infection, toxins, or trauma need to be eliminated. • If demyelination caused by lead toxicosis becomes permanent, then prognosis for return to function is poor. Lead toxicosis, if chronic, may need to be treated repeatedly due to the deposition of lead in bones.

SURGICAL CONSIDERATIONS
• Placement of a ventricular feeding tube via ingluviotomy may be indicated for greater ease of assisted feeding during prolonged recoveries. • Removal of heavy metal foreign bodies via gastric lavage, endoscopy or surgery.

MEDICATIONS

DRUG(S) OF CHOICE
• **For controlling seizures:** ○ Midazolam 0.5–1.0 mg/kg IM. May be repeated up to three times if necessary. May also be administered intracloacally—use double the dose for this. ○ Diazepam may also be administered via continuous intravenous or intraosseous infusion at 1 mg/kg/h if further seizure control is needed. Once seizures have been controlled for 12–24 hours, the diazepam is tapered to zero over the next 12–24 hours. If further seizure control is needed, phenobarbital may be infused at 2–10 mg/kg/h. Once the seizures have been stopped, phenobarbital may be used for long term seizure control at 2–7 mg/kg PO q12h. ○ Levetiracetam (Keppra) 60 mg/kg IM; can use for long-term seizure control at dosing up to 100 mg/kg PO q8h. • **Cerebral edema:** ○ Mannitol 0.5–1.0 grams/kg IV over 20 minutes. This may be repeated up to three times in a 24-hour period. • **Chlamydiosis:** ○ Doxycycline mg/kg PO q or 75–100 mg/kg IM q5–7 days for a minimum of 45 days, or 25 50 mg/kg PO q24h for 45 days. • **PDD:** ○ Currently, treatment with an NSAID has shown some success at reducing clinical signs.

Meloxicam 0.5 mg/kg PO q12–24h. Treatment is usually life-long. • **Lead toxicity:** see "Heavy metal toxicosis". • **Hypocalcemia:** see "Hypocalcemia/hypomagnesemia". • **Dyslipidemia syndrome/lipid emboli:** see "Dyslipidemia/Hyperlipidemia". ○ Dilution of the vascular sludge with crystalloid diuresis is a mainstay. ○ Cholesterol lowering statin drugs may be compounded and tried in birds. These include: ▪ Atorvastatin 3 mg/kg q24h. ▪ Simvastatin 1 mg/kg q24h. ▪ Lovastatin 6 mg/kg q24h.

CONTRAINDICATIONS
Avoid steroid use in birds with neurologic disease. Birds seem to be more sensitive to the immunosuppressive effects of steroids, and this might worsen clinical signs, particularly if the cause is infectious. Steroid use may also lead to hyperglycemia, which is shown to have a poorer prognosis in head trauma.

PRECAUTIONS
Phenobarbital may not reach therapeutic levels and may be ineffective to control seizures in some avian species (i.e., African greys).

POSSIBLE INTERACTIONS
N/A

ALTERNATIVE DRUGS
• For seizure control, gabapentin has not been investigated scientifically, but has been used at a dose of 20 mg/kg PO q12h.
• Zonisamide may be helpful (use concurrently with levetiracetam) at a dose of 10–20 mg/kg PO q12h. • For PDD, some clinicians prefer celecoxib 10 mg/kg PO q24h.

 FOLLOW-UP

PATIENT MONITORING
• Successive neurologic exams essential to monitor response to treatment and prognosis.
• If seizures are being controlled with phenobarbital, blood levels should be checked 2–3 weeks after starting therapy. • If seizures are being controlled with levetiracetam and zonisamide, blood levels can be monitored to verify therapeutic (based on humans) dosing.
• If lead toxicity has been chronic, or if environmental exposure cannot be reliably limited, periodic testing of blood lead levels is recommended.

PREVENTION/AVOIDANCE
Eliminating or limiting exposure to toxins, sources of trauma (e.g., ceiling fans) or infectious agents that cause neurologic disease. Screening birds for *Chlamydia psittaci* prior to introducing them into flocks can reduce the risk.

POSSIBLE COMPLICATIONS
Some neurologic signs may be permanent, and may persist even after treatment.

EXPECTED COURSE AND PROGNOSIS
Varies with the cause. PDD and many other viral disease carry a guarded to poor prognosis. Treating lead toxicosis often carries a fair to good prognosis. Trauma is variable.

 MISCELLANEOUS

ASSOCIATED CONDITIONS
N/A

AGE-RELATED FACTORS
N/A

ZOONOTIC POTENTIAL
Chlamydiosis and Newcastle disease are reportable diseases. They are also zoonotic. Chlamydiosis can cause a serious respiratory disease, and Newcastle disease can cause conjunctivitis in people.

FERTILITY/BREEDING
N/A

SYNONYMS
N/A

SEE ALSO
Atherosclerosis
Botulism
Heavy metal toxicity
Dyslipidemia/hyperlipidemia
Hypocalcemia/hypomagnesemia
Liver disease
Mycotoxicosis
Organophosphate/carbamate toxicity
Otitis
Paramyxoviruses
Pituitary/brain tumors
Proventricular dilatation disease
Seizure
Sick-bird syndrome
Trauma
Tumors
Viral disease
Appendix 7, Algorithm 9. Neurologic signs

ABREVIATIONS
AST—aspartate aminotransferase
ATP—adenosine triphosphate
CN—cranial nerve
CT—computed tomography
EMG—electromyography
GI—gastrointestinal
IM—intramuscular
IV—intravenous
MRI—magnetic resonance imaging
NSAID—nonsteroidal anti-inflammatory drug
PCR—polymerase chain reaction
PDD—proventricular dilatation disease
PO—per os, i.e. by mouth

Suggested Reading:
Antinoff, N., Orosz, SE. (2012). Don't be nervous! A Clinician's approach to the avian neurologic system. *Proceedings of the Association of Avian Veterinarians Annual Conference*, Louisville, KT.
Clippinger, T.L., Bennett, R.A., Platt, S.R. (2007). The avian neurologic examination and ancillary neurodiagnostic techniques: A review update. *The Veterinary Clinics of North America. Exotic Animal Practice*, **10**(3):803–836.
Johnson-Delaney, C.A., Reavill, D. (2009). Toxicoses in birds: Ante- and postmortem findings for practitioners. *Proceedings of the Association of Avian Veterinarians Annual Conference*, Milwaukee, WI.
Orosz, S.E., Bradshaw, G.A. (2007). Avian neuroanatomy revisited: From clinical principles to avian cognition. *The Veterinary Clinics of North America. Exotic Animal Practice*, **10**(3):775–802.

Author Stephen Dyer, DVM, DABVP (Avian)
Acknowledgement Erika Cervasio, DVM, DABVP (Avian)

Nutritional Imbalances

BASICS

DEFINITION
Nutritional deficiency can be defined as an insufficient intake of a bioavailable nutrient essential to support a given physiological state.

There are approximately 30 essential nutrients in birds; however, the requirements vary greatly between species. Theoretically, a deficiency state is possible for any essential nutrient; however, some deficiencies have only been experimentally produced, whereas others are commonly seen in practice. Deficiencies are more likely to have a clinical manifestation during times of growth, reproduction, stress, or disease; however, birds that appear healthy to the owner may have subclinical deficiencies that may increase the risk of morbidity or they may have mild signs. A classic example of a deficient diet in pet birds is a seed-only diet. This type of diet is deficient in many essential nutrients including total protein, vitamin A, vitamin D, calcium, and phosphorous.

PATHOPHYSIOLOGY
The predominant cause of nutritional deficiencies is poor diet. Less commonly primary malabsorption may also cause nutritional deficiencies. An all seed diet, unbalanced home-prepared diets, and poor quality commercial feeds have all been demonstrated as causes of nutritional imbalances. As many nutrients are required for energy metabolism, a high fat or calorically dense diet may lead to relative deficiencies. Each deficiency may cause characteristic pathology; for instance, iodine deficiency will affect thyroid function and may cause goiter, whereas vitamin A deficiency may cause squamous metaplasia affecting the epithelium and mucus membranes.

SYSTEMS AFFECTED
Any system may be affected depending on the type of deficiency or imbalance, common examples include: • Behavioral—changes in vocalization patterns have been observed in cockatiels fed a vitamin A/carotenoid deficient diet. • Endocrine/Metabolic—iodine deficiency may cause goiter in budgerigars; however this is not usually accompanied by decreased thyroid hormones in circulation.
• Gastrointestinal—vitamin A deficiency may cause squamous metaplasia which may manifest as oral plaques or a decrease in number or appearance of the choanal papillae (species dependent). • Hemic/Lymphatic/Immune—vitamin A and vitamin E are known in many species to have a role as immune modulators and deficiencies in these vitamins may exacerbate other morbidities.
• Hepatobiliary—biotin deficiency may cause hepatic lipidosis in cockatiels.
• Musculoskeletal—vitamin D, calcium, and phosphorus deficiencies or imbalances may cause metabolic bone disease and developmental pathology. • Nervous ○ Thiamin deficiency may cause neurological signs. Most commonly seen in pescivorous birds that have ingested thiaminase-containing fish. ○ Vitamin E deficiency may cause encephalomalacia and generalized neurological signs. • Neuromuscular ○ Calcium deficiency may cause seizures or seizure-like muscle tremors; African grey parrots appear to be particularly susceptible to this. ○ Magnesium deficiency may exacerbate clinical signs associated with calcium deficiency. ○ Vitamin E deficiency may cause torticollis. • Ophthalmic—vitamin A deficiency may cause pathologic keratinization of the conjunctival mucus membranes, swelling, irritation, severe pain, and in advanced cases inability to open the eyes. • Renal/Urologic—vitamin A deficiency may cause squamous metaplasia affecting the renal tubules and subsequently renal disease. • Reproductive ○ Vitamin D and/or calcium deficiencies may predispose females to dystocia or egg binding and abnormal shell production. ○ Vitamin A deficiency may cause embryonic death or congenital defects.
• Respiratory ○ Vitamin A deficiency may cause crusting around the nares or cere. ○ Vitamin A deficiency may predispose birds to secondary respiratory infection. • Skin/Exocrine ○ Vitamin A deficiency may cause poor plumage, decreased coloring (as carotenoids play a part in synthesis of some pigments). ○ Vitamin A deficiency may cause uropygial gland (preen gland) disorders such as inflammation and impaction. ○ Protein deficiency may cause an abnormal and thin plumage. Specifically methionine deficiency has been associated with streaks across the feathers known as stress bars.

GENETICS
N/A

INCIDENCE/PREVALENCE
The prevalence of nutritional imbalance is still considered high, despite increased awareness to the importance of incorporation of formulated pelleted diet in pet bird nutrition. Wild birds nursed by good Samaritans may also present with nutritional deficiencies.

GEOGRAPHIC DISTRIBUTION
N/A

SIGNALMENT
• **Species:** Any species may be affected; however deficiencies are less likely in birds that are allowed to roam freely and select food on their own (such as roaming wild birds, backyard poultry, or waterfowl). • **Mean age and range:** Any age could be affected, growing animals and reproducing animals are more vulnerable. • **Predominant sex:** Females are more likely to be affected during reproduction. Otherwise, both sexes are equally at risk.

SIGNS
Historical Findings
Comprehensive diet history should always be included as part of the medical history. This should include type of diet fed, amounts offered and consumed, and frequency and type of treats provided. Other pertinent information regarding previous diets, length of time fed current diet and any information regarding body weight fluctuations should also be noted.

Physical Examination Findings
• Obesity. • Failure to thrive. • Blunted choanal papillae. • Poor body condition.
• Poor plumage. • Neurological signs.
• Weakness/tremors. • Limb deformity.

CAUSES
Nutritional deficiencies are the result of an imbalanced diet, most commonly an all-seed diet or an imbalanced home-prepared diet that consists of human food items. Vitamin E and thiamin deficiencies have been described in pescivorous birds as fish may contain thiaminase that breaks down thiamin and fatty fish contain high levels of lipids which may become oxidized and reduce availability of vitamin E.

RISK FACTORS
History of an imbalanced diet; all seed diets, unbalanced home-cooked diets, and inappropriate diets for a reared wild bird are all possible risk factors. High fat may increase the risk for nutritional deficiencies if diets are not balanced.

DIAGNOSIS

DIFFERENTIAL DIAGNOSIS
• Dermatological pathologies resulting from vitamin A deficiency may need to be differentiated from neoplastic, traumatic, or infectious processes depending on appearance and location. • Seizures and neurological abnormalities caused by calcium deficiency or thiamin deficiency need to be differentiated from neurological disease associated with trauma, infection, inflammation, or idiopathic processes. • Skeletal pathology caused by calcium, phosphorous or vitamin D deficiency should be differentiated from congenital abnormalities, infectious diseases causing abnormal growth, protein/energy malnutrition, or trauma.

CBC/BIOCHEMISTRY/URINALYSIS
CBC
• Deficiencies of folate or iron may cause nonregenerative anemias; however, these are not commonly seen. • Riboflavin and pyridoxine deficiencies may cause an increase in heterophil counts and a decrease in lymphocyte counts.

NUTRITIONAL IMBALANCES (CONTINUED)

Biochemistry
- Hypocalcemia may be seen in birds on a calcium deficient diet; however, normal total calcium does not rule out hypocalcemia and ionized calcium measurement may be indicated to assess the bird's calcium status.
- Hypomagnesemia may accompany hypocalcaemia. • Hyperuricemia may result from kidney disease secondary to vitamin A deficiency.

OTHER LABORATORY TESTS
Specific tests are available to diagnose specific nutritional deficiencies—for example, thiamin deficiency may be diagnosed by decreased serum transketolase activity. However, common deficiencies such as vitamin A deficiency cannot be easily diagnosed using blood tests alone.

IMAGING
Skeletal pathology resulting from vitamin D, phosphorous, or calcium deficiencies may be viewed on radiographs. Coelomic ultrasound may be beneficial in evaluating parenchymatic changes associated with hepatic lipidosis due to biotin deficiency or for the diagnosis of hepatopathies associated with vitamin A, copper, or iron accumulation.

DIAGNOSTIC PROCEDURES
- Hepatic biopsies are used for quantification and diagnosis of vitamin A deficiency or toxicity. • Skin biopsies may be useful in diagnosing vitamin A deficiency, which characteristically manifests as squamous metaplasia. • Deficiencies in B vitamins may be detected with high performance liquid chromatography (HPLC), for instance for detection of thiamin deficiency. • Thiamin deficiency may also be diagnosed with an erythrocyte transketolase assay. • Low or borderline-low plasma ionized calcium may be a finding in birds with calcium or severe vitamin D deficiency. • Response to therapy may also be the diagnostic method employed in some cases where the index of suspicion for a nutritional deficiency is high—for example, resolution of goiter in response to correction of iodine intake or improvement in ophthalmic or respiratory signs in response to an improved diet with sufficient vitamin A.

PATHOLOGIC FINDINGS
Vitamin A deficiency may cause pathology affecting epithelial tissues throughout the body; specifically, epithelial turn over and keratinization. These nonspecific changes are called squamous metaplasia. Pathology associated with calcium, phosphorous or vitamin D deficiency includes skeletal changes, which are covered under metabolic bone disease in this book. Postmortem findings in birds with severe nutritional imbalances are beyond the scope of this section and may be found elsewhere (please refer to the suggested reading list).

TREATMENT

APPROPRIATE HEALTH CARE
Most often patients diagnosed with nutritional deficiencies may be managed as outpatients with recommendations to correct the diet and additional supplementation if needed. Vitamin A deficiency causing respiratory disease, seizures due to hypocalcaemia, severe respiratory distress due to goiter, and dystocia due to calcium deficiency are some examples of nutritional deficiencies that need to be addressed urgently with supportive care in parallel with dietary correction and supplementation.

NURSING CARE
Nursing care is required in cases of birds presenting with acute clinical signs as described above. Oxygen support, medical seizure management, and fluid therapy may be needed in cases of acute respiratory distress or seizures. Calcium by injection may help with seizure control in birds presenting with acute hypocalcaemia and may also assist in dystocia with a possible nutritional background. Budgerigars with goiter due to iodine deficiency presenting with acute respiratory distress may need oxygen support. Injectable glucocorticoids may be used as emergency treatment to relieve goiter associated dyspnea.

ACTIVITY
There is no need for activity alteration in most cases.

DIET
Unbalanced diet should be addressed to resolve most nutritional deficiencies. Current recommendations for psittacine diets are that 50–75% of the daily caloric intake to be provided by commercial pelleted diets and the rest from a selection of fruits, vegetables, and small amounts of seeds, grains and nuts. The nutritional content of common food items may be found on nutrient databases online such as the USDA database.

CLIENT EDUCATION
Client education is key for the management of poor nutrition. The importance of balanced nutrition should be discussed as part of a wellness examination or a new bird examination. For pet psittacines, directions should be provided to allow the owner to select from good quality commercial formulated pelleted diets. Providing the bird several diets as options may be a good way to find a preferred diet and make dietary transition an easier process. An appropriate treat/produce allowance should also be included.

SURGICAL CONSIDERATIONS
N/A

MEDICATIONS

DRUG(S) OF CHOICE
- Vitamin and mineral supplementation may be included in the management of nutritional deficiencies especially in acute cases or short term while the patient is transitioned to an appropriate diet. • Calcium gluconate (10%): 5–100 mg/kg IM/slow IV, may be given to psittacines or raptors in acute hypocalcaemia. May cause arrhythmias; therefore, monitor heart rate while administering. • Magnesium sulfate: 20 mg/kg IM single treatment (or as needed) for hypomagnesemia. • Iodine (sodium iodide 20%): 2 mg (0.01 mL)/bird IM as needed. • Vitamin A: 20,000–30,000 IU/kg IM. • Thiamine: 1–3 mg/kg IM every seven days. • Vitamin D3: 3300 U/kg IM every seven days as needed. Beware of excessive use leading to hypervitaminosis D. Moderate sunlight exposure (11–30 minutes per day, avoiding overheating) may be used in conjunction or as replacement.

CONTRAINDICATIONS
None.

PRECAUTIONS
None.

POSSIBLE INTERACTIONS
N/A

ALTERNATIVE DRUGS
N/A

FOLLOW-UP

PATIENT MONITORING
Monitoring of response to treatment and dietary modification is important to assess accuracy of diagnosis and owner compliance. Diet transition may be challenging in many cases and owners can find the process exasperating without proper guidance.

PREVENTION/AVOIDANCE
Providing a balanced diet and avoiding all-seed or home-prepared unbalanced diets.

POSSIBLE COMPLICATIONS
Changes to the respiratory and oral epithelium may cause ulcerations and secondary infections.

EXPECTED COURSE AND PROGNOSIS
Most clinical signs described above are expected to resolve with appropriate nutrition. Some anatomic changes associated with chronic deficiencies may be permanent.

MISCELLANEOUS

ASSOCIATED CONDITIONS
N/A

Nutritional Imbalances (Continued)

AGE-RELATED FACTORS
Growing animals and reproducing females are generally more susceptible to clinical manifestation. It is also hypothesized that geriatric animals in general have altered nutritional requirements and may be more prone to clinical deficiencies.

ZOONOTIC POTENTIAL
None.

FERTILITY/BREEDING
Reproducing females have higher energy and nutrient demands. Calcium metabolism is paramount for egg shell deposition, embryonic tissues, and contractions during egg laying; therefore, deficiencies in either calcium or vitamin D may predispose a bird to egg binding or dystocia. Vitamin A is associated with successful breeding; therefore, it is necessary to ensure that the reproductive female is fed an appropriate diet even prior to breeding.

SYNONYMS
Nutritional deficiencies, nutritional insufficiency, dietary imbalance

SEE ALSO
Chronic egg laying
Dystocia/egg binding
Hypocalcemia/hypomagnesemia
Infertility
Lameness
Iron storage disease
Metabolic bone disease
Obesity
Oral plaques
Overgrown beak and nails
Pododermatitis
Rhinitis/sinusitis
Seizure
Thyroid diseases
Trauma
Vitamin D toxicosis

ABBREVIATIONS
N/A

INTERNET RESOURCES
http://ndb.nal.usda.gov/ndb/search/list (USDA Nutrient Database for Standard Reference).
http://nagonline.net/ (Zoo and Aquariums Nutrition Advisory group).

Suggested Reading
Brightsmith, D.J. (2012). Nutritional levels of diets fed to captive Amazon parrots: does mixing seed, produce, and pellets provide a healthy diet?. *Journal of Avian Medicine and Surgery*, **26**(3):149–160.
Fidgett, A.L., Gardner, L. (2014). Advancing avian nutrition through best feeding practice. *International Zoo Yearbook*, **48**(1) 116–127.
Hess, L., Mauldin, G., Rosenthal, K. (2002). Estimated nutrient content of diets commonly fed to pet birds. *The Veterinary Record*, **150**:399–404.
Koutsos, E.A., Matson, K.D., Klasing, K.C. (2001). Nutrition of birds in the order Psittaciformes: a review. *Journal of Avian Medicine and Surgery*, **15**(4):257–275.
Schmidt, R.E., Reavill, D.R., Phalen, D.N. (2003). *Pathology of Pet and Aviary Birds*. 2nd edn. Ames, IA: Blackwell Publishing.
Author Jonathan Stockman, DVM, DACVN

 Client Education Handout available online

Obesity

BASICS

DEFINITION
Birds whose weight exceeds their optimal weight by more than 20% are considered obese. Obesity is the most common nutritional disease of pet birds.

PATHOPHYSIOLOGY
Obesity occurs when energy intake chronically exceeds energy expenditure. This may occur due to lower metabolic requirement of some species, increased efficiency of energy absorption, decreased energy expenditure, or increased intake usually due to high calorie diets, poor nutritional content, or poor biofeedback of satiation.

SYSTEMS AFFECTED
• Dermatological—lipomas, xanthomatosis, pododermatitis. • Reproductive—infertility, dystocia. • Cardiovascular—atherosclerosis, heart failure. • Hepatic—hepatic lipidosis. • Respiratory—air sac impingement by visceral fat. • Gastrointestinal—lipomas and fatty deposits can cause crop impingement leading to regurgitation. • Endocrine—hypothyroidism.

GENETICS
There may be a genetic basis for obesity.

INCIDENCE/PREVALENCE
Common in captive species with high caloric intake.

GEOGRAPHIC DISTRIBUTION
Worldwide among captive avian species.

SIGNALMENT
• **Species:** ○ Common in many cage birds. ○ Amazons, budgerigars, quaker parrots, rose-breasted and sulfur-crested cockatoos, pigeons, and waterfowl are particularly susceptible. ○ Risk may increase with age and subsequent decrease in metabolic rate. • **Mean age and range:** N/A • **Predominant sex:** Both sexes equally susceptible.

SIGNS

Historical Findings
• Exercise intolerance. • Seed-based diet (especially sunflower, safflower, hemp, rape, niger, spray millet). • Other high fat diet (peanuts, peanut butter, cheese, eggs). • Limited time out of cage or sedentary behavior outside cage.

Physical Examination Findings
• Pectoral soft tissue may bulge bilaterally to keel. • Excessive fat in abdomen, axillae/flank, and subcutis. • Excessive caudal coelomic fat can cause abdominal distention. • Skin may have yellowish tint either from underlying fat or xanthomatosis. • Lipomas may be present, particularly over cranial sternum and crop area. Skin over lipomas may be xanthomatous, ulcerated, or focal necrosis may occur. • Keel bone may be difficult to palpate. • Plumage may appear to have bald patches due to separation of featherless tracts due to subcutaneous fat. • A wide based stance may be present. • Dyspnea due to compression of abdominal and caudal thoracic air sacs by visceral fat (especially Amazon parrots, budgerigars, cockatiels, Pionus parrots).

CAUSES
• Excess consumption of foods which are high in simple starches or fats. • Excess consumption of deficient diets to satisfy nutritional demands. • Behavioral polyphagia secondary to boredom. • Lower energy requirements in some species (budgerigars, quakers, and rose-breasted cockatoos are suspected examples). • Lack of sufficient exercise either through behavioral tendencies (Amazon parrots) or husbandry deficiency. • Hypothyroidism is sometimes speculated to be the underlying cause of obesity, but is difficult to confirm using laboratory testing, and may be over-diagnosed.

RISK FACTORS
See "Signalment and Species".

DIAGNOSIS

DIFFERENTIAL DIAGNOSIS

Differentiating Similar Signs
• Sick birds may present with their plumage fluffed in such a way as to appear to have more bulk than normal. Palpation of the keel and pectoral muscles will distinguish the obese bird from the sick bird, which is often underweight. • Caudal coelomic distention can also be caused by organomegaly, masses, or ascites.

Differentiating Causes
• An exhaustive history of diet both offered and consumed should be taken, including all treats, as well as frequency and amounts fed, in order to elucidate if and how excess calories are provided. • A thorough history of opportunities for activity should be taken, including cage size, time and activities outside cage, and access to swings and ladders.

CBC/BIOCHEMISTRY/URINALYSIS
• Complete blood count is typically normal. • Biochemistry panel may be normal or may show elevations in liver enzymes or hypercholesterolemia. Lipemia may be present, and may indicate hepatic lipidosis even without elevations in liver enzymes.

OTHER LABORATORY TESTS
• Cholesterol may need to be added separately to some biochemical panels. • Persistently elevated cholesterol may warrant a lipoprotein electrophoresis. • Serum thyroxine levels and TSH stimulation testing is needed to confirm suspected hypothyroidism.

IMAGING
• Obesity may complicate radiography due to loss of caudal coelomic detail, and increased soft tissue density overlying the air sacs. • The liver silhouette is frequently enlarged, and the cardiohepatic "waist" may be obliterated.

TREATMENT

APPROPRIATE HEALTH CARE
• Once a patient is determined to be in adequate health by physical examination and CBC/biochemistry panel, decreasing dietary caloric content and increasing activity is usually the most effective treatment for obesity. • Liver disease, kidney disease, or concurrent infection should be addressed prior to initiating changes in diet or activity.

NURSING CARE
• Weight loss should be monitored closely so as not to exceed 3% of body weight per week. • More rapid weight loss in obese birds may result in hepatic lipidosis.

ACTIVITY
• Activity may be increased by rearranging perching and food and water dishes to maximize flight, ambulation, and effort. • Cage furniture such as swings and ladders can encourage physical activity. • Increasing cage size may benefit some birds. • Time outside the cage can be increased to encourage flying, climbing, and play. • Many birds (particularly Amazons, budgerigars, cockatiels, and cockatoos) enjoy a variety of music, and can be encouraged to bob and "dance" with their owners to music the bird likes. This may be of most use for very overweight and sedentary birds to begin increasing activity without over-stressing extremely unfit individuals. • Increased frequency of bathing can add flapping, feather shaking, and active grooming. • Increased diversity of activities in addition to providing foraging activities may help decrease polyphagia secondary to boredom.

DIET
• Foraging toys and techniques can help delay rate of calorie intake which is known to be more rapid in captivity than in the wild. • Although dietary improvements may be needed, changes should be gradual to prevent anorexia due to aversion or non-recognition of the novel food item. • Reductions may be needed in amount of food, fat content of food, and overall dietary composition. • An ideal diet is predominantly a commercial pellet, approximately 20% fruit and vegetables, and no greater than two tablespoons of seed and nuts. (Exceptions: macaws requiring a greater amount of oil in their diet, and cockatiels consuming greater than 50% pellets may be at risk for renal dysfunction.) • Transitioning a bird to a

(CONTINUED)

 OBESITY

pelleted diet can take many months, particularly if the pelleted diet is a completely novel food to the bird. Many repeated offerings of a novel food are required before it will be tasted and eventually accepted, similar to mammalian species. 10–13 months of consistently offering a food may be required in very novelty-phobic individuals. • During the transition phase, birds will often fling their pellets with a sharp head flick. Although this resembles a human expression of disgust, it may simply signify the end of that particular tasting experience, and should in no way discourage the owner from continuing to present the desired new diet. • Pellets can be offered near a preferred perch or in a preferred food dish or foraging toy to increase appeal. • Small birds are more likely to pick at pellets spread on a flat surface on plain white paper or on a mirrored surface. • For highly bonded birds, the owner may pretend to eat the food to generate interest. • For birds that already eat one nonseed food item, only that food and pellets may be offered in the morning, when appetite is highest. Seeds may be withheld and presented later in the day. (Note: it is *very* important that this technique only be attempted in birds that have been assessed as healthy enough to endure a brief fast if they refuse this meal, preferably via a thorough physical examination and minimal database of CBC and chemistry panel.)

SURGICAL CONSIDERATIONS
Lipomas may require surgical resection if correction of diet and activity do not cause reduction, particularly in cases of significant crop impingement or necrosis. Some lipomas are very vascular, and hemostatic techniques may be needed.

 MEDICATIONS

DRUG(S) OF CHOICE
• L-carnitine (1000 mg/kg of feed) for lipomas. • L-thyroxine (0.1 mg per 4 oz drinking water for 1–4 months) to be used for documented hypothyroidism only.

CONTRAINDICATIONS
The use of thyroid hormone treatment is not recommended in obesity of nonthyroid origin, as dangerous cardiovascular complications can ensue.

PRECAUTIONS
See "Appropriate Health Care" and "Nursing Care".

POSSIBLE INTERACTIONS
N/A

ALTERNATIVE DRUGS
N/A

 FOLLOW-UP

PATIENT MONITORING
See "Appropriate Health Care" and "Nursing Care".

PREVENTION/AVOIDANCE
• Birds should be fed a species-appropriate high-quality, nutritionally sound diet.
• Ample opportunity for exercise must be provided, as well as encouragement for sedentary individuals.

POSSIBLE COMPLICATIONS
• Associated hepatic lipidosis can affect drug metabolism, particularly of anesthetics.
• Hypercholesterolemia, atherosclerosis, and heart failure have all been associated with chronic obesity. • Visceral fat deposits can cause air sac impingement, compromising respiratory effort. • Fatty or lipomatous deposits over the crop can cause regurgitation.
• Incidence of dystocia may be increased.
• Both male and female fertility may be impaired, although the mechanism of this is not completely understood. • Pododermatitis and arthritis may be exacerbated.

EXPECTED COURSE AND PROGNOSIS
• Dietary improvements and increased activity can be difficult to achieve, as they usually require changes in both owner and patient behavior. • Due to their high metabolic rate, husbandry corrections often reverse obesity and prevent complications.

☑ MISCELLANEOUS

ASSOCIATED CONDITIONS
See "Possible Complications".

AGE-RELATED FACTORS
See "Signalment".

ZOONOTIC POTENTIAL
N/A

FERTILITY/BREEDING
See "Possible Complications".

SEE ALSO
Atherosclerosis
Coelomic distention
Hepatic lipidosis
Dyslipidemia/hyperlipidemia
Infertility
Lipoma
Nutritional deficiencies
Pododermatitis
Regurgitation/vomiting
Thyroid diseases
Tumors
Xanthoma

Suggested Reading
Macwhirter, P. (1994). Malnutrition. In: Ritchie, B.W., Harrison, G.J., Harrison, L.R. (eds), *Avian Medicine: Principles and Application*. Lake Worth, FL: Wingers Publishing, Inc., pp. 842–861.
McDonald, D. (2006). Nutritional considerations: Section I. In: Harrison, G.J., Lightfoot, T.L. (eds), *Clinical Avian Medicine*, Vol. I. Palm Beach, FL: Spix Publishing, Inc., pp. 86–107.
Rupley, A.E. (1997). Nutritional Diseases. In: Rupley, A.E., *Manual of Avian Practice*. Philadelphia, PA: WB Saunders, pp. 298–301.
Smith, J.M, Roudybush, T.E. (1997). Nutritional disorders. In: Altman, R.B., Clubb, S.L., Dorrestein, G.M., Quesenberry, K. (eds), *Avian Medicine and Surgery*. Philadelphia, PA: WB Saunders, pp. 501–516.
Author Elisabeth Simone-Freilicher, DVM, DABVP (Avian)

Client Education Handout available online

OCULAR LESIONS

BASICS

DEFINITION
Any abnormality of the eye or associated structures. An extensive discussion of all ocular lesions is beyond the scope of this text. The most common ophthalmologic lesions including cataracts, conjunctivitis, periocular swelling, uveitis and retinal detachment are addressed.

PATHOPHYSIOLOGY
• *Cataracts*: A cataract is opacity of any part of the lens or its capsule. Cataracts may be associated with malformation, genetic disorders, nutritional deficiency, trauma, senescence, toxicity, or other ocular disorders.
• *Conjunctivitis*: Conjunctivitis is inflammation of the conjunctiva and is one of the most common ocular lesions in pet birds. Common causes of conjunctivitis include vitamin A deficiency, infection, and toxins. Inflammation can be secondary to trauma, such as that experienced during handling, shipping, self-induced, or from a conspecific.
• *Periocular swelling*: Periocular swelling involves focal or diffuse disorders of the eyelids, infraorbital sinus, or nasal gland. Common causes of periocular swelling include trauma, infections, neoplasia, and nutritional deficiency. • *Uveitis*: Uveitis is the term for inflammation of the uvea and can be divided into categories based on location: anterior, intermediate, posterior, and panuveitic. Causes of uveitis can include trauma, infection, immune-mediated, and neoplasia. • *Retinal disease*: Retinal detachment can be either retinal separation or a tear. Retinal separation occurs between the retinal pigment epithelium and neural retinal, resulting in subretinal fluid, exudate, etc. A retinal tear is most often due to trauma in raptors.

SYSTEMS AFFECTED
• Ophthalmic—this is the primary system affected. • Behavioral—birds may present for anorexia or behavior changes if vision is compromised. • Gastrointestinal—birds may present for anorexia or nonspecific signs of illness. • Hemic/Lymphatic/Immune—in immune-mediated conditions.
• Reproductive—reproduction may be affected if vision is compromised.
• Respiratory—birds may have abnormal respiration if disease extends into the sinuses or upper airway. • Skin/Exocrine—birds may present for visible swellings or skin discoloration.

GENETICS
Hereditary autosomal recessive cataract has been reported in Norwich and Yorkshire canaries. Cataract and optic nerve hypoplasia has been reported in turkeys.

INCIDENCE/PREVALENCE
Conjunctivitis is one of the most common ocular lesions in pet birds. *Mycoplasma gallisepticum* has caused outbreaks of contagious conjunctivitis in wild house finches. Hyphema was the most frequent ocular finding in a retrospective survey of ocular examinations in raptors. 15.4% of 24 birds in quarantine that died, mostly cockatiels, Amazon parrots, and budgerigars, were reported to have cataracts on necropsy.

GEOGRAPHIC DISTRIBUTION
Worldwide distribution.

SIGNALMENT
• **Species:** All avian species are susceptible to these conditions. • **Species predilections:** Pet psittacine birds and passerines (conjunctivitis), raptors secondary to trauma (uveitis), canaries (cataracts). • **Mean age and range:** Aged birds may be more likely to have cataracts. All age ranges can be affected by these conditions. • **Predominant sex:** N/A

SIGNS
Signs can be nonspecific or birds may present with visible lesions.

Historical Findings
Owners may report gradual or sudden changes in behavior or appearance based on the condition.

Physical Examination Findings
• *Periocular swelling*: Eyelids and/or periocular areas may appear generally or focally swollen, erythematous, and/or scaly; palpebral fissure width may be reduced; nictitans may be prolapsed; eyelid and periocular feather loss; discharge to the point of eyelids matted together; conjunctival hyperemia or swelling and/or facial swelling; facial rubbing.
• *Corneal pathology*: Ulceration detected with application of fluorescein stain. Cornea scars are opaque, but fluorescein negative.
• *Conjunctivitis*: Hyperemia, blepharospasm, variable photophobia, chemosis, serous to mucopurulent discharge; nictitans mobility may be impaired; eyelid margins may be sealed together; periocular feather loss and self-trauma may occur. • *Cataracts*: Lens opacities identified by direct or retroillumination may be focal to diffuse, dyscoria and posterior synechiae in cases with previous or current uveitis, lens luxation or subluxation in chronic cataracts, vision loss depending on extent of disease, globe size reduced in cases of congenital microphthalmos or acquired phthisis.
• *Uveitis*: Photophobia, blepharospasm, corneal edema, aqueous flare, hypopyon, vitreous opacity, hypotony or secondary glaucoma, miosis, dyscoria, iris thickening or discoloration, rubeosis irides, and or anterior or posterior synechiae are signs of active uveitis. Aqueous flare may be detected with a bright light source focused to the smallest diameter possible, in a darkened room. Focal to diffuse retinal edema, hemorrhage, retinal detachment, and vitreal opacity may be present in cases of active posterior uveitis. Diffuse corneal edema, posterior synechiae, anterior synechiae, secondary glaucoma, cataract, retinal atrophy, chronic detachment, and blindness can be sequelae of chronic uveitis. • *Retinal detachment*: This occurs most often as the result of trauma, especially in raptors. Retinal detachment may be serous, with fluid accumulation between the retina and retinal pigmented epithelium. Detachment may also be the result of a tear in the retina, often visible on ophthalmoscopy.

CAUSES
• *Cataracts*: Hereditary autosomal recessive cataract in canaries, in association with crooked toe in Brahma chickens, maternal vitamin E deficiency in turkeys can cause cataract in offspring, blunt or perforating trauma, aging change, dinitrophenol fed to chicks, avian encephalomyelitis, and chronic uveitis and retinal degeneration.
• *Conjunctivitis*: Trauma, vitamin A deficiency, infectious causes (bacteria, mollicutes including *Mycoplasma, Ureaplasma, Acholeplasma*, viruses, nematodes, and trematodes). Bacterial causes can include *Pseudomonas, Staphylococcus, Pasteurella, Citrobacter, Escherichia coli, Actinobacter, Actinobacillus, Erysipelothrix, Clostridium botulinum, Mycobacteria sp, Bordetella avium, Chlamydia psittaci, Haemophilus, Salmonella sp, Aeromonas*, Fungal causes can include *Candida* and *Aspergillus sp*. Viral causes include: poxvirus, herpesviruses, avian influenzavirus, paramyxovirus, adenovirus, cytomegalovirus, and others. Parasitic causes can include *Ceratospira, Oxyspirura, Philophthalmus, Thelazia, Serratia*, and cryptosporidium. Photosensitization from certain plants, chemical/ammonia burn, and neoplasia has also been associated with conjunctivitis. • *Periocular swelling*: Many of the same causes of conjunctivitis listed above including trauma, infections (bacterial, parasitic, viral), nutritional deficiency, and neoplasia. In addition to the above differentials, *Knemidokoptes pilae*, infraorbital sinusitis, and nasal or salt gland inflammation can result in periocular swelling. • *Uveitis*: Blunt or perforating trauma (often involving hemorrhage), corneal ulceration, viral (including herpesvirus-induced lymphomatosis and West Nile virus), bacterial (*Pasteurella*, salmonellosis, *Mycoplasma*, others), mycotic (*Aspergillus*), protozoal infection (toxoplasmosis), and neoplasia can also cause uveitis. • *Retinal detachment*: Common sequela of trauma especially in raptors.

RISK FACTORS
Birds on inadequate diets may be predisposed to ocular lesions related to hypovitaminosis A. Birds housed in unsanitary conditions may be predisposed to ammonia or other irritant-induced conjunctivitis. Inappropriate housing

OCULAR LESIONS (CONTINUED)

may predispose to traumatic injury. Vitamin E deficiency in turkey hens results in cataracts of offspring. Brahma chickens with crooked toes have been reported to develop spontaneous cataracts.

DIAGNOSIS

DIFFERENTIAL DIAGNOSIS
Subcutaneous emphysema from air sac trauma or leakage can give the appearance of pathology around the neck and face. Ink from newsprint or toys and even owner lipstick can cause discoloration around the eyes that could be mistaken for pathology. Upper respiratory disease or regurgitation can cause matting of the feathers around the face and eyes that could be mistaken for ocular disease.

CBC/BIOCHEMISTRY/URINALYSIS
Blood work may or may not be helpful depending on etiology of disease. CBC changes may be present with infectious or chronic disease. A minimum diagnostic database is helpful to rule out underlying disease.

OTHER LABORATORY TESTS
Chlamydia, Mycoplasma, viral, fungal, and parasitic screening tests are indicated if signs warrant. Bacterial or fungal cultures may be indicated. Paranasal flush or aspirates can be submitted for cytologic examination and/or culture.

IMAGING
Radiographs of the head and whole body may be indicated as part of a complete work-up and to rule out underlying respiratory disease. CT may be more helpful for evaluating sinuses than plain radiography. Contrast radiography has been reported. Ultrasound can be helpful to assess the globe and surrounding tissues.

DIAGNOSTIC PROCEDURES
Complete ophthalmologic examination is crucial for any ocular or orbital problem. This includes fluorescein stain and tonometry (either TonoPen or TonoVet). An otoscope without a cone provides magnification for examination of the anterior chamber. The direct ophthalmoscope set on the slit beam in a darkened room may be used to detect aqueous flare. Posterior segment examination may be performed with either a direct ophthalmoscope, or with a Finoff transilluminator and lens (28–40 diopters). Consider consultation with a board-certified ophthalmologist when feasible. Topical mydriatics used for mammals are not useful in birds due to the presence of skeletal muscle in the iris.

Additional diagnostic procedures that may be warranted include exfoliative conjunctival or corneal cytologic examination (including Gram's, acid-fast, Giminez, periodic acid–Schiff stains if warranted), biopsy, (rarely—aqueous or vitreous paracentesis), and other procedures as indicated. Skin scraping or tape preps can help rule out parasitic causes such as *Knemidokoptes*.

PATHOLOGIC FINDINGS
The presence of intracellular bacteria in heterophils or epithelial cells can be seen with bacterial infections. Epidermal hyperplasia with ballooning degeneration, intraepithelial vesicles, and eosinophilic intracytoplasmic inclusion are seen on histopathologic examination of skin biopsies in cases of poxvirus. Conjunctival cytologic examination of birds with chlamydia may reveal conjunctival epithelial cell hyperplasia, inflammatory cell infiltrate, intracytoplasmic chlamydial elementary bodies on Giemsa- or Gimenez-stained samples, and antigen may be demonstrated by IFA (immunofluorescent assay). *Mycoplasma* inclusions are small basophilic dots on the surface of epithelial cells. Neoplastic cells may be seen on cytologic or biopsy samples in cases of neoplasia.

TREATMENT

APPROPRIATE HEALTH CARE
Birds with ocular lesions can generally be managed on an outpatient basis unless significant pathology is present.

NURSING CARE
Supportive care as indicated including fluid therapy and gavage feeding in anorexic patients. Warm compresses and ocular flushing may help loosen up caked debris resulting in eyelid matting. In cases of suspected infectious disease, birds should be isolated. Protective gear should be worn by caretakers if zoonotic disease is suspected. Elizabethan or protective collars may be helpful in cases of self-trauma. Supportive care and management of secondary bacterial and fungal infections may be the only treatment option for some viral diseases.

ACTIVITY
No activity restriction is indicated unless patient is systemically ill or if vision is impaired. Confining the patient to a smaller area may help facilitate treatment.

DIET
Birds on an inadequate diet should be converted to an appropriate diet. Assisted feeding may be indicated if vision is impaired or the bird is not eating.

CLIENT EDUCATION
Owners should be educated on proper nutrition for their bird. If the event was caused by trauma or excessive restraint, owners should be counseled on ways to avoid future trauma and proper restraint techniques.

SURGICAL CONSIDERATIONS
Foreign bodies, including conjunctival nematodes and trematodes, should be removed. Cataract surgery (lensectomy by needle discission and aspiration or conventional extracapsular extraction or phacoemulsification) can be considered in cases of cataracts; this should be performed with a veterinary ophthalmologist. Tarsorrhaphy is an option for protection of the cornea in cases of severe ulceration. Enucleation or modified evisceration may be indicated in cases of neoplasia or painful conditions refractory to therapy. Eyelid repair may be indicated in cases of eyelid laceration.

MEDICATIONS

DRUG(S) OF CHOICE
Systemic antibiotic, antifungal, antiparasitic, or antiviral therapies as warranted. Broad-spectrum topical ophthalmic antibiotics (e.g., bacitracin-polymyxin B-neomycin, tetracycline, chloramphenicol, gentamycin, fluoroquinolone) are indicated in some cases of infectious conjunctivitis, keratoconjunctivitis, and uveitis. Systemic NSAIDs are warranted in cases of severe inflammation or uveitis. Topical NSAIDs including flurbiprofen or diclofenac can help control local inflammation. Mydriasis cannot realistically be obtained with topical therapy in birds so pupil preservation is best controlled by reducing inflammation. Lubricating therapies should be instituted if lagophthalmos is present or associated with eyelid swelling. Parenteral vitamin A (10,000–25,000 IU/300g body weight) is helpful in vitamin A deficiency.

CONTRAINDICATIONS
Topical steroids or NSAIDs should not be used in cases of corneal ulceration. Follow FARAD guidelines in food animal species including poultry and waterfowl.

PRECAUTIONS
Systemic or topical medication (especially corticosteroid) therapy is not without risk and potential side effects or complications should be discussed with owners prior to treatment. Corticosteroid therapy may exacerbate underlying viral or bacterial disease. Oil/petrolatum-based ointments can cause greasy residue on feathers; water-based formulations preferred whenever feasible.

POSSIBLE INTERACTIONS
Topical and/or systemic steroid therapy could result in generalized immunosuppression so judicious use is recommended.

ALTERNATIVE DRUGS
TPA (tissue plasminogen activator) has been administered intravitreally in cases of intraocular hemorrhage. Safety and efficacy of topical and oral medications to control

glaucoma in mammals have not been extensively investigated in birds but can be considered in cases of glaucoma.

FOLLOW-UP

PATIENT MONITORING
Follow-up ophthalmologic examination will help determine response to therapy. More frequent visits are warranted in cases of corneal ulceration or severe disease.

PREVENTION/AVOIDANCE
Vaccination of poultry for Marek's disease and avian encephalomyelitis may be warranted. Vaccines for other viral diseases including poxvirus and West Nile virus can be considered. Identify and treat minor corneal ulceration early to prevent progression. Treat traumatic injury and uveitides aggressively and early. Quarantine new additions to the aviary or flock and isolate affected birds early.

POSSIBLE COMPLICATIONS
Severe corneal pathology may result in globe rupture. Blindness can result from severe or untreated disease. Corticosteroid therapy may exacerbate underlying viral or bacterial disease.

EXPECTED COURSE AND PROGNOSIS
Course and prognosis depends on the disease condition. Minor disease can be self-resolving or responsive to therapy and associated with a good outcome. Severe disease can result in progressive pathology refractory to therapy.

MISCELLANEOUS

ASSOCIATED CONDITIONS
Other lesions including fractures may be present in cases of trauma so thorough physical examination is warranted. Any of these ocular conditions can be related to systemic disease and/or secondary complications.

AGE-RELATED FACTORS
Aged birds may have higher morbidity and mortality than younger birds. Cataracts are more common in aged birds.

ZOONOTIC POTENTIAL
Infectious disease causes including *Chlamydia, Salmonella, Mycobacteria*, influenza, and paramyxoviruses are potentially zoonotic.

FERTILITY/BREEDING
Birds with vision impairment or systemic disease may have reduced fertility or difficulty breeding.

SYNONYMS
Chlamydia—parrot fever
Knemidokoptes pilae—scaly face and leg mite

SEE ALSO
Adenoviruses
Airborne toxicosis
Air sac rupture
Aspergillosis
Bordetellosis
Cryptosporidiosis
Mycobacteriosis
Mycoplasmosis
Nutritional deficiencies
Paramyxoviruses
Respiratory distress
Rhinitis and Sinusitis
Salmonellosis
Trauma
Tumors
Viral disease
West Nile virus

ABBREVIATIONS
CT—computed tomography
IFA—immune fluorescence assay

INTERNET RESOURCES
http://lafeber.com/vet/raptor-ophthalmology-ocular-lesions/

Suggested Reading
Kern, T. (1997). Disorders of the special senses. In: Altman, R.B., Clubb, S.L., Dorrestein, G.M., Quesenberry, K. (eds), *Avian Medicine and Surgery*. Philadelphia, PA: WB Saunders, pp. 563–580.
Abrams, G., Paul-Murphy, J., Murphy, C. (2002). Conjunctivitis in birds. *The Veterinary Clinics of North America. Exotic Animal Practice.* **5**(2):287–309.
Murphy, C., Kern, T., McKeever, K, *et al.* (1982). Ocular lesions in free-living raptors. *Journal of the American Veterinary Medical Association*, **181**:1302–1304.

Authors Jennifer Graham, DVM, DABVP (Avian/ECM), DACZM
Ruth M. Marrion, DVM, DACVO, PhD

Oil Exposure

BASICS

DEFINITION
Oil exposure in birds is defined as a contamination of external structures and/or internal ingestion or inhalation of petroleum-based or other types of oil. This exposure can occur on an individual basis or can affect large populations of wild birds. In both cases, the type of oil will affect the outcome of exposure and can lead to both short and long term effects on a wide variety of body systems.

PATHOPHYSIOLOGY
• Environmental sources of oil include natural seeps, pipeline breaks, drilling platform or well accidents, tanker collisions, illegal/deliberate releases and dumping of waste oil by the general public. In addition, the individual bird can encounter cooking or waste oil within a household setting. • Birds can be exposed to oil during any of their daily activities, including feeding, loafing, flying, breeding, nesting, and raising young.
• External contamination leads to disruption of feather alignment and possible skin and corneal irritation/burns. • Internal contamination can occur from ingestion of oil through feeding in contaminated waters or from preening of oiled feathers. Exposure to highly refined fuels, such as jet or diesel fuels, can result in inhalation of highly volatile fumes that lead to respiratory and neurological abnormalities.

SYSTEMS AFFECTED
• Skin/Exocrine—oil disrupts the microscopic alignment of hooks and barbules on feathers, resulting in the inability of birds to maintain waterproofing. This, in turn, can lead to loss of buoyancy in the water, exposure to environmental elements resulting in hypo- or hyperthermia, and loss of flight capability. Highly volatile fuels can also cause varying degrees of skin irritation or burns through direct contact with exposed areas.
• Gastrointestinal—ingestion of oil causes direct inflammatory effects on the gastrointestinal tract as well as an alteration in electrolyte and water transport across the intestinal mucosa. Inability to feed normally in the wild or in captivity during rehabilitation activities can result in malnutrition and weight loss.
• Hepatobiliary—liver abnormalities result from exposure to polyaromatic hydrocarbons and other toxic components of oil. • Hemic/Lymphatic/Immune—ingestion of crude oil has been linked to development of hemolytic anemia in different avian species. Anemia of chronic disease and malnutrition is also seen. Petroleum exposure results in direct suppression of immune system function and this is exacerbated by immunosuppression resulting from the stress of oiling and captivity during rehabilitation.
• Reproductive—exposure to oil can result in failed reproduction in the short-term due to loss of breeding activity, contamination of eggs or young with oil and disruption of normal reproductive hormonal levels. Long-term population effects on reproductive success have been documented following large-scale oil spills. • Musculoskeletal—malnutrition often results in pectoral muscle atrophy seen on admission to rehabilitation. Secondary problems related to captive care include keel, hock and foot sores resulting from abnormal pressure points on the bird's body when it is held out of the water for prolonged periods of time. • Endocrine/Metabolic—internal exposure to oil may result in impairment of salt gland function, resulting in osmoregulatory dysfunction. • Behavioral—any wild bird contaminated with oil that is subsequently captured and then brought into rehabilitation experiences a large amount of stress during the entire process. This results in behavior consistent with that of a wild animal that believes it is in constant danger until the moment of release back to the wild.
• Respiratory—inhalation of highly volatile fumes from refined fuels can result in direct damage to exposed areas of the respiratory tract. • Nervous—exposure to highly volatile fumes from refined fuels can result in neurological abnormalities, including ataxia and seizure activity. • Ophthalmic—highly volatile fuels can cause corneal irritation or burns through direct contact or via exposure to fumes from the oil.

GENETICS
N/A

INCIDENCE/PREVALENCE
• Birds can be exposed to oil almost anywhere in their habitat though there is increased incidence of oiling in areas where oil is extracted, transported or refined. • Incidence may rise in winter months due to the increased possibility of storms leading to transport spills. In addition, many seabirds congregate in large numbers on the open ocean during this time period, making them more vulnerable to oiling.

GEOGRAPHIC DISTRIBUTION
Worldwide—especially in developed countries which utilize large quantities of petroleum products.

SIGNALMENT
All avian species, no specific age or sex predilection though seabirds and waterfowl are especially vulnerable.

SIGNS

General Comments
• Some types of refined oils will be clear in coloration and difficult to discern on feathers.
• In some instances, wild birds will die from problems associated with oil exposure before they reach land and are never found during search and collection efforts. • More heavily oiled birds are often found first and brought for rehabilitation because they are more incapacitated from the oil. Ironically, trace to lightly oiled birds are often in poorer shape by the time they are weakened enough from oil exposure to be captured.

Historical Findings
• In the USA, the National Response Center (telephone: 800-424-8802) will advise the responsible agencies (Coast Guard, EPA, state and Federal natural resource agencies) about the event. An incident command system is then instituted for the spill response. • If an individual oiled bird is found with possible petroleum contamination, it is recommended that the state wildlife agency be contacted as this bird may be part of a larger spill event. • Trained wildlife response personnel are engaged by the spiller (the responsible party) to conduct wildlife operations. • The general public should be discouraged from search and collection of wild birds during an oil spill incident. • An individual oiled bird, whether wild or owned, may be brought to a veterinary practice for care after being found oiled. • Outside the USA, Oil Spill Response Limited (OSRL) (telephone: +44 (0) 2380331551)—the largest international industry-funded cooperative—is available 24 hours a day, 365 days of the year to attend spills of oil, chemicals and other hazardous substances worldwide (http://www.oilspillresponse.com/services/member-response-services/).

Physical Examination Findings
• Oil contaminated feathers and skin—in most instances, this is fairly obvious though highly refined fuels may be colorless and vapors may have dissipated by the time the bird is found. Feathers may be matted, wet, feel oily and smell like the petroleum product.
• Oil may occlude the nares or the oropharyngeal area—bird may be in respiratory distress. • Increased respiratory effort, rate, abnormal respiratory tract sounds—with inhalation of toxic fumes.
• Neurological signs including ataxia, seizures—with exposure to toxic fumes. • If the cornea or periocular area is irritated or burned—epiphora, conjunctival redness/swelling, corneal edema or ulceration.
• Hypo- or hyperthermia depending on environmental factors. • Thin to emaciated body condition unless found within the first 24–48 hours of oiling. • Dehydration.
• Evidence of regurgitation and diarrhea—

OIL EXPOSURE (CONTINUED)

often bloody. • Anemia. • If the bird has been out of the water for prolonged periods, it may show evidence of pressure sores on the keel, posterior aspects of the hocks and the feet.

CAUSES
• Cooking oils. • Motor oils. • Petroleum products—crude, gasoline, intermediate distillates (diesel, jet fuels), heavy distillates (lubricants, waxes), residues (heavy fuel oils, asphalt, tar, coke).

RISK FACTORS
• Birds of the open ocean (shearwaters, petrels, fulmars, albatrosses) are likely to encounter petroleum released with drilling platform accidents, tanker accidents or illegal discharges. They are especially vulnerable because they live far away from land and often drown when their waterproofing is compromised with oil exposure. • Birds that inhabit coastal areas and only come to land to breed (loons, grebes, auks, gannets/boobies)—similar to open ocean birds and often live in areas with heavy tanker traffic. • Birds that inhabit near shore environments (pelicans, wading birds, gulls, terns, shorebirds)—less vulnerable because they can get to land more easily but often live in environments with multiple sources of potential oil release. • Waterfowl (ducks, geese, swans)—in general, more hardy than birds in the above categories.

DIAGNOSIS

DIFFERENTIAL DIAGNOSIS
• External contamination with another substance, for example, dispersants (used during oil spill responses), household solvents, glues. • Internal exposure to other inhaled or ingested toxins, for example, Teflon®, poisonous plants, heavy metals. • Infectious diseases leading to multisystemic signs—polyomavirus, herpesvirus, mycobacteriosis, chlamydiosis.

CBC/BIOCHEMISTRY/URINALYSIS
• CBC may reveal evidence of anemia, increased white blood cell count due to direct effects of oil or secondary bacterial and fungal infections. • Biochemistry panels may show low total protein levels (malnutrition, hepatic dysfunction), hypoglycemia, hepatic and renal enzyme abnormalities, electrolyte disturbances resulting from malnutrition, dehydration and the multisystemic effects of the toxic components in oil.

OTHER LABORATORY TESTS
• Contaminated feathers can be "fingerprinted" to identify the specific composition of the oil that is present. This is a process using gas chromatography/mass spectrometry and is only available through certain laboratories. • ELISA test to detect polyaromatic hydrocarbons in oil from feather analysis.

IMAGING
• Radiographs may be useful in discerning respiratory abnormalities (inhalant toxicity, aspergillosis infections secondary to immune suppression and captive environment), organomegaly such as an enlarged hepatic silhouette or adrenal hyperplasia, and thickened gastrointestinal tract walls. • Not a primary method of diagnosis for oil exposure.

DIAGNOSTIC PROCEDURES
• Can place a few feathers in a pan of water and watch for an oil sheen to appear—not considered a specific test of oil contamination, since the sheen may be from other substances. • Fluorescein tests to reveal possible corneal ulceration. • Endoscopy may show ingested oil within the gastrointestinal tract. • Laparoscopy may reveal evidence of secondary bacterial or fungal infections.

PATHOLOGIC FINDINGS
• Necropsies of oiled birds cannot be performed in large oil spill events, as the bodies are considered legal evidence, unless permission is obtained from the proper authorities. • Gross necropsy lesions highly variable. • Oil may be present on feathers, in oropharyngeal area, trachea, or in the gastrointestinal tract. • Lining of gastrointestinal tract may be inflamed or hemorrhagic. • Lungs may appear hemorrhagic with exposure to volatile fumes. • Salt glands and adrenal glands may be enlarged. • No specific or consistent lesions in birds exposed to oil.

TREATMENT

APPROPRIATE HEALTH CARE
• All personnel working with oiled birds during a spill response should wear appropriate personal protective equipment and have received hazardous materials training. • Search and collection of oiled birds during large-scale oil spills is only conducted by trained personnel. • When birds are found during search and collection, they are initially stabilized in the field. This involves removing any oil occluding eyes, nares and glottis and impairing movement of wings and legs as well as administration of a balanced electrolyte solution via gavage. Birds may also need to be treated for hypo-or hyperthermia with the use of warm or cold packs appropriate for field use. • Because birds affected during large-scale oil spills are considered legal evidence for use in damage assessment proceedings against the responsible party, a strict chain of custody procedure is followed during their time in captivity. The location where birds are found is documented with GPS readings, and an individual leg band is placed on the bird (or other means of identification) so that it can be tracked throughout the rehabilitation process. • When a bird reaches the rehabilitation center, it is given a thorough physical examination with documentation of the extent of oiling, is photographed and a small sample of oiled feathers is taken for legal purposes. • Any abnormalities noted on physical examination require appropriate medical care. Because a large number of birds are often seen during spill responses, treatment is often given on a "herd health" basis administered for the "average" bird rather than tailored to each individual. • Triage is an essential part of the rehabilitation process during an oil spill response. Decisions are made based on the resources available, health status of the bird, and other general factors, for example, the historical success of rehabilitation of the species. This means that euthanasia plays a significant role in oiled wildlife response efforts.

NURSING CARE
• Scrupulous attention should be paid to disinfection and cleanliness during all procedures in oiled bird care. Because these birds are often immunocompromised, every step should be taken to decrease the incidence of secondary bacterial or fungal infections. Aspergillosis is the most common infectious disease seen in an oil spill response and prophylactic antifungal medications are often administered to decrease the possibility of this infection. • Initial care for dehydrated oiled birds consists of fluid administration given either IV or orally via gavage. Hetastarch therapy for hypoproteinemic birds may be used in individual cases at 15 ml/kg divided 3–4 times per day followed by maintenance fluids. • Oil is gently removed from eyes with an ophthalmic irrigation solution, and from the nares and/or glottis using dampened gauze sponges. • Body temperature abnormalities (normal 102–105°F) should be corrected. • For seabirds that are not used to being out of the water, padded wraps to decrease the incidence of pressure sores over hocks and feet are helpful. Keel wraps are sometimes placed to decrease pressure sores over this area. • Feeding—high calorie nutritional slurries, for example, Lafeber Emeraid Piscivore® slurry at 10% fat, are also administered via gavage, usually alternating with oral electrolyte fluids. Birds are encouraged to self-feed by offering appropriate food items, for example, whitebait or night smelt for piscivores. • Obligate marine species are given additional salt in their feedings while housed on land or in fresh water. • Housing—a decision must be made whether to house a bird individually vs. in a group pen. In general, birds that are gregarious in the wild, for example, pelicans or common murres, can be housed together. Birds are housed in indoor pens until they are cleaned and then

OIL EXPOSURE

are moved to outdoor aviaries or pools appropriate to the species once cleared for prerelease conditioning. • Washing—A 1–2% Dawn® dishwashing solution at 104–106°F is used to bathe oiled birds. The water is placed into appropriately sized tubs and the bird is moved from tub to tub until the bath water is clean. • Rinsing—All soap must then be rinsed off thoroughly in order to ensure proper waterproofing. This requires rinsing with water at 104–106°F utilizing a high pressure nozzle and making sure to get every portion of the bird's feathers clean. When a bird is properly rinsed, the water will bead off the feathers and the bird's feathers will appear dry. • Drying—birds are dried off utilizing incubators, heat lamps or warm air pet dryers set on "low". They should be closely monitored to make sure that they are not overheating. Most birds will preen their feathers back into alignment in this time period.

ACTIVITY
• Oiled birds are housed in indoor pens that restrict their activity before cleaning. • After cleaning, these birds can be moved to outdoor aviaries or pools in order to condition them prior to release.

DIET
See above. If the bird is regurgitating, losing weight and/or has diarrhea, the veterinarian may choose to focus on rehydration methods and back off gavage feeding of nutritional slurries.

CLIENT EDUCATION
If an owned bird becomes oiled within the household, clients should be instructed in methods to limit further access to sources of oil in the future. Appropriate education should also be given to address individual problems associated with the oiling incident, such as proper wound care or dietary considerations.

SURGICAL CONSIDERATIONS
N/A

MEDICATIONS

DRUG(S) OF CHOICE
• Antibiotics ○ Amoxicillin/clavulinic acid 125 mg/kg PO q12h. ○ Enrofloxacin 10–20 mg/kg PO, SC, IM q12–24h. • Antifungals ○ Itraconazaole 20 mg/kg PO q24h • Analgesics ○ Meloxicam 0.5–2.0 mg/kg PO q12–24h. • Ophthalmic medications ○ Avoid use of petroleum based ointments. ○ Ciprofloxacin drops—one drop in affected eye(s) q4–6h. • Salt 250 mg/kg PO every other day. • Sucralfate 25 mg/kg PO q8h before food or other drugs. • Vitamin supplementation ○ Vitamin B1 for those birds eating thawed frozen fish (1–2 mg/kg PO q24h). ○ Commercial vitamin for fish-eating birds (Mazuri Auklet Vitamins®).

CONTRAINDICATIONS
Repeated injections of enrofloxacin are associated with severe muscle necrosis.

PRECAUTIONS
Avoid use of antibiotics and glucocorticoids, unless clinically necessary, as these medications may increase the risk of aspergillosis and other infectious diseases in immunocompromised oiled birds.

POSSIBLE INTERACTIONS
N/A

ALTERNATIVE DRUGS
Alternative antibiotics and antifungal medications can be used based on culture and sensitivity results.

FOLLOW-UP

PATIENT MONITORING
• Prerelease conditioning: Once birds are cleaned and medical issues such as malnutrition and anemia are resolved, they are moved into outdoor aviaries or pools for prerelease conditioning and observation of waterproofing, appetite and behavior. • Release evaluation: Criteria used to determine suitability for release include behavior, for example, diving appropriately in pools, body weight/condition, waterproofing, normal blood values, and resolution of all other abnormalities noted during the rehabilitation process, for example, wounds healed and so on. • Postrelease: Monitoring via radio and satellite telemetry has been utilized in certain species following large-scale oil spill incidents.

PREVENTION/AVOIDANCE
• Recognize potential for oil exposure in the household and take steps to cover and safely dispose of any cooking or waste oil in this setting. • The Oil Pollution Act of 1990 mandates that each state in the United States have a contingency plan for oiled wildlife response. It also requires every company that moves petroleum in the United States to have a "plan to prevent spills that may occur" and have a "detailed containment and cleanup plan" for oil spills.

POSSIBLE COMPLICATIONS
Many oiled birds will succumb or must be euthanized due to the effects of oil or as a result of the secondary effects of captive care. These complications are discussed under "Systems Affected".

EXPECTED COURSE AND PROGNOSIS
The success of treatment for oil exposure varies widely depending on factors such as the type of oil spilled, the amount of oiling on the bird, the species of bird affected, preplanning for spill response, the expertise of response personnel and many others considerations.

MISCELLANEOUS

ASSOCIATED CONDITIONS
N/A

AGE-RELATED FACTORS
N/A

ZOONOTIC POTENTIAL
During an oil spill response, there is potential for a large number of captive wild birds to be housed together indoors. Some of these birds may become ill with infectious diseases such as aspergillosis or chlamydiosis, which are potentially zoonotic. In addition, stressed birds may be shedding gastrointestinal pathogens such as Campylobacter or Salmonella. As stated before, it is imperative that personnel working with oiled birds be familiar with and follow protocols designed to minimize exposure to zoonotic agents.

FERTILITY/BREEDING
See Reproduction under "Systems Affected".

SYNONYMS
N/A

SEE ALSO
Anemia
Aspiration
Coagulopathies
Dehydration
Diarrhea
Emaciation
Gastritis/enteritis
Hemorrhage
Infertility
Liver disease
Neurologic conditions
Ocular lesions
Pneumonia
Regurgitation/vomiting
Respiratory distress
Seizure
Trauma

ABBREVIATIONS
N/A

INTERNET RESOURCES
International Bird Rescue
http://www.bird-rescue.org/our-work/aquatic-bird-rehabilitation/our-process-for-helping-oiled-birds.aspx
Rescue and Rehabilitation of Oiled Birds—USGS National Wildlife Health Center
http://www.bird-rescue.org/our-work/aquatic-bird-rehabilitation/our-process-for-helping-oiled-birds.aspx
Oiled Wildlife Care Network (CA)
http://www.vetmed.ucdavis.edu/owcn/
Tristate Bird Rescue & Research:
https://tristatebird.org/oil-spill-response-and-services/

Oil Exposure (Continued)

Suggested Reading

Jessup, D.A., Leighton, F.A. (1996). Oil pollution and petroleum toxicity to wildlife. In: Fairbrother, A., Locke, L.N., Hoff, G.L. (eds), *Noninfectious Diseases of Wildlife*. Ames, IA: Iowa State University Press, pp. 141–156.

Leighton, F.A., Peakall, D.B., Butler, R.G. (1983). Heinz-body hemolytic anemia from the ingestion of crude oil: a primary toxic effect in marine birds. *Science*. **220**:871–873.

Mazet, J.A.K., Newman, S.H., Gilardi, K.V.K., *et al.* (2002). Advances in oiled bird emergency medicine and management. *Journal of Avian Medicine and Surgery*, **16**(2):146–149

Tseng, F.S. (1999). Considerations in care for birds affected by oil spills. *Seminars in Avian and Exotic Pet Medicine*, **8**(1):21–31.

USGS (1999). Oil. In: Friend, M., Franson, J.C. (eds), *Field Manual of Wildlife Diseases*. Information and Technology Report 1999-001, Biological Resources Division, United States Geological Survey, pp. 309–315.

Author Florina S. Tseng, DVM

Oral Plaques

BASICS

DEFINITION
Oral plaques are generally yellow, creamy, or white raised lesions on the mucosa of the oropharynx. In birds, oral plaques are most commonly caused by hypovitaminosis A, or infections with avian poxvirus, herpesvirus (pigeons, doves, canaries, Amazons, Bourke's parakeets), bacteria, *Candida* (most commonly *C. albicans*), *Trichomonas* (*T. gallinae, T. columbae*) or *Capillaria* spp.

PATHOPHYSIOLOGY
• Hypovitaminosis A causes squamous metaplasia resulting in hyperkeratotic squamous cells which plug the mucous and salivary glands, causing swelling and nodules. The mucous itself is thicker and tenacious, and when combined with exfoliated cornified squamous cells, will appear as mucosal plaques. • Candidiasis is an opportunistic disease most commonly caused by *Candida albicans*, although *C. parpasilosis* and rarely other species have been reported. Generally seen in very young patients or adults with underlying disease causing immunocompromise. Infection may be superficial or invade deeply into mucosa, particularly in the crop. Suppression of normal oral or ingluvial flora by antibiotics (typically tetracyclines) can predispose patients. May also be a secondary pathogen following lesions caused by bacteria, viruses, or hypovitaminosis A. • Avian poxvirus is transmitted through biting insects or conspecific aggression resulting in skin trauma. The virus stimulates epithelial cell DNA synthesis, resulting in hyperplasia. The exudative lesions often become secondarily infected, resulting in plaques, swellings, or caseous plugs throughout the mucosa of the oropharynx. Amazon parrots are considered especially susceptible to a diphtheritic form of avian poxvirus, although this is seen much less now that fewer birds remain from the imported population. • Herpesvirus results in fibrinonecrotic pseudomembranes secondary to inflammatory and necrotic debris. A mutation of the herpesvirus which causes infectious laryngotracheitis virus in chickens is believed to be responsible for a severe upper respiratory disease in Amazon parrots and Bourke's parakeets, also called Amazon tracheitis virus.

SYSTEMS AFFECTED
• Gastrointestinal—oral plaques can cause dysphagia, pharyngitis. *Candida* and *Trichomonas* commonly affect the crop, and sometimes lower gastrointestinal tract. *Mycobacteria* spp. can form granulomas throughout the lower gastrointestinal tract. *Capillaria* may be found in the oropharynx, esophagus, crop, and small intestine. Avian poxvirus lesions may found in the crop.
• Renal—hypovitaminosis A-associated squamous metaplasia of renal tubular epithelium, resulting in kidney dysfunction.
• Respiratory—lesions associated with the glottis or proximal trachea can cause cough or dyspnea. Hypovitaminosis A-associated squamous metaplasia of upper respiratory epithelium, predisposing patient to sinusitis, rhinitis. *Trichomonas* and herpesvirus can affect the larynx and trachea. Avian poxvirus can affect the trachea, air sacs and lungs.
• Skin/exocrine—avian poxvirus often results in skin lesions, particularly of the face and feet. Herpesvirus may result in foot lesions. *Mycobacteria* may cause granulomatous swellings at any location.
• Ophthalmic—avian poxvirus can cause blepharitis, chemosis, conjunctivitis, and ulcerative lesions of the eyelids, or secondary keratitis, corneal ulcers, and corneal perforation.

SIGNALMENT
• **Species:** ○ Young birds may be more predisposed to infectious causes. ○ Candidiasis is seen more frequently in neonatal cockatiels and pigeons, and may be more common in climates with increased humidity and warmer temperatures. In older birds, candidiasis is more commonly secondary to delayed crop emptying or underlying disease. ○ Pigeons and raptors are more susceptible to trichomoniasis (called "canker" and "frounce," respectively). In pet birds, trichomoniasis occurs most frequently in canaries, zebra finches, budgerigars, and cockatiels. ○ Avian poxvirus has been seen in canaries, psittacines (Amazons, pionus parrots, lovebirds), passerines, raptors, pigeons and doves, starlings, mynahs, waterfowl, pheasants, quail, and ostrich chicks. ○ Herpesvirus is usually a disease of the young and immunosuppressed, particularly in pigeons. ○ *Capillaria* is most common in budgerigars, macaws, canaries, pigeons, and gallinaceous birds. • **Mean age and range:** N/A • **Predominant sex:** N/A

SIGNS
Historical Findings
• Inappetance or anorexia. • Dysphagia.
• Gaping, yawning, repeated tonguing of oral mucosa. • Regurgitation or vomiting, increased upper gastrointestinal transit time (especially with crop involvement).
• Dyspnea, cough (especially when glottis and proximal trachea affected). • Halitosis (especially candidiasis).

Physical Examination Findings
May include dehydration, weight loss, and weakness secondary to negative energy balance. Oral plaque lesions on the oropharyngeal mucosa may be:
• Diphtheritic or caseous: (avian poxvirus, herpesvirus, *Candida, Trichomonas, Capillaria*). • Granulomas: (mycobacterial, other bacteria, fungal). • Hyperkeratotic: (hypovitaminosis A). • Easily removed: (hypovitaminosis A, *Trichomonas*). • Difficult to remove: (bacterial, avian poxvirus). • Grey or brown, fibrinous and friable: ("wet" form of avian poxvirus).

CAUSES
• Nutritional deficiency (generally seed-based diet): hypovitaminosis A. • Bacterial: often Gram-negative (*E. coli, Klebsiella* spp., *Pseudomonas aeruginosa, Pasteurella sp.*), more rarely Gram-positive (*Staphylococcus* spp.), anaerobic, or mycobacterial. Choanal bacterial lesions are sometimes seen with sinusitis. • Fungal: *Candida* spp. • Parasitic: *Trichomonas*. • Viral: avian poxvirus, herpesvirus, which may occur concurrently in pigeons with trichomoniasis, causing more severe lesions.

RISK FACTORS
See "Signalment".

DIAGNOSIS

DIFFERENTIAL DIAGNOSIS
Differentiating Similar Signs
• Dried food can accumulate in the mucosa and resemble plaques; this is often dependent on diet and eating habits. • Hypovitaminosis A can cause a pustular glossitis along the lateral surfaces of the tongue and around the glottis. Although at first glance these may resemble plaques, careful inspection will reveal that these lesions are not superficial like oral plaques, but instead are comprised of purulent or caseous material within the salivary pores lining the tongue. These may require debridement in addition to vitamin A supplementation to resolve. • Pigeons may present with sialolithiasis, concretions in the caudal pharynx originating from the salivary glands.

Differentiating Causes
• On physical examination, hypovitaminosis A also presents with blunting of the choanal papillae. This blunting usually appears as a very even, symmetrical scalloping, rather than the irregular erosion seen secondary to chronic sinusitis. Poxvirus usually presents with crusty or nodular lesions of the head, legs, and feet.
• Lesions may be swabbed for cytology, direct saline microscopy, and Gram-stain.
• Additional diagnostic testing when indicated (e.g., acid fast stain, mycobacterial or viral PCR, bacterial culture and sensitivity, mycobacterial culture, and fungal culture).
• Lesions refractory to treatment should be biopsied to rule out underlying neoplasia.

CBC/BIOCHEMISTRY/URINALYSIS
• In nonsystemic cases the hemogram and biochemical profile may be normal.
• Leukocytosis may occur with secondary or concurrent systemic infection. • Anemia seen with severe gastrointestinal ulceration and

Oral Plaques (Continued)

bleeding. • Hypoglycemia seen with prolonged anorexia. • Hyperuricemia seen secondary to hypovitaminosis A squamous metaplasia of renal tubule epithelium.

CYTOLOGY
• May see squamous cells with abundant basophilic keratin granules, often together in rafts or sheets with hypovitaminosis A. Secondary bacterial infection may or may not be present. Inflammatory cells are usually not a component. • May see bacteria, or "ghost rods" with mycobacterial infections. • May see oval darkly basophilic budding yeast with *Candida*. The presence of pseudohyphae (slender unbranched chains of tubular cells) is suggestive of tissue invasion. • May see basophilic piriform flagellated protozoa with an undulating membrane with *Trichomonas* • Intracytoplasmic (Bollinger) or intranuclear inclusion bodies with avian poxvirus. • Intranuclear (Cowdry) bodies with herpesvirus.

OTHER LABORATORY TESTS
For direct microscopic examination for motile *Trichomonas*, an esophageal swab may be performed with a wet cotton-tipped applicator. A water droplet from the swab is then expressed onto a slide. Warming may improve motility and detection of the organism.

IMAGING
• Survey radiographs are often normal, but delayed motility and gastrointestinal dilation may be noted with lower GI involvement.
• Contrast radiography may be used to distinguish dilated intestinal loops from intestinal thickening seen in mycobacteriosis.

TREATMENT

APPROPRIATE HEALTH CARE
The degree of nursing care needed is dependent on severity of disease.

NURSING CARE
• Nondebilitated patients with mild lesions and signs may be treated on an outpatient basis if hydrated and eating well on their own.
• Supportive care may be needed for anorexic, dehydrated, and/or debilitated patients, including fluids (SQ, IV, or IO), easily digestible nutrition, including gavage feeding in cases of inappetance, and heat support for birds experiencing weight loss or negative energy balance.

ACTIVITY
Most birds with this condition may be allowed to determine their own level of activity.

DIET
Although dietary improvements may be needed, changes should be gradual and deferred until the patient has recovered.

CLIENT EDUCATION
The overall predisposing factors that led to the disease need to be addressed. Clients should be educated on nutritional requirements for their particular species. Any underlying husbandry deficiencies should be addressed.

SURGICAL CONSIDERATIONS
In very severe cases, debridement of the lesions under sedation may be needed.

MEDICATIONS

DRUG(S) OF CHOICE
• May use analgesia when inappetance or dysphagia is present, including meloxicam (0.5–1 mg/kg PO q12h). • Hypovitaminosis A—vitamin A (5000–20,000 IU/kg IM).
• *Candida*—for superficial infections: nystatin (300,000 IU/kg PO q12h); however, dose volume is large and requires direct contact with lesions to be effective. For severe infections: fluconazole (2–5 mg/kg PO q24h), itraconazole (10 mg/kg PO q12h for 21 days), or ketoconazole (20–30 mg/kg PO q12h). • *Trichomonas*—metronidazole (20–50 mg/kg PO q12h for 3–5 days), ronidazole (6–10 mg/kg PO q24h for 7–14 days), or carnidazole (20–30 mg/kg PO q12h for 3–5 days). • Bacterial—appropriate antibiotic therapy based on culture and sensitivity. • Pigeon herpesvirus—treat topically and systemically with acyclovir (80 mg/kg PO q8h for 7–10 days, or 40 mg/kg IV/SQ q8h). • Amazon tracheitis (herpes virus)—successful specific treatment has not been reported. • Avian poxvirus—vitamin A, treatment of secondary bacterial infections if present. • *Capillaria*—fenbendazole (100 mg/kg PO), or oxfendazole (10–40 mg/kg PO), repeat in 14 days.

CONTRAINDICATIONS
Metoclopramide and cisapride are not effective for improving crop motility.

PRECAUTIONS
Fenbendazole can cause toxicity in a number of species, including pigeons and doves, vultures, storks, cockatiels and lories. Nystatin should not be given in cases of suspected gastrointestinal ulceration, as this may result in absorption-dependent toxicity. A fecal cytology or fecal occult blood test may help identify this, particularly in patients with anemia.

POSSIBLE INTERACTIONS
N/A

ALTERNATIVE DRUGS
N/A

FOLLOW-UP

PATIENT MONITORING
If inappetance, anorexia, dehydration, or significant weight loss occurs: hospitalize patient for supportive care and treatment, and obtain appropriate diagnostics.

PREVENTION/AVOIDANCE
• Birds should be fed a species-appropriate high-quality, nutritionally sound diet.
• Owners should observe good hygiene, and housing, nest boxes, food, and water supply should be carefully kept as clean as possible.
• Flock biosecurity measures should be observed. • Routine treatment of pigeons for trichomoniasis is recommended to reduce shedding. • Pigeons should not be fed to raptors, to prevent the transmission of *Trichomonas*. • A vaccine for avian poxvirus is available for pigeons, doves, and canaries. The vaccine recommended only for healthy flocks to prevent viral recombination.

POSSIBLE COMPLICATIONS
See "Pathophysiology" and "Systems Affected".

EXPECTED COURSE AND PROGNOSIS
• Following treatment, prognosis is good for bacterial and nutritional plaques or candidiasis in patients who are not seriously debilitated. Recovery from nutritional plaques is usually within 2–3 weeks. • Recovery from the nonsepticemic form of avian poxvirus is usually 3–4 weeks, if uncomplicated by bacterial or fungal secondary infections. Some strains in passerines and columbids can leave survivors prone to tumor formation. The diphtheritic ("wet") and septic forms are associated with high mortality. Survivor immunity may be life-long. • Amazon tracheitis virus is usually lethal, and recovery may be as long as nine months. • Acyclovir may reduce mortality in pigeon herpesvirus.
• Psittacines are prone to relapses of trichomoniasis. Significant morbidity and mortality can be seen. Recovery following treatment may be as rapid as 1–2 days.
• Mycobacterial lesions carry a guarded prognosis and treatment is controversial due to zoonotic potential, particularly in susceptible individuals.

MISCELLANEOUS

ASSOCIATED CONDITIONS
Chronic and recurrent trichomoniasis in pigeons is associated with concurrent herpesvirus.

AGE-RELATED FACTORS
See "Signalment".

(CONTINUED)

ZOONOTIC POTENTIAL
Mycobacteria spp. can be infectious to humans. The potential varies with species within this genus, and the degree to which it occurs is controversial.

FERTILITY/BREEDING
N/A

SYNONYMS
N/A

SEE ALSO
Aspergillosis
Aspiration
Candidiasis
Ingluvial stasis/hypomotility
Flagellate enteritis
Herpesviruses

Intestinal helminthiasis
Mycobacteriosis
Nutritional deficiencies
Poxvirus
Regurgitation/vomiting
Respiratory distress
Rhinitis/sinusitis
Squamous cell carcinoma
Tracheal/syringeal diseases
Trichomoniasis
Viral disease
Appendix 7, Algorithm 8. Oropharyngeal lesions

ABBREVIATIONS
N/A

INTERNET RESOURCES
N/A

Suggested Reading
Gelis, S. (2006). Evaluating and treating the gastrointestinal system. In: Harrison, G.J., Lightfoot, T.L. (eds), *Clinical Avian Medicine*, Vol. I. Palm Beach, FL: Spix Publishing, Inc., pp. 411–440.
Hoefer, H.L. (1997). Diseases of the gastrointestinal tract. In: Altman, R.B., Clubb, S.L., Dorrestein, G.M., Quesenberry, K. (eds), *Avian Medicine and Surgery*. Philadelphia, PA: WB Saunders, pp. 419–453.
Lumeij, J.T. (1994). Gastroenterology. In: Ritchie, B.W., Harrison, G.J., Harrison, L.R. (eds), *Avian Medicine: Principles and Application*. Lake Worth, FL: Wingers Publishing, Inc., pp. 482–521.
Author Elisabeth Simone-Freilicher, DVM, DABVP (Avian)

ORGANOPHOSPHATE AND CARBAMATE TOXICITY

BASICS

DEFINITION
Organophosphates (OP) and carbamates (CA) are common components of many pesticides used in a variety of agricultural, urban, and suburban applications. They are used for lawn and crop pest control, topical parasite control in livestock, and as avicides for control of perceived bird pests. OP and CA use has largely replaced the use of organochlorine compounds, such as DDT, as the former do not accumulate in the environment and tend to break down rapidly. Because OPs and CAs do not persist in the environment, intoxications seen clinically are typically acute and localized to a specific area where pesticide application has recently occurred.

PATHOPHYSIOLOGY
The activities of OP and CA are as cholinesterase inhibitors. Toxic effects are a result of the toxin binding the enzyme acetylcholinesterase (AChE) in the peripheral and central nervous system. AChE breaks down the neurotransmitter acetylcholine (ACh) at synaptic junctions. Therefore, inhibition of the enzyme that breaks down the neurotransmitter leads to potentiation of that neurotransmitter's effects. Increased ACh activity results in hyperstimulation of the parasympathetic nervous system and resulting clinical signs can classified as muscarinic, nicotinic, or central. Muscarinic signs are usually the first to appear and include ptyalism, decreased upper GI motility/crop stasis, diarrhea, and dyspnea caused by contriction of airways and increased airway secretions. Nicotinic and central signs are primarily neurologic and include generalized weakness, tremors, and seizures.

The most clinically significant effects are typically seen in the cardiovascular and respiratory systems. This persistent stimulation leads ultimately to paralysis of the muscles involved in respiration, which is the cause of death in intoxicated birds.

SYSTEMS AFFECTED
- Gastrointestinal signs resulting from OP/CA intoxication include increased saliva production, crop stasis, delayed gastric emptying, vomiting, and diarrhea.
- Neuromuscular effects of OP/CA intoxication include muscle tremors, seizures, and generalized weakness • The primary cardiovascular effect of OP/CA intoxication is bradycardia, which can be pronounced, and associated hypotension. • Respiratory distress due to increased secretions within the respiratory tract is a significant sequela of OP/CA intoxication. Continued cholinergic stimulation will ultimately lead to respiratory failure from exhaustion of respiratory muscles.
- OP/CA exposure may inhibit aquatic birds' abilities to regulate electrolytes by decreasing the activity of Na,K-ATPase in salt glands.

GENETICS
N/A

INCIDENCE/PREVALENCE
OP/CA intoxications occur across the United States and the world. In a retrospective study of the National Wildlife Health Center's mortality database, it was found that 335 avian mortatility events occurred in 42 states between 1980 and 2000, with the greatest number of events occurring in Washington, Virginia, and Ohio. Areas with a high percentage of agricultural activity may have a higher incidence of intoxication events.

GEOGRAPHIC DISTRIBUTION
OP/CA intoxications occur across the United States and the world, whereever OP/CA toxins are are used for pest control.

SIGNALMENT
Waterfowl, raptors, passerines are the groups most commonly affected by anticholinesterase intoxication. Geese and dabbling ducks, which congregate in large numbers in agricultural areas, may be affected by OP/CA by either ingesting treated vegetation or invertebrate prey items or from contact with agricultural run-off. Because these birds often feed together in large flocks, large numbers of waterfowl may be affected at one site. Raptors, especially eagles and hawks, which scavenge as a normal part of their feeding strategy, are most commonly affected by feeding on carcasses that have been treated with OP/CA. However, raptor species that feed on insect prey (such as kestrels and Swainson's hawks) may become intoxicated via this route. Passerine species most commonly affected are those which are commonly found in agricultural areas, such as blackbirds, grackles, and starlings. These species may also be targeted for intentional poisoning due to their perceived nuisance to agricultural sites.

SIGNS
• Affected birds typically in good body condition • Ataxia • Convulsions/tremors • Lethargy • Nictitans prolapse • Respiratory distress • Clenched feet • Ptyalism • Upper GI hypomotility • Bradycardia.

CAUSES
Birds may become intoxicated by OP/CA by the following routes: • **Ingestion-prey items** that have been treated with these toxins ○ Invertebrate prey ○ Carcasses of larger vertebrates that are scavenged ○ Consumption of treated seed ▪ Intentional baiting of nuisance birds ○ Vegetation w/pesticide residue • **Water runoff** • **Inhalation** • **Absorption through skin.**

RISK FACTORS
Risk factors for OP/CA intoxication are primarily related to feeding strategy and environment. Birds who congregate in agricultural areas to feed, such as dabbling ducks and geese, are more likely to be affected in greater numbers. Bird species that feed on carrion or grain also have a higher likelihood of intoxication that other species.

DIAGNOSIS

DIFFERENTIAL DIAGNOSES
• Heavy metal toxicosis. • Hypocalcemia. • Algal toxicosis/Avian Vacuolar Myelinopathy. • Botulism.

CBC/CHEMISTRY/URINALYSIS
Usually unremarkable.

OTHER LABORATORY TESTS
N/A

IMAGING
Usually unremarkable, although there may be ingesta present in the GI tract in acutely intoxicated birds or increased density of the lung tissue associated with increased respiratory secretions.

DIAGNOSTIC PROCEDURES
Definitive diagnosis of OP/CA intoxication is by assessment of AChE activity in the plasma (sublethal intoxications) or brain tissue (lethal intoxications). A decrease of greater than 20% AChE activity is indicative of exposure to OP/CA toxins; a decrease greater than 50% decrease is indicative of lethal exposure. As there are not established normal values for brain or plasma AChE activity for most species of birds, the AChE activity of the patient must be compared with the AChE activity of a normal conspecific. Assessment of AChE activity will confirm exposure to OP/CA toxins, but will not determine a specifc toxin.

Gastrointestinal contents may be assayed for specific toxins. Fat, liver, and kidney samples may also be analyzed for the presence of toxins. However, because OP/CA break down quickly, this is often unrewarding.

PATHOLOGIC FINDINGS
There are no specific gross lesions associaetd with OP/CA intoxication. Affected birds are typically in good body condition and may have food material present in the upper GI tract, indicatative of acute intoxication. Fluid accumulation in lungs consistent with respiratory failure may be found on post-mortem examination.

TREATMENT

APPROPRIATE HEALTH CARE
• Evacuate crop contents to prevent further absorption of ingested toxin. • Respiratory support, including supplemental oxygen therapy and/or assisted ventilation may be necessary in patients demonstrating severe

ORGANOPHOSPHATE AND CARBAMATE TOXICITY (CONTINUED)

respiratory signs. Prognosis in these cases is poor.

NURSING CARE
- Nutritional and fluid support should be provided to patients too debilitated to eat and drink on their own.
- Thermal support may also be necessary, but should be monitored closely as patients having muscle fasiculations may have a tendency to overheat.
- Basic cleanliness should be maintained in patients who are recumbent and unable to preen.
- Large waterfowl, such as geese, who are sternally recumbent, should be housed on heavily padded substrate or on a cushion that elevates the keel off of the substrate to avoid development of decubital ulcers along the keel.

ACTIVITY
N/A

DIET
N/A

CLIENT EDUCATION
N/A

SURGICAL CONSIDERATIONS
N/A

MEDICATIONS

DRUG(S) OF CHOICE
- Atropine 0.5mg/kg IM or SQ TID-QID until symptoms subside.
- Pralidoxime iodide (2-PAM) 20–100mg/kg IM.
- Activated charcoal 2–8mg/kg PO BID until GI tract cleared.

CONTRAINDICATIONS
Pralidoxime iodide treatment may be contraindicated in cases of CA intoxication as some CA inhibit AChE activity.

PRECAUTIONS
N/A

POSSIBLE INTERACTIONS
N/A

ALTERNATIVE DRUGS
N/A

FOLLOW-UP

PATIENT MONITORING
N/A

PREVENTION/AVOIDANCE
N/A

POSSIBLE COMPLICATIONS
Sublethal doses of OP/CA have been shown to result in an inability to thermoregulate, changes in breeding and nesting behavior, and increased susceptibility to traumatic injury.

EXPECTED COURSE AND PROGNOSIS
Prognosis is generally poor without therapy and supportive care. Course and prognosis are related to degree of intoxication.

MISCELLANEOUS

ASSOCIATED CONDITIONS
Sublethal intoxications may result in impairments in mentation and reaction ability that could predispose affected birds to traumatic injuries, such as vehicular trauma or predation.

AGE-RELATED FACTORS
Immature birds are more severely affected by the activities of anticholinesterase toxins.

ZOONOTIC POTENTIAL
N/A

FERTILITY/BREEDING
Direct effects of OP/CA exposure on breeding are not well described. However, loss of invertebrate prey items in an area treated with these toxins may result in increased instances in nest abandonment and changes in incubation behavior. Decreased nest attentiveness and changes in song character have been observed in passerines.

SYNONYMNS
Anticholinesterase toxicity

SEE ALSO
Anticoagulant rodenticide
Botulism
Ingluvial stasis/hypomotility
Diarrhea
Heavy metal toxicity
Hypocalcemia/hypomagnesemia
Neurologic conditions
Respiratory distress
Seizure
Trauma

ABBREVIATIONS
ACh–acetylcholine
AChE–acetylcholinesterase
CA–carbamate
GI–gastrointestinal
OP–organophosphate

INTERNET RESOURCES
http://www.merckmanuals.com/vet/toxicology/insecticide_and_acaricide_organic_toxicity/carbamate_insecticides_toxicity.html

Suggested Reading
Fleischli, M.A., Franson, J.C., Thomas, N.J., et al. (2004). Avian mortality events in the United States caused by anticholinestersase pesticides: A retrospective summary of National Wildlife Health Center records from 1980 to 2000. *Archives of Environmental Contamination and Toxicology*, **46**:542–550.

Franson, J.C., Friend, M. (1999). Organophosphorous and carbamate pesticides. In: Franson, J.C., Friend, M. (eds), *USGS Field Manual of Wildlife Diseases: General Field Procedures and Diseases of Birds*, pp. 287–293.

Grue, C.E., Hart, A.D.M., Mineau, P. (1991). Biological consequences of depressed brain cholinesterase activity in wildlife. In: Mineau, P. (ed), *Cholinesterase-inhibiting Insecticides—Their Impact on Wildlife and the Environment*. Amsterdam, The Netherlands: Elsevier Science, pp. 151–209.

Henry, C.J., Kolby, E.J., Hill, E.F., et al. (1987). Case history of bald eagles and other raptors killed by organophosphorous insecticides toppically applied to livestock. *Journal of Wildlife Diseases*, **23**:292–295.

Hill, E.F. (1995). Organophosphorous and carbamate pesticides. In: Hoffman, D.J., Rattner, B.A., Burton, G.A. Jr., Cairnes, J. Jr. (eds), *Handbook of Ecotoxicology*. Boca Raton, FL: Lewis, pp. 243–274.

Author Shannon M. Riggs, DVM

Osteomyelosclerosis and Polyostotic Hyperostosis

BASICS

DEFINITION
A term describing the radiographic appearance of the avian skeleton as it relates to reproductive/hormonal state. It is characterized by generalized increased medullary opacity to some or all bones that is typically seen 10–14 days before egg formation.

PATHOPHYSIOLOGY
- Polyostotic hyperostosis is a commonly observed finding on radiographs of the female avian patient. • The pathogenesis remains unknown, as the condition is seen in birds with reproductive-associated activity and birds suffering from reproductive disorders.
- Bone marrow ossification occurs secondary to rising estrogen levels and is a physiologic change related to normal egg laying. • Bone marrow ossification may also relate to a pathologic change such as gonadal tumors.
- The increased opacity relates to calcium mobilization and storage for use in eggshell formation.

SYSTEMS AFFECTED
- Musculoskeletal—increased bone opacity from calcium storage in the avian skeleton.
- Reproductive—a finding in reproductively active female birds.

GENETICS
No genetic predisposition as it most often occurs as a normal physiologic change.

INCIDENCE/PREVALENCE
The finding is common in reproductively active hens but is not generally considered a pathologic condition.

GEOGRAPHIC DISTRIBUTION
N/A

SIGNALMENT
- **Species:** No species predilection, as it is seen in all species of birds but is most commonly seen in budgerigars and cockatiels. • **Mean age and range:** Sexually mature hens (sexual and hormonal maturity varies widely among species). • **Predominant sex:** Females of egg-laying age. Although rare, this condition can occasionally occur in male birds in association with gonadal tumors/disease.

SIGNS

General Comments
The skeleton may increase in weight by up to 25% by the replacement of hematopoietic tissue with medullary bone.

Historical Findings
- Behavioral changes secondary to hormonal changes. • Aggressive behaviors such as cage or nest protection. • Broodiness and nesting behaviors, hiding, paper shredding. • Egg laying, may be chronic and uncontrolled.
- Fecal retention secondary to normal nesting behavior. • Lethargy, weakness, or decreased appetite may be observed in birds with chronic egg laying.

Physical Examination Findings
- No directly attributable physical exam findings. • Body weight may increase in relation to reproductive state. • An egg may be palpable in the coelomic cavity.

CAUSES
- Reproductive activity and hormonal changes. • Egg laying. • Calcium mobilization for egg shell formation.

RISK FACTORS
A poor diet without appropriate vitamins and calcium supplementation may lead to pathology.

DIAGNOSIS

DIFFERENTIAL DIAGNOSIS
- Osteomyelosclerosis—disease occurs in non-laying hens and cocks as a result of specific pathology. • Osteopetrosis—induced by avian leucosis virus in chickens.
- Neoplasia including metastatic disease—uncommon, but may represent alterations secondary to reproductive tract neoplasia.
- Metabolic bone disease—a decreased overall opacity to bones that can result in weakness, lethargy, and pathologic fractures. • Coelomic distension—may indicate a reproductively active hen with reproductive tract enlargement.

CBC/BIOCHEMISTRY/URINALYSIS
- Hypercalcemia is commonly seen as a normal physiologic response to egg laying.
- Hypercholesterolemia and hypertriglyceridemia may be seen as a physiologic response to egg laying in the ovulating hen. • Hyperuricemia occurs during egg laying from compression of renal parenchyma.

OTHER LABORATORY TESTS
Protein electrophoresis—hyperglobulinemia may be seen in egg-laying birds.

IMAGING
- Standard radiography is the primary method of identification. • Generalized increased medullary opacity of some or all bones (primarily long bones) is observed.
- Radiography is also useful for evaluation of concurrent or associated medical conditions.

DIAGNOSTIC PROCEDURES
N/A

PATHOLOGIC FINDINGS
- Polyostotic hyperostosis is not considered a pathologic finding alone. • It can be seen in birds that have gonadal tumors.

TREATMENT

APPROPRIATE HEALTH CARE
Required of hens that present with radiographic finding of polyostotic hyperostosis and concurrent medical conditions.

NURSING CARE
As appropriate for other medical conditions that may be diagnosed, no direct care.

ACTIVITY
No restriction of activity level is required in a healthy egg laying hen.

DIET
Hens should be on appropriate calcium supplementation such as egg-shell powder, mineral blocks, or cuttlebone.

CLIENT EDUCATION
- May indicate future egg laying. • May indicate reasons for behavior changes.
- Clients should be educated about chronic egg laying, egg binding, yolk coelomitis, and other reproductive disease of pet birds.
- Leuprolide acetate can be administered if the hen starts to exhibit reproductive behaviors.

SURGICAL CONSIDERATIONS
Surgery is only indicated with concurrent reproductive disease and may include salpingohysterectomy.

MEDICATIONS

DRUG(S) OF CHOICE
- No direct treatment for the radiographic finding. • Leuprolide Acetate, a long acting GnRH analog used to prevent ovulation and egg production. Doses vary: 700–800 mcg/kg IM for birds <300 grams, 500 mcg/kg IM for birds >300 grams q21–30d.

CONTRAINDICATIONS
N/A

PRECAUTIONS
N/A

POSSIBLE INTERACTIONS
N/A

ALTERNATIVE DRUGS
N/A

FOLLOW-UP

PATIENT MONITORING
Routine radiography is useful to determine presence of increased ossification and to determine reproductive status.

Osteomyelosclerosis and Polyostotic Hyperostosis

PREVENTION /AVOIDANCE
- Discourage excessive egg laying by reducing the photoperiod to no more than 10 hours per day by covering the cage at night, keep the hen separate from any mate or perceived mates, reduce foods high in fat and overall quantity of food, and remove nests and nesting material. • Leuprolide acetate can be administered monthly or seasonally or when the hen is exhibiting reproductive behavior to prevent ovulation and egg laying. • A deslorelin implant can also be placed and may prevent egg laying for several months.

POSSIBLE COMPLICATIONS
N/A

EXPECTED COURSE AND PROGNOSIS
N/A

MISCELLANEOUS

ASSOCIATED CONDITIONS
- Reproductive activity. • Egg laying.
- Hormonal behaviors.

AGE-RELATED FACTORS
N/A

ZOONOTIC POTENTIAL
N/A

FERTILITY/BREEDING
- Since elected spaying/sterilization is not routine, any female bird can lay eggs.
- Breeding of pet birds should be discouraged.

SYNONYMS
"egg-laying bones"

SEE ALSO
Cere and skin, color changes
Chronic egg laying
Cystic ovaries / ovarian neoplasia
Dystocia / egg binding
Tumors
Vitamin D toxicosis

ABBREVIATIONS
EPH—protein electrophoresis

INTERNET RESOURCES
http://www.lafebervet.com/avian-medicine-2/avian-emergency-medicine/reproductive-emergencies/

Suggested Reading
Baumgartner, R., Hatt, J.-M., Dobeli, M., Hauser, B. (1995). Endocrinologic and pathologic findings in birds with polyostotic hyperostosis. *Journal of Avian Medicine and Surgery*, 9(4) 251–254.
Bowles, H.L. (2006). Evaluating and treating the reproductive system. In: Harrison, G.J., Lightfoot, T.L. (eds), *Clinical Avian Medicine*, Vol. 2. Palm Beach, FL: Spix Publishing, Inc., pp. 519–540.
Pollock, C.G., Orosz, S.E. (2002). Avian reproductive anatomy, physiology, and endocrinology. *The Veterinary Clinics of North America. Exotic Animal Practice*, **5**(3):441–474.
Stauber, E., Papageorges, M., Sande, R., Ward, L. (1990). Polyostotic hyperostosis associated with oviductal tumor in a cockatiel. *Journal of the American Veterinary Medical Association*, **196**(6):939–940.

Author Anthony A Pilny, DVM, DABVP (Avian)

Otitis

BASICS

DEFINITION
Otitis is defined as inflammation of any of the structures of the ear. It can be further localized based on one of three areas of the ear:
• Otitis externa includes inflammation of the structures surrounding the ear opening and any structures leading to the tympanic membrane such as the external ear canal or external acoustic meatus. • Otitis media is inflammation of the tympanic membrane and structures in the tympanic cavity. Structures in the tympanic cavity include muscles, ligaments, the cochlea, and columella.
• Otitis interna is inflammation of tissues associated with the membranous labyrinth. This includes the chochlear organ and a vestibular organ. In most instances, disease of the middle ear and inner ear occur together and some will refer to otitis media/interna rather than trying to separate these two areas.

PATHOPHYSIOLOGY
Otitis is an inflammatory reaction to some event in one of three areas of the ear. Infection is the most common cause for the inflammatory reaction. Bacterial infection can be the result of invasion of organisms from other areas of the head such as the Eustachian tube, pharynx, sinus, and nasal cavity. Otitis externa may result from opportunistic bacteria on the skin near the ear opening. Connections to the respiratory system, such as the cervicocephalic air sacs and infraorbital sinus, in the middle ear can lead to extension of disease of the upper respiratory system causing otitis media/interna. Inflammation due to infectious disease processes can lead to purulent and serous discharges that may not be able to drain from the ear canal leading to further disease and damage of the ear structures. Neoplastic processes can invade the ear canal from other areas of the skull or originate in the ear canal. One class of antibiotics, aminoglycosides, are a well documented cause of otitis by mechanisms that are not entirely clear but may cause damage to the cochlea.

SYSTEMS AFFECTED
• Nervous—birds may lose the ability to hear. Birds may show signs of imbalance, proprioceptive deficits, head tilt, torticollis.
• Ophthalmic—as a continuation of an upper respiratory or sinus infection, there may be conjunctivitis also present. • Respiratory—if otitis, especially interna/media, can be a secondary result of a sinus or upper respiratory infection, there may be signs of upper respiratory disease present. • Skin/Exocrine—if otitis externa is present, there may be discharge around the ear opening and some feather loss in the area.
• Gastrointestinal—severely affected birds maybe anorexic.

GENETICS
Macaws may hatch with a thin membrane covering the ear canal that should be gone by day 35 posthatch. If this membrane remains intact, it can lead to inflammation of the outer ear canal.

INCIDENCE/PREVALENCE
Otitis externa, otitis media, and otitis interna are all uncommon to rare conditions in all orders of birds. Of the three locations for disease, otitis externa is the most common. And bacterial infection is the most common cause of otitis externa. Otitis media and interna remain rare disease diagnoses in birds.

GEOGRAPHIC DISTRIBUTION
N/A

SIGNALMENT
• **Species:** All orders of birds are susceptible to otitis. In companion animal practice, lovebirds are more frequently seen with otitis externa than other species of birds. • **Mean age and range:** Any age bird is susceptible.
• **Predominant sex:** No predilection for male or female.

SIGNS
Historical Findings
• Otitis externa—owners may report increased scratching and preening of the skin and feathers near the ear opening. The owners may report feather loss near the ear opening. Crusted feathers and discharge around the ear opening may be seen. If bilateral disease, this will be seen on both sides of the head. Observant owners may also report hearing loss. • Otitis media/interna—owners may report lack of balance as evidenced by falling off the perch, not able to perch, unable to fly, or a head tilt. Signs of respiratory disease may also be seen.

Physical Examination Findings
• Otitis externa: ○ The skin around external ear openings is scaly with absence of feathers. ○ Purulent or crusted discharge is around ear opening. ○ The opening may become swollen and is no longer clearly visible. ○ Exudate, swelling, purulent discharge seen in ear canal with otoscope. ○ Patient does not respond to loud noises if hearing loss. ○ Growths around the ear canal. ○ Swelling under the skin near the tympanum. • Associated signs: Nasal discharge, conjunctivitis, sneezing. • Otitis media/interna: Head tilt, torticollis, stargazing, proprioceptive loss, ataxia, nystagmus.

CAUSES
• Bacterial: Salmonella, Corynebacterium, *Escherichia coli*, *Pseudomonas aeruginosa*, *Staphylococcus aureus*, *Pasteurella multocida*, *Klebsiella oxytoca*, *Proteus mirabilis*, *Kocuria kristinae*, *Ornithobacterium rhinotracheale*, *Enterococcus* species, *Mycobacteria* spp. • Fungi: *Candida* spp, *Microsporum gallinae*, *Aspergillus* sp. • Parasitic: *Knemidocoptes*, harvest mites, fleas, *Cryptosporidium baileyi*. • Viral: Paramyxovirus, poxvirus. • Foreign body: Possible foreign object, such as a piece of grass or hay, can lodge in an ear canal. • Neoplasia: Carcinoma of the ceruminous gland of the ear, squamous cell carcinoma.

RISK FACTORS
N/A

DIAGNOSIS

DIFFERENTIAL DIAGNOSIS
• Otitis externa: No other disease condition will result in skin and feather changes around the external ear opening. • Otitis media/interna: Neurologic signs associated with this condition can also be caused by central nervous system disease, any disease condition that causes whole body weakness.

CBC/BIOCHEMISTRY/URINALYSIS
• CBC: Possibly increased white blood cell count due to an absolute heterophilia. Chronic disease may lead to an increase in monocytes. • Plasma biochemistry: Not associated with abnormalities. • Urinalysis: N/A

OTHER LABORATORY TESTS
• Culture and sensitivity testing: Bacterial and fungal infections can be diagnosed in this manner. • Cytology and skin scrapings: Can be used to diagnose parasitic disease.
• Biopsy: If neoplasia is suspected, this will be necessary to rule out chronic infection versus neoplasia. • Viral testing as necessary.

IMAGING
Diagnostic imaging, such as skull radiographs, is usually normal. If otitis includes skeletal structures, disease may be present on plain films. In larger species of birds with the use of CT or MRI machines of high resolution, disease may be apparent in otitis media/interna.

DIAGNOSTIC PROCEDURES
In larger species, small diameter endoscopes maybe useful to explore the external ear canal and remove tissue and debris for cytology and culture.

PATHOLOGIC FINDINGS
Inflammation and infectious disease organisms can be seen on tissue samples from area affected by otitis.

TREATMENT

APPROPRIATE HEALTH CARE
Most of these patients are stable and do not require in-hospital treatment. Most patients are treated as outpatient. In some cases,

(CONTINUED)

debridement of discharge in or around the external ear canal may be needed.

NURSING CARE
Nursing care is necessary in cases where otitis media/externa is present and the patient has severe vestibular disease and cannot perch nor stand properly. In those cases, most patients also need to be fed as they are anorexic. Necessary care may include a padded enclosure with no perches until the patient is more stable. Assist feedings and parenteral fluids may be necessary until the patient is stable and able to drink and eat on its own.

ACTIVITY
N/A

DIET
N/A

CLIENT EDUCATION
If the patient had torticollis or a head tilt as a result of otitis media/interna, those conditions may not fully resolve once the disease condition has been treated. Clients should be instructed on how to set up an enclosure for a patient with chronic head tilt or torticollis.

SURGICAL CONSIDERATIONS
N/A

MEDICATIONS

DRUG(S) OF CHOICE
• Antibiotic selection is based on culture and sensitivity results. Antibiotics can be given orally. They can also be applied to the area around the opening of the ear canal and into the ear canal. Topical water-based or ophthalmic antibiotic solutions are preferable to oil-based preparations. • NSAIDS: Meloxicam 0.5 mg/kg q12h, PO. • Possible saline ear canal flushes. • Analgesics as necessary.

CONTRAINDICATIONS
• Aminoglycosides should be avoided as they are toxic to inner ear structures. • Topical and systemic steroids are contraindicated as they may exacerbate signs associated with infection.

PRECAUTIONS
Avoid vigorous flushing around the opening of the ear canal.

POSSIBLE INTERACTIONS
N/A

ALTERNATIVE DRUGS
Acetic and boric acid commercial solution if a bacterial infection is suspected.

FOLLOW-UP

PATIENT MONITORING
Follow up within a week of diagnosis and treatment to observe effectiveness of antibiotics in case of bacterial or fungal disease.

PREVENTION/AVOIDANCE
N/A

POSSIBLE COMPLICATIONS
Long-term hearing loss. Long-term difficulty balancing on perch or flying.

EXPECTED COURSE AND PROGNOSIS
Uncomplicated infection that is susceptible to drug treatment should resolve in 10–14 days. More complicated disease, that is associated with osteomyelitis, may take much longer to resolve. Disease involving structures associated with balance and proprioception may not recover to a normal state.

MISCELLANEOUS

ASSOCIATED CONDITIONS
Upper respiratory disease.

AGE-RELATED FACTORS
N/A

ZOONOTIC POTENTIAL
N/A

FERTILITY/BREEDING
N/A

SYNONYMS
N/A

SEE ALSO
Heavy metal toxicity
Hypocalcemia/hypomagnesemia
Lameness
Mycoplasmosis
Neurologic conditions
Organophosphate/carbamate toxicity
Pituitary/brain tumors
Rhinitis and sinusitis
Seizure
Trauma

ABBREVIATIONS
N/A

INTERNET RESOURCES
N/A

Suggested Reading
Doneley, B. (2010). *Avian Medicine and Surgery in Practice: Companion and Aviary Birds*. London: Manson Publishing.
Martel, A., Haesebrouck, F., Hellebuyck, T., Pasmans, F. (2009). Treatment of otitis externa associated with *Corynebacterium kroppenstedtii* in a peach-faced lovebird (*Agapornis roseicollis*) with an acetic and boric acid commercial solution. *Journal of Avian Medicine and Surgery*, **23**:141–150.
Rival, F. (2005). Auricular diseases in birds. *Proceedings of the 8th European Association of Avian Veterinarians Conference*, Arles, France, April 24–30.
Shivaprasad, H. (2007). The avian ear: anatomy and diseases. *Proceedings of the Association of Avian Veterinarians Annual Conference*.
Author Karen Rosenthal, DVM, MS

Ovarian Cysts, Neoplasia and Cystic Ovaries

BASICS

DEFINITION
Cystic ovarian disease is somewhat common in a variety of pet bird species, but occurs in higher frequency in budgerigars, cockatiels, canaries and pheasants. The exact cause is unknown, although an endocrine imbalance is suspected. The disease is characterized by a single or multiple cystic structures on the ovary.

Ovarian neoplasia can occur in any species and predominantly affects mature or older avian subjects. Cockatiels, budgerigars and chickens are more commonly affected amongst all commonly known species. The ovary is usually consumed by a proliferative, firm, multinodular mass when first identified or visually examined. The physical appearance may either be a flattened mass with a grainy surface or an enlarged ovary with a multitude of varying sized follicles. Ovarian neoplasms are typically classified by tissue of origin.

PATHOPHYSIOLOGY
Cystic ovarian disease is suspected to be an endocrine related disease affecting the ovary of a variety of avian species. A secondary condition is hyperostosis of long bones, noted both on radiographic appearance and histopathology sections of bones of the extremities.

The cause of ovarian neoplastic diseases is usually thought to be either a genetic predisposition (especially in psittacine birds) or of a herpesvirus induced disease, also known as Marek's Disease, in gallinaceous birds.

SYSTEMS AFFECTED
- Reproductive—one or more cystic structures may be noted on the ovary. Coelomic distention is a common secondary clinical sign. Reproductive performance may be affected negatively. Infertility is common in both disease states. Oviductal neoplasms may occur concomitantly with ovarian neoplasms. • Musculoskeletal—hyperostosis of the bones of the extremities is commonly noted on radiographs in both diseases. Pectoral muscle mass loss is common in advanced stages of both disease states.
- Hepatobiliary—metastasis of ovarian carcinomas to the liver, spleen or other internal organs may occur.
- Gastrointestinal—body wall herniation with incorporated loops of bowel may occur with advanced cases of either disease. Feces may collect on the vent or even pile up due to coelomic distension and dorsal displacement of the vent by the swollen coelom or by distension from body wall herniation.
- Neuromuscular—left leg lameness may occur if the neoplasm applies pressure to the ischiatic nerve. • Behavioral—affected birds may show nesting behavior. If coelomic distention is severe, patients may show lethargy, ataxia, or respiratory distress.

GENETICS
Suspected to be genetically inherited in budgerigars and cockatiels.

INCIDENCE/PREVALENCE
- Cystic ovarian disease: Fairly common condition in cockatiels and budgerigars. Can be noted in a variety of other species. Eclectus parrots seem to have a higher incidence amongst the larger psittacines. • Ovarian neoplasia: In psittacine birds, budgerigars and cockatiels are overrepresented, suspected to be from a genetically inherited trait. In chickens that have not been vaccinated for Marek's Disease and/or have been exposed to carrier birds or birds symptomatic of Marek's Disease, the incidence is fairly high.

GEOGRAPHIC DISTRIBUTION
N/A

SIGNALMENT
- **Species:** Can occur in any avian species.
- **Species predilections:** Budgerigar, cockatiel, eclectus parrot, poultry. • **Mean age and range:** Affected birds are sexually mature and have usually had several seasons of production or breeding. • **Predominant sex:** Female.

SIGNS

General Comments
- Initially, the signs of ovarian cysts are hidden from human inspection until the patient becomes clinically affected. • Large ovarian cysts and ovarian neoplastic masses often create coelomic distention from just the sheer size of the cyst/neoplasm or from ascites. • Infertility may be noted in breeder birds and initiate the reason for an endoscopic view of the ovary. • Radiographs often reveal bilateral hyperostosis of long bones (radius, ulna, femur and tibiotarsus). In severely affected birds, all bones may show a bright homogeneous, mineral opacity. Large cysts or neoplasms will often reveal a soft tissue dense mass immediately cranial to the anterior renal pole on a true lateral radiographic projection.
- Some birds exhibit nesting behavior for long periods of time with no egg production.

Historical Findings
Many cases are presented as 'Egg bound' due to the coelomic distention with no recent egg production. Breeders note infertility as a reason for the examination.

Physical Examination Finding
Most cases of both cystic ovarian disease and ovarian neoplasia present with some degree of coelomic distention. Coelomic palpation often reveals a fluid filled coelom with no egg present. In severe cases of coelomic distention caused by large ovarian cysts or ascites, dyspnea and exercise intolerance may be noted.

CAUSES
- Endocrine related metabolic imbalances are assumed to be the initiating factor for cystic ovarian disease in psittacines and passerines.
- Ovarian neoplasia may be an end stage disease of cystic ovarian disease. Lymphoma of the ovary secondary to Marek's Disease is common in chickens. Other ovarian neoplasias may occur without a specific etiologic agent.

RISK FACTORS
- Allowing chronic hormonal stimulation in species known to be prone to excessive egg laying (budgerigars, cockatiels, and eclectus) may predispose to cystic ovarian disease.
- Marek's related ovarian lymphoma is common in backyard flocks and/or small breeder setups where baby chicks are exposed to carrier or unvaccinated adults.

DIAGNOSIS

DIFFERENTIAL DIAGNOSIS
- Coelomic swelling due to ascites may be secondary to peritonitis, liver disease, cardiac disease or oophoritis. • Coelomic swelling with a firm intracoelomic mass may be caused by a retained egg, ectopic egg or a neoplastic disease of an abdominal organ. • Dyspnea due to coelomic distention may be also noted with ascites or neoplasia of an organ that has displaced or compressed abdominal air sacs.
- Any infectious, inflammatory, fungal, metabolic or neoplastic disease affecting the ovary or oviduct may be implicated with respect to infertility issues.

CBC/BIOCHEMISTRY/URINALYSIS
- Complete blood counts in cases of cystic ovarian disease will vary from normal to a picture of an inflammatory hemogram where ascites is a part of the clinical picture.
- Ovarian neoplastic diseases often reveal normal hemograms unless lymphoid neoplasia is involved. Lymphoma of the ovary is a common sequelae in Marek's cases. As with other lymphoid neoplastic disease cases, circulating anaplastic or neoplastic lymphocytes may be apparent in white blood cell differential counts. The PCV may be low if there is metastasis to the bone marrow.
- Blood chemistry profiles generally show elevations of AST and CK, indicative of muscle loss and/or internal inflammatory disease.

IMAGING
- Whole body survey radiographs in pure cystic ovarian diseases may show a fluid or soft tissue dense, spherical space occupying mass or masses juxtaposed to the cranial pole of the renal silhouette on the lateral view. • Cases involving ascites will produce a homogeneous fluid density encompassing the vast majority or the entire space of the abdominal coelom.

Ovarian Cysts, Neoplasia and Cystic Ovaries (Continued)

- Homogenous osseous density of long bones termed hyperostosis is a common finding in ovarian cystic disease and many cases of ovarian neoplastic disease. This infers an endocrine imbalance involving calcium mobilization. • Ovarian neoplasias produce a soft tissue dense space occupying mass cranial to the cranial division of the kidney in a lateral view. The shape may be lobulated to spherical. • Barium contrast radiography has a great benefit in relation to space occupying coelomic masses. On a lateral projection, barium filled loops of intestines are displaced ventrally in cases of ovarian cysts and ovarian neoplasia. Ventro-dorsal (V/D) radiographic projections of both ovarian cystic disease and ovarian neoplasia generally reveal displacement of barium filled ventriculus and/or intestinal loops to the center or right side of the coelomic cavity. • Ultrasonography of the coelom is extremely valuable if performed and read by a radiologist or someone with experience and training in identifying anatomical abnormalities of the avian ovary. Ultrasonography of the normal ovarian ducts and ovaries is challenging and visualization of the ovaries depends on their size and quality of the ultrasound equipment and user experience. Ovarian follicles appear as rounded, variably echogenic structures, where the echogenicity changes with maturation, increasing in echogenicity as the yolk develops. Ovarian cysts can be single or multiple. Occasionally, effusion of an undetermined source may be the only finding when ovarian cysts are present. They appear sonographically as clearly defined, rounded anechoic, thin walled compartments with distal acoustic enhancement. Ovarian neoplasms typically appear as large masses. They are well defined structures, usually easily detected, large, rounded masses of mixed echogenicity, seen sonographically as marked focal or diffuse inhomogenous echotexture.
- Coelomic CT scans have the potential to be valuable in mid and large size psittacines, galliformes and anseriformes. There are several limiting factors in reference to the ability of the scan to deliver an accurate diagnostic image. An extremely critical factor is the capability of the CT unit. Many machines used in the veterinary market are not advanced enough to provide images on small patients with small size pathological issues. Slices smaller than 2 mm are generally needed to diagnose masses or cystic structures in our avian patients. Secondly and as equally important, the size of the patient and the corresponding size of the mass must be of adequate size for the specific machine being utilized for the CT scan to identify the pathologic abnormality on the ovary.

DIAGNOSTIC PROCEDURES
Coelomic endoscopy is diagnostic in cases of both cystic ovarian disease and ovarian neoplasia where ascites is not the predominant clinical presentation. Cystic masses on the ovary may be an incidental finding in cases of infertility investigation or when radiographs indicate potential ovarian disease. Solid, nodular soft tissue masses both small and large can be visualized on endoscopy of the left abdominal coelom in cases of ovarian neoplastic diseases.

PATHOLOGIC FINDINGS
- Depending on the severity of the ovarian cystic disease, surgical or necropsy visual findings reveal one of the following: one large cystic structure attached to the surface of the ovary that may be several millimeters to several centimeters in diameter depending on the species involved or numerous medium size ovarian cystic structures that fill up a large area in the abdominal coelom. The cysts are well encapsulated and filled with either clear fluid or yolk material. In cockatiel species, the cysts often take up 50–75% of the coelom. These cysts are easy to aspirate with a 25–26 gauge needle or winged infusion set attached to a 3–12cc syringe. Care must be taken in small species not to suction the entire amount of the fluid contents as this may create a metabolic imbalance by drawing off fluid that contains protein and electrolytes that the patient uses to create a homeostatic metabolic environment. These cystic structures are loosely attached to the ovarian parenchyma yet often have a vascular supply at the attachment site. • Ovarian neoplastic disease may take on several shapes and sizes in avian species. These cancerous masses generally involve the entire ovarian parenchyma. Ovarian neoplasms may create dozens of medium size follicles that take up a large portion of the coelom. Certain types of ovarian neoplasms may produce a large solid mass that engulfs the entire ovary. The following types of ovarian tumors have been reported: lymphoma, adenocarcinoma, leiomyosarcoma, adenoma, and granulosa cell tumors. In poultry, Marek's induced lymphoma is very common. Granulosa cell tumors are yellow lobulated irregular masses that are friable. They may become quite large and fill the abdominal cavity. Ovarian carcinomas are variably sized, firm, grey-white, and multilobular. They may implant on serosal surfaces of adjacent organs, the mesentery, and body wall. They may also metastasize. Arrhenoblastomas are grey-white lobulated masses comprised of structures resembling seminiferous tubules. Carcinomatosis has been reported in poultry as sequelae to ovarian neoplastic disease.
- Histopathology of ovarian cystic disease reveals cysts lined by flattened cells caused by trapped surface mesothelium. Granulosa cells line follicular cysts. • In reference to histopathology of ovarian neoplastic disease, the findings will vary greatly depending on the specific cell type of ovarian neoplasm present. Granulosa cell tumors are comprised of nests and trabeculae of slightly pleomorphic cells with eosinophilic cytoplasm and vesicular nuclei separated by variable amounts of stroma. Histologically, carcinomas are comprised of poorly differentiated epithelial cells forming cords, acini, and papillary structures.

TREATMENT

APPROPRIATE HEALTH CARE
Patients exhibiting symptoms of lethargy, coelomic distention and/or dyspnea will benefit by being hospitalized in a quiet, warm environment. Dyspneic patients should be placed in an oxygen cage. Cases presenting with a large fluid distended coelom, will need coelomocentesis to remove up to 50% of the suspected ascitic fluid to relieve some of the pressure and the air sac displacement.

NURSING CARE
- Placing the patient in a warmed hospital cage that is secluded and has dimmed lighting will help the bird's homeostasis mechanisms. Anorectic patients should be gavage fed with an avian nutrient formula for sick birds. Intravenous fluids may be required for critically depressed patients. Patients with ovarian neoplasia need supportive care of heat, fluids and nutrition while medical or surgical management is contemplated.
- Removal of the female from any conspecific male birds, nesting material or hiding places and removal of mirrors and other self-stimulating objects from the cage or home environment is advised for female birds with cystic ovarian disease. Birds with ascites or large fluid filled cysts should have the fluid aspirated to relieve the physical pressure of the fluid or cyst on the respiratory system.

ACTIVITY
If the patient is totally nonsymptomatic, no restriction on activity will be necessary. Weak or lethargic patients need to be hospitalized in a quiet, warm environment. Restricted activity may be beneficial in both cases to conserve the patient's strength and keep respiratory efforts at a baseline level.

CLIENT EDUCATION
- Cystic ovarian disease is a chronic recurring condition. Life-long changes will be necessary to help control the disease. Manipulation of the environment is essential to diminish hormonal stimulation that may lead to ovarian cyst formation. Changes to the home environment must include removal of all nest boxes, nesting materials and mirrors. Preferably the female should be kept in a different room from any conspecific male and owners should refrain from stroking or petting the bird on the back and not let the bird hide under blankets, sheets or towels. Owners must watch for signs of coelomic

Ovarian Cysts, Neoplasia and Cystic Ovaries (Continued)

swelling, respiratory distress or nesting type behavior. Recurring symptoms must be attended to quickly. • Avian patients with ovarian cancer are in need of frequent re-assessments. Even with surgical removal or debulking, chemotherapy or radiation therapy, neoplasms of the ovary may grow at an extraordinary rate. Coelomic swelling or pressure on surrounding renal tissue needs to be re-checked on a routine basis, the timing of which will depend on the species, size of the original mass and overall health of the patient. A re-check examination by endoscopy, radiology or ultrasonography is required to re-assess and evaluate true changes in size of the neoplasm. • Prevention is key in reference to Marek's induced disease in chickens—advise owners to purchase vaccinated chicks and maintain closed flocks.

SURGICAL CONSIDERATIONS
Surgery for correction of ovarian cystic disease is risky and should only be attempted by an accomplished avian surgeon and only after coelomocentesis and medical therapy have failed. Exploratory surgery will be a challenge on these patients. Their respiratory status is challenged by the sheer size of the mass. They are generally in a catabolic state and if the abdominal air sacs are incised and the cystic fluid enters this space, the patient may drown. The patient should be placed on a surgical board with head elevated. The surgical approach may be with the patient in dorsal recumbency for a left lateral approach or in right lateral recumbency for a left lateral approach. Extreme care must be taken on the surgical approach. Inside the coelom, the intestinal loops are often displaced and may be in a location that will interfere with the usual entry location. Once in the coelom, the cystic structure is identified and often aspirated to collapse the structure so the ovary and surrounding structures may be identified. Once all vital structures (kidney, intestinal loops, ureter) have been identified, the surgeon may either ligate the cystic structure with Hemoclips or a surgical ligature. Radiosurgical excision with and without endoscopy has been performed. Extreme care must be taken not to ligate the anterior pole of the kidney, the renal artery or the ureter. The larger the species, the easier it is to visualize the vital structures around the ovary.

Ovarian neoplastic diseases are generally very difficult to resolve surgically. In large species (macaws, raptors, chickens, ducks) there exists the potential for a successful surgical outcome. In general, ovarian neoplasms encompass the entire ovary and are closely attached to the cranial pole of the left kidney, making surgical excision difficult and risky.

MEDICATIONS

DRUG(S) OF CHOICE
• There are several therapeutic options for treating ovarian cysts in the avian patient. Human Chorionic Gonadotropin (HCG) has been used for decades in avian patients to control excessive layers. Leuprolide acetate, a gonadotropin-releasing hormone agonist, has been used for over two decades to treat excessive egg laying in psittacines.
• Deslorelin acetate, an implant formulation of a gonadotropin-releasing hormone agonist, has recently been used anecdotally to treat a variety of ovarian diseases in avian species. Reports from several practitioners reveal some cases have shown diminishment of cyst size with the use of deslorelin acetate.
• Cabergoline, a potent dopamine receptor agonist, may be used to treat cystic ovary patients. It works in humans by reducing prolactin release from the pituitary gland.
• Ovarian neoplastic masses have shown size reduction with the use of deslorelin acetate.
• Dosages for the various pharmaceutical agents mentioned above are: ○ Cabergoline: 10–20 ug/kg q24h or as needed. ○ Deslorelin acetate (implant): 4.7 mg SQ. ○ HCG: 500–1000 units/kg IM on days 1,3,7 q3–6 weeks. ○ Leuprolide acetate: 200–800 ug/kg IM q14 days for three treatments (average dosage recommendation is 500 ug/kg).

CONTRAINDICATIONS
N/A

PRECAUTIONS
• Depo-Provera, medroxyprogesterone has been used in years past to treat excessive egg laying and ovarian cyst formation. This drug has been proven to cause fatal hepatic lipidosis in avian species. It is not recommended for use since other safer alternatives are available. • The use of Deslorelin acetate in "nonferret" species is considered extra-label and is prohibited. Use in chickens, a food animal as categorized by the FDA, may actually be illegal.

POSSIBLE INTERACTIONS
N/A

ALTERNATIVE DRUGS
• Releaves, a holistic supplement consisting of palm fruit and raspberries, has been suggested to diminish egg laying in psittacines and therefore may be of benefit to shrink ovarian cysts. • Alternative treatments or therapies have not been documented to provide a benefit for ovarian neoplastic diseases.

FOLLOW-UP

PATIENT MONITORING
Weekly, monthly or quarterly re-check appointments are recommended to assess shrinkage of the cystic or neoplastic ovarian mass. Referral to a facility that has endoscopy, ultrasound or CT capability will provide more accurate assessment of changes in size of the mass.

PREVENTION/AVOIDANCE
• Hormone stimulation is the main at home avoidance therapy that will be useful to slow or prevent recurrence of cystic structures. • In specific reference to chickens, vaccinating day old chicks against Marek's Disease or purchasing Marek's vaccinated young chicks and maintenance of a closed flock is the best way to reduce the potential for Marek's associated ovarian lymphosarcoma.

POSSIBLE COMPLICATIONS
• In reference to ovarian cystic disease, recurrence is likely. Close attention needs to be paid to coelomic swelling, dyspnea, lethargy, depression and/or anorexia.
• Patients with ovarian neoplastic masses may show acute symptoms of depression, lethargy, anemia and/or dyspnea at any time frame post diagnosis or initiation of therapy.

EXPECTED COURSE AND PROGNOSIS
Both diseases have a high potential for progression. Death or elective euthanasia may be possible outcomes for both ovarian cystic disease and ovarian neoplastic diseases.

MISCELLANEOUS

ASSOCIATED CONDITIONS
Both ovarian cystic disease and ovarian neoplastic diseases may cause coelomic distension, dyspnea and/or ascites.

AGE-RELATED FACTORS
Both conditions occur in mature female birds.

ZOONOTIC POTENTIAL
N/A

FERTILITY/BREEDING
Both diseases often result in an infertile hen.

SYNONYMS
Ovarian cysts
Carcinomatosis

OVARIAN CYSTS, NEOPLASIA AND CYSTIC OVARIES (CONTINUED)

SEE ALSO
Ascites
Cere and skin, color changes
Chronic egg laying
Coelomic distention
Dystocia/egg binding
Egg yolk coelomitis / reproductive coelomitis
Lymphoid neoplasia
Infertility
Lameness
Salpingitis / uterine disorders
Tumors

ABBREVIATIONS
N/A

INTERNET RESOURCES
N/A

Suggested Reading
Bowles, H.L. (2002). Reproductive diseases of pet bird species. *The Veterinary Clinics of North America. Exotic Animal Practice*, **5**:489–506.
Hofbauer, H., Krautwald-Junghanns, M.E. (1999). Transcutaneous ultrasonography of the avian urogenital tract. *Veterinary Radiology & Ultrasound*, **40**(1):58–64.
Latimer, K.S. (1994). Oncology. In: Ritchie, B.W., Harrison, G.J., Harrison, L.R. (eds), *Avian Medicine: Principles and Application*. Lake Worth, FL: Wingers Publishing, Inc., pp. 640–672.

Orosz, S., Dorrestein, G.M., Speer, B.L. (1997). Urogenital disorders. In: Altman, R.B., Clubb, S.L., Dorrestein, G.M., Quesenberry, K. (eds), *Avian Medicine and Surgery*. Philadelphia, PA: WB Saunders, pp. 614–644.
Schmidt, R.E, Reavill, D.R, Phalen, D.N. (2003). Reproductive diseases. In: Schmidt, R.E., Reavill, D.R., Phalen, D.N., *Pathology of Pet and Aviary Birds*, 2nd edn. Ames, IA: Iowa State Press, pp. 109–120.

Author Gregory Rich, DVM, BS Medical Technology

Client Education Handout available online

Overgrown Beak and Nails

BASICS

DEFINITION
Overgrowth of the avian beak and/or nails is a fairly common clinical presentation for captive avian species, the majority of which are psittacine pets. Both the beak and nails of birds continually grow, are worn down daily through environmental interactions, and are continuously being replaced. The avian beak can be considered overgrown when the length or shape interferes with normal activities including feeding and gripping. What is considered an overgrown beak may vary with the individual bird as well as the specific species of bird. Nails can be considered overgrown when the arc extends more than 180 degrees, extends parallel to the plane of the metatarsus, grow in abnormal directions, become caught on inanimate objects such as the cage or a towel, or become sharpened such that the human caretaker experiences discomfort when handling the bird.

PATHOPHYSIOLOGY
The underlying mechanisms that result in overgrowth of the avian beak specifically involve the horny sheaths that cover the bones of the beak, the rhamphotheca. The maxillary rhamphotheca (also known as the rhinotheca) and the mandibular rhamphotheca (also known as the gnathotheca) are essentially thick, keratinized integuments that are comprised of a dermis and a modified epidermis. For overgrowth to occur, either external and/or internal changes occur to the tissue such that the normal relationship of building up, wear down and replacement of tissue is disrupted.

The mechanisms that result in overgrowth of the bird's nails similarly involve the horny or keratinized integument that extends from the distal digits. The normal mechanism of growth is such that the heavily keratinized dorsal plate which covers the dorsal and lateral surfaces grows faster than the softer ventral plate occurring on the ventrum of the digits. When this normal relationship is disrupted, either from external and/or internal changes to the tissue and growth pattern, nail overgrowth can occur.

SYSTEMS AFFECTED
• Skin/Exocrine. • Behavioral—may result in reluctance to move normally (with nail overgrowth) and reluctance to use the beak as a tool (with beak overgrowth) such as with preening and feeding. • Musculoskeletal—may result in injury and straining to the musculature and supporting soft tissue structures of the leg if the nails become entangled such as can occur with a towel or blanket. • Nervous—nerve damage may result in a situation such as is described above in which the bird is found hanging for some time by an entangled overgrown nail. • Respiratory—damage to the beak where the nares are located can result in chronic or temporary respiratory issues such as chronic rhinitis.

GENETIC
Beak overgrowth such as occurs in scissor beak deformity and mandibular prognathism in young birds, may in part have a genetic basis for these beak growth abnormalities.

INCIDENCE/PREVALENCE
Common among captive avian species, particularly in psittacine birds.

GEOGRAPHIC DISTRIBUTION
N/A

SIGNALMENT
• **Species:** Primarily psittacine birds. • **Species predilections:** Beak growth abnormalities such as scissor beak deformity and mandibular prognathism have been most commonly observed in young cockatoos and macaws. • **Age:** ○ Most commonly occur once the bird has fledged and the keratinized tissues of the beak and nails have hardened. ○ Psittacine beak and feather disease (PBFD) commonly occurs in birds less than two years of age, although older birds that survived the virus may be presented intermittently for abnormal beak overgrowth. • **Predominant sex:** N/A

SIGNS
Abnormal appearance and/or shape to the beak or nails.

Physical Examination Findings
Overgrown beaks:
• May notice changes resulting from trauma such as bleeding, punctures, fractures or avulsion. • Scissor beak deformity such that the premaxilla/maxilla/rhinotheca laterally deviate to the left or right, while the mandible/gnathotheca overgrown in the opposing direction. • Complete or partial separation or splitting of the mandible into two sections. • Mandibular prognathism such that the mandible abnormally extends over the maxilla. • Presence of small honeycomb proliferative lesion involving the unfeathered areas of the bird, such as the beak, feet and legs, resulting in an abnormally growing beak and appearance to the legs and feet.
• Overgrown, abnormally shaped, easily fractured beak such as can be seen with infection by psittacine circovirus (PCV) in psittacine beak and feather disease. • Nodular growth on the beak from a tumor can present as an overgrown or abnormally growing beak.

Overgrown nails:
• Nails that are overgrown such that they no longer maintain a typical 180 degree arch. • Nails that induce pain when the bird is handled. • Nails that become entangled in a towel when the bird is handled.

CAUSES
Overgrown beaks:
• Trauma—including attacks by other birds or pets, attacks from wild animals such as raccoons, blunt trauma such as collision with a solid surface and beak fractures. • Trauma specifically to the growth areas of the beak sometime in the bird's past. • Malnutrition or nutritional imbalances. Includes nutritional secondary hyperparathyroidism. • Underlying medical conditions. • Malocclusion from multiple etiologies. • Infection with *Knemidokoptes pilae* mites (scaly face mite). • Infection of the beak with bacterial agents. • Infection of the beak with fungal agents, including aspergillosis. • Systemic infection with psittacine circovirus or PCV resulting in abnormal beak growth. • Presence of neoplasia in the beak such as a squamous cell carcinoma. • Possible undetermined genetic causes.

Overgrown nails:
• Trauma—including attacks by other birds or pets, getting nails caught or tangled in something, nail fractures, and traumas of all types that damage the nail growth beds. • Infection with *Knemidokoptes pilae* mites. • Systemic infection with psittacine circovirus or PCV resulting in abnormal nail growth. • Neoplasia involving the digits and/or nails. • Abnormal wear due to decreased movement with age or chronic illness. • Inappropriate perch sizes and surfaces resulting in decreased wear of the nails.

RISK FACTORS
• Poor husbandry. • Poor or inappropriate management. • Malnutrition. • Exposure to other birds, pets and animals where the bird is not protected. • Free flying birds such that collision traumas occur.

DIAGNOSIS

DIFFERENTIAL DIAGNOSIS
Beak that appears overgrown yet is normal for a particular species, such as with a long-billed corella (*Cacatua tenuirostris*).

CBC/BIOCHEMISTRY/URINALYSIS
Some clinicians suggest there is a link between liver disease and beak overgrowth in certain species including budgerigar, cockatiels and Amazon parrots. In this situation, liver inflammation and function tests may be elevated.

Performing a CBC with biochemistries is always good medicine for an abnormal presentation.

OVERGROWN BEAK AND NAILS

(CONTINUED)

OTHER LABORATORY TESTS
• Fungal and bacterial cultures and cytology if a beak infection is suspected. • Cytology if suspect *Knemidokoptes pilae* infection. • PCR testing for PBFD if suspected. • Needle aspirate or tissue biopsy of suspected neoplasia.

IMAGING
• Radiographs maybe indicated to further characterize beak involvement in certain cases. • MRI or CT scans may be indicated to further characterize beak involvement in certain cases.

DIAGNOSITC PROCEDURES
• Skin scraping if suspect *Knemidokoptes pilae* mites. • Biopsy of abnormal beak or nail lesions.

PATHOLOGIC FINDINGS
Gross and histopathologic findings will differ depending upon the underlying condition.

TREATMENT

APPROPRIATE HEALTH CARE
• It is important to understand that the vast majority of birds presented for routine "grooming" will have a normal beak that does not require attention. The majority of these birds may only require nail or wing trims per client request. • Less frequently, a bird will present with a recurrent overgrown beak and/or nails that will need more involved trimming. • Prior to trimming a beak, it is important to mentally visualize what a normal beak looks like for that specific species. • It is often helpful to manually hold the bird's beak closed prior to trimming and shaping. • Some clinicians prefer to use a Dremel for shaping of both the beak and the nails, while others may prefer toe nail clippers, handheld cautery, or other instruments. If a Dremel is utilized, the clinician may elect to wear protective eyewear. • Styptic powder or other appropriate clotting agents should be prepared and available prior to commencing these procedures. Both the nails and beak do have a blood supply such that trimming either too aggressively can result in bleeding. If bleeding occurs, styptic powder should be applied to halt the bleeding and the client informed that the bird may be sensitive in the area where the bleeding occurs. In the case of beak involvement, this could cause the bird to be reluctant to eat or prefer to eat soft food until the discomfort resolves. • Birds do not need to be sedated or anesthetized for these procedures to occur. • For birds presented with beak trauma, the patient should be examined quickly to assess the extent of the injuries as well as assessing its general condition, brief wound cleaning may be indicated, any bleeding should be stopped immediately, and then the patient should be placed in a warm, quiet environment, prior to further treatment.

NURSING CARE
• In the case of beak trauma, subcutaneous (SQ), intravenous (IV), or intraosseous (IO) fluids often are not indicated. If the clinician elects to administer fluids, warmed Lactated Ringers solution or other isotonic crystalloid solutions can be administered up to the shock fluid doses of 90–100 mL/kg of body weight through a catheter for IO or IV administration or no more than 10 mL/kg/site for SQ administration. • If a toe nail continues to bleed, application of a pressure bandage on that toe or the bleeding toe and the adjacent toe, may be indicted.

ACTIVITY
N/A

DIET
If the bird is not on a formulated pelleted diet with supplementation of fruits and vegetables as indicated per the specific species, conversion to an appropriate improved plane of nutrition should be discussed and ideally gradually implemented by the client.

CLIENT EDUCATION
Once the nails or beak become overgrown from whatever the cause, intermittent trimming will most likely be required, at various intervals, for the remainder of the bird's life.

SURGICAL CONSIDERATIONS
• If a neoplasia is present, surgical removal may be indicated. • If the bird has a beak growth abnormality (scissor beak deformity or mandibular prognathism), that was not managed successfully or at all when the beak was still pliable by use of physical therapy, then surgical correction can be considered if the owner elects not to have intermittent beak trimmings performed. • Trauma to the beak may require a variety of surgical intervention techniques depending on the nature of the trauma. Hence, each individual situation needs to be dealt with on an individual basis.

MEDICATIONS

DRUG(S) OF CHOICE
Ivermectin, dosed at 0.2 mg/kg PO, SQ or IM, is recommended for treatment of scaly faced mite caused by *Knemidokoptes pilae*. One to three weekly treatments may be required.

CONTRAINDICATIONS
N/A

PRECAUTIONS
N/A

POSSIBLE INTERACTIONS
N/A

ALTERNATIVE DRUGS
N/A

FOLLOW-UP

PATIENT MONITORING
Repeat trims as needed.

PREVENTION/AVOIDANCE
Nutritionally balanced diet.

POSSIBLE COMPLICATIONS
Weight loss or death can occur due to an overgrown beak that is left unaddressed such that the bird can no longer or not easily eat.

EXPECTED COURSE AND PROGNOSIS
Prognosis with a beak injury resulting in an overgrown beak is usually guarded to good but depends on the specific injury and situation.

MISCELLANEOUS

ASSOCIATED CONDITIONS
N/A

AGE-RELATED FACTORS
N/A

ZOONOTIC POTENTIAL
N/A

FERTILITY/BREEDING
N/A

SYNONYMS
N/A

SEE ALSO
Beak fracture
Beak malocclusion
Circoviruses
Liver disease
Nutritional deficiencies
Rhinitis and sinusitis
Trauma

ABBREVIATIONS
IO—Intraosseous
IV—Intravenous
PBFD—Psittacine beak and feather disease
PCV—Psittacine circovirus
SQ—Subcutaneous

INTERNET RESOURCES
N/A

Suggested Reading
Cooper, J.E., Harrison, G.J. (1994). Dermatology. In: Ritchie, B., Harrison, G., Harrison, L. (eds), *Avian Medicine Principles and Applications*. Lake Worth, FL: Wingers Publishing, pp. 607–639.

King, A.S., McLelland, J. (1984). Skeletomuscular system. In: King, A.S., McLelland, J. (eds), *Birds: Their Structure and Function*. Philadelphia, PA: Bailliere Tindall, pp. 23–51.

Lintner, M.S. (2013). The beak trim. In: *Proceedings of the Association of Avian Veterinarians Annual Conference*, pp. 189–193.

Olsen, G.H. (2003). Oral biology and beak disorders of birds. In: Rupley, A. (ed.), *The Veterinary Clinics of North America. Exotic Animal Practice*, **6**:505–521.

Speer, B., Echols, S. (2013). Surgical procedures of psittacine skull. In: *Proceedings of the Association of Avian Veterinarians Annual Conference*, pp. 99–109.

Author Amy B. Worell, DVM, ABVP (Avian)

 Client Education Handout available online

Pancreatic Diseases

BASICS

DEFINITION
Any disease process that directly affects the structure and/or function of a bird's pancreas. These include: • Pancreatitis • Exocrine pancreatic insufficiency (EPI) • Diabetes mellitus (DM) • Neoplasia • Pancreatic atrophy • Pancreatic necrosis • Cystic pancreas • Amyloidosis.

PATHOPHYSIOLOGY
• Pancreatitis is an inflammatory condition of the pancreas. Damage to the acinar cell walls results in the leakage of digestive enzymes (primarily protease, trypsin, and phospholipase) within the pancreatic parenchyma resulting in auto-digestion.
• Exocrine pancreatic insufficiency is associated with insufficient synthesis and/or secretion of pancreatic enzymes from the tubuloacinar glands. This can be associated with inadequate enzyme production by the pancreatic acinar cells or by obstruction of the pancreatic ducts. • Diabetes mellitus is discussed elsewhere in this book. • Primary neoplasia of the pancreas includes papilloma (secondary to PsHV-3), adenoma, adenocarcinoma, lymphoma, pancreatic duct carcinoma, and islet cell carcinoma. Metastatic neoplasia can also invade the avian pancreas. • Pancreatic atrophy occurs secondary to the loss of functional pancreatic tissues, often with subsequent pancreatic fibrosis. • Pancreatic necrosis is typically an acute disease process where the pancreatic tissue rapidly degenerates. • Cystic pancreas and amyloidosis are rarely reported and are typically considered incidental findings on necropsy.

SYSTEMS AFFECTED
• Behavioral—the bird may exhibit signs of pain, depression, or feather destructive behaviors (particularly over the ventral coelom). • Endocrine—diabetes mellitus.
• Gastrointestinal—altered GI motility secondary to local chemical coelomitis. Enhanced vascular permeability may also lead to focal or generalized coelomitis.
• Hemic/Lymphatic/Immune—leukocytosis.
• Hepatobiliary—hepatic lesions may result from fatty infiltrates, cholestasis, leukocytic infiltrates, or focal damage by pancreatic enzyme release. • Respiratory—inflammation, organomegaly, and coelomic effusion may compromise air sac volume.
• Skin—feather loss if bird is exhibiting feather destructive behaviors.

GENETICS
Quaker parrots have a high instance of idiopathic acute pancreatic necrosis. *Neophema* spp. are more likely to suffer from chronic pancreatitis secondary to paramyxovirus.

INCIDENCE/PREVALENCE
Unknown. Due to difficult antemortem diagnosis, most pancreatic diseases are identified postmortem.

SIGNALMENT
• **Species:** ○ Pancreatitis has been documented in a multitude of avian species. ○ Exocrine pancreatic insufficiency has been reported in a sulfur-crested cockatoo, yellow-naped Amazon parrot, cockatiel, macaw, chickens, pigeons, quail, and an Indian Hill mynah. ○ Primary pancreatic neoplasia has been documented in ducks, guinea fowl, chickens, cockatiels, Amazon parrots, and macaws. Metastatic pancreatic neoplasia has been reported in ducks, a great tit, a golden eagle, and an Adelie penguin.
○ Pancreatic atrophy has been reported in chickens, ostriches, budgerigars, a blue and gold macaw, and a peregrine falcon.
○ Pancreatic necrosis occurs most commonly in Quaker parakeets, but has also been documented in chickens, turkeys, magpies, cockatoos, a kiwi, and a red-tailed hawk.
○ Pancreatic cysts have been documented in chickens. ○ Pancreatic amyloidosis has been reported in Pekin ducks. • **Mean age and range:** Not documented. • **Predominant sex:** N/A

SIGNS
General Comments
Pancreatic diseases are often undiagnosed antemortem because symptoms can be vague and nonspecific.

Historical Findings
• Lethargy/weakness. • Anorexia/poor appetite, although an increased appetite can be seen with EPI and DM.
• Depression/recumbency.
• Vomiting/regurgitation.
• Diarrhea/voluminous foul-smelling droppings. • Undigested food in the feces/oily feces. • Polyuria/polydipsia (PU/PD).
• Weight loss/failure to thrive.
• Feather-damaging behaviors/aggression.
• Labored breathing.

Physical Examination Findings
• Poor body condition/emaciation.
• Dehydration. • Coelomic distension (fluid or masses), with subsequent decreased or absent abdominal air sac sounds. • Coelomic pain. • Feather loss, particularly over the ventral coelom. • Pasted vent. • Dried vomitus on head feathers.

CAUSES
• Pancreatitis can be idiopathic, or associated with infection of the pancreas with bacteria (including *Mycobacteria* and *Chlamydia*), viruses (West Nile virus, paramyxovirus [PMV-3], reovirus, adenovirus, herpesvirus, avian influenza, polyomavirus, poxvirus, duck hepatitis B virus, avian encephalomyelitis virus, and infectious bronchitis virus), fungi (*Aspergillus*), or parasites (nematodes, trematodes, coccidia, cryptosporidia). It can also be caused by obesity, toxins (primarily zinc, mycotoxins, and selenium), neoplasia, trauma, or secondary to egg-yolk coelomitis or medroxyprogesterone therapy. • Exocrine pancreatic insufficiency can be idiopathic, congenital, or secondary to any disease process that causes damage to the pancreatic acinar cells or obstruction of the pancreatic ducts (pancreatitis, pancreatic atrophy, pancreatic necrosis, or neoplasia). • Pancreatic neoplasia is often idiopathic, but infection with herpesvirus (PsHV-3) in psittacines and retrovirus (avian osteoperosis virus strain Pts-56) in guinea fowl have been linked to pancreatic neoplasia. • Pancreatic atrophy can be caused by pancreatitis, pancreatic necrosis, prolonged caloric deficiency, zinc toxicosis, pancreatic duct obstruction, atherosclerosis of the pancreatic vasculature, or selenium deficiency. • Pancreatic necrosis is often idiopathic, but other potential causes include acute pancreatitis, high-fat diets, zinc or levamisole toxicity, and infection (adenovirus, avian influenza, herpesvirus, polyomavirus, poxvirus, West Nile virus, and *Chlamydia*).

RISK FACTORS
• Obesity. • High-fat diets.

DIAGNOSIS

DIFFERENTIAL DIAGNOSIS
• Primary gastrointestinal disease (ingluvitis, gastroenteritis, GI ulceration, foreign body, obstruction, proventricular dilatation disease [PDD]). • Hepatobiliary disease (cholangiohepatitis, biliary outflow obstruction, toxic injury). • Secondary causes of weight loss and chronic diarrhea (hepatic failure, kidney failure, PDD). • Secondary causes of coelomic effusion (heart disease, liver disease, neoplasia, egg yolk coelomitis). • Secondary causes of coelomic organomegaly (liver disease, reproductive tract disease, neoplasia). • Secondary causes of PU/PD (primary renal disease, reproductive tract disease, endocrinopathy, diabetes insipidus).

CBC/BIOCHEMISTRY/URINALYSIS
• Lipemia. • Elevated triglycerides and cholesterol. • Elevated amylase (>1100 IU/dl) and/or lipase. • Elevated AST, ALT, ALP, CK. • Hyperglycemia (typically >1000 mg/dL with DM). • Leukocytosis.
• Glucosuria/ketonuria.

OTHER LABORATORY TESTS
• Lower increases of amylase and lipase may be associated with extra-pancreatic causes, such as renal disease, GI disease, or glucocorticoid or medroxyprogesterone administration. • Fecal amylase and trypsin levels may be reduced, whereas fecal starch and fat levels may be elevated. • Trypsin-like immunoreactivity (TLI) is species-specific and

PANCREATIC DISEASES (CONTINUED)

has not been studied in birds. Cobalamin and folate levels have also not been assessed in birds. • Blood triglyceride levels can be assessed after administration of corn oil alone or mixed with pancreatic enzymes. • Assessing blood insulin, glucagon, or fructosamine levels may be beneficial in the diagnosis of DM. • Blood zinc levels may be indicated if zinc intoxication is suspected. • PCR or antibody testing for suspected viral, bacterial, chlamydial, or fungal infection. • Protein electrophoresis may show decreased protein levels if EPI is present, or increased globulins if inflammation is present.

IMAGING
• Coelomic ultrasound may demonstrate coelomic effusion or masses. The pancreas can be partially visualized between the descending and ascending duodenum, but unless there is profound enlargement or a mass present, it may be difficult to image. • Although the pancreas is not visible radiographically, coelomic visceral enlargement, ascites, hepatomegaly, or other systemic problems that may be correlated with pancreatic disease may be noted.

DIAGNOSTIC PROCEDURES
Biopsy of the pancreas via coeleoscopy or coeleotomy can be diagnostic, but focal lesions may be missed.

PATHOLOGIC FINDINGS
• Pancreatitis: The pancreas may be grossly firm, irregular, or hemorrhagic, and may contain mycobacterial or fungal granulomas, or suppurative lesions. A purulent exudate or coelomic effusion may be present. Infiltration of the pancreas with lymphocytes, heterophils, or histiocytes may occur, and fibrin deposition may be noted. Bacteria, parasites, or viral inclusion bodies may be evident. Fibrosis can be seen with chronic pancreatitis. • Exocrine pancreatitic insufficiency: EPI may be caused by pancreatitis or pancreatic atrophy, so pathologic changes associated with these diseases may be present. Occlusion of the pancreatic ducts may also be present, due to hyperplasia, neoplasia, inflammatory exudates, or the presence of nematodes or trematodes. • Diabetes mellitus: Degenerative changes may be seen in the pancreatic islets, including hypoplasia, atrophy, and/or vacuolation of islet cells. • Neoplasia: Masses and coelomic effusion may be grossly present. Primary or metastatic neoplastic cells are noted within the pancreatic parenchyma. • Pancreatic atrophy: The pancreas is small and possibly irregular, with shrinkage of acinar cells and loss of zymogen granules. Acini are irregular, and the cells usually contain clear cytoplasm. Fibrosis may be present. • Pancreatic necrosis: The pancreas is firm and pale, with variable hemorrhage and adjacent fat necrosis. Histological abnormalities include coagulative necrosis of pancreatic acini, hemorrhage within the pancreatic lobules, and multifocal necrosis of adjacent mesenteric fat. Coelomic effusion may be present. • Zinc toxicity: Parenchymal mottling may be noted. The primary histological lesion is degranulation and vacuolation of the acinar cells. Cellular necrosis, mononuclear inflammatory infiltrates, and interstitial fibrosis may also be present.

TREATMENT

APPROPRIATE HEALTH CARE
• Aggressive in-patient treatment is indicated with pancreatitis (especially acute pancreatitis), pancreatic necrosis, and unregulated DM. • Uncomplicated EPI and pancreatic atrophy may be treatable on an outpatient basis.

NURSING CARE
• Aggressive fluid therapy is indicated to correct hypovolemia and dehydration, and to maintain adequate pancreatic microcirculation. Fluids can be administered IV, IO, SC, or orally. A balanced electrolyte solution, such as lactated Ringer's solution, is typically the first choice. Addition of colloids may be indicated if the bird is in shock or is profoundly hypovolemic. • Pain management is typically indicated. Both narcotic analgesics and NSAIDs may be utilized. • Antibiotics are indicated if a bacterial etiology is suspected. • Antiemetic medications may be warranted if the bird is actively vomiting or regurgitating, or if GI ileus is present. • The use of pancreatic enzymes may benefit pancreatitis patients, as well as birds with EPI. • In critical cases, a plasma transfusion might be indicated, in order to replace protease inhibitors and decreasing further pancreatic damage.

ACTIVITY
Limited activity may be beneficial with active disease and during recovery.

DIET
• Continue to feed normally unless intractable vomiting or regurgitation is present. Due to their high metabolic rates, birds should ideally not be fasted. • Syringe-feeding, gavage-feeding, and enteral feeding via an esophagostomy or duodenostomy tube may be indicated in anorectic animals. • Low-fat diets are indicated. A formulated diet would be optimal. • Seed-based diets should be avoided due to high fat content and lower digestibility. • The addition of pancreatic enzymes to the diet is indicated with EPI.

CLIENT EDUCATION
• Discuss the need for aggressive treatment in critically ill patients. • Discuss the difficulty of obtaining a diagnosis antemortem, and the possible need for more invasive diagnostics. • Discuss the need for lifelong medical therapy for animals with EPI or DM. • Make sure that the client is aware birds with pancreatic diseases often come with a guarded-to-poor prognosis.

SURGICAL CONSIDERATIONS
• Pancreatic biopsy obtained either endoscopically or via coeleotomy is often necessary to confirm the diagnosis. • The placement of an esophagostomy or duodenostomy feeding tube may be indicated. • Extrahepatic biliary obstruction secondary to pancreatic disease may require surgical correction. • In the case of pancreatic neoplasia, surgery may be indicated to obtain biopsies, and/or remove or debulk the tumor(s). • The presence of coelomic effusion, organomegaly, or coelomic masses may compromise the air sacs, increasing the risk for anesthetic complications.

MEDICATIONS

DRUG(S) OF CHOICE
• Antiemetic drugs and GI motility stimulants may be indicated if vomiting, regurgitation, or ileus is present. ○ Metoclopramide: 0.3–0.5 mg/kg IV, IM, PO q8–12h ○ Cisapride: 0.5–1.5 mg/kg PO q8h. ○ Maropitant: 1–2 mg/kg SC, PO (questionable efficacy in birds). • Antibiotics if evidence of bacterial or chlamydial infection is present. Therapy should be based on bacterial culture and sensitivity results, if possible. • Pain management is indicated with pancreatitis, pancreatic necrosis, and neoplasia. ○ Butorphanol: 0.5–4 mg/kg IM, IV q1–4h. ○ Tramadol: 11–30 mg/kg q6–12h (questionable efficacy in some species). ○ Meloxicam: 0.5–1 mg/kg IM, IV, PO q12–24h. ○ Carprofen: 2–10 mg/kg IM, IV, PO q12–24h. • Powdered pancreatic enzymes are indicated in cases of EPI, and may also benefit some patients with pancreatitis. • The use of Omega 6 and Omega 3 fatty acids may help to reduce hyperlipidemia and pancreatic inflammation. • The use of insulin and/or oral hypoglycemic medications may be indicated if DM is present. • Chelation therapy is indicated if zinc toxicity is present. • Chemotherapy protocols may be indicated in cases of primary or metastatic pancreatic neoplasia.

CONTRAINDICATIONS
• Anticholinergics. • Corticosteroids (unless part of a chemotherapy protocol). • Diuretics.

PRECAUTIONS
Tetracyclines may increase fibrosis in cases of chronic pancreatitis.

ALTERNATIVE DRUGS
Plant-based organic enzyme supplements have been marketed for birds. The efficacy of these products in cases of EPI is questionable.

PANCREATIC DISEASES (CONTINUED)

FOLLOW-UP

PATIENT MONITORING
• Amylase and lipase levels should be rechecked 1–2 weeks after the patient is discharged. • Patients that survive pancreatitis or pancreatic necrosis should be monitored closely for the development of EPI and DM.

PREVENTION/AVOIDANCE
• Reduce weight if obese. • Avoid high-fat diets, and attempt to convert birds from a seed-based diet to a formulated diet.

POSSIBLE COMPLICATIONS
• Failed response to medical therapy. • Other life-threatening disease may be linked to these diseases.

EXPECTED COURSE AND PROGNOSIS
• The prognosis is good for patients with mild pancreatitis, EPI, and DM. • The prognosis is poor to grave for patients with necrotizing pancreatitis, pancreatic necrosis, or malignant neoplasia.

MISCELLANEOUS

ASSOCIATED CONDITIONS

Life-threatening
• Coelomitis. • Coelomic effusion.
• Hepatobiliary damage from cholestasis or due to leakage of pancreatic enzymes.
• Metastasis of malignant pancreatic neoplasia.

Nonlife-Threatening
• Diabetes mellitus. • Exocrine pancreatic insufficiency.

ZOONOTIC POTENTIAL
Some causative agents of pancreatitis are potentially zoonotic (e.g. *Chlamydia, Mycobacteria,* Avian Influenza virus).

FERTILITY/BREEDING
Birds with congenital pancreatic abnormalities should not be bred.

SEE ALSO
Chlamydiosis
Coelomic distention
Diabetes mellitus
Egg yolk coelomitis
Heavy metal toxicosis
Ingluvial hypomotility/stasis
Mycobacteriosis
Paramyxoviruses
Regurgitation/vomiting
Undigested food in droppings

ABBREVIATIONS
ALP—alkaline phosphotase
ALT—alanine aminotransferase
AST—aspartate aminotransferase
CK—creatine kinase
DM—diabetes mellitus
EPI—exocrine pancreatic insufficiency
GI—gastrointestinal
PMV-3—paramyxovirus type III
PsHV-3—psittacine herpesvirus type III
PU/PD—polyuria/polydipsia
TLI—trypsin-like immunoreactivity

INTERNET RESOURCES
http://www.vin.com/VIN.plx

Suggested Reading
Doneley, R. (2001). Acute pancreatitis in parrots. *Australian Veterinary Journal,* **79**(6):409–411.
Doneley, R. (2010). Disorders of the avian pancreas. In: Doneley, R., Avian Medicine and Surgery in Practice: Companion and Aviary Birds. London: Manson Publishing, pp. 182–184.
Hudelson, K.S., Hudelson, P.M. (2006). Endocrine considerations. In: Harrison, G.J., Lightfoot, T.L. (eds), *Clinical Avian Medicine* Palm Beach, FL: Spix Publishing, Inc., pp. 541–557.
Pilny, A.A. (2008). The avian pancreas in health and disease. *The Veterinary Clinics of North America. Exotic Animal Practice,* **11**(1):25–34.
Schmidt, R.E., Reavill, D.R. (2014). Lesions of the avian pancreas. *The Veterinary Clinics of North America. Exotic Animal Practice,* **17**(1):1–11.

Author David E. Hannon, DVM, DABVP (Avian)

Papillomas, Cutaneous

BASICS

DEFINITION
Cutaneous papillomas are considered benign neoplasms, characterized by hyperplasia of the epithelium. Papillomas are colloquially referred to as cutaneous or skin warts, and are typically of either viral (Papilloma viruses papillomaviruses or herpesviruses) or unknown origin.

PATHOPHYSIOLOGY
- Papillomaviruses are considered host specific, and typically affect the featherless skin, most commonly associated with the skin of the eyelids, at the junction of the beak and face, and on the feet and legs. • Transmission is thought to be by direct contact with infected birds, but this has not been proven. • Infection with papillomaviruses of the epidermal basal cells can result in wart-like lesions of the skin or a persistent asymptomatic infection.

SYSTEMS AFFECTED
Skin.

GENETICS
There is no known genetic basis for cutaneous papillomas.

INCIDENCE/PREVALENCE
Incidence and prevalence of cutaneous papillomatosis is unknown. One study from the early 1970s of wild Chaffinches demonstrated a prevalence of 1.8%. Cutaneous papillomatosis in psittacine species is considered rare.

GEOGRAPHIC DISTRIBUTION
Cutaneous papillomatosis can potentially affect birds anywhere in the world, with no clear seasonal pattern.

SIGNALMENT
- **Species:** All avian species susceptible; in psittacines, cutaneous papillomatosis was observed in several African grey parrots, a Quaker parrot, an Amazon parrot, and several budgerigars. • **Mean age and range:** All ages susceptible. • **Predominant sex:** Both genders sexes susceptible.

SIGNS
Skin lesions include papules, nodules, or discrete masses on the feet, legs, or around the eyes and commissures of the beak. Masses may be focal to multifocal, and are hyperplastic with a cauliflower-like or wart-like appearance (e.g., papilliferous).

CAUSES
Papillomaviruses are suspected as the cause of most cutaneous papillomas, but not all cutaneous papillomas contain viral particles. A novel herpesvirus (Psittacid Herpesvirus 2) was found to be associated with a cutaneous papilloma and a cloacal papilloma in an African grey parrot.

RISK FACTORS
A potential risk factor is exposure of any avian species to a bird(s) with cutaneous papillomas.

DIAGNOSIS

DIFFERENTIAL DIAGNOSIS
- Poxvirus. • Bacterial or fungal cutaneous abscesses abscesses/granulomas.
- *Knemidokoptes* mites. • Soft tissue sarcoma or other epithelial tumors.

CBC/BIOCHEMISTRY/URINALYSIS
No specific blood parameter changes.

OTHER LABORATORY TESTS
N/A

IMAGING
N/A

DIAGNOSTIC PROCEDURES
The diagnosis of skin disease has traditionally relied on biopsy and histopathologic examination of skin lesions. Electron microscopy of tissue preparations from cutaneous papillomas was used to confirm the presence of papilloma viral particles. More recently, molecular tools, such as PCR and/or immunohistochemistry, are increasingly used to analyze tissue samples.

PATHOLOGIC FINDINGS
- Gross lesions: White to gray, fleshy, or cauliflower-like growths, most typically on a short stalk. • Histopathologic lesions: Papillary mass with pedunculated fibrovascular connective tissue stalk, stratified squamous epithelium covered with a hyperkeratotic layer, with or without the presence of basophilic intranuclear inclusion bodies at junction of keratinized and nonkeratinized epithelium. In the dermis, lymphocytes and heterophils are occasionally present.

TREATMENT

APPROPRIATE HEALTH CARE
Outpatient supportive care.

NURSING CARE
Supportive care may include fluid support, anti-inflammatory medications (e.g., meloxicam), and appropriate antimicrobials.

ACTIVITY
N/A

DIET
N/A

CLIENT EDUCATION
Warn client that little is known about transmission, but contact transmission may be important, so other birds in the household may be susceptible.

SURGICAL CONSIDERATIONS
Surgical excision of papillomatous lesions may be an option, especially if function (e.g., flight, vision, ability to eat) is affected. The lesions can be surgically excised using scalpel blade, cautery, laser, or cryosurgery, but the papillomas may recur.

MEDICATIONS

DRUG(S) OF CHOICE
N/A

CONTRAINDICATIONS
N/A

PRECAUTIONS
N/A

POSSIBLE INTERACTIONS
N/A

ALTERNATIVE DRUGS
N/A

FOLLOW-UP

PATIENT MONITORING
Monitor areas of excised papillomas for recurrence.

PREVENTION/AVOIDANCE
N/A

POSSIBLE COMPLICATIONS
Recurrence of the papilloma after excision or secondary microbial (e.g., bacterial, fungal) infection at the surgical site.

EXPECTED COURSE AND PROGNOSIS
Cutaneous papillomas may spontaneously regress, or may never resolve on their own and continue to proliferate. If functional capacity, such as vision or ability to eat, is affected, the prognosis may be guarded to poor.

MISCELLANEOUS

ASSOCIATED CONDITIONS
N/A

AGE-RELATED FACTORS
N/A

ZOONOTIC POTENTIAL
N/A

FERTILITY/BREEDING
N/A

SYNONYMS
N/A

SEE ALSO
Cere and skin, color changes
Dermatitis
Ectoparasites
Feather disorders

Papillomas, Cutaneous

(Continued)

Herpesviruses
Lameness
Mycobacteriosis
Nutritional deficiencies
Papillomatosis
Pododermatitis
Poxvirus
Tumors

ABBREVIATIONS
N/A

INTERNET RESOURCES
N/A

Suggested Reading
Gill, J.H. (2001). Avian skin diseases. *The Veterinary Clinics of North America. Exotic Animal Practice*, **4**:463–492.
Perez-Tris, J., Williams, R.A.J., Abel-Fernandez, E., et al. (2011). A multiplex PCR for detection of poxvirus and papillomavirus in cutaneous warts from live birds and museum skins. *Avian Diseases*, **55**:545–553.
Phalen, D. (1997). Viruses. In: Altman, R.B., Clubb, S.L., Dorrestein, G.M., Quesenberry, K. (eds), *Avian Medicine and Surgery*. Philadelphia, PA: WB Saunders, pp. 281–322.
Styles, D.K., Tomaszewsk,i E.K., Phalen D.N. (2005). A novel psittacid herpesvirus found in African grey parrots (*Psittacus erithacus erithacus*). *Avian Pathology*, **34**:150–154.
Worell, A.B. (2013). Dermatological conditions affecting the beak, claws, and feet of captive avian species. *The Veterinary Clinics of North America. Exotic Animal Practice*, **16**:777–799.

Author Kurt K. Sladky, MS, DVM, DACZM, DECZM (Herpetology)

PARAMYXOVIRUSES

BASICS

DEFINITION
- Paramyxoviruses are enveloped, single-stranded, negative-sense RNA viruses. There are two subfamilies: Paramyxovirinae, which contains the avian paramyxoviruses (APMV), and Pneumovirinae, which contains the avian metapneumoviruses (aMPV).
- There are 12 serotypes of APMV. A wide variety of avian species may be infected.
- Newcastle disease (ND) is caused by highly virulent isolates of APMV-1, known as virulent Newcastle disease virus (vNDV). Low virulence isolates of APMV-1 only occasionally cause clinical disease. • ND is pathotyped by ability to cause disease in naive chickens (*Gallus gallus*). Pathotypes in increasing order of virulence: asymptomatic enteric, lentogenic, mesogenic, neurotropic velogenic, and viscerotropic velogenic. Pathotypes may not be applicable to ND infection in species other than domestic fowl, as clinical signs may vary widely. • Pigeon paramyxovirus 1 (PPMV-1; variant of vNDV) is commonly isolated from domestic pigeons (*Columba livia domestica*). May infect non-*Columbidae* species and cause severe disease.
- aMPV has four subtypes (A, B, C, D). Primarily a disease of turkeys (*Meleagris gallopavo*) and chickens (*Gallus gallus*), but has been isolated from wild waterfowl.

PATHOPHYSIOLOGY
- APMV: Virus attaches to the host cell membrane. The viral nucleocapsid is released into the cytoplasm where replication takes place. Virions are released by budding. Virus is shed in feces and oropharyngeal secretions. Transmission is fecal-oral, aerosol, or contact with mucous membranes (especially conjunctiva). Pests (flies, mice), humans, and vehicles may act as fomites. • aMPV: Primarily targets respiratory epithelium, may target respiratory macrophages. aMPV attaches to the host cell and enters the cytoplasm where replication takes place. Virions are released by budding. Transmission is unpredictable, only contact confirmed to spread virus. *E. coli* coinfection is believed to worsen clinical disease.

SYSTEMS AFFECTED
- APMV—variable by host species and serotype. • aMPV—primarily respiratory, variable by host species.

GENETICS
N/A

INCIDENCE/PREVALENCE
- APMV-1: Poultry considered main reservoir. Asymptomatic-enteric pathotype common in wild birds. vNDV may be endemic in Phalacrocoracidae (cormorants and shags). ND common in Africa, Asia, Central America, and parts of South America; sporadic in other parts of the world.
- PPMV-1: Common in domestic pigeons and doves. • APMV 2-12: Unknown, likely variable by species. • aMPV: Common in poultry, variable in wild birds.

GEOGRAPHIC DISTRIBUTION
- APMV: Worldwide. • aMPV: Worldwide.

SIGNALMENT
- **Species:**
  - APMV infect a wide variety of avian species. Common subtype-species associations are:
    - APMV-1 & vNDV: Any avian species. Chickens considered most susceptible.
    - PPMV-1: Pigeons and doves, may infect other species (especially poultry).
    - APMV-2: Turkeys, passerines, chickens, psittacines, rails (*Rallidae*), raptors, waterfowl.
    - APMV-3: Turkeys, psittacines (*Neophema* spp., budgerigars [*Melopsittacus undulatus*], parrotlets [*Forpus* spp.]), passerines (finches [*Fringillidae*]), ostriches (*Struthio* spp.).
    - APMV-4: Ducks, geese.
    - APMV-5: Budgerigars.
    - APMV-6: Ducks, geese, rails, turkeys.
    - APMV-7: Pigeons, doves, turkeys, ostriches.
    - APMV-8: Ducks, geese.
    - APMV-9: Ducks.
    - APMV-10: Rock hopper penguins (*Eudyptes chrysochome*), Magellanic peguins (*Spheniscus magellanicus*).
    - APMV-11: Common snipe (*Gallinago gallinago*).
    - APMV-12: Eurasian wigeon (*Anas penelope*).
  - aMPV: Turkeys, chickens, numerous wild birds.
- **Mean age and range:** N/A
- **Predominant sex:** N/A

SIGNS

General Comments
ND: Incubation period 2–21 days (average 2–6 days).

Historical Findings
- **ND:**
  - Poultry: Marked, peracute increase in mortality, sudden drop in egg production, misshapen eggs. Neurologic, GI, and/or respiratory signs. Viscerotropic velogenic ND carries up to 100% mortality and is associated with GI signs. Neurotropic velogenic ND carries ~50% mortality, respiratory and neurologic signs predominate. Other pathotypes show low to no mortality.
  - Psittacines: Species with reported ND infections include *Neophema* spp., budgerigars, yellow-headed Amazons (*Amazona oratrix*), yellow-naped Amazons (*Amazona auropalliata*). Malaise, respiratory signs may be reported.
  - Raptors: ND has been reported in falcons (*Falco* spp.). Neurologic +/− GI signs are seen.
  - Variable in other species.
- **PPMV-1**
  - Pigeons
    - Variable morbidity (30–100%) and mortality (≤77% adults, ≤95% juveniles).
    - GI and neurologic signs.
    - Poor feather quality.
  - Poultry: as for ND.
- **APMV 2–11:** Variable by serotype, host species. Generally a combination of GI, respiratory, and/or neurologic signs.
- **APMV 8–12:** Asymptomatic.
- **aMPV**
  - Turkeys
    - Severe morbidity (up to 100%).
    - Variable mortality (up to 50%).
    - Severely decreased egg production.
    - Respiratory signs.
  - Chickens
    - Variable morbidity (up to 80%).
    - Low mortality.
    - Decreased egg production.
    - Swollen head.
    - Respiratory signs.

Physical Examination Findings
- **ND**
  - Vary by host species & pathotype of virus. Most often a combination of respiratory, GI, and neurologic signs.
  - Chickens and turkeys
    - Viscerotropic
      - Lethargy
      - Facial swelling
      - Conjunctivitis
      - Mucoid oral discharge
      - Dyspnea
      - Cyanosis (esp. comb)
      - Diphtheritic mucous membranes.
    - Neurotropic
      - Hypermetria
      - Head & muscle tremors
      - Head tilt
      - Unilateral paralysis.
    - Mesogenic
      - Respiratory signs
      - Neurologic signs: head tremors, head tilt, paralysis.
    - Lentogenic
      - Respiratory signs (mild).
  - Psittacines
    - Lethargy
    - Oculonasal discharge
    - Conjunctivitis
    - Dyspnea
    - Neurologic signs.
  - Falcons
    - Neurotropic
      - Hyperesthesia
      - Clonic spasms
      - Ataxia
      - Head tremors
      - Dysphagia
      - Tongue paresis

PARAMYXOVIRUSES

(CONTINUED)

- Ptyalism
- Amaurosis
- 3rd eyelid paresis/paralysis
- Progressive pelvic limb paralysis
- Convulsions.
■ Viscerotropic
- Severe depression
- Mucoid diarrhea and hematochezia
- Constant vocalization.
• **PPMV-1**
  ○ Pigeons
    ■ Ataxia
    ■ Head tilt
    ■ Limb paresis
    ■ Pecking aside seeds
    ■ Diarrhea
    ■ Poorly developed feathers if infected during molt.
• **APMV-2**
  ○ Turkeys
    ■ Respiratory signs (mild to severe)
    ■ Sinusitis
    ■ Decreased egg production.
• **APMV-3**
  ○ Psittacines and passerines
    ■ Death
    ■ Circling
    ■ Head tilt
    ■ Worsening CNS signs with excitement
    ■ Ataxia
    ■ Opisthotonus
    ■ Steatorrhea
    ■ Weight loss
    ■ Dyspnea.
• **APMV-4:** asymptomatic
• **APMV-5**
  ○ Budgerigars
    ■ Death
    ■ Depressed mentation
    ■ Dyspnea
    ■ Diarrhea
    ■ Head tilt.
• **APMV-6**
  ○ Turkeys
    ■ Respiratory signs (mild)
    ■ Decreased egg production.
• **APMV-7**
  ○ Turkeys
    ■ Rhinitis
    ■ Dyspnea.
• **APMV 8–12:** asymptomatic.
• **aMPV**
  ○ Turkeys
    ■ Snicking
    ■ Rales
    ■ Sneezing
    ■ Nasal discharge
    ■ Coughing
    ■ Head shaking
    ■ Foamy ocular discharge
    ■ Swollen infraorbital sinuses
    ■ Submandibular edema
    ■ Uterine prolapse (secondary to cough).
  ○ Chickens
    ■ Nasal discharge
    ■ Foamy ocular discharge

■ Peri/infra-orbital sinus swelling
■ Rales
■ Head tilt
■ Ataxia
■ Opisthotonus.

CAUSES
• ND: vNDV, a virulent form of APMV-1
• PPMV-1: A variant of vNDV
• APMV 2–12, aMPV: Relevant virus

RISK FACTORS
• ND:
  ○ Exposure to migratory birds, illegally acquired birds, or unvaccinated birds.
  ○ Raptors: feeding on pigeons and quail, use of pigeons in training.
• PPMV-1: Exposure to pigeon/dove racing or trade events.
• aMPV & many APMV: Exposure to wild birds.

DIAGNOSIS

DIFFERENTIAL DIAGNOSIS
• ND:
  ○ Poultry: High pathogenicity avian influenza, infectious bronchitis, infectious laryngotracheitis, mycoplasmosis, fowl cholera, aspergillosis, avian metapneumovirus, other bacterial infectious.
  ○ Other host species: Variable by clinical signs.
• PPMV-1:
  ○ Pigeons: Salmonellosis, adenovirus type 1.
  ○ Other species: Variable by clinical signs.
• APMV 2–12: Variable by clinical signs.
• aMPV: Low pathogenicity avian influenza, infectious laryngotracheitis, fowl cholera, mesogenic vNDV, other APMV, other acute infections.

CBC/BIOCHEMISTRY/URINALYSIS
Variable results depending on systems affected and severity of infection.

OTHER LABORATORY TESTS
• **ND:**
  ○ Contact regulatory authorities for guidance in sample acquisition from suspect cases.
  ○ Viral DNA identification
    ■ VI: Tracheal, oropharyngeal, or cloacal swabs; organ tissue samples—used to inoculate embryonated chicken eggs. May take up to seven days.
    ■ HA: Performed on samples after virus isolation.
    ■ HI: Performed with APMV-1 antiserum on samples with a positive HA.
      - APMV-1 antiserum may cross-react with APMV-3 and 7, rarely 2 and 4. Monoclonal antibodies may reduce cross-reactions.
    ■ Molecular diagnostics: rRT-PCR most commonly used. Genetic variation between viruses may reduce sensitivity. Once outbreak strain is determined, specific primers may be generated.
  ○ Serology
    ■ Generally not diagnostically useful due to widespread vaccination.
    ■ May be used to monitor flock immunity.
    ■ ELISA usually specific to host species.
• **PPMV-1**
  ○ VI: Brain samples ideal (virus persists 5 weeks), cloacal swabs (virus persists 3 weeks).
  ○ RT-PCR: More sensitive than VI.
• **APMV 2-12:** Combination of VI and HA.
• **aMPV**
  ○ Viral DNA identification:
    ■ VI: Sample oculonasal discharge, choana, sinus or intranasal scrapings. Best done as early as possible. Successful VI more difficult in severely affected birds.
    ■ RT-PCR: Generally highly sensitive. Some kits only test for certain subtypes.
  ○ Serology:
    ■ ELISA: Variable sensitivity depending on subtype of virus.
    ■ Virus neutralization and indirect immunofluorescence also available.

IMAGING
Variable results depending on systems affected, severity of infection.

DIAGNOSTIC PROCEDURES
N/A

PATHOLOGIC FINDINGS
• APMV (all): Variable depending on systems affected.
• aMPV
  ○ Turkeys: Tracheal deciliation, mucoid exudate in turbinates and trachea, reproductive-associated coelomitis, misshapen eggs, airsacculitis, pericarditis, pneumonia, perihepatitis
  ○ Chickens: Yellow gelatinous to purulent edema of subcutaneous tissues of head.

TREATMENT

APPROPRIATE HEALTH CARE
• ND
  ○ Cull infected birds.
• On-farm care for flocks. Individuals with severe disease may require hospitalization.
• Proper biosecurity measures (see "Internet Resources").
• APMV and aMPV inactivated by most disinfectants, including bleach, ethanol, quaternary ammonia, iodophors. Removal of organic matter before application of disinfectants is critical.
• Avoid high stocking density and multi-age stock.
• Provide adequate ventilation and high quality litter.
• Remove manure frequently.

PARAMYXOVIRUSES (CONTINUED)

NURSING CARE
Symptomatic care indicated by patient, situation. Fluid therapy, oxygen therapy, hand-feeding, supplemental heat may all be indicated.

ACTIVITY
N/A

DIET
N/A

CLIENT EDUCATION
Proper biosecurity measures (see "Internet Resources").

SURGICAL CONSIDERATIONS
N/A

MEDICATIONS

DRUG(S) OF CHOICE
• No antiviral drugs are known to treat APMV or aMPV. • Secondary infections common, especially respiratory. Select antimicrobials based on pathogen, sensitivity, location of infection, and patient species.

CONTRAINDICATIONS
N/A

PRECAUTIONS
Drug selection in food-producing animals should be made in compliance with regulatory statutes.

POSSIBLE INTERACTIONS
N/A

ALTERNATIVE DRUGS
N/A

FOLLOW-UP

PATIENT MONITORING
• ND: Alert authorities—notifiable disease to the World Organization for Animal Health. • Monitor clinical status of both sick and asymptomatic exposed animals. • Infected patients should be isolated for 14 days (aMPV) or 30 days (APMV) beyond resolution of clinical signs. • APMV: Repeat testing (VI, RT-PCR) of recovered patients to determine if virus being shed.

PREVENTION/AVOIDANCE
• Proper biosecurity measures (see Internet Resources)
• Avoid contact with birds of unknown health status–30-day quarantine for all new birds.
• ND
  ○ Vaccines available, used in numerous species.
  ○ Vaccination may substantially reduce clinical disease, cannot completely prevent virus replication/shedding.
  ○ Protocols for poultry vary and are informed by husbandry and geographic location.
  ○ Vaccination of falcons is recommended immediately after importation or as soon as moult is finished and training begins.
  ○ Unvaccinated "sentinel" birds may be used.
• APMV 2-12
  ○ Vaccines may be available against certain virus subtypes in specific species.
• aMPV
  ○ Vaccines available for turkeys and chickens. Protection, especially in chickens, may not be complete.

POSSIBLE COMPLICATIONS
Secondary infections (especially respiratory).

EXPECTED COURSE AND PROGNOSIS
• ND:
  ○ Falcons: Grave; death in 3–7 days.
  ○ Psittacines: Guarded to grave. Infected psittacines may intermittently shed virus for extended periods (>1 year).
  ○ Chickens:
    ■ Viscerotropic: Grave; up to 100% mortality.
    ■ Neurotropic: Guarded; ~50% mortality.
    ■ Mesogenic: Good; low to no mortality.
• PPMV-1
  ○ Pigeons: Guarded, highly variable mortality
  ○ Poultry: As for ND
• APMV-2: Good
• APMV-3:
  ○ Turkeys: Good
  ○ Psittacines and passerines: Guarded to grave, up to 100% mortality (may be species-dependent).
• APMV-5:
  ○ Budgerigars: Grave, up to 100% mortality.
• APMV-4, 8–12: Unknown, likely species dependent.
• aMPV
  ○ Turkeys: Variable mortality (0.4–50%) depending on age, vaccination status. Uncomplicated infections resolve in 10–14 days.
  ○ Chickens: Good-to-variable morbidity, low mortality.

MISCELLANEOUS

ASSOCIATED CONDITIONS
Secondary respiratory infections.

AGE-RELATED FACTORS
Juveniles may be more susceptible.

ZOONOTIC POTENTIAL
• ND: Zoonotic. Transient infections in humans, ≤2 day duration. Common symptoms involve the eyes (conjunctivitis, epiphora, blepharitis); headaches and chills less common. • aMPV: Clinical disease not reported, but turkey production workers have tested seropositive.

FERTILITY/BREEDING
APMV and aMPV may cause severe reproductive losses.

SYNONYMS
• ND: Exotic Newcastle disease, pseudo-fowl pest, pseudo-fowl plague, avian pest, avian distemper, Tetelo disease, Ranikhet disease, Korean fowl plague • APMV-2: Yucaipa virus • APMV-5: Kunitachi virus • aMPV: Avian pneumovirus, turkey rhinotracheitis, swollen head syndrome, avian rhinotracheitis

SEE ALSO
Aspergillosis
Avian influenza
Circoviruses
Herpesviruses
Mycoplasmosis
Neurologic conditions
Pasteurellosis
Pneumonia
Respiratory distress
Rhinitis/sinusitis
Salmonellosis
Seizure
West Nile virus
Viral disease

ABBREVIATIONS
aMPV—avian metapneumovirus(es)
APMV—avian paramyxovirus(es)
APMV-[X]—avian paramyxovirus serotype-[X] (e.g. APMV-1—avian paramyxovirus serotype-1)
GI—gastrointestinal
HA—hemagglutination assay
HN—hemagglutinin-neuraminidase
ND—Newcastle disease
PPMV-1—pigeon paramyxovirus serotype-1
RT-PCR—reverse transcriptase PCR
rRT-PCR—real time reverse transcriptase PCR
VI—virus isolation
vNDV—virulent Newcastle disease virus

INTERNET RESOURCES
USDA—Biosecurity for Birds: http://www.aphis.usda.gov/animal_health/birdbiosecurity/
Center for Food Safety & Public Health at Iowa State University: www.cfsph.iastate.edu

Suggested Reading
Capua, I., Alexander, D.J. (eds) (2009). *Avian Influenza and Newcastle Disease: A Field and Laboratory Manual*. Milan, Italy: Springer-Verlag Italia.
Suarez, D., Miller, P.J., Koch, G., et al. (2013). Newcastle disease, other avian paramyxoviruses, and avian metapneumovirus infections. In: Swayne, D.E. (ed.), *Diseases of Poultry*, 13th edn. Hoboken, NJ: John Wiley & Sons, Iinc, pp. 89–138.

Author Andrew D. Bean, DVM, MPH, CPH

PASTEURELLOSIS

BASICS

DEFINITION
Pasteurellosis (also called Avian Cholera) is a bacterial infection of free-ranging birds caused by *Pasteurella multocida*, a Gram-negative, aerobic/facultative anaerobic species of bacteria. *P. multocidida* causes disease in a wide variety of mammalian and avian species, and has been reported in more than 100 species of bird. At least 16 serotypes of *P. multocida* exist that cause disease in wild bird populations, which are differentiated from those causing disease in domestic fowl. Serotype 1 is most frequently identified in waterfowl outbreaks along the North American migratory flyways, with the exception of the Atlantic flyway.

Pasteurellosis is primarily a disease of wild waterfowl, with outbreaks occurring on wintering and breeding grounds where birds are in close proximity and in high concentrations. Frequency and incidence of outbreaks have increased since the disease was first described in the mid-1940s, due in large part to the loss of wetland habitat, forcing more and more birds to congregate in smaller areas. Avian cholera has been called the most significant infectious disease of North American waterfowl.

PATHOPHYSIOLOGY
Acute pasteurellosis is an extremely rapidly progressing disease, resulting in death within 6–48 hours following exposure. Exposure to *P. multocida* can occur via mucous membranes of the upper respiratory and digestive tracts. Infection can occur through ingestion of contaminated food or water, direct contact between infected and non-infected birds, aerosolization of organism from the surface of contaminated water (bathing, landing on/taking off from water), or through wounds in the skin. Deceased birds may shed very large numbers of the organism into the environment. *P. multocida* can persist in water for weeks to months, prolonging outbreaks if proper site management is not implemented.

Once the organism enters the body, the disease process progresses rapidly to septicemia, endotoxemia, and death. Severity of signs and rapidity of progression are related to the specific strain the virulence of that strain, species affected, sex, age, immune status due to previous exposure, concurrent injury/illness, route of exposure, and exposure dose. While a chronic carrier state has been defined for *P. multocida* infections in poultry, this has not been established as a potential source of infection in wild birds.

SYSTEMS AFFECTED
• Cardiovascular—death from *P. multocida* usually occurs as a result of circulatory collapse due to hypovolemic shock +/− consumptive coagulopathy. • Respiratory—infected birds may have severe upper respiratory discharge and respiratory distress. • Neuromuscular—infected birds may display a variety of neurologic signs ranging from lethargy to erratic flight, sometimes attempting to land on water significantly above the water's surface.

GENETICS
N/A

INCIDENCE/PREVALENCE
Pasteurellosis has been called a "disease for all seasons". Related to seasonal migration patterns: Fall into spring.

GEOGRAPHIC DISTRIBUTION
• Major outbreaks among waterfowl tend to occur along the major migratory flyways, excepting the Atlantic flyway. • Pasteurellosis gained more recognition as a significant disease of free-ranging waterfowl in North America in the 1970s. Locations that experience the most frequent and significant pasteurella outbreaks are located along major migratory flyways and include California's Central Valley, Klamath Basin at the California/Oregon border, the Texas Panhandle, and the Platte River Basin in Nebraska. • Pasteurella outbreaks in wild waterfowl are related to loss of habitat by increasing concentrations of birds.

SIGNALMENT
Waterfowl, such as ducks and geese, and coots are most frequently affected in large outbreaks of *P. multocida*. Scavenger species, such as gulls and corvids, are also commonly affected when attracted to large numbers of carcasses at an outbreak site, becoming infected by consuming infected tissues.

SIGNS
Often affected birds are found dead without clinical signs. However, when clinical signs are present they can include:
• Lethargy. • Birds do not try to evade capture and may die within a matter of minutes after capture. • Convulsions. • Swimming in circles. • Opisthotonus. • Erratic flight; attempting to land before water's suface. • Mucous discharge from mouth. • Bloody nasal discharge. • Soiled vent. • Pasty, yellow droppings/bloody droppings.

CAUSES
Multiple serotypes of *P. multocida* cause pasteurellosis in free-ranging birds. Serotypes that cause disease in wild birds are different from those that cause disease in captive poultry. There is also a differentiation of serotypes on the Atlantic flyway of the United Stated compared to the rest of the country.

RISK FACTORS
High concentrations of birds in close proximity in wintering and breeding grounds

DIAGNOSIS

DIFFERENTIAL DIAGNOSES
• Botulism. • Heavy metal toxicosis.
• Organophosphate/carbamate toxicity.
• Duck viral enteritis.

CBC/CHEMISTRY/URINALYSIS
Due to acute progression of the disease, finding for these diagnostics are usually unremarkable.

OTHER LABORATORY TESTS
• Culture of heart blood, liver, bone marrow.
• Can persist for weeks in bone marrow, so scavenged portions can be used for diagnostics. • Coccobacilli on smear of heart blood.

IMAGING
Due to acute progression of the disease imaging is generally not feasible.

DIAGNOSTIC PROCEDURES
Pasteurella serotyping.

PATHOLOGIC FINDINGS
• Histopathology findings nonspecific in acute deaths. • Affected birds typically in good body condition. • Food in upper GI, indicating acute death. • Hemorrhages on the surface of the heart muscle and along the coronary band. • Small white-yellow foci within the hepatic parenchyma. ○ May also see darkening, copper tone, swelling, and fracture of the liver upon handling.
• Hemorrhages on the surface of the ventriculus. • Ropey nasal discharge.
• Intestines contain thick, yellowish fluid or clear mucous. • Intestinal contents and nasal discharge contain very large numbers of *P. multocida* organisms.

TREATMENT

APPROPRIATE HEALTH CARE
The degree of health care provided is based on the severity of the clinical signs.

NURSING CARE
Nutritional and fluid support should be provided to patients too debilitated to eat and drink on their own. Thermal support should also be provided as needed. Basic cleanliness should be maintained in patients who are recumbent and unable to preen. Large waterfowl, such as geese, who are sternally recumbent, should be housed on heavily padded substrate or on a cushion that elevates the keel off of the substrate to avoid development of decubital ulcers along the keel.

ACTIVITY
N/A

PASTEURELLOSIS (CONTINUED)

DIET
N/A

CLIENT EDUCATION
N/A

SURGICAL CONSIDERATIONS
N/A

MEDICATIONS

DRUG(S) OF CHOICE
• Antibiotic therapy should be based on culture and sensitivity in individual birds, but is impractical for treatment of flocks of wild birds. • Canada geese have been treated successfully with intramuscular injections of oxytetracycline, followed by a 30 day treatment course of tetracycline-infused feed.

CONTRAINDICATIONS
N/A

PRECAUTIONS
N/A

POSSIBLE INTERACTIONS
N/A

ALTERNATIVE DRUGS
P. multocida tends to be susceptible to a variety of antibiotics commonly used in veterinary medicine (e.g., enrofloxacin, amoxicillin-clavulanate).

FOLLOW-UP

PATIENT MONITORING
N/A

PREVENTION/AVOIDANCE
• Surveillance of sites where waterfowl congregate is the most important defense against large outbreaks for *P. multocida*. At sites where an outbreak has occurred, or suspected outbreak is occurring, collection of carcasses will mitigate further contamination of the environment. Removal of carcasses, which can act as decoys, will also decrease attraction of other birds to a contaminated area and prevent scavenging of infected carcasses.

Scavenger species, such as gulls and corvids, do not succumb to *P. multocida* infection as rapidly as do waterfowl, and may therefore spread the organism far from the original outbreak site. • Habitat management is another facet of the prevention/management of *P. multocida* outbreaks. Contaminated areas may be drained, and water diverted to another, uncontaminated location, or flooded to decrease the overall concentration of the organism in the environment. Hazing practices may also be employed to prevent congregation of birds at a contaminated site. • A killed vaccine has been developed which has been shown to be effective in preventing disease in Canada geese. However, immunity is short-lived, only lasting for approximately 12 months, making its application for *P. multocida* management in migratory waterfowl limited. Canada geese have also been treated successfully post-exposure with oxytetracycline injections, followed by administration of tetracycline in feed, but again, application to large numbers of free-ranging birds would be difficult.

POSSIBLE COMPLICATIONS
N/A

EXPECTED COURSE AND PROGNOSIS
P. multocida infection in wild birds is typically a peracute to acute disease process, usually resulting in death within a matter of hours. Due to the rapid progression of this septicemic disease, prognosis for survival and recovery is poor.

MISCELLANEOUS

ASSOCIATED CONDITIONS
N/A

AGE-RELATED FACTORS
• Chickens more susceptible as they reach maturity. • Ducks >4 weeks. • Turkeys, young–mature.

ZOONOTIC POTENTIAL
Infection with *P. multocida* is not considered a high human health risk due to differences in species supsceptibilities to different strains. Most pasteurella infections in humans are the result of domestic animal bites; however, individuals working at outbreak sites or performing necropsies on suspected *P. multocida* infected birds should take appropriate precautions and wear appropriate personal protective equipment (PPE) to prevent passage of disease.

FERTILITY/BREEDING
N/A

SYNONYMNS
Fowl cholera
Avian cholera
Avian hemorrhagic septicemia

SEE ALSO
Avian influenza
Bite wounds
Botulism
Heavy metal toxicosis
Herpesviruses
Organophosphate/carbamate toxicity
Salmonellosis
Viral disease

ABBREVIATIONS
N/A

INTERNET RESOURCES
N/A

Suggested Reading
Franson, J.C., Friend, M. (1999). Avian cholera. In: Franson, J.C., Friend, M. (eds), *USGS Field Manual of Wildlife Diseases: General Field Procedures and Diseases of Birds*. USGS, pp. 75–92.
Samuel, M.D., Boltzer, R.G., Wobeser, G.A. (2007). Avian cholera. In: Thomas, N.J., Hunter, B., Atkinson, C.T. (eds), *Infectious Diseases of Wild Birds*. Ames, IA: Blackwell Publishing, pp. 239–269.
Wobeser, G.A. (1997). Avian cholera. In: Wobeser, G.A. (ed.), *Diseases of Wild Waterfowl*, 2nd Edition. New York: Plenum Press, pp. 57–69.
Author Shannon M. Riggs, DVM

Phallus Prolapse

BASICS

DEFINITION
Only 3% of bird species have a phallus: an intromittent (protruding) form in ratites, tinamous, kiwis, and Anseriformes; a nonintromittent form in Galliformes (chickens, turkeys) and Emberizidae (towhees, juncos, buntings, New World sparrows). Only in males, the structure is derived from the internal cloacal wall, which is generally associated with sperm competition. In the Argentine lake duck, the corkscrew-shaped phallus can approach 16 inches in length. This condition can cause concern for a client or Good Samaritan because the phallus is visibly protruding from the cloaca.

PATHOPHYSIOLOGY
• With the intromittent phallus, it normally lies on the cloacal floor and becomes erect due to lymphatic engorgement produced by the left and right lymphatic bodies. Semen then travels in a groove, the phallic sulcus. In all intromittent phallic species, except the ostrich, the phallus has a blind-ended long hollow tube that with lymphatic engorgement, inverts out like a latex glove finger. The left lymphatic body is larger than the right, and this incongruity causes the erect phallus's spiral twist. • With the nonintromittent phallus, two lateral folds on the ventral edge of the vent become engorged, protruding that edge, which the male quickly contacts to the female's protruding oviduct, with semen transferring down the groove and into the oviduct. • Vasa parrots (*Coracopsis vasa, C. nigra*) have an intermediate form, with a fleshy bag-like protrusion (4 × 5 cm in size) from the male's cloaca inserted into the female's cloaca, with copulation lasting up to 100 minutes. This protrusion can be physiologically prolapsed in the species during breeding season and should not automatically be considered a pathological prolapse.
• Excepting the Vasa parrot example, if the phallus does not return fully into the cloaca, then one would consider that a pathological prolapse requiring veterinary assistance.

SYSTEMS AFFECTED
• Reproductive—may limit reproductive success, trauma to phallus and/or cloacal tissue. • Behavioral—usually occurs during breeding season in seasonal species, can include self-mutilation of phallus if causing discomfort, also may be associated with chronic masturbation on inanimate objects. • Gastrointestinal—caking may lead to inability to empty GI system in timely manner. • Renal/Urologic—urate caking may block outflow of ureters, leading to primary renal disease. • Skin/Exocrine—chronic prolapse may lead to urate and fecal caking of skin and feathers around the vent. • General—generalized weakness may lead to prolapse.

GENETICS
No.

INCIDENCE/PREVALENCE
Unknown.

GEOGRAPHIC DISTRIBUTION
Likely worldwide.

SIGNALMENT
• **Species:** Ostrich, rhea, emu, tinamou, kiwi, ducks, geese, swans. • **Mean age and range:** Mature males and immature males attempting to breed. • **Predominant sex:** Male.

SIGNS

Historical Findings
Breeding season, usually female present, may have been competing males, possible diarrhea/polyuria of explosive nature or chronic.

Physical Examination Findings
Phallus hanging partway or completely out of cloaca, may exteriorize during defecation or urination. Tissue may appear normal (usually a pale white/yellow) or may be necrotic and inflamed. May be fecal/urate caking around vent on feathers. Diarrhea and/or polyuria may be noted. In some cases, phallus may only extrude when bird becomes sexually stimulated by presence of female, male competitors, or even a human owner (other animal) to which it has "mate bonded." Bird may have generalized weakness or have systemic neurologic disease.

CAUSES
Trauma during successful breeding or breeding fatigue are the most likely causes. Gastrointestinal/cloacal disease (cryptosporidiosis and histomoniasis have both been documented causes in ostrich chicks) should be considered secondarily, especially if more than one bird is affected. Generalized weakness may lead to prolapse and inability to retract, likewise any severe systemic neurologic disease such as Eastern Equine Encephalitis or West Nile Virus should be on the differential list.

RISK FACTORS
Breeding season, poor male: female ratio, male competitors, unreceptive females, immature males around females.

DIAGNOSIS

DIFFERENTIAL DIAGNOSIS
Cloacal prolapse, intestinal prolapse, oviductal prolapse, gastrointestinal foreign body protruding from cloaca, foreign body adhered to feathers around vent, neoplasia.

CBC/BIOCHEMISTRY/URINALYSIS
Possibly leukocytosis if site infected or inflamed, possible polycythemia or hyperproteinemia.

OTHER LABORATORY TESTS
Fecal floats, direct smears, fecal Acid Fast stains, fecal cultures if evidence of concurrent diarrhea or if there is a flock issue.

IMAGING
N/A

PATHOLOGIC FINDINGS
Grossly, the phallus would be engorged, possibly necrotic or traumatized. Other lesions noted in the cloacal or phallic tissue may indicate an underlying etiology such as cryptosporidiosis.

TREATMENT

NURSING CARE
Keep phallus lubricated and moist, ideally with a water soluble lubricant.

ACTIVITY
Separate the affected individual away from females and potentially other males. Best to keep in quiet, dark area to encourage limited activity and excitement until the issue is resolved.

DIET
Only if dietary issues causing GI disease lead to prolapse.

CLIENT EDUCATION
N/A

SURGICAL CONSIDERATIONS
• Ideally, the phallus can be replaced into the cloaca. However, in most cases, the bird strains to push it back out immediately or the phallus reprolapses with defecation/urination. In these cases, placing two simple interrupted sutures to decrease the size of the cloacal opening but still allow defecation/urination may buy some time for the tissue to detumensce and be kept within the cloaca. Suture selection should depend on the size of the bird, but often suture strength required is greater than expected. • If those sutures are not holding, the tissue is necrotic/inflamed, or client does not want to attempt salvage, then a phallectomy can be performed. As the structure is only used for copulation, not urination, there is no urethra to rebuild. In some cases, the use of a lidocaine/bupivicaine/sodium bicarbonate local ring and deeper block may suffice for analgesia without needing to resort to general anesthesia. The block is applied proximal to where the amputation is planned. It can be a vascular organ, so expect moderate site hemorrhage with a local block. Depending on the size of the phallus, hemoclips, LigaSeal, or transfixation/circumferential sutures may be used for ligation. The author usually places the

Phallus Prolapse (Continued)

ligation just proximal to a crushing hemostat clamp, then cuts distal to that clamp, slowly removing the clamp. How far proximal on the phallus one attempts amputation may depend on access, equipment, and surgical expertise. Often small "stumps" that are left can be retained by the bird within the cloaca with no apparent long term repercussions. Too aggressive an amputation may lead to damage of the cloaca, slippage of ligation and secondary hemorrhage during recovery, or deformation of the cloaca, leading to defecation and urination issues. • In most cases, the bird's body appears to "self-amputate" the ligation site over time and sutures/hemoclips are shed out unnoticed over the next few weeks to months. • While some birds may still be able to successfully father offspring, phallectomy may lead to inability to naturally successfully copulate, an important consideration for valuable breeders. Ostriches seem to have been reported to have a higher risk of unsuccessful siring.

MEDICATIONS

DRUG(S) OF CHOICE
• Application of a hypertonic sugar solution may be of assistance. If sutures are placed on the cloaca, the application of a DMSO gel or infusion of the cloaca with DMSO may help reduce inflammation and lead to phallus erection reduction. • Antibiotics should be considered if the tissue appears necrotic or infected. If amputation is performed, consider prophylactic antibiotics because of the location (cloaca) of the surgery. • Analgesia should be considered. NSAIDs and opioids may be warranted, depending on species and studies supporting analgesic effects of a particular drug and dose with that species. • An underlying cause may warrant direct treatment of that disease to manage the animal holistically and also prevent recurrence.

CONTRAINDICATIONS
Remember that if bird is considered a food animal, all applicable laws should be considered and severely limit the use of medications.

PRECAUTIONS
N/A

POSSIBLE INTERACTIONS
N/A

ALTERNATIVE DRUGS
N/A

FOLLOW-UP

PATIENT MONITORING
Visual reassessment of the phallus or amputation site should be performed daily to weekly until considered resolved. Experienced veterinarians may be able to manually palpate the phallus or stalk.

PREVENTION/AVOIDANCE
Avoid breeding situations. Neutering or sterilizing birds is still extremely difficult and has high mortality risks; therefore, this should only be contemplated by an experienced surgeon or as a supplemental procedure of last resort.

POSSIBLE COMPLICATIONS
Inability to breed, cloacal fecal/urine leakage, and partial permanent mild prolapse of phallus/stalk/cloaca.

EXPECTED COURSE AND PROGNOSIS
Medical management generally has a guarded prognosis since most cases present in an advanced state to a veterinarian. While amputation may lead to inability to sire offspring (or not), if done correctly most birds have a good prognosis, unless underlying health issues led to the problem and were uncorrectable.

MISCELLANEOUS

ASSOCIATED CONDITIONS
N/A

AGE-RELATED FACTORS
Usually only in mature males during breeding season.

ZOONOTIC POTENTIAL
No.

FERTILITY/BREEDING
While many male birds can still successful father offspring, permanent damage or removal of the phallus likely does decrease reproductive success.

SYNONYMS
N/A

SEE ALSO
Cloacal disease
Cryptosporidiosis
Endoparasites
Gastrointestinal foreign body
Infertility
Trauma
Viral disease

ABBREVIATIONS
N/A

INTERNET RESOURCES
N/A

Suggested Reading
Bezuidenhout, A.J., Penrith, M.L., Burger, W.P. (1993). Prolapse of the phallus and cloaca in the ostrich (*Struthio camelus*). *Journal of the South African Veterinary Association*, **64**(4):156–158.
Coles, B. (2008). Diversity in anatomy and physiology: Clinical significance. In: Coles, B. (ed.), *Essentials of Avian Medicine and Surgery* 3rd edn. Ames, IA: Blackwell, pp. 19–20.
Penrith, M.L., Bezuidenhout, A.J., Burger, W.P., Putterill, J.F. (1994). Evidence for cryptosporidial infection as a cause of prolapse of the phallus and cloaca in ostrich chicks (*Struthio camelus*). *Onderstepoort Journal of Veterinary Research*, **61**(4):283–289.

Author Eric Klaphake, DVM, DABVP (Avian), DACZM, DABVP (Reptile-Amphibian)

Plant and Avocado Toxins

BASICS

DEFINITION
Avocado toxicosis is associated with a wide range of clinical signs in birds. Weakness and depression are usually noted initially, followed by respiratory distress secondary to heart failure. Many plants commonly found in and around a home environment are potentially toxic to birds and listed in Table 1. *Common Plant Toxins, Sources and Effects.* However, there are many other plants considered to be toxic for birds and this list should not be considered all inclusive.

PATHOPHYSIOLOGY
Avocado toxicosis is caused by persin, a fungicidal toxin found in all parts of the avocado. Cardiac glycosides affect heart contractility and rhythm. Oxalate crystals cause inflammation of the oropharyngeal mucosa. Protoanemonin and phorbol esters are GI irritants. Nitrotoxins are believed to inhibit succinate dehydrogenase and fumarase, resulting in failure of the tricarboxilic acid cycle. Grayanotoxins are alkaloids inhibiting inactivation of calcium channels, resulting in prolonged depolarization and excitation. They are also significant GI irritants. Sulfur-containing alkaloids are oxidizing agents activated by mechanical manipulation (cutting and crushing of the plant) and affecting erythrocytes resulting in hemolytic anemia.

SYSTEMS AFFECTED
- Cardiovascular—myocardial necrosis (avocado); altered heart contractility and rhythm (cardiac glycosides, grayanotoxin); tachycardia, pale mucous membranes, collapse (*Allium*). • Respiratory—related to heart failure (avocado, cardiac glycosides).
- GI—swelling/edema of oral mucous membranes (oxalate crystals), GI irritation (protoanemonin, phorbol ester).
- Nervous—weakness, ataxia, paralysis, seizures and coma (cardiac glycosides, nitrotoxin, grayanotoxin).
- Hemic—hemolytic anemia (*Allium*).
- Renal/Urologic—renal failure (rhubarb, *Allium*), hemoglobinuria and hemoglobinuric nephrosis (*Allium*).

GENETICS
N/A

INCIDENCE/PREVALENCE
Between January 2006 and December 2011, the ASPCA Animal Poison Control Center managed 1866 calls regarding exotic birds. Avocado exposure represented 2.6% of those calls.

GEOGRAPHIC DISTRIBUTION
N/A

SIGNALMENT
No specific species, age or sex predilection.

SIGNS

General Comments
- Not all species of birds are equally affected by persin, the toxic principle of avocado. Median lethal dose (LD_{50}) in budgerigars is approximately 2 g. • The adverse effects of avocado may occur as quickly as 15–30 minutes following ingestion, but may also be delayed up to 30 hours. • Clinical signs associated with toxic plant ingestion are variable depending on the plant species and quantity consumed.

Table 1

	Common plant toxins, sources and effects.		
Toxins	*Plant sources*	*Systems affected*	*Clinical signs*
Persin	*Persea Americana* (avocado)	Cardiovascular	Weakness, depression, dyspnea, shock
Cardiac glycosides	*Convallaria majalis* (lily of the valley) *Digitalis purpurea* (foxgloves) Rhododendron *Nerium oleander* (oleander) *Taxus media* (yew) *Kalanchoe* sp	Cardiovascular Nervous	Arrhythmias, cardiac arrest Tremors, ataxia, seizures, coma
Oxalate crystals	*Schefflera* sp (umbrella plant) *Spathephyllum* sp (peace lily) *Dieffenbachia* sp (dumb cane) *Philodendron* sp *Epiprenum* sp (pothos) *Alocasia* sp (Elephant's ear)	GI	Swelling/edema of oral mucous membranes, dysphagia, ptyalism, regurgitation, inappetence
	Rheum (rhubarb)	Renal/urinary GI	Vomiting, swelling/edema of oral mucous membranes, clinical signs of renal failure such as lethargy, polyuria, polydipsia, anorexia, anuria, coma
Protoanemonin	*Montana rubens* (clematis)	GI	Diarrhea, ptyalism, vomiting
Phorbol ester	*Euphorbia* sp (poinsettia)	GI	Diarrhea, vomiting
Nitrotoxin	*Coronilla varia* (crown vetch)	Nervous	Tachypnea, weakness, ataxia, tremors, collapse
Grayanotoxin	Ericaceae family: Rhododendron, Pieris, Menziesia, Leucothoe, Ledum, Kalmia	Nervous Cardiovascular GI	Weakness, ataxia, paralysis, coma Bradycardia, hypotension Mucosal irritation, ptyalism, vomiting
Sulfur-containing alkaloids	*Allium cepa* (domesticated onion) *Allium porrum* (leek) *Allium schoenoprasum* (chives) *Allium sativum* (garlic)	Hemic Renal/urinary	Lethargy, weakness, tachycardia, pale mucous membranes, collapse, death

Plant and Avocado Toxins (Continued)

Historical Findings
• Potential exposure to avocado or toxic plants. • Time spent outside of the cage without supervision. • Healthy bird in good body condition becoming acutely ill.

Physical Examination Findings
• Weakness, depression, lethargy. • Cardiac arrhythmias, tachycardia, bradycardia, increased capillary refill time, pale mucous membranes. • Tachypnea, dyspnea. • Erythematous oropharyngeal mucous membranes, ptyalism, repetitive head-shaking, regurgitation/vomiting, diarrhea. • Ataxia, paralysis, tremors, seizures, coma. • Discolored urine and urates.

CAUSES
• Toxins affecting the cardiovascular system—Persin (avocado), cardiac glycosides (lily of the valley, foxgloves, rhododendron, oleander, yew, *Kalanchoe* sp), and, to a lesser degree, grayanotoxin. • Toxins affecting the GI system—oxalate crystals (umbrella plant, peace lily, dumb cane, *Philodendron* sp, pothos, elephant's ear, rhubarb), protoanemonin (clematis), phorbol ester (poinsettia). • Toxins affecting the nervous system—nitrotoxin (crown vetch), grayanotoxin (*Ericaceae family: Rhododendron, Pieris, Menziesia, Leucothoe, Ledum, Kalmia*), and, to a lesser degree, cardiac glycosides. • Toxins affecting the hemic system—sulfur-containing alkaloids (*Allium*). • Toxins affecting the renal/urinary system—oxalate crystals (rhubarb), sulfur-containing alkaloids (*Allium*).

RISK FACTORS
The presence of avocado or toxic plants in and around the house renders toxic exposure possible. Birds allowed to roam freely out of their cage in the house without supervision or birds that are taken outside to the garden without being directly supervised are at higher risk of plant toxicity.

DIAGNOSIS

DIFFERENTIAL DIAGNOSIS
• Degenerative (cardiomyopathy)—history of exercise intolerance, distended coelom. • Metabolic (hepatic lipidosis, atherosclerosis)—inappropriate diet, obesity, distended abdomen, palpable liver lobes, liver lobes visualized through the abdominal skin by wetting the feathers, yellow-discolored urates, biliverdinuria, intermittent lameness. • Infectious (*Chlamydia*, herpesvirus, reovirus, adenovirus, paramyxovirus)—history of recent exposure to other birds (*Trichomonas, Helicobacter* sp)—stomatitis in cockatiels. • Immune–mediated anemia. • Traumatic (head trauma)—ecchymosis, skin wound, ocular lesions, ear bleeding, palpable fracture. • Toxic (heavy metal toxicosis)—history of foreign body ingestion potentially containing zinc, lead or copper.

CBC/BIOCHEMISTRY/URINALYSIS
• No specific abnormalities reported with most toxins. • Electrolyte imbalances and hemoconcentration due to vomiting and diarrhea. • Hemolysis, hemolytic anemia, occasional Heinz bodies and hemoglobinuria with *Allium* toxicosis. • Indicated to assess evidence of infection, inflammation, and organ function especially if toxin exposure is not documented.

OTHER LABORATORY TESTS
None.

IMAGING
• Whole body radiographs—to evaluate internal organs for abnormalities. • Echocardiography—to evaluate the heart for cardiac abnormalities.

DIAGNOSTIC PROCEDURES
• ECG—to evaluate cardiac arrhythmias.

PATHOLOGIC FINDINGS
• Gross and pathological findings will differ depending upon the toxin ingested. • Avocado—myocardial necrosis, subcutaneous edema, hydropericardium, generalized congestion. • Oxalate crystals—oropharyngeal erythema and edema, renal tubular damage. • Allium—hemoglobinuric nephrosis, hepatosplenic erythrophagocytosis.

TREATMENT

APPROPRIATE HEALTH CARE
• Emergency inpatient intensive care management—patient exhibiting severe depression, cardiac/neurologic/respiratory abnormalities, anemia and shock. • Inpatient medical management—patient vomiting with/without moderate to severe oropharyngeal erythema and edema. • Outpatient medical management—patient otherwise normal; therapeutic approach may require brief hospitalization.

NURSING CARE
• Crop lavage—if early decontamination is possible, that is, within two hours of ingestion. If clinical signs have already started, decontamination is no longer indicated. Sedation or general anesthesia with endotracheal intubation may be considered to perform crop lavage. Stainless steel ball-tipped feeding tube or red rubber feeding tube may be used. The crop is instilled with 20 ml/kg of warm saline solution, massaged gently and emptied by aspirating its content. This process is repeated as needed 3–4 times until the crop content has been removed. • Fluid therapy—may be provided subcutaneously in birds exhibiting mild dehydration (5–7%) or intraosseously/intravenously in birds with moderate to severe dehydration (>7%) or critically ill birds exhibiting clinical signs consistent with shock. Oral fluids are contraindicated in birds severely depressed or birds exhibiting GI/neurological signs. In most patients, Lactated Ringers Solution or Normosol crystalloid fluids are appropriate. For birds with hypotensive shock, a colloidal solution may be administered concurrently with crystalloids. Blood transfusion or hemoglobin replacement product should be considered with severe anemia. • Oxygen therapy—indicated if respiratory distress or anemia is present; beneficial for any sick avian patient. • Warmth (85–90°F/27–30°C)—to minimize energy spent to maintain body temperature. • Pericardiocentesis—may be considered if pericardial tamponade is diagnosed on ECG or echocardiography.

ACTIVITY
Patient activity level should be adjusted according to its general status. Birds exhibiting severe illness should be kept in a small, warm environment. Birds showing neurological deficits may require an adapted environment with lower perches and padded surfaces to prevent injury.

DIET
Birds reluctant to eat may require assisted feeding. Neurological impairment and severe depression increase the risk of aspiration pneumonia. Enteral nutritional preparations may be administered initially at 20–30 ml/kg directly in the crop and the volume administered may be progressively increased according to the patient's tolerance.

CLIENT EDUCATION
• The prognosis of birds following avocado ingestion depends on how quickly treatment is administered. Death may occur if no treatment is provided. This is also true for birds exposed to cardiac glycosides, neurotoxins and sulfur-containing alkaloids. • Birds suffering from erythema/edema following ingestion of clematis, poinsettia, or a plant containing calcium oxalates typically recover with symptomatic and supportive therapy.

SURGICAL CONSIDERATIONS
N/A

MEDICATIONS

DRUG(S) OF CHOICE
• Activated charcoal (2000–8000 mg/kg of body weight); may be repeated every 6–8 hours on a case-by-case basis for avocado toxicosis because it is currently unknown whether persin undergoes enterohepatic recirculation. • Cathartics—magnesium hydroxide (Milk of magnesia, Roxane) may be mixed with activated charcoal (10–12 ml MgOH with 5 ml of activated charcoal PO

PLANT AND AVOCADO TOXINS

(CONTINUED)

once); sodium sulfate (Golytely, 2000 mg/kg PO q24h) × 2 days. • Bulking cathartics—psyllium (Metamucil, 2.5 ml mixed with 60 ml of water or enteral feeding formula q12–24h) or peanut butter and mineral oil (2:1) added to the diet daily until clinical signs resolve. • Sucralfate (25 mg/kg PO q8h)—indicated to treat oropharyngeal irritation from calcium oxalate ingestion. • Diuretic (furosemide 0.15–2.0 mg/kg PO, IM, IV q8h) may be indicated for pulmonary edema and pericardial effusion. Critically dyspneic birds often require higher dosages (4–8 mg/kg) to achieve initial stabilization. • ACE inhibitor (enalapril 1.25 mg/kg PO q 8–12h) may be indicated for heart failure. • Bronchodilator (aminophylline 10 mg/kg IV q3h then 4 mg/kg PO q6–12h; 3 mg/ml water or sterile saline for nebulization) may be indicated if respiratory compromise is present. • Nonsteroidal anti-inflammatory medication (meloxicam 0.5 mg/kg PO/IM q12h) may be considered to alleviate pain from oropharyngeal inflammation. • Antioxidants (vitamin E 0.06 mg/kg IM q7d; vitamin C 20–50 mg/kg IM q1–7d) have been suggested with *Allium* toxicity to reduce oxidative erythrocyte damage.

CONTRAINDICATIONS
Emesis—emetic medications typically used in mammals are ineffective in birds.

PRECAUTIONS
• Fluid deficits should be addressed prior to using cathartics, diuretic and nonsteroidal anti-inflammatory medication. Cathartics may lead to diarrhea and worsen dehydration. • All medications used to treat avocado/toxic plants ingestion are off-label use.

POSSIBLE INTERACTIONS
• Activated charcoal may affect drug absorption. • Combination of high-dose diuretics and ACE inhibitors may alter renal perfusion.

ALTERNATIVE DRUGS
None.

FOLLOW-UP

PATIENT MONITORING
Varies with the toxin ingested and the severity of clinical signs.

PREVENTION/AVOIDANCE
Birds should never have access to avocado and toxic plants in their surroundings.

POSSIBLE COMPLICATIONS
• If the ingested toxin is already absorbed from the gastrointestinal tract, treatment may be unrewarding. • Aspiration pneumonia may occur if a bird vomits and is unable to hold its head upright or swallow properly. • Depending on the organ system affected and the toxin involved, chronic changes (cardiac disease, renal disease, liver disease) may persist following recovery.

EXPECTED COURSE AND PROGNOSIS
• Varies with the toxin ingested, the quantity consumed by the bird and the delay between the ingestion and presentation to a veterinarian. • Once respiratory signs develop following avocado ingestion, death usually follows quickly. • Ingestion of calcium oxalate containing plants is generally associated with a good prognosis with no reported deaths in the avian literature.

MISCELLANEOUS

ASSOCIATED CONDITIONS
• Heart disease. • GI disease. • Neurologic disease. • Renal/urinary disease.

AGE-RELATED FACTORS
N/A

ZOONOTIC POTENTIAL
N/A

FERTILITY/BREEDING
N/A

SYNONYMS
N/A

SEE ALSO
Airborne toxicosis
Cardiac disease
Dehydration
Heavy metal toxicity
Organophosphate/carbamate toxicity

ABBREVIATIONS
GI—gastrointestinal
ACE—angiotensin-converting enzyme

INTERNET RESOURCES
https://www.vetlearn.com/veterinary-technician/toxicology-brief-avian-avocado-toxicosis

Suggested Reading
Hargis, A.M., Stauber, E., Casteel, S., et al. (1989). Avocado (*Persea americana*) intoxication in caged birds. *Journal of the American Veterinary Medical Association*, **194**(1):64–66.

Lightfoot, T.L., Yeager, J.M. (2008). Pet bird toxicity and related environmental concerns. *The Veterinary Clinics of North America. Exotic Animal Practice*, **11**(2):229–259.

Richardson, J. (2006). Implications of toxic substances in clinical disorders. In: Harrison, G.J., Lightfoot, T. (eds), *Clinical Avian Medicine*. Palm Beach, FL: Spix Publishing, pp. 711–719.

Wade, L.L., Newman, S.J. (2004). Hemoglobinuris nephrosis and hepatosplenic erythrophagocytosis in a dusky-headed conure (*Aratinga weddelli*) after ingestion of garlic (*Allium sativum*). *Journal of Avian Medicine and Surgery*, **18**(3):155–161.

Author Isabelle Langlois, DMV, DABVP (Avian)

 Client Education Handout available online

PNEUMONIA

BASICS

DEFINITION
Inflammation of the lung parenchyma secondary to infectious etiologies (bacterial, fungal, viural, parasitic), toxin exposure, or foreign material inhalation. Commonly associated with inflammation in the airsacs (airsacculitis).

PATHOPHYSIOLOGY
Birds have a very specialized, efficient respiratory system with lungs that are adhered to the dorsolateral body wall. Gas exchange occurs in air capillaries, which communicate with other airways (as opposed to the blind sacs of mammalian alveoli). Birds have a system of air sacs, most of which are associated with ventilation. The air sacs connected to the lungs have very little blood supply and are not lined with cilia, thus are not able to clear out items via the mucociliary apparatus. Most of them are dependent, leading to deposition of inhaled particles, which cannot exit the body and thus create a source of inflammation as the body tries to wall them off.

Pathogens may enter the respiratory tract by inhalation or hematogenous spread. Infections in the upper respiratory tract may spread into the lower tract. The avian immune system is effective in encapsulating sources of infection/inflammation and over time. These encapsulated areas may not allow exposure of the etiologic agents to the immune system and blood supply, making identification challenging and treatment more difficult.

SYSTEMS AFFECTED
Respiratory.

GENETICS
N/A

INCIDENCE/PREVALENCE
Common in pet birds and poultry.

GEOGRAPHIC DISTRIBUTION
Unlimited distribution.

SIGNALMENT
- **Species:** ○ All species and ages, both sexes. ○ Hand-fed birds (especially young) are more susceptible to aspiration pneumonia.
- **Species predilections:** African Grey parrots may be predisposed to aspergillosis/fungal pneumonia.

SIGNS

General Comments
If a bird presents for respiratory distress, discuss with the owner the possible risk of death from handling. Place in oxygen before and during examination and between diagnostic tests. Birds have a very efficient respiratory system, and oxygen may not help.

Historical Findings
Owners may note mild to severe signs, including general sick bird signs (lethargy, depression, inappetance, weight loss, decreased activity, fluffed, sitting on the bottom of the cage), progressive exercise intolerance, voice change/loss of voice (if associated with lesions in syrinx/caudal trachea), open-beak breathing, increased respiratory effort (tail bob, exaggerated keel movement), coughing (may need to differentiate from mimicking a human coughing sound).

Physical Examination Findings
Respiratory rate and effort should be evaluated before birds are handled. Assess for tail bob, exaggerated keel movement, open-beak breathing, and/or extended neck. If dyspnea is noted, place the patient in oxygen cage for 5–10 minutes before the examination. Take examination in stages if dyspnea worsens during handling. Carefully hold all dyspneic birds upright (especially if coelomic distention is noted). Signs of pneumonia on physical examination include:
- Increased respiratory rate. • Increased respiratory effort (tail bob). • Increased or rough lung sounds may be heard on auscultation of dorsal body wall. • Open-beak breathing. • Thin body condition.

CAUSES

Infectious
- Bacterial—may be inhaled/aspirated or spread hematogenously (sepsis); Gram-negative pathogens commonly noted although Gram-positive organisms have been reported as well (*Streptococcus spp.*, *Staphylococcus spp.*); specific etiologies include *Nocardia*, *Chlamydia psittaci*, *Mycobacterium* (*M. avium* or *M. genavense*), *Mycoplasma spp.*
- Fungal—including *Aspergillus spp.*, *Penicillium spp.*, Zygomycetes (including *Mucor*, *Rhizopus*, *Absidia*), *Trichosporon spp.*; Cryptococcal pneumonia may be associated with cryptococcal sinusitis. • Viral—including *Polyomavirus*, *Herpesvirus*, *Poxvirus*, *Paramyxovirus* (including Exotic Newcastle Disease [END]), avian influenza (AI).
- Parasitic—including *Sarcocystis spp.* (*S. falcatula* most significant) especially in Old World Psittacines; *Sternostoma tracheolum* (air sac mites in canaries and finches).

Noninfectious
- Aspiration (especially hand fed birds).
- Airborne toxins—most commonly related to exposure to pyrolysis products from overheating pans coated with polytetrafluoroethylene (PTFE); also includes exposure to aerosol sprays, paint fumes, cigarette smoke, fumes from burned food and self cleaning ovens; most often causes acute death but birds that survive may demonstrate signs of pneumonia.

RISK FACTORS
- Malnutrition—birds on an all-seed (Vitamin A deficient) diet are more prone to hypovitaminosis A, which can promote squamous metaplasia within the respiratory tract. • Hand feeding. • Systemic disease/immune suppression. • Exposure to airborne irritants (cigarette smoke, dust, burning food or coated pans). • Poor husbandry (inadequate diet, unsanitary conditions).
- Stress.

DIAGNOSIS

DIFFERENTIAL DIAGNOSIS
- Upper airway disease (including nasal, glottis and tracheal disease)—differentiate with radiographs and physical examination.
- Heart failure—differentiate with physical examination (auscultation), radiographs and ultrasound. • Pulmonary hypersensitivity—most common in macaws exposed to dust producing birds, differentiate with history, radiographs and hematocrit (polycythemia is often present). • Pulmonary neoplasia—differentiate with biopsy and CBC.
- Trauma—differentiate with history. • Any source of coelomic distention which compresses airsacs (fluid accumulation from coelomitis, heart or liver failure; organomegaly; neoplasia, etc.)—differentiate with physical examination and radiographs.

CBC/BIOCHEMISTRY/URINALYSIS
- CBC—a leukocytosis may be present (if the lesions are localized to the air sacs or are well walled off in the lung parenchyma, there may be little to no elevation in the white blood cell count). The leukogram may vary depending on the etiologic agent. Bacterial infections are more likely to cause heterophilia with or without a left shift. Fungal, chlamydial, or mycobacterial infections may produce a monocytosis, which may only be a relative monocytosis with an otherwise normal white blood cell count. If gas exchange is impaired for a long enough time, polycythemia may occur. Conversely, anemia may be present related to chronic disease. • Chemistry panel—creatine kinase may be elevated if the infection is adjacent to or invading the muscle surrounding the lungs, but otherwise the chemistry panel helps identify other concurrent disease processes which may affect the treatment plan.

OTHER LABORATORY TESTS
- Protein electrophoresis (EPH) may help confirm the presence of inflammation in the absence of a leukocytosis. • Molecular diagnostics and serology may be available for specific etiologic agents (see the sections for each etiologic agent). • Bacterial and fungal culture and sensitivity are very important—submit tracheal wash samples or tissue from lung and airsac biopsies. • Cytology should be performed if a tracheal wash sample is obtained. • Histopathology should be performed on any biopsies taken.

PNEUMONIA (CONTINUED)

IMAGING
Whole body radiographs are important to identify abnormalities in the lung fields and air sacs. On the lateral view, patchiness to the normal "honeycomb" appearance may suggest pulmonary infiltrates or granulomas; thickening of the air sac membranes ("air sac lines") and opacities within the air sacs may be recognized. On the ventro-dorsal view, asymmetry of the lung fields and lateral air sacs may be noted. The pectoral muscles overlap the lung tissue, making interpretation of lungs on the ventro-dorsal view challenging.

DIAGNOSTIC PROCEDURES
- Endotracheal wash is a technique that allows sampling without surgery. Use 0.5–1.0 mL/kg saline in an anesthetized patient; submit sample for cytology as well as aerobic and fungal cultures. • Celioscopy is a minimally invasive technique to visualize the air sacs and caudo-ventral border of the lungs. The air sacs should be completely clear and the lungs should be bright pink without any exudate or hemorrhage. Biopsies should be obtained of any lesions and lung tissue and submitted for histopathology and aerobic and fungal culture. Tissue should be saved in the freezer for possible molecular diagnostics as directed by the histopathologic findings.

PATHOLOGIC FINDINGS
- The lungs may appear hemorrhagic or have white or yellow patches on the surface when visualized. Air sacs may have increased vascularity or may contain white to yellow caseous plaques (fungal plaques may appear fluffy). • Histology will demonstrate inflammation characteristic of the etiologic agent. • *Polyomavirus*—mononuclear infiltration. • *Herpesvirus*—edema and congestion. • *Poxvirus*—epithelial hyperplasia of the airways. • Inhaled bacteria—intraluminal hemorrhage, a small amount of suppurative inflammation and fibrin deposition. • Hematogenous bacteria—more severe effacement of parenchyma with hemorrhage, congestion, fibrinopurulent exudate and intralesional bacteria. • Nocardiosis and mycobacteriosis—parenchymal granulomatous lesions. • *Mycoplasma* pneumonia—lymphocytic infiltrates which may develop into lymphoid nodules. • Fungal infections—granulomatous lesions in the lungs and air sacs (acute infections may associated with severe hemorrhage, fibrin deposition, and edema; chronic infections lead to the development of granulomatous nodules which consist of necrotic cores surrounded by degenerate inflammatory cells, multinucleated giant cells and epithelioid macrophages, which contain hyphae). • *Sarcocystis*—severe pulmonary congestion, edema, or hemorrhage. • Air sac mites—when present in the parabronchi cause infiltration of lymphocytes, plasma cells and macrophages. • Toxin exposure—edema and hemorrhage within the gas exchange tissues. • Foreign body inhalation—initially characterized by hemorrhage, congestion, edema and fibrin deposition; chronic lesions are characterized by accumulations of lymphocytes, macrophages and multinucleated giant cells.

TREATMENT
APPROPRIATE HEALTH CARE
- Inpatient care is required for birds that require oxygen or supportive care (fluids/feeding). • Outpatient care is feasible for birds that are stable off of oxygen and are eating and maintaining hydration.

NURSING CARE
- Fluid therapy as needed based on deficits (subcutaneous or intravenous [IV]/intraosseus [IO]) to maintain hydration. • Oxygen may be required (measure respiratory rate on and off of oxygen to determine if it is helping).

ACTIVITY
Keep quiet and warm, minimize activity. Birds may not tolerate exercise well with respiratory compromise.

DIET
Offer patient's normal diet. Gavage may be needed if decreased appetite noted.

CLIENT EDUCATION
Respiratory disease carries a guarded prognosis and may require very long term and aggressive therapy (granulomas are hard to penetrate with medications and some may need to be surgically debrided). Aggressive care as quickly as possible will increase the chances of recovery.

SURGICAL CONSIDERATIONS
Birds with pneumonia will likely require ventilatory support during anesthesia. They may not respond as well to gas anesthetics if too much functional tissue is affected. Be prepared to place an air sac cannula if there is suspicion that the trachea or syrinx is involved. However, if there is severe lung parenchymal disease (affecting air exchange tissues), the air sac cannula may not help.

MEDICATIONS
DRUG(S) OF CHOICE
- Base on culture/sensitivity, when possible. If no culture is possible, use a broad spectrum combination of antibiotics (i.e., not just enrofloxacin alone). Doses are for psittacines, unless otherwise noted. Treatment should persist for at least one month, and repeat radiographs, CBC, +/− endoscopy should be used to determine if treatment may be discontinued. • Antibiotics (very debilitated or septic birds may require injectable antibiotics for a few days then switched to oral versions). ○ Ceftazidime—50–100 mg/kg IM or IV q 4–8h. ○ Piperacillin/tazobactam—100 mg/kg IM or IV q6–12h. ○ Ticarcillin/clavulanic acid—200 mg/kg IM q12h. ○ Doxycycline (Vibravenos injection, drug of choice for *Mycoplasma* and *Chlamydia*)—75–100 mg/kg IM q5-7d for 4-6 weeks. ○ Enrofloxacin—wide dose range; historically 5–15 mg/kg BID was recommended and may still be efficacious; currently 20–25 mg/kg q24h accepted for SQ, IM, PO (IM dose should be used only once to minimize muscle trauma, dilute SQ dose in fluids to minimize tissue trauma), no anaerobic coverage, *Not approved in food producing birds. ○ Amoxicillin/Clavulanic acid—125 mg/kg PO BID. ○ Trimethoprim/Sulfamethoxazole—30–100 mg/kg PO q12h. ○ Doxycycline (drug of choice for *Mycoplasma* and *Chlamydia*)—25–50 mg/kg PO q12–24h. ○ Azithromycin (used for primarily intracellular infections, like *Chlamydia psittaci*—(10–20 mg/mg PO q48h for at least five treatments for nonintracellular infections (macaws); 40 mg/kg PO q24h for 30 days for intracellular infections (macaws); 40 mg/kg PO q48h for 30 days for intracellular infections (cockatiels)). • Antifungals (see Aspergillosis chapter). • Nebulized medications—Accepted as beneficial. Twice daily treatment is recommended for at least 15 minute sessions. Create a nebulization chamber out of a large plastic storage container or a carrier wrapped in a clear plastic garbage bag. For treatment of the lung and air sacs, the nebulized particles should be <3–5 μm in diameter (this information is available in nebulizer package inserts). ○ Saline (0.9%) and sterile water—may reduce viscosity of secretions without any additives. ○ Enrofloxacin—10 mg/mL saline. ○ Amikacin—5–6 mg/mL saline. ○ Piperacillin—10 mg/mL saline. ○ Terbutaline—0.01 mg/kg with 9 mL saline—bronchodilation. ○ N-acetyl-L-cysteine (Mucomyst)—22 mg/mL in sterile water—mucolytic agent; tracheal irritation/reflex bronchoconstriction noted in mammals, use preceded by bronchodilators in mammals. • Other medication: Ivermectin is the drug of choice for tracheal mites—0.2 mg/kg topically over the right jugular once every two weeks for three doses.

CONTRAINDICATIONS
N/A

PRECAUTIONS
Some birds may not tolerate medications as well as others and adverse reactions are possible despite all precautions. In general, hypovolemia/dehydration should be corrected before starting most therapeutic agents. Trimethoprim/Sulfa and enrofloxacin may cause vomiting. Doxycycline may cause regurgitation and should be used at a lower

PNEUMONIA (CONTINUED)

dose range in macaws and cockatoos. Alterations in liver or kidney function might affect medication choices (i.e., imidazoles should be used with caution in patients with altered liver function, azithromycin should not be used in birds with liver or renal disease, doxycycline should be used with caution in birds with liver disease). Sulfa antibiotics should not be used in birds with dehydration, liver disease or bone marrow suppression. Nebulized amikacin is unlikely to cause renal dysfunction, but should be used with caution in patients with existing renal disease.

POSSIBLE INTERACTIONS
- Do not use amoxicillin/clavulanic acid with allopurinol. • Foods containing calcium, aluminum, magnesium, or iron may interfere with absorption of oral doxycycline and enrofloxacin.

ALTERNATIVE DRUGS
Omega 3 fatty acid supplementation may help as an adjunct treatment (for its anti-inflammatory effects), but does not replace the need for antimicrobials if indicated.

FOLLOW-UP

PATIENT MONITORING
- CBC should be performed regularly (weekly to start with, then monthly to every six months for fungal or other granulomatous disease). • Radiographs should be performed at least before cessation of treatment to confirm absence of disease. Long-term infections should be monitored with radiographs every 3–6 months. Follow up endoscopy to visualize previously diseased lungs and air sac spaces may be necessary to determine if radiographic changes are related to active disease or scar tissue.

PREVENTION/AVOIDANCE
Ensure good ventilation, clean housing, regular exposure to direct sunlight, nutritionally balanced diet.

POSSIBLE COMPLICATIONS
- Permanent damage may lead to residual exercise intolerance or potentially lack of response to treatment. • Initial improvement then recrudescence if treatment discontinued too early.

EXPECTED COURSE AND PROGNOSIS
Prognosis is variable depending on etiology. Simple bacterial infections may clear up in weeks, but fungal and other granulomatous infections may take months (to years) of treatment.

MISCELLANEOUS

ASSOCIATED CONDITIONS
Upper airway disease.

AGE-RELATED FACTORS
Young birds with aspiration pneumonia are more likely to develop infections of multiple etiologies due to immune suppression. Older birds, especially after being fed a lifetime of seeds or being exposed to long-term second hand smoke, may be more at risk due to immune suppression.

ZOONOTIC POTENTIAL
Chlamydia psittaci is zoonotic and reportable. Mycobacterial infections are potentially zoonotic and careful consideration should be taken when determining if treatment is warranted. *Paramyxovirus* that causes END may cause mild disease in humans. Avian influenza is zoonotic.

FERTILITY/BREEDING
Note regulations for use of therapeutics in poultry. No antibiotics are approved for laying hens but withdrawal times can be determined. Fluoroquinolones, nitroimidazoles, glycopeptides, and cephalosporins are all prohibited for use in food animals. Access the Food Animal Residue Avoidance Databank for egg withdrawal times for nonprohibited drugs (http://www.farad.org).

SYNONYMS
Lower respiratory tract disease

SEE ALSO
Air sac mites
Aspergillosis
Aspiration
Avian Influenza
Airborne toxicosis
Chlamydiosis
Coelomic distention
Herpesviruses
Mycobacteriosis
Mycoplasmosis
Paramyxoviruses
Polyomavirus
Poxvirus
Respiratory distress
Sarcocystis
Tracheal/Syringeal Disease
Trauma

ABBREVIATIONS
N/A

INTERNET RESOURCES
Compendium of Measures To Control *Chlamydophila psittaci* Infection Among Humans (Psittacosis) and Pet Birds (Avian Chlamydiosis), 2010, National Association of State Public Health Veterinarians (NASPHV): http://www.nasphv.org/Documents/Psittacosis.pdf.
Food Animal Residue Avoidance Databank: http://www.farad.org (use to help determine withdrawal times for poultry cases).

Suggested Reading
Hawkins, M.G., Barron, H.W., Speer, B.L., et al. (2013). Birds. In: Carpenter, J.W., Marion, C.J. (eds), *Exotic Animal Formulary*. St. Louis, MO: Elsevier, pp. 183–437.
Hillyer, E.V., Orosz, S., Dorrestein, G.M. (1997). Respiratory system. In: Altman, R.B., Clubb, S.L., Dorrestein, G.M., Quesenberry, K. (eds), *Avian Medicine and Surgery*. Philadelphia, PA: WB Saunders, pp. 387–411.
Orosz, S.E., Lichtenberger, M. (2011). Avian respiratory distress. *The Veterinary Clinics of North America. Exotic Animal Practice*, **14**(2):241–256.
Schmidt, R.E., Reavill, D.R., Phalen, D.N. (2003). *Pathology of Pet and Aviary Birds*. Ames, IA: Iowa State Press.
Tully, T.N., Harrison G.J. (1994). Pneumonology, In: Ritchie, B.W., Harrison, G.J., Harrison, L.R. (eds), *Avian Medicine: Principles and Application*. Lake Worth, FL: Wingers Publishing, Inc., pp. 556—581.
Author Anneliese Strunk, DVM, Diplomate ABVP (Avian)

Pododermatitis

BASICS

DEFINITION
Avian bumblefoot is a catch-all term for inflammatory, infectious or degenerative conditions of the avian foot. The condition can range from mild swelling and redness of the plantar surfaces of the feet and toes to ulceration and potentially cellulitis and osteomyelitis. Strictly-speaking, pododermatitis is inflammation of the dermal structures of the feet; however, this term is often used interchangeably with bumblefoot in avian species.
   Etiology varies according to species and specific environmental conditions and management systems.

PATHOPHYSIOLOGY
- Regardless of the inciting cause, ischemic conditions and/or vascular congestion are often central to the development of bumblefoot. • Poor hygiene can cause pododermatitis with extension of infectious disease to deeper tissues resulting in cellulitis, tenosynovitis and osteomyelitis.
- Malnutrition can also cause or exacerbate pododermatitis. Hypovitaminosis A with squamous metaplasia of the plantar epithelium can predispose birds to plantar erosions and damage to deeper tissues.

SYSTEMS AFFECTED
Typically, bumblefoot begins as a disorder of the skin, but the disease can progress to affect all of the anatomic structures of the feet.

GENETICS
No genetic predisposition is appreciated.

INCIDENCE/PREVALENCE
Bumblefoot is a disease of captive animals and is almost always related to deficiencies in husbandry. Incidence and prevalence are therefore variable and dependent on specific husbandry conditions.

GEOGRAPHIC DISTRIBUTION
N/A

SIGNALMENT
- **Species:** Heavy bodied species, such as birds of prey (*Accipitridae*), parrots (*Psittacidae*), penguins (*Speniscidae*), storks (*Ciconiidae*), cranes (*Gruidae*), and waterfowl/shorebirds (*Anseriformes*), are predisposed to pododermatitis, though all species can be affected. • **Mean age and range:** All ages are susceptible. • **Predominant sex:** Both sexes are susceptible, though males of some species are more frequently affected due to relatively increased body weight.

SIGNS

General Comments
- Pododermatitis can assume a variety of presentations depending on the inciting cause and chronicity. A number of classification systems are described in various publications. In one classification scheme, three types are described: ○ In type I dermatitis, lesions are mild, localized and characterized as either degenerative or proliferative. In degenerative conditions the normally pebbled skin on the plantar surface of the feet becomes smooth and thin. In proliferative conditions, hyperkeratosis occurs resulting in "corns" and cracking on the plantar surfaces. ○ In type II, bacterial or fungal infections begin to play a role in the disease process. *Staphyloccocus aureus, Eschericia coli, Pseudomonas* spp., and sometimes yeast or fungi are often encountered. There is typically a noticeable inflammatory component to type II pododermatitis. ○ In type III pododermatitis, the changes are typically chronic, infected and involve deeper soft and bony tissues of the feet. • In some chronic cases, a static condition can develop in which a heavy eschar develops over an ulcer on the plantar aspect of the foot. In these cases, over time excessive fibrous connective tissue can develop under the eschar which may eventually cause significant discomfort and lameness.

Historical Findings
Typically, predisposing factors can be identified in the history or physical examination. Often there is history of an injury to the pelvic limbs resulting in direct injury to the feet or resulting in uneven weight-bearing. Persistently wet or unhygienic substrate conditions often result in pododermatitis. Changes in substrate or perching (inappropriate size, shape, and surface) also can result in acute pododermatitis. Birds that are overweight or malnourished (hypovitaminosis A) are predisposed to pododermatitis. Birds that have been previously affected by pododermatitis are predisposed to recrudescence of disease.

Physical Examination Findings
Pododermatitis can affect one or both pelvic limbs resulting in varying degrees of lameness. With mild cases, clinical signs can be as seemingly insignificant as minor smoothing of the keratinized skin on the plantar surfaces of the toes/feet. More advanced cases may present with swelling of the toes and feet, ulcerations with eschar formation on the plantar surfaces. Joint effusion and abscessation of the soft tissues and joints may also be encountered.

CAUSES
Causes of pododermatitis include both acute and chronic trauma, and poor hygiene. Lacerations, puncture wounds and abrasions can become secondarily infected leading to pododermatitis. Abnormal wear on weight-bearing surfaces as may occur with birds suffering from degenerative joint disease, obesity, and so on, can also lead to pododermatitis.

RISK FACTORS
The risk factors for bumblefoot are conditions that result in abnormal trauma to the structures of the feet, or change the skin and other tissue's ability to withstand normal wear. Risk factors include, but are not limited to obesity, inactivity, inappropriate substrate and/or perches, poor hygiene and/or chronically wet substrate. Traumatic injuries to the feet and limbs can create portals for ingress of infective organisms can result in bumblefoot. Degenerative conditions of the pelvic limbs can also lead to bumblefoot as abnormal forces are applied to the structures of the feet. When injuries or degenerative conditions affect a pelvic limb it is not uncommon for bumblefoot to develop in the contralateral foot.

DIAGNOSIS
The diagnosis of pododermatitis is relatively straightforward; however, diagnostics beyond a physical examination are often required to characterize the diagnosis and direct therapy.

DIFFERENTIAL DIAGNOSIS
Neoplastic conditions affecting the structures of the feet can sometimes mimic bumblefoot. Articular gout can also create conditions that may be confused with bumblefoot.

CBC/BIOCHEMISTRY/URINALYSIS
CBC often shows signs of inflammation. Changes in the differential may reflect the duration of the disease process. A chemistry panel is indicated to evaluate organ function; amyloidosis is frequently encountered as a sequela to chronic cases of bumblefoot.

OTHER LABORATORY TESTS
- Fine-needle aspiration and impression smears of affected tissue for cytology and potentially culture/sensitivity are often useful in diagnosis and direction of therapy. If joints are noted to be distended, arthrocentesis may be indicated. • Biopsies and histopathology of affected tissue can provide extremely important information, especially in cases that are proliferative and/or do not respond to therapy as expected.

IMAGING
- High-detail radiographs of the feet are indicated to rule out osteomyelitis of the bony structures of the feet. • Fistulograms utilizing iodinated contrast material can also be useful. • Angiography utilizing iodinated contrast materials suitable for IV administration may be a worthwhile diagnostic to characterize the blood supply to the feet. • Ultrasonography with a high-frequency transducer is an effective imaging modality in larger species, and can be especially helpful in identifying pockets of fluid and evaluating the vasculature structures of the feet. Stand-off pads can facilitate effective imaging of structures just

under the skin. • Computed tomography with or without contrast may also prove useful if the machine is capable of small enough slices to provide adequate detail. • Thermography has been described as useful in detecting subclinical bumblefoot in poultry. This modality may be useful in evaluation of cases of lameness with no external lesions or in the monitoring of cases under treatment.

DIAGNOSTIC PROCEDURES
Pathogen-specific PCR of aspirates/fluid if indicated.

PATHOLOGIC FINDINGS
Dermatitis, cellulitis, tenosynovitis and osteomyelitis are common findings in bumblefoot cases. Vasculitis with thrombosis and loss of blood supply to affected tissue is also frequently encountered. Lack of adequate vascular supply is a very important finding as it will dramatically affect the bird's ability to recover and resolve bumblefoot lesions.

TREATMENT
• Correction of husbandry deficiencies and other predisposing factors is critical to addressing bumblefoot in avian patients. In very mild cases, correction of these predisposing factors may be adequate for complete resolution of the lesions. • In mild cases in which the plantar skin is intact, protective bandaging and nonsteroidal anti-inflammatory medications may be sufficient to reverse the process and allow the skin to heal and resume its normal conformation. • In more severe cases where ulceration, abscessation and necrotic tissue are present, surgical debridement is necessary. Debridement needs to be precise and delicate to preserve vital structures if possible. Magnification can be helpful. In cases where the resultant surgical defect is small enough and the site is considered clean, primary closure may be possible. Large or contaminated lesions that cannot be closed need to be managed until they are considered candidates for surgical closure. Wet-to-dry bandages can be a very effective debridement technique, especially in severely infected and necrotic lesions. • Bandaging is an exceedingly important aspect of treating bumblefoot in avian patients. Many different bandaging schemes are described, but all share the common goal of relieving pressure on the lesion and encourage circulation of blood to the plantar structures. Donut-shaped pads are often fashioned out of various materials and incorporated into the bandage. Great care needs to be taken to avoid creating excessive pressure at the contact points of any pad incorporated in to the bandage. Wrapping the toes with soft cast padding or other similar materials at the contact points can help avoid the development of iatrogenic lesions. • The use of vacuum-assisted closure (VAC) devices is often discussed as a possible option in treating severe bumblefoot lesions in birds, but implementation of this modality is challenging to say the least. In terrestrial birds that are not inclined to chew vacuum lines, VAC therapy can be considered as a possible option for treatment in the early stages of treatment. The benefit of VAC therapy should be considered in light of the limitations it will obviously have on activity.

APPROPRIATE HEALTH CARE
Patients that are self-sustaining and ambulatory are typically treated as outpatients, but brief hospitalization may be appropriate for initiation of analgesic therapy and collection of baseline diagnostic data (including screening radiographs). Debilitated patients may need extended periods of hospitalization for supportive care until stable.

NURSING CARE
Bandage maintenance is critical in management of bumblefoot cases. Quarters for convalescing birds should be managed such that the bandages stay clean and dry until scheduled changes. If a bandage is noted to appear soiled, wet, and so on, it should be changed immediately. It is recommended that bandages should be changed every 24–72 hours depending on the severity. The need for sedation/anesthesia and/or analgesia for bandage changes can be decided on a case by case basis. If the bird vigorously resists bandaging and appears overly stressed by the bandage change procedure, anesthesia is warranted.

ACTIVITY
In most cases, activity should be encouraged and maximized. Birds with bumblefoot should be kept in lean body condition in order to avoid excessive pressure on the structures of the feet. Exercise will also encourage circulation of blood and lymph in the feet. Flighted birds should be given as much opportunity to fly as possible.

DIET
Birds suffering from bumblefoot should have their vitamin A status and body condition critically evaluated with adjustments made as necessary. Vitamin A supplementation should be pursued carefully as over-supplementation can easily occur. In most cases oral supplementation is preferred over parenteral delivery. Often, providing a balanced diet is sufficient and specific vitamin A administration is not necessary.

Body condition should be kept as lean as possible to minimize pressure on the bird's feet. Providing a healthy diet in appropriate proportions combined with adequate exercise will help keep birds as lean as possible in captive situations.

CLIENT EDUCATION
Bumblefoot is a disorder of captive animals and is directly related to deficiencies in husbandry and diet. Most cases of bumblefoot could be avoided with proper education of keepers and implementation of a minimum standard of husbandry. Clients working with high-risk species such as waterfowl and falcons should be aware of the risk factors for development of bumblefoot and should monitor their charges closely. Falcons can develop severe bumblefoot within days of the introduction of inappropriate perching. Thus, keepers must be properly educated regarding appropriate perching for the species they work with.

SURGICAL CONSIDERATIONS
• Surgical intervention is often indicated in cases of bumblefoot. Surgical procedures should be carefully planned as clumsy and over-aggressive debridement can result in damage to vital structures of the feet and affect the long-term outcome of the case. • Necrotic tissue should be carefully removed with efforts to identify and preserve tendons, ligaments, joint capsules, vasculature and nerves if possible. Resultant surgical defects can be closed primarily or allowed to heal via second intention depending on the level of residual contamination/necrotic tissue present. In cases where joints or tendon sheaths are affected or abscessed, indwelling catheters can be placed to allow repetitive local therapy via flushing or delivery of medications. • There are procedures described in which pedicle advancement flaps are used to close defects on the plantar surfaces of the feet. These procedures can be considered in chronic cases with persistent open plantar lesions. Following the application of advancement flaps, careful bandaging and antisepsis needs to be practiced following the procedure.

MEDICATIONS
• Oral and parenteral nonsteroidal anti-inflammatory medications are frequently indicated in cases of bumblefoot. Meloxicam, ketoprofen, and carprofen are commonly used in avian patients and are typically well tolerated and seemingly effective in avian patients. • Antimicrobial therapy is indicated when open lesions are present or there is obvious evidence of infection. Drug choice should ideally be based on culture results though *Staphylococcus aureus* and *E. coli* are commonly encountered isolates. • Since in many cases of bumblefoot blood flow to the lesion(s) is impaired, efforts to promote perfusion are indicated. Pentoxifylline at a dose of 15 mg/kg orally q8–12h for 6–8 weeks has been used with seemingly positive effect. • Various topical medications are used in the treatment of bumblefoot. Antimicrobial

(CONTINUED)

creams or ointments are used when open lesions are present. Silver sulfadiazine cream is very frequently used when open lesions are present. Pluronic gel compounded to contain an appropriate antibiotic can be very effective in local treatment of infectious processes.
• Mixtures containing dimethyl sulfoxide, various corticosteroids, nonsteroidal anti-inflammatory medications, and antibiotics are described for use in treatment of bumblefoot and labeled with names such as "Bumble-Be-Gone". These mixtures should be used with caution, especially when containing corticosteroids as birds tend to be quite sensitive to the adverse effects of corticosteroids. • Topical application of materials such as granulated sugar, raw/medical honey, and papaya can be safe and effective as part of the therapeutic approach for open bumblefoot lesions. These materials can be incorporated into wet-to-dry bandages and aid dramatically with antisepsis, debridement and encouragement of granulation tissue formation. • There are many topical products available that may have some application in the treatment of avian bumblefoot. Various topical antimicrobial products as well as collagen-containing products, platelet-derived growth factor and platelet-rich plasma may all have applications depending on the specific features of the case.

DRUG(S) OF CHOICE
There are no medications that are uniformly indicated in cases of bumblefoot; selection of medication(s) should be based on the specifics of individual cases.

CONTRAINDICATIONS
• Nonsteroidal anti-inflammatory medications should be used with caution in birds with existing or suspected renal disease.
• If topical medications containing corticosteroids are used, care should be taken to monitor for signs of adverse effects. Even topical corticosteroid administration can result in systemic effects, including immunosuppression.

PRECAUTIONS
N/A

POSSIBLE INTERACTIONS
N/A

ALTERNATIVE DRUGS
N/A

FOLLOW-UP
Therapy for bumblefoot is often necessarily protracted, requiring weeks to months depending on the severity of disease. Birds that have previously been affected by bumblefoot are predisposed to recurrence of disease in the future, even after many months to years. Cases that appear to be going well will often abruptly worsen, so frequent rechecks are recommended. The follow-up schedule will, of course, depend on the individual case, but in severe cases rechecks every one to two weeks may be warranted until the lesions appear to be improving. Following apparent resolution of lesions, the bird should be rechecked at monthly intervals for 2–3 months and then every six months until the bird is determined to be stable.

PATIENT MONITORING
Serial CBCs, chemistry panels and cultures of open lesions are often indicated to provide optimal care for birds suffering from bumblefoot. Serial imaging via radiography, computed tomography and/or ultrasonography can also be useful, especially in cases that are not responding as expected to therapy.

PREVENTION/AVOIDANCE
Optimal nutrition, maintenance of healthy body condition and adequate exercise are all central to prevention of bumblefoot in captive birds. Appropriate substrate and perching, including hygiene, are also important in prevention of bumblefoot. Immediate and aggressive therapy of traumatic lesions to the feet or toes is also important.

POSSIBLE COMPLICATIONS
• In cases of unilateral bumblefoot, it is not uncommon to see lesions develop in the contralateral foot due to uneven weight bearing. Chronic bumblefoot can also lead to degenerative conditions of the joints of the pelvic limbs. • Systemic amyloidosis is frequently encountered in birds with long-standing bumblefoot lesions. The liver, spleen, kidneys and heart are commonly affected by amyloidosis in these cases. Biopsies of affected tissues are necessary to diagnose the condition. • Birds with chronically infected bumblefoot lesions are also at risk for sepsis and bacterial seeding of visceral organs. As a result hepatic abscessation and development of endocarditis is occasionally encountered.

EXPECTED COURSE AND PROGNOSIS
Mild cases of bumblefoot hold a good-to-fair prognosis based on causative factors and can be expected to make fast and complete recoveries with appropriate therapy and husbandry modification. More severe or chronic cases carry a guarded prognosis. Treatment of severe cases can take many months and recrudescence of disease following apparently successful therapy is common. In very severe cases in which septic arthritis, tenosynovitis and/or osteomyelitis is present, the prognosis is poor for complete recovery and euthanasia may even be considered.

MISCELLANEOUS

ASSOCIATED CONDITIONS
Amyloidosis, sepsis, endocarditis, degenerative joint disease.

AGE-RELATED FACTORS
Older, chronically affected birds are likely to be affected by associated conditions as listed above.

ZOONOTIC POTENTIAL
N/A

FERTILITY/BREEDING
N/A

SYNONYMS
Pododermatitis

SEE ALSO
Arthritis
Cere and skin, color changes
Dermatitis
Ectoparasites
Fracture/luxation
Lameness
Nutritional deficiencies
Obesity
Overgrown beak and nails
Papilloma (cutaneous)
Poxvirus
Splay leg/slipped tendon
Trauma

ABBREVIATIONS
N/A

INTERNET RESOURCES
http://www.themodernapprentice.com/perches.htm

Suggested Reading
Blair, J. (2013). Bumblefoot: A comparison of clinical presentation and treatment of pododermatitis in rabbits, rodents, and birds. *Veterinary Clinics of North America: Exotic Animal Practice*, **16**(3):715–735.
Harcourt-Brown, N.H. (1996). Foot and leg problems. In: Benyon, P.H., Forbes, N.A., Harcourt-Brown, N.H. (eds), *BSAVA Manual of Raptors, Pigeons and Waterfowl*. Ames, IA: Iowa State University Press, pp. 163–167.
Remple, J.D. (2006). A multi-faceted approach to the treatment of bumblefoot in raptors. *Journal of Exotic Pet Medicine*, **15**(1):49–55.

Author Ryan De Voe, DVM, Dipl. ACZM, Dipl. ABVP (Reptile & Amphibian)

 Client Education Handout available online

Polydipsia

BASICS

DEFINITION
Abnormal or excessive water consumption (thirst), typically greater than 100 mL/kg/day.

Many birds, such as carnivores and frugivores, obtain their water intake requirements through their diet, and some small xerophilic species can survive on metabolic water produced from a dry diet in the absence of water intake. However, the water consumption rate of most species follows a regular relation with body mass. Birds weighing 100 g or more drink approximately 5% of their body mass per day, but as the body mass decreases, water consumption rates rise, up to about 50% for birds weighing 10–20 g. Birds with salt glands drinking salt water may drink substantially more than those without salt glands.

PATHOPHYSIOLOGY
• Water consumption is controlled by interactions between the kidneys, pituitary gland, and hypothalamus. • Polydipsia (PD) usually occurs as a compensatory response to polyuria (PU) to maintain hydration. Occasionally PD may be the primary process and PU is the compensatory response. Then, the patient's plasma becomes relatively hypotonic because of excessive water intake, and ADH secretion is reduced, resulting in PU.

SYSTEMS AFFECTED
• Cardiovascular—alterations in circulating fluid volume. • Endocrine/Metabolic—the hypothalamus and pituitary gland play a role in compensation to PD. • Renal/Urologic—kidneys.

GENETICS
Strains of Japanese quail and leghorn chickens with hereditary DI have been described.

SIGNALMENT
• Congenital causes of PD may be seen in younger animals. • Most of the conditions that cause PD are more common in middle-aged to older birds. • Psychogenic PD is often associated with juvenile hand-reared cockatoos.

SIGNS
N/A

CAUSES
• **Primary PD:** Behavioral problems, pyrexia, pain, hypotension, and any disease process that directly affects the anterior hypothalamic thirst center (psychogenic polydipsia). It has also been reported secondary to lithium administration. ○ Mild hepatoencephalopathy and altered function of portal osmoreceptors can cause PD in mammals and may have a similar effect in birds. Hepatoencephalopathy, due to increased ammonia, has not been proven to occur in birds at this time. ○ Nutritional secondary hyperparathyroidism (NSHP), due to calcium utilization exceeding dietary intake and subsequent stimulation of the parathyroid gland, may have an effect on the thirst center in the brain, resulting in PD.
• **Secondary PD:** Can occur with any disease process that also causes polyuria: ○ Renal disease: ▪ Nephrogenic diabetes insipidus (NDI). ▪ Inflammatory—glomerulonephritis (polyoma virus, paramyxovirus, pox virus, reovirus, adenovirus, aspergillosis), hypercalcemic nephropathy, renal gout, bacterial pyelonephritis, yolk coelomitis. ▪ Degenerative—hypovitaminosis A, obstructive uropathy (urolith, cloacolith, neoplasia, egg binding), polycystic kidney disease, amyloidosis, mineralization, nephrosis. ▪ Toxic—heavy metals (lead, zinc, mercury), acetone, aflatoxin, ethylene glycol, drug-induced (allopurinol, aminoglycosides, cephalosporins, sulfonamides, tetracyclines), hypervitaminosis D. ▪ Renal ischemia—hypoperfusion (dehydration, anesthesia, hypovolemia, atherosclerosis), neoplasia (adenocarcinoma, nephroblastoma). ○ Systemic or metabolic disease: ▪ Central diabetes insipidus (CDI). ▪ Liver disease. ▪ Pancreatitis. ▪ Gastrointestinal disease. ▪ Septicemia. ▪ Localized abscess, salpingitis. ▪ Neoplasia (pituitary or pineal adenoma/adenocarcinoma). ▪ Metabolic: □ Diabetes mellitus (DM). □ Disseminated intravascular coagulopathy (DIC). □ Hypercalcemia. □ Hyperthyroidism. □ Hyperadrenocorticism (HAC). ○ Dietary: ▪ Breeding/feeding offspring. ▪ Excessive dietary sodium. ○ Stress/excitement. ○ Congenital/genetic.

RISK FACTORS
• Renal disease or liver disease. • Administration of diuretics, corticosteroids, or anticonvulsants.

DIAGNOSIS

DIFFERENTIAL DIAGNOSIS
Differentiating Causes
• If associated with excessive weight loss, consider renal failure, pyelonephritis, liver failure, DM, neoplasia, hyperthyroidism. • If associated with polyphagia, consider DM, hyperthyroidism. • If associated with hypercalcemia, consider hypervitaminosis D, normal egg production, reproductive tract disease, neoplasia. • If associated with coelomic distention, consider liver disease, reproductive disease, neoplasia, normal egg production. • If associated with a behavioral or neurological disorder, consider hepatic failure, primary PD, CDI, heavy metal or salt intoxication.

CBC/BIOCHEMISTRY/URINALYSIS
• Serum sodium levels may help differentiate primary PD from primary PU. Hyponatremia or decreased serum osmolarity suggests primary PD, whereas hypernatremia or increased serum osmolarity indicates primary PU. • Increased uric acid level may be indicative of renal failure, but dehydration, lipemia, high-protein diet, or muscle catabolism may also cause hyperuricemia. Persistent hyperuricemia after rehydration and fasting is indicative of renal disease. • Decreased uric acid levels are indicative of liver dysfunction. • Elevated levels of AST and bile acids may indicate liver disease. • Persistent hyperglycemia is indicative of DM. • Hypercalcemia may be present with normal egg production, reproductive tract disease, neoplasia, or hypervitaminosis D. • Hyperphosphatemia and hyperkalemia may also occur with renal failure. • Hyperkalemia with concurrent hyponatremia are suggestive of HAC in mammals, but this has yet to be determined in birds. • Hypoalbuminemia (as determined by plasma EPH) supports renal or hepatic causes of PU/PD. • Heterophilia suggests infectious or inflammatory disease. • Alterations in urine specific gravity are not as reliable as an indicator of renal tubular disease in birds, as is seen with mammals, due to the bird's limited ability to concentrate urine. • Glucosuria and ketonuria are indicative of DM, but urine glucose may be increased with fecal contamination. • Proteinuria may indicate renal disease, but avian urine normally contains much higher protein levels than mammalian urine. Proteinuria may also be associated with fecal contamination. • Avian urine is typically isosmotic because the predominant reptilian-type nephrons cannot concentrate urine beyond plasma osmolarity, making assessment of urine specific gravity difficult. Hyposthenuria in a dehydrated bird may indicate an inability to concentrate urine.

OTHER LABORATORY TESTS
• Serum EPH may show hypoproteinemia secondary to a protein-losing nephropathy, or elevated globulin levels indicating an inflammatory or infectious disease process. • Since urine samples are typically contaminated with GI flora, urine cultures tend to be non-diagnostic, unless an uncontaminated urine sample can be obtained directly from the ureter. • PCR testing for suspected pathogens (e.g. *Aspergillus*, polyomavirus, paramyxovirus, etc.) that can potentially cause PU/PD might be rewarding. • Assessment and comparison of plasma and urine osmolality may be beneficial in determining if a bird is able to concentrate its urine. • Assessment of circulating levels of AVT can be used to rule out DI, but normal values have not been established for most species.

IMAGING
Plain and contrast radiography, ultrasound, CT, and MRI can all be used to image coelomic structures, such as the kidneys, liver,

POLYDIPSIA
(CONTINUED)

and reproductive tract, and may also demonstrate coelomic masses or effusion.

DIAGNOSTIC PROCEDURES
• Water deprivation testing is useful in ruling out DI and psychogenic polydipsia. This test should only be considered after all other possible causes of PU/PD have been ruled out. ○ *Traditional water deprivation test*: The bird is weighed, and blood and urine samples are collected for assessment of PCV, total solids and osmolality of blood, and specific gravity and osmolality of urine. The bird is then placed in a cage with no food or water for the duration of the test. Blood and urine parameters are evaluated every 3–4 hours for 12–48 hours, depending on the species and physical condition of the bird (smaller birds should be evaluated more frequently). The test should be discontinued when the patient demonstrates the ability to concentrate its urine or loses 4–5% of body weight. Birds with psychogenic polydipsia should develop more concentrated urine and an increase in PCV, total solids, and plasma osmolality, consistent with dehydration. Birds with DI become dehydrated but maintain dilute urine. ○ *Gradual water deprivation test*: This test is preferred over the traditional water deprivation test because it allows the kidneys and cloaca to respond gradually to increasing plasma osmolality. The bird's average water intake is determined, and then reduced by 10% over 3–5 days. Final deprivation should be done in the hospital. At this point, the traditional water deprivation test protocol should be followed.
• *Vasopressin/desmopressin response test*: This test is utilized to distinguish between CDI and NDI. Vasopressin and desmopressin are ADH analogues. Administration to birds with CDI should result in increased urine specific gravity and osmolality. Birds with NDI will not respond.

TREATMENT
APPROPRIATE HEALTH CARE
Treat the underlying cause of polydipsia whenever possible.

NURSING CARE
• Serious medical consequences are rare if the patient has free access to water and is willing and able to drink. Until the mechanism of PD is understood, discourage owners from limiting access to water. Treatment should be directed at the underlying cause. • PD patients should be provided with free access to water unless they are vomiting. If the patient is vomiting, replacement maintenance fluids should be administered after the patient has been assessed and appropriate diagnostic samples have been obtained. Parental fluids should also be administered when other conditions limit oral intake or dehydration persists despite PD. • Base fluid selection on knowledge of the underlying cause for fluid loss. In most avian patients, Normosol-R, Plasmalyte-R, Plasmalyte-A, and 0.9% NaCl are acceptable replacement fluid. • Primary PD should be treated by limiting water intake to a normal daily volume. The patient should be monitored closely to avoid iatrogenic dehydration. • Monitor body weight, hydration status (moistness of MM, skin turgor, eye position), perfusion parameters (mentation, MM color, CRT, pulse quality, HR, temperature of extremities), droppings.

ACTIVITY
N/A

DIET
• Nutritional support. ○ Provide adequate nutrition and calories. ○ Gradually transition the bird to a healthy, balanced diet.

CLIENT EDUCATION
• Monitor appetite, body weight, droppings.
• Stress the importance of keeping the patient hydrated.

SURGICAL CONSIDERATIONS
N/A

MEDICATIONS
DRUG(S) OF CHOICE
Varies with underlying cause.

CONTRAINDICATIONS
Do not administer ADH or any of its synthetic analogs (desmopressin, vasopressin) to patients with primary PD because of the risk of inducing water intoxication.

PRECAUTIONS
Until renal and hepatic failure have been excluded as potential causes for PU/PD, use caution in administering any drug that is eliminated via these pathways.

POSSIBLE INTERACTIONS
N/A

ALTERNATIVE DRUGS
N/A

FOLLOW-UP
PATIENT MONITORING
• Hydration status by clinical assessment of hydration and serial evaluation of body weight. • Fluid intake and urine output provide a useful baseline for assessing adequacy of hydration therapy.

PREVENTION/AVOIDANCE
Complete water deprivation.

POSSIBLE COMPLICATIONS
Dehydration.

EXPECTED COURSE AND PROGNOSIS
Depends on the etiology of polydipsia.

MISCELLANEOUS
ASSOCIATED CONDITIONS
N/A

AGE-RELATED FACTORS
N/A

ZOONOTIC POTENTIAL
N/A

FERTILITY/BREEDING
N/A

SYNONYMS
N/A

SEE ALSO
Dehydration
Diabetes Insipidus
Hyperglycemia/Diabetes mellitus
Liver disease
Pituitary tumor
Polyuria
Renal disease
Thyroid diseases
Vitamin D toxicosis
Appendix 7, Algorithm 10. Polydipsia

ABBREVIATIONS
ADH—antidiuretic hormone
AST—aspartate aminotransferase
AVT—arginine vasotocin
CDI—central diabetes insipidus
CT—computerized tomography
DI—diabetes insipidus
DIC—disseminated intravascular coagulopathy
DM—diabetes mellitus
EPH—electrophoresis
GI—gastrointestinal
HAC—hyperadrenocorticism
MRI—magnetic resonance imaging
NDI—nephrogenic diabetes insipidus
PCR—polymerase chain reaction
PU/PD—polyuria/polydipsia

INTERNET RESOURCES
N/A

Suggested Reading
Echols, M.S. (2006). Evaluating and treating the kidneys. In: Harrison, G.J., Lightfoot, T.L. (eds), *Clinical Avian Medicine*, Vol. I. Palm Beach, FL: Spix Publishing, Inc., pp. 451–492.
Hudelson, K.S., Hudelson, P.M. (2006). Endocrine considerations. In: Harrison, G.J., Lightfoot, T.L. (eds), *Clinical Avian Medicine*, Vol. I. Palm Beach, FL: Spix Publishing, Inc., pp. 541–557.
Lennox, A.M., Doneley, R. (2004). Working up polyuria and polydipsia in a parrot. *Proceedings of the Association of Avian Veterinarians Annual Conference*, pp. 59–66.

POLYDIPSIA (CONTINUED)

Oglesbee, B. (2003). Approach to the polydipsic/polyuric bird. *Proceedings of the Western Veterinary Conference Annual Conference.*

Portions Adapted From
Polzin, D.J. (2011). Polyuria and polydipsia. In: Tilley, L.P., Smith, F.W.K., Jr. (eds), *The 5-Minute Veterinary Consult: Canine and Feline*, 5th edn. Chichester, UK: John Wiley & Sons, Ltd.

Author David E. Hannon, DVM, DABVP (Avian)

POLYOMAVIRUS

BASICS

DEFINITION
Avian polyomavirus infection (APV) is a nonenveloped virus in the family *Polyomaviridae* that has been documented to affect avian psittacine neonates as well as adults. Clinical signs range from feather dystrophy and depression to slow crop emptying and death in budgerigar species. Nonbudgerigar psittacines may show slow crop emptying, diarrhea, subcutaneous hemorrhages and/or acute death in neonates to sub-clinical infections in adult parrots. Canaries and finches have been shown to be clinically susceptible to APV.

PATHOPHYSIOLOGY
Infection with APV generally takes place 7–14 days after direct exposure to an infected or carrier bird. Dander or body fluids from an infected or carrier bird may also transmit viral particles to a naïve bird by human hands, articles of clothing or feeding utensils. On a cellular level, APV causes massive internal hemorrhages, hepatic necrosis, myocardial cell necrosis and/or atrophy of lymphoid tissue in all species infected. In budgerigars, APV can be detected in skin and feather follicles.

SYSTEMS AFFECTED
- Budgerigar species: ○ Feather—follicle cellular lysis and necrosis causing lack of down feather formation and generalized feather dystrophy. ○ Skin—hemorrhages can be detected on the skin in a variety of locations. ○ Hepatobiliary—hepatocellular swelling and necrosis. ○ Spleen—splenomegaly, cellular swelling and necrosis. ○ Cardiovascular—pericardial effusion, karyomegaly and myocardial cell necrosis. ○ Renal—karyomegaly and cellular necrosis.
- Nonbudgerigar psittacines: ○ Hepatobiliary—hepatocellular necrosis, generally severe. ○ Spleen—karyomegaly and cellular necrosis. ○ Renal—karyomegaly and cellular necrosis. ○ Skin—subcutaneous hemorrhages. ○ Cardiac—karyomegaly and pericardial effusion. ○ Hemic/Lymphatic—karyomegaly of vessel walls, atrophy of lymphoid tissue in the spleen and bursa.

RISK FACTORS
The biggest risk factor is not quarantining and testing new birds coming into a collection.

GENETICS
None.

INCIDENCE/PREVALANCE
- Higher incidence is noted in the spring and summer. This increase coincides with hatching season. • Prevalence is higher in aviaries that do not practice proper a "Closed Aviary Concept".

GEOGRAPHIC DISTRIBUTION
APV can be found in all geographic regions where budgerigar and nonbudgerigar psittacine parrots are being raised in an avicultural environment.

SIGNALMENT
- **Species**: Budgerigar, nonbudgerigar psittacine and passerine. • **Species predilection:** ○ Eclectus Parrots, Caiques and Hawkhead parrots are more susceptible to fatal Polyomavirus infections. ○ All neonatal and juvenile psittacines are susceptible.
- **Mean age and range:** ○ Sudden death in birds less than three weeks of age post exposure to carrier or clinically infected birds. ○ Adult budgerigars and nonbudgerigar species may become sub-clinical carriers.

SIGNS
- Sudden to peracute death in birds less than three weeks of age is common. Subcutaneous hemorrhages, delayed crop emptying, regurgitation and diarrhea are common in nonbudgerigar psittacines. Feather dystrophy is common in budgerigars that survive the initial infection. • Owners usually report poor crop emptying, regurgitation, diarrhea or acute death in neonatal psittacines. Breeders/owners of budgerigars usually notice feather abnormalities or acute death in neonates and fledglings. • On physical examination of budgerigar babies, subcutaneous hemorrhages and/or feather abnormalities may be noted prior to death or in older survivors. • Nonbudgerigar species often show lethargy, slow crop emptying, diarrhea and subcutaneous hemorrhages.

CAUSES
- Exposure to carrier birds or birds clinically infected with Polyomavirus is the most common cause of infection. The virus may also be carried on the owner's/breeder's hands from one bird to the next. Contaminated feeding utensils, water/food bowls and caging material may also serve as fomites. • Housing multiple species of non-vaccinated baby psittacines together increases the risk of infection. Stress of travel, inadequate environmental temperature and/or over-crowding may lower an individual's ability to fight off infection.

DIAGNOSIS

DIFFERENTIAL DIAGNOSIS
Any cause of acute death in psittacine neonates should be considered until testing or histopathology accurately identifies the causative agent. Viral diseases such as Pacheco's, PBFD and adenovirus must be on a rule-out list. Systemic infections such as bacterial sepsis or chlamydiosis may create similar internal organ pathology.

CBC/BIOCHEMISTRY/URINALYSIS
Patients that are infected, but not near death, may show elevated white blood cell counts, low PCV and generally have extremely elevated AST, CK enzymes.

OTHER LABORATORY TESTS
Specific antemortem diagnostic tests involve oral, cloacal or fecal DNA probe testing for viral particles. Paired serum samples to show increases in antibody titers are useful in survivors.

IMAGING
N/A

PATHOLOGIC FINDINGS
Gross necropsy findings in fatal APV infections are fairly classic and generally involve noting subcutaneous hemorrhages on the crop, abdominal skin, patagium and/or torso. Most cases have a pale oral cavity. A variety of feather abnormalities may be noted, specifically in young budgerigar cases. Internal gross lesions often reveal serosal hemorrhages on many organs, hydropericardium, hepatomegaly, splenomegaly, ascites and myocardial hemorrhages.

Histopathological lesions in both groups often reveal hepatic, splenic and myocardial necrosis. Intranuclear inclusions may be noted in heart, liver, splenic and renal tissues. Inclusions in feather follicles may also be seen in budgerigar species.

TREATMENT

APPROPRIATE HEALTH CARE
Supportive care with nutritional support, intravenous or intraosseous fluids and warmth are helpful but may not slow the progression of signs. Interferon and other immune stimulation may provide help in isolated cases.

NURSING CARE
Patients suspected of having a polyomavirus infection should be under strict quarantine in either the veterinary clinic/hospital or at the owner's home/aviary. All humans coming in contact with infected and clinical birds should wear protective clothing that can be discarded after handling known positive cases. These birds should be treated or fed last.

ACTIVITY
N/A

DIET
N/A

CLIENT EDUCATION
Clients with diseased or infected carrier birds should be advised to stay away from pet stores, bird fairs or other breeder's facilities until their home/aviary has been properly disinfected. They need to be advised that bird dander from infected birds can transmit to naive birds without direct contact. Environmental testing for APV is highly recommended.

POLYOMAVIRUS (CONTINUED)

Feeding utensils, nest boxes, cages, counter tops, floors and walls of the home/aviary should all have DNA probe testing to identify APV positive zones that need disinfection.

Prevention is the key to APV. Polyomavirus vaccination has been approved for all psittacines over five weeks of age and repeated two to four weeks later. Vaccinated APV negative birds are resistant to infection and hence, eliminate spread to other birds.

SURGICAL CONSIDERATIONS
N/A

MEDICATIONS

DRUG(S) OF CHOICE
No antiviral medications are effective. Interferon has been used with mixed results.

CONTRAIDICATIONS
N/A

PRECAUTIONS
N/A

POSSIBLE INTERACTIONS
N/A

ALTERNATIVE DRUGS
N/A

FOLLOW-UP

PATIENT MONITORING
Serologic testing of the affected bird(s) to determine their seroconversion is helpful to determine if they are still infected. Environmental disinfection and DNA probe testing is critical to make sure the locale where the infected bird(s) originated from is critical to protect other birds and prevent the spread of the virus to other locations.

PREVENTION/AVOIDANCE
• Vaccination of psittacine birds can help protect naive birds and eliminate spread to other birds/facilities. Neonates and adult psittacines can be vaccinated, especially those that will be exposed to non-vaccinated birds from other breeders or homes. • A large percentage of infected budgerigar and nonbudgerigar psittacine neonatal patients often expire within two weeks of exposure. Vaccination will not be effective in neonates that have recently been infected with the virus. • Surviving budgerigar and nonbudgerigar patients may take one to two weeks to begin acting normal. Feather growth abnormalities are common in surviving budgerigar species. • Polyomavirus is very stable in the environment for long periods of time. Environmental disinfection with 10% sodium hypochlorite, 70% ethanol and Avinol-3 have all been shown to effectively inactivate the virus.

POSSIBLE COMPLICATIONS
Increased susceptibility to other infectious diseases.

EXPECTED COURSE AND PROGNOSIS
Most neonate and juvenile birds with clinical disease will die within two weeks of exposure. Surviving budgerigar and nonbudgerigar patients may take 1–2 weeks to begin acting normal. Feather growth abnormalities are common in surviving budgerigar species.

MISCELLANEOUS

ASSOCIATED CONDITIONS
See above.

AGE-RELATED FACTORS
Sudden death in birds less than three weeks of age postexposure to carrier or clinically infected birds. Adult budgerigars and nonbudgerigar species may become subclinical carriers.

ZOONOTIC POTENTIAL
N/A

FERTILITY/BREEDING
Decreased egg hatchability and early embryonic deaths may occur in budgerigar species.

SYNONYMS
Papovavirus (previous classification), APV, budgerigar fledgling disease

SEE ALSO
Anemia
Coagulopathies
Feather disorders
Heavy metal toxicity
Hemorrhage
Infertility
Liver disease
Sick bird syndrome

Suggested Reading
Garcia, A.P., Latimer, K.S., Niagro, F.D., et al. (1994). Diagnosis of polyomavirus-induced hepatic necrosis in psittacine birds using DNA probes. *Journal of Veterinary Diagnostic Investigation*, **6**:308–314.
Ritchie, B.W. (1995). Papovaviridae. In: Ritchie, B.W. (ed.), *Avian Viruses: Function and Control*. Lake Worth, FL: Wingers Publishing, pp. 127–170.
Ritchie, B.W., Latimer, K.S., Leonard, J., et al. (1999). Safety, immunogenicity and efficacy of an inactivated avian polyomavirus vaccine. *American Journal of Veterinary Research*, **59**(2):143–148.
Ritchie, B.W., Niagro, F.D., Latimer, K.S., et al. (1991). Polyomavirus infections in adult psittacine birds. *Journal of the Association of Avian Veterinarians*, **5**:202–206.
Ritchie, B.W., Niagro, F.D., Latimer, K.S., et al. (1993). Efficacy of an inactivated polyomavirus vaccine. *Journal of the Association of Avian Veterinarians*, **7**:187–192.

Author Gregory Rich, DVM, BS Medical Technology

 Client Education Handout available online

Polyuria

BASICS

DEFINITION
Polyuria (PU) is defined as a true and persistent increase in urinary output and water intake. PU is a fairly nonspecific sign.

PATHOPHYSIOLOGY
• Normal urine output and water intake vary immensely among different bird species, age, and physiologic state. • Urine output is dependent on GFR, tubular reabsorption of solutes and water, and patency of the urinary tract. • The medullary or mammalian nephron contains a loop of Henle, which produces urine. A countercurrent multiplier present within the medullary cone is capable of producing concentrated urine however the degree of concentration is much less in birds when compared to mammals. • Urine may flow via retrograde peristalsis from the urodeum into the coprodeum and distal portion of the colon, where water and sodium are reabsorbed in the ceca (when present) and rectum. ○ This can create considerable differences between the osmolality of ureteral urine and that of voided urine, particularly in dehydrated birds. ○ In a study of dehydrated pigeons, voided urine was 30% more concentrated than that obtained via ureteral cannulation.

SYSTEMS AFFECTED
• Endocrine/Metabolic. • Gastrointestinal system—ceca, rectum, cloaca play a role in reabsorption of ureteral water.
• Behavioral—polydipsia.

GENETICS
N/A

INCIDENCE/PREVALENCE
N/A

GEOGRAPHIC DISTRIBUTION
N/A

SIGNALMENT
• **Species:** Polyuria can be observed in any avian species: ○ Neophema—PMV-3, a potential cause of PU, is most commonly reported in Genus Neophema. ○ Guinea fowl (*Numida meleagris*), chickens, African grey parrots (*Psittacus erithacus*)—adenovirus causing pancreatic disease. ○ Finches and canaries—bacterial nephritis.
○ Pigeons—Salmonella infections.
○ Raptors—lead poisoning. • **Mean age and range:** N/A • **Predominant sex:** N/A

SIGNS
General Comments
• Is PU truly present? Birds with PU are commonly presented for "diarrhea"
○ Evaluate the volume and color of urine and urate components. ○ Is the urine output normal for the species and age of the bird?

Historical Findings
• Is PU a transient or persistent finding? • Is there polydipsia (PD)? • Is there a history of exposure to renal disease risk factors such as heavy metals, nephrotoxic drugs, supplementation with excessive levels of vitamin D3, etc. • Does the bird consume high moisture food items like fruit, gruel, water-laden vegetables, nectar? • Is there a history of breeding or feeding offspring?

Physical Examination Findings
• Measure body weight. • Evaluate hydration status (moistness of mucous membranes, skin turgor, eye position). • Evaluate perfusion parameters (mentation, mucous membrane color, capillary refill time, pulse quality, heart rate, temperature of the extremities).
• Perform a complete physical examination to identify additional clinical signs that may further characterize underlying disease.

CAUSES
• **Prerenal disease** ○ Increased intake of fluids ▪ Psychogenic polydipsia ▪ Fluid administration (overhydration) ○ Dietary imbalances ▪ Hypervitaminosis D ▪ Excessive dietary sodium ○ Drugs ▪ Diuretics
▪ Glucocorticoids, endogenous (stress) or exogenous ○ *E. coli* endotoxin ○ Electrolyte abnormalities ▪ Hypercalcemia ○ Hormonal conditions ▪ Diabetes insipidus ▪ Diabetes mellitus ▪ Hyperadrenocorticism
▪ Hypoadrenocorticism ▪ Hyperthyroidism?
▪ Pituitary adenoma/carcinoma • **Liver disease (severe hepatopathy)** • **Pancreatitis**
• **Salpingitis or localized abscess** • **Sepsis**
• **Renal disease or gout:** PU/PD is an uncommon sign of renal disease in the bird:
○ Important causes of infectious nephritis associated with PU include salmonellosis, adenoviral disease affecting the pancreas, psittacid herpesvirus-1, paramyxovirus-3, avian polyomavirus, West Nile virus, sarcoystosis, psittacosis. ○ Toxic exposure: heavy metals (e.g. zinc, lead), nephrotoxic drugs (e.g. aminoglycosides, sulfonamides), selenium poisoning, oak toxicosis.
• **Postrenal disease:** Postobstructive diuresis (urolithiasis, cloacolith, neoplasia, egg binding).

RISK FACTORS
• **Medical conditions** ○ Glomerular filtration rate, conditions that reduce ▪ Blood loss
▪ Severe dehydration ▪ Shock
○ Inflammation, chronic
○ Immunocompromise ○ Old age ○ Systemic hypertension ○ Uncontrolled diabetes mellitus
• **Medications** ○ Nephrotoxins: NSAIDs, intravenous contrast media • **Environmental factors** ○ Poor husbandry ▪ Adverse conditions that might lead to dehydration
▪ High moisture conditions that promote the development of moldy feed (mycotoxicosis)
▪ Improper nutrition.

DIAGNOSIS

DIFFERENTIAL DIAGNOSES
• Diarrhea (increased fluid in feces).
• Normal droppings in nectar-eating birds like lories and lorikeets. • Normal droppings in birds with diets high in fruits, water-laden vegetables. • Normal droppings in birds on gruels, tube feeding, or hand feeding formulas. • Normal droppings in breeding birds or birds hand feeding young.
• Transient polyuria as a stress response.

CBC/BIOCHEMISTRY/URINALYSIS
• Complete blood count or at least hematocrit/total protein. • Biochemistry panel including: ○ Blood glucose ○ Blood urea nitrogen, creatinine (useful indicator of prerenal dehydration) ○ Uric acid ○ Albumin, total protein ○ Sodium. • Urinalysis.

OTHER LABORATORY TESTS
• Amylase: Increased concentrations have been reported in some pancreatitis cases.
• Lactate. • Blood lead/zinc levels. • Serology as indicated. • Urine osmolality.

IMAGING
• Survey radiographs. • Contrast studies of the gastrointestinal tract can help isolate the location of the kidneys and the source of organomegaly. • Sonographic evaluation of coelomic organs. • Alternate imaging techniques include: ○ Nuclear scintigraphy to evaluate renal function. ○ MRI, CT. ○ Do not perform intravenous excretory urography in birds with severe renal compromise.

DIAGNOSTIC PROCEDURES
• Water deprivation testing. • Laparoscopic examination of the kidneys and renal biopsy.

PATHOLOGIC FINDINGS
Depending on the underlying site of disease, signs of inflammation, organomegaly, edema, etc. may be visible in:
• Adrenal glands • Brain • Gastrointestinal tract • Kidneys • Liver • Pancreas
• Reproductive tract.

TREATMENT

APPROPRIATE HEALTH CARE
Treat the underlying cause of polyuria whenever possible.

NURSING CARE
• Unless PU is caused by overhydration, careful attention to fluid therapy is required to prevent significant abnormalities in hydration. • Monitor body weight, hydration status (moistness of MM, skin turgor, eye position), perfusion parameters (mentation, MM color, CRT, pulse quality, HR, temp of extremities), droppings.

POLYURIA (CONTINUED)

ACTIVITY
N/A

DIET
• Nutritional support ∘ Provide adequate nutrition and calories ∘ Gradually transition the bird to a healthy, balanced diet.

CLIENT EDUCATION
• Monitor appetite, body weight, droppings. • Stress the importance of keeping the patient hydrated. • Long-term prognosis.

SURGICAL CONSIDERATIONS
N/A

MEDICATIONS

DRUG(S) OF CHOICE
Although PU is an uncommon sign of renal disease, consider antibiotic therapy if nephritis is suspected, since 50% of avian nephritis cases are associated with bacterial infection.

CONTRAINDICATIONS
• Diuretics could exacerbate fluid loss and dehydration. • Stop or avoid all nephrotoxic drugs that could cause or aggravate renal disease such as aminoglycosides or sulfonamides.

PRECAUTIONS
Drugs primarily cleared by renal tubular secretion or filtration (e.g., penicillin) may demonstrate a shorter half-life in the polyuric patient.

POSSIBLE INTERACTIONS
• Dehydration can result in adverse responses to NSAIDs, especially cyclooxygenase (COX) inhibitory drugs like meloxicam.

ALTERNATIVE DRUGS
Medical management of diabetes mellitus can include insulin, glipizide.

FOLLOW-UP

PATIENT MONITORING
• Activity level. • Appetite and water intake. • Droppings. • Body weight. • Hydration status (moistness of MM, skin turgor, eye position). • Perfusion parameters (mentation, MM color, CRT, pulse quality, HR, temperature of extremities).

PREVENTION/AVOIDANCE
Water deprivation.

POSSIBLE COMPLICATIONS
Dehydration.

EXPECTED COURSE AND PROGNOSIS
N/A

MISCELLANEOUS

ASSOCIATED CONDITIONS
Dehydration.

AGE-RELATED FACTORS
N/A

ZOONOTIC POTENTIAL
N/A

FERTILITY/BREEDING
Polyuria can be a normal finding in breeding birds or birds feeding offspring.

SYNONYMS
N/A

SEE ALSO
Adenoviruses
Avocado/plant toxins
Chlamydiosis
CNS tumors
Dehydration
Diabetes insipidus
Diabetes mellitus
Heavy metal toxicity
Herpesviruses
Hyperuricemia
Liver disease
Pancreatitis
Paramyxoviruses
Polydipsia
Polyomavirus
Proventricular dilatation disease
Renal disease
Salmonellosis
Sarcocystis
Thyroid diseases
Urate/fecal discoloration
Viral disease
Vitamin D toxicosis
West Nile Virus
Appendix 7, Algorithm 11. Polyuria

ABBREVIATIONS
PMV—paramyxovirus

INTERNET RESOURCES
N/A

Suggested Reading
Smarick, S. (2009). Urine output. In: Silverstein, D.C., Hopper, K. (eds), *Small Animal Critical Care Medicine*. St. Louis, MO: Saunders Elsevier, pp. 865–868.
Starkey, S.R., Morrisey, J.K., Stewart, J.E., Buckles, E.L. (2008). Pituitary-dependent hyperadrenocorticism in a cockatoo. *Journal of the American Veterinary Medical Association*, **232**:394–398.
Starkey, S.R., Wood, C., de Matos, R., et al. (2010). Central diabetes insipidus in an African Grey parrot. *Journal of the American Veterinary Medical Association*, **237**:415–419.
Author Christal Pollock, DVM, Dipl. ABVP (Avian)

POXVIRUS

BASICS

DEFINITION
Avian poxviruses (APVs) are large (150–250 × 265–350 nm) double stranded, enveloped DNA viruses. APVs are in the *Chordopoxvirinae* subfamily of the *Poxviridae* family. Disease from infection is most commonly categorized as cutaneous or "dry" (nodules and crusting lesions on unfeathered skin) and diphtheroid or "wet" (fibronecrotic membranes of the upper gastrointestinal or respiratory systems). Septicemic disease is also possible; animals become acutely depressed and die. Mortality rates are higher with diphtheroid and septicemic disease.

PATHOPHYSIOLOGY
APVs replicate in the cytoplasm of epithelial cells and the virus induces a hyperplastic host response. Infection may be limited to the cells around the sight of inoculation, but in some cases viremia develops. Viral replication in the liver and bone marrow then lead to a secondary viremia and replication in multiple organs. Disease is more severe in viremic birds. APVs cannot penetrate intact epithelium and infections depend on disruption of epithelial integrity. Direct, indirect, ingestion, and inhalation may be sources of transmission. Biting insects are considered a major vector. In domestic species, the incubation period is typically 5–10 days and flock outbreaks last 2–3 months; some flocks may be chronically infected.

Cutaneous (dry) pox starts as papule and vesicle formation in areas of unfeathered skin. Vesicles rupture, leaving erosions that scab. Secondary bacterial or fungal infection is common. Lesions typically take about a month to heal in domestic species and up to 3 months or more in wild species. Secondary infections can complicate healing and lead to chronic wounds. Occasionally, skin lesions are dark, discolored areas without papule/vesicle formation. Scarring, digit loss, and hypopigmentation may be present after healing. Cutaneous pox frequently develops in psittaciformes, falconiformes, passeriformes, galliformes, and waterfowl and is the most common form identified in free-ranging, wild birds. Mortality rates tend to be low unless lesions are severe and limit basic behaviors (e.g., feeding).

Diphtheroid (wet) pox occurs on mucosa. Erosions that fill with fibronecrotic material develop on the tongue, hard palate, larynx, trachea, bronchi, esophagus, or crop. Secondary bacterial or fungal infections are common. Diphtheroid pox frequently develops in psittaciformes, pheasants, quail, columbiformes, and galliformes. It is more common in domestic or captive wild species and is infrequently identified in free-ranging wild birds. It is the form historically associated with Amazon parrots (*Amazona* spp.) at import stations. Mortality rates are high.

Septicemic pox causes severe disease. Animals become acutely depressed and dyspneic and frequently die. Multiple organs may be impacted. Necrosis of the heart and liver as well as air sacculitis, pneumonia, peritonitis, and necrotic debris in the alimentary tract have been described. Septicemic pox is frequently reported in captive canaries (*Serina* spp.).

The form of disease (cutaneous, diphtheroid, or septicemic) and severity of infection varies with route of inoculation, viral strain, viral load, and immune status of the bird. Flocks may have multiple forms in a single outbreak and individual birds within a flock may have more than one form. Birds that survive infection typically have a protective immune response to that strain. A carrier state after recovery appears possible.

APVs have been described in over 230 bird species from 70 families in 23 orders. Strains are named for the species in which they were first identified; molecular evaluation is not complete for all strains identified. There are 10 strains identified by the International Committee on Taxonomy of Viruses; canarypox, fowlpox, juncopox, mynahpox, pigeonpox, psittacinepox, quailpox, sparrowpox, starlingpox, turkeypox (http://www.ictvonline.org/virusTaxonomy.asp). Fowlpox from chickens (*Gallus gallus domesticus*) is the type species. Strains are typically specific to individual species and disease tends to be more severe in adapted hosts, though this is not universal.

SYSTEMS AFFECTED
• Skin • Respiratory • Gastrointestinal
• Ophthalmic • Hemic/Lymphatic/Immune.

GENETICS
N/A

INCIDENCE/PREVALENCE
The incidents rate ranges from 0–50% across various species in captive and free-range settings, often falling in the 5–20% range. In captive psittacines in North America, disease is uncommon. Many canary and chicken flocks are infected and animals may be vaccinated as part of routine management; clinical disease in pet canaries is uncommon.

GEOGRAPHIC DISTRIBUTION
Worldwide except Arctic and Antarctic.

SIGNALMENT
• **Species:** While strains appear specific for individual species, some strains can infect nonadapted hosts. Disease may be more or less severe in the aberrant host. All species should be considered susceptible to potential infection to a species-specific strain. Captive-bred pet parrots, birds largely held indoors, or held outdoors with good protection from insect pests are less commonly affected. In psittacines, disease was most often associated with importation of birds from the wild. South and Central American Amazon and pionus (*Pionus* spp.) parrots, macaws (e.g., Ara spp.), lovebirds (*Agapornis* spp.), Quaker parrots (*Myiopsitta monachus*), and conures (e.g., *Aratinga* spp.) were considered most susceptible. During this period, disease in African species, cockatoos, and cockatiels was less common. APV is commonly reported in domestic chickens and pigeons (*Columba livia domestica*) and disease can be severe. • **Mean age and range:** Any age, disease may be more common and/or severe in juvenile birds. • **Predominant sex:** N/A

SIGNS

Historical Findings
There may be a history of recent additions to the flock. In outdoor-housed flocks, weather changes that increase insect populations (e.g., warm, moist conditions) may be reported. In cutaneous pox, caregivers typically note lesion development around the face or digits. Signs may be chronic and multiple birds in a flock may be affected. Birds with severe lesions may have trouble eating or finding food (e.g., proliferations around the mouth or eyelids) or have reduced mobility (e.g., severe digit lesions). In diphtheroid pox, caregivers may appreciate lethargy, anorexia, increased breathing rate or effort, open mouth breathing, or reduced appetite. Septicemic pox may be reported as acute depression, increased respiratory rate or effort, and death.

Physical Examination Findings
Cutaneous pox typically presents as proliferative, crusted lesions around the palpebral margins, oral commissure, wattles, digits, and other unfeathered regions. In some species (e.g., penguins), lesions may be noted on the wings and cloaca. Mobility issues may be noted due to impingement of digit motion or pain. Diphtheroid pox presents with a moist, fibronecrotic membrane in the oral cavity that is tightly adhered to underlying tissue. Stridor, dyspnea, and/or tachypnea may be present if lesions are in the respiratory system. Cutaneous and diphtheroid lesions hemorrhage readily if disturbed. Animals may be thin or in good body condition depending on duration of signs and if lesions impact eating and drinking. Animals may be dehydrated. Septicemic pox may present with severe depression and respiratory changes, animals may be in appropriate body condition but dehydration and other signs of severe disease are expected.

CAUSES
Clinical signs are related to epithelial hyperplasia and secondary bacterial or fungal infection.

RISK FACTORS
Conditions that increase population density (e.g., fall flocking behavior, high-density

Poxvirus (Continued)

commercial farming), reduce immune function (e.g., transportation stress), and increase insect vectors (e.g., warm, moist weather) can increase disease.

DIAGNOSIS

DIFFERENTIAL DIAGNOSIS
- Cutaneous—bacterial infection, papillomavirus, *Knemidokoptes* mite infection, *Trichophyton gallinae* infection (Favus).
- Diphtheroid—herpesvirus (infectious laryngotracheitis virus), fungal infection (e.g., candidiasis, aspergillosis), hypovitaminosis A, trichomoniasis. • Septicemic—bacterial sepsis, pneumonia, toxin, any cause of acute mortality.

CBC/BIOCHEMISTRY/URINALYSIS
Changes may be consistent with secondary disease conditions (e.g. leukocytosis with secondary bacterial infection).

OTHER LABORATORY TESTS
N/A

IMAGING
N/A

DIAGNOSTIC PROCEDURES
Clinical signs are suggestive. Cytology and/or histology of lesions demonstrate large intracytoplasmic vacuolar inclusions (Bollinger bodies) that push the nucleus to the margin of the cell and contain small, round, pale eosinophilic inclusions, consistent with pox infection. PCR with sequencing or viral isolation from diseased tissue or blood (in septicemic cases) should complement histology or cytology for definitive diagnosis. For clinical cases in commonly infected captive species (e.g., chickens), physical examination and possibly cytology are generally sufficient for management needs. Secondary infections are common and lesions are often inflammatory. Serology may support exposure to APV.

PATHOLOGIC FINDINGS
Gross findings include lesions as noted above. In addition, intestinal thickening and splenic, liver, and pulmonary congestion may be noted. Histopathology changes in cutaneous pox include epidermal hyperplasia and hypertrophy. Large, granular eosinophilic inclusions are noted in epithelial cells above the basal layer. While inclusions are typically intracytoplasmic, in some infections, intranuclear inclusions have also been reported. Necrosis and secondary bacterial or fungal infection near the surface of the lesion is common. In the diphtheroid form, serous membranes have severe fibrinous inflammation and liver necrosis may be present. In the septicemic form, pulmonary changes such as edema and fibrinous pneumonitis develop.

TREATMENT

APPROPRIATE HEALTH CARE
There is no direct treatment for poxvirus. Management is based on appropriate supportive care in individual animals and population management strategies for flocks. Cutaneous lesions are often minor and self-limiting. If severe proliferation or secondary infection is present, standard treatments should be implemented. Crusts may be removed (e.g., moistened and slowly broken down, surgically debulked). Topical antimicrobials, systemic antibiotics and antifungals, analgesics, and anti-inflammatory medication may be indicated. Ocular medications may be appropriate if palpebral integrity is compromised. Diphtheroid cases may need hospitalization for intensive supportive care, if individual patient management is desired. If patients are admitted, care must be taken to prevent transmission to other in-hospital patients.

NURSING CARE
If animals are eating and ambulating normally, little supportive care is needed with cutaneous disease. If lesions are severe and animals have been unable to eat or drink, fluid and nutrition support is appropriate. In diphtheroid and septicemic forms of the disease, oxygen therapy and more aggressive fluid and nutritional support may be indicated. Nebulizing may be considered in birds with severe respiratory compromise.

ACTIVITY
N/A, except as appropriate for managing general condition.

DIET
N/A, unless base diet is considered inappropriate for species.

CLIENT EDUCATION
Clients should be instructed that APVs are long-lasting in the environment. Appropriate cleaning and other forms of environmental management (e.g., dust control) should be implemented. Pest control methods (e.g., mosquito reduction or netting, treatment for mites) should be reviewed. Enhanced biosecurity measures should be implemented (e.g., isolation of clinically affected and exposed birds, work with clinically affected birds last, separate tools for clinically affected birds). Vaccination for certain species should be discussed (e.g., domestic chicken, turkey, quail).

SURGICAL CONSIDERATIONS
If lesions are significant, debridement may be appropriate.

MEDICATIONS

DRUG(S) OF CHOICE
There is no treatment for poxvirus but topical antimicrobials, systemic antibiotics and antifungals, analgesics, and anti-inflammatory medication may be indicated. Ocular medications may be appropriate if palpebral integrity is compromised.

CONTRAINDICATIONS
N/A

PRECAUTIONS
N/A

POSSIBLE INTERACTIONS
N/A

ALTERNATIVE DRUGS
N/A

FOLLOW-UP

PATIENT MONITORING
Typically, once lesions are healing, birds will recover. Follow-up schedules should be based on severity of clinical disease. Clinicians should discuss flock management options with owners.

PREVENTION/AVOIDANCE
Prevention is primarily through appropriate biosecurity and management of insect vectors to reduce exposure. Unlike many enveloped viruses, virions remain viable in the environment for longs periods of time and are resistant to disinfection. Contaminated perches, nest boxes, bedding, and so on should be destroyed. Virions can be inactivated with 1% potassium hydroxide, 2% NaOH, 5% phenol, and steam or exposure to 50°C for 30 minutes or 60°C for 8 minutes. Virkon appears to be effective. Infected and exposed birds should be isolated. Vaccines for some strains are available (e.g., fowl pox) and may be appropriate. Many commercially produced chickens are vaccinated as juveniles. Protection is generally strain specific. The vaccine may not protect fully against some wild strains.

POSSIBLE INTERACTIONS
Complications from APV infection are primarily related to secondary infections, as noted above. Permanent scarring (e.g., digit loss, palpebral anatomy) may develop in some cutaneous cases. *Poxviridea* may have oncogenic properties; skin tumors in Columbiformes and pulmonary tumors in canaries have been reported.

EXPECTED COURSE AND PROGNOSIS
Prognosis for cutaneous form is often good. Diphtheroid and septicemic forms are often fatal.

(CONTINUED)

 MISCELLANEOUS

ASSOCIATED CONDITIONS
N/A

AGE-RELATED FACTORS
N/A

ZOONOTIC POTENTIAL
N/A

FERTILITY/BREEDING
May reduce breeding success; disease in chicks may be more severe than in adults.

SYNONYMS
N/A

SEE ALSO
Aspergillosis
Candidiasis
Ectoparasites
Herpesvirusus
Nutritional deficiencies
Oral plaques
Papilloma
Pneumonia
Respiratory distress
Trichomoniasis

ABBREVIATIONS
APV–avipoxvirus

INTERNET RESOURCES
Vaccine information: http://www.hyline.com/aspx/redbook/redbook.aspx?s=5&p=35
Disinfection techniques: http://www.fao.org/docrep/004/y0660e/y0660e03.htm

Suggested Reading
Ritchie, B.W. (1990). *Poxviridae. Avian Viruses: Function and Control.* Lake Worth, FL: Wingers Publishing, pp. 285–311.
van Riper C., III, Forrester, D.J. (2007). Avian pox. In: Thomas, N.J., Hunter, D.B., Atkins, C.T. (eds), *Infectious Diseases of Wild Birds.* Ames, IA: Blackwell Publishing, pp. 131–176.
Weli, S.C., Tryland, M. (2011). Avipoxiruses: Infection biology and their use as vaccine vectors. *Virology Journal*, **8**:49.

Author Leigh Ann Clayton, DVM, DABVP (Avian, Reptile-Amphibian)

PROBLEM BEHAVIORS: AGGRESSION, BITING AND SCREAMING

BASICS

DEFINITION
There are different scientific levels of analysis relevant to understanding and solving behavior problems. Each level addresses different sources of influence and strategies for change. In this section, the behavioral level of analysis is presented, also known as behavior analysis. Behavior analysis is the scientific investigation of environmental influences on behavior. Applied behavior analysis (ABA) is the behavior-change technology based on this science. The essential focus of ABA is learning, that is, behavior change due to contact with the environment. The term environment includes all the stimuli, events, and conditions that the animal experiences, including the interactions between caregivers and their birds. As such, the appropriate unit of analysis in ABA is the individual learner, client, and ever-changing environment.

From the behavioral perspective, a problem behavior is defined as any activity that is the result of learning processes and occurs at a strength that is 1. problematic to the client, 2. negatively impacts the bird's health and welfare, or 3. interferes with the bird's ability to live successfully in human care. Strength refers to any measurable characteristic of behavior, for example, frequency, rate, duration, and intensity.

PATHOPHYSIOLOGY
In this context, the problem behaviors of interest are neither a result nor a symptom of underlying pathophysiology. Rather, problem behaviors are the result of functionally related antecedents (events that occur before the behavior) and functionally related consequences (events that occur after the behavior). These correlates comprise the three-term contingency: antecedent-behavior-consequence (ABC). As behavior never occurs in the absence of antecedents and consequences, the three-term contingency is the smallest, meaningful unit of behavior analysis.

Antecedents signal the behavior-consequence contingency ahead and consequences provide the feedback about the adequacy of past behavior. Consequences are also the purpose for behaving, that is, that which we behave to gain or escape.

For example, given too little to do (antecedent), many birds learn to scream (observable behavior), to gain attention (consequence). Given an imposing hand (antecedent), many birds learn to bite (observable behavior), to escape being touched or picked up (consequences). In both cases, it is reasonable to predict that unless something changes, the birds will continue to scream and bite because doing so is purposive, neither willy-nilly nor the result of alleged character traits (e.g., vicious, neurotic). Biting and screaming persist because they are functional behaviors.

SYSTEMS AFFECTED
Behavioral.

GENETICS
At both the population level (evolutionary history) and individual level (personality), inherited behavioral tendencies influence and are influenced by learning outcomes. Like a Gordian knot, the co-influences of biology and the environment are inextricably intertwined. However, most commonly reported problems with pet birds (screaming, biting, inactivity) have not been associated with genetic markers.

INCIDENCE/PREVALENCE
No conclusive data are currently available. However, problem behaviors are commonly reported by clients and are likely the main reason clients relinquish birds.

GEOGRAPHIC DISTRIBUTION
Problem behaviors are ubiquitous.

SIGNALMENT
• **Species:** No conclusive data currently available. • **Mean age and range:** All • **Predominant sex:** No conclusive data currently available.

SIGNS

General Comments
The behavior of interest is dependent on certain environmental conditions, that is, predicted by specific antecedents and maintained by functional consequences. In discussions with veterinarians, clients may present behavior problems with hypothetical constructs and intangible labels such as vicious, aggressive, territorial, dominant, attention seeking, jealous, hormonal, and fearful instead of describing behavior. Behavioral descriptions include patterns such as screaming, biting, lunging, flying at people or other pets, chewing unapproved household items, and inactivity.

Historical Findings
Insufficiencies in the environment account, at least partly, for most common problem behaviors. Changes in the environment can improve these problems and may ameliorate them entirely.

Physical Examination Findings
Physical examination is often normal; comorbid medical problems are possible (e.g., poor skin and feather condition, overweight, early pododermatitis).

CAUSES
Problem behaviors are caused by a general lack of knowledge about how behavior works, which translates to insufficient environmental arrangement and training. Specifically, clients inadvertently reinforce the "wrong" behaviors, create persistent wrong behaviors with intermittent reinforcement, and insufficiently reinforce the "right" behaviors. Understanding the behavior ABCs (antecedent-behavior-consequences) is a powerful tool to reveal the ways in which functional misbehavior is shaped. By asking, "What sets the occasion for the behavior?" and, "What does the bird get out of doing the behavior?" causes can be determined.

For example, lunging and biting is often maintained by escape of aversive stimuli (e.g., the clients themselves, aversive locations); screaming is often maintained by intermittent attention (good or bad, obvious or subtle); and chewing and inactivity are often the result of a lack of stimulus diversity or experience using enrichment devices.

RISK FACTORS
Rate of reinforcement: too low, too high, or intermittent; clients who force and coerce parrots to behave; insensitive or insufficient skill at observing and responding appropriately to the birds body language signaling comfort or discomfort.

DIAGNOSIS

DIFFERENTIAL DIAGNOSIS
Physical and ethological correlates for problem behavior should be considered. However, many clients and veterinarians underestimate learning as the reason for problem behavior.

CBC/BIOCHEMISTRY/URINALYSIS
N/A

OTHER LABORATORY TESTS
N/A

IMAGING
N/A

DIAGNOSTIC PROCEDURES
The diagnostic protocol in the ABA model is centered on identifying the ways in which antecedents and consequences set the occasion for the behavior and reinforce or punish the behavior, thereby increasing or decreasing the probability of the behavior occurring again. The following key questions facilitate diagnostic ABCs:
1. What does the behavior look like? Use observable descriptions not hypothetical constructs or intangible labels.
2. What antecedent conditions predict when the behavior will occur? Consider when the behavior is most likely.
3. What is the immediate consequence of the behavior? Consider what change in the environment occurs immediately following the behavior.
4. Will the behavior likely increase (maintain) or decrease if nothing changes? Additional information may be obtained by also asking, "Under what conditions does the bird not exhibit the behavior, that is, when is the bird successful?"

Problem Behaviors: Aggression, Biting and Screaming

PATHOLOGIC FINDINGS
N/A

TREATMENT

APPROPRIATE HEALTH CARE
An effective treatment protocol requires a direct link between diagnosis and treatment (Table 1). Once the ABCs are identified, treatment consists of systematically redesigning the environment to replace the problem behavior with an existing alternative and to teach new skills. For each problem behavior, answering the following questions will inform the treatment plans:
1. What acceptable alternative behavior can the bird do to get the same outcome?
2. What do you (the client) want the bird to do instead?

Essential strategies include differential selection of alternative (DRA) or incompatible (DRI) behavior, whereby reinforcers are contingently delivered for acceptable behaviors. New skills are taught with shaping, the process of incrementally reinforcing small approximations to the final goal. Data should be taken to ensure that the intervention is heading in the right direction and trigger revisions.

NURSING CARE
N/A

ACTIVITY
N/A

DIET
If the bird is on an inappropriate diet for the species, dietary modification should be implemented as part of a holistic wellness program. Increased foraging and reserving favored food items for positive reinforcement training is important.

CLIENT EDUCATION
Educating clients and providing ongoing support are critical to effective intervention. As clients are responsible for implementing the plan, it must be created in collaboration with them.

SURGICAL CONSIDERATIONS
N/A

MEDICATIONS

DRUG(S) OF CHOICE
If abnormal neurophysiology is suspected based on lack of response to appropriate plans, medication such as selective serotonin reuptake inhibitors (e.g., fluoxetine), tricyclic antidepressants (e.g., clomipramine) and benzodiazepams (e.g., diazepam) may be considered. It is inappropriate to expect medication to compensate for a poor environment. Reproductive cycles may exacerbate behavior problems and temporary administration of medications that reduce reproductive hormones, such as GnRH agonists (e.g., leuprolide acetate for depot suspension), may be appropriate as part of a comprehensive management strategy.

CONTRAINDICATIONS
N/A

PRECAUTIONS
N/A

POSSIBLE INTERACTIONS
N/A

ALTERNATIVE DRUGS
N/A

FOLLOW-UP

PATIENT MONITORING
Varies with the case. A single meeting is often sufficient to develop a plan that clients can implement. Routine (typically weekly)

Table 1

	Examples of how to use key questions to prompt functional evaluation of problem behavior and identify treatment options.	
Question	*Purpose of Question*	*Example*
What does the behavior look like?	Describe it in observable (operational) terms, without labels.	Typical Label: Behavior A: The bird is aggressive. Behavior B: The bird is territorial. Operationalized behaviors: Behavior A: The bird bites hands. Behavior B: The bird lunges and bites other bird's leg
What conditions predict when the behavior will occur?	Antecedent evaluation. Identify the relevant environmental stimuli that cue or set the stage for the behavior.	Behavior A happens when the bird is in the cage, the client's hand is within 6". Behavior B happens when both birds are on the play gym and food is in one bowl.
What does the bird get for get away from by doing the behavior?	Consequence evaluation. Identify the relevant environmental stimuli that are reinforcing (maintaining) the behavior.	After behavior A happens, the hand is pulled away and the bird is not touched. After behavior B happens, the other bird leaves the area and the bird has access to the food.
Under what conditions does the bird not exhibit the behavior?	Identify the environment when the bird is most successful. This step helps clients realize the bird is successful under certain conditions.	Behavior A is least likely to happen when the bird is on a table or the hand is still and bird has walked toward the hand. Behavior B is least likely to happen when food is put in multiple food bowls on the play-gym or there is no food present.
What do you want the bird to do instead?	Treatment. Identify another behavior the bird can do in place of the problem behavior for the same outcomes.	For Behavior A, the bird can lean away to remove hand. The bird can learn with shaping to step up for valued reinforcers. For behavior B, the birds can eat from separate bowls once two bowls are installed. Or, food can be omitted from the play area.

PROBLEM BEHAVIORS: AGGRESSION, BITING AND SCREAMING

communication may help support the clients through the initial implementation phase.

PREVENTION/AVOIDANCE
To build successful behavior repertoires, make the "right" behaviors easier and more reinforcing than the "wrong" behaviors. Allow behavior to be effective through empowerment, choices, and environments rich with stimulus diversity.

POSSIBLE COMPLICATIONS
Fluent behavior can be suppressed but not eliminated. Thus, problem behaviors can recover as a result of inadvertent reinforcement or when the context cues in which the behavior was learned reappear. Care must be taken to avoid either event. The treatment plan should be faithfully implemented again if this occurs.

When force and coercion is a lifestyle, rather than a very occasional event, animals have too little control over their own outcomes. This lack of empowerment is correlated with physiologic changes consistent with lower immunity and higher inflammatory parameters and suboptimal welfare.

EXPECTED COURSE AND PROGNOSIS
A well-designed intervention typically produces initial responsiveness quickly, within a week. However, the longer the problem behavior has been rehearsed the longer it may take to change. Also, behavior problems can become more persistent with repeated failed interventions.

MISCELLANEOUS

ASSOCIATED CONDITIONS
N/A

AGE-RELATED FACTORS
N/A

ZOONOTIC POTENTIAL
N/A

FERTILITY/BREEDING
N/A

SYNONYMS
As noted above, labels are often used in place of describing the problem behavior. Some common labels used with pet parrots that bite or scream include viscous, aggressive, hormonal, territorial, dominant, and fearful.

SEE ALSO
Feather destructive behavior

ABBREVIATIONS
ABA—applied behavior analysis
DRA—differential reinforcement of alternative behavior

INTERNET RESOURCES
behaviorworks.org
clickertraining.com
goodbirdmagazine.com
thegabrielfoundation.org/education/
parrotenrichment.com

Suggested Reading
Clayton, L.A., Friedman, S.G., Evans, L.A. (2012). Management of specific and excessive posturing behavior in a hyacinth macaw (*Anodorhynchus hyacinthinus*) by using applied behavior analysis. *Journal of Avian Medicine and Surgery*, **26**(2):107–110.

Friedman, S.G. (2007). A framework for solving behavior problems: Functional assessment and intervention planning. *Journal of Exotic Pet Medicine*, **16**(1):6–10.

Friedman, S.G., Edling, T.M., Cheney, C.D. (2006). Concepts in behavior: Section I. In: Harrison, G.J., Lightfoot, T.L. (eds), *Clinical Avian Medicine*, Vol. I. Palm Beach, FL: Spix Publishing, Inc., pp. 46–59.

Friedman, S.G., Haug, L.I. (2010). From parrots to pigs to pythons: Universal principles and procedures of learning. In: Tynes, V.V. (ed.), *Behavior of Exotic Pets*. Chichester, UK: John Wiley & Sons Ltd, pp. 190–205.

Heidenreich, B. (2005). *The Parrot Problem Solver*. Neptune City, NJ: T.F.H. Publication.

Authors Leigh Ann Clayton, DVM, DABVP (Avian, Reptile-Amphibian)
Susan G. Friedman, PhD

 Client Education Handout available online

Proventricular Dilatation Disease

BASICS

DEFINITION
Proventricular dilatation disease (PDD) is a neurological disease caused by avian bornavirus. One of the characteristic syndromes can be dilation of the proventriculus, ventriculus and intestines resulting in maldigestion.

PATHOPHYSIOLOGY
PDD is a progressive disease of the central and peripheral nervous systems, characterized by central nervous system dysfunction as well as dysfunction of other organ systems including the gastrointestinal tract because of inflammatory lesions in the nerves which innervate those organs. Lesions may also be found in the heart and adrenal glands. The nerve dysfunction is cause by invasion of inflammatory cells into neurons. More accurately described as neuropathic ganglioneuritis or lymphoplasmocytic ganglioneuritis. Often equated to an auto-immune disease.

SYSTEMS AFFECTED
• Nervous • Gastrointestinal • Cardiovascular • Ophthalmic • Endocrine/Metabolic.

GENETICS
N/A

INCIDENCE/PREVALENCE
• ABV infection may be endemic. Some authors report 33–60% infection rates in some homes or aviaries. Other laboratories report much lower infection rates. • Clinical disease is uncommon, except in infrequent outbreaks.

GEOGRAPHIC DISTRIBUTION
Worldwide in captive birds.

SIGNALMENT
PDD has been diagnosed in over 50 psittacine (parrot) species as well as canaries, honeycreepers, weaver finches, waterfowl, toucans, and birds of prey. Sexes are affected equally. All ages are affected.

SIGNS
• Weight loss, muscle wasting, emaciation. • Dilation of the proventriculus and/or ventriculus. • Vomiting, diarrhea, undigested seeds in stools. • CNS disorders, seizures, ataxia, proprioceptive deficits, paresis/paralysis, blindness, abnormal head movements. • Polyuria. • Cardiac abnormalities.

CAUSES
• Avian bornavirus (ABV). • There are 10 known genotypes of ABV and seven of these are known to affect psittacine birds. The virulence of these genotypes can vary. There is also evidence that an asymptomatic bird harboring one genotype can sometimes become ill after exposure to a second novel genotype. • There seems to be an unidentified triggering mechanism which can cause clinical expression of ABV infection.

RISK FACTORS
• Exposure to infected birds shedding the virus in urine and feces. • Indoor housing with other infected birds. • Birds infected at a young age may have increased risk.

DIAGNOSIS

DIFFERENTIAL DIAGNOSIS
• Foreign body ingestion (such as leather from toys). • Heavy metal toxicosis. • Macrorhabdus ornithogaster. • Neoplasia. • General emaciation. • Severe parasitic infection. • Severe internal papillomatosis. • Traumatic injury. • Hydrocephalus. • Bacterial, fungal, or viral CNS infections. • Nutritional deficiencies.

CBC/BIOCHEMISTRY/URINALYSIS
• Elevation of creatine phosphokinase. • Anemia. • Hypoproteinemia. • Leukocytosis can occur early in the disease. • Hyperuricemia if neurologic signs result in difficulty drinking.

OTHER LABORATORY TESTS
• Serological tests used include indirect immunoflouorescence, ELISA, western blot assay—the sensitivity and specificity of any of these tests is uncertain. • PCR: Reverse transcriptase (RT) polymerase chain reaction (PCR) has been used to detect ABV in droppings, feathers, blood and tissues on postmortem. Positive ABV PCR test results do not mean that the bird has or ever will develop clinical disease. • Antiganglioside antibody. • There is no generally accepted reliable laboratory screening test. Results vary greatly between laboratories. Standardization of testing procedures, as well as specimen type and handling will be important to establish confidence in laboratory testing. • Few laboratories offer genotype testing. • Bacterial or yeast overgrowth can be seen in the case of gastrointestinal stasis; fecal Gram's stains or cultures may be helpful in diagnosing these conditions.

IMAGING
• Radiography can be used to identify an enlarged, dilated proventriculus and/or ventriculus. The use of contrast media is helpful. • Fluoroscopy can be used to demonstrate disruptions of ventricular contractions, keeping in mind PDD is not the only disease that causes disruption of ventricular contractions.

DIAGNOSTIC PROCEDURES
• Biopsy of crop tissue "crop biopsy" is a procedure where a full thickness biopsy of crop tissue is taken ensuring that a blood vessel and its accompanying nerve is included. Histopathology can reveal affected peripheral neurons exhibiting lymphoplasmocytic infiltrates. At least three ganglia should be examined. • Partial thickness biopsy of the ventriculus or adrenal biopsy could be performed but are more invasive than crop biopsy.

PATHOLOGIC FINDINGS
• Gross pathology lesions can include profound emaciation, dilation of the proventriculus and ventriculus, thin walled proventriculus, atrophy of ventricular muscles and dilation of the intestinal tract. • Some cases have only CNS lesions, adrenal gland dysfunction, or myocardial lesions or arrythmias and cannot be observed grossly. • Histopathologic lesions are characterized by lymphoplasmacytic-ganglioneuritis. Lesions may be seen in the CNS as well as peripheral nerves. • ABV can be demonstrated in the tissue by PCR or immunohistochemistry techniques.

TREATMENT

APPROPRIATE HEALTH CARE
• Highly digestible diet. • Nutritional anti-oxidants. • Nonsteroidal anti-inflammatory drugs such as celecoxib or meloxicam can reduce and or reverse inflammatory cell infiltration into neurons. • Cyclosporine may be used in advanced cases. • Amantadine can help with CNS signs. • Metoclopramide to assist with intestinal motility.

NURSING CARE
• Nursing care will typically be prolonged for months or years, consisting of NSAID therapy and nutritional management. • Birds with PDD will often ingest foreign bodies, care must be taken to avoid opportunities to ingest materials. • Probiotics or antibiotics may help be helpful to prevent/treat secondary enteric infections.

ACTIVITY
Activity should not be limited if the patient is physically capable.

DIET
• The diet should be highly digestible and preferably contain higher than normal fiber foods to aid in intestinal motility. • Birds on a seed diet may have more difficulty digesting foods. Conversion to a formulated diet may be very difficult in an ill bird. Supplemental feeding of highly digestible foods will be beneficial. • Supplementation with foods high in anti-oxidants may help reduce inflammation.

CLIENT EDUCATION
The potential of exposure and disease in contact birds in the home or aviary is of concern. Infected or clinical birds should ideally be isolated from other birds. However, not all birds that test positive will develop

disease, and if they do it could potentially take years or decades. Unfortunately, there is no general accepted screening process, especially to declare a bird free of infection. Ideally birds with different genotypes of ABV will not be mixed.

SURGICAL CONSIDERATIONS
N/A

MEDICATIONS

DRUG(S) OF CHOICE
• Celecoxib—20 mg/kg PO SID, or in divided doses. • Meloxicam. • Amantadine (to improve CNS function) 10 mg/kg SID.

CONTRAINDICATIONS
• Gastrointestinal ulcerations.
• Hypersensitivity to NSAIDs. • Renal disease.

PRECAUTIONS
• Renal disease. • Hepatic disease. • Cardiac disease. • Hemorrhagic disorders.

POSSIBLE INTERACTIONS
• Possibly interactions with NSAIDS include furosemide and fluconazole. • Combination with antidepressants such as fluoxetine may cause bleeding or bruising.

ALTERNATIVE DRUGS
N/A

FOLLOW-UP

PATIENT MONITORING
• Body weight as well as food types and consumption should be monitored at home. Weight gains can be misleading if the intestinal tract becomes dilated and filled with ingesta. • Owners should be aware of signs which may indicate secondary infections, especially with spore-forming bacterial infections or yeast due to prolonged intestinal transit times. • If treating with NSAIDS, feces should be monitored for melena or fresh blood which may indicate proventricular ulceration. • Monitor dilation of the proventriculus by periodic radiography.

PREVENTION/AVOIDANCE
Birds which test positive should ideally be isolated. With very high infection rates reported by some laboratories and in some facilities, this may be impractical.

POSSIBLE COMPLICATIONS
• Foreign body ingestion. • Secondary aspiration pneumonia.

EXPECTED COURSE AND PROGNOSIS
• ABV infection does not equate to clinical PDD. • Birds which test positive for ABV may not develop clinical disease for years, or may never develop clinical PDD. • There is a poor prognosis for birds exhibiting severe clinical illness.

MISCELLANEOUS

ASSOCIATED CONDITIONS
N/A

AGE RELATED FACTORS
Birds that are affected at a very young age may have more rapid and severe progression of disease.

ZOONOTIC POTENTIAL
There are abundant references to bornaviruses and psychiatric disorders in humans; however, a definitive connection has not been proven. Avian bornaviruses are divergent from those of mammals.

FERTILITY/BREEDING
• Avian bornavirus can be transmitted through the egg. Artificial incubation and handrearing will not eliminate transmission.
• Infected birds can breed successfully.

SYNONYMS
Macaw wasting disease

SEE ALSO
Emaciation
Coelomic distention
Enteritis/gastritis
Gastrointestinal foreign body
Heavy metal toxicity
Macrorhabdus orithogaster
Neurologic conditions
Polyuria
Regurgitation/vomiting
Sick bird syndrome
Undigested food in droppings

ABBREVIATIONS
PDD—proventricular dilatation disease

INTERNET RESOURCES
Living with PDD, a Yahoo support group for owners of birds with PDD: https://groups.yahoo.com/neo/groups/livingwithpdd/info.

Suggested Reading
Hoppes, S., Tizard, I., Shivaprasad, H.L. (2013). Avian bornavirus and proventricular dilatation disease: Diagnostics, pathology, prevalence, and control. *The Veterinary Clinics of North America. Exotic Animal Practice*, **16**(2):339–355.

Author Susan L Clubb, DVM, DABVP (Avian)

 Client Education Handout available online

REGURGITATION AND VOMITING

BASICS

DEFINITION
Regurgitation is the retrograde expulsion of ingesta from the esophagus, including the crop. Vomiting is the retrograde expulsion of ingesta from the proventriculus, ventriculus, or intestine. Because of the anatomic arrangement of the ventriculus, vomiting from the intestine is rare in birds. Both vomiting and regurgitation (V/R) are usually active processes in birds. The rule that regurgitation occurs passively and vomiting is active is not valid in birds. The origin of the expelled ingesta can be determined based on the pH. The pH of the proventiculus should be very acidic, and the pH of the crop and esophagus should be close to neutral.

PATHOPHYSIOLOGY
Following stimulation of the gastrointestinal (GI) tract, vestibular system, or central nervous system, signals are sent to the GI tract and a coordinated contraction of the smooth and skeletal muscles propel ingesta back into and subsequently out of the mouth.

SYSTEMS AFFECTED
• Gastrointestinal. • The respiratory system may develop aspiration pneumonia or rhinitis.

GENETICS
There does not appear to be a genetic predilection.

INCIDENCE/PREVALENCE
Regurgitation and vomiting are common clinical signs.

GEOGRAPHIC DISTRIBUTION
None

SIGNALMENT
• **Species:** These signs can occur in birds of any species, age, or sex. • **Mean age and range:** None. • **Predominant sex:** Behavioral regurgitation is somewhat more common in male birds.

SIGNS
General Comments
These are clinical signs rather than a disease. However some findings are common.

Historical Findings
Owners may witness the expulsion of material from the mouth. Undigested, mucus-coated food may be found in the enclosure.

Physical Examination Findings
Dried food or mucus may be matting the feathers of the head (often referred to in lay press as head-sweating). Sour or fetid breath odor can occur, especially when there is infection in the crop. Distended, fluctuant, or thickened crop can be noted during examination.

CAUSES
• **This can be a behavioral condition,** usually related to misplaced courtship or sexual behavior toward objects or people. • **Normal casting behavior** in raptors involves vomiting up the indigestible components of their prey. The casts are formed in the ventriculus. • **Crop or esophageal disorders:** ○ Bacterial, yeast or trichomonas infections are common in the crop. Bacterial infections often result in sour, mucous exudate in the crop. Yeast (*Candida spp.*) and trichomonas infections often cause thickened crop walls. ○ Foreign bodies in the crop are usually quite large, or they will pass down to the ventriculus. They are generally readily palpable. ○ Ingluvioliths are usually urate stones and result from coprophagia. They are readily palpable on examination. ○ Burns or fistulae occur commonly in handfed baby birds from overheated or unevenly heated formula. These birds may have severe infections. • **Proventricular or ventricular diseases:** ○ Infectious ▪ Avian bornavirus (ABV) (proventricular dilatation disease: PDD) ▪ *Macrorhabdus ornithogaster* ▪ *Candida spp.* ▪ *Cryptosproridium spp.* ▪ Other parasites ▪ *Mycobacterium spp.* ○ Foreign bodies, including heavy metals, usually will be trapped in the ventriculus. The outflow is particle size dependent. Heavy metals have the additional risk of inhibiting motility and eroding the mucosa. ○ Neoplasia in the gastrointestinal tract occurs most commonly at the isthmus between the proventriculus and ventriculus. Most commonly, the birds die from rupture and subsequent hemorrhage of the mass. ○ Ulcers can occur in the stomach or intestine. *Macrorhabdus* infection is a major cause in small species. Stress, nonsteriodal anti-inflammatory drugs (NSAIDs), and heavy metals are risk factors for any species. ○ Koilin dysplasia is a rare condition, but can result in obstruction of the ventricular outflow. • **Obstructive conditions** of the intestinal tract are uncommon. The ventriculus strains out most foreign bodies before they reach the intestine. When they do occur, they are severe and acute. Intussusception occasionally occurs. • **Motion sickness** occurs in many birds. Sometimes there is a significant stress component involved. The birds are generally healthy otherwise. • **Extreme polydipsia** can result in regurgitation, especially if the bird is handled. The overfilled, distended crop readily refluxes the contents. • **Iatrogenic** (drugs, anesthetics) regurgitation is relatively common. Macaws appear particularly prone to this condition.

RISK FACTORS
Social and imprinting factors can affect behavioral regurgitation. Poor sanitation of environment, water, and food can result in bacterial overgrowth and infection of the upper gastrointestinal tract. Immunosuppression or immunologic immaturity can allow infections to take hold more easily. Exposure to infectious agents is necessary for primary pathogens to affect the bird. Some, such as ABV have very long incubation periods. Access to foreign material is a risk factor for birds with a tendency to consume them.

DIAGNOSIS

DIFFERENTIAL DIAGNOSIS
Regurgitation and vomiting are signs. Differentials include upper GI diseases and behavioral alterations.

CBC/BIOCHEMISTRY/URINALYSIS
There are no specific changes peculiar to this clinical sign. Any changes very much depend on the specific cause and the severity of the disease. Infectious causes of vomiting and regurgitation may show inflammatory responses on the CBC. Electrolyte alterations may occasionally occur. If severe, biochemical alterations may occur due to catabolic effects.

OTHER LABORATORY TESTS
Cytology, culture, molecular diagnostics from fecal, crop or gastric (proventricular, or ventricular) samples can help identify infections. Special stains may be needed to identify some organisms. Acid-fast stains are useful for identifying *Cryptosporidium spp.* or *Mycobacterium spp.* Fecal flotation and direct (wet mount) examination should be used to help identify parasites. Direct examination also allows visualization of *Macrorhabdus ornithogaster.* Avian bornavirus PCR may have some utility; however, this should be interpreted carefully. Not all infected birds test positively, and many asymptomatic birds test positively.

IMAGING
Radiography, possibly including positive contrast can help determine if there is gastric or intestinal involvement. Fluoroscopy with contrast can be useful to determine motility.

DIAGNOSTIC PROCEDURES
• Crop biopsy with histopathology is still the most commonly recommended diagnostic test for proventricular dilatation disease. A full thickness biopsy, incorporating visible vessels (and the associated, but difficult to see nerves) is needed. • Endoscope-assisted gastric lavage can be used to inspect the lining of the stomach. This is often done in conjunction with the crop biopsy, using the crop incision as an access point to the coelomic esophagus.

PATHOLOGIC FINDINGS
This is highly variable and depends on the specific cause of the V/R.

REGURGITATION AND VOMITING (CONTINUED)

TREATMENT

APPROPRIATE HEALTH CARE
Treatment of identified cause of the problem is necessary to resolve the problem long term. Hospitalization may be required during the initial phase of treatment.

NURSING CARE
Fluid and nutritional support is needed. Nutritional support may be difficult to achieve unless the regurgitation/vomiting can be controlled. Inhibition of vomiting and regurgitation using anti-emetic drugs is indicated.

ACTIVITY
No activity restrictions should be needed, unless required by fluid lines or other supportive care. Patients with motion sickness may need premedication before traveling.

DIET
Prolonged fasting is not practical in most birds. A very bland, easily digested diet should be given. Gavage feeding may be required if the patient is anorectic. If the patient is severely polydipsic, water may have to be restricted or offered intermittently.

CLIENT EDUCATION
Client education will depend on the specific diagnosis.

SURGICAL CONSIDERATIONS
Surgery or endoscopy may be indicated for removal of foreign material from the gastrointestinal tract. Crop biopsy is indicated for diagnosis of PDD.

MEDICATIONS

DRUG(S) OF CHOICE
• Metoclopramide may be used as an anti-emetic. • Appropriate antimicrobial therapy should be used to treat any infections identified. • Celecoxib or other COX-2 selective drug should be used to control inflammation for ABV infections. • Sucralfate or histamine-2 blockers should be used if endoscopy, surgery, or gastric bleeding occurs.

CONTRAINDICATIONS
None.

PRECAUTIONS
Drugs that have any GI side effects may complicate monitoring of the patient's progress.

POSSIBLE INTERACTIONS
None.

ALTERNATIVE DRUGS
N/A

FOLLOW-UP

PATIENT MONITORING
The frequency and severity of V/R should be monitored during the course of therapy. Other monitoring depends on the specific diagnosis and any complications encountered during the course of the disease.

PREVENTION/AVOIDANCE
This is dependent on the specific etiology.

POSSIBLE COMPLICATIONS
Dehydration and weight loss are the primary concerns for birds with regurgitation and vomiting. Aspiration pneumonia or aspiration rhinitis may occur if the regurgitated material is inhaled or extends into the choana. The various causes of this sign have their own potential complications. ABV often results in neurologic signs. *Macrorhabdus* infection often results in gastric bleeding and may predispose to gastric neoplasia.

EXPECTED COURSE AND PROGNOSIS
The prognosis is highly dependent on the underlying cause of the problem:
• Prognosis is excellent for behavioral regurgitation and motion sickness.
• Prognosis is good for simple upper gastrointestinal infections. • Prognosis is fair for ulcers, Macrorhabdus infections, and foreign bodies. • Prognosis is guarded-to-poor for ABV, cryptosporidiosis, and koilin dysplasia. • Prognosis is grave for neoplasia.

MISCELLANEOUS

ASSOCIATED CONDITIONS
N/A

AGE-RELATED FACTORS
N/A

ZOONOTIC POTENTIAL
Most etiologies of V/R have no zoonotic potential.

FERTILITY/BREEDING
N/A

SYNONYMS
N/A

SEE ALSO
Aspiration
Candidiasis
Cryptosporidiosis
Dehydration
Emaciation
Enteritis/gastritis
Gastrointestinal foreign bodies
Ingluvial hypomotility/ileus
Macrorhabdus ornithogaster
Polydipsea
Proventricular Dilatation Disease
Rhinitis and sinusitis
Sick-bird syndrome
Appendix 7, Algorithm 3. Regurgitation and vomiting

ABBREVIATIONS
V/R—vomiting and/or regurgitation

INTERNET RESOURCES
N/A

Suggested Reading
Denbow, D.M. (2000). Gastrointestinal anatomy and physiology. In: Whittow, G.C. (ed.), *Sturkie's Avian Physiology*, 5th edn. San Diego, CA: Academic Press, pp. 299–325.
Fudge, A.M. (2000). Avian liver and gastrointestinal testing In: Fudge, A.M. (ed.), *Laboratory Medicine: Avian and Exotic Pets*. Philadelphia, PA: WB Saunders, pp. 47–55.
Girling, S. (2004). Diseases of the digestive tract of psittacine birds. *In Practice*, **26**:146–153.
Hadley, T.L. (2005). Disorders of the psittacine gastrointestinal tract. *The Veterinary Clinics of North America. Exotic Animal Practice*, **8**:329–349.
Author Kenneth R. Welle, DVM, DABVP (Avian)

Client Education Handout available online

RENAL DISEASE

BASICS
OVERVIEW
• Renal disease describes any disorder affecting the kidneys. • Renal failure is end-stage renal disease severe enough to negatively impact renal function.

PATHOPHYSIOLOGY
• As renal tissue loses function, the remaining nephrons initially increase their performance as a renal functional adaptation however a significant loss of renal tissue causes a fall in the glomerular filtration rate. This cumulative loss of nephrons with subsequent loss of renal function can be a slow, insidious process (chronic renal failure) or rapid (acute renal failure). • The cause of renal disease can be anything that compromises renal perfusion (prerenal), structure and function of the nephrons (primary renal disease), or anything that prevents normal renal excretion (postrenal disease).

SYSTEMS AFFECTED
• Nervous—renomegaly can cause compression or impingement on the branches of the lumbosacral plexus that pass through or near the kidneys. • Gastrointestinal—ceca, rectum, cloaca play a role in reabsorption of ureteral water. • Cardiovascular—affected when fluid overload occurs. • Respiratory—affected when fluid overload occurs. • Musculoskeletal—rare reports of osteopenia of long bones secondary in renal tumor case report. • Behavioral—rare instances of feather destructive behavior/self-mutilation overlying the synsacrum described.

GENETICS
Genetic problems are poorly described in birds but presumably play a role in some conditions like renal neoplasia in budgerigar parakeets.

INCIDENCE/PREVALENCE
Based on retrospective avian pathology studies, renal disease is relatively common in birds.

GEOGRAPHIC DISTRIBUTION
N/A

SIGNALMENT
• **Species:** Renal disease can be seen in any avian species; commonly reported conditions include: ◦ Diclofenac nephrotoxicity (Old World *Gyps* vultures). ◦ Primary renal tumors (budgerigar parakeets). ◦ Renal amyloidosis (captive waterfowl, shorebirds, gulls). ◦ Renal parasites (waterfowl, marine birds). ◦ Urolithiasis (poultry layers). • **Mean age and range:** ◦ Juvenile birds: Vitamin D3 toxicosis (psittacine chicks), renal coccidiosis (aquatic birds), "perirenal hemorrhage syndrome" (turkey poults), urolithiasis (pullets). ◦ Adult birds: Primary renal tumors (young to middle-aged adult budgerigar parakeets). • **Predominant sex:** ◦ Urolithiasis in hens. ◦ Primary renal tumors and "perirenal hemorrhage syndrome" in male birds.

SIGNS
General Comments
• Appreciable changes in urinary output are uncommon with renal disease, but polyuria, anuria, oliguria, or even stranguria and hematuria can be observed. • Although many clinical signs are similar to those seen in mammals, there are also signs of renal disease unique to birds: ◦ Unilateral or bilateral ataxia, lameness, neurologic deficits, paralysis secondary to direct compression or impingement on the branches of the lumbosacral plexus that pass through or near the kidneys. ◦ Constipation due to compression of the rectum.

Historical Findings
• Nonspecific clinical signs of illness: lethargy, anorexia, fluffed and ruffled appearance, unwillingness or inability to fly. • Lameness or paralysis with no history of trauma. • Change in droppings: volume, color of urine or urates, diarrhea. • Regurgitation or vomiting. • Inadequate husbandry: nutritional imbalances, poor sanitation. • In addition to signs of nephritis, infectious bronchitis virus is characterized by respiratory signs in broilers and decreased egg production, poor egg quality in layers.

Physical Examination Findings
• Visual examination: ◦ The affected leg may be held feebly, resting the plantar surface of the tarsometatarsus on the perch or cage floor. ◦ Dyspnea may be observed secondary to coelomic distension, ascites. ◦ Carefully observe droppings for changes in volume or color of urine and/or urates. • Hands-on examination: ◦ Evidence of weight loss, emaciation. ◦ Evidence of dehydration. ◦ Crop stasis. ◦ Coelomic distention due to a mass lesion, renomegaly, and/or ascites. ◦ In larger birds like raptors, the caudal renal division can be palpated with a lubricated gloved finger inserted into the cloaca. ◦ Compare reflexes and sensation in birds with evidence of ataxia or lameness. With renal tumors, reduced sensation may be detected distal to the stifle. ◦ Feather destructive behavior, self-mutilation over the synsacrum.

CAUSES
Anatomic
• Renal agenesis is the most frequently described inherited defect. • Renal hypoplasia, dilated ureter, remnant ureters. • Solitary or multiple renal cysts.

Metabolic
• Dehydration. • Diabetes mellitus. • Lipidosis.

Neoplasia
• Renal carcinoma, nephroblastoma are the most common tumors of the avian kidney. • Other reported renal tumors include: renal adenoma, cystadenoma, fibrosarcoma, embryonal nephroma, and metastatic tumors like multicentric lymphoma or seminoma.

Nutrition
• High dietary calcium. • High dietary protein. • High dietary calcium + high dietary protein. • Low dietary protein: Renal lipidosis has been associated with starvation, low protein diets, and biotin deficiency. • Vitamin A deficiency: Metaplastic changes can lead to ureteral obstruction and post-renal failure. • Excessive dietary vitamin D3. • High cholesterol diets.

Infection
• An estimated 50% of avian nephritis patients have bacterial disease. ◦ Most cases are associated with multisystemic disease or sepsis. ◦ Less commonly, nephritis can develop secondary to infection ascending from the cloaca or colon. • Gram-negative bacteria ◦ *E. coli* ◦ *Klebsiella* ◦ *Salmonella* spp. ◦ *Yersinia* spp. ◦ *Proteus* ◦ *Pseudomonas* ◦ *Pasteurella multocida* ◦ *Brachyspira alvinipulli*. • Gram-positive bacteria ◦ *Staphylococcus* spp. ◦ *Steptococcus* spp. ◦ *Listeria* spp. ◦ *Erysiplelothrix rhusiopathiae* (Coturnix quail). • *Mycobacterium* spp. • *Chlamydophila psittaci* • Viruses known to infect the kidneys are generally part of systemic infection, and include: ◦ Adenovirus: can cause renomegaly, but usually an incidental finding ◦ Astrovirus ◦ Bornavirus ◦ Coronavirus ◦ Enterovirus ◦ Herpesvirus (Marek's disease) ◦ Infectious bronchitis virus ◦ Influenza virus ◦ Nephritis virus ◦ Newcastle disease virus ◦ Orthovirus ◦ Paramyxovirus (pigeons) ◦ Polyomavirus: Best known virus affecting kidneys of caged birds ◦ Poxvirus ◦ Reovirus ◦ Reticuloendotheliosis virus ◦ Retrovirus (avian leukosis/lymphoid leukosis) ◦ Togavirus: West Nile virus, EEE, WEE. • Fungal nephritis is relatively uncommon but can result from local extension of fungal air sacculitis (*Aspergillus* spp.) or less commonly via fungal thrombosis. • Parasites ◦ Renal coccidiosis (*Eimeria* spp., *Isospora* spp.) ◦ Renal trematodes (scattered reports) ◦ Protozoa: Cryptosporidium spp., Encephalitozoon, Microsporidia, Plasmodium (avian malaria), Sarcocystis (rare to sporadic cases).

Inflammation
• Amyloidosis: associated with chronic, systemic inflammatory conditions (e.g., pododermatitis).

Idiopathic
• Glomerular lipidosis. • Sudden death syndrome or "perirenal hemorrhage syndrome" is the main cause of acute death in large turkey flocks from 8-14 weeks of age.

Toxic exposure
• **Avicide (3-chloro-p-toluidine).** • **Ethylene glycol.** • **Heavy metals: Lead,**

RENAL DISEASE (CONTINUED)

zinc. • **Mycotoxins.** • **Nephrotoxic medications** ◦ Allopurinol. ◦ Antibiotics: Aminoglycosides (e.g., gentamicin), sulfonamides. ◦ Nonsteroidal anti-inflammatory drugs (NSAIDs). ■ Birds display a highly variable response to NSAIDs possibly because of differences in protein binding. ■ Nephrotoxicity have been reported with: ▫ Carprofen in Harris's hawk (*Parabuteo unicinctus*), northern saw-whet owl (*Aegolius acadicus*), Maribou stork (*Leptoptilos crumeniferus*). ▫ Diclofenac in Old World vultures (*Gyps* spp.), Japanese quail (*Coturnix japonica*), mynah birds (Sturnidae), pigeons (*Columba livia*), broiler chickens. ▫ Flunixin meglumine in flamingos, cranes, northern bobwhite quail (*Colinus virginianus*), eiders (*Somateria spectabilis, S. fischeri*). ▫ Ketoprofen in eiders, broiler chickens. • **Oxalate**—ingestion of plants containing toxic levels. • **Rodenticides (vitamin D3 analogs).** • **Salt toxicosis.**

Traumatic
• Capture myopathy leading to muscle damage and myoglobinuric nephrosis.
• Direct trauma is a very rare cause of renal hemorrhage because the kidneys are protected within the synsacrum.

RISK FACTORS
• **Medical conditions** ◦ Glomerular filtration rate, conditions that reduce ■ Blood loss ■ Severe dehydration ■ Shock. ◦ Inflammation, chronic ◦ Immunocompromise ◦ Old age ◦ Systemic hypertension ◦ Uncontrolled diabetes mellitus. • **Medications** ◦ Diuretics, administration of. ◦ Nephrotoxins: NSAIDs, intravenous contrast media. • **Environmental factors** ◦ Poor husbandry ■ Adverse conditions that might lead to dehydration. ■ High moisture conditions that promote the development of moldy feed (mycotoxicosis). ■ Improper nutrition. ■ Overcrowding (renal coccidiosis). ◦ Stress, prolonged: ■ Amyloidosis is typically associated with prolonged periods of stress and chronic inflammatory disease.
• **Shipping stress** is considered a risk factor for urolithiasis in poultry.

DIAGNOSIS

DIFFERENTIAL DIAGNOSIS
• Differentials for early renal disease are quite broad since this includes anything that can cause nonspecific signs of illness.
• Hematuria: The source can be the kidneys, reproductive tract, cloaca, or distal gastrointestinal tract. • Coelomic distention: In addition to renomegaly, differentials include gonadal enlargement, egg binding, as well as other mass lesions. • Differentials for lameness include orthopedic disease or metabolic disorders.

CBC/BIOCHEMISTRY/URINALYSIS
• Complete blood count: nonspecific, stress-related changes ◦ Mild nonregenerative anemia. ◦ Mild-to-marked heterophilia, monocytosis, lymphopenia. • Biochemistry panel ◦ Uric acid is an insensitive renal function test. Normal uric acid values do not mean kidney function is normal; however, uric acid levels can rise when glomerular filtration decreases by more than 70–80%. ◦ Elevations in blood urea nitrogen (BUN) can be seen with dehydration in birds, or after high-protein meals in carnivorous birds. ◦ Creatinine levels may be elevated in pet birds fed high-protein diets (e.g., parrots on dog chow). Creatinine also increases in significantly dehydrated pigeons. ◦ Urea:Creatinine and Urea:Uric acid ratios can be used to better define pre-and postrenal azotemia as both ratios should be high during dehydration and postrenal obstruction. ◦ Low serum protein, hypoalbuminemia has been reported in chickens with advanced tubular nephrosis and interstitial nephritis. ◦ Avian kidneys usually cannot concentrate sodium or electrolytes much above normal blood levels, and significant changes are not generally reported. • Urinalysis. ◦ The presence of renal casts can indicate renal pathology. ◦ Persistent hematuria has been associated with renal neoplasia, avian polyomavirus, bacterial and viral nephritis, and some forms of toxic nephropathy. ◦ Myoglobinuria can cause false positive hematuria. ◦ Porphyrinuria, as seen in lead-poisoned Amazon parrots, can result in urine that mimics hemoglobinuria, hematuria. ◦ Although proteinuria is the hallmark sign of glomerulonephritis in mammals, voided urine samples are "normally" positive for protein due to fecal contamination in the cloaca.

OTHER LABORATORY TESTS
• Protein electrophoresis. • Blood heavy metal levels. • Serology, PCR testing as indicated based on history, signalment. • Mammalian tests with implications for the future in avian medicine: ◦ Urinary cystatin C as a specific marker for renal tubular dysfunction. ◦ N-Acetyl-beta-D-glucosaminidase (NAG) is an exoglycolytic enzyme used as a marker for renal tubular damage.

IMAGING
• **Survey radiograph** ◦ Evaluate renal divisions ■ The cranial lobe of the kidneys is superimposed by the gonads in some birds. ■ Middle and the caudal renal lobes are superimposed by the bony pelvis. ◦ Evaluate renal size ◦ Loss or compression of the abdominal air sac diverticula. ◦ Renal enlargement leads to displacement of the proventriculus, grit-filled ventriculus, and intestines on the lateral view. • **Radiopacity** ◦ Dehydration ◦ Fibrosis ◦ Gout ◦ Mineralization • **Contrast studies** of the gastrointestinal tract can help determine the location of the kidneys and the source of organomegaly. • **Sonographic evaluation:** Evaluate renal size, determine location of an abdominal mass, scan for the presence of an obstruction, and evaluate renal vascular status.
• **Alternate imaging techniques:** ◦ Nuclear scintigraphy to evaluate renal function. ◦ MRI, CT. ◦ Do not perform intravenous excretory urography in birds with severe renal compromise.

DIAGNOSTIC PROCEDURES
• Renal biopsy ◦ Indication: Definitively diagnose primary renal disease and specific pathologic tissue damage when considering specific therapy. ◦ Contraindications: Small kidneys, advanced disease, coagulopathy. ◦ Usually performed via endoscopy; however, a dorsal pelvic surgical approach is also described. • Culture: Identification of bacteria within renal tissue can be difficult; blood culture preferred. • Measure systolic blood pressure. • Water deprivation test useful to rule out unknown causes of polyuria/polydipsia including diabetes insipidus and psychogenic polydipsia.
• Cytology: Murexide text to identify uric acid crystals. • Histopathology: Congo red staining method to reveal amyloid. • GFR can be measured in a laboratory setting by measuring clearance of an exogenous marker like H-inulin. This test requires extensive anesthesia and cannulation of the ureters for extended periods of time to obtain uncontaminated urine.

PATHOLOGIC FINDINGS

Gross findings
• Enlarged kidneys may bulge out of the renal fossa. • Masses: ◦ Adenomas and adenocarcinomas are large, somewhat friable, usually focal, nodular swellings that vary from white, tan to red-brown in color. ◦ Renal masses are often located in the cranial pole of the affected kidney. • Discolorations: ◦ Pallor. ◦ Chalky white urate deposits, tubules and ureters distended with urates. ◦ Renal coccidiosis: Kidneys are often enlarged with white-to-yellowish nodules containing urates and/or oocysts. ◦ Amyloid-infiltrated kidneys may be enlarged, firm, pale, or waxy.

Histopathology
• Renal histologic lesions are rarely pathognomonic for a specific disease process, many different diseases cause similar lesions.
• Amyloid-infiltrated tissues stain brown with iodine, brown color changes to blue with addition of sulfuric acid. • Bacterial or viral nephritis: granulomas, interstitial nephritis, degeneration or nephrosis, glomerulopathy, tubular dilation and impaction with inflammatory cells. As nephritis becomes chronic, tubular necrosis, cyst formation, distortion and interstitial fibrosis with mononuclear cell infiltration become evident.
• Renal coccidiosis: cytologic smears of renal tissue and ureters often contain different

RENAL DISEASE

(CONTINUED)

stages of coccidian oocysts. • Toxic nephropathy: proximal tubular necrosis, visceral gout, nephrosis.

TREATMENT

APPROPRIATE HEALTH CARE
• Treat the underlying cause of renal disease whenever possible. • Management follows the same principles as for mammals.

NURSING CARE
• Fluid therapy to maintain hydration, treat dehydration (which can rapidly exacerbate renal disease), and/or manage hyperuricemia. Monitor the patient closely for signs of fluid overload. • Nutritional support.

ACTIVITY
N/A

DIET
Gradually transition the bird to a healthy, balanced diet.

CLIENT EDUCATION
• Monitor appetite, body weight, droppings. • Stress the importance of keeping the patient hydrated. • Long-term prognosis.

SURGICAL CONSIDERATIONS
• Primary renal tumors are extremely difficult to remove successfully: ○ Location within renal fossae makes access difficult. ○ Extensive vascular network poses a significant risk of hemorrhage. ○ Close association with the lumbosacral plexuses means there is a risk of neurologic damage. • Removal of postrenal obstruction (e.g., ureterotomy).

MEDICATIONS

DRUG(S) OF CHOICE
• Antibiotic therapy ○ Base on culture and sensitivity whenever possible or select a broad-spectrum agent (e.g., beta-lactams). ○ A course of 4–6 weeks minimum is recommended. • Analgesia (e.g., meloxicam, opioids) for renal tumor patients. • Corticosteroids (e.g., methylprednisolone) may reduce peritumoral inflammation and edema. • Parenteral vitamin A for oliguric and anuric renal patients with hyperuricemia. • Diuretics (e.g., furosemide, mannitol) to reduce fluid overload and increase urinary output in oliguric and anuric patients.

CONTRAINDICATIONS
• Avoid all nephrotoxic drugs that could exacerbate renal disease. • Steroids can cause general immunosuppression and predispose to development of secondary infection.

PRECAUTIONS
• Calcium EDTA ○ Associated with nephrotoxicosis in mammals, producing an acute but reversible necrotizing nephrosis of proximal convoluted tubules. ○ Calcium EDTA has been used for weeks with no deleterious effects in birds.

POSSIBLE INTERACTIONS
Dehydration and renal disease may result in adverse responses to NSAIDs, especially cyclooxygenase (COX) inhibitors like meloxicam.

ALTERNATIVE DRUGS
• Chelation therapy to manage heavy metal toxicosis. • Colchicine to reduce renal fibrosis. • Chemotherapy: The use of carboplatin has been described in the management of a budgerigar parakeet with renal adenocarcinoma. • ACE inhibitors to correct hypertension. • Omega fatty acids have been used as an adjunct therapy for their anti-inflammatory, lipid-stabilizing, and renal protective properties.

FOLLOW-UP

PATIENT MONITORING
• Body condition, body weight. • Appetite, water intake. • Droppings including color and volume of urine and urates. • Hydration status. • CBC/biochemistry panel to evaluate treatment response and disease progression.

PREVENTION/AVOIDANCE
Ensure adequate water intake.

POSSIBLE COMPLICATIONS
• Birds can develop articular or visceral gout with any severe renal dysfunction that results in chronic, moderate to severe hyperuricemia. • Gastrointestinal ulcers are rarely reported in birds with renal disease. Ventricular erosions have been associated with naturally occurring urolithiasis in chickens. • Hemostatic abnormalities like fibrinous renal vessel thrombi are noted with some forms of renal disease but are rare in birds when compared to mammals.

EXPECTED COURSE AND PROGNOSIS
• Renal disease can be a chronic condition that progresses over months to years eventually leading to renal failure. • Poor long-term prognosis.

MISCELLANEOUS

ASSOCIATED CONDITIONS
• Gout. • Sepsis.

AGE-RELATED FACTORS
• Some causes of renal disease, like renal tumor and amyloidosis, are more common in adult birds. • Young birds are more commonly affected by some infectious agents and some nutritional causes of renal disease.

ZOONOTIC POTENTIAL
Some of the viruses known to infect the kidneys have zoonotic potential: Influenza virus, Newcastle disease virus, togaviruses.

FERTILITY/BREEDING
In addition to signs of nephritis, infectious bronchitis virus is characterized by decreased egg production and poor egg quality in layers.

SYNONYMS
Renal insufficiency, renal failure, nephritis, nephrosis

SEE ALSO
Dehydration
Diabetes insipidus
Diarrhea
Feather damaging behavior/self-injurious behavior
Heavy metal toxicity
Hyperuricemia
Hyperglycemia/diabetes mellitus
Polydipsia
Polyuria
Lameness
Nutritional deficiencies
Proventricular dilatation disease
Tumors
Urate/fecal discoloration
Viral disease
Vitamin D toxicosis
Viral neoplasms

ABBREVIATIONS
GFR—glomerular filtration rate

INTERNET RESOURCES
N/A

Suggested Reading
Knoopska, B., Gburek, J., Golab, K., Warwas, M. (2013). Influence of aminoglycoside antibiotics on chicken cystatin binding to renal brush-border membranes. *Journal of Pharmacy and Pharmacology*, **65**(7):988–994.

MacWhirter, P. (2002). Use of carboplatin in the treatment of renal adenocarcinoma in a budgerigar. *Exotic DVM*, **4**(2):11–12.

Naidoo, V., Duncan, N., Bekker, L., Swan, G. (2007). Validating the domestic fowl as a model to investigate the pathophysiology of diclofenac in Gyps vultures. *Environmental Toxicology and Pharmacology*, **24**:260–266.

Simova-Curd, S., Nitzl, D., Mayer, J., Hatt, J.M. (2006). Clinical approach to renal neoplasia in budgerigars (Melopsittacus undulatus). *Journal of Small Animal Practice*, **47**:504–51.

Wimsatt, J., Canon, N., Pearce, R., et al. (2009). Assessment of novel avian renal disease markers for the detection of experimental nephrotoxicosis in pigeons (Columba livia). *Journal of Zoo and Wildlife Medicine*, **40**(3):487–494.

Author Christal Pollock, DVM, Dipl. ABVP (Avian)

 Client Education Handout available online

RESPIRATORY DISTRESS

BASICS

DEFINITION
Respiratory distress refers to difficulty in breathing, but also the psychological experience associated with such difficulty.

PATHOPHYSIOLOGY
Respiratory distress most commonly occurs secondary to lower respiratory tract disease. Upper respiratory disease may induce respiratory distress with complete and bilateral nasal obstruction. Nonrespiratory diseases (coelomic cavity disease) leading to compression of the air sacs may induce respiratory distress.

SYSTEMS AFFECTED
- Respiratory—lower or upper respiratory diseases. • Reproductive—organomegaly, dystocia, ovarian cyst, egg-related peritonitis. • Renal/Urologic—organomegaly, gout. • Hepatobiliary—organomegaly, metabolic disease, ascites. • Gastrointestinal—organomegaly. • Cardiovascular—cardiac disease with pulmonary edema.

GENETICS
None.

INCIDENCE/PREVALENCE
Common.

GEOGRAPHIC DISTRIBUTION
N/A

SIGNALMENT
Species, age or sex predilection vary with the cause of respiratory distress:
- Tracheal obstruction—small bird species, that is, lovebirds, cockatiels;
- Hypersensitivity syndrome—macaw species;
- Aspergillosis – African grey parrots, Amazons and *Pionus* sp; • Dystocia – female;
- *Sternostoma* sp – canaries, finches.

SIGNS

General Comments
• The veterinarian must quickly evaluate if the patient's condition requires immediate emergency intervention. • Physical examination may have to be postponed until the bird is more stable; patient should be held upright without impairing keel movements; consider oxygen administration. • Large airway disease—inspiratory stridor, increased rate and effort with open-beak breathing exacerbated by exertion. Gasping occurs with almost complete obstruction. Tracheal transillumination may be helpful to identify *Sternostoma* sp in passerines and to locate a tracheal foreign body. • Small airway disease (hypersensitivity syndrome)—soft expiratory wheezes, severe respiratory distress with open-beak breathing, and sometimes collapse with gasping. • Parenchymal disease (lungs, air sacs)—increased respiratory rate and/or effort. Poor body condition with a history of lethargy/anorexia is usually present. Moist expiratory sounds may be auscultated with pulmonary disease while clicking noises suggest air sac disease. • Coelomic diseases—increased respiratory rate and effort at rest; becomes severely dyspneic with open-beak breathing on handling; history of lethargy/anorexia prior to the development of respiratory signs. Coelomic distension is present in most cases.

Historical Findings
• Exposure to toxic aerosols. • Acute onset of respiratory distress while eating. • Malnutrition, seeds in the diet. • Suboptimal hygiene, corn cob litter. • Macaw with periodic respiratory distress; housed with cockatiel, cockatoo or African grey species. • Hand-fed juvenile bird, severely depressed bird, vomiting/regurgitation. • Recent contact with sick bird. • Old world parrot kept outdoors with access to dirt ground in Gulf coast area. • Recent anesthetic event with tracheal intubation.

Physical Examination Findings
• Lower respiratory involvement—depression, exercise intolerance, open-beak breathing, neck extension, obvious sternal movements, wings held away from the body, inspiratory (large airways) or expiratory (small airways) noise, gasping, coughing, tail bobbing, head shaking, coelomic distension, moving black specks in trachea. • Upper respiratory involvement—facial or nasal swelling/asymmetry, nasal discharge, matted feathers around the nares, malformed rhinotheca, sneezing, wheezing, nasal sounds with open-beak breathing, repetitive yawning, nasal or oral cavity scratching, rubbing beak on perches, tachypnea without increased respiratory efforts, papilloma.

CAUSES
• Infectious—bacteria (*Chlamydia* sp, Gram-negative bacteria, *Mycobacterium* sp), fungi (*Aspergillus* sp), virus (herpesvirus, poxvirus, paramyxovirus, reovirus), parasites (*Sarcocystis* sp, *Sternostoma* sp, *Trichomonas* sp, *Syngamus* sp). • Metabolic—ascites (liver/heart disease), organomegaly. • Inflammatory—egg-related peritonitis. • Toxic—aerosols (polytetrafluoroethylene (Teflon®), self-cleaning ovens, bleach, ammonia, etc.), avocado. • Immune mediated—hypersensitivity syndrome. • Iatrogenic—aspiration pneumonia, membranous stenosis following tracheal intubation. • Traumatic—collision, ruptured air sac, bite wounds. • Degenerative—congestive heart failure, atherosclerosis. • Accidental—foreign body aspiration. • Neoplastic—respiratory or extra-respiratory tumors impairing breathing.

RISK FACTORS
• Medical conditions—malnutrition may predispose to respiratory diseases. • Environmental factors—exposure to toxic aerosols, suboptimal environment (poor hygiene, etc.).

DIAGNOSIS

DIFFERENTIAL DIAGNOSIS
See "Causes".

CBC/BIOCHEMISTRY/URINALYSIS
• CBC—to identify evidence of inflammation/infection. • Biochemistry—to evaluate organ function.

OTHER LABORATORY TESTS
• Serum protein electrophoresis—to identify and characterize inflammation. • Specific pathogen testing—to determine if *Chlamydia* sp, *Aspergillus* sp, *Mycobacterium* sp, herpes virus may be present. Sensitivity and specificity vary depending on testing modalities and which pathogen is targeted.

IMAGING
• Skull radiographs—to identify upper respiratory disease. Subtle lesions may not be apparent. • Whole body radiographs—to identify lower respiratory disease or pathology compressing the air sacs. Subtle or peracute (aspiration pneumonia) lower respiratory disease may not be apparent. • Contrast radiographs—to outline upper respiratory lesion or tracheal foreign body. • Fluoroscopy—to outline tracheal foreign body after contrast administration. • Pre- and postcontrast computed tomographic examination—to identify upper or lower respiratory disease. • Magnetic resonance imaging—to identify upper or lower respiratory disease. • Coelomic ultrasonography—to identify and characterize coelomic disease compressing the air sacs; to evaluate large respiratory lesions adjacent to body wall. • Echocardiography—to identify cardiac disease.

DIAGNOSTIC PROCEDURES
• Endoscopy—to directly assess upper and lower respiratory system; to collect cytology/culture/histopathologic samples. • Cytology—to evaluate cells and microbial population. Choanal swab; nasal, sinus, and trans-tracheal lavage; and impression smears from biopsy samples may be submitted for evaluation. • Culture—to evaluate microbial population. See cytology for sample types. • Histopathology—to characterize lesions in diseased tissue. • ECG—to identify cardiac diseases.

PATHOLOGIC FINDINGS
Gross and histopathologic findings will differ depending upon the underlying cause of respiratory distress.

RESPIRATORY DISTRESS (CONTINUED)

TREATMENT

APPROPRIATE HEALTH CARE
• Emergency inpatient intensive care management: All birds in respiratory distress require immediate stabilization in a warmed oxygen-enriched incubator. • Surgical management once patient is stable: placement of air sac cannula, tracheotomy, tracheal resection and anastomosis may be performed following initial stabilization.

NURSING CARE
Supportive care
• Oxygen therapy—indicated for all cases of respiratory distress. Ideal method of oxygen delivery is within a cage. Humidification is required if oxygen supplementation is performed for more than a few hours. Humidification is achieved by bubbling the oxygen through a container with sterile saline or distilled water. • Fluid therapy—may be provided SQ in birds exhibiting mild dehydration (5–7%) or IO/IV in birds with moderate to severe dehydration (>7%) or critically ill birds exhibiting signs consistent with shock. Avian daily maintenance fluid requirements are 50 mL/kg. In most patients, Lactated Ringers Solution or Normosol crystalloid fluids are appropriate. • Warmth (85–90°F/27–30°C)—to minimize energy spent to maintain body temperature.

Therapeutic care
• Nasal flush—to remove mucous/discharge and improve breathing. • Endoscopy—to remove tracheal foreign bodies (medium and large birds) and debulk tracheal granulomatous lesions. Prior placement of an air sac cannula recommended. • Suction—to retrieve tracheal foreign bodies in small birds where the endoscope is too large to be inserted within the small trachea. Prior placement of an air sac cannula is required. An appropriately sized urinary catheter is inserted in the trachea until it reaches the foreign body and suction is applied.
• Coelomocentesis—to aspirate coelomic effusion/drain cystic structures and decrease air sac compression. Best performed guided by ultrasound under mild sedation. May be performed blindly on the right lateral coelom just cranial to the cloaca. Patient should be maintained in an upright position to decrease the likelihood of fluid entering the lungs. Concurrent oxygen therapy recommended.

ACTIVITY
Patient activity should be minimal to avoid worsening of clinical signs.

DIET
Birds presenting in respiratory distress may require assisted feeding—must be performed rapidly to minimize stress and worsening of respiratory signs. A smaller volume is typically administered. Enteral nutritional preparation may be administered initially at 20 mL/kg directly in the crop. Volume administered may be progressively increased according to the patient tolerance.

CLIENT EDUCATION
Balancing the need to perform diagnostic tests and administer treatment in light of the patient's condition is critical. Patient clinical condition may worsen despite appropriate care.

SURGICAL CONSIDERATIONS
Surgical considerations will differ depending upon the underlying cause of respiratory distress.

Air sac cannula placement—to provide an alternate airway. Emergency placement of an air sac cannula is indicated with large airway obstruction. An air sac tube may partially improve respiration in birds with primary lung and air sac disease. Various types of tubing or standard endotracheal tubing can be modified for cannulation purposes. The diameter of the cannula should approximate the bird's tracheal diameter. Consider sedation or inhalant anesthesia to facilitate placement of the cannula. The patient is positioned in lateral recumbency with the upper leg pulled caudally. Skin incision is performed over the triangle created by the cranial muscle mass of the femur, ventral to the synsacrum and caudal to the last rib. Using a pair of hemostats, the caudal air sac is entered making a popping noise. The air sac cannula is inserted between the last two ribs or just caudal to the last rib. The bird should immediately begin to breathe through the tube if placed correctly. Patency may be evaluated using the end of a laryngoscope (fogging) or a feather may be held in front of the cannula (breath-associated movements). The cannula is sutured into place using a Chinese finger trap technique. Consider placing a filter over the end of the tube to prevent entry of particulate matter.

MEDICATIONS

DRUG(S) OF CHOICE
• Medication will differ depending upon the underlying cause of respiratory distress.
• Nebulization therapy – within an oxygen cage, through an endotracheal tube or face mask. Antibiotics (enrofloxacin 10 mg/mL saline), antifungals (amphotericin B 7-10 mg/mL sterile water), bronchodilators (aminophylline 3 mg/mL sterile saline) may be nebulized. Ultrasonic nebulization is preferred. Particle size needs to be less than five microns to allow for optimal dispersion. Nebulization sessions typically last 15–30 minutes and may be repeated several times a day depending on the medication delivered and the patient's clinical condition.
• Bronchodilators (terbutaline 0.01 mg/kg IM q6–8h; aminophylline 10 mg/kg IV q3h then 4 mg/kg PO q6–12h) to address bronchoconstriction. Corner stone treatment for patient with hypersensitivity syndrome.
• Benzodiazepine (midazolam 0.5–2mg/kg IM)—to decrease anxiety associated with being unable to breathe normally. • Analgesic (butorphanol 1–4 mg/kg IM q2–6h) is indicated to address pain. Adverse effects in light of the patient's overall condition need to be considered. • Diuretic (furosemide 0.15–2.0 mg/kg PO, IM, IV q8h) to treat pulmonary edema. Critically dyspneic birds often require higher dosages (4–8 mg/kg) to achieve initial stabilization. • Nonsteroidal anti-inflammatory (meloxicam 0.5 mg/kg PO/IM q12h) to decrease inflammation that may be associated with the disease condition. Adverse effects in light of the patient's overall condition need to be considered.
• Antiparasitic therapy—varies with the specific cause. • Antibiotic therapy – to treat primary and secondary bacterial infection that may be associated with the disease condition. Antibiotic selection varies with the specific cause of respiratory distress. • Antifungal therapy—to treat aspiration pneumonia and fungal disease. Antifungal selection varies with the causative agent, the severity of clinical signs, the location of infection and patient species.

CONTRAINDICATIONS
None.

PRECAUTIONS
Fluid deficits should be addressed prior to using nonsteroidal anti-inflammatory medication.

POSSIBLE INTERACTIONS
None.

ALTERNATIVE DRUGS
N/A

FOLLOW-UP

PATIENT MONITORING
Varies with the cause of respiratory distress.

PREVENTION/AVOIDANCE
• Avoid exposure to toxic aerosols. • Offer a balanced diet according to the bird species.

POSSIBLE COMPLICATIONS
• Varies with the cause of respiratory distress.
• Suffocation with tracheal obstruction.

EXPECTED COURSE AND PROGNOSIS
• Varies with the cause of respiratory distress.
• Respiratory distress secondary to upper airway disease carries a better prognosis than respiratory distress secondary to lower airway, coelomic or cardiac disease. • The delay between the onset of respiratory distress and time to seeking veterinary care influences prognosis. Death may occur if treatment is

RESPIRATORY DISTRESS (CONTINUED)

delayed. • Teflon exposure and severe fungal infections carry a grave prognosis.

MISCELLANEOUS

ASSOCIATED CONDITIONS
Varies with the cause of respiratory distress.

AGE-RELATED FACTORS
Aspiration pneumonia should be ruled out in handfed juvenile birds with respiratory distress.

ZOONOTIC POTENTIAL
Chlamydia psittaci is zoonotic.

FERTILITY/BREEDING
N/A

SYNONYMS
Dyspnea

SEE ALSO
Airborne toxicosis
Air sac mites
Ascites
Aspergillosis
Aspiration
Cardiac disease
Chlamydiosis
Coelomic distention
Cystic ovarian disease and ovarian neoplasia
Dehydration
Egg binding/dystocia
Egg yolk coelomitis/reproductive coelomitis
Paramyxovirus
Pneumonia
Pox virus
Rhinitis/sinusitis
Tracheal/syringeal diseases
Viral diseases
Appendix 7, Algorithm 4. Respiratory distress

ABBREVIATIONS
N/A

INTERNET RESOURCES
N/A

Suggested Reading
Graham, J. (2006). The dyspneic avian patient. *Seminars in Avian and Exotic Pet Medicine*, **13**(3):154–159.
Phalen, D. (2000). Respiratory medicine of cage and aviary birds. *The Veterinary Clinics of North America. Exotic Animal Practice*, **3**(2):423–452.
Westerhof, I. (1995). Treatment of tracheal obstruction in psittacine birds using a suction technique: A retrospective study of 19 birds. *Journal of Avian Medicine and Surgery*, **9**(1):45–49.
Hillyer, E.V., Orosz, S., Dorrestein, G.M. (1997). Respiratory system. In: Altman, R.B., Clubb, S.L., Dorrestein, G.M., Quesenberry, K. (eds), *Avian Medicine and Surgery*. Philadelphia, PA: WB Saunders, pp. 387–411.

Author Isabelle Langlois, DMV, DABVP (Avian)

RHINITIS AND SINUSITIS

BASICS

DEFINITION
Rhinitis is inflammation of one or both nasal cavities. Sinusitis is inflammation of one or more diverticulae of the infraorbital paranasal sinus.

PATHOPHYSIOLOGY
The upper respiratory system in birds consists of the paired nasal cavities and paired infraorbital paranasal sinuses. Air flows into the nasal cavity through bilateral nares (nostrils). The nasal cavity is divided by the nasal septum, which is incomplete in psittacine birds. Three nasal conchae are present (rostral, middle, and caudal) that serve to warm and moisten inspired air. The middle and caudal conchae communicate with the infraorbital sinus through small dorsal openings. Air exits the nasal cavity ventrally through the choana, a slit-like opening in the palate. The infraorbital sinus is the only true paranasal sinus in birds. It is located ventromedial to the orbit and consists of several diverticulae—rostral, preorbital, infraorbital, postorbital, and mandibular. The right and left paranasal sinuses communicate in psittacine birds but not in passerine birds. The infraorbital sinus communicates with the cervicocephalic air sac.

Most bacterial and fungal infections of the nasal cavity and infraorbital sinus are secondary to other diseases of the nasal cavity or infraorbital sinus that predispose these structures to infection, such as hypovitaminosis A, foreign bodies, airborne toxins and irritants, developmental abnormalities, and neoplasia.

SYSTEMS AFFECTED
• Gastrointestinal • Ophthalmic • Respiratory.

GENETICS
Choanal atresia and imperforate naris may have a genetic basis.

INCIDENCE/PREVALENCE
Unknown.

GEOGRAPHIC DISTRIBUTION
None.

SIGNALMENT
• Species predilections: Variable. • Mean age and range: Variable. Congenital diseases such as choanal atresia and imperforate naris are seen in young birds. Neoplastic diseases are generally seen in middle aged to older birds. • Predominant sex: None

SIGNS
Historical Findings
Wheezing, sneezing, nasal snuffling, infraorbital sinus flaring, head shaking, open-mouth breathing, lethargy, reduced appetite.

Physical Examination Findings
Rhinitis
Serous to mucopurulent ocular or nasal discharge, matted feathers surrounding the affected naris or nares, redness or swelling of the affected naris or nares, distortion or erosion of the nasal opening and cavity or distortion along upper beak extending rostrally from affected naris or nares, nasal obstruction, blunting or loss of the choanal papillae, choanal discharge, open mouth breathing.

Sinusitis
Periorbital swelling, sneezing, infraorbital sinus flaring with breathing, sunken eyed appearance (reported with gram negative bacterial infections in macaws, termed "sunken sinus syndrome"), serous to mucopurulent ocular or nasal discharge.

CAUSES
• **Congenital** ○ Choanal atresia: most commonly reported in African grey parrots. ○ Imperforate naris or nares. • **Nutritional** ○ Hypovitaminosis A. • **Neoplasia** ○ Nasal or sinus adenocarcinoma. ○ Squamous cell carcinoma. ○ Lymphoma. ○ Fibroma, fibrosarcoma. ○ Basal cell carcinoma. ○ Malignant melanoma. • **Immune-mediated** (allergic) rhinitis and sinusitis. • **Infectious** ○ *Viral*: rarely the underlying cause of rhinitis or sinusitis in companion birds: ■ Avian influenza virus ■ Herpesvirus ■ Paramyxovirus-1 (e.g., Newcastle disease virus) ■ Poxvirus ■ Reovirus. ○ *Primary bacterial agents* (may also be considered as secondary or opportunistic agents in some cases): ■ *Bordetella avium* ■ *Chlamydia psittaci* ■ *Enterobacter* spp. (eg. canaries and finches) ■ *Helicobacter* sp. ("spiral bacteria") in cockatiels ■ *Mycoplasma* spp. (e.g., finches, budgerigars, cockatiels) ■ *Mycobacterium* spp. ■ *Salmonella* spp. ■ *Yersinia pseudotuberculosis* (eg. canaries and finches) ○ *Secondary (opportunistic) bacterial agents*: may also be considered primary agents in certain cases: ■ *Escherichia coli* (*E. coli*) ■ *Haemophilus* sp. ■ *Klebsiella* spp. ■ *Nocardia asteroides* ■ *Pasteurella* spp. ■ *Pseudomonas aeruginosa* ■ *Staphylococcus* spp. ■ *Streptococcus* spp. ○ *Fungal agents*: fungal infections are often considered opportunistic and noncontagious: ■ *Aspergillus* spp. (particularly in African grey parrots) ■ *Candida albicans* ■ *Cryptococcus neoformans* ■ *Mucor* sp. ○ *Parasitic agents* ■ *Knemidocoptes pilae* (e.g., canaries) ■ *Trichomonas* spp. (e.g., canaries). • **Trauma.** • **Foreign bodies:** food particles, wood shavings, and so on. • **Rhinoliths:** proliferative nasal granulomas—most frequently reported in African grey parrots.

RISK FACTORS
• Poor air quality (e.g., exposure to cigarette smoke, cooking fumes, etc.). • Inadequate humidity. • Inadequate bathing. • Hypovitaminosis A (results in squamous metaplasia of the epithelial lining to the upper respiratory tract).

DIAGNOSIS

DIFFERENTIAL DIAGNOSIS
• Lower respiratory disease. • Diseases of the oral cavity. • Ocular disease.

CBC/BIOCHEMISTRY/URINALYSIS
Hematologic and biochemistry findings are generally nonspecific. Infectious and inflammatory diseases may be associated with leukocytosis and heterophilia or monocytosis.

OTHER LABORATORY TESTS
• Cytology: Useful to evaluate fluid and tissue samples. • Special stains (e.g., Gram, acid fast): Useful to further evaluate micro-organisms such as bacteria and fungi. • Microbiologic culture and sensitivity: Useful for identification of micro-organisms and for determining suitable antibiotics for treatment. • Infectious disease screening: ○ Serology—serologic tests are available for a number of infectious agents, such as *C. psittaci* and *Aspergillus* spp. Additional serologic assays such as galactomannan evaluation are also available for further screening for aspergillosis in birds. ○ PCR – PCR assays are available for a number of infectious agents, such as *C. psittaci* and *Mycobacterium* spp. ○ Other—additional testing methods are available and be useful in select cases, such as immunohistochemistry (IHC) and *in situ* hybridization (ISH) on tissue samples. • Plasma protein electrophoresis (EPH): Potentially useful for further evaluating for inflammatory response to disease. • Histopathology: Indicated for tissue samples obtained through biopsy of affected structures.

IMAGING
• Survey radiography: Survey radiographs of the skull can be helpful in evaluating bony and soft tissue structures. General anesthesia should be used for imaging. Superimposition of bony structures can make interpretation of findings difficult. Additional views such as oblique or skyline views may provide additional diagnostic information. • Contrast radiography: Contrast radiographs may be especially helpful in the diagnosis of choanal atresia and other obstructive diseases. Iodinated contrast such as iohexal (Omnipaque, Nyecomed Inc., Princeton, NJ) is diluted to 15–29% in sterile saline and instilled into the naris. Absence of flow through the choanal slit into the oral cavity can be diagnostic for choanal atresia. • Computed tomography (CT): CT with or without contrast enhancement can be very helpful in further evaluation of the skull. • Magnetic resonance imaging (MRI): Like CT, MRI can be helpful in further evaluating

RHINITIS AND SINUSITIS (CONTINUED)

the skull. MRI is most suitable for diseases affecting soft tissue structures. CT and MRI allow for image reconstruction matrices, eliminating problems with superimposition of anatomic structures by conventional radiography.

DIAGNOSTIC PROCEDURES
Sample collection methods
- **Nasal swab:** A sterile mini-tip culturette is placed into the naris for sample collection. This method is best suited for serous to mucopurulent nasal discharge. Environmental contamination is common.
- **Nasal aspiration:** Syringe aspiration of contents of the nasal cavity through the naris may be useful for collection of fluid samples. Environmental contamination is common.
- **Nasal lavage (nasal irrigation, nasal flush):** The awake bird is positioned with the head down and approximately 5–10 mL/kg of warm sterile saline or water is instilled into the nasal cavity through the naris. The fluid is collected as it flushes through the choanal slit and out the oral cavity. The fluid sample is representative of the nasal cavity but not the infraorbital sinus except possibly for the preorbital diverticulum. Contamination with oral flora is unavoidable.
- **Sinus aspiration and lavage:** Sinus aspiration and lavage can safely be performed in the awake bird. The preorbital diverticulum can be sampled through a site half way between the eye and the external naris under (or over) the zygomatic bone. The needle is directed perpendicular to the skin. A 25-gauge hypodermic or butterfly needle is suitable in most cases. Fluid within the preorbital diverticulum can be aspirated using this method. The infraorbital diverticulum of the infraorbital sinus can be aspirated directly ventral to the orbit dorsal to the zygomatic arch. Great care must be taken to avoid the ocular globe. If no fluid is immediately aspirated, sterile saline can be instilled and aspirated.
- **Choanal swab:** A moistened sterile swab is inserted into the rostral diverticulum of the choana. The sample only represents a small portion of the upper respiratory tract. Contamination with oral flora is common.
- **Endoscopy:** Endoscopy can be used to visualize and obtain diagnostic samples from the upper respiratory tract. Anatomic areas amenable to endoscopy include the rostral choanal sulcus, external naris (in larger birds), and the infraorbital sinus through a surgical approach.

PATHOLOGIC FINDINGS
Variable.

TREATMENT
APPROPRIATE HEALTH CARE
Birds with rhinitis or sinusitis are most often managed on an outpatient basis unless surgery is done.

Medical Treatment
- **Nasal and sinus lavage:** Lavage of the nasal cavity or infraorbital sinus using the methods described earlier can be useful to flush these cavities to remove inflammatory and foreign debris and to instill medications. For chronic conditions, such as rhinolithiasis with nasal cavity erosion, clients can be instructed on performing nasal irrigation at home.
- **Rhinolith debridement:** Rhinoliths can be manually removed using bent-tipped hypodermic needles, ear curettes, and other devices. Topical or general anesthesia is advised.
- **Nebulization:** Nebulization can be used to deliver medications to the nasal cavity and infraorbital sinus, provided that there is patent air flow through these anatomic structures.

SURGICAL TREATMENT
- **Sinusotomy:** Surgical access to the preorbital diverticulum of the infraorbital sinus is midway between the medial canthus of the eye and external naris, below the zygomatic arch. Through this site, the diverticulum can be visually or endoscopically explored. Purulent or caseous debris can be debrided. The incision can be left open to heal, or drains or stents can be placed, allowing continued drug delivery and lavage.
- **Sinus trephination:** If the rostral or preorbital diverticulum within the frontal bone is involved, bony trephination may be necessary to gain surgical access to the sinus cavity. The bony defect can be left open for continued lavage and drug delivery.
- **Choanal atresia and imperforate naris repair:** Several surgical techniques have been described for the repair of choanal atresia and imperforate naris.

NURSING CARE
Extensive nursing care is generally not required in cases of rhinitis or sinusitis.

ACTIVITY
Restrictions in activity are generally not required in cases of rhinitis or sinusitis.

DIET
If rhinitis or sinusitis is believed associated with hypovitaminosis A, dietary changes should be advised.

CLIENT EDUCATION
Clients should avoid exposure of their birds to airborne toxins and irritants such as cigarette smoke and cooking fumes.

SURGICAL CONSIDERATIONS
Patients should be assessed for lower respiratory disease, which would increase the risks of general anesthesia.

MEDICATIONS
DRUG(S) OF CHOICE
Topical antimicrobials
- **Antibiotics:** Broad spectrum ophthalmic antibiotic solutions are often prescribed for topical application. ◦ Fluoroquinolones ▪ Ciprofloxacin HCl 3 mg/mL (0.3%) (Ciloxan, Alcon) ▪ Ofloxacin 3 mg/mL (0.3%) (Ocuflox, Allergan). ◦ Aminoglycosides ▪ Gentamicin sulfate 3 mg/mL (0.3%). • **Antifungals** ◦ Clotrimazole 1% solution ◦ Amphotericin B, at dilutions no greater than 0.05 mg/mL. Sinus irrigation using a 50 mg/mL solution was associated with severe granulomatous inflammation and death in an African grey parrot.

Systemic antimicrobials
- **Antibiotics:** Indicated for more severe or invasive bacterial infections, based on culture and sensitivity results. Bactericidal antibiotics are generally preferred over bacteriostatic antibiotics. • **Antifungals:** Indicated for more severe or invasive mycotic infections. ◦ Azole antifungals ▪ Itraconazole ▪ Voriconazole. ◦ Amphotericin B.

Antimicrobials for nebulization
- Antibiotics ◦ Enrofloxacin ◦ Gentamicin.
- Antifungals ◦ Clotrimazole 1% solution ◦ Terbinafine. • Disinfectants ◦ F10®SC (1:250 dilution) (Health and Hygiene, South Africa).

CONTRAINDICATIONS
None.

PRECAUTIONS
None.

POSSIBLE INTERACTIONS
None.

ALTERNATIVE DRUGS
None.

RHINITIS AND SINUSITIS (CONTINUED)

FOLLOW-UP

PATIENT MONITORING
Birds with rhinitis or sinusitis should be periodically evaluated until the condition is resolved. For chronic cases such as rhinolithiasis with nasal cavity erosion, birds should be re-evaluated every few months.

PREVENTION/AVOIDANCE
Airborne toxins and irritants such as cigarette smoke and cooking fumes should be avoided. Diets should be balanced and adequate in vitamin A and other nutrients.

POSSIBLE COMPLICATIONS
Systemic spread of bacterial or fungal infections (sepsis) can occur with advanced or refractory rhinitis or sinusitis. Neoplasms can be locally aggressive or metastasize to regional or distant sites.

EXPECTED COURSE AND PROGNOSIS
Complete resolution is achievable in most cases of uncomplicated bacterial or mycotic rhinitis or sinusitis, and in cases involving foreign bodies or trauma. However, resolution of underlying conditions such as congenital choanal atresia or neoplasia is often challenging at best.

MISCELLANEOUS

ASSOCIATED CONDITIONS
Hypovitaminosis A is often associated with other conditions.

AGE-RELATED FACTORS
None.

ZOONOTIC POTENTIAL
C. psittaci, one cause of rhinitis and sinusitis in birds, is a zoonotic and reportable disease.

FERTILITY/BREEDING
None.

SYNONYMS
None.

SEE ALSO
Airborne toxicosis
Aspergillosis
Bordetellosis
Mycoplasmosis
Nutritional deficiencies
Respiratory distress
Tumors
Viral disease

ABBREVIATIONS
None.

INTERNET RESOURCES
None.

Suggested Reading
Noonan, B.P., de Matos, R., Butler, B.P., et al. (2014). Nasal adenocarcinoma and secondary chronic sinusitis in a hyacinth macaw (*Anodorhynchus hyacinthinus*). *Journal of Avian Medicine and Surgery*, **28**(2):143–150.
Schmidt, R.E., Reavill, D.R., Phalen, D.N. (2003). *Pathology of Pet and Aviary Birds.* Ames, IA: Iowa State Press, pp. 17–40.
Morrisey, J.K. (1997). Diseases of the upper respiratory tract of companion birds. *Seminars in Avian and Exotic Pet Medicine*, **6**(4):195–200.
Tully, T.N., Harrison, G.J. (1994). Pneumonology. In: Ritchie, B.W., Harrison, G.J., Harrison, L.R. (eds), *Avian Medicine: Principles and Application.* Lake Worth, FL: Wingers Publishing, Inc., pp. 556–580.
King, A.S., McLelland, J. (1984). *Birds, Their Structure and Function.* Philadelphia, PA: Baillière Tindall, pp. 110–144.
Author Lauren V. Powers, DVM, DABVP (Avian/ECM)

Client Education Handout available online

SALMONELLOSIS

BASICS

DEFINITION
Salmonellosis is due to infection with an organism from the genus *Salmonella*, a Gram-negative bacterium. There are two species of *Salmonella*, with *S. enterica* subsp. *enterica* being responsible for most infections in birds. There are over 1000 serotypes of *S. enterica enterica*. Infections can lead to clinical signs but are often asymptomatic in many animals.

PATHOPHYSIOLOGY
Salmonella organisms are most commonly transmitted between animals and/or people by the fecal-oral route, but some serotypes can be transmitted transovarially. Although systemic signs do occur, gastrointestinal signs such as diarrhea are most common.

SYSTEMS AFFECTED
- Gastrointestinal—diarrhea due to enteritis.
- Musculoskeletal—polyarthritis and osteomyelitis have been reported.
- Conjunctivitis in pigeons. • Acute sepsis, leading to death.

GENETICS
N/A

INCIDENCE/PREVALENCE
The prevalence of asymptomatic infections in poultry and waterfowl is high; these types of birds often act as carriers (infected, asymptomatic shedders). The prevalence is lower in other avian species and may be more likely to lead to clinical disease.

GEOGRAPHIC DISTRIBUTION
Salmonella sp. are distributed across the entire world. Wildlife species, such as waterfowl, and poultry often act as asymptomatic reservoirs.

SIGNALMENT
- **Species:** All avian species are susceptible to *Salmonella* infections. • **Mean age and range:** Young, old, or immunocompromised animals are more likely to develop clinical disease.

SIGNS

General Comments
Many animals infected with *Salmonella* will not show any clinical signs.

Historical Findings
- Exposure to another animal, which was either asymptomatic or sick, that was shedding the organism in its feces.
- Inappropriate husbandry leading to increased pathogen exposure or increased susceptibility due to immunosuppression.

Physical Examination Findings
- Diarrhea, which can be hemorrhagic.
- Non-specific signs, such as lethargy, anorexia, polyuria, and/or conjunctivitis.
- Swelling of joints.

CAUSES
Salmonellosis is caused by ingestion of *Salmonella* bacteria.

RISK FACTORS
Poor husbandry or exposure to asymptomatic birds may increase exposure to *Salmonella*. Immunosuppression may increase risk of developing clinical disease from infection.

DIAGNOSIS

DIFFERENTIAL DIAGNOSIS
Other causes of diarrhea including *E. coli*, *Yersinia*, *Pasteurella*, influenza, *Giardia*, coccidia, toxins, or bacterial septicemias.

CBC/BIOCHEMISTRY/URINALYSIS
Evidence of dehydration or inflammation, such as increased PCV or total protein or a leukocytosis, may be present, but these tests are not specific for salmonellosis.

OTHER LABORATORY TESTS
A sample of feces should be submitted for culture to isolate the organism. The diagnostic laboratory should be notified of suspected *Salmonella* as an etiologic agent because special procedures are needed for optimum isolation. Serology can be used as a screening tool in a poultry flock situation but is not typically performed for individual birds. Pulsed field gel electrophoresis is used in investigations to determine the source of an outbreak.

IMAGING
Radiographs can be obtained to examine changes in joints or bones, but a fine needle aspirate and culture will be needed to isolate the organism.

DIAGNOSTIC PROCEDURES
Collection of feces or a tissue sample for culture will be needed to definitively diagnose salmonellosis. A fine needle aspirate may be needed to collect a sample from a joint or bone. Samples from affected tissues can be collected at necropsy.

PATHOLOGIC FINDINGS
Gross and histopathologic findings will vary with severity and duration of disease but can include dehydration, enteritis, muscle necrosis, hepatomegaly, splenomegaly, pericarditis, epicarditis, granuloma formation in various organs, and diffuse inflammation.

TREATMENT

APPROPRIATE HEALTH CARE
Asymptomatic cases are typically not recognized or treated. Symptomatic cases are treated as described below.

NURSING CARE
Most cases of salmonellosis can be treated only with fluids (PO, SQ, IV) and appropriate nutrition.

ACTIVITY
Movement of the animal should be limited to reduce fecal contamination and reduce exposure of other animals or people.

DIET
Although appetite may be decreased, oral intake of normal food and fluids is acceptable.

CLIENT EDUCATION
Salmonella is zoonotic, and owners must take precautions to eliminate exposure to themselves, other humans, and other animals. Hand washing and cleaning and disinfection of contaminated areas and equipment are essential.

SURGICAL CONSIDERATIONS
N/A

MEDICATIONS

DRUG(S) OF CHOICE
Salmonellosis is not typically treated with antibiotics because infections are typically self-resolving with supportive care and because of the possibility of the development of antibiotic-resistant strains of *Salmonella*. Antibiotics can be used in severe cases, but drug choice must be based on culture and sensitivity results.

CONTRAINDICATIONS
Antibiotics that have not been shown to be appropriate based on culture and sensitivity.

PRECAUTIONS
Salmonella can develop antibiotic resistance.

POSSIBLE INTERACTIONS
N/A

ALTERNATIVE DRUGS
N/A

FOLLOW-UP

PATIENT MONITORING
Fecal output and consistency should be monitored and will typically return to normal within a week. Attitude should also improve with fluid administration. Repeat cultures can be performed, but shedding of the organism can occur for days to weeks after clinical signs resolve.

PREVENTION/AVOIDANCE
Avoid exposure to *Salmonella* by keeping environments clean and disinfected and purchasing food from reputable sources.

POSSIBLE COMPLICATIONS
People can become sick with salmonellosis.

SALMONELLOSIS

EXPECTED COURSE AND PROGNOSIS
Most patients that have predominantly gastrointestinal signs will recover without sequelae. More severe or systemic infections that lead to osteomyelitis or sepsis have a poorer prognosis.

MISCELLANEOUS

ASSOCIATED CONDITIONS
N/A

AGE-RELATED FACTORS
Young and old animals are more susceptible to disease.

ZOONOTIC POTENTIAL
There are approximately 1.3 million cases of human salmonellosis in the United States annually. Most of these cases are foodborne, but a portion is related to animal exposure.

FERTILITY/BREEDING
Salmonella can infect the testes or ovaries, which could affect breeding potential.

SYNONYMS
Pullorum disease, fowl typhoid, and paratyphoid are distinct diseases in poultry and are associated with particular serotypes of *Salmonella*. Typhoid and paratyphoid refers to infections in humans with *Salmonella* serotype Typhi and *Salmonella* Paratyphi, respectively, but these serotypes do not affect animals.

SEE ALSO
Cloacal disease
Colibacillosis
Diarrhea
Enteritis and gastritis
Regurgitation/vomiting
Urate/fecal discoloration

ABBREVIATIONS
None

INTERNET RESOURCES
Centers for Disease Control and Prevention (CDC): http://www.cdc.gov/salmonella/
CDC Healthy Pets: http://www.cdc.gov/healthypets/
U.S. Department of Agriculture Healthy Birds: http://healthybirds.aphis.usda.gov/
U.S. Geological Survey Field Manual: http://www.nwhc.usgs.gov/publications/field_manual/chapter_9.pdf
Merck Manuals: http://www.merckmanuals.com/vet/poultry/salmonelloses/overview_of_salmonelloses_in_poultry.html

Suggested Reading
Hernandez, S.M., Keel, K., Sanchez, S., I. (2012). Epidemiology of a *Salmonella enterica* subsp. Enterica serovar Typhimurium strain associated with a songbird outbreak. *Applied and Environmental Microbiology*, **78**:7290–7298.
Marietto-Goncalves, G.A., de Almeida, S.M., de Lima, E.T., *et al.* (2010). Isolation of *Salmonella enterica* serovar Enteriditis in blue-fronted Amazon Parrot (*Amazona aestiva*). *Avian Diseases*, **54**:151–155.

Author Marcy J. Souza, DVM, MPH, DABVP (Avian), DACVPM

Salpingitis, Oviductal Disease and Uterine Disorders

BASICS

DEFINITION
The oviduct is the structure that handles the ova following ovulation. Various segments of the oviduct are responsible for specific aspects of egg development. The oviductal segments from cranial to caudal are the infundibulum, magnum, isthmus, uterus and vagina. There is a discrete sphincter between the uterus and vagina. The vagina opens into the urodeum of the cloaca. Most birds only have a left ovary and oviduct. Various disease processes can affect the oviduct and can result in salpingitis and/or impaction of the oviduct, dystocia, and so on. Neoplasia is also occasionally found to affect the oviduct.

PATHOPHYSIOLOGY
Any disorder that creates inflammation in the oviduct can affect normal egg formation leading to the development of abnormal eggs and/or impaction of the oviduct. Nutritional disease that results in hypocalcemia can affect normal muscle function of the oviduct as well as impact egg shell formation. Ascending or systemic bacterial or fungal infections can lead to salpingitis, and some pathogens have tropism for the oviduct. Neoplastic conditions of the oviduct can become secondarily infected or lead to abnormal egg development and impaction as well.

SYSTEMS AFFECTED
Reproductive tract.

GENETICS
There does not appear to be a genetic predisposition of oviductal disease in birds.

INCIDENCE/PREVALENCE
Incidence and prevalence are unknown across all avian species. Anecdotally, prevalence appears to be quite high in captive female birds.

GEOGRAPHIC DISTRIBUTION
N/A

SIGNALMENT
• **Species:** Female birds of any species can be affected by oviductal disease; however, prolific, heavy-laying species seem pre-disposed. Domestic chickens and ducks, cockatiels and various finch species are often diagnosed with oviductal disorders. • **Mean age and range:** Older female birds that have been reproductively productive are predisposed to oviductal disease.
• **Predominant sex:** Female.

SIGNS

General Comments
Clinical signs associated with oviductal disease can be subtle and nonspecific. Occasionally birds with diseased oviducts may strain or have some discharge from the vent. Often though, the symptoms are vague; lethargy, anorexia/weight loss, and fluffed feathers.

Historical Findings
A common history for a bird with oviductal disease is an acute change in egg laying behavior and/or the observation of changes in the appearance of the eggs being passed. A classic presentation is a bird that has been laying heavily for a period of time will begin passing soft-shelled or deformed eggs, followed by complete cessation of laying. Occasionally cases present in more chronic stages where laying ceased many months to years prior to presentation. It is important to understand that birds that have never laid an egg can also be affected by oviductal disease.

Physical Examination Findings
Usually, physical examination findings are nonspecific (poor body condition, dehydration, etc.). Examination findings specific to oviductal disease typically manifest as coelomic distention. Oviductal disease can result in coelomic effusion resulting in grossly apparent coelomic distention. Occasionally the oviduct may be apparent during coelomic palpation. Changes in the oviduct near the cloaca can cause deformation and dysfunction of the vent, resulting in abnormal conformation and/or accumulation of feces and urates around the vent.

If eggs are available for examination, they may have soft shells, defects in the shells, be streaked with blood and/or purulent material or be foul-smelling.

CAUSES
• Infectious organisms that can cause primary salpingitis in birds include bacteria (*Gallibacterium anatis, Mycoplasma sp., Riemerella anatipestifer, E. coli*, etc.), viruses (avian influenza, adenovirus, paramyxovirus, coronavirus, etc.) and fungi (*Aspergillus* sp., etc.). • Uterine carcinoma or adenocarcinomas are the most common neoplasms associated with the avian oviduct.

RISK FACTORS
Chronic egg-laying, excessive body condition/coelomic fat, malnutrition, and egg binding can predispose birds to developing oviductal disorders. Long-term exposure to estrogenic compounds has been associated with oviductal disease in domestic poultry.

DIAGNOSIS

Diagnosis of oviductal disease is usually straightforward and involves a careful history, physical examination and imaging. Further support and/or characterization of the disease is possible via advanced imaging, clinical pathology and laparoscopy.

DIFFERENTIAL DIAGNOSIS
Ovarian disease, egg-yolk peritonitis/coelomitis, dystocia.

CBC/BIOCHEMISTRY/URINALYSIS
• A leukocytosis is sometimes present, being primarily heterophilic or monocytic depending on the chronicity of the disease process. A nonregenerative anemia may also be noted secondary to chronic disease. • The chemistry panel may show hypercalcemia and hyperphosphatemia as well as hyperproteinemia (particularly hyperglobulinemia) if reproductive activity is still occurring. Other chemistry changes associated with injury or dysfunction of other coelomic organs may be observed if the oviductal disease has created coelomitis.

OTHER LABORATORY TESTS
• Cytology of impression smears from oviductal tissue or discharge, if available, can provide useful information. Culture and sensitivity of tissue or discharge can also be useful in diagnosis and direction of therapy.
• If biopsies are collected or salpingectomy performed, histopathology is indicated. Definitive diagnosis as well as identification of healthy margins will help direct further therapy. • If coelomic fluid or oviductal fluid is present, it can be aspirated and cytology and culture and sensitivity performed on it.

IMAGING
• Plain radiography with often show a mass effect in the coelomic region if the oviduct is enlarged. Fluid accumulation within the coelom can also be appreciated with plain radiographs. Further imaging is usually required to definitively identify disease of the oviduct however. • Computed tomography can be extremely effective in identifying the oviduct and characterizing disease of the organ. If there is significant expansion of the coelomic visceral mass or coelomic effusion, an adequate acoustic window may exist for meaningful ultrasonographic examination of the coelom.

DIAGNOSTIC PROCEDURES
• Laparoscopic examination is often utilized to definitively diagnose oviductal disease. Laparoscopy can be challenging if air sac space is limited for manipulation of the scope. Also, great care should be utilized when there is coelomic effusion present as perforation of the coelomic membrane can result in fluid entering the air sacs/lungs and effectively drown the patient. If the oviduct is accessible, it is possible to collect biopsies from the serosal surface. However, perforation of the oviduct can result in introduction of material from the lumen into the coelom so caution is recommended. • Intraluminal examination of the oviduct via endoscopy is possible as well. Accessing the oviduct through the cloaca can be challenging or impossible depending on how disease has affected the anatomy.

PATHOLOGIC FINDINGS
• The pathologic findings affecting the oviduct can range from acute salpingitis of varying severity with or without a suppurative

(CONTINUED) SALPINGITIS, OVIDUCTAL DISEASE AND UTERINE DISORDERS

component to a more chronic granulomatous inflammatory condition. The presence of material within the oviduct is variable, but often there is intraluminal inspissated yolk material or purulent material present.
• Adenocarcinoma is occasionally noted to affect the avian oviduct and can be difficult to impossible to differentiate from an inflammatory condition grossly. Benign neoplastic change of the oviduct is possible, but not commonly encountered.

TREATMENT
APPROPRIATE HEALTH CARE
• Cessation of further reproductive activity is typically indicated in cases of oviductal disease. Treatment with gonadotropin releasing hormone agonist medications such as leuprolide and deslorelin has proven effective in curtailing ovarian activity and subsequent egg laying in avian species. Other hormones have been used in avian patients with variable results. • In cases of infectious salpingitis, nonsteroidal anti-inflammatory medications are typically indicated. Specific antimicrobial agents should be employed as indicated. • Other nonspecific treatments such as fluid therapy and nutritional support should be employed as needed. • Attempts can be made to treat cases of salpingitis conservatively and preserve the reproductive capabilities of the bird. In cases in which there is inspissated material within the lumen of the oviduct, administration of prostaglandin E_2 to relax the ureterovaginal spincter and prostaglandin $F_{2\alpha}$ to stimulate oviductal contraction may be effective in clearing the oviduct of unwanted material. Alternatively, salpingotomy followed by flushing of the oviduct can be attempted. Conservative therapy is not often successful.

Salpingohysterectomy is often necessary for definitive therapy of oviductal disease.

NURSING CARE
Many cases of oviductal disease are chronic and patients are relatively stable. However, the occasional bird will present in the terminal stages of disease or suffering from sepsis, organ failure, and so on and will require intensive nursing care for effective therapy. Fluid and nutritional therapy can be critical in addition to more specific therapies to address the oviductal disease.

ACTIVITY
No activity restrictions accompany the treatment of avian oviductal disease.

DIET
Disease of the avian oviduct can sometimes be secondary to malnutrition. Diets deficient in vitamin A, vitamin D, calcium and other minerals can lead to malformed eggs and potential dystocia. When treating cases of oviductal disease, in addition to providing adequate caloric intake, complete diets should be offered to support normal oviductal function.

CLIENT EDUCATION
Clients working with high risk species should be well-versed regarding nutritional requirements and management tools to help limit excessive egg laying. Recognition of the signs of oviductal disease is also important.

SURGICAL CONSIDERATIONS
Salpingohysterectomy is often necessary to definitively treat oviductal disease in birds. A left flank approach is usually employed to facilitate removal of the oviduct. Often transection of the last left rib is necessary to achieve proper exposure. Salpingohysterectomy is typically a fairly straightforward procedure, even with a diseased oviduct. Ligation of the oviduct's blood supply in the mesosalpinx is made much easier and safer with the use of hemostatic clips. The oviduct should be ligated and removed as close to the cloaca as possible. When disease is limited to the oviduct, salpingohysterectomy can be curative. Surgical cure is possible even in cases of neoplasia if disease-free margins are attainable.

MEDICATIONS
DRUG(S) OF CHOICE
• Nonsteroidal anti-inflammatory medications, antibiotic and antifungal medications and hormone agonists/analogues are frequently indicated in treatment of oviductal disease in birds. Addition of other medications is usually based on the unique needs of individual cases. • Empirically chosen antibiotics should reflect the common occurrence of *E. coli* and other Gram-negative isolates and *Mycoplasma* sp. as causative agents of salpingitis. • The gonadotropin releasing hormone analogs leuprolide acetate and deslorelin are frequently used to discontinue reproductive activity in avian patients. It should be understood that these therapeutics are more effective in preventing reproductive activity than halting reproductive activity once it has begun. The use of leuprolide is well described in the available literature and a number of different protocols are described. A dose of 250–750 ug/kg IM q2–6 weeks is recommended. • If prostaglandin therapy is attempted, prostaglandin E_2 is dosed at 0.02–0.1 mg/kg applied topically to the uterovaginal sphincter. Prostaglandin $F_{2\alpha}$ is dosed at 0.02–0.1 mg/kg IM or intracloacal once.

CONTRAINDICATIONS
N/A

PRECAUTIONS
N/A

POSSIBLE INTERACTIONS
N/A

ALTERNATIVE DRUGS
N/A

FOLLOW-UP
Recommendations depend on the exact diagnosis and chosen therapy.

PATIENT MONITORING
• Physical examination to determine resolution of disease or recurrence pattern.
• Coelomocentesis may be required if effusion is present.

PREVENTION/AVOIDANCE
• Proper diet (including maintenance of normal body condition) and avoidance of management systems that encourage over-production of eggs can help avoid oviductal disease. Allowing birds to incubate eggs or use of dummy eggs can help limit egg production, as can manipulation of light cycles and caloric intake in some species.
• Proper hygiene of substrate and nesting material may help to avoid ascending infections.

POSSIBLE COMPLICATIONS
• If medical treatment is pursued, irreversible scarring of the oviduct may occur rendering it permanently nonfunctional. Therefore, the next time the bird ovulates the ova will not be handled appropriately and disease may recur.
• If salpingohysterectomy is performed and ovarian activity continues, internal ovulation may occur resulting in egg-yolk peritonitis/coelomitis.

EXPECTED COURSE AND PROGNOSIS
Recurrence of disease is common in birds that are treated medically. Birds that have salpingectomies usually have no further problems, but should be monitored for future ovarian function.

MISCELLANEOUS
ASSOCIATED CONDITIONS
Misovulated ova ("internal laying"), egg yolk peritonitis.

AGE-RELATED FACTORS
Older female birds that have been reproductively productive are predisposed to oviductal disease.

ZOONOTIC POTENTIAL
N/A

FERTILITY/BREEDING
Birds suffering from oviductal disease are often rendered infertile. Cases of mild

SALPINGITIS, OVIDUCTAL DISEASE AND UTERINE DISORDERS (CONTINUED)

salpingitis or dystocia that are treated medically have the best chance of retaining their reproductive capabilities.

SYNONYMS
N/A

SEE ALSO
Adenoviruses
Ascites
Aspergillosis
Avian influenza
Chronic egg laying
Cloacal disease
Coelomic distention
Colibacillosis
Ovarian cysts/neoplasia
Dystocia/egg binding
Egg yolk coelomitis/reproductive coelomitis
Hypocalcemia/hypomagnesemia
Infertility
Mycoplasmosis
Nutritional deficiencies
Paramyxoviruses
Tumors

ABBREVIATIONS
N/A

INTERNET RESOURCES
N/A

Suggested Reading

Altman, R.B. (1997). Soft tissue surgical procedures. In: Altman, R.B., Clubb, S.L., Dorrestein, G.M., Quesenberry, K. (eds), *Avian Medicine and Surgery*. Philadelphia, PA: WB Saunders, pp. 704–732.

Joyner, K.L. (1994). Theriogenology. In: Ritchie, B.W., Harrison, G.J., Harrison, L.R. (eds), *Avian Medicine: Principles and Application*. Lake Worth, FL: Wingers Publishing, Inc., pp. 748–788.

Divers, S.J. (2010). Avian endosurgery. *The Veterinary Clinics of North America. Exotic Animal Practice*, **13**(2):203–216.

Author Ryan DeVoe, DVM, Dipl. ACZM, Dipl. ABVP (Reptile & Amphibian)

SEIZURES

BASICS

DEFINITION
A seizure is a sudden disruption of the forebrain's normal neurotransmission that can result in altered consciousness and/or other neurological and behavioral manifestations. When the seizures are recurrent, this is described as epilepsy. Generalized seizures involve both cerebral hemispheres and can have several phases, including tonic (sustained muscle contraction), myoclonic (brief muscle contraction), clonic (rhythmic muscle contraction) or atonic (loss of muscle tone). Focal seizures result in initial activation of only one part of one cerebral hemisphere and can take many forms, including focal motor, focal sensory or focal autonomic seizures.

PATHOPHYSIOLOGY
Seizures are caused by increased excitation or decreased inhibition of the electrical activity of the neurons in the forebrain; although the exact events that create and terminate a seizure are unknown.

SYSTEMS AFFECTED
Nervous.

GENETICS
N/A

INCIDENCE/PREVALENCE
N/A

GEOGRAPHIC DISTRIBUTION
• West Nile Virus: Reported in parrots in California and Louisiana. • Lead toxicity is more prevalent in urban regions that have older housing with lead paint.

SIGNALMENT
• **Species prediliction**: African grey parrots—seizures related to hypocalcemia ages 2–15 years old.

SIGNS
Historical Findings
Since the seizure activity is not usually witnessed by the veterinarian, the owner's historical findings are most important to differentiate a seizure from other events (see differential diagnoses below). Having the owner video the seizure can be very helpful in differentiation. There are four stages the owner may recognize and describe:
1. Prodromal or pre-ictal period: Behavioral change that precedes a seizure by hours to days—can include restless, agitation, decreased interaction/vocalization with owner, fearful behavior, "clinging" to owner.
2. Aura period: Subjective sensation at onset of seizure—usually not noted in animals but can include regurgitation/vomiting and voiding feces/urates.
3. Ictal period or seizure event: Usually lasts 60–90 seconds; signs in birds include loss of consciousness, vocalization, flapping of wings, contraction of limbs, abnormal placement of the head with it thrown backward and falling. Focal seizures can start in one part of the body and progress to generalized seizures or can stay as a brief cluster of signs.
4. Postictal period: Abnormal behavior immediately following the seizure: restlessness, aggression, lethargy, confusion, vision loss, hunger, thirst, disorientation, ataxia.

Once it has been determined that the patient had a seizure, other historical findings can include such as head trauma, exposure to other sick birds, diet changes, housed outdoors in area with endemic West Nile Virus, current medications, exposure to toxins, any recent illnesses or change in the droppings.

Physical Examination Findings
• Unless the episode is observed, most patients with intracranial disease have normal physical examination findings and are normal between seizures. Signs of trauma such as bruising, bleeding, fractures, corneal ulceration can be seen if the patient fell during the seizure. Rarely, other signs of forebrain disease can be seen such as altered behavior, central blindness, decreased consciousness, circling and/or a head turn.
• Animals with extrahepatic causes of seizures can have other signs of disease, such as depression, ascites, biliverdinuria and weight loss in birds with hepatic encephalopathy.
• Animals with systemic infectious disease such as *Chlamydia psittaci* will have signs of respiratory and GI disease and those with systemic proventricular dilatation disease (PDD) can have multiple systems affected.

CAUSES
Causes of seizures are categorized as extracranial and intracranial with intracranial further categorized as structural or functional disorders: • **Extracranial** (toxicities and metabolic disturbances that interfere with CNS function or are directly neurotoxic): ◦ Metabolic: Hepatic encephalopathy, hypoglycemia, hypocalcemia, hypomagnesemia. ◦ Nutritional: Thiamine deficiency (in piscivorous birds eating frozen fish). ◦ Trauma: Head or neck. ◦ Toxic: Lead, zinc, organic mercury, pyrethroids, organophosphates, carbamates, methylxanthines. • **Intracranial:**
◦ Structural disorders (there is gross structural cause within the brain): ▪ Degenerative: Neuronal ceroid lipfuscinosis—reported in one lovebird and one mallard duck.
▪ Anatomic: Hydrocephalus. ▪ Neoplastic: Primary (astrocytoma, primitive neuroectodermal tumor, glial cell tumors, choroid plexus papillomas, lymphosarcoma) or metastatic neoplasia. ▪ Infectious/inflammatory/auto-immune: *Chlamydia psittaci*, West Nile Virus, *Clostridium tertium* and PDD associated with avian bornavirus.
▪ Vascular: Secondary to atherosclerosis.
◦ Functional disorders (no gross structural changes are evident in the brain and the cause is probably neurochemical dysfunction): Idiopathic epilepsy (diagnosis of exclusion when all other causes have been ruled out).

RISK FACTORS
Exposure to lead containing products (i.e., paint, toys).

DIAGNOSIS

DIFFERENTIAL DIAGNOSIS
• Differential diagnoses include tremors, involuntary movements, syncope, metabolic collapse (i.e., hypoglycemia), vestibular disease, intermittent claudication like syndrome, exercise intolerance, peripheral nerve disease (i.e., chronic lead toxicity, sciatic compression from renal tumor), neuromuscular disease. • Historical findings:
◦ If the episode occurs after exercise/activity: Suspect syncope, metabolic collapse, intermittent claudication like syndrome, exercise intolerance, hypoglycemia or neuromuscular disease. ◦ If the animal is unconscious during the episode: Suspect seizure or severe syncope. ◦ If the animal has atypical behavior after the episode ("postictal period"): Suspect seizure although short episodes of confusion can be seen with severe syncope. • Physical examination findings:
◦ Peripheral nerve disease or neuromuscular disease: CP deficits, decreased grip/movement of affected limb(s) and/or muscle atrophy of affected limbs. ◦ Vestibular disease: Nystagmus, ataxia, head tilt. ◦ Syncope: Arrhythmias, murmur, pulse deficits.

CBC/BIOCHEMISTRY/URINALYSIS
CBC
• Leukocytosis if there is an infectious disease causing seizures. • Heterophilia with or without a left shift can be seen with bacterial or fungal infections. • Eosinophilia can be seen with parasitic diseases. • As with any chronic disease, anemia of chronic inflammation can be seen.

BIOCHEMISTRY
• Elevation in creatinine phosphokinase (CPK) and AST is the most common finding as these are released from damaged muscles during rhythmic muscular contraction or muscle trauma if the bird falls.
• Hypoglycemia can be seen as a cause of seizures or a result of glucose utilization during prolonged seizure activity.
• Hypocalcemia and hypomagnesemia have been reported specifically in African grey parrots. • Elevation in bile acids and decrease in albumin can be seen with hepatic encephalopathy.

SEIZURES (CONTINUED)

OTHER LABORATORY TESTS
- Blood lead and zinc levels. • West Nile virus (WNV): Reverse transcriptase PCR. • Avian bornavirus: Real time PCR of urine, Western blot assay. • *Chlamydia psittaci*: Paired antibody titers, culture and/or PCR.

IMAGING
- Metallic densities can be seen in the gastrointestinal tract if lead toxicity has caused the seizures. Absence of metal does not preclude lead toxicity, as only 25% of birds with lead toxicity have metal noted radiographically. • Whole body radiographs including the skull should be performed to rule out any trauma that could have caused the seizures or as a result of falling during the seizure. • Echocardiography should be performed in any bird with evidence of heart disease (ie arrhythmia, murmur, pulse deficits) to rule out syncope as a cause of collapse vs. seizures.

DIAGNOSTIC PROCEDURES
- Advanced imaging such as MRI and CT has been used in birds to look for intracranial disease. • Electroencephalography (EEG) can be performed to confirm seizure activity but is usually nonspecific. May only be effective in larger birds so that the electrodes can be properly placed. • Cerebrospinal (CSF) fluid analysis can be obtained to look for elevations in leukocyte and protein levels although there are no established reference ranges for avian species.

PATHOLOGIC FINDINGS
- Pathologic findings depend on the underlying cause of the seizures. If there is an intracranial functional cause for the seizures (i.e., idiopathic epilepsy) then no abnormalities will be noted pathologically. Intracranial structural causes could include gross lesions such as hydrocephalus or only those noted on histopathology such as PDD. Likewise, extracranial causes such as head trauma may be grossly noted. • Samples may also be collected for specific etiology such as virus isolation or immunohistochemistry staining for West Nile Virus, PCR for avian bornavirus and/or PCR or culture for *Chlamydia psittaci*.

TREATMENT

APPROPRIATE HEALTH CARE
- **Emergency management:** ○ Systemic stabilization of airway, breathing and circulation: Ensure patent airway and administer 100% oxygen therapy, obtain intravenous or intraosseous access. ○ Stop the seizure activity: ■ Benzodiazepines: Diazepam IV/IO/cloacal (0.5–2.0 mg/kg) or midazolam administered IV/IO/IM/IN (0.1–2 mg/kg)—can repeat 2–3 times; can use a CRI of diazepam (0.1–0.5 mg/kg/hr IV or IO); if 1–2 doses fail to control seizures consider adding longer-acting anticonvulsant such as phenobarbital or levetiracetam (if phenobarbital is inappropriate).
  ■ Phenobarbital (2 mg/kg IV)—can repeat up to two times (contraindicated with liver disease). ■ Levetiracetam (20–60 mg/kg IV).
  ■ If seizures have still not stopped, consider propofol boluses (1–2 mg/kg IV/IO), ketamine boluses (5 mg/kg IV/IO) or inhalant isoflurane. ○ Correct the underlying condition: ■ If hypocalcemic: 10% calcium gluconate IV/IO (0.5–1.5 ml/kg) over 10 minutes while monitoring heart rate and rhythm. ■ If hypoglycemic: 25% dextrose IV or IO to effect over 15 minutes or oral glucose solution WITH CAUTION to prevent aspiration. ■ If head trauma and suspect increased intracranial pressure: Hypertonic saline (7.5% NaCl:4 ml/kg, 3% NaCl:5.4 ml/kg over 15–20 min IV or IO) or mannitol (0.5–1.0 g/kg over 15–20 min IV/IO).
- **Outpatient medical management:** ○ See "Medications" section.

NURSING CARE
- A well-padded cage with no perches should be provided to prevent further trauma if another seizure occurs. • Eye lubrication should be applied if the patient is obtunded and not blinking properly to prevent ocular ulceration. • The patient's body temperature and cage temperature should be monitored as patients with prolonged seizure activity can be hyperthermic and those that are obtunded may not be able to maintain body temperature and will require supplemental heat therapy. • Fluid therapy (SC, IC or IO) is required if the patient is not able to eat and drink properly. If the patient is hypoglycemic, 2.5% dextrose should be added to the IV or IO fluids until the patient is able to eat on its own. • Oxygen therapy should be administered as needed.

ACTIVITY
Patients with severe neurologic dysfunction should not be allowed to fly.

DIET
- Food and water should be removed from the patients' cage if they are depressed or obtunded to prevent aspiration and/or drowning in the water dish.
- Supplementation with oral calcium and/or magnesium as needed if hypocalcemic or hypomagnesemic. • Support feeding as needed for patients with hypoglycemia.
- Specialized low protein diets may be formulated for birds with hepatic encephalopathy. • Dietary thiamine supplementation can be added to piscivore diets lacking vitamin B1.

CLIENT EDUCATION
- Clients need to understand the goals of seizure treatment, which are to reduce the frequency and severity of seizures, to minimize the potential side effects and to maximize the patients' quality of life. It is important for owners to know that the bird may still have seizures even when on the medications. • Compliance is very important and clients need to be prepared that their bird may require lifelong medications 1–3 times daily and that doses cannot ever be skipped.
- Clients should be made aware that anticonvulsant drug therapy in avian patients is not well studied and, therefore, frequent serum drug monitoring will be required.

SURGICAL CONSIDERATIONS
Ketamine and inhalant anesthetics should be used with caution if any increase in intracranial pressure is suspected.

MEDICATIONS

DRUG(S) OF CHOICE
- Underlying diseases causing seizures should be treated appropriately. For example, lead toxicity should be treated with chelating agents and hypocalcemia with oral calcium supplementation. • There is one published paper about the pharmacokinetics of levetiracetam in Hispaniolan Amazon parrots, but there are no published serum therapeutic levels of anticonvulsants in birds. Therefore guidelines are based on those for canine patients: ○ Phenobarbital: Canine dose is 2–3 mg/kg PO q12h with suggested serum therapeutic range of 20–30 mg/dl; 2 mg/kg PO q12h in an African grey resulted in nondetectable serum phenobarbital levels. ○ Potassium bromide: Canine dose is 20–40 mg/kg PO q24h with suggested serum therapeutic range of 1–3 mg/ml; 80 mg/kg PO q24h in a cockatoo kept serum levels between 1.7–2.2 mg/ml but in an African grey 50 mg/kg PO q12h and 100 mg/kg PO q12h resulted in serum levels of 0.6 and 0.7 mg/ml, respectively, and was ineffective at controlling seizures. ○ Levetiracetam: Canine dose is 10–20 mg/kg PO q 8 hr with suggested serum therapeutic range of 5.5–21 mcg/ml; 50 mg/kg and 100 mg/kg PO q 8h reached this range in an African grey; 50 mg/kg PO q 8 hr and 100 mg/kg PO q 12 hrs has been suggested based on pharmacokinetic studies in Hispaniolan Amazons. ○ Zonisamide: Canine dose is 2.5–10 mg/kg PO q12h with suggested serum therapeutic range of 10–40 mcg/ml; 20 mg/kg PO q8h reached this range in an African grey. ○ Gabapentin: Canine dose 10–20 mg/kg PO q8h with suggested serum therapeutic range of 4–16 mg/l; 20 mg/kg PO q12h was used in an African grey.

CONTRAINDICATIONS
- Behavioral medications such as selective serotonin reuptake inhibitors (SSRIs), tricyclic antidepressants (TCAs) and antipsychotics can lower seizure threshold and should not be used with anticonvulsants. Specifically in birds, extrapyramidal side

(CONTINUED) SEIZURES

effects have been reported in a macaw with clomipramine and haloperidol therapy.
• Steroids are contraindicated in cases of traumatic brain injury as methylprednisolone use is associated with an increase in mortality two weeks and six months postinjury in humans.

PRECAUTIONS
• Possible side effects of anticonvulsant medications include sedation and sometimes ataxia so patients should be closely monitored when starting these medications.
• Phenobarbital should not be used in patients with liver disease.

POSSIBLE INTERACTIONS
There are several drug interactions with phenobarbital; therefore no other medication should be administered without checking first with a veterinarian.

ALTERNATIVE DRUGS
N/A

FOLLOW-UP
PATIENT MONITORING
• Because there are few published doses of anticonvulsant therapies in avian patients, serum drug levels should be checked. If the serum levels are too low, samples should be collected for serum peak and trough measurements so that serum half-life can be calculated. • Depending on the half-life of the medication this should be initiated 1–3 weeks after initiating therapy until adequate seizure control is reached.

PREVENTION/AVOIDANCE
• Avoid contact with heavy metals or other toxic substances. • Controversial use of off label WNV vaccine in birds.

POSSIBLE COMPLICATIONS
Any patient with seizures may progress to status epilepticus, which is an emergency.

EXPECTED COURSE AND PROGNOSIS
Prognosis depends on the underlying cause. If the underlying cause can be diagnosed and cured, permanent CNS damage may have already occurred and may be irreversible or may improve with time. Progressive diseases such as multifocal PDD are almost always fatal.

MISCELLANEOUS
ASSOCIATED CONDITIONS
N/A

AGE-RELATED FACTORS
Neoplasia and ischemic stroke are seen more frequently in older animals, whereas degenerative and anatomical causes are seen more frequently in younger animals.

ZOONOTIC POTENTIAL
If seizures are associated with *Chlamydia psittaci*.

FERTILITY/BREEDING
N/A

SYNONYMS
N/A

SEE ALSO
Heavy metal toxicity
Hypocalcemia/hypomagnesemia
Neoplasia
Neurologic conditions
Nutritional deficiencies
Otitis
Pituitary/brain tumors
Proventricular dilatation disease (PDD)
Sick-bird syndrome
Trauma
West Nile virus

ABBREVIATIONS
N/A

INTERNET RESOURCES
N/A

Suggested Reading
Beaufrère, H., Nevarez, J., Gaschen, L., et al. (2011). Diagnosis of presumed acute ischemic stroke and associated seizure management in a Congo African grey parrot. *Journal of the American Veterinary Medical Association*, **239**:122–128.
Delk, K. (2012). Clinical management of seizures in avian patients. *Journal of Exotic Pet Medicine*, **21**:132–139.
DiFazio, J., Fletcher, D.J. (2013). Updates in the small animal patient with neurologic trauma. *The Veterinary Clinics of North America. Small Animal Practice*, **43**:915–940.
Platt, S. (2012). Seizures. In: Platt, S., Garosi, L. (eds), *Small Animal Neurologic Emgergencies*. London, UK: Manson Publishing Ltd, pp. 155–172.
Schnellbacher, R., et al. (2014). Pharmacokinetics of levetiracetam in healthy hispaniolan Amazon parrots (Amazona ventralis) after oral administration of a single dose. *Journal of Avian Medicine and Surgery*, **28**(3):193–200.
Author Nicole R. Wyre, DVM, DABVP (Avian)

Sick Bird Syndrome

BASICS

DEFINITION
Birds are regularly presented to the veterinarian with an apparent (owner-perceived) acute onset of illness that is represented by an array of highly nonspecific clinical signs. This phenomenon is termed "sick bird syndrome", but also could be called "acute nonspecific illness". There are two main reasons for this. First, birds tend to hide (mask) their illnesses until they are too ill to continue doing so. Secondly, owners miss many clinical signs, either due to lack of knowledge or inadequate observational skills. The result, quite often, is that a critically ill avian patient with vague signs is brought to the veterinarian for examination and treatment.

PATHOPHYSIOLOGY
There are numerous potential underlying disease processes and organ system failures that could cause a client to finally notice that their bird is sick, but the reason that the patient is being presented is because it is too weak to continue hiding more obvious signs of illness. Underlying pathophysiological mechanisms that could lead to this include septicemia, toxemia, cardiovascular collapse, anemia, dehydration, hypoxemia, electrolyte abnormalities, acid–base imbalances and caloric deficits/hypoglycemia.

SYSTEMS AFFECTED
Any of the organ systems could be affected when presented in this state. Disease of any one organ system could be the primary problem, but as the disease process progresses, multiple organ systems could become involved.

GENETICS
N/A

INCIDENCE/PREVALENCE
This syndrome is very common for reasons already outlined.

GEOGRAPHIC DISTRIBUTION
N/A

SIGNALMENT
- **Species:** All species of birds can become ill and show nonspecific clinical signs. Smaller species might present later in the course of illness since owners do not necessarily observe their clinical signs as readily or handle them as much as the larger species. • **Mean age and range:** All ages are susceptible.
- **Predominant sex:** Both sexes are susceptible.

SIGNS

General Comments
When presented with an acutely ill bird with nonspecific clinical signs, it is very important to take a thorough history, as this and the physical examination will be instrumental in determining a course of action. This should be done before instituting an exhaustive diagnostic testing plan. The patient can be placed in a warmed incubator in oxygen while the history is being taken.

Historical Findings
History is not necessarily specific. Often the owner's impression is that the bird was healthy the day before. With careful history-taking, however, once the owner is questioned about specific details, it might be determined that there were clinical signs earlier that had not been deemed relevant.

Physical Examination Findings
These findings and clinical signs are generally not specific and could be seen in association with a wide variety of ailments. They simply mean that the bird is sick:
- Decrease in appetite or thirst. • In some cases, polydipsia. • Depression/droopy/lethargy/listless. • Not moving—sitting on perch or at bottom of cage. • Eyes closed or partially closed. • Minimal response to stimuli. • Fluffed—feathers fluffed so the bird looks bigger. • "Diarrhea"—wet droppings; usually perceived by the owner as diarrhea but more often than not it is polyuria. • Not talking or vocalizing.

CAUSES
Many avian illnesses could result in nonspecific clinical signs and eventual presentation in an advanced state. An incomplete list includes:
- Bacterial infections—Gram-negatives and a few Gram-positives. • Chlamydiosis. • Mycobacteriosis. • Viral infections—herpesvirus, polyomavirus, circovirus, bornavirus, other. • Fungal infections—candidiasis, aspergillosis, macrorhabdosis. • Parasitic infections—flagellates, cryptosporidiosis, helminthiasis. • Toxicologic conditions—lead, zinc, inhaled, other. • Neoplastic conditions. • Metabolic disorders. • Malnutrition and related/secondary diseases. • Other noninfectious disorders—reproductive diseases, atherosclerosis.

RISK FACTORS
Smaller patients are often more difficult to monitor and therefore they are more likely to be presented in an advanced state of illness.

DIAGNOSIS

DIFFERENTIAL DIAGNOSIS
These are listed under "Causes".

CBC/BIOCHEMISTRY/URINALYSIS
Note: If it is considered to be too risky to take blood from a critically ill bird, then this should be delayed. The patient needs to be stabilized first. • Blood work, if done on an acutely ill bird, should be performed as soon as possible (in-house as opposed to sending to a laboratory). It pays to have an in-house blood chemistry analyzer and capabilities to perform a quick CBC. • Findings will vary depending on the pathologic process causing the clinical signs.

OTHER LABORATORY TESTS
- This will vary also, depending upon the specific historical and physical examination findings • The veterinarian should not fall into the trap of "more is better". Simply not knowing what is wrong with a patient does not necessitate the need for doing as many tests as one can think of. Logic and common sense should prevail. One finding might lead to the need for one more specific test. For example, a radiograph showing metallic densities in the ventriculus combined with potential signs of lead or zinc toxicosis, would prompt the clinician to screen heavy metal levels in the blood. • In critically ill patients, it is reasonable to perform the least invasive tests first, such as direct fecal examination, fecal Gram's staining, and sampling and analysis of any identifiable lesions.

IMAGING
- Radiographs might be part of the diagnostic plan, depending on the patient, the client's finances and the risk of performing the procedure and the likelihood of getting good positioning. If radiographs can be taken quickly and safely, there is often valuable information that can be gained. • There is considerable debate in the veterinary community as to how to best perform radiographs in a critically ill bird. Some advocate for the use of, and routinely administer gas anesthesia; others will manually restrain without it; a third option would be to consider butorphanol (IM) and/or midazolam (IM or intranasal) for these instances. • A quick DV standing radiograph can be done if looking for metallic foreign bodies (heavy metal poisoning). This minimizes the risk in the critically ill patient, but would otherwise be considered a nondiagnostic radiograph.

DIAGNOSTIC PROCEDURES
Depending on historical or physical examination specific findings, certain diagnostic procedures could be performed in order to arrive at a diagnosis. Obtaining of samples through swabbing, needle aspirates or biopsy techniques could yield valuable information in cases where accessible lesions are identified.

PATHOLOGIC FINDINGS
There could be a wide variety of lesions associated with nonspecific illness clinical signs, depending upon the underlying condition and the organ system(s) involved. These are too numerous and varied to list here.

Sick Bird Syndrome (Continued)

TREATMENT
APPROPRIATE HEALTH CARE
Essentially all birds presented in critical condition should be hospitalized and treated aggressively. Diagnostic and therapeutic plans need to be determined at the outset, and it is important to understand that sometimes it is best to delay any invasive or hands-on diagnostic tests until the patient is more stable. This cannot be overemphasized.

NURSING CARE
In essentially all hospitalized cases there is a need for stabilization, which will include:
• Thermal support (incubate at 80–90°F).
• Oxygen therapy. • Fluid therapy—fluids can be given by IO, IV or SQ routes; occasionally by gavage. • Nutritional support—depending on the hydration status and other factors, gavage feeding can be done. It is best to warm the patient and ensure hydration before gavage feeding.

ACTIVITY
Sick birds should be limited in their activity—birds in the hospital are usually incubated; sick birds in the home environment should not be allowed to fly around the house; often a "hospital" cage needs to be set up in the owner's home for a few days after hospitalization.

DIET
Should be tailored to each individual situation.

CLIENT EDUCATION
• In all cases, it is important to outline the course of action with the client, including the diagnostic and treatment plan and the prognosis. • It should be stressed with all clients, during annual or postpurchase examinations that they need to look for changes in their bird's behavior, demeanor, appetite, droppings, breathing or anything else at all times. The key word is changes. If it is different from the usual state of the bird, then it is worth investigating as opposed to taking a "wait and see" approach.

SURGICAL CONSIDERATIONS
N/A

MEDICATIONS
DRUG(S) OF CHOICE
The knee-jerk tendency in the avian veterinary hospital is to "shotgun" treat every sick bird that is presented with nonspecific signs. While this may work in a percentage of cases, it does not justify this type of approach. The goal is to have an evidence-based rationale when confronted with any decision that needs to be made concerning treatment. The patient needs to be stabilized above and beyond all other considerations. If diagnostic testing results will be available shortly, it is best to wait for the results of those tests in order to initiate specific treatment.
• If there is a legitimate argument for bacterial disease, then empirical antibiotic therapy may be initiated. • If there is a legitimate argument for fungal disease, then antifungal medication may be initiated. • If there is a legitimate argument for chelation therapy, then calcium EDTA therapy may be initiated. • The previous arguments follow for all types of medication.

CONTRAINDICATIONS
N/A

PRECAUTIONS
N/A

POSSIBLE INTERACTIONS
N/A

ALTERNATIVE DRUGS
N/A

FOLLOW-UP
PATIENT MONITORING
Trained veterinary staff should closely monitor hospitalized patients. Once released from the hospital, each patient should be monitored as needed, depending on the severity of the condition and the specific disorder. This should include examinations as well as specific laboratory testing as indicated.

PREVENTION/AVOIDANCE
Education of every client is essential in order to prevent birds from being presented in an advanced state of illness.

POSSIBLE COMPLICATIONS
N/A

EXPECTED COURSE AND PROGNOSIS
Prognosis is extremely variable, depending upon the specific diagnosis.

MISCELLANEOUS
ASSOCIATED CONDITIONS
N/A

AGE-RELATED FACTORS
N/A

ZOONOTIC POTENTIAL
N/A

FERTILITY/BREEDING
N/A

SYNONYMS
N/A

SEE ALSO
Airborne toxicosis
Anemia
Anorexia
Dehydration
Dystocia/egg binding
Nutritional deficiencies
Regurgitation/vomiting
Respiratory distress
Trauma
Appendix 7, Algorithm 1. Sick bird syndrome

ABBREVIATIONS
N/A

INTERNET RESOURCES
N/A

Suggested Reading
Doneley, B., Harrison, G., Lightfoot, T. (2006). Maximizing information from the physical examination. In: Harrison, G.J., Lightfoot, T.L. (eds), *Clinical Avian Medicine*. Palm Beach, FL: Spix Publishing, Inc., pp. 153–211.
Author George A. Messenger, DVM, DABVP (Avian)

Client Education Handout available online

SPLAY LEG AND SLIPPED TENDON

BASICS

DEFINITION
• Avian slipped tendon is a condition where the gastrocnemius tendon displaces laterally or medially from the caudal/plantar aspect of the intertarsal joint. • Avian splay leg is a condition of leg malposition resulting from an orthopedic change/rotation in the femur or tibiotarsus.

PATHOPHYSIOLOGY
The root mechanisms of orthopedic conditions of avian chicks are not fully understood and are extrapolated to early research in poultry. Several dietary factors may play a role, but manganese deficiency either through inadequate diet or excess calcium supplementation is a possible primary predisposing factor. The deviation of a slipped gastrocnemius tendon may occur secondary to valgus or varus deformities of the legs. Once deviated the contracture of the tendon perpetuates the condition accelerating the limb deformity and inflammation of the tarsal joint. Splay leg may be the result of any deviation along the limb leading to a negative cascade affecting the other bones and joints of the limb. Other factors, such as deficiencies in calcium and vitamin D_3, may lead to weaker bones and growth plates that are affected by outside forces, such as slipping on smooth surfaces, trauma from handling, and so on.

SYSTEMS AFFECTED
Slipped tendon
• Musculoskeletal—affecting the tibiotarsus, tarsal joint, tarsometatarsus, and/or the gastrocnemius tendon.

Splay leg
• Musculoskeletal—affecting the coxofemoral joint, femur, stifle joint, tibiotarsus, tarsal joint, and/or tarsometatarsus.

GENETICS
None definitively known.

INCIDENCE/PREVALENCE
Rare in adults, uncommon in chicks with proper diet, common in chicks with improper diet or growth/weight monitoring.

GEOGRAPHIC DISTRIBUTION
None.

SIGNALMENT
• **Species:** All avian species. • **Species predilections:** Seen in waterfowl and parrot species. • **Mean age and range**: Young chicks at or around fledging. • **Predominant sex:** None.

SIGNS
Historical Findings
Signs reported by the owner.

Physical Examination Findings
• Acute lameness and inability to stand and/or walk. Most commonly diagnosed in young chicks and those species with relatively long legs and those with a combination of larger body and shorter legs. • Upon palpation, the "slipped" tendon may be visibly deviated laterally or more commonly medially. In the acute phase, then tendon may be manually positioned caudally upon extension of the tarsal joint but deviates when the joint is flexed. Edema is present in the acute phase leading to progressive joint inflammation, tendonitis, and bone bruising. For splay leg, the legs will be malpositioned when the patient is placed in a flat surface.

CAUSES
Slipped tendon
• Trauma. • Manganese deficiency. • Vitamin B6 deficiency. • Vitamin B12 deficiency. • Biotin deficiency. • Choline deficiency. • Methionine deficiency. • Valgus/Varus malposition.

Splay leg
• Trauma. • Vitamin E deficiency. • Selenium deficiency. • Valgus/Varus malposition.

RISK FACTORS
Typically seen in chicks prior and around fledging due to rapid physical development and growth. Species with longer lengths of legs in relation to body size may also be at increased risk. Heavy bodied species with relatively shorter legs (i.e. waterfowl) are also at an increased risk.

DIAGNOSIS

DIFFERENTIAL DIAGNOSIS
Angular limb deformity, joint infection, trauma, degenerative joint disease, stifle abnormalities.

CBC/BIOCHEMISTRY/URINALYSIS
Often will be within normal limits unless secondary inflammation or infection has developed to a point where the white blood cell (WBC) count begins to increase.

OTHER LABORATORY TESTS
None.

IMAGING
Radiographs should be performed to rule out any underlying pathology to the joint or bones.

DIAGNOSTIC PROCEDURES
None.

PATHOLOGIC FINDINGS
Inflammation of the tissues around the tarsal joint and calcaneal tendon.

TREATMENT

APPROPRIATE HEALTH CARE
Initially, the patient needs to be assessed and the affected limb(s) stabilized. Once diagnostics are performed, the patient may be managed as an outpatient in mild cases. More severe cases should be managed inpatient initially until the limb(s) appear to be properly aligned.

NURSING CARE
• Mild-to-early cases of slipped tendon may be managed medically with bandaging and/or splinting of the affected leg. • Hobbles may be used to correct splay leg if caught early enough for 2–4 days. In some cases, housing the chick in a deep nesting cup may facilitate proper positioning of the limbs for corrective growth.

ACTIVITY
• Slick surfaces in the brooder/nest box should be avoided with the proper use of substrates, such as shaving, carpet, and so on. • Waterfowl should be encouraged to exercise through swimming.

DIET
A well-balanced diet should be provided, avoiding excessive protein/calorie intake. Since excessive caloric intake is a risk factor, monitoring daily weight gain is vital in prevention and therapeutics. The daily weight gain should be appropriate for the species but on average is around 10% daily. Leg and tendon issues may be exacerbated with excessive calcium supplementation leading to decreased manganese absorption.

CLIENT EDUCATION
Due to the varying levels of prognosis, client education about expectations of outcome is very important.

SURGICAL CONSIDERATIONS
• Several techniques have been described for re-positioning the gastrocnemius tendon and holding it in place for proper limb function. One technique involves a skin incision alongside the tendon opposite of the luxation. The soft tissues are dissected and the tendon sheath is sutured to the periosteum along the trochlear ridge. In addition, deepening the trochlear groove has been used in conjunction with this technique or as a single technique with variable success. • Another technique describes the placement of staples into each distal tibiotarsal condyle with sterile cable meshed between and over the tendon to hold it in place. A less invasive approach with variable success would be the external use of sutures to hold the tendon over the caudal position. Unfortunately, the nature of forces exerted on this area often lead to tearing of tissues by the suture material.

SPLAY LEG AND SLIPPED TENDON (CONTINUED)

MEDICATIONS

DRUG(S) OF CHOICE
NSAIDs may be used in management of the pain and inflammation associated with the stresses on the joints/bones.

CONTRAINDICATIONS
None.

PRECAUTIONS
None.

POSSIBLE INTERACTIONS
None.

ALTERNATIVE DRUGS
None.

FOLLOW-UP

PATIENT MONITORING
Patients should be monitored daily for progress of therapy. Weekly radiographs may allow visual monitoring of return to normal positioning of bones or the development of additional lesions.

PREVENTION/AVOIDANCE
• Most of the nutritional deficiencies can be prevented by placing chicks on a balanced and appropriate diet for the species. • Daily monitoring of weight (same time of day before feeding) is extremely important to adjust the caloric input to regulate a slow steady weight gain,

POSSIBLE COMPLICATIONS
• Use of bandages and splints may lead to abnormal angles of pressures on the joints or constrictive blood flow which may lead to edema and/or discomfort. • Surgical corrections may breakdown due to the high forces and mobility of the tarsal joint.

EXPECTED COURSE AND PROGNOSIS
Depending on the severity presented during the initial examination. Mild cases may be managed medically with a fair prognosis but more advanced cases that require surgical correction have a guarded to poor prognosis.

MISCELLANEOUS

ASSOCIATED CONDITIONS
• Valgus/Varus deformity. • Angel wing.
• Curled toes.

AGE-RELATED FACTORS
• Most commonly diagnosed in young chicks.
• Seen most commonly in species with relatively long legs and those with a combination of larger body and shorter legs.

ZOONOTIC POTENTIAL
No.

FERTILITY/BREEDING
Suspected in poultry to have some genetic relationships but not definitively proven.

SYNONYMS
Slipped tendon—Perosis.
Splay leg—Spraddle leg, bow leg, valgus deformity.

SEE ALSO
Valgus/Varus deformity
Angel wing
Curled toes

ABBREVIATIONS
None.

INTERNET RESOURCES
None

Suggested Reading
Samour, J. (2008). Management-related Diseases. In: Samour, J. (ed.), *Avian Medicine*, 2nd edn. Mosby/Elsevier Inc., pp. 260–261.
Tully, T.N., Dorrenstein, G.M., Jones, A. (eds) (2009). *Handbook of Avian Medicine*, 2nd edn. WB Saunders/Elsevier Inc.
Author Rob L. Coke, DVM, DACZM, DABVP (Reptile & Amphibian), CVA

Squamous Cell Carcinoma

BASICS

DEFINITION
Squamous cell carcinoma (SCC) is a tumor of squamous epithelial cells that is commonly diagnosed in psittacine birds. Cutaneous and upper gastrointestinal SCC are most common. This type of tumor may be located in any organ that has a squamous epithelial surface.

PATHOPHYSIOLOGY
• In humans and other animals, predisposing factors may include UV damage from sun exposure or chronic inflammation or infection. • In birds, hypovitaminosis A due to a seed-based diet may be a predisposing factor as the resultant squamous metaplasia may progress to squamous cell carcinoma, particularly in the oral cavity. • Other suspected predisposing factors include chronic infections or chronic feather-damaging behavior (FDB), but additional research is needed to determine their role in disease.

SYSTEMS AFFECTED
• Skin and cutaneous glands (uropygial). • Gastrointestinal tract (particularly upper—oropharynx/choana, esophagus, proventriculus). • Respiratory tract.

GENETICS
No significant genetic predisposition has been identified.

INCIDENCE/PREVALENCE
• Represents 10% of psittacine submissions in one pathology service retrospective study (Garner, 2006). Relatively common in companion birds. • Older surveys estimate incidence at 1.8% (Reece, 1992) and 1.1% (Leach, 1992).

GEOGRAPHIC DISTRIBUTION
N/A

SIGNALMENT
Predisposed species include budgerigars, cockatiels and Amazon sp. Additional species may include Conure sp., African grey and lovebird, although tumors in many different species have been reported. There does not appear to be a sex predilection.

SIGNS
• Variable depending on lesion location. • Oral cavity: Voice change, dyspnea or stridor, foul odor from mouth from caseous debris or concurrent bacterial infections, bleeding or a mass, anorexia. • Upper esophageal: Anorexia, regurgitation, foul odor or caseous debris. • Beak: Mass lesion evident, thickened or altered keratin with erosions, secondary infections. • Dermal: Mass is normally evident, may also present as a nonhealing wound, crusting or ulceration. Birds may feather-pick over the area or lesions may arise in areas of chronic feather destructive behavior or chronic infections. • Uropygial gland: May appear as an abscess or nonhealing wound with ulceration and/or crusting.

CAUSES
• Underlying causes or genetic alterations are unknown. • May be related to concurrent infections or chronic inflammation (see "Risk Factors"). • Poor flooring or caging may cause lesions in certain species (flamingos).

RISK FACTORS
• SCC in birds does not appear to be related to UV exposure based on lesion locations. • Lesions may be more likely in areas with chronic inflammation or infection. • Birds with oral squamous metaplasia secondary to hypovitaminosis A may be predisposed to developing oral SCC lesions. • Inappropriate husbandry leading to foot lesions or other chronic skin lesions/irritation.

DIAGNOSIS

• Based on histopathology from mass excision or biopsy. • Extent of lesions may be determined by imaging (radiograph, CT, MRI), particularly for those lesions involving the oral cavity, beak, upper esophagus or uropygial gland. • Contrast imaging may be useful for determining the extent/depth of lesions.

DIFFERENTIAL DIAGNOSIS
• Dermal: Chronic bacterial or fungal infections/abscesses, nonhealing wound or chronic inflammation. Self-mutilation secondary to feather-damaging behavior (FDB). • Oral cavity: Squamous metaplasia with secondary infection, bacterial or fungal abscesses, chronic wound. • Beak: Fungal or bacterial infections. • Uropygial gland: Impaction or abscessation.

CBC/BIOCHEMISTRY/URINALYSIS
Depending on location and extent of lesions, may see evidence of inflammation or chronic infection (leukocytosis with heterophilia or monocytosis).

OTHER LABORATORY TESTS
Frequently, there may be concurrent infections at the neoplastic site. Bacterial or fungal cultures may be indicated to dictate appropriate antibiotic or antifungal therapy.

IMAGING
• Radiographs +/– CT are indicated to delineate the degree of bony involvement of lesions. If lesions have significant bony involvement, this will alter the treatment strategy. • If CT images are obtained, contrast can be included to help identify the extent of neoplastic involvement.

DIAGNOSTIC PROCEDURES
• Biopsy is the gold standard for diagnosis. Where feasible, complete excision with a margin should be pursued. For lesions that cannot be completely removed, debulking is indicated followed by adjuvant therapy. • Histological margins should be inspected to determine the possibility of residual disease.

PATHOLOGIC FINDINGS
• Histologically, lesions are comprised of infiltrating nests and cords of moderately differentiated to poorly differentiated squamous cells. "Keratin pearls" may be observed, which are composed of central cores of compressed, laminated keratin. • Lesions that do not extend beyond the basement membrane of the epithelium are diagnosed as *in situ* lesions. • There is not sufficient evidence to indicate that more poorly differentiated lesions have a worsened prognosis or shorter survival times. • Tumors are generally locally invasive and only rarely are reported to metastasize.

TREATMENT

Consultation with an oncology specialist is strongly recommended when deciding on appropriate therapy for a patient and may be required for more advanced therapies (i.e., radiation).

APPROPRIATE HEALTH CARE
• Treatment of concurrent diseases (ex: bacterial or fungal infections) should be provided based on diagnostics (biochemical and hematologic values, culture where appropriate). • Birds that are generally malnourished or dehydrated should have appropriate supportive care prior to surgery or other neoplastic treatments.

NURSING CARE
• If the patient is underweight or not eating due to pain or inflammation, supportive feeding and fluid support may be indicated. • Altered caging or perching may be required for patients postsurgery, particularly after amputations, as their balance may be altered. This is particularly true for large or heavy birds. • Owners should be instructed as to appropriate precautions for patient's postchemotherapy administration: wear chemotherapy gloves to change caging for 1–3 days, launder any towels separately and keep feces, urine and urates away from children or immunocompromised adults.

ACTIVITY
No activity restrictions are required unless the animal is post-surgical and may require a collar to prevent picking at bandages or surgical sites.

DIET
No special diets are indicated. Patients may have mild-to-moderate anorexia following treatment with chemotherapeutics and supplemental nutrition may be indicated to prevent weight loss.

Squamous Cell Carcinoma (Continued)

CLIENT EDUCATION
• Follow appropriate precautions following chemotherapy (when given). • Monitor patients closely for changes in attitude, appetite and droppings during therapy. • Watch conspecifics closely if they attempt to bite lesions or bandages. • It may be recommended for owners to weigh their birds at home to more closely monitor their progress and response to therapies, particularly if anorectic.

SURGICAL CONSIDERATIONS
• Complete excision is the treatment of choice as adjunctive therapies for residual disease are rarely successful. • Imaging and histopathologic examination of the margins from previous excisions should be used to guide therapy. • Amputation is indicated if imaging indicates the lesion may affect underlying bony structures or to obtain complete excision at distal locations.

MEDICATIONS

DRUG(S) OF CHOICE

Analgesia
• Analgesic support is very important postsurgery as well as for lesions that appear to be causing pain. Options include meloxicam (1–2 mg/kg SC or PO q12h, although the safety of long-term dosing at this dose range has not been established), tramadol (5–30 mg/kg PO q8–12h), butorphanol (1–5 mg/kg IM or SC q12h). Sedation may be seen at higher doses and dose intervals of tramadol.

Chemotherapy
• Cisplatin has been used systemically and intralesionally, with varying degrees of success. Carboplatin has been dosed systemically. The dosing for these agents is extrapolated from doses from dogs and cats and from a small number of pharmacokinetic studies in sulfur-crested cockatoos (Filippich and Charles, 2004). Reported doses in the literature for systemic carboplatin range from 5–27 mg/kg IV. Typical dosing interval is 3–4 weeks; however, this may need to be adjusted based on hematologic monitoring if leukopenia is severe. Appropriate body surface area algorithms have not been validated for chemotherapy administration in birds.

Topical therapies
• Aldara™ (Imiquimod) is an immune response modifier that has been used for carcinoma in-situ in lesions and Bowen's disease in cats and for cervical papillomas in humans. It is applied topically to lesions every other day for 30 days.

Other Therapies

Cryotherapy
• This involves the application of a cryoprobe to the site of the lesion, with the goal of achieving necrosis of locally invasive lesions. • Use has been reported in several case reports for birds with localized lesions. • Multiple episodes are generally required and the success has been mixed.

Radiation
• External beam radiotherapy: This is indicated for lesions that are not surgically resectable and are too deep to use more superficial regional therapies (cryotherapy, strontium-90). The goal may be curative or palliative and this will determine the number of fractions provided. A typical full course will consist of 12 fractions at 4Gy/fraction three times weekly, while a typical palliative protocol would be 4 fractions of 8Gy/fraction once weekly. • Strontium-90: This is a radioactive probe that delivers a very high dose of radiation with limited penetration (≤2 mm) and can be very effective for lesions with minimal tissue penetration. One treatment may be sufficient, but multiple treatments can be used.

Phototherapy
• Phototherapy may be indicated for superficial lesions. Two case reports document its use in birds, however, neither case resulted in tumor remission. The photosensitizer Hexylether pyropheophorbide-a was used in both cases.

CONTRAINDICATIONS
• If regional or distant metastasis has occurred, the use of extensive local therapies is not recommended. Local therapy/surgical debulking for palliation may still be indicated. • Glucocorticoids may have significant immunosuppressive effects in birds and may potentiate concurrent infections and are not indicated. • If neoplastic lesions are immediately adjacent to critical structures (ie: glottis in the oral cavity), extreme caution must be observed when using therapies that may cause inflammation or potential resultant fibrosis (intralesional chemotherapy, cryotherapy, external beam radiotherapy).

PRECAUTIONS
• Systemic chemotherapies have been reported to cause a decrease in heterophil count in multiple avian species. WBC should be monitored in patients approximately one week postadministration (depending on the agent used) and again just prior to therapy. If the heterophil count is decreased to <25% of the normal range for the species being treated, a treatment break should be strongly considered. Administration of broad spectrum antibiotics should also be considered based on the clinical situation. Patients should be monitored closely for evidence of secondary infections and not exposed to situations that may make them more likely to contract infections. • Patients with pre-existing hepatic or renal compromise may need dose adjustments prior to administration of chemotherapy and should be monitored closely for alterations in uric acid, AST, ALT or bile acids. • Due to the anatomy of the avian renal portal system, systemic drug administration is recommended in the cranial half of the body (brachial vein, jugular vein) as opposed to the metatarsal vein. It is not known how the routes of administration of chemotherapeutics may affect uptake.

POSSIBLE INTERACTIONS
Use caution if patient is on antifungal therapy or antibacterial therapy that may be expected to cause renal or hepatic toxicity if patient is also on chemotherapeutics. Dose adjustments and careful monitoring are necessary.

ALTERNATIVE DRUGS
N/A

FOLLOW-UP

PATIENT MONITORING
• Careful patient monitoring for side effects is important. Different treatment protocols require different types of monitoring:
◦ Surgical excision: Patients should be monitored closely for evidence that they are damaging surgical sites as this is common. Surgical sites should be bandaged appropriately and a soft collar may be indicated for the first 7–10 days postsurgery to allow sites to fully heal. Patients with amputations that may affect balance or how they perch (full wing amputation or limb amputation) should have their feet closely monitored for pododermatitis. Surgical sites should also be monitored on a regular basis for evidence of tumor recurrence, even if margins appear clean on histopathology.
◦ Chemotherapy: Most common side effects are changes in total WBC, heterophil count, decreased appetite, increased regurgitation and general lethargy. It is recommended that patients have a CBC performed one week posttreatment and also immediately prior to the next treatment to ensure there is a sufficient heterophil count to prevent infection. If patient's total WBC or heterophil count is <25% of normal, a treatment delay and broad spectrum antibiotics should be considered. ◦ Radiation: Treatment sites should be monitored for skin irritation or redness.

PREVENTION/AVOIDANCE
• Appropriate pelleted diets may prevent hypovitaminosis A, which may cause squamous metaplasia; these lesions may predispose patients to developing SCC.
• Appropriate treatment of bacterial and fungal infections is important as chronic infections may also be a predisposing factor.
• It is also important to manage chronic feather pickers as it is reported that some SCC lesions occur in areas of chronic feather damaging behavior.

SQUAMOUS CELL CARCINOMA (CONTINUED)

POSSIBLE COMPLICATIONS
- Postsurgery, animals may pick at surgical sites or bandages and have the potential to cause surgical site dehiscence. Owners should monitor sites closely. • Chemotherapeutics may cause side effects, including anorexia, regurgitation, changes in stool quantity and character, and may make patients more predisposed to secondary infections.
- Radiation therapy may cause superficial changes to the skin (redness) that should be monitored. • Therapies that are expected to cause local necrosis (ex. cryotherapy, intralesional chemotherapy) may affect surrounding normal tissues and should be used with caution around vital structures (ex. glottis). • Topical immunomodulators (ex Aldara) are expected to cause skin crusting and flaking, patients should be watched carefully so they do not aggravate lesions.

EXPECTED COURSE AND PROGNOSIS
- If complete surgical excision is not possible, the overall prognosis for complete response is poor. • Survival times may vary significantly depending on the size and location of the tumor. Some patients may live over a year if lesions are small and can be managed locally.
- Lesions that are larger or impair crucial structures, like the glottis, are expected to cause earlier significant morbidity. Oral tumors frequently result in difficulty breathing and dysphagia, which will shorten life expectancy.

MISCELLANEOUS

ASSOCIATED CONDITIONS:
- Feather damaging behavior.
- Hypovitaminosis A. • Chronic bacterial/fungal infections.

AGE-RELATED FACTORS
- SCC is diagnosed more commonly in older patients. • Older patients may also have more frequency concurrent conditions that may affect treatment decisions (renal or hepatic disease).

ZOONOTIC POTENTIAL
None.

FERTILITY/BREEDING
- As genetic predispositions have not been identified, there are no contraindications for breeding affected animals. • Effects of chemotherapy or radiation on fertility in birds have not been well studied.

SYNONYMS
None.

SEE ALSO
Beak fracture
Beak malocclusion
Candidiasis
Feather damaging behavior
Lipoma
Nutritional deficiencies
Oral plaques
Rhinitis and sinusitis
Tracheal/syringeal diseases
Trichomoniasis
Tumors
Uropygial gland disease
Xanthoma

ABBREVIATIONS
SCC–squamous cell carcinoma

INTERNET RESOURCES
N/A

Suggested Reading
Filippich, L.J., Charles, B.G. (2004). Current research in avian chemotherapy. *The Veterinary Clinics of North America. Exotic Animal Practice*, **7**:821–831.

Garner, M. (2006). Overview of tumors: Section II. Harrison, G.J., Lightfoot, T.L. (eds), *Clinical Avian Medicine*. Palm Beach, FL: Spix Publishing, Inc., pp. 566–571.

Kent, M.K. (2004). The use of chemotherapy in exotic animals. *The Veterinary Clinics of North America. Exotic Animal Practice*, **7**:807–820.

Leach, M.W. (1992). A survey of neoplasia in pet birds. *Seminars in Avian and Exotic Pet Medicine*, **1**(2):52–64.

Reavill, D.R. (2004). Tumors of pet birds. *The Veterinary Clinics of North America. Exotic Animal Practice*, **7**:537–560.

Reece, R.L. (1992). Observations on naturally occurring neoplasms in birds in the state of Victoria, Australia. *Avian Pathology*, **21**:3–32.

Author Ashley Zehnder, DVM, PhD, DABVP (Avian)

Thyroid Diseases

Diseases of the thyroids are uncommon in birds. They may include endocrine changes, or the thyroids may enlarge, creating a space-occupying mass which is responsible for the clinical signs. The thyroids produce thyroxine, which has numerous effects on metabolism.

BASICS

DEFINITION
Any disease state of the thyroids is included here. They include hypothyroidism, thyroid hyperplasia (goiter), and thyroid neoplasia.

PATHOPHYSIOLOGY
Hypothyroidism in pet birds is uncommon enough in the literature that the mechanism responsible for its development is not known. In chickens, humans, and dogs, autoimmune thyroiditis is often responsible for the destruction of the thyroid cells. This has not been documented in other birds. Iodine deficiency can result in thyroid hyperplasia. Without iodine, functional thyroid hormone cannot be produced. This in turn fails to activate the negative feedback to the pituitary and excessive thyroid stimulating hormone (TSH) is produced, resulting in hyperplasia of the thyroid glands. The enlarged glands may compress the esophagus and trachea. The etiology of thyroid neoplasia is not known.

SYSTEMS AFFECTED
• Gastrointestinal changes may occur if the coelomic esophagus is compressed by enlarged thyroid glands. • Hemic changes may include mild nonregenerative anemia. • Respiratory effects of thyroid disease are common. With goiter or thyroid neoplasia, the trachea may be displaced or compressed leading to vocal changes, wheezing, or dyspnea. • Skin conditions are the most prominent aspect of hypothyroidism. Poor feather development, delayed molt, excess subcutaneous fat deposition, and hyperkeratosis.

GENETICS
These conditions are not known to be genetic.

INCIDENCE/PREVALENCE
Thyroid hyperplasia was once very common in budgerigars. With improved diets, it is now uncommonly seen. It occurs most commonly in macaws and budgerigars. Thyroid neoplasia is an uncommonly encountered tumor in pet birds. There is only a single confirmed report of hypothyroidism in parrots. Many others are suspected but not confirmed by appropriate diagnostics.

GEOGRAPHIC DISTRIBUTION
There appears to be no specific geographic distribution.

SIGNALMENT
• **Species:** This can occur in any species. Macaws and budgerigars are predisposed to goiter. • **Mean age and range** N/A • **Predominant sex** N/A

SIGNS

General Comments
Thyroid hyperplasia usually presents as a respiratory or gastrointestinal problem.

Historical Findings
Owners may note skin and feather problems, vocal changes or breathing abnormalities.

Physical Examination Findings
Hypothyroidism is associated with feather hypoplasia, feather loss, and epidermal atrophy.
Obesity may be seen in hypothyroid birds. Hypothyroidism may be associated with lipemia, nonregenerative anemia, and hypercholesterolemia. Recurrent skin infections may be encountered in hypothyroid birds. Birds with goiter or thyroid neoplasia may exhibit respiratory wheezes or squeaks, due to the pressure placed on the syrinx. Severe cases may result in overt dyspnea. The thyroid may also compress the coelomic esophagus, resulting in crop stasis.

CAUSES
• The cause of hypothyroidism in pet birds is not usually determined. In chickens, there is a genetically influenced autoimmune thyroiditis that results in hypothyroidism. However, this has not been demonstrated in other birds.
• Iodine deficiency is usually thought to be the cause of goiter. The ingestion of goitrogenic food items may also contribute.
• The cause of thyroid neoplasia is not generally determined. It could be speculated that they may result from chronic hyperplasia, but this has not been documented.

RISK FACTORS
Dietary iodine deficiency is the main risk factor for thyroid hyperplasia. Ingestion of goitrogenic foods such as cruciferous vegetables (kale, Brussels sprouts, mustard greens), soy-based foods, and some fruits can exacerbate the problem by inhibiting iodine metabolism.

DIAGNOSIS

DIFFERENTIAL DIAGNOSIS
Hypothyroidism should be distinguished from viral feather diseases such as psittacine beak and feather disease (PBFD) and polyomavirus, as well as other skin disorders. Thyroid neoplasia and hyperplasia should be differentiated from respiratory or upper gastrointestinal disorders as well as other masses in the coelomic inlet.

CBC/BIOCHEMISTRY/URINALYSIS
Mild anemia, gross lipemia, and hypercholesterolemia may be seen with hypothyroidism.

OTHER LABORATORY TESTS
Hypothyroidism must be diagnosed using a TSH stimulation test. The thyroxine (T4) level should at least double 4–6 hours after administration of TSH.

IMAGING
Very large thyroids may be seen radiographically. However, the thyroids are located just inside the coelomic inlet in birds. Definitive diagnosis of thyroid hyperplasia or neoplasia may require advanced imaging such as computed tomography (CT).

DIAGNOSTIC PROCEDURES
Endoscopy using a coelomic inlet approach will allow visualization of the thyroid. Biopsy techniques should be carefully applied, since the thyroids are very closely associated with the vasculature.

PATHOLOGIC FINDINGS
Gross enlargement of both thyroids may be seen with thyroid hyperplasia. Enlargement of one thyroid occurs with neoplasia. Most are not functional tumors, so atrophy of the contralateral thyroid is not common. Pathology of hypothyroidism may include skin changes such as ortho- and parakaratotic hyperkeratosis of the epidermis and moderate, widespread vacuolar degeneration and necrosis in the follicular epithelium.

TREATMENT

APPROPRIATE HEALTH CARE
Most patients would be treated as outpatients. Exceptions would be severely enlarged thyroids resulting in dyspnea or esophageal compression with nutritional deficits resulting from crop stasis.

NURSING CARE
Oxygen therapy is indicated for dyspneic patients. Fluid and nutritional support should be provided as needed. A liquid diet may pass through easier than a solid diet in patients with external esophageal compression.

ACTIVITY
Unless dyspneic, activity levels need not be altered.

DIET
As many hypothyroid birds are obese, a review of the diet to assure a balanced diet is provided is warranted. Additional caloric restriction may be indicated, but should be carefully monitored during thyroid supplementation to prevent excessive weight loss. In addition to iodine supplementation, goitrogenic items should be restricted in the diet of birds with thyroid hyperplasia.

CLIENT EDUCATION
Dietary counseling.

THYROID DISEASES (CONTINUED)

SURGICAL CONSIDERATIONS
Although not reported, surgical removal of a neoplastic thyroid gland may be possible. The close association of the gland with the carotid artery makes this a very delicate and risky procedure. Access to the site is very restricted as well.

MEDICATIONS

DRUG(S) OF CHOICE
- Hypothyroidism: Synthetic thyroid hormone is used as replacement therapy in cases of hypothyroidism. A starting dose of 0.2 µg/kg can be used, and then adjusted based on monitoring blood levels. • Thyroid hyperplasia: Lugol's iodine solution can be used by making a stock solution of 1 mL per 30 mL of water. A single drop (0.05 mL) can be added to 250 mL of drinking water.
- Radioactive iodine therapy could be considered in cases of thyroid neoplasia.

CONTRAINDICATIONS
None.

PRECAUTIONS
None.

POSSIBLE INTERACTIONS
None known.

ALTERNATIVE DRUGS
There are some commercial supplements containing iodine for birds. These may be used as alternatives to Lugol's iodine.

FOLLOW-UP

PATIENT MONITORING
Patients receiving thyroid supplementation should be monitored every 3–6 months. A general examination, hematology, chemistries, and thyroid levels should be evaluated at these visits.

PREVENTION/AVOIDANCE
Balanced nutrition may prevent thyroid hyperplasia.

POSSIBLE COMPLICATIONS
- Iatrogenic hyperthyroidism could occur with excess supplementation of thyroxine.
- Dyspnea, weight loss, or regurgitation may occur with very large thyroid glands.

EXPECTED COURSE AND PROGNOSIS
- Hypothyroid birds can be managed long term with supplementation. • Birds with goiter generally respond well to therapy.
- Thyroid neoplasia carries a poor to grave prognosis.

MISCELLANEOUS

ASSOCIATED CONDITIONS
None known.

AGE-RELATED FACTORS
None known.

ZOONOTIC POTENTIAL
None

FERTILITY/BREEDING
Hypothyroidism could affect breeding success, although there is little data supporting this.

SYNONYMS
None.

SEE ALSO
Circoviruses
Feather damaging behavior
Feather disorders
Nutritional deficiencies
Obesity
Polyomavirus
Tracheal/syringeal diseases
Tumors

ABBREVIATIONS
N/A

INTERNET RESOURCES
N/A

Suggested Reading
Schmidt, R.E., Reavill, D.R. (2008). The avian thyroid gland. *The Veterinary Clinics of North America. Exotic Animal Practice*, 11:15–23.

Author Kenneth R. Welle, DVM, DABVP (Avian)

Toxoplasmosis

BASICS

DEFINITION
Toxoplasma gondii is a coccidian parasite which rarely infects avian species.

PATHOPHYSIOLOGY
• Cats are the definitive host. They become infected from the feces of other cats or from eating birds or rodents. Sexual reproduction occurs within the intestinal mucosa and the infective oocysts are shed in the feces. • After the ingestion of oocysts, *T. gondii* matures into bradyzoites and sporozoites, which then infect the intestine. After several rounds of replication, tachyzoites spread throughout the body via the bloodstream and lymphatic system. Tachyzoites then infect various tissues throughout the body. Further replication occurs intracellularly until the cell bursts, causing tissue necrosis. • Clinical signs depend on the number of tachyzoites released, the immune system's ability to prevent the spread of the tachyzoites, and the organs damaged by the tachyzoites.

SYSTEMS AFFECTED
• Central nervous system • Ophthalmic
• Renal • Musculoskeletal • Hepatic • Splenic
• Respiratory.

GENETICS/INCIDENCE/PREVALENCE
Worldwide.

SIGNALMENT
• Asymptomatic infections are common in chickens, ducks and many wild birds.
• Outbreaks have been reported in chickens, passerines, guinea fowl and zoo collections.
• Canaries and mynahs show acute respiratory signs. • Canaries (*Serinus canaria*) that do not exhibit the acute respiratory phase of the disease will first present with the ocular and neurologic form of the disease (i.e., torticollis, blindness). • Canaries and mynahs in the acute phase may show the following: hepatosplenomegaly, catarrhal pneumonia, myositis. • Partridges (*Perdix perdix*) are more susceptible than other gallinaceous birds.
• Experimentally infected pigeons show rapidly progressive renal disease. • The disease has also been reported in mynahs, finches and lorikeets.

SIGNS
Historical Findings
• Lethargy • Anorexia • Weight loss
• Depression • Vague respiratory signs
• Torticollis • Acute death • Acute blindness.

Physical Examination Findings
• Respiratory signs—tachypnea, dyspnea
• Anorexia and weight loss • General debilitation • Blindness • Conjunctivitis
• Head tilt • Torticollis • Circling • Ataxia
• Seizure • Fever.

CAUSES
Ingestion of infected oocysts.

RISK FACTORS
• Feral cats with access to aviaries or zoos.
• Pet birds housed with cats. • Free range birds.

DIAGNOSIS

DIFFERENTIAL DIAGNOSIS
• Infectious: PMV, mycobacteriosis, listeriosis, *Apergillus sp.*, sarcocystis, *Baylisascaris*, West Nile Virus, EEE, duck plague, leukocytozoon infection, avian malaria, botulism. • Trauma. • Toxins and medications side effects: organophosphates, dimetridazole, heavy metals. • CNS disease: hepatoencephalopathy, CNS neoplasia, CNS abscess, epilepsy, avian vacuolar myelinopathy.
• Nutritional: hypovitaminosis B, vitamin E or selenium deficiency, hypocalcemia.
• Vascular: arteriosclerosis, infarct.

CBC/BIOCHEMISTRY/URINALYSIS
Hemogram
• Nonspecific changes. • Leukopenia or leukocytosis.

Biochemistry Panel: Results are variable based on the organ affected.
• Liver: Elevated bile acids, AST. • Muscle: Elevated AST, CK. • Renal: Elevated UA.
• Elevated albumin and TP.

OTHER LABORATORY TESTS
Indirect fluorescent antibody and the modified agglutination tests have been performed experimentally

IMAGING
Survey radiographs
• Hepatomegaly • Splenomegaly
• Pneumonia.

CT Scan
• Hepatomegaly • Splenomegaly
• Pneumonia • Encephalitis.

DIAGNOSTIC PROCEDURES
• Note: antemortem diagnosis is difficult.
• Immunohistochemistry on brain tissue slides. • PCR on brain tissue slides or other infected tissues. • CSF tap to diagnose encephalitis.

PATHOLOGIC FINDINGS
• **Gross lesions include:** Dark red, mottled, edematous and/or emphysematous lungs, and pale and swollen liver, spleen and kidneys.
• *Toxoplasma* trophzoites within affected tissues. Microscopically seen as 3–6 μm basophilic bodies, which are found free in the tissues or within the cytoplasm of many epithelial cells and macrophages. Multifocal necrosis can also be noted. • **Muscular:** Myositis especially of the pectoral muscle.
• **Neurologic:** ○ Tissue cysts with bradyzoites within the meninges and neurophil of cerebrum and cerebellum. ○ Degeneration of blood vessel walls and edema. ○ Subsequent tissue necrosis with/without hemorrhage.
○ Nonsuppurative inflammation.
• **Pneumonia:** ○ Gross changes: Hyperemia, fibrin and exudate. ○ Severe, focal necrotizing interstitial pneumonia with proliferation of type II pneumonocytes. • **Hepatic**
○ Hepatomegaly. ○ Hepatocellular degeneration and/or necrosis. ■ Gross findings seen as discrete pale or (less often) dark red foci, sharply delineated from adjacent parenchyma. Size ranging from <1mm to several centimeters ■ Either single cell or multifocal aggregates of necrotic hepatocytes.
• **Splenomegaly** ○ Experimentally infected quail had hemosiderin laden macrophages.
• **Myositis** especially of the pectoral muscle. Multifocal necrotizing myosites with infiltration of lymphocytes, plasma cells and neutrophils. • **Eosinophilic myocarditis.**
• **Hemovascular:** ○ Arteriolar dilation resulting in hyperemia. ○ Increased permeability resulting in edema. • **Ocular:**
○ Nonsuppurative chorioretinitis with macrophages that contain the tachyzoite form of *T gondii* in the subretinal space.
○ Iridocyclitis. ○ Aggregates of tachyzoites within the nerve fiber layer of the retina +/− necrosis.

TREATMENT

APPROPRIATE HEALTH CARE
• Outpatient care is possible if symptoms are minimal. • Inpatient care is necessary for symptomatic birds.

NURSING CARE
• Nutritional support via gavage feeding.
• Dyspneic birds placed in incubator with 78–85% oxygen supplementation at 5 L/min.
• Minimize stress by placing in a quiet, dark cage. • Place in warm, humid environment (85°F and 70% humidity), unless there is evidence of head trauma. • Fluid therapy if dehydrated and there is no pulmonary edema:
○ SQ, IO or IV depending on level of dehydration and vascular access. ○ 60–150 mL/day maintenance depending on species plus any additional fluids to correct for dehydration and ongoing losses.

ACTIVITY
If clinical for disease, activity should be restricted until the patient has made a full recovery.

DIET
No diet modification is necessary unless anorexic. All anorexic patients should receive assisted feedings.

TOXOPLASMOSIS (CONTINUED)

SURGICAL CONSIDERATIONS
• Pulmonary edema makes patients high risk for anesthetic complications. • Consider furosemide prior to biopsy.

CLIENT EDUCATION
Eliminate direct and indirect contact with cats and cat feces.

MEDICATIONS

DRUG(S) OF CHOICE
• Pyrimethamine: 0.5mg/kg PO every 12 hours and Trimethoprim-sulfamethoxazole (30 mg/kg PO q8h) for 14–28 days.
• Diclazuril: 10 mg/kg PO every 24 hours on day 0, 1, 2, 6, 8 and 10. • Furosemide 0.1–2 mg/kg PO, SQ, IM or IV q6–24h to counteract edema. • Meloxicam 0.35–1.0 mg/kg PO q12–24h to treat inflammation.

FOLLOW-UP

PATIENT MONITORING
• Monitor for response to treatment and resolution of clinical signs. • Monitor the patient's weight once to twice daily to assess nutritional status. • Hemocytology: Monitor for resolution of leukocytosis or leukopenia. • Neurologic cases: Perform serial neurologic exams to asses patients response to treatment.

PREVENTION/AVOIDANCE
• Prevent access to cat feces. • Prevent feral cat access to aviary/zoo collections.

POSSIBLE COMPLICATIONS
• Ocular Atrophy. • Death.

EXPECTED COURSE AND PROGNOSIS
Experimentally infected pigeons show rapidly progressive renal disease. Asymptomatic infections are common in chickens, ducks and many wild birds. Course and prognosis depends on the number of tachyzoites released, the immune system's ability to prevent the spread of the tachyzoites, and the organs damaged by the tachyzoites.

MISCELLANEOUS

ASSOCIATED CONDITIONS
Secondary infections from parasites, bacteria, viruses, and/or fungi may be seen.

AGE-RELATED FACTORS
N/A

ZOONOTIC POTENTIAL
None

FERTILITY/BREEDING
Avoid the use of sulfonamides in reproductive animals as they may be teratogenic.

SYNONYMS
N/A

SEE ALSO
Aspergillosis
Atherosclerosis
Botulism
Hemoparasites
(Systemic) coccidiosis
Mycobacteriosis
Nutritional deficiencies
Organophosphate/carbamate toxicity
Paramyxoviruses
Pneumonia
Seizure
Viral disease
West Nile virus

ABBREVIATIONS
AST—aspartate aminotransferase
CK—creatine kinase
UA—uric acid
TP—total protein

INTERNET RESOURCES
N/A

Suggested Readiings
Calton, W.W., McGavin, M.D. (1995). *Thomson's Special Veterinary Pathology*, 2nd edn. St. Louis, MO: Mosby.
Doneley, R.J.T. (2009). Bacterial and parasitic diseases of parrots. *The Veterinary Clinics of North America. Exotic Animal Practice*, **12**:3.
Dorrenstein, G.M. (2002). Avian pathology challenge. *Journal of Avian Medicine and Surgery*, **16**(3):240–244.
Harrison, G.J., Lightfoot, T.L. (eds) (2006). *Clinical Avian Medicine*. Palm Beach, FL: Spix Publishing, Inc., pp. 462–463, 509–510, 903.
Author Erika Cervasio, DVM, DABVP (Avian)

Tracheal Disease and Syringeal Disease

BASICS

DEFINITION
Condition/disease affecting the larynx, trachea, or the syrinx. The syrinx is an organ located at the caudal bifurcation of the trachea, responsible for production of the voice in birds, which involves the vibration of tympaniform membranes within the syringeal tympanum, during the expiratory phase of respiration (in most birds). The syrinx is a particularly complex organ in singing passerines.

PATHOPHYSIOLOGY
- Clinical signs are usually caused by the reduction of the tracheal or syringeal lumen caused by tissue proliferation, exudate, foreign bodies, traumatic tracheal collapse, or external compression. In viral diseases (e.g., herpesviruses, poxviruses), tracheal lesions may just be part of a systemic disease process.
- Coughing is a reflex associated with tracheal irritation. Birds rarely cough with cardiomegaly.

SYSTEMS AFFECTED
Respiratory.

GENETICS
N/A

INCIDENCE/PREVALENCE
N/A

GEOGRAPHIC DISTRIBUTION
N/A

SIGNALMENT
- **Species:** ○ Among psittaciformes, blue and gold macaws are suspected to be at greater risk of postintubation tracheal stenosis. A variety of other species have been reported with this condition. ○ Tracheal foreign body: Encountered in small psittaciformes, especially cockatiels. • **Mean age and range:** Any age. • **Predominant sex:** Both.

SIGNS

Historical Findings
- Increased amplitude of respiratory movements. • Tail bobbing. • Respiratory distress, open-mouth breathing. • Respiratory noises. • Coughing. • Voice changes. • Voice loss. • Lethargy. • Anorexia, weight loss.
- Decreased flight and hunting performance in falconry birds.

Physical Examination Findings
- Inspiratory dyspnea. • Increased inspiratory noises upon auscultation. • Open-mouth breathing. • Rhythmic overinflation of the infraorbital sinus. • Cyanosis. • Abnormal body posture. • Inflamed glottis. • Mass on the glottis. • Foreign body, mites or tracheal lesions may be visible upon tracheal transillumination.

CAUSES
- **Tracheal stenosis following intubation:** ○ Occurs about 1–2 weeks after an intubation event. ○ The exact causative event is unknown: possible physical or chemical irritation of the mucosa by the endotracheal tube itself, constant flow of air responsible for focal mucosal desiccation may cause inflammation and promote infection locally in the trachea, which progresses in stenosis of the trachea due to excessive fibrous tissue or granuloma formation. • **Foreign body** (e.g., millet seeds, insect part). • **Infectious causes:** ○ Bacterial: Various bacterial agents have been implicated in tracheitis, which may lead to tracheal stenosis either by fibrous tissue or granuloma formation: ▪ *Chlamydia psittaci* in turkeys. ▪ *Bordetella avium* in turkeys. ▪ *Mycoplasma* spp. are commonly found in tracheas of birds of prey and are considered commensals. ▪ *Enterococcus faecalis,* associated with chronic tracheitis in canaries. ▪ *Mycobacterium genavense*, reported in an Amazon parrot with granulomatous tracheitis. ○ Viral: ▪ Herpesvirus infections: Infectious laryngotracheitis in chickens, Amazon laryngotracheitis in Amazon parrots and a few other psittacine species (in recently imported birds), Psittacid herpesvirus 1 (glottal lesions, mainly in green-winged macaws), cytomegalovirus in Australian finches, and psittacid herpesvirus 3 infection in Eclectus parrots and Bourke's parakeets. ▪ Poxvirus (diphtheric or wet form) in passerine birds, various parakeets, lovebirds, and mynahs. ○ Fungal: Aspergillus infection can be responsible for granulomas in the trachea. The syrinx seems to be a site of predilection. Commonly diagnosed in Amazon parrots, African grey parrots, macaws, and falcons. ○ Parasitic: *Syngamus* spp. (mainly *S. trachea*), *Cyathostoma* spp. (mainly *C. bronchialis*), and acarids such as *Sternostoma tracheacolum* (tracheal, air sac mites). Tracheal worms infest mainly anseriformes, birds of prey, and poultry whereas tracheal mites infest primarily canaries and Australian finches. • **Toxic:** Inhaled toxins and smoke-inhalation injuries may cause severe necrotizing tracheitis.
- **Trauma:** Attack by a predator or a conspecific may lead to tracheal trauma and collapse. Damage to the *crista ventralis* during intubation in species with this anatomical feature in the glottis, damage to the bronchial bifurcation in certain species (e.g., *Spheniscus* spp. penguins). • **Nutritional deficiencies:** ○ Hypovitaminosis A has been associated with epithelial changes in the trachea and notably in the syrinx. It is also a risk factor for infectious diseases of the trachea by altering the mucosal defenses and increasing air turbulence. ○ Goiter has been associated with respiratory signs in budgerigars. • **Neoplasia:** Uncommon. Tracheal osteochondroma has been reported in parrots. Masses in the thoracic inlets have been associated with syringeal and bronchial compressions.

RISK FACTORS
- Postintubation tracheal stenosis and tracheal trauma: ○ Repeated intubation. ○ Intubation too distal in the trachea, where it narrows significantly. ○ Use of cuffed endotracheal tubes. • Poor diet, especially lacking in vitamin A or iodine (budgerigars).
- Infectious tracheal diseases may originate from a conspecific (lack of quarantine, persistently infected bird, especially with herpesviruses). Tracheal mites are transmitted from contaminated feces. • Exposure to inhalants (polytetrafluoroethylene, smoke, air fresheners, other toxic aerosols). • Exposure to intermediate or paratenic hosts for tracheal worms (e.g., earthworms). Exposure to contaminated feces for species with direct life cycle.

DIAGNOSIS

DIFFERENTIAL DIAGNOSIS
- Nasal and sinus diseases. • Lower respiratory diseases (expiratory dyspnea).
- Cardiovascular diseases with respiratory signs. • Iodine deficiency (frequent in budgies) can be responsible for goiter disease, that may cause a partial external obstruction of the trachea. • Overheating, may cause panting and open mouth breathing. • Cough mimicking.

CBC/BIOCHEMISTRY/URINALYSIS
- Inflammatory leukogram: ○ Heterophilic leukocytosis and/or monocytosis may be present if an infectious or inflammatory process is ongoing. ○ Marked leukocytosis is usually encountered with avian chlamydiosis and aspergillosis, although some cases of aspergillosis may show a normal leukogram.
- Polycythemia may be seen with chronic hypoxemia (more common in chronic lower respiratory diseases).

OTHER LABORATORY TESTS
- Arterial blood gas: Gold standard to assess oxygenation and ventilation. $PaO_2 \leq 80$ mmHg indicates hypoxemia. • Cytology of tracheal/syringeal lesion: May help identify Aspergillus hyphae. • Bacterial culture and sensitivity of a tracheal swab. • Biopsy of lesions if possible: more sensitive for bacterial culture and fungal culture, histopathology.
- PCR: Commercially available for *Chlamydia psittaci* (conjunctival/choanal/cloacal swab), *Mycobacterium genavense/avium/intracellulare*, *Mycoplasma* spp, Psittacid herpesvirus, other viruses. • Serology: ○ Available for *Chlamydia psittaci* (elementary agglutination antibody test (IgM), indirect fluorescent antibody, complement fixation, or ELISA tests (IgY)). Recommended to pair with a PCR test.

TRACHEAL DISEASE AND SYRINGEAL DISEASE (CONTINUED)

◦ Serology for aspergillosis is also available but may be of low sensitivity and specificity.
• Serum galactomannan assay to test for Aspergillosis seems to be associated with low sensitivity.

IMAGING

Radiography
• Radiographs: Can be useful in the diagnosis of tracheal stenosis. The affected part of the trachea becomes more radio-opaque, and the tracheal rings, readily visualized, can be irregular, or collapsed. The syrinx is more difficult to visualize, due to surperimposition of structures, but syringeal granulomas can usually be identified. Overinflation of caudal air sacs may be appreciated with syringeal diseases causing a valve-like effect resulting in air trapping into these air sacs. The tracheal and syringeal cartilages become calcified with aging and a radiographic conspicuous syrinx is not unusual in older birds. In many male ducks, a large syringeal bulla is also present.

Tracheoscopy
• The gold standard to investigate tracheal and syringeal diseases. The procedure must be performed with care not to induce further lesions of the trachea. This is facilitated by the use of a 0° scope. The diameter of the scope will depend on the size of the bird (commonly used are 1.9 and 2.7 mm). Limits of this examination include tracheal length and size of the bird. Psittaciformes lack a pessulus (median syringeal cartilage) in the syrinx.

Endoscopy
• Endoscopy of the interclavicular air sac may be useful to visualize the outside of the trachea, syrinx, and possible lesions or masses in the thoracic inlet.

DIAGNOSTIC PROCEDURES
• Pulse oximetry: Measures the hemoglobin saturation in oxygen (likely underestimated by mammalian pulse oximeters).
• Capnometry: Measures the end-tidal CO_2 in the expired gas: should not be affected in upper respiratory diseases.

PATHOLOGIC FINDINGS
Pathologic findings will depend on the cause of the disease

TREATMENT

APPROPRIATE HEALTH CARE
• An air sac tube should be placed either in the caudal thoracic or abdominal air sac if tracheal obstruction is suspected (see air sac cannulation). It is also indicated during tracheal surgery for gas anesthesia administration. • A tracheoscopy can be performed under anesthesia if the bird is stable or when the air sac tube has been placed. If a lesion partially or completely obstructing the trachea is visualized, endoscopically guided surgical debriding can be performed (e.g., aspergilloma). An endoscopic radiosurgical electrode or CO_2 laser may be used. Samples may be taken for diagnostic tests during this same procedure for cytology and culture. Topical administration of antimicrobial may be performed during the tracheoscopy and is indicated for fungal tracheitis. • Removal of foreign body can also be attempted during a tracheoscopy. • Endoscopic removal of tracheal worms may be indicated if tracheal obstruction is an issue. • Nebulization allows high local concentration of therapeutics, maximizing its efficacy while minimizing systemic absorption, reducing potential for toxicity, and reducing drug biotransformation. Humidification of the mucociliary escalator may also improve its efficiency. Because of the trachea and bronchi large diameters, even large particle size (e.g., humidification, vaporization) should be deposited in these locations. Nebulization time is typically 15–30 minutes and frequency is typically once to twice daily. In birds, the respiratory surface and functional efficiency of avian lungs may lead to greater systemic absorption of the drugs being nebulized than in other species.

NURSING CARE
• Oxygen therapy if the patient is dyspneic.
• Gavage feeding in case of anorexia. • Fluid replacement or maintenance therapy.
• Sedation may be useful to reduce the distress associated with severe dyspnea. Midazolam or midazolam-butorphanol are usually the drugs of choice for mild to marked sedation.

ACTIVITY
N/A

DIET
Provide a well-balanced diet.

CLIENT EDUCATION
Inform the clients of the risk of tracheal stenosis secondary to intubation.

SURGICAL CONSIDERATIONS
• Seed tracheal foreign bodies may be removed with the help of a needle inserted in the trachea just below the seed. The seed is slightly dislodged cranially by the needle's bevel. Then, a syringe should be connected to the needle and air pushed in an attempt to force the seed out through the glottis.
• Tracheotomy: Indicated for removal of foreign objects or surgical debriding for which endoscopic treatment was unsuccessful. In small birds (e.g., cockatiels), tracheoscopy may be challenging and a tracheotomy may be the only option. Tracheotomy of approximately 50% of tracheal circumference is performed between tracheal rings. Once the procedure is completed, the tracheotomy is repaired with absorbable simple sutures comprising at least one ring on each side of the incision with the knots outside of the tracheal lumen. • Tracheostomy: Has been described in parrots and birds of prey. Long-term survival may be poor. • Tracheal resection and anastomosis: Surgical treatment is necessary in almost all cases of tracheal stenosis. Five to ten tracheal rings or 10% of the tracheal length can usually be removed safely but up to 12–15 rings have been removed in extreme cases. The anastomosis is performed with absorbable simple sutures comprising at least one ring on each side of the incision. Two additional sutures to counteract tension may be placed when more than 10% of the trachea is removed.
• Stenting of the trachea has been described using a custom-made nitinol stent in an Eclectus parrot, placed by tracheoscopy and fluoroscopy.

MEDICATIONS

DRUG(S) OF CHOICE
• Tracheal stenosis: ◦ Dexamethasone: 0.05–0.1 mg/kg locally on the affected area in the trachea during tracheoscopy to reduce inflammation. ◦ Antibiotics: According to results of bacterial culture and sensitivity. A broad spectrum antibiotic may be used while waiting for the results or if the owner declines the culture (e.g., amoxicilline/clavulanic acid 125mg/kg PO q12h, enrofloxacin, 10–15 mg/kg q12–24h). • Avian chlamydiosis: ◦ Doxycycline: 25–50 mg/kg PO q12h for 21–45 days, long-lasting injectable may be given at 70 mg/kg IM q1week, water-based medication in cockatiels: 200–400 mg/L. ◦ Azithromycin: 10–40 mg/kg PO q24–48h for 21–45 days. • Aspergillosis: A combination of antifungal is recommended in confirmed cases: ◦ Amphotericin B 1.5 mg/kg IV q12h or voriconazole 10–20 mg/kg IV q12h during 1–3 days. ◦ Amphotericin B 0.1–5 mg/ml of sterile water intratracheal. ◦ Voriconazole 10–20 mg/kg PO q12h. ◦ Terbinafine 15–25 mg/kg PO q12h mg/kg. ◦ Antifungals frequently need to be given for several months. • Parasites: ◦ Ivermectin: 0.2–0.4 mg/kg IM, SC, PO, topical. Repeat 10–14 days in most cases. ◦ Fenbendazole: 20–50 mg/kg PO q24h for five days.
• Nebulization: ◦ Antibiotics: Amikacin 5–6 mg/mL sterile water, enrofloxacin 10 mg/mL of sterile water, piperacillin 10 mg/mL of sterile water. ◦ Antifungals: Amphotericin B 0.1–5 mg/mL of sterile water, terbinafine 0.01 mg/kg in 5–10 mL of sterile saline, voriconazole 10 mg/mL of sterile saline. ◦ Mucolytic: N-acetyl-cysteine 22 mg/ml of sterile water; may induce bronchospasm and may be combined with bronchodilators.

CONTRAINDICATIONS
• Although corticosteroids can be beneficial in the treatment of tracheal stenosis, they are contraindicated if an infection is suspected.

TRACHEAL DISEASE AND SYRINGEAL DISEASE (CONTINUED)

Birds are corticoid-sensitive species so the overall bird condition should be carefully evaluated before using steroid medications. Stenotic lesions also frequently harbor secondary pathogens. Systemic adverse effects may also be encountered in birds. • Systemic enilconazole and ketoconazole are highly toxic in birds and should not be employed through this route. • Triazoles may cause hepatotoxicity in cytochrome P450 sensitive species such as African grey parrots. Itraconazole frequently causes anorexia in birds of prey.

PRECAUTIONS
N/A

POSSIBLE INTERACTIONS
N/A

ALTERNATIVE DRUGS
N/A

FOLLOW-UP

PATIENT MONITORING
• Progression of clinical signs. • Regular tracheoscopy to monitor progression of tracheal stenosis and resolution of fungal granuloma or postoperatively to monitor for recurrence (can happen up to several years after the procedure).

PREVENTION/AVOIDANCE
• Intubate with great care, and ensure the endotracheal tube is not inserted too deep inside the trachea. • Move the head and neck of intubated birds carefully. • Use well lubricated uncuffed endotracheal tubes. • Humidification of oxygen may be performed under anesthesia using humidifiers for inhalant gas. • Provide a well-balanced diet. • Quarantine any new bird to limit the risk of contagious infectious disease. • Parasites: Reducing access to intermediate host such as limiting access to the ground (earthworms). • Limit exposure to household toxic inhalants.

POSSIBLE COMPLICATIONS
• Severe dyspnea, cyanosis, death if not treated. • Complication of surgery: Bleeding, dehiscence, subcutaneous emphysema, recurrence of stenosis over time. Voice changes may also be encountered due to neurologic and muscular trauma on syringeal muscles and modification of tracheal airflow.

EXPECTED COURSE AND PROGNOSIS
• Medical therapy has not been associated with good results in tracheal stenosis and recurrence/postoperative complications have been reported commonly after tracheal resection and anastomosis. The prognosis for tracheal stenosis is therefore guarded.
• Aspergillosis is associated with a guarded prognosis as treatment is long and difficult.
• Other causes of tracheal disease are usually associated with a good prognosis after adequate treatment is given.

MISCELLANEOUS

ASSOCIATED CONDITIONS
Associated conditions may be associated with the etiology of the tracheal disease; for example:
• Aspergillosis granulomas, if present in the trachea, can also be found elsewhere; therefore, imaging of the whole body is recommended. • *Chlamydia psittaci* can also be responsible for hepatitis, pneumonia and airsacculitis. • Mycoplasma infection is often responsible for conjunctivitis and sinusitis in columbiformes, passeriformes, and galliformes.

AGE-RELATED FACTORS
None.

ZOONOTIC POTENTIAL
Avian chlamydiosis is a zoonotic disease.

FERTILITY/BREEDING
N/A

SYNONYMS
N/A

SEE ALSO
Aspergillosis
Airborne toxicosis
Air sac mites
Air sac rupture
Aspergillosis
Aspiration
Avian influenza
Chlamydiosis
Herpesviruses poxvirus
Mycoplasmosis
Nutritional deficiencies
Paramyxoviruses
Poxvirus
Respiratory distress
Thyroid diseases
Trauma
Viral disease

ABBREVIATIONS
N/A

INTERNET RESOURCES
N/A

Suggested Reading
Clippinger, T.L., Bennett, R.A. (1998). Successful treatment of a traumatic tracheal stenosis in a goose by surgical resection and anastomosis. *Journal of Avian Medicine and Surgery*, **12**:243–247.
Dennis, P.M., Avery Bennett, R.A., Newell, S.M., Heard, D.J. (1999). Diagnosis and treatment of tracheal obstruction in a cockatiel (*Nymphicus hollandicus*). *Journal of Avian Medicine and Surgery*, **13**(4):275–278.
Jankowski, G., Nevarez, J.G., Beaufrère, H., et al. (2010). Multiple tracheal resections and anastomoses in a blue and gold macaw (*Ara ararauna*). *Journal of Avian Medicine and Surgery*, **24**(4):322–329.
McLelland, J. (1965). The anatomy of the rings and muscles of the trachea of Gallus domesticus. *Journal of Anatomy, London*, **99**(3):651–656.
Sanchez-Migallon Guzman, D., Mitchell, M., Hedlund, C.S., et al. (2007). Tracheal resection and anastomosis in a mallard duck (*Anas platyrhynchos*) with traumatic segmental tracheal collapse. *Journal of Avian Medicine and Surgery*, **21**:1500–1157.

Authors Delphine Laniesse, Dr.Med.Vet, IPSAV
Hugues Beaufrère, Dr.Med.Vet., PhD, Dipl ABVP (Avian), Dipl ECZM (Avian), Dipl ACZM

Trauma

BASICS

DEFINITION
A wound or injury to any body system.

PATHOPHYSIOLOGY
• Blunt force producing injury to the body or head. • Laceration, abrasion or puncture wound to the skin or musculature. • Self-mutilation secondary to behavioral abnormalities, pain or discomfort. • Edema or subcutaneous hemorrhage secondary to electrical shock. • Head trauma results in direct damage to the intracranial structures, disturbance of auto-regulatory mechanisms (depletion of ATP, electrical gradient damage and glutamine release), failure of pressure auto-regulation of arteries and damage to the cerebral ischemic response.

SYSTEMS AFFECTED
• Skin/Exocrine • Musculoskeletal • Cardiovascular • Behavioral—as primary cause of self-induced trauma • Neuromuscular • Neurologic • Ophthalmic • Respiratory.

GENETICS/INCIDENCE/PREVALENCE
• Trauma is a common avian emergency. • There is no geographic distribution.

SIGNALMENT
• Paired cockatoos: The male often attacks the female mate. • Paired eclectus: The female often attacks the male mate. • Wildlife. • Birds with flight or allowed unsupervised activity outside the cage. • Cockatoos have a higher incidence of self-mutilation. • Quaker parakeets have a severe and sometimes fatal self-mutilation syndrome.

GEOGRAPHIC DISTRIBUTION
None.

SIGNS

Historical Findings
• Recent blunt force trauma. • Fight with another animal. • Pair-bonded birds. • Active bleeding. • History of behavioral abnormalities/self-mutilation. • History of night fright. • Unsupervised activity.

Physical Examination Findings
• Laceration, abrasion or puncture wound to the body or head. • Hemorrhage. • Subcutaneous bruising. • Fracture(s) or malposition of a limb. • Damaged or missing feathers. • Bleeding or broken blood feather. • Lethargy. • Recumbency. • Coma. • Shock. • Seizures. • Neurologic disease: Cranial nerve or peripheral nerve deficits.

CAUSES
• Gunshot wound. • Vehicular impact. • Flying into a window or other stationary objects. • Interaction with another pet or predatory species. • Behavioral abnormalities leading to self-mutilation. • Internal pain or discomfort. • Electrical shock. • Chemical exposure. • Recent fall. • Neuromusculoskeletal disease that causes weakness. • Night fright.

RISK FACTORS
• Bird with free flight in an unsafe environment. • Housed with a cage mate. • Unsupervised activity.

DIAGNOSIS

DIFFERENTIAL DIAGNOSIS
• Coagulopathies • Dermatopathy • Nutritional deficiencies • Infection • Systemic disease • Neoplasia.

CBC/BIOCHEMISTRY/URINALYSIS

Hemogram
• Anemia secondary to loss and/or thrombocytopenia due to platelet consumption. • Polychromasia is noted in mallard ducks 12 hours posthemorrhagic event.

Biochemistry panel
• Muscle/liver enzyme abnormalities. *Note*: Caution must be taken in birds with excessive hemorrhage. Only 1% of the bird's weight in grams is recommended for blood work (i.e., 1 mL of blood for every 100 g of body weight). Therefore, all blood loss from hemorrhage and bruising must be taken into account.

IMAGING
• Survey radiographs: May identify fractures, luxations, air sac rupture, SQ emphysema, visceral damage or ascites. • CT scan: Better identifies fractures, luxations, and visceral damage. In addition, spinal injury and brain injury can be assessed.

DIAGNOSTIC PROCEDURES
Blood pressure measurement can be considered to access cardiovascular status.

PATHOLOGIC FINDINGS
• Hemorrhage, bruising and tissue necrosis from a laceration or puncture wound. • Three phases of wound healing. ○ Inflammatory phase occurs in first 0–36 hours where there is leukocyte, monocyte, plasma cell and macrophage infiltration. ○ Collagen phase occurs from 3–4 days in which the microfibril aggregates form and capillaries bud into the wound area. ○ Maturation phase occurs from weeks to months in which there is remodeling of collagen along tension of the wound. • Edema and subcutaneous hemorrhage as a result of electrical shock. • Fractures or malposition of any bone. • Damage and missing feathers as a result of feather damaging behavior. • Head trauma. ○ Hemorrhage in the epidural, subdural and/or leptomeningeal areas and in the brain parenchyma. ○ Cerebral edema. ○ Extra cerebral hemorrhage. • Spinal trauma. ○ Vertebral fracture. ○ Laceration, transection or compressive lesions of the spinal cord. *Note*: In animals, the margin of error in a concussive injury that results in unconsciousness or death is small due to the small size of the patient's brain.

TREATMENT

APPROPRIATE HEALTH CARE
• Outpatient medical management is possible if the hemorrhage is minimal, the laceration/abrasion is small or in broken blood feathers with minimal blood loss. • Inpatient medical management is required in cases of excessive blood loss.

NURSING CARE
• Nutritional support via gavage feeding. • Dyspneic birds place in incubator with 78–85% oxygen supplementation at 5 L/min. • Stop active hemorrhage with compression, pressure wrap application (except around keel), silver nitrate application on a bleeding nail, the removal of the broken blood feather, hemostatic matrix or emergency surgery. • *Note*: Although pulling a blood feather is commonly recommended, this can result in damage to the feather follicle. • Minimize stress by placing in a quiet, dark cage. • Stabilize all fractures/malposition first with a splint or figure-of-eight bandage. Once the patient is stable, long term repair can be performed by surgical correction or splinting. • Stop access to feathers in self/mutilation cases with an E-collar designed for birds or bandaging. Next work up the patient for self-mutilation (See "Feather Damaging Behavior"). • Place in warm, humid environment (85°F and 70% humidity), unless there is evidence of head trauma. • Fluid therapy: ○ SQ, IO or IV depending on level of dehydration and vascular access. ○ 60–150 mL/day maintenance depending on species, plus any additional fluids to correct for dehydration and ongoing losses.

ACTIVITY
Activity should be restricted until the wound has healed or the patient has made a full recovery.

DIET
No diet modification is necessary unless anorexic. All anorexic patients should receive assisted feedings.

CLIENT EDUCATION
• Debilitated birds have a poor prognosis despite treatment. • Discuss the importance of avian enrichment. • Discuss the potential risks in the bird's every day environment.

SURGICAL CONSIDERATIONS
• Hypotension should be corrected before surgery. • If the patient is clinical for its anemia or excessive blood loss, stabilization

(CONTINUED)

should occur before surgery. • Electrocautery should be used to minimize blood loss.

MEDICATIONS
DRUG(S) OF CHOICE
Fluid therapy
• Subcutaneous fluids can be given for mild levels of dehydration (<4–5%) • IO/IV fluids should be given for moderate levels of dehydration (approximately 5–8% dehydration). 80% of fluid deficit should be replaced over 6–8 hours in acute losses and over 12–24 hours in chronic losses.

Management of hypovolemic shock (>10% dehydration)
• First determine that phase by assessing the BP, HR, mm and CRT. • For treatment of all phases, warm the fluids to 100–103°F.
• **Compensatory phase**—the first stage of shock. ○ Typical examination findings—tachycardia and hypertension. ○ This phase is typically seen with blood loss of less than 20% of total blood volume. ○ Treatment includes volume replacement of the deficit with crystalloids IV/IO over a 12-hour period. Where maintenance equals 60–150 mL/day (depending on the species) and deficit equals % dehydration $\times$ BW$_{kg}$ $\times$ 1000. • Early decompensatory phase—second stage of shock in which there has been decreased blood flow to the kidneys, GI tract, skin and muscles. ○ Typical examination findings include tachycardia, +/– hypothermia and normal to decreased blood pressure, pale mm, prolonged CRT, cool limbs and depression. ○ This phase typically occurs with blood loss of greater than 25–30% of total blood volume. ○ Treatment: Crystalloid bolus (10 mg/kg) and HES bolus (3–5 mL/kg) or Oxyglobin (5 mL/kg) repeated until the blood pressure is greater than 90 mmHg (approximately 3–4 total) IV/IO.
• **Decompensatory phase**—final stage of shock: ○ The LD$_{50}$ of ducks (Anas platyrhynchas) was 60% of their total blood volume. ○ Typical examination findings include bradycardia, hypothermia, hypotension, pale mm and absent CRT. ○ Treat in the following order: ○ HSS (3–5 mL/kg) over 10 minutes +/– HES (3 mL/kg) IV/IO. ○ Warm patient with supplemental heat such as a Bair Hugger. ○ Give crystalloids (10 mLkg) and HES (3–5 mL/kg) bolus IV/IO. Repeat until blood pressure is greater than 90 mm/Hg (3–4 boluses total). ○ Once stable, place on crystalloids at maintenance rate + deficits and ongoing losses. • If there is no response to above treatments, check the BG, PCV, TP and ECG. • If hypoglycemic give 50% dextrose 50–100 mg/kg IV slow to effect. Dilute 1:1 with 0.9% saline. • If abnormal cardiac contractility give nitroglycerin—place a 1/8 inch/2.5 kg strip on the skin. Wear gloves. • If PCV <20%, consider a blood transfusion or use of a blood alternative. • *Note*: blood transfusions are rarely used in emergency situations due to lack of availability. • Multiple transfusions carry increased risk of a fatal reaction. ○ Calculate the patient's total blood volume (8% of body weight) and replace 10% of the blood volume IV/IO. ○ Alternatively, give Oxyglobin (if available) at 5 mL/kg boluses IV slowly over one minute, every 15 minutes. • Pain management ○ Butorphanol: Dosing varies on species; 1–5mg/kg IM, IV q1-4h. ○ Metacam: 0.35–1.0 mg/kg PO q12–24h.
• Nutritional support via gavage feeding.
• Antibiotics: ○ Chose broad spectrum antibiotics for lacerations and abrasions. ○ For bites chose piperacillin or a fluoroquinolone.

Head trauma
No specific research has been conducted in birds. Guidelines have been adapted from other species. • Treat hypovolemia with low volume resuscitation with HSS (3 mL/kg) over 10 minutes +/– HES at 3 mLkg IV/IO. • Treat hemorrhage as described above. • Address respiratory compromise. • Keep at normal body temperature. Avoid any additional heat. • Mannitol can be given at 0.25-1.0 g/kg every 4-6 hours for a total of 3 doses in 24 hours. • Maintain normoglycemia. • Jugular venipuncture is contraindicated. • Keep quiet and in low light. • Oxygen supplementation. • Elevate head 25-30° if recumbent. • Perform serial neurological examinations every 30–60 minutes. • Pain management: ○ Use caution with opiods due to respiratory depression. ○ Avoid benzodiazepines due to neurologic depression. ○ Control seizures with a propofol CRI.

Wound care
• Clean and remove feathers 2–3 cm from wound. • Remove devitalized tissue and foreign material. • Lavage with 0.9% saline, LRS, dilute chlorhexidene or povidone iodine 1% or less. • Wound closure with primary or secondary intention. • Give broad spectrum antibiotics. • Open wounds: Apply topical ointments and occlusive dressing with 90% water (1% silver sulfadiazine cream, triple antibiotic cream). • Prevent access with a wrap or E-collar.

CONTRAINDICATIONS
• Corticosteroids complicate head trauma due to fluid retention. • Jugular phlebotomy and supplemental heat in patients with head trauma. • Phlebotomy if greater than 1% of blood loss has occurred.

FOLLOW-UP
PATIENT MONITORING
• Monitor for response to treatment and resolution of clinical signs (HR, CRT, mm quality). • Blood pressure—perform serial measurements to evaluate trends until normalized. • Head trauma—perform serial neurologic exams to asses patients response to treatment. • Anemia and thrombocytopenia—patient follow up should occur until the platelet count and PCV/TS have normalized. • Monitor the patient's weight once to twice daily to assess nutritional status. • Fractures—splints should be assessed weekly until removal and surgical stabilization should be monitored once to twice weekly. • Hemocytology—polychromasia will be noted within 12 hours after blood loss.

PREVENTION/AVOIDANCE
• Provide proper enclosures for birds to prevent injury. • Perform routine wing trimming if a flighted bird cannot be kept safely in the environment. • Provide enrichment and behavior modification for self-mutilating patients. • Discuss risk of unsupervised interaction with another animal or unsupervised activity. • Avoid airborne perfumes/odors and topical products around birds with feather damaging behavior. • Use appropriate wing trimming techniques.

POSSIBLE COMPLICATIONS
• Death. • Depending on wound the bird may be left with inability to fly, a permanent deformity or a permanent peripheral nerve deficit. • Head trauma may result in seizures or permanent cranial nerve deficits.

EXPECTED COURSE AND PROGNOSIS
Varies with the severity of traumatic injuries. Nerve and vascular damage – may have poor prognosis for healing and return to function, especially if the extremities are involved

MISCELLANEOUS
ASSOCIATED CONDITIONS
None.
AGE-RELATED FACTORS
N/A
ZOONOTIC POTENTIAL
N/A
FERTILITY/BREEDING
Avoid teratogenic antibiotics in laying hen. Traumatic injury can result from a conspecific, particularly with breeding pairs of cockatoos and eclectus parrots.
SYNONYMS
Injury, wounds, contusion, abrasion
SEE ALSO
Anemia
Beak fracture
Feather damaging behavior
Fracture/luxation
Hemorrhage
Lameness
Neurologic conditions

TRAUMA (CONTINUED)

Problem behaviors
Seizure

ABBREVIATIONS
PCV—packed cell volume
TS—total solids
SC—subcutaneous
IV—intravenous
IO—intraosseous
mm—mucus membranes
BG—blood glucose
ECG—electrocardiogram
CRT—capillary refill time
HSS—hypertonic saline
HES—hetastarch
LRS—Lactated Ringer's Solution
GI—gastrointestinal

INTERNET RESOURCES
N/A

Suggested Reading
Carpenter, J.W. (2013). *Exotic Animal Formulary*, 4th edn. St. Louis, MO: Elsevier.
Harrison, G.J., Lightfoot T.L. (eds) (2006). *Clinical Avian Medicine*. Palm Beach FL: Spix Publishing.
Lennox, A. (2013). Avian critical care. Proceedings BSAVC.
Lichtenberger, M. (2004). Response to fluid resuscitation after acute blood loss in Peking ducks. *Proceedings of the Association of Avian Veterinarians Annual Conference.*
Lichtenberger, M. (2005). Avian shock: Recognition and treatment. Presented at the 11th International Veterinary Emergency and Critical Care Symposium (IVECCS 2005), September 7-11, Atlanta, GA.
Lichtenberger, M. (2005). Normal Indirect blood pressure in different species of birds. *Proceedings of the Association of Avian Veterinarians Annual Conference.*

Author Erika Cervasio, DVM, DABVP (Avian)
Acknowledgement Stephen Dyer, DVM, DABVP (Avian)
Marla Lichtenberger, DVM, DACVECC
Jorg Mayer, DVM, MS, DABVP (ECM), DECZM (Small Mammal)

 Client Education Handout available online

TRICHOMONIASIS

BASICS

DEFINITION
Trichomoniasis is the disease due to infection with the protozoan *Trichomonas gallinae* or *T. columbae* (usually affecting the upper digestive tract), or *T. gallinarum* (affecting the lower digestive tract).

PATHOPHYSIOLOGY
Infection occurs with a virulent organism from a contaminated environment or direct contact with infected birds, after a 4–18 day incubation period; *T. gallinae* and *T. columbae* damage and cause caseated material and adherent plaque buildup on the mucosa of the oropharynx, the digestive tract down to and including the proventriculus, and to the upper and sometimes lower respiratory system (sinuses, trachea and even air sacs). There is also pathology from invading visceral tissue especially the liver. *T. gallinarum* affects the intestines and the ceca. There are less virulent strains and infection with these may provide cross protection from more virulent strains.

SYSTEMS AFFECTED
- Gastrointestinal—via damage by the organism. • Respiratory—blockage with caseated material. • Behavioral—especially in reaction to the general irritation from the parasite, and overall lethargy.
- Hepatobiliary—due to invasion into liver.

GENETICS
N/A

INCIDENCE/PREVALENCE
Common

GEOGRAPHIC DISTRIBUTION
Worldwide, found in captive and wild species.

SIGNALMENT
- **Species:** *T. gallinae, T. columbae*: pigeons (and other doves), raptors, canaries, budgerigars, cockatiels, other parrots, mynahs, poultry. *T gallinarum:* poultry.
- **Mean age and range:** Found in all age ranges. **Predominant sex:** None.

SIGNS

Historical Findings
- May be reported as normal. • Decreased activity/lethargy/weak flight or being unwilling to fly. • Decreased appetite and weight loss, and increased water consumption. • Gagging, neck stretching, regurgitation, head flicking (ejecting bits of food as well), difficulty swallowing. • Nasal discharge and difficulty breathing, "blowing bubbles." • Green diarrhea as well as polyuria and green urates.

Physical Examination Findings
- May be asymptomatic. • Lethargy, reduced body condition, or failure to thrive.
- Caseated plaques (white, yellow, tan or brown) at commissure of beak, oral cavity, oropharynx, laryngeal mound, sinuses and choana. • Plaques can progress to large proliferative/invasive masses. • Plaques can be easily removed without bleeding. • Greenish fluid accumulation within oropharynx.
- Regurgitation or crop and gut atony.
- Dyspnea, nasal discharge, sneezing.
- Thickened esophagus and crop wall.
- Hepatomegaly. • Polyuria and/or biliverdinuria. • Diarrhea (may be sour-smelling). • Umbilical swelling in juveniles.

CAUSES
Trichomonas organisms are labile and do not have a stable cyst. There is no intermediate host. They can be transmitted via contamination of food and water; parental feeding of chicks; and eating infected birds.

RISK FACTORS
- Overcrowding, lack of quarantine of newly introduced birds, poor hygiene, compromised immune system. • Infected parents. • Feeding on potentially infected species (in the case of raptors feeding on pigeons for example).
- Pigeon herpesvirus lesions may make the bird more susceptible to worse lesions.

DIAGNOSIS

DIFFERENTIAL DIAGNOSIS
- Malnutrition: Squamous metaplasia from hypovitaminosis A. • Candidiasis. • Poxvirus. • Herpesvirus. • Bacterial stomatitis. • Neoplasia. • Capillaria. • Nodules in the pigeon choanal area can be due to harmless sialolytes. • Digestive tract malfunction caused by: hepatic disease, GIT foreign bodies, neoplasia, other parasites, viruses, yeast and bacteria, proventricular dilatation disease, and heavy metal toxicity. • Causes of respiratory distress: other parasites (tracheal mites), viral, fungal (aspergillosis) or bacterial disease, respiratory foreign bodies, or toxic/irritating inhalants (PTFE). • Diarrhea: Other enteritis-causing organisms, toxicants, diet sensitivities, and others. • Liver lesions can resemble those of bacterial infections (including mycobacterium), *Aspergillus*, neoplasia, or histomoniasis. • Plaque lesions due to pigeon herpesvirus are generally more voluminous, located more rostrally, and the sites bleed when the plaques are removed.

CBC/BIOCHEMISTRY/URINALYSIS
The hemogram would likely be affected with a leukocytosis to a varying extent depending on progression of disease. Biochemistry might show hypoalbuminemia and might show changes with the liver-associated values if the organism has invaded the liver.

OTHER LABORATORY TESTS
For *T. gallinae* and *T. columbae*, wet mount of an esophageal swab or of the plaques might show flagellated protozoa (teardrop to round shape), three to five long anterior flagella and a tailing flagella attached to an undulating membrane (8–14 μm in size). It may slowly turn in one spot, or have irregular back and forth forward motion. For *T. gallinarum* a wet mount of fecal material can show a similar organism. The organism dies very quickly after the sample cools down so the direct evaluation must be immediate. In the absence of a wet mount the organisms can be seen on a sample smear stained with a "quick" cytological stain or trichrome stain.

IMAGING
Radiology might show a thickened crop/esophageal wall. The hepatic silhouette may be widened and irregular radioopacities present within. Contrast GI radiology might show ingluvial or gastric stasis. There might be opacities noted in the air sacs as well. The lesions if adjacent to bone can cause lytic change.

DIAGNOSTIC PROCEDURES
Upper GI endoscopy can allow evaluation of mucosal surfaces for lesions. Laparoscopy might be required to visibly evaluate the serosa of liver and other viscera, and air sacs (biopsies as needed) might be useful in cases where lesions are not present in more accessible locations.

PATHOLOGIC FINDINGS
Ulcerative and proliferative lesions, plaques and caseated material found in potentially many places, from sinuses down through the proventriculus and the trachea and the air sacs. The crop wall may be thickened and opaque otherwise. The liver may have discolored lesions present. Organisms may not be visible with histopathology.

TREATMENT

APPROPRIATE HEALTH CARE
For milder cases the patient can get home care assuming the caretaker is able to properly medicate and monitor food and fluid intake. For seriously ill birds, hospital care may be required to provide supportive care and debriding as needed of caseous material.

NURSING CARE
Supportive care needed may include fluid therapy (subcutaneous or intravascular crystalloids; with sufficient debilitation and/or hypoalbuminemia intravascular colloidal fluids may be indicated); supplemental warmth and humidity; oxygen therapy if patient is dyspneic; assisted feeding.

ACTIVITY
Activity should be reduced until signs resolve.

DIET
There may be maldigestion and significant weight loss; hyperalimentation may be required during recovery once digestive

TRICHOMONIASIS (CONTINUED)

function resumes. The use of liquified or other easily digested foods and added digestive enzymes to assist assimilation can also be helpful.

CLIENT EDUCATION
The treatment will likely be successful if there isn't sufficient damage caused by the organism however in some cases the prognosis will be poor. The overall predisposing factors that led to the disease need to be addressed including new bird quarantine and testing. Some protocols advocate regular flock treatments to reduce the burden of organisms before it causes too many birds in a flock to become ill.

SURGICAL CONSIDERATIONS
Surgical curettage of larger pockets such as in the sinuses can help limit the damage and resolve infection. With sufficient plaque buildup blocking the tracheal lumen, surgical placement of an air sac cannula might be helpful. Otherwise this is a medically managed condition.

MEDICATIONS

DRUG(S) OF CHOICE
Members of the nitroimidazole class are generally effective. Drugs include metronidazole at 25–50 mg/kg q12–24h, as well as ronidazole, and carnidazole.

CONTRAINDICATIONS
Some references report metronidazole toxicity in some types of birds such as finches. At this time the USA prohibits the use of nitroimidazoles in food producing poultry; confirm any legal restrictions at your location.

PRECAUTIONS
The nitroimidazoles are known to have neurologic side effects and dose may vary between individuals so observe for signs of reactions.

POSSIBLE INTERACTIONS
None.

ALTERNATIVE DRUGS
None.

FOLLOW-UP

PATIENT MONITORING
The patient should be continually monitored to make sure food intake is adequate, that any indication of a secondary infection is not present, and to make sure plaques are fully resolved within weeks. A follow-up esophageal swab and wet mount may be required to confirm success of therapy.

PREVENTION/AVOIDANCE
Quarantine and testing of new birds into groups, and regular testing of birds during annual or semiannual health examinations via esophageal swab and wet mount. In the case of raptors, not feeding prey of susceptible species. The organisms are sensitive to drying and die soon after elimination from the host, so control of the organism in the environment is straightforward.

POSSIBLE COMPLICATIONS
Permanent damage to the oropharyngeal tissues, sinuses and liver may occur. A plaque may slough and obstruct the airway.

EXPECTED COURSE AND PROGNOSIS
Prognosis is good for mild-to-moderately affected cases. Improvement of general clinical signs can start occurring within a few days and all lesions will likely disappear within weeks. Cases with deep tissue or visceral migration have a more guarded to poor prognosis and may take longer to return to normal.

MISCELLANEOUS

ASSOCIATED CONDITIONS
The many other diseases and infectious agents that are associated with the risk factors of this disease (lack of quarantine, overcrowding, immune system compromise) often exacerbate this condition. Pigeon herpesvirus lesions are often made worse with Trichomonas.

AGE-RELATED FACTORS
Nestlings are often more severely affected than adults. Adults are more likely to be infected without signs.

ZOONOTIC POTENTIAL
None.

FERTILITY/BREEDING
Birds should be cleared of this organism before breeding as chicks can be severely affected.

SYNONYMS
Canker, trich, roup, frounce

SEE ALSO
Air sac mites
Aspergillosis
Candidiasis
Diarrhea
Enteritis
Flagellate enteritis
Macrorhabdus ornithogaster
Nutritional deficiencies
Oral plaques
Poxvirus

ABBREVIATIONS
N/A

INTERNET RESOURCES
N/A

Suggested Reading
Altman, R.B., Clubb, S.L., Dorrestein, G.M., Quesenberry, K. (eds) (1997). *Avian Medicine and Surgery*. Philadelphia, PA: WB Saunders,
Ritchie, B.W., Harrison, G.J., Harrison, L.R. (eds) (1994) *Avian Medicine: Principles and Application*. Lake Worth, FL: Wingers Publishing, Inc.
Rosskopf, W., Woerpel, R. (1996). *Diseases of Cage and Aviary Birds*, 3rd edn. Baltimore, MD: Williams and Wilkins.
Samour, J. (ed.) (2000). *Avian Medicine*. London, UK: Mosby Publishing.
Tully, T., Lawton, M., Dorrestein, G. (2000). *Avian Medicine*. Woburn, MA: Butterworth-Heinemann.
Author Vanessa Rolfe, DVM, ABVP (Avian)

TUMORS

AVIAN

BASICS

DEFINITION
Tumors are a collection of cells that show abnormal growth patterns and are no longer under control. They are classified into benign tumors that grow locally and do not metastasize, in situ tumors that arise in the epithelium and do not invade the basement membrane and malignant tumors that can invade both locally or distantly by metastasis.

PATHOPHYSIOLOGY
Tumors occur when normal cells have uncontrolled proliferation or loss of appropriate cell death. The loss of normal cellular homeostasis is caused by a variety of genetic, chemical, physical, viral and hormonal factors.

SYSTEMS AFFECTED (STUDY CITED IN FILIPPICH, 2004)
- Tumors have been reported in many organ systems in birds but in 1992 the most common system affected was the skin (31.7%), urinary (25.1%) and genital (17.3%) systems based on necropsies of 1539 birds. • The following is a list of tumors that have been described in various organ systems in birds ○ Skin and Uropygial gland—basal cell tumor, basal cell carcinoma, fibroma, fibrocarcinoma, hemangioma, hemangiosarcoma, lipoma, liposarcoma, myelolipoma, hemangiolipoma, osseous metaplasia, hematopoietic neoplasms, cutaneous pseudolymphoma, malignant melanoma, cutaneous papilloma, squamous cell carcinoma, granular cell tumor, xanthoma, mast cell tumor with mastocytosis, hamartoma, cutaneous lymphoma, myxoma. ○ Renal—nephroblastoma, adenocarcinoma, adenoma. ○ Reproductive—ovarian/oviductal adenocarcinoma, carcinoma, cystadenocarcinoma, granulosa cell tumor; testicular seminoma, Sertoli cell and interstitial cell tumor, ovarian hemangiosarcoma. ○ Hemic/Vascular/Lymphatic/Immune—thymoma, hemangiosarcoma, hemangioma, lymphosarcoma (see "Lymphoid Neoplasia"). ○ Gastrointestinal—leiomyosarcoma, papilloma, squamous cell carcinoma, proventricular adenocarcinoma, proventricular hemangiosarcoma, ventricular adenocarcinoma, teratoma attached to ventriculus, pancreatic adenoma, pancreatic adenocarcinoma and pancreatic carcinoma. ○ Hepatobiliary—cholangiocarcinoma, hepatocellular carcinoma, lymphosarcoma, hepatic lipoma. ○ Ophthalmic—conjunctival xanthoma, intraocular osteosarcoma, conjunctival lipoma, undifferentiated carcinoma, multicystic adenoma, pigmented iridociliary adenoma, teratoid medulloepithelioma (presumptive), ganglioneuroma (presumptive), lymphoma, metastatic papillary cystic carcinoma, conjunctival squamous cell carcinoma, teratoma. ○ Musculoskeletal—chondroma, chondrosarcoma, osteoma, osteosarcoma, rhabdomyosarcoma, rhabdomyoma, synovial cell sarcoma, teratoma. ○ Endocrine—adrenal carcinoma, thyroid carcinoma, thyroid adenoma. ○ Respiratory tract—sinus carcinoma, sinus adenocarcinoma, primary pulmonary carcinoma, air sac carcinomas, undifferentiated pulmonary tumors of cockatiels, teratoma. ○ Nervous—pituitary adenocarcinoma, pituitary carcinoma, chromophobe pituitary tumors, astrocytoma, primitive neuroectodermal tumor, glial cell tumors, choroid plexus papillomas, lymphosarcoma.

GENETICS
Cancer is a genetic disease. Although not proven in avian species, there is suspicion that inbreeding of captive psittacine species results in a closed gene pool can lead to increased incidence of neoplasia.

INCIDENCE/PREVALENCE
(INFORMATION BASED ON FILIPPICH, 2004)
- More frequently reported in captive vs. wild birds. • Incidence increased over time from 1.17% in 1933 to 2.6% in 1959 based on necropsies at the Philadelphia Zoo.
- Incidence in birds in Australia (10,000 commercial, zoological, aviculture, pet and free-flying wild birds) in 1992 was 3.8%.
- Incidence in birds in the United States (15,000 birds) in 2002 was 6.95%.

GEOGRAPHIC DISTRIBUTION
N/A

SIGNALMENT
- **Species/Order predilections (information based on Filippich, 2004 and Reavill, 2004):** ○ Most commonly reported in Psittaciformes (3.6%), Galliformes (1.41%) and Anseriformes (0.89%) with the lowest incidence in Passeriformes (0.46%). ○ In Psittaciformes, budgerigars (*Melopsittacus undulatus*) have the highest incidence of 15.8–24.2%. They have a higher incidence of ovarian, oviductal, testicular, renal, chromophobe pituitary tumors and lipomas. ○ Gray-cheeked parakeet (*Brotogeris pyrrhopterus*) has a higher incidence of proventricular adenocarcinoma. ○ Cockatoos have a higher incidence of lipomas. ○ Cockatiels (*Nymphicus hollandicus*) have a higher incidence of ovarian, oviductal and testicular tumors. ○ New World Psittaciformes such as macaws (*Ara spp*), Amazon parrots (*Amazona spp*) and Hawk-headed parrots (*Deroptyus accipitrinus accipitrinus*) have a higher incidence of papillomas associated with herpesvirus. • **Mean age and range:** As with all species, incidence of neoplasia increases with age but should not be ruled out in juvenile patients. • **Predominant sex:** Primary renal tumors more frequently reported in young-middle aged male budgerigars, chromophobe pituitary tumors are more frequently reported in budgerigars less than 30 months of age.

SIGNS
- Historical and physical examination findings are often nonspecific, progressive and depend on the organ system affected. Specifically the owner may report swelling in the affected region that may also be palpated or visualized on clinical examination.
- Specific historical signs and physical examination findings can point to the organ system affected. For instance, regurgitation, melena or hematochezia may be noted with gastrointestinal tumors and dyspnea, exercise intolerance and tachypnea may be noted with respiratory tumors.

CAUSES
- Viral infections: Papillomatosis is caused by herpesvirus which has been linked with bile duct carcinoma, retroviruses have been detected in renal carcinomas of budgerigars.
- Environmental: Exposure to carcinogens.
- Genetic: Inbreeding.

RISK FACTORS
See above "Causes" and "Species/Order Predilections"

DIAGNOSIS

DIFFERENTIAL DIAGNOSIS
In general, swellings associated with tumors tend to be nonpainful, progressive and more chronic in nature. Specific differentials for swellings associated with the coelom, skin and musculoskeletal system are:
- Coelom: Abscess/granuloma, egg (either mineralized or unmineralized), hepatosplenomegaly due to systemic infection. • Skin: Feather cysts, cutaneous avipoxvirus infection, feathers cysts, granuloma/abscess, foreign body reaction.
- Musculoskeletal: Hematoma, soft tissue injury, callus from previous fracture, granuloma/osteomyelitis.

CBC/BIOCHEMISTRY/URINALYSIS
- CBC ○ Nonregenerative anemia secondary to chronic inflammation/disease.
○ Leukocytosis if associated with secondary infection or tumor necrosis. ○ Leukopenia if associated with immunosuppression.
- Biochemistry: Depending on the system involved, changes may be seen on the serum biochemistry. Note that the absence of elevations in specific enzymes does not preclude severe involvement of that organ system. • Hepatic involvement: Elevations in a spartate transaminase (AST), gamma glutamyl transferase (GGT) and/or bile acids.
- Renal involvement: Elevations in uric acid,

phosphorus and/or potassium. • Muscloskeletal involvement: Elevations in AST and creatinine phosphokinase (CPK). • Skeletal involvement: Elevations in calcium and/or phosphorus if bone is being destroyed. • Generalized nonspecific abnormalities: Elevations in AST and CPK in if the bird is weak and/or anorexic and has muscle break down.

OTHER LABORATORY TESTS
Herpesviral PCR and serological testing can be performed in patients with suspected herpesviral papillomatosis. Positive PCR or serology indicating exposure does not definitively mean the lesion is associated with herpesvirus, therefore biopsy (incisional or excisional) should still be performed for definitive diagnosis.

IMAGING
Diagnostic imaging is used to not only diagnose tumors but also help for staging, surgical and radiation treatment planning and response to therapy. • Radiographs: Used as a screening test for skeletal and coelomic tumors which is followed by other imaging to better differentiate and define tumor extent and evaluate for metastasis. In birds, whole body radiographs are often obtained. This may be followed with a barium series if intracoelomic tumors are noted and ultrasound is not available. • Ultrasound: Used to evaluate coelomic tumors and is most helpful if effusion is present. This is helpful to determine which organ is affected and can be used for ultrasound guided biopsy, fine needle aspiration or collection of coelomic fluid to help with diagnosis. • Computed tomography (CT): Used for better evaluation of the lungs, ribs and bones for primary and metastatic lesions (more reliable than radiographs). When combined with IV contrast, CT helps to delineate the tumor extension for surgical and radiation therapy. • Magnetic resonance imaging (MRI): Although in its infancy in avian medicine, this modality is used to evaluate tumors of the central nervous system. • Scintigraphy: Rarely reported in avian medicine, this modality is used to detect early bone metastases but can also be used for renal, thyroid, lung and liver evaluation. • Positron emission tomography (PET): This modality is used to detect distant metastases. Use for detection of cancer in avian patients has not been reported, although results of whole body PET scans have been published in healthy Hispaniolan Amazon parrots.

DIAGNOSTIC PROCEDURES
• Diagnostic cytology via fine needle aspiration, scrapings and impression smears (i.e., superficial cutaneous lesions) may be helpful for initially differentiating between cyst vs. granuloma vs. tumor but can be nondiagnostic or inconclusive. • Biopsy with histologic examination and immunohistochemistry is necessary for definitive diagnoses. ○ Methods for biopsy collection include needle core, punch, incisional or excisional. These can be obtained surgically, endoscopically, ultrasound or CT guided. ○ The goal is to determine if the lesion is a tumor, if it is benign vs. malignant, the histologic type and margins (if applicable). ○ If a viral etiology is suspected, such as herpesvirus with papillomatosis and retroviruses in renal tumors, samples should also be submitted for PCR testing or special stains. • Coelomic fluid analysis can sometimes aid in diagnosis of a tumor if neoplastic cells are present in the effusion. • Bone marrow biopsy or aspiration is warranted if abnormalities are noted in the thrombocytes, white or red blood cells.

PATHOLOGIC FINDINGS
Gross pathologic findings
• Solid tumors can be noted in any organ system. Grossly they usually appear pale in color but can be red/purple if affecting the vascular system. Tumors can appear papillomatous, encapsulated and smooth or cystic and nodular with obvious extension into surrounding tissue. Lesions may be singular or multiple and in cases of metastasis can also be found in the lungs, liver and spleen.

Histopathologic findings
• Histology is necessary for definitive diagnosis. At this time, there is no standardized histopathologic grading for avian tumors. Additional biopsies may be warranted for immunohistochemistry, special stains and viral PCR.

TREATMENT
APPROPRIATE HEALTH CARE
Before initiating treatment in a patient with a tumor, the goals, side effects, and potential outcome of the treatment must be discussed. It first must be established if the goal is palliative or curative as this will help to tailor the treatment plan. • Surgical oncology: Surgery is the treatment of choice for curing solitary tumors with limited potential for metastasis that can be removed with adequate margins. Surgery may also be used for diagnostic biopsy, palliation of symptoms by removing painful tumors or surgical debulking as an adjunct to radiation therapy and/or chemotherapy. Cryosurgery or hyperthermia can be used with specific, small, solitary, solid tumors. • Chemotherapy: Successful use of chemotherapy in birds has been reported. It is indicated in tumors that are known to be sensitive to chemotherapy (i.e., round cell tumors), as an adjuvant therapy to eradicate micrometastasis, as an adjuvant therapy to prevent recurrence after incomplete surgical excision, to decrease tumor size before surgery/radiation therapy or as palliative treatment when a definitive cure is not an option. Specific doses are not well established in birds so many doses are extrapolated from canine and feline medicine. Reported chemotherapeutic agents used in birds include asparaginase (*Note*: Elspar is no longer manufactured at the time of publication but can be compounded), carboplatin, chlorambucil, cisplatin, cyclophosphamide, doxorubicin, prednisone and vincristine sulfate. Pharmacodynamic studies have been published for cisplatin and carboplatin in sulfur-crested cockatoos. • Radiation therapy: Use of radiation therapy in birds has been reported with variable success. It is used for control of local, solid tumors either alone or in combination with surgery, and can be administered before surgery to decrease the tumor size or after incomplete surgical excision. It is also used as a palliative treatment when surgery is not an option. Extrapolating doses used in canine and feline medicine seems to be inadequate for treatment of tumors in birds. Additionally, birds seem to have a high normal tissue tolerance as doses as high as 72 gy in 4 gy fractions showed minimal epidermal changes and no chronic changes in ring-necked parakeets. • Photodynamic therapy: Alone or combined with photosensitizers, this modality has been successfully used in avian patients with squamous cell carcinomas. Photosensitizing agents that have been used in birds in combination with photodynamic therapy include hexyl ether pyropheophorbide-a and porfimer sodium. • Immunotherapy, molecular and targeted therapies are on the forefront of tumor treatment in humans and domestic species. There has been some initial success with immunotherapy in avian patients and as more research is done in this field, hopefully this will be a novel modality for avian tumor treatment in the future.

NURSING CARE
• If the neoplasia is causing pain or discomfort appropriate analgesia should be administered. Opioids, NSAIDs and local analgesic blocks are often used together although the animal must be appropriately hydrated and have normal renal and hepatic biochemistry values before NSAIDs are administered. • If the patient is dehydrated or hypovolemic secondary to systemic illness or anorexia/regurgitation, fluid therapy should be initiated. Subcutaneous, intravenous, intraosseous and oral routes can be utilized. • Oxygen support is necessary if the patient is dyspneic due to primary pulmonary neoplasia or air sac compression secondary to organomegaly or ascites. • If ascites is present and causing respiratory distress, therapeutic coelomocentesis can be performed. • If there is a secondary fungal or bacterial infection, appropriate antimicrobials should be

TUMORS (CONTINUED)

administered based on culture and sensitivity results.

ACTIVITY
If the patient is weak or systemically ill, activity should be limited. A cage can be set up with low/no perches to conserve energy and the patient should not be permitted to fly.

DIET
The patient should be encouraged to eat. Supportive feeding may be necessarily in the form of gavage or syringe feeding in anorexic patients or those that are losing weight.

CLIENT EDUCATION
• It is important to discuss the owners' goals in treatment of the neoplasia—whether it be curative or palliative. • Owners should be aware that prognosis is difficult to assess in each patient due to the paucity of reported cases in avian medicine.

SURGICAL CONSIDERATIONS
Any patient with cancer is at a higher risk for anesthetic and surgical complications.

MEDICATIONS

DRUG(S) OF CHOICE
See chemotherapy above.

CONTRAINDICATIONS
Cisplatin may cause nephrotoxicity in sulfur-crested cockatoos and should not be used in patients with pre-existing renal disease.

PRECAUTIONS
Because chemotherapy can cause myelosupression, vomiting/regurgitation, diarrhea, anorexia, neuropathy, hepatotoxicity, hypersensitivity and/or nephrotoxicity, side effects of any other medication must be thoroughly investigated before initiation to prevent worsening of these toxicities.

POSSIBLE INTERACTIONS
Drug interactions can occur with chemotherapeutic agents. Therefore, no medications or supplements should be started without checking with a veterinarian first.

ALTERNATIVE DRUGS
• Prednisone or prednisolone can be used as a single chemotherapeutic agent for palliation. It must be used judiciously in avian patients as significant side effects such as immunosuppression, diabetes mellitus, hyperadrenocorticism and hepatic damage have been reported following administration. • COX-2 inhibitors have been shown to be a successful adjuvant or palliative single agent for carcinomas treatment in other species. It is unknown if there is COX-2 overexpression in avian carcinomas but it has been used in a cockatiel with pancreatic adenocarcinoma. • GnRH agonists and analogues have been helpful in palliative management of ovarian adenocarcinoma in cockatiels.

FOLLOW-UP

PATIENT MONITORING
• In cases of solitary tumors that are not surgically removed, the size of the tumor should be monitored for response to treatment. Frequency depends on the treatment modality, type of tumor and if the goals are palliative vs. curative. • Follow-up often includes diagnostic imaging if the tumor is not readily visible or if metastasis is a concern.

PREVENTION /AVOIDANCE
N/A

POSSIBLE COMPLICATIONS
• Recurrence or metastasis of the neoplasia can occur. • Side effects of chemotherapeutic agents can be severe and can include toxicity as well as infections secondary to immunosuppression. Birds seem to be resistant to side effects of radiation therapy but focal dermal irritation can occur.

EXPECTED COURSE AND PROGNOSIS
To date, prognosis in avian patients with tumors is difficult to determine due to the low number of published cases although successful treatment of has been achieved with chemotherapy, surgical, radiation and photodynamic therapy.

MISCELLANEOUS

ASSOCIATED CONDITIONS
• Immunosuppression as a result of the primary tumor or the treatment can occur, leading to secondary bacterial and/or fungal infections. • As with other species, paraneoplastic syndromes have been reported in avian species.

AGE-RELATED FACTORS
As with all species, incidence of neoplasia increases with age but should not be ruled out in juvenile patients.

ZOONOTIC POTENTIAL
N/A

FERTILITY/BREEDING
If the reproductive tract (i.e., oviduct or testes) is involved this can decrease fertility.

SYNONYMS
Neoplasia, cancer, cancerous growth, malignant growth, malignancy.

SEE ALSO
CNS tumors, Brain tumors, Pituitary tumors
Lipomas
Lymphoid neoplasia
Ovarian cysts, Neoplasia and Cystic ovaries)
Squamous cell carcinoma
Viral neoplasms: Marek's disease, Lymphoid leukosis and Reticuloendotheliosis
Xanthoma

ABBREVIATIONS
N/A

INTERNET RESOURCES
N/A

Suggested Reading
Barron, H.W., Roberts, R.E., Latimer, K.S., et al. (2009). Tolerance doses of cutaneous and mucosal tissues in ring-necked Parakeets (*Psittacula krameri*) for external beam megavoltage radiation. *Journal of the American Veterinary Medical Association*, **23**(1):6–9.
Farese, J.P., Withrow, S.J., Gustafson, D.L., et al. (2013). Therapeutic modalities for the cancer patient. In: Withrow, S.J., MacEwen, E.G. (eds). *Small Animal Clinical Oncology*, 5th edn. Philadelphia, PA: WB Saunders Co., pp. 149–304.
Filippich, L.J. (2004). Tumor control in birds. *Seminars in Avian and Exotic Pet Medicine*, **13**(1):25–43.
Graham, J., Reavill, D., Zehnder, A. (2014). Avian oncology master class. *Proceedings of the Association of Avian Veterinarians Annual Conference*, pp. 169–181.
Reavill, D.R. (2004). Tumors of pet birds. *The Veterinary Clinics of North America. Exotic Animal Practice*, **7**:537–5604.
Author Nicole R. Wyre, DVM, DABVP (Avian)

Undigested Food in Droppings

 BASICS

DEFINITION
Certain avian disorders can result in the passing of undigested food in the droppings; this is most often noticed in birds that are eating seed. It may be more difficult to detect the presence of undigested food in the stool of a bird that is eating a formulated diet or other nonseed items; in those cases there might be bulky or loose stools without obvious undigested food. Diarrhea is defined as an increased frequency and liquidity of the fecal discharges; in the avian patient, one can also add fecal consistency and the presence of undigested food.

PATHOPHYSIOLOGY
Formation of a proper fecal portion is dependent on proper intestinal, hepatic and pancreatic function as well as the mechanical functions of the coprodeum, urodeum and vent. Alterations in any of these organ systems or structures may alter the avian patient's ability to create a formed, consistent fecal movement. There are a few mechanisms by which a bird can develop the presence of undigested food in the stool. In the normal bird, there is back and forth movement of ingesta between the proventriculus and the ventriculus. Disease of either or both of these organs can interfere with this process, resulting in the passage of poorly prepared digesta into the small intestine and eventually into the stool. Disease of the small intestines can result in this clinical sign due to malabsorption or hypermotility. Pancreatic disease with resultant exocrine deficiency can result in inadequate digestion. Coelomic disorders can interfere with gastrointestinal function, resulting in clinical signs of such.

SYSTEMS AFFECTED
Note: while in most cases the cause of undigested food in the droppings is going to be related to disease of the gastrointestinal (GI) tract or the pancreas, there might be other organ systems affected, either secondarily or concomitantly with the GI tract. • Gastrointestinal—disease of the proventriculus, ventriculus, small intestine, pancreas. • Hepatobiliary—any liver disorder in theory, but undigested food is not commonly seen with liver disease. • Musculoskeletal—muscle wasting due to inability to properly digest nutrients. • Nervous—CNS signs – muscle weakness and seizures seen with lead toxicosis, bornaviral disease. • Renal/urologic—renal disease seen with zinc and lead, kidneys secondarily affected with many disorders. • Reproductive—disorders involving the female reproductive tract could secondarily affect the GI tract and pancreas. • Skin—may see feather picking secondary to giardiasis, poor general feather condition with severe GI or any chronic disease.

GENETICS
N/A

INCIDENCE/PREVALENCE
While the incidence and prevalence of this clinical sign are not known, there are many disorders that can cause; although many of them might be fairly uncommon, when they are all added up, this clinical sign is regularly seen. It is important to note that there are a number of etiologies and pathophysiologies that could be involved, and while some of them may be fairly rare, if all else is ruled out, they should be included in the differential diagnosis and ultimately ruled out.

GEOGRAPHIC DISTRIBUTION
This clinical sign has no geographic boundaries. Some of the disorders might be more common in certain parts of the world.

SIGNALMENT
• **Species:** ○ Any species of birds could develop this clinical sign. Some species are known for certain disorders: ▪ *Macrorhabdus ornithogaster* (avian gastric yeast): Budgerigars, cockatiels, lovebirds, finches, parrotlets. ▪ Giardiasis: Budgerigars, cockatiels, lovebirds, rarely other parrots, finches. ▪ *Spironucleus* spp: Pigeons (*S. columbae*), turkeys (*S. meleagridis*), budgerigars, cockatiels, Australian King Parrots. ▪ Cochlosomiasis: Cockatiels, finches, esp. Gouldians. ▪ Candidiasis: Cockatiels, finches. ▪ Gizzard worms – *Dyspharynx* spp, *Acuaria* spp – finches and galliformes. ▪ PDD: African grey parrots, Amazon parrots, cockatoos, macaws and others. ▪ Proventricular/ventricular neoplasia: Budgerigars, grey-cheeked parakeets, lovebirds and Amazon parrots. ▪ Mycobacteriosis: Grey-cheeked parakeets, Amazon parrots, waterfowl, and zoo collections. ▪ Internal papillomatosis: Macaws, Amazon parrots, and conures.
• **Mean age and range:** Many infectious disorders such as flagellate enteritis, bacterial infection and macrorhabdosis are seen more commonly in the young; neoplastic processes are more likely in the elderly patient.
• **Predominant sex:** Both sexes are susceptible.

SIGNS

General Comments
Signs and findings include those that may be seen in conjunction with this clinical sign and those that occur concurrently as a result of the inciting problem.

Historical Findings
There are many possible historical findings in association with this clinical sign, depending on the specific disorder.

Physical Examination Findings
• Undigested food in droppings—seeds or other food items seen; if not eating seeds, often seen as bulky or voluminous feces. • Malodorous feces. • Diarrhea (not to be confused with polyuria). • Anorexia or, in some cases, increased appetite. • Poor body condition with loss of body fat and muscle wasting. • Distended crop; delayed crop emptying. • Regurgitation/vomiting. • Weakness, depression due to lack of nutrient assimilation. • True diarrhea—liquid frequent stools. • Biliverdinuria due to hepatic disease. • Hemoglobinuria—in lead poisoning cases. • Hematochezia and/or melena—in some cases of enteritis, zinc toxicosis. • Gastrointestinal stasis—lead poisoning, grit impaction. • Seizures and other neurologic signs—lead poisoning. • Coelomic distention—due to fluid, enlarged liver, mass, reproductive disease.

CAUSES
A comprehensive list of potential causes is listed below. In general, disorders of the proventriculus, ventriculus and pancreas are arguably the most common causes for this clinical sign, yet a number of small intestinal, hepatic and systemic disorders should be included for the sake of thoroughness.

Proventricular disease
• Gastric foreign bodies or impaction. • Proventricular dilatation disease (PDD)—bornavirus-related. • Macrorhabdosis. • Internal poxvirus. • Herpesvirus. • Bacterial infections—primary or secondary. • Mycobacterial disease. • Candidiasis. • Zygomycete fungal infection. • Cryptosporidiosis. • Nematodes: *Spiroptera sp.*, *Dyspharynx sp. Tetrameres* sp., others. • Internal papillomatosis. • Proventricular carcinomas—at the isthmus.

Ventricular disease
• Foreign bodies – damaged koilin – can perforate and cause coelomitis. • Grit impaction. • Mineralization. • Zinc toxicosis—ventricular erosion and pancreatic damage. • PDD. • Adenoviral infection. • Bacterial infection. • Mycobacterial infection. • Ventricular mycosis – *Candida*; occasional zygomycetes – esp. finches. • Nematodes. • Neoplasia—carcinomas.

Small intestinal disease
• Candidiasis. • Coccidial infections. • Flagellate enteritis—giardiasis, spironucleosis, and cochlosomiasis. • Bacterial enteritis—*Escherichia coli*, *Klebsiella*, *Salmonella*, and *Enterobacter* spp. *Clostridium*, *Campylobacter*, *Enterococcus* spp. others. • Mycobacterial infection. • Chlamydiosis. • Nematodes. • Intestinal neoplasia. • Viral disease—PDD, paramyxovirus, herpesvirus, adenovirus, reovirus, rotavirus.

Pancreatic diseases
• Pancreatitis. • Pancreatic atrophy/fibrosis. • Chlamydiosis. • Zinc toxicosis.

Hepatic disease
• Numerous etiologies/disorders but probably unlikely to result in undigested food in droppings.

UNDIGESTED FOOD IN DROPPINGS (CONTINUED)

Coelomic disease
• Aspergillosis. • Coelomitis—perforated GI tract, egg-related. • Coelomic neoplasia.

RISK FACTORS
Poor sanitation, poor food quality, malnutrition, stress, overcrowding, exposure to other birds with contagious diseases, age, concurrent diseases, long term antibiotic use (candidiasis), allowing birds to roam throughout the home (lead, zinc, other toxins) are predisposing factors.

DIAGNOSIS

DIFFERENTIAL DIAGNOSIS
Covered under causes of this clinical sign.

CBC/BIOCHEMISTRY/URINALYSIS
• CBC may show anemia in lead poisoning cases. • Leukocytosis may be seen in cases of significant inflammation/infection (aspergillosis, mycobacteriosis, chlamydiosis, other). • Eosinophilia seen sometimes with giardiasis and other flagellates.
• Hypoproteinemia is seen in cases where there is significant malabsorption. • There may be other electrophoretic changes associated with infection, protein loss, liver diseases, egg-laying disorders and other conditions. • Elevated bile acids and liver enzymes will be seen in cases of liver disease.

OTHER LABORATORY TESTS
• Direct fecal smears—can see yeasts, several types of parasites including flagellates, avian gastric yeast. • Fecal parasite testing – zinc sulfate centrifugation – helminths. • Fecal Gram's staining (trying to match with culture results and to not overinterpret results).
• Fecal cytology – several stains – Dif-Quik, fungal stains, other. • Fecal acid fast staining—*Cryptosporidium* spp. and *Mycobacterium* spp. • Fecal culture—aerobic and anaerobic. • Fecal occult blood testing.
• Lead and zinc blood testing. • Fecal PCR testing—*Chlamydia*, *Cryptosporidium* sp., *Macrhorhabdus ornithogaster*, Bornavirus and other viral testing.

IMAGING
Radiography can be very useful in determining the cause of this clinical sign. Contrast radiographs might be helpful to delineate lesions such as tumors or to properly outline the gastrointestinal tract and in order to see the lining and the lumina of the viscous organs. Heavy metal foreign bodies might be visible in the ventriculus; the proventriculus could be dilated in cases of PDD and in lead toxicosis, internal papillomatosis and other conditions – it is best to do a contrast study in order to further clarify. A dilated, thin-walled proventriculus and ventriculus is suggestive of PDD but not pathognomonic. Fluoroscopy could also be done to further examine the gastrointestinal tract and especially its motility.

DIAGNOSTIC PROCEDURES
• Endoscopy—of the proventriculus or ventriculus (not often done) or the coelomic cavity - in order to obtain mucosal samples for cytology and culture. • Laparoscopy—to obtain biopsy specimens of internal organs such as liver, pancreas, small intestine.
• Biopsy of the crop (or less commonly, the proventriculus, ventriculus or small intestine)—generally in an attempt to confirm PDD.

PATHOLOGIC FINDINGS
Gross pathological findings for the more common causes of this clinical sign are:
• PDD— enlarged, dilated thin walled proventriculus and/or ventriculus with potential dilation of the small intestine; histologically, lymphoplasmacytic myenteric ganglioneuropathy; lesions also in CNS, heart, adrenals. • Macrorhabdosis – minimal gross findings – excess mucous production; histologically, organisms seen in the proventricular mucosa, mononuclear inflammation. • Lead toxicosis—hemoglobinuria, may see enlarged dilated proventriculus or other areas of the GI tract; histologically, renal and hepatic pathology.
• Zinc toxicosis—grossly may see melena due to erosive ventriculitis; histologically, pancreatic lesions-vacuolation and degranulation of acinar cells. • Proventricular/ventricular carcinomas—infiltrative to proliferative mass, usually located at the isthmus. • Flagellate enteritis—distended intestines with excess mucous or gas or fluid, or no lesion; histologic evidence of organisms along the villi. • Proventricular/ventricular/intestinal bacterial or fungal or yeast infection—redness, exudation, erosions, degeneration of koilin layer, organisms present in koilin or deeper into the tissues.
• Clostridial overgrowth in intestines—focal to diffuse hemorrhage, necrosis and fibrin deposition; presence of large gram-positive spore forming rods. • Mycobacteriosis—diffuse and/or nodular thickening and opacification of the intestinal wall; histologically, diffuse infiltration of lamina propria with large macrophages that contain acid-fast bacteria. • Pancreatic atrophy/fibrosis—pale, small firm pancreas; histologically normal acini replaced by interstitial fibrosis.

TREATMENT

APPROPRIATE HEALTH CARE
• Patients with mild-to-moderate disease without other, more serious clinical signs can be treated as outpatients, pending results of testing. • Patients with severe disease need to be hospitalized in order to stabilize while a thorough diagnostic work-up is performed.
• For inpatients with foreign bodies, including lead and zinc and other heavy metals, surgical intervention might be necessary.

NURSING CARE
Nursing care must be done in the sicker patients. This might include incubation, gavage feeding, fluid therapy and possibly oxygen therapy.

ACTIVITY
N/A

DIET
It makes sense that dietary manipulation could be beneficial for these patients, especially while diagnostic testing results are pending and also if the patient is weakened or otherwise doing poorly. Formulated diets should be more digestible for these patients.

CLIENT EDUCATION
Proper administration of medications should be explained. Clients should be advised to watch for worsening of the patient's condition and to return to the hospital as needed. Monitoring of the droppings by the owner should be emphasized.

SURGICAL CONSIDERATIONS
N/A

MEDICATIONS

DRUG(S) OF CHOICE
• For candidiasis, nystatin (300,000 IU/kg PO q12h) may be used for mild infection. Azole antifungals (itraconazole 5–10 mg/kg QD, terbinafine, voriconazole) are indicated for severely invasive and systemic infection.
• For *Macrorhabdus* infection, amphotericin B (compounded) at 100 mg/kg PO or gavage BID for up to 30 days; recent literature suggests doses as low as 25 mg/kg BID for 14 days might be effective. Treatment with sodium benzoate in drinking water at 1 tbs/L.
• For giardiasis, cochlosomiasis and spironucleosis, various nitroimidazoles can be used, including metronidazole, ronidazole, carnidazole and dimetridazole. • For cryptosporidiosis, ponazuril might be successful. • For lead and zinc toxicosis, 35 mg/kg calcium disodium EDTA IM BID – usually done five days on and 2–5 days off, then repeated; it is unclear if this is a necessary precaution. • For PDD, Cox-2 inhibitors such as celecoxib at 10 mg/kg QD PO. • For bacterial infections, selection of the appropriate antibiotic should be based upon the results of culture and sensitivity testing. Note that there is a degree of clinical judgment needed when faced with the results of a fecal or cloacal culture – the presence of a Gram-negative organism does not equate with disease of the gastrointestinal tract. • For

UNDIGESTED FOOD IN DROPPINGS (CONTINUED)

clostridial infections, clindamycin (100 mg/kg PO QD) and amoxicillin/clavulanate (125 mg/kg PO BID) have been suggested.
• For mycobacteriosis, several multiple-drug protocols have been recommended. • For pancreatic insufficiency, pancreatic enzymes can be supplemented.

CONTRAINDICATIONS
N/A

PRECAUTIONS
N/A

POSSIBLE INTERACTIONS
N/A

ALTERNATIVE DRUGS
N/A

FOLLOW-UP

PATIENT MONITORING
The owner at home should monitor all patients, either during the diagnostic work-up phase, and during and after treatment, or by the veterinary team in the hospital. Follow-up examinations, including laboratory testing – as needed – should be done.

PREVENTION/AVOIDANCE
Prevent further access to heavy metals and foreign bodies. Avoid exposure/housing with individuals that have been proven to have infectious, contagious diseases.

POSSIBLE COMPLICATIONS
N/A

EXPECTED COURSE AND PROGNOSIS
The presence of seed in the droppings is generally a fairly grave sign. Determining a prognosis is dependent upon arriving at a proper diagnosis. In theory the sooner the diagnosis is achieved, the better the prognosis. For bacterial, flagellate, fungal and chlamydial infection the prognosis with treatment should be good. Cryptosporidiosis is difficult to treat and the prognosis is fairly poor. Birds with clinical PDD generally don't survive very long, but there are exceptions. Lead and zinc toxicosis cases can be cured with chelation therapy for the proper length of time. Mycobacterial infections are very tedious and difficult to treat and treatment is controversial.

MISCELLANEOUS

ASSOCIATED CONDITIONS
Conditions associated with this clinical sign have all been listed.

AGE-RELATED FACTORS
The very young are most affected in all of these protozoal diseases; parents/adults are often asymptomatic carriers.

ZOONOTIC POTENTIAL
There is some question about the zoonotic potential of cryptosporidiosis, although it seems to be very rare or unlikely; mycobacterial infections are potentially zoonotic but rarely reported; Giardia might be zoonotic but not well documented; some bacterial infections such as chlamydiosis and salmonellosis could cause disease in people.

FERTILITY/BREEDING
Several of these conditions could adversely affect future fertility, especially egg-related coelomitis, neoplasia, aspergillosis, PDD, mycobacteriosis.

SYNONYMS
N/A

SEE ALSO
Aspergillosis
Chlamydiosis
Cryptosporidiosis
Diarrhea
Emaciation
Enteritis/gastritis
Flagellate enteritis
Gastrointestinal foreign bodies
Heavy metal toxicity
Ingluvial hypomotility/ileus
Macrorhabdus ornithogaster
Mycobateriosis
Pancreatic diseases
Proventricular dilatation disease
Regurgitation/vomiting
Sick-bird syndrome
Tumors
Urate/fecal discoloration
Viral disease

ABBREVIATIONS
AGY—avian gastric yeast
PDD—proventricular dilatation disease

INTERNET RESOURCES
VIN searches and PubMed searches are useful to learn more detail about any of these individual topics.

Suggested Reading
Brandão, J., Beaufrère, H. (2013). Clinical update and treatment of selected infectious gastrointestinal diseases in avian species. *Journal of Exotic Pet Medicine*, **22**:101–117.
Clyde, C.I., Patton, S. (2000). Parasitism of caged birds. In: Olsen, G.H., Orosz, S.E. (eds), *Manual of Avian Medicine*. St. Louis, MO: Mosby, Inc, pp. 424–448.
Doneley, R.J.T. (2009). Bacterial and parasitic disease of parrots. *The Veterinary Clinics of North America. Exotic Animal Practice*, **12**(3):423–431.
Patton, S. (2000). Avian parasite testing. In: Fudge, A.M. (ed.), *Laboratory Medicine–Avian and Exotic Pets*. Philadelphia, PA: WB Saunders, pp. 147–156.
Reavill, D.R., Schmidt, R.E., Phalen D.N. (2003). Gastrointestinal system and pancreas. In: Schmidt, R.E., Reavill, D.R., Phalen D.N. (eds), *Pathology of Pet and Aviary Birds*. Ames, IA: Iowa State Press, pp. 41–65.

Author George A. Messenger, DVM, DABVP (Avian)

URATE AND FECAL DISCOLORATION

BASICS
DEFINITION
The avian dropping consists of three portions: the stool, the urates and liquid urine. The normal color of the stool of most birds on a seed diet is a rich dark green due to the presence of biliverdin. The urates are normally a very pure chalky white color. Birds may be presented for evaluation of abnormal urate/urine or fecal color, or these abnormalities may be discovered during the examination of the patient. The evaluation of bird droppings is an essential component of the overall exam. Abnormal coloration of the fecal or urate/urine portion of the dropping can help to reveal very valuable information about the health of the avian patient.

PATHOPHYSIOLOGY
• Fecal color can be strongly influenced by diet, bile pigments, and potentially blood from the GI tract. Bile pigments are a strong dark green color. This color presents in comparatively increased amount when there is inadequate food intake, but also can be often visible in the droppings from patients who eat primarily seed. • Birds with lead poisoning have increased destruction of red blood cells, resulting in intravascular hemolysis. This, combined with nephrosis, contributes to hemoglobinuria and/or hematuria. • Birds with upper gastrointestinal lesions or bleeding diathesis (due to vitamin K antagonist toxins and other causes) can have black stools indicative of melena. • In the presence of hepatic disease, biliverdinuria may occur, causing the urates to turn green or yellow. This occurs because of the accumulation of biliverdin, which is normally removed by the liver. Birds produce small amounts of bilirubin because of the decreased production of an enzyme (biliverdin reductase) that converts biliverdin into bilirubin.

SYSTEMS AFFECTED
• Gastrointestinal—upper gastrointestinal bleeding (black stools – melena). • Hemic/Lymphatic/Immune—intravascular hemolysis with lead toxicosis (red, pink, brown urine or urates). • Hepatobiliary—liver disease of all types (yellow, golden, mustard or green urates). • Renal/urinary—lead toxicosis – renal tubular necrosis, hemoglobinuria (red, pink, brown urates/urine); renal disease can result in hematuria also. • Musculoskeletal—muscle damage after intramuscular injection.

GENETICS
N/A

INCIDENCE/PREVALENCE
Discoloration of the urates or feces is a fairly common finding in birds.

GEOGRAPHIC DISTRIBUTION
Any of these diseases can be seen anywhere in the world.

SIGNALMENT
• **Species:** ◦ Amazon parrots, possibly galahs and eclectus parrots – more likely to develop hemoglobinuria secondary to lead poisoning. ◦ All birds otherwise susceptible to most disease processes that cause any of these clinical signs. • **Mean age and range:** All ages are susceptible. • **Predominant sex:** Both sexes are susceptible.

SIGNS
General Comments
Historical findings and physical findings will vary depending on the specific problem.
Historical Findings
It is important to get a good dietary history as well as to question the introduction of new toys into the bird's cage.
Physical Examination Findings
Lead toxicosis
• General nonspecific signs. • PU/PD. • Hemoglobinuria. • Weakness. • Seizures; occasionally other neurologic signs. • Anemia.
Liver diseases
• General nonspecific signs. • Vomiting/regurgitation. • Diarrhea. • Biliverdinuria. • Yellow to orange urates. • Enlarged liver. • Possible abdominal pain.
Intestinal diseases
• Nonspecific signs. • Diarrhea. • Undigested food in droppings. • Possible melena or frank blood (rare - usually cloacal) or clay colored stools.

CAUSES
Note: Leaching of dyes from colored newspaper can retroactively discolor the stool or urine part of a bird's dropping (when it is dropped onto the paper) and this should not be misconstrued as an abnormality associated with the patient.

Urate and/or urine discoloration
• **Green**—biliverdinura due to liver disease or (rarely) hemolytic anemia; GI stasis in anorexic birds or in stools that are not fresh, bile pigments may leach into urine/urates. • **Yellow**—liver disease; anorexia; muscle injury such as IM injection; use of B complex vitamins. • **Brown or reddish brown**—hemoglobinuria secondary to lead poisoning (see causes for red). • **Red**—hematuria/hemoglobinuria most often due to lead toxicosis (also describe as "tomato soup" or port wine or chocolate) and rarely mercury or zinc toxicosis; hematuria can also result from kidney, cloacal or reproductive disease. • **Pink**—hematuria or hemoglobinuria (same causes as for red). • **Orange**—vitamin B injection in last few hours. • **Blue, yellow, red, green, purple, orange – bright colors**—from blueberries, blackberries, beets, pomegranates, or other fruits or vegetables or from artificial dyes or food coloring in some toys or foods (including some formulated diets). This is usually going to be seen in the clear urine and perhaps in the stool and generally will not discolor the urates.

Fecal discoloration
• **Brown**—often normal in birds eating mainly formulated diet; normal in finches, canaries, quail and others. Coliform infections may produce a distinct brown color change and liquidity in certain species; may have an odor; as will viral diseases with secondary bacterial infection. • **Green** (more than normal)—this color can often present in comparatively increased amount when there is inadequate food intake. Budgerigars with macrorhabdosis will classically have a slightly paler green and looser stool. • **Clay**—malabsorption/maldigestion; intestinal disease, pancreatic and possibly liver disease. • **Black** – melena—caused by upper GI bleeding due to a variety of stomach (including ulcerative ventriculitis due to zinc toxicosis) or intestinal lesions, or a bleeding disorder. Anorexia or moribund states commonly result in this change, especially in smaller birds. Budgerigar stools often look black after they have dried out, but upon close exam of a fresh stool, they are dark green. • **Red**—hematuria (frank blood) due to lower GI disease (often cloacal), but also possibly from reproductive or urinary disease. • **Orange, yellow, blue, red, orange, purple**—as with urine/urate discoloration, these can be due to ingestion of beets, berries and other fruits and vegetables as well as artificial dyes or food coloring in food or toys.

RISK FACTORS
• Birds allowed to roam the house are more likely to develop heavy metal toxicosis • Poor diets and inactivity can very often lead to hepatic lipidosis in some species. • Exposure to other ill birds (herpesvirus, *Chlamydia*, etc.) increases the likelihood of acquiring infections

DIAGNOSIS
DIFFERENTIAL DIAGNOSIS
Differentials for all forms of urate and fecal discoloration are listed under causes.

CBC/BIOCHEMISTRY/URINALYSIS
• Findings will vary depending on the pathologic process causing the clinical sign. • Hypochromic regenerative anemia in lead toxicosis cases; basophilic stippling and ballooning degeneration in RBCs. • Liver disease—elevated bile acids; often see AST elevations, but this is not specific for liver.

OTHER LABORATORY TESTS
In most cases, the nature of the urine or fecal color change will determine the direction of the diagnostic work-up. Some additional tests that might be necessary include:
• Gross examination of droppings—make sure to carefully and critically examine and evaluate each portion. Roll out feces on a piece of white paper to look for undigested

URATE AND FECAL DISCOLORATION (CONTINUED)

food and blood clots and to differentiate urates from feces. • Fecal direct smear in the case of potential flagellates or helminths or avian gastric yeasts. • Fecal cytology—WBCs, RBCs and occasionally other findings. • Fecal flotation for parasite ova and *Giardia* cysts. • Fecal Gram's stains—if patient has diarrhea. Beware of overinterpreting the results. • Fecal acid-fast stain—mycobacteria and cryptosporidia. • Fecal culture and sensitivity—in cases of suspected bacterial infection, or to corroborate Gram's stain results. • Blood lead (and occasionally zinc) testing.

IMAGING
• Hepatomegaly or decreased liver size with some liver diseases. • Heavy metal objects can sometimes be seen in the ventriculus or other parts of the GI tract with lead and/or zinc poisoning.

DIAGNOSTIC PROCEDURES
• Laparoscopic or surgical biopsy of the liver or pancreas in some cases. • Exploratory laparotomy to evaluate organs in cases of suspected reproductive and other conditions. • Cloacoscopy in the case of lower GI bleeding (frank blood).

PATHOLOGIC FINDINGS
There could be a wide variety of lesions associated with discolored feces and urates. Some of the more common ones include:
• Any form of liver disease and associated pathology—hepatic lipidosis, chlamydiosis, mycobacteriosis, viral disease (herpesvirus and other), biliary carcinomas, bacterial disease, inflammatory and immune mediated conditions, aspergillosis, and cirrhosis. • In macrorhabdosis cases, increased mucous or mild changes in the proventriculus near or at the isthmus, presence of organisms microscopically. • In zinc toxicosis cases, erosive ventriculitis and pancreatic lesions may be seen. • In cases of small intestinal or pancreatic disease resulting in malabsorption/maldigestion or melena, numerous pathologic processes could be present, including chronic pancreatitis with fibrosis, pancreatic neoplasia, chronic zinc poisoning (pancreas), GI parasites (helminths or flagellates), bacterial enteritis, mycobacteriosis, chlamydiosis, neoplasia and other lesions.

TREATMENT

APPROPRIATE HEALTH CARE
During the examination, the severity of the bird's condition is determined. Birds that are seriously ill need to be hospitalized and subjected to an appropriate diagnostic workup and treatments as deemed necessary.

NURSING CARE
Fluid, thermal and nutritional support may be necessary in birds that are hospitalized.

ACTIVITY
N/A

DIET
N/A

CLIENT EDUCATION
In all cases, it is important to outline the course of action with the client, including the diagnostic and treatment plan and the prognosis. Explanation of the normal and abnormal appearance of avian droppings is always useful with all bird owners.

SURGICAL CONSIDERATIONS
N/A

MEDICATIONS

DRUG(S) OF CHOICE
• For lead and zinc toxicosis – calcium EDTA 35 mg/kg IM BID for 5–10 days; might need to be repeated or continued for longer periods. • For liver diseases—many possible therapies depending on the underlying cause. If underlying cause is not known then milk thistle products are often used; lactulose has been historically used but this might be of questionable efficacy. • For bacterial infections, antibiotics should be based upon the results of culture and sensitivity testing. • For macrorhabdosis, amphotericin B at 25–100 mg/kg PO BID for 10–30 days. The lower dosage might be efficacious.

CONTRAINDICATIONS
N/A

PRECAUTIONS
N/A

POSSIBLE INTERACTIONS
N/A

ALTERNATIVE DRUGS
N/A

FOLLOW-UP

PATIENT MONITORING
Each patient should be monitored as needed, depending on the severity of the condition and the specific disorder. This should include examinations as well as specific laboratory testing and examination of the droppings.

PREVENTION/AVOIDANCE
N/A

POSSIBLE COMPLICATIONS
N/A

EXPECTED COURSE AND PROGNOSIS
Prognosis is extremely variable, depending upon the specific diagnosis. Cases of lead poisoning carry a favorable prognosis if the condition is diagnosed and treated promptly.

MISCELLANEOUS

ASSOCIATED CONDITIONS
N/A

AGE-RELATED FACTORS
N/A

ZOONOTIC POTENTIAL
N/A

FERTILITY/BREEDING
N/A

SYNONYMS
N/A

SEE ALSO
Anorexia
Campylobacteriosis
Chlamydiosis
Cloacal disease
Clostridiosis
Coccidiosis (intestinal)
Colibacillosis
Cryptosporidiosis
Diarrhea
Enteritis/gastritis
Flagellate enteritis
Heavy metal toxicity
Herpesviruses
Intestinal helminthiasis
Liver disease
Macrorhabdus ornithogaster
Pancreatic diseases
Polyuria
Proventricular dilatation disease
Salmonellosis
Sick bird syndrome
Undigested food in droppings
Viral disease

ABBREVIATIONS
N/A

INTERNET RESOURCES
N/A

Suggested Reading
Bauck, L. (2000). Abnormal droppings. In: Olsen, G.H., Orosz, S.E. (eds), *Manual of Avian Medicine*. St. Louis, MO: Mosby, Inc., pp. 62–69.
Doneley, B., Harrison, G., Lightfoot, T. (2006). Maximizing information from the physical examination. In: Harrison, G.J., Lightfoot, T.L. (eds), *Clinical Avian Medicine*. Palm Beach, FL: Spix Publishing, Inc., pp. 153–211.
Fudge, A.M. (2000). Avian liver and gastrointestinal testing. In: Fudge A.M. (ed.), *Laboratory Medicine–Avian and Exotic Pets*. Philadelphia, PA: WB Saunders, pp. 47–55.
Author George A. Messenger, DVM, DABVP (Avian)

UROPYGIAL GLAND DISEASE

BASICS
DEFINITION
Disease or condition affecting the uropygial gland. The uropygial gland is a bilobed gland found dorsally at the base of the tail, secreting a lipoid sebaceous material that is suspected to have several roles: plumage waterproofing in some species, antimicrobial activity, protection against certain ectoparasites (e.g., feather lice), anti-abrasive effects during preening, conversion of provitamin D to vitamin D which is then ingested during preening, production of pheromones, sex-related changes in the UV appearance of the plumage, and excretion of some pesticides and pollutants.

PATHOPHYSIOLOGY
Trauma, neoplasia, impaction or infection of the gland, responsible for its increase in size, possible loss of function, and further infiltration/damage to adjacent uropygial structures.

SYSTEMS AFFECTED
Skin/Exocrine.

GENETICS
N/A

INCIDENCE/PREVALENCE
N/A

GEOGRAPHIC DISTRIBUTION
N/A

SIGNALMENT
- **Species:** All species of birds that have a uropygial gland (Amazon parrots, Hyacinth macaws, Pionus parrots, some columbiformes, ratites, bustards, frogmouths, and woodpeckers do not have a uropygial gland). • **Mean age and range:** Any age. • **Predominant sex:** Both.

SIGNS
Historical Findings
- Poor feather quality. • Mass dorsally at the base of the tail. • Overpreening at the base of the tail. • Bleeding dorsally at the base of the tail. • Water birds: disruption of feather waterproofing in some instances.

Physical Examination Findings
- Increased size of the uropygial gland. • Asymmetry of the gland. • Ulceration/necrosis of the gland, that can be associated with localized hemorrhage. • Feather loss around the gland.

CAUSES
- Nutritional: hypovitaminosis A responsible for glandular metaplasia and hyperkeratosis of the gland and possible impaction. • Neoplasia: adenoma, adenocarcinoma, squamous cell carcinoma, papilloma. • Inflammation: adenitis, chronic dermatitis. • Infection: bacterial abscess, fungal granuloma. • Impaction. • Foreign body. • Trauma: rupture of the gland causing inflammation and scar tissue formation.

RISK FACTORS
- Trauma: self-mutilation or trauma to the area of the gland. • Improper diet (hypovitaminosis A).

DIAGNOSIS
DIFFERENTIAL DIAGNOSIS
See "Causes".

CBC/BIOCHEMISTRY/URINALYSIS
Unremarkable in most cases. A mild leukocytosis and monocytosis may be present in inflammatory/infectious processes.

OTHER LABORATORY TESTS
- Cytology (on gland secretion or impression smear of ulcers): May be useful in differentiating between an infectious process (bacteria and inflammatory cells may be visible) and neoplasia (neoplastic cells may be observed) • Histopathology: In case of poor response to initial therapy or an obvious mass, a biopsy may be collected. • Bacteriology: Bacterial culture and sensitivity is recommended in suspected cases of infection of the gland (a swab of the secretions or ulcers can be performed). • Fungal culture.

IMAGING
Not indicated as a diagnostic tool for this condition.

DIAGNOSTIC PROCEDURES
Biopsy for histology (essential to confirm a neoplastic process).

PATHOLOGIC FINDINGS
Findings specific to the various causes (abscess, neoplasia, squamous metaplasia).

TREATMENT
APPROPRIATE HEALTH CARE
- Digital pressure may be applied on the gland to express its content in cases of impaction. Hot compresses can be applied on the gland prior to this procedure to soften the content and facilitate the expulsion. • If ulcerations or wounds secondary to trauma are noted, wound care should be performed: disinfection (e.g., diluted chlorhexidine), debridement of necrotic tissues if necessary, topical ointment if indicated (e.g., medical honey, silver sulfadiazine cream), dressing. • If digital pressure cannot resolve the impaction, an incision over the affected lobe of the gland can be performed. Expulsion of the material should follow, and the gland may be flushed with sterile saline or diluted chlorhexidine. The wound should then be allowed to heal by secondary intention. A culture swab or the impacted material may be submitted for bacterial culture and sensitivity. Cytology should be performed on the removed materials. • Surgery: Excision of the uropygial gland can be performed without adverse consequences in most species of birds (may cause feather abnormalities in water birds). The surgery is recommended in suspected cases of neoplasia, and may be recommended in extensive infections or severe trauma. Recurrent impaction or infiltrative processes may also justify an excisional surgery that may involve the entire uropygium.

NURSING CARE
ACTIVITY
Same as usual.

DIET
Provide a well-balanced diet. In case of birds reluctant to switch to a pelletized diet, all-seed diet should be supplemented with multivitamin powder.

CLIENT EDUCATION
The client should be informed of the localization and normal aspect of the uropygial gland. Normal behavior (preening in the area of the gland) can be discussed.

SURGICAL CONSIDERATIONS
- Uropygial gland excision: The uropygial gland is very vascular and, therefore, surgical excision should be performed with care as to limit the risk of hemorrhage. A fusiform incision is made over the uropygial gland and the gland is dissected with blunt and radiosurgical dissection. The gland may greatly extend cranially. The blood supply is typically located at different levels along the cranial, middle, and caudal portions of the gland. The follicles of the rectrices should be spared during dissection and coagulation if possible. The skin closure is routine, however if significant tension is encountered secondary intention healing may be selected using bandages. • Uropygium amputation: In case of significant infiltration of cancerous tissue, excision of the whole uropygium may be indicated to obtain clear margins. Computed tomography scan of the affected area is indicated to better delineate the tumor. Significant hemorrhage may be encountered during the surgery.

MEDICATIONS
DRUG(S) OF CHOICE
- Neoplasia: ○ Systemic chemotherapy. ○ Topical chemotherapy (e.g., 5-fluorouracil). ○ Intratumoral chemotherapy (e.g., carboplatin, cisplatin). ○ Radiation therapy (Strontium-90 therapy). ○ Cryotherapy. • Bacterial infection: antibiotic therapy according to the results of culture and sensitivity. • Anti-inflammatory/analgesics may be indicated (e.g., meloxicam).

UROPYGIAL GLAND DISEASE (CONTINUED)

CONTRAINDICATIONS
None.

PRECAUTIONS
Chemotherapy and radiation therapy should be performed by specialists in oncology as they represent a serious health hazard.

POSSIBLE INTERACTIONS
None.

ALTERNATIVE DRUGS
N/A

FOLLOW-UP

PATIENT MONITORING
• Monitor the size of the gland/area of surgery (using calipers). • Regular CBC and biochemistry panels are indicated in birds receiving chemotherapy.

PREVENTION/AVOIDANCE
• Annual health checks during which the uropygial gland should be evaluated.
• Provide a well-balanced diet.

POSSIBLE COMPLICATIONS
• Lack of response to treatment. • Recurrence after treatment. • Bleeding during treatment/surgery. • Permanent injury to the gland.

EXPECTED COURSE AND PROGNOSIS
• Impaction: Good, although recurrence is possible. • Abscess, granuloma, adenitis: Usually good after treatment. • Trauma: Good except if severe trauma. • Neoplasia: Guarded, may not respond to treatment or may recur.

MISCELLANEOUS

ASSOCIATED CONDITIONS
• Poor feather condition. • Disruption of the waterproof property of the plumage in some aquatic bird species. • Self-mutilation in the uropygial area.

AGE-RELATED FACTORS
None.

ZOONOTIC POTENTIAL
None.

FERTILITY/BREEDING
N/A

SYNONYMS
N/A

SEE ALSO
Tumors

ABBREVIATIONS
N/A

INTERNET RESOURCES
N/A

Suggested Reading
Atlman, R.B. (1997). Soft tissue surgical procedures. In: Altman, R.B., Clubb, S.L., Dorrestein, G.M., Quesenberry, K. (eds), *Avian Medicine and Surgery*. Philadelphia, PA: WB Saunders, pp. 704–732.
Lucas, A.M., Stettenheim, P.R. (1972). *Avian Anatomy: Integument, Part II*. Agriculture Handbook 362, Washington, DC: U.S. Government Printing Office, pp. 613–626.
Nemetz, L.P., Broome, M. (2004). Strontium-90 therapy for uropygial neoplasia. *Proceedings of the Association of Avian Veterinarians Annual Conference*, pp. 15-20.

Authors Delphine Laniesse, Dr.Med.Vet, IPSAV
Hugues Beaufrère, Dr.Med.Vet., PhD, Dipl ABVP (Avian), Dipl ECZM (Avian), Dipl ACZM

Viral Disease

BASICS

DEFINITION
Viruses are defines as microorganisms that consist of nucleic acids (DNA or RNA) and a protein coat. Viruses are dependent on living cells by which it inserts the nucleic acid into the host cells genome to replicate itself and the protein coating.

PATHOPHYSIOLOGY
The act of viral replication within the cell results in cellular injury that leads to a disease state. The type and location of the viral infection determines the disease course. The same species of virus may have variable genetic differences called strains that can result in clinical states ranging from asymptomatic to death. The structure of the virus has surface receptors that are specific to the target cells/tissue. Once inside the cell, the virus particle "opens" up releasing its nucleic acid into the cytoplasm that then is transcribed into the host cells genome. Once in place, the cellular metabolism is adjusted to make more viral nucleic acids and proteins. Once the cell has been fully utilized the viral particles bud through the cellular membrane (enveloped) or lyse the cell to release the virus (non-enveloped).

Clinical signs widely vary depending on the location of infection, family of virus, and strain of virus within the genus. The host organism may have innate defenses (i.e. non-pathogenicity in non-target species). Other variables of the host susceptibility may include the functionality of the immune system, nutritional levels, environmental stress, reproductive status, etc.

SYSTEMS AFFECTED
• Behavioral—some of the encephalitis viruses (WNV/EEE/WEE) will directly affect the cognitive state presenting as depression or lethargy. • Cardiovascular—WNV has been shown to cause myocarditis, polyomavirus may present as subcutaneous ecchymosis along the trunk and limbs. • Endocrine/Metabolic—infectious bursal disease resides in the bursa. • Gastrointestinal—PDD often presents with passage of undigested food or regurgitation, internal papillomatosis affecting the oral cavity or cloaca. • Hemic/Lymphatic/Immune—circoviruses often result in leukopenia, anemia of chronic disease may result in highly stimulated WBC production. • Hepatobiliary—internal papillomatosis resulting in neoplasia of the pancreas or liver, herpesvirus, adenovirus, picornavirus. • Nervous—encephalitis viruses (WNV/EEE/WEE), paramyxovirus, influenza virus. • Renal/Urologic—retrovirus, adenovirus, polyomavirus. • Respiratory—influenza viruses. • Skin/Exocrine—external papillomatosis, psittacine beak and feather disease and polyomavirus often result in feather abnormalities, poxvirus lesions on the head or legs.

GENETICS
Many avian viruses are genetically structured to affect commonly related species.
• Psittaciformes—polyomavirus, psittacine beak and feather disease (circovirus), proventricular dilatation disease (bornavirus), Pacheco's disease (herpesvirus), adenovirus, Amazon tracheitis virus (retrovirus), papillomatosis (herpesvirus, papovavirus), picornavirus, poxvirus. • Passeriformes—poxvirus, herpesvirus, polyomavirus, paramyxovirus. • Columbiformes—paramyxovirus, herpesvirus, adenovirus, poxvirus, circovirus. • Galliformes—Marek's disease (herpesvirus), infectious laryngotracheitis (herpesvirus), Newcastle's disease (paramyxovirus), avian leucosis (retrovirus), reticuloendotheliosis (REV, retrovirus), egg drop syndrome (adenovirus), infectious bursal disease (birnavirus), marble spleen disease (adenovirus), rotavirus. • Anseriformes—Derzsy's disease (parvovirus), duck plague virus (herpesvirus), duck viral hepatitis (picornavirus), parvovirus, paramyxovirus, orthomyxovirus. • Falconiformes—herpesvirus, poxvirus, adenovirus, flavivirus. • Multiple species—Eastern/Western equine encephalitis virus (EEE/WEE, togavirus), West Nile virus (WNV, flavivirus), avian influenza (multiple strains, orthomyxovirus).

INCIDENCE/PREVALENCE
N/A

GEOGRAPHIC DISTRIBUTION
Many viral pathogens are worldwide but some such as West Nile virus was introduced from Africa and rapidly spread across the United States starting in 1999 to the mid-2000s. Psittacine beak and feather disease is endemic in Australia. Viruses spread my mosquitos are more common in sub-topical/tropical environments especially in the spring and fall.

SIGNALMENT
All species may be susceptible to viral pathogens of any age or sex. Juveniles with less developed immune systems may be more vulnerable to viral infections (i.e., polyomavirus, West Nile virus, etc.)

SIGNS
Clinical signs are often nonspecific for many of the viral infections with generic presentations of lethargy, anorexia, weakness, neurologic changes, regurgitation, and dyspnea. See the systems review above for more details.

CAUSES
Types of infections
• Subclinical—no overt signs of infection, immune system actively controlling infection.
• Peracute—very rapid progression of infection with either eventual recovery and antibody response or death within hours to days. • Acute—rapid progression of infection with either eventual recovery and antibody response or death within days to weeks.
• Persistent ○ Chronic infections—long-term infection with shedding of viral pathogens for months to years, may or may not develop antibodies. ○ Latent infections—long-term infection with intermittent shedding of viral pathogens for months to years, may have intermittent to no antibody titers, infection may re-activate. ○ Slow infections—progressive infection over the course of months to years, variable antibody response, no clinical signs early.

Routes of transmission
• Direct skin-to-skin contact. • Fecal/oral ingestion. • Inhalation. • Mechanical or arthropod vector.

RISK FACTORS
• Improper/inadequate diet. • Poor sanitation/lack of routine disinfection.
• Mixing of different species. • No quarantine of new individuals to an existing group.
• Increased stress in the environment.

DIAGNOSIS

DIFFERENTIAL DIAGNOSIS
• Bacterial infection. • Fungal infection.
• Parasitic infection. • System disease—neurologic lesions, renal failure, and so on.

CBC/BIOCHEMISTRY/URINALYSIS
Initial laboratory results may be within normal limits but may reveal a generalized increase in white blood cells (WBCs) with varying peaks in heterophils, lymphocytes, and monocytes. In certain severe cases, the infection may result in cellular depletion, anemia and/or leukopenia. Plasma enzyme levels may not be altered unless the virus is targeting that particular organ, that is, liver and AST.

OTHER LABORATORY TESTS
Antigen detection
• Immunofluorescent assay (IFA): Sample prepared from blood smears or tissue/cell preps, requires specific antibody conjugates related to the species and pathogen.
• Enzyme-linked immunosorbent assay (ELISA): Sample generally prepared from plasma or serum, requires specific antibody conjugates related to the species and pathogen. • Hemagglutination inhibition assay (HA): Sample generally prepared from plasma or serum, utilizes immune complex formation of avian red blood cells (West Nile virus, Newcastle's disease, influenza viruses).
• Immunohistochemistry (IHC): Similar to ELISA but is used to detect the virus in tissue.

Genetic detection
• Polymerase chain reaction (PCR): Extracts the nucleic acids from a sample and amplifies

VIRAL DISEASE (CONTINUED)

the potential target genetic material then compares it to the target virus to determine its presence in the sample. • *In situ* hybridization (ISH): A genetic probe detecting target nucleic acids within a cell or tissue sample. May be used on formalin fixed tissue sample.

Virus detection
• Electron microscopy (EM): Direct means of identification of the virus in situ in the tissue of cell culture, size and morphology of the virus may lead to descriptive etiology. • Virus Isolation (VI): Gold standard for identification, needs a fresh or frozen sample to propagate the virus in cell culture, further IFA or PCR to determine viral species, time consuming and expensive, false negatives due to transportation issues.

IMAGING
Radiographs may reveal organomegaly (liver, kidney, spleen, etc.). Ultrasonography may be helpful in evaluating internal structures to aid in differentials and in determining and removing aberrant fluid accumulation.

DIAGNOSTIC PROCEDURES
• Cytology: Fine needle aspirates or touch preps of lesions may reveal cellular changes with inclusion bodies. • Biopsy: Laparoscopic or surgical excisional/incisions samples may be submitted for histopathology for visual assessment of lesions and/or submitted for additional diagnostics as described above.

PATHOLOGIC FINDINGS
A full gross necropsy of any deceased patient is invaluable to determine the extent of any infectious process and the evaluation and sampling of any lesions will aid in determining an etiology. A pathologist is able to detect and visualize any inclusion bodies and characterize them as intracytoplasmic or intranuclear and with the source tissue may be able to determine a working diagnosis. These tissues may be submitted for additional diagnostics as described above.

TREATMENT
APPROPRIATE HEALTH CARE/NURSING CARE
• Patients should be isolated separately from other susceptible avian species. • Birds with respiratory viral infections initially will benefit from an intensive care unit (ICU) with oxygen supplemented. • Debilitated patients may not be able to thermoregulate and would benefit from placement into a heated ICU or in larger patients a cage with added subfloor heating or heat lamp. • If the patient is dehydrated, supplemental fluid administration is warranted. ○ Warm fluids to 100–103°F. ○ Lactated Ringer's Solution (LRS) given 50–60 ml/kg q24h subcutaneously. ○ IV or IO catheter may be placed in severely debilitated patients with 10 mL/kg slow bolus over 5–10 minutes. • Patients who are not eating need to have caloric supplementation with one of the commercial powdered/liquid enteric formulations per label instructions.

ACTIVITY
The patient should not have to have any changes in activity, though if isolated might be moved into a smaller enclosure limiting activity.

DIET
The diet should be evaluated to correct any nutritional deficiencies (i.e., seed based) and supplemented with fresh food items. In the convalescent patient, syringe or tube feeding with one of the commercially available enteric diets is necessary to provide enough calories to support the metabolism and the immune system.

CLIENT EDUCATION
See "Prevention".

SURGICAL CONSIDERATIONS
None initially.

MEDICATIONS
DRUG(S) OF CHOICE
• Antibiotics may be chosen to combat secondary infections ○ Enrofloxacin: 15 mg/kg PO q24h. ○ Trimethoprim Sulfa: 30–60 mg/kg PO q12–24h. ○ Doxycycline: 25–50 mg/kg PO q 12–24h. • Nonsteroidal anti-inflammatories may be used to reduce the secondary inflammation from infection ○ Meloxicam : 0.5–1.0 mg/kg PO q12–24h. • Antivirals may be used but need to be specially compounded for the size of patient • Acyclovir and Famciclovir has been used in patients with herpes viral infections. • Interferon α has been used as a general immunomodulation agent and antiviral with limited success. • Amantadine has been used against influenza A viruses.

CONTRAINDICATIONS
Many of the human antiviral agents have not been evaluated pharmacokinetically or pharmacodynamically in avian species.

PRECAUTIONS/POSSIBLE INTERACTIONS
Since many of the human antiviral drugs have not been used/evaluated, side-effects are unknown but may be extrapolated from the human literature. Since use of these drugs would be considered "experimental and off-label," the veterinarian should utilize a release form of liability after discussing these options with the owner.

ALTERNATIVE DRUGS
Use of acupuncture and herbal therapy would be considered alternatives to traditional drug therapy. Although not directly targeting the virus, acupuncture utilizes the body's own mechanisms to create a homeostatic balance thus improving the immune system. Herbal therapy also utilizes this mechanism, though many of the current pharmacologic agents were once used in the raw form as rural herbal therapies.

FOLLOW-UP
PATIENT MONITORING
Routine physical examinations and blood screening should be performed on a routine basis at least once to twice a year.

PREVENTION/AVOIDANCE
• Isolation of any ill bird. ○ To provide individualized medical care. ○ To prevent spread of any potential pathogens. • Quarantine of any new birds into an existing collection. ○ At least 30 days. ○ Coupled with diagnostic screening for any underlying health problems and pathogen surveillance. • Provide a proper enclosure with adequate size for the species with behavioral enrichment provided. • Offer a balanced, healthy diet. • Regularly clean and disinfect the housing as well as "cage furniture" and bowls. • Vaccinations are available for certain diseases. • West Nile virus, polyomavirus, poxvirus, etc. • Titer responses have been variable between species. • Some vaccines are not approved in avian species (WNV—equine) but have been used extensively in zoos to prevent disease in highly susceptible species: corvids, jays, raptors, flamingos, etc.

POSSIBLE COMPLICATIONS
As antiviral use in avian species is relatively sparse, most of the management of a viral infection is dependent on early detection and symptomatic care. Latent/chronic carriers may shed virus into the environment and infect others.

EXPECTED COURSE AND PROGNOSIS
Depending on which virus in which species as avian species may be latent carriers shedding the virus to acute death.

MISCELLANEOUS
ASSOCIATED CONDITIONS
Some viral agents may produce neoplastic to neoplastic-like conditions (internal papillomatosis) that may lead to a false or delayed diagnosis.

AGE-RELATED FACTORS
Avian patients with other underlying medical conditions may lead to a suppressed immune system as well as juvenile birds with a less developed immune system may be more susceptible to vial pathogens.

Viral Disease (Continued)

ZOONOTIC POTENTIAL
Many of the avian viruses are specific to birds but some such as highly pathogenic avian influenza A (H5N1) has been shown to be zoonotic and potentially pandemic (see "Avian Influenza" for additional details). West Nile virus as well as the other encephalitis viruses are also zoonotic (see "West Nile virus" for additional details).

FERTILITY/BREEDING
None.

SYNONYMS
None.

SEE ALSO
Adenoviruses
Avian influenza
Circoviruses
Herpesviruses
Lymphoid neoplasia
Papilloma (cutaneous)
Paramyxoviruses
Polyomavirus
Poxvirus
Proventricular dilatation disease (PDD)
Viral neoplasms
West Nile virus

ABBREVIATIONS
None.

INTERNET RESOURCES
None.

Suggested Reading
Greenacre, C.B. (2005). Viral diseases of companion birds. *The Veterinary Clinics of North America. Exotic Animal Practice*, **8**(1):85–105.
Ritchie, B.W. (1995) *Avian Viruses, Function and Control*. Lake Worth, FL: Wingers Publishing.

Author Rob L. Coke, DVM, DACZM, DABVP (Reptile & Amphibian), CVA

Viral Neoplasms: MD, LL, and RE

BASICS

DEFINITION
Lymphoid neoplasia in chickens and related species is most often due to Marek's disease (MD), lymphoid leukosis (LL), or reticuloendotheliosis (RE), all of which are caused by viruses. These viruses are common in chicken flocks (especially MD virus), and it can be difficult to differentiate between these viral etiologies due to their similar clinical signs.

PATHOPHYSIOLOGY
- MD—caused by Marek's disease virus (MDV), an alphaherpesvirus initially affecting B lymphocytes but later predominantly involving T lymphocytes, horizontally transmitted by the respiratory route from the inhalation of infected dust or skin/feather dander. Numerous strains of serotype 1 exist and their virulence varies, with recent emergence of more virulent strains. Serotypes 2 and 3 are not oncogenic. Incubation period from time of infection to time of clinical signs can range from a few weeks to several months.
- LL—caused by lymphoid leukosis virus (LLV), an alpharetrovirus predominantly involving B lymphocytes, vertically (through the egg) or horizontally transmitted (oculonasal, oral, respiratory or skin, via feces, saliva or skin dander). Disease is most commonly due to a virus of subgroup A or B, less commonly subgroup J. Also can be transmitted as a contaminant of live vaccines (MDV, fowl pox) produced in chicken embryo cells or tissues. Incubation period from time of infection to time of clinical signs can range from a few weeks to several months.
- RE—caused by reticuloendotheliosis virus (REV), a gammaretrovirus involving B or T lymphocytes, vertically or horizontally transmitted. Mosquitoes can transmit REV. Also can be transmitted as a contaminant of live vaccines (MDV, fowl pox) produced in chicken embryo cells or tissues. Incubation period from time of infection to time of clinical signs can range from two weeks to several months.

SYSTEMS AFFECTED
- Cardiovascular
  ○ MD—atherosclerosis.
- Gastrointestinal
  ○ Diarrhea, hepatomegaly.
- Hemic/Lymphatic/Immune
  ○ MD—immunosuppression due to thymic and bursal atrophy.
  ○ LL—leukemia is not typically seen, thus the term "leukosis" is used. Commonly affects the bursa of Fabricius.
  ○ RE—immunosuppression is considered one of the most important effects of infection; stunted growth can be due to immunosuppression.
- Musculoskeletal
  ○ MD, RE—Stunted growth.
- Neuromuscular
  ○ MD can include paralysis due to T cell infiltration and demyelination of peripheral nerves, typically the sciatic nerves and less often the nerves of the wings or neck. MD less commonly causes a transient paralysis of 1–2 days due to encephalitis.
- Ophthalmic
  ○ MD—iris abnormalities due to lymphocyte infiltration (gray discoloration, unequal PLRs, misshapen iris); blindness.
- Renal/Urologic
  ○ MD—glomerulopathy due to immune complex deposition.
- Reproductive
  ○ Decreased egg production and quality, reproductive tract tumors.
- Skin/Endocrine
  ○ MD—enlarged/swollen feather follicles visible in skin.
  ○ RE—abnormal feathering ("nakanuke").

GENETICS
MD—genetic resistance to MD is present in some lines of birds.

INCIDENCE/PREVALENCE
- MDV and LLV are considered to be ubiquitous in chicken flocks, and REV is considered common.
- LL—incidence of neoplasms in infected flocks is usually only 1–2%, although losses of up to 20% can occur.
- RE—clinical disease is rare, but losses from mortality or condemnation at slaughter in affected flocks can be as high as 20%.

GEOGRAPHIC DISTRIBUTION
Worldwide.

SIGNALMENT
- **Species**
  ○ Chickens: MD, RE, LL
  ○ Turkeys: MD, RE
  ○ Pheasants: MD, RE
  ○ Partridges: RE
  ○ Prairie chickens: RE
  ○ Peafowl: RE
  ○ Quail: MD, RE
  ○ Ducks: RE
  ○ Geese: MD, RE
- **Strain/age predilections**
  ○ MD: Genetic resistance to MD is present in some lines of birds.
- Mean age and range
  ○ MD: ≥4 weeks old, most commonly at 10–24 weeks of age.
  ○ LL: ≥14 weeks old, with most mortality at 24–40 weeks of age. Generally, resistance to infection increases with age.
  ○ RE: stunted growth can be noticed as early as 1 month of age; lymphomas typically occur at ≥15 weeks of age, depending on bird species.
- **Predominant sex**
  ○ MD, LL: females are more likely to develop tumors than males.

SIGNS
General Comments
Any of the three viruses can cause lethargy, diarrhea, inappetence, emaciation, dehydration, depressed egg laying, stunted growth.

Historical Findings
- Lameness or weakness: MD.
- Regurgitation: MD.
- Blindness: MD.
- Head tilt: MD.
- Labored respirations: MD.
- Stunted growth: ME, RE.

Physical Examination Findings
- Unilateral leg paralysis or crop stasis due to lymphocyte infiltration into peripheral nerves (sciatic nerve plexus, sciatic nerve, GI tract), sometimes called "fowl paralysis" or "range paralysis"—MD.
- Iris abnormalities due to lymphocyte infiltration (gray discoloration, unequal PLR's, misshapen iris)—MD.
- Enlarged/swollen feather follicles visible in skin—MD.
- Abnormal feather development, with adhesion of the barbs to a localized section of the shaft ("nakanuke")—RE.
- Abdominal distention—MD, LL.

CAUSES
- Inhalation of skin/feather dander from infected birds—MD.
- Vertical transmission from hen—LL, RE.
- Vertical transmission from rooster—RE.
- Horizontal transmission from infected birds—MD, LL, RE.

RISK FACTORS
- LL: Incidence may be reduced by the presence of infectious bursal disease virus.
- MD: High-protein diets or selection for fast growth rate may increase susceptibility.
- MD: Concurrent infection with other immunosuppressive viruses will usually exacerbate disease (infectious bursal disease virus, chicken infectious anemia virus, REV).

DIAGNOSIS

DIFFERENTIAL DIAGNOSIS
All three viruses should be considered in the differential diagnosis list, as well as other neurological or visceral diseases, for example, ovarian adenocarcinoma.

CBC/BIOCHEMISTRY/URINALYSIS
- RE: Anemia is sometimes seen.
- LL, MD: leukemia is rarely seen.

OTHER LABORATORY TESTS
- Antibody titer measurement (ELISA or virus neutralization) is available for all three viruses, but since the viruses all occur commonly, evidence of exposure for a particular virus does not necessarily confirm the etiology of observed clinical signs. Viral

Viral Neoplasms: MD, LL, and RE (Continued)

antibodies present in birds <4 weeks old are likely to have been maternally derived. Viral antibodies against LLV in day-old chicks indicate the presence of exposure in the hen; therefore, the chick may be congenitally infected and could spread the infection to other chicks. Individual birds without antibodies in a known infected flock may have "tolerant infection" and be viremic shedders of LLV or REV. Samples for antibody detection include plasma, serum, or egg yolk (LLV).
• Viral detection by virus isolation, PCR tests or ELISA is available for all three viruses but, as above, evidence of infection does not confirm etiology. Samples for testing include buffy coat cells from heparinized whole blood (MDV, LLV), oviduct swabs (LLV), cloacal swabs (LLV, REV), egg albumen (LLV, REV), embryo tissue (LLV), meconium (LLV), feces (LLV, REV), oral swabs (LLV), semen (LLV, REV), or suspensions of splenic, feather tip or lymphomatous tissue (MDV, LLV).

IMAGING
N/A

DIAGNOSTIC PROCEDURES
Antemortem diagnostic tests are unlikely to be informative as to a specific etiologic diagnosis.

PATHOLOGIC FINDINGS
• Various gross findings can be seen, depending on the phase of disease:
  ○ Swelling and loss of striations in peripheral nerves, especially the sciatic nerves, less frequent in adult birds than in young birds—MD, less striking in RE.
  ○ Hepatomegaly—LL, MD, RE.
  ○ Diffuse, miliary or nodular tumors in liver—LL, MD, RE.
  ○ Splenomegaly—LL, MD, RE.
  ○ Splenic atrophy—MD.
  ○ Diffuse, military or nodular tumors in spleen—LL, MD (usually diffuse), RE.
  ○ Spleen is soft in texture—LL.
  ○ Bursal enlargement—LL, MD.
  ○ Bursal atrophy—MD, RE.
  ○ Nodular tumors in bursa—LL, RE.
  ○ Thymus atrophy—RE.
  ○ Diffuse or focal tumors in bone marrow—LL.
  ○ Normal bone marrow—MD.
  ○ Intestinal thickening and annular lesions—RE.
  ○ Vessel thickening (atherosclerosis)—MD.
  ○ Kidney involvement—LL, MD.
  ○ Ovarian involvement—LL, MD.
  ○ Proventricular involvement—MD.
  ○ Heart involvement—MD.
  ○ Muscle involvement—MD.
  ○ Feather follicle involvement—MD.
  ○ Skin lesions on head and mouth—RE.
  ○ Abnormal feathering ("nakanuke")—RE.
  ○ Iris involvement—MD.
  ○ Leukemia, although uncommon—LL (lymphoblastic), MD (lymphocytic).
  ○ If lymphomas are found in ducks, geese, pheasants or quail, REV is a likely cause rather than MDV or LLV.
• Various microscopic findings can be seen in cytologic or histopathologic examination, depending on the phase of disease:
  ○ Extravascular infiltrations of lymphoblasts—LL, RE.
  ○ Perivascular infiltrations of pleomorphic lymphocytes, sometimes blastic—MD.
  ○ Brain edema—MD.
  ○ Glomerulopathy due to immune complexes—MD.
  ○ Immunoproliferative lesions in feather pulp—MD.

TREATMENT

APPROPRIATE HEALTH CARE
Treatment has been unsuccessful for birds clinically affected by any of these viruses; symptomatic infections become fatal.

NURSING CARE
N/A

ACTIVITY
N/A

DIET
N/A

CLIENT EDUCATION
When obtaining new chickens, clients should choose those that were vaccinated against MDV as day-old chicks.

SURGICAL CONSIDERATIONS
N/A

MEDICATIONS

DRUG(S) OF CHOICE
Treatment has been unsuccessful for birds clinically affected by any of these viruses; symptomatic infections become fatal. However, in one study, chickens fed a diet containing the cortisol-reducing drug metyrapone showed regression or lack of Marek's disease tumors compared to a control group. Nonsteroidal anti-inflammatory drugs may temporarily improve the quality of life of affected birds.

CONTRAINDICATIONS
N/A

PRECAUTIONS
N/A

POSSIBLE INTERACTIONS
N/A

ALTERNATIVE DRUGS
See "Drug(s) of choice" section above.

FOLLOW-UP

PATIENT MONITORING
N/A

PREVENTION/AVOIDANCE
• MD: Vaccination is 90% effective, either *in ovo* (day 18 of incubation) or on the day of hatch, using a product containing turkey herpesvirus as well as another MD virus (often CV1988-Rispens). Vaccination can prevent some lymphoma formation and clinical disease, but does not prevent superinfection by especially virulent MDV strains. Vaccine-induced immunity takes two weeks to develop, so vaccinated chicks should be kept away from infection sources the first two weeks of life. Re-vaccination of an adult bird does not cause harm, but may be unnecessary since nearly all birds will have become naturally exposed by this time.
• LL, RE: Vaccines are not commercially available, therefore eradication from a flock depends on breaking the vertical transmission cycle from dam to chicks (eliminating dams and roosters that are infected), and prevention of reinfection of chicks. Although roosters do not transmit LLV via semen to embryos, they can be a venereal source of infection for hens. Hatched chicks can be reared in isolation in small groups and tested for viremia and viral antibodies from approximately 8 weeks of age to verify virus-free status.
• MD: Genetic resistance to MD is present in some lines of birds and can be a useful component of prevention.
• RE: Reduction of insect vectors (mosquitoes).
• Hygiene, including removal of used litter and disinfection. MDV is long-lived outside a host, retaining infectivity for 4–8 months at room temperature or for at least 10 years at 4°C. LLV and REV only survive for a few hours outside of a host. The most effective types of disinfectants for these viruses are chlorine-releasing agents and iodophors; chlorhexidine is ineffective. Freezing and thawing will degrade LLV and REV, as will high temperatures (>50°C.) or pH extremes (<5 or >9). MDV requires a more extreme pH (<3 or >11) or temperature (>60°C.) for inactivation than LLV or REV.
• Biosecurity, preventing introduction of new viral strains to a bird enclosure (especially for MDV).
• Minimization of stress, especially in newly-hatched chicks (to encourage development of immunity) and around the time of onset of egg production.

POSSIBLE COMPLICATIONS
N/A

EXPECTED COURSE AND PROGNOSIS
Symptomatic infections with any of these three viruses invariably become fatal.

VIRAL NEOPLASMS: MD, LL, AND RE (CONTINUED)

MISCELLANEOUS

ASSOCIATED CONDITIONS
- LL—fowl glioma.
- MD—atherosclerosis.

AGE-RELATED FACTORS
LL—congenitally infected chicks are an important source of infection for other chicks in the hatchery and during the brooding period.

ZOONOTIC POTENTIAL
Although seropositivity has been seen in humans, there is no direct evidence of disease potential in humans for MDV, LLV or REV.

FERTILITY/BREEDING
Poor egg production, egg size, fertility, hatchability and chick growth rate.

SYNONYMS
- MD—fowl paralysis, range paralysis, polyneuritis, neurolymphomatosis gallinarum, acute leukosis, early mortality syndrome, Alabama redleg, gray eye.
- LL—big liver disease, lymphatic leukosis, visceral lymphoma, lymphocytoma, lymphomatosis, visceral lymphomatosis.

SEE ALSO
Atherosclerosis
Dermatitis
Lameness
Lymphoid neoplasia

Neurologic conditions
Ocular lesions
Regurgitation/vomiting
Respiratory distress
Tumors
Viral disease

ABBREVIATIONS
LL—lymphoid leukosis
LLV—lymphoid leukosis virus
MD—Marek's disease
MDV—Marek's disease virus
RE—reticuloendotheliosis
REV—reticuloendotheliosis virus

INTERNET RESOURCES
Dinev, I. Virus-induced neoplastic diseases: Marek's disease. http://www.thepoultrysite.com/publications/6/diseases-of-poultry/201/virusinduced-neoplastic-diseases-mareks-disease (last accessed October 6, 2015).
Dinev, I. Lymphoid leucosis. http://www.thepoultrysite.com/publications/6/diseases-of-poultry/202/lymphoid-leukosis (last accessed October 6, 2015).
Dunn, J. (2013). Merck Veterinary Manual: Overview of neoplasms in poultry. http://www.merckmanuals.com/vet/poultry/neoplasms/overview_of_neoplasms_in_poultry.html (last accessed October 6, 2015).
Morishita, T.Y., Gordon, J.C. (2013). Cleaning and disinfection of poultry facilities. http://ohioline.osu.edu/vme-fact/pdf/0013.pdf (last accessed October 6, 2015).
OIE Terrestrial Manual (2010). Marek's Disease. http://www.oie.int/fileadmin/Home/eng/Health_standards/tahm/2.03.13_MAREK_DIS.pdf (last accessed October 6, 2015).
Payne, L.N., Venugopal, K. (2000). Neoplastic diseases: Marek's disease, avian leukosis and reticuloendotheliosis. *Scientific and Technical Review of the Office International des Epizooties (Paris)*, **19**(2):544–564. http://www.oie.int/doc/ged/D9316.PDF (last accessed October 6, 2015).

Suggested Reading
Dunn, J.R., Gimeno, I.M. (2013). Current status of Marek's disease in the United States and worldwide based on a questionnaire survey. *Avian Diseases*, **57**:483–490.
Gross, W.B., Siegel, P.B. (2003). Effect of metyrapone on host defense against Marek disease lymphoid tumors in chickens. *Journal of Avian Medicine and Surgery*, **17**(3):144–146.
Nair, V., Schat, K.A., Fadly, A.M., Zavala, g. (2013). Neoplastic Diseases. In: Swayne, D.E., Glisson, J.R., McDougal, L.R., et al. (eds.), *Diseases of Poultry*, 13th edn. Hoboken, NJ: John Wiley & Sons, Inc., pp. 513–604.
Payne, L.N. (1998). Retrovirus-induced disease in poultry. *Poultry Science*, **77**:1204–1212.
Ritchie, B,W. (1995). *Avian Viruses: Function and Control*. Lake Worth, FL: Wingers Publishing, pp. 200–204, 365–376.

Author Lisa Harrenstien, DVM, DACZM

Vitamin D Toxicosis

BASICS

DEFINITION
Excessive levels of vitamin D_3 in the body resulting in hypercalcinosis and associated clinical signs.

PATHOPHYSIOLOGY
Excessive oral supplementation or parenteral administration of vitamin D, resulting in increased vitamin D levels in the body, which is then converted in the liver to 25-hydroxycholecalciferol and further hydroxylated to 1,25-dihydroxycholecalciferol (cholecalciferol, vitamin D_3) in the kidney. Hypervitaminosis D results in excessive uptake of calcium (and phosphate) in the gastrointestinal tract, thereby mimicking toxic changes seen in hypercalcinosis.

SYSTEMS AFFECTED
• Renal/Urologic—calcium deposits form in the kidney (nephrocalcinosis), resulting in polyuria and polydipsia as a result of renal failure and gout. • Endocrine/Metabolic—disruption of the calcium homeostasis as the excessive levels of vitamin D result in increased calcium levels in the blood (hypercalcemia). • Behavioral —birds may become disoriented. • Cardiovascular—calcium and uric acid deposits may be formed in the pericardium and/or arterial walls, thereby resulting in acute death. Hypercalcemia may also result in cardiac arrhythmias. • Gastrointestinal—anorexia, nausea and vomiting may be seen resulting from hypercalcemia and/or hyperuricemia. • Musculoskeletal—painful joints (due to articular gout) and/or muscle weakness may be seen; demineralization of bones may also occur. • Skin/Exocrine—accumulation of whitish colored material (calcium deposits) may be present in the skin. • Reproductive—reduced productivity and embryonic mortality may occur in hens that are fed high levels of vitamin D.

GENETICS
There are currently no indications for an underlying genetic basis.

INCIDENCE/PREVALENCE
An exact incidence is not known, but the disease is nowadays considered rare due to the increased knowledge regarding commercial hand-feeding formulas for chicks.

GEOGRAPHIC DISTRIBUTION
N/A

SIGNALMENT
• **Species predilections**: Macaws, particularly Blue and gold (*Ara ararauna*) and Hyacinth macaws (*Anodorrhynchus hyacinthinus*) appear highly sensitive, but the condition may also be found in other species (e.g., cockatiels, Grey parrots, cockatoos). • **Mean age and range**: The condition is particularly common in young, developing chicks. • **Predominant sex**: No sex predilection is known.

SIGNS

Historical Findings
Polyuria/polydipsia, anorexia, depression, crop stasis and weight loss (or poor weight gain); most often involves growing psittacine neonates which are fed a home-made diet or commercially prepared diet to which a vitamin and/or mineral supplement has been added. Single or multiple members of the same clutch may be affected.

Physical Examination Findings
Clinical examination may reveal birds in poor condition that appear weak and/or disoriented. Birds may also present with a distended crop, poly- and/or hematuria. Upon physical examination swollen, painful joints (articular gout) and accumulation of whitish colored calcium deposits in the subcutis may also be noted.

CAUSES
• Vitamin D toxicosis may occur when feeding developing birds vitamin D containing supplements and/or by parenteral administration of vitamin D containing supplements. • Toxicity is suggested to occur at levels as low as 4–10 times the recommended dose and may be exacerbated by high dietary levels of calcium and/or phosphorus. • Toxicity can vary highly between different species (e.g., chickens can tolerate up to 100 times the recommended dose), suggesting different needs and sensitivities to vitamin D_3 in the different bird species.

RISK FACTORS
Neonatal birds fed home-made diets or regular diets supplemented with vitamin D and/or calcium supplements.

DIAGNOSIS

DIFFERENTIAL DIAGNOSIS
• Pathologic hypercalcemia resulting from other causes, predominantly prolonged and excessive dietary calcium intake; other causes of hypercalcemia may include primary hyperparathyroidism, pseudohyperparathyroidism (i.e., hypercalcemia associated with neoplasia), and cholecalciferol rodenticide ingestion, but these have not or rarely been reported in birds. • Other causes of polyuria and/or polydipsia (see Polyuria and Polydipsia chapters). • Other causes of visceral and/or articular gout, including kidney failure and dehydration (see Hyperuricemia). • Diagnosis is usually based on the (dietary) history combined with typical radiographic and biochemical changes.

CBC/BIOCHEMISTRY/URINALYSIS
• Blood biochemistry will usually reveal hypercalcemia, hyperphosphatemia and/or hyperuricemia; eucalcemia, however, does not necessarily rule out the condition. • Elevated plasma creatine kinase levels have also been noted in affected birds.

OTHER LABORATORY TESTS
Measurement of vitamin D metabolites in the blood may potentially aid in the diagnosis, although this is not routinely done.

IMAGING
Radiographic findings: Renomegaly and (extensive) mineralization of the kidneys and other organs (e.g., proventriculus, lungs).

DIAGNOSTIC PROCEDURES
N/A

PATHOLOGIC FINDINGS
• Upon gross pathology, widespread soft tissue calcification and damage, especially to the kidneys is the most noticeable finding. Visceral and/or articular gout may also be observed. • Special calcium stains are needed to clearly separate urate deposition from calcium deposition.

TREATMENT

APPROPRIATE HEALTH CARE
• Treatment includes diuresis with crystalloid fluids (e.g., 0.9% NaCl) via the SC, IV or IO route, in combination with correction of the diet. • Diuretics (e.g., furosemide) are added to the treatment in cases of persistent hypercalcemia following diuresis with crystalloid fluids.

NURSING CARE
Treatment is mostly symptomatic and aimed at promoting forced diuresis, combined with nutritional support (force-feeding) in anorectic patients.

ACTIVITY
N/A

DIET
Provision of a well-balanced diet and discontinue use of all supplements immediately. Vitamin D_3 concentrations present in various hand-feeding formulas and formulated diets for adults range from 400 to 2970 IU/kg, thereby following the current recommendations for parrot species (500–2000 IU/kg feed) and, therefore, need no further supplementation.

CLIENT EDUCATION
Clients should be aware of the potential risks associated with feeding home-made diets and/or supplementation with vitamins and/or minerals, particularly in breeds such as macaws, as this may result in deficiencies and/or over dosages.

VITAMIN D TOXICOSIS (CONTINUED)

SURGICAL CONSIDERATIONS
Hypercalcemia may be associated with cardiac arrhythmias. Therefore, close anesthetic monitoring is warranted. In addition, provide sufficient fluid support to prevent further renal damage.

MEDICATIONS

DRUG(S) OF CHOICE
- Furosemide is recommended in cases of persistent hypercalcemia and can be administered in doses of up to 10 mg/kg q12h PO / SC / IM / IV. • In mammals, treatment with calcitonin is advocated for cases of severe and/or persistent hypercalcemia to rapidly decrease the magnitude of hypercalcemia and reduce the resulting tissue damage. No reports, however, exist on the use of calcitonin in birds. • Allopurinol can be used in patients with elevated plasma uric acid concentrations, although its use is controversial. • In case of nausea, regurgitation and/or crop stasis, the motility-enhancing and anti-emetic drug metoclopramide (0.5 mg/kg q8–12 h PO / SC / IM) may be administered.

CONTRAINDICATIONS
NSAIDs and other drugs that may impair renal function should be used with caution and their use omitted, if possible. Also omit the use of calcium- and/or vitamin D containing drugs, as these may exacerbate the condition.

PRECAUTIONS
Caution is warranted when administering drugs for which the kidneys form the main route of excretion, since renal function may be impaired in these patients, thereby resulting in a decreased elimination.

POSSIBLE INTERACTIONS
N/A

ALTERNATIVE DRUGS
N/A

FOLLOW-UP

PATIENT MONITORING
Closely monitor calcium and phosphate levels as well as renal function (uric acid levels) throughout the therapy.

PREVENTION/AVOIDANCE
Avoid the use of home-made diets (particularly in neonates) as well as the use of vitamin and/or mineral supplements (including those containing vitamin D precursors) in birds that are fed commercial diets, especially in species prone to hypervitaminosis D.

POSSIBLE COMPLICATIONS
Renal failure with subsequent hyperuricemia and visceral or articular gout are common.

EXPECTED COURSE AND PROGNOSIS
Prognosis is usually poor in patients that demonstrate signs of (extensive) tissue calcification and visceral or articular gout.

MISCELLANEOUS

ASSOCIATED CONDITIONS
Vitamin D toxicosis is often accompanied by hypercalcemia and hyperuricemia, the latter resulting from renal failure due to nephrocalcinosis.

AGE-RELATED FACTORS
N/A

ZOONOTIC POTENTIAL
N/A

FERTILITY/BREEDING
Hypervitaminosis D in a hen may result in poor reproductive performance and/or embryonic mortality (high levels of vitamin D_3 and calcium are transferred to the embryo).

SYNONYMS
Hypervitaminosis D, vitamin D toxicity

SEE ALSO
Anticoagulant rodenticide
Dehydration
Hyperuricemia
Polydipsia
Polyuria
Renal disease
Thyroid diseases

ABBREVIATIONS
N/A

INTERNET RESOURCES
N/A

Suggested Reading
Brue, R.N. (1994). Nutrition. In Ritchie, B.W., Harrison, G.J., Harrison, L.R. (eds), *Avian Medicine: Principles and Application*. Lake Worth, FL: Wingers Publishing, Inc., pp. 63–95.
De Matos, R. (2008). Calcium metabolism in birds. *The Veterinary Clinics of North America. Exotic Animal Practice*, **11**(1):59–82.
Schoemaker, N.J., Lumeij, J.T., Beynen, A.C. (1997). Polyuria and polydipsia due to vitamin and mineral oversupplementation of the diet of a salmon crested cockatoo (*Cacatua moluccensis*) and a blue and gold macaw (*Ara ararauna*), *Avian Pathology*, **26**(1):201–209.
Takeshita, K., Graham, D.L., Silverman, S. (1986). Hypervitaminosis D in baby macaws. In *Proceedings of the Annual Meeting of the Association of Avian Veterinarians*, Miami, FL, pp. 341–346.
Authors Yvonne van Zeeland, DVM, MVR, PhD, Dip. ECZM (Avian, Small Mammal)
Nico Schoemaker, DVM, PhD, Dip. ECZM (Small mammal, Avian), Dipl. ABVP (Avian)

West Nile Virus (WNV)

BASICS

DEFINITION
West Nile virus (WNV) is an arbovirus, transmitted by female mosquitoes, that causes neurological disease and potentially death in multiple bird species, horses, humans, and certain other individual mammal and reptile species. As of 2015, new lineages are still being identified and debated as to whether genetically distinct enough to warrant being identified as a separate lineage, with anywhere from 5–9 distinct lineages being considered. Lineage 1 contains those emerging and most concerning for disease–Clade 1a (Europe, Middle East, Asia, North America) and Clade 1b (Australia). Lineage 2 clades have been less concerning and were previously found in South Africa and Madagascar, though variants in Europe have recently caused human and avian disease. Lineages 3–5 are found in Eurasia, are poorly understood, and are not associated with disease. A concerning finding in a 2010 Arizona outbreak, was cocirculation of three distinct genetic variants, including strains with novel envelope protein mutations, which adds to the complexity and perhaps explanation as to why this virus continues to do so much damage in endemic areas.

PATHOPHYSIOLOGY
Usually the bird is bitten by an infected mosquito; however, other arthropods such as ticks, hippoboscid flies, and other fly species may be sources of transmission. *Culex spp.* (*C. pipens*) and *Aedes spp.* are primary vector species. In some cases, less efficient methods of acquiring WNV include ingestion of infected carcasses (reported in alligators fed WNV-infected horsemeat) or contaminated fluids (blood, feces, urine, saliva, crop milk). In mammals, maternal transmission has been suggested.

SYSTEMS AFFECTED
• Nervous—torticollis, opisthotonos, nystagmus, ataxia, clenched feet, upper motor neuron signs, head tilt, uncoordinated flight, paralysis, tremors, and seizures. • Behavioral—depression, ruffled feathers, decreased activity to lethargy. • Ophthalmic—chorioretinitis, cortical blindness. • Musculoskeletal—drooping wings, clenched feet, neck muscle rigidity. • Cardiovascular—bradycardia, murmurs, cardiac arrest (death). • Gastrointestinal—anorexia, rapid weight loss, diarrhea, regurgitation. • Endocrine/Metabolic—hyperthermia or hypothermia. • Renal/Urologic—polyuria to hypouria, green staining of urates. • Respiratory—nasal discharge, labored breathing, aspiration. • Skin/Exocrine—urate/fecal matting by vent, wounds over pressure points from thrashing, possible keel region feather loss. • Immune—in humans, immunosuppression is reported.

GENETICS
Since only 1/150 infected humans develops meningitis or encephalitis, this is suggestive for a genetic susceptibility in humans. It is unknown at this time for any avian species, but may also be the case.

INCIDENCE/PREVALENCE
In many places, WNV is considered endemic, but seasonal patterns occur based upon mosquito activity (late summer). There also appear to be periodic "spikes." For humans in the USA: 2003—9868 cases; 2008—1356 cases; 2012—5674 cases. Likely similar patterns emerge based upon a myriad of factors occur in birds as well. However, isolated cases outside mosquito season have been documented in birds. Incubation period is 4–14 days in birds. Some birds become detectably viremic by one day postinoculation.

GEOGRAPHIC DISTRIBUTION
WNV is firmly established in avian and mosquito populations worldwide. The virus is endemic and transmission is reinitiated annually in the summer within temperate areas of North America and Europe.

SIGNALMENT
• **Species:** ○ All bird species are likely susceptible and serve as reservoirs; most cases likely subclinical. Species most likely to be clinical include: corvids (crows, ravens, jays, magpies), raptors (owls, eagles, falcons, hawks). ○ Endangered or threatened species of special concern in the USA are California condors, Florida scrub jays, greater sage grouse, and native Hawaiian birds. ○ High rates of death were observed in free-ranging, juvenile American white pelicans in nesting colonies, captive lesser scaup ducklings, and experimentally-infected and free-ranging greater sage grouse. ○ Other birds have been documented with WNV infection, including flamingos, penguins, emus, wild turkeys, cormorants, kori bustards, bronze-winged ducks, sandhill cranes, common coots, and red-legged partridges, and others. ○ A variety of psittacine species housed in outdoor aviaries, many of which were of Australian origin, had clinical WNV disease. ○ American robins and house sparrows may serve as significant, nonclinical reservoirs for WNV.
• **Mean age and range:** Young of the year seem more susceptible. • **Predominant sex:** N/A

SIGNS

Historical Findings
Usually summer, drought seems to contribute to severity, also areas where there is not aggressive mosquito mitigation

Physical Examination Findings
See "Systems affected".

CAUSES
West Nile virus.

RISK FACTORS
Immunosuppression, species predilection, lack of previous exposure, mutation of virus, drought, poor regional mosquito mitigation, +/– lack of vaccination

DIAGNOSIS

DIFFERENTIAL DIAGNOSIS
Other arboviruses (EEE, WEE, VEE, St. Louis Encephalitis, LaCrosse, Powassan, Buggy Creek), trauma, heavy metal toxicity, bacterial meningitis, fungal meningitis, protozoal meningitis, *Baylisascaris procyonis*, hepatoencephalopathy, neoplasia, other toxins, vascular disease (stroke, aneurysm, atherosclerosis), avian vacuolar myelinopathy, nutritional deficiency or excess (Vitamin E, B6, B12 deficiencies).

CBC/BIOCHEMISTRY/URINALYSIS
Generally unremarkable, most abnormalities secondary to issues like dehydration, anorexia. Elevated CPK is not uncommon.

OTHER LABORATORY TESTS
• Twofold to fourfold increase in paired WNV-specific antibodies in acute and convalescent sera, IgM in CSF, or IgM in serum (suggestive). ELISA, with confirmation of results by plaque reduction neutralization test (Cornell University Diagnostic Laboratory). Virus isolation from serum, CSF, tissues (brain, heart, kidneys and spleen), oral/cloacal swabs, and/or urine of some animals. The period in which virus can be detected in live animals is limited, and can be especially difficult in animals with low viremia titers. RT-PCR can be more sensitive than virus isolation. Bodily fluids such as blood (centrifuged for separation of serum or plasma), CSF, urine, saliva, or swabs of body cavities (oropharyngeal or cloacal cavities, rectum), or tissues (heart, kidney, and spleen have been consistently useful for virus isolation and PCR testing in birds and can also be useful for immunohistochemistry [IHC] in birds); feather pulp, nonvascular feathers, brain, eye, spinal cord, liver, and others; tissues can be pooled to possibly increase sensitivity. Testing maggots from carcasses for RNA may be useful in decomposed birds. • Most state public health laboratories conduct WNV testing; however, virus isolation and plaque reduction neutralization tests are time and labor intensive and require BSL-3 conditions.

IMAGING
Only useful to rule other etiologies.

DIAGNOSTIC PROCEDURES
N/A

PATHOLOGIC FINDINGS
Gross lesions are often absent, but can be nonspecific, including white–tan mottling or streaking of the myocardium, splenomegaly,

WEST NILE VIRUS (WNV) (CONTINUED)

congested cerebral vessels, and poor nutritional condition. Histologic lesions can be minimal to severe, and can include heterophilic to lymphoplasmacytic myocarditis, encephalitis, ganglionitis, hepatitis, and nephritis. Vasculitis can also occur.

TREATMENT

APPROPRIATE HEALTH CARE
- Patients should be isolated separately from other susceptible avian species. • Birds with respiratory viral infections initially will benefit from an intensive care unit (ICU) with oxygen supplemented. • Debilitated patients may not be able to thermoregulate and would benefit from placement into a heated ICU or in larger patients a cage with added subfloor heating or heat lamp. • If the patient is dehydrated, supplemental fluid administration is warranted. ○ Warm fluids to 100–103°F. ○ Lactated Ringer's Solution (LRS) given 50–60 mL/kg q24h subcutaneously. ○ IV or IO catheter may be placed in severely debilitated patients with 10 mL/kg slow bolus over 5–10 minutes. • Patients who are not eating need to have caloric supplementation with one of the commercial powdered/liquid enteric formulations per label instructions.

NURSING CARE
• The bird may need to be stabilized to keep from thrashing and self-traumatizing, regurgitating and aspirating. Large birds may benefit from hay bales or a sling. Smaller birds may require towels and removal from a wire cage to prevent falls. Padding may be needed to minimize pressure sores or rub wounds. Consider a tail feather protector. • Most severe cases require nutritional support and fluid support—oral, subcutaneous, or intravenous—depending on blood results. Some may swallow with orally placed food; others require supportive feeding or esophageal tube placement. • Regular cleaning/plucking of fecal/urate contaminated feathers should be assessed. • Cold laser therapy and ultrasound to minimize edema and ligament contracture may be warranted. • Heat supplementation or hyperthermia management varies from case to case.

ACTIVITY
Minimize exposure to sound, light, and tactile stimulation early on. However, over time increasing exposure to stimulus seems to encourage the bird to help with management. If not standing on own initiate physical therapy.

DIET
Assisted nutritional support is usually required early on and essential for the extended temporal component of recovery from this disease. Watch closely, especially early on, for aspiration from regurgitation.

CLIENT EDUCATION
Birds clinically affected by WNV may recover over a few weeks, but most severely affected cases carry a guarded prognosis and require a commitment of supportive care for 45–90 days. There may also be permanent damage to the neurologic, ophthalmic, and cardiac systems.

SURGICAL CONSIDERATIONS
Potential placement of feeding tubes.

MEDICATIONS

DRUG(S) OF CHOICE
Most treatments are supplemental as mentioned in "Nursing care". The use of broad spectrum or synergistic use of antimicrobials may be helpful but do not address the primary problem. The use of anti-inflammatories such as NSAIDs, corticosteroids, and/or IV DMSO may also assist. In cases of seizures or disorientation, IM midazolam may help in the short term. For birds on a good balanced diet, vitamin E injections to prevent captive capture myopathy are generally not needed or helpful.

CONTRAINDICATIONS
N/A

PRECAUTIONS
Despite previous mention, it is important to remember that corticosteroids are 4–8 times more potent in birds vs. mammals; so repeated use should be carefully evaluated. They also likely suppress the immune system.

POSSIBLE INTERACTIONS
N/A

ALTERNATIVE DRUGS
N/A

FOLLOW-UP

PATIENT MONITORING
Follow-up WNV plaque neutralization serology may help track; repeat CBCs and chemistries may help monitor for organ failure, hydration issues, and secondary infections.

PREVENTION/AVOIDANCE
Mosquito control measures should be implemented: screened housing, fans, avoiding stagnant water, larvicides, and stocking mosquito fish in ponds. Insect repellants may create toxicity concerns with birds and are not recommended for birds at this time. Isolation of infected individuals and quarantine of new animals is recommended. Avoid feeding potentially contaminated meat/carcasses.

Four vaccines were developed for use in horses: a killed vaccine (West Nile-Innovator® DNA vaccine, Fort Dodge Animal Health), a recombinant vaccine in a canarypox vector (Recombitek®, Merial Animal Health), a flavivirus chimera vaccine (Equi-Nile™, Intervet), and a recombinant DNA plasmid-pCBWN (CDC/Fort Dodge Animal Health-not yet licensed). Many zoological facilities vaccinate sensitive avian species with available vaccines. Extra-label use of vaccines or use of vaccines that have not been adequately assessed in the target animal (i.e., controlled challenge studies) should be used with caution and not assumed to be protective. *While not common, anaphylactic reactions to these vaccines, up to and including death have been seen in various avian species including lorikeets and roseate spoonbills.* Numerous vaccines have been tested to various degrees in birds (some without challenge) with varied responses. Flamingoes failed to seroconvert after a single vaccination with the killed product. A modified live vaccine was tested in domestic geese in Israel with 75–94% protection. The killed equine vaccine, DNA plasmid vaccine, and recombinant equine vaccine provided partial protection in island scrub jays. Some red-tailed hawks vaccinated with a DNA-plasmid vaccine had partial protection while American robins and California condors vaccinated with the same vaccine seroconverted. Results were variable among adult and juvenile thick-billed parrots vaccinated with the killed equine vaccine. Seroconversion occurred in some penguins following administration of DNA plasmid and killed vaccines. A DNA plasmid vaccine failed to protect greater sage grouse from mortality. Oral vaccines in fish crows were ineffective.

In adult falcons, trials with two vaccines, one with a killed vaccine, while the second represented the canarypox recombinant live virus vector vaccine. The falcons were challenged with a WNV lineage 1 strain. The best results for both vaccines required three doses, that is alleviation of clinical signs, absence of fatalities and reduction of virus shedding and viremia; the recombinant vaccine conveyed slightly better protection than the killed vaccine; although side effects at the vaccination sites were noted. A recombinant avian adenovirus vaccine was given to Coturnix quail and showed promising antibody responses, however, there was no challenge with virus, nor is that vaccine commercially available.

WNV does not persist for long periods in the environment. 70% ethanol and bleach are sufficient for general cleaning. Viricides such as Virkon® are highly effective when concern is high but can be damaging to skin and mucus membranes.

West Nile Virus (WNV)

(Continued)

WNV has been spread horizontally shortly after experimental inoculation in some birds that were housed in close captive quarters, as well as in the American alligator. Infected individuals should be isolated. Viremia usually wanes after 5–10 days in birds. Infectious virus persisted in tissues of house sparrows for up to 43 days. Antibodies persist in some previously infected birds for years to life-long. Recent studies suggest antibody titer decay is faster in juvenile birds. In Northern cardinals (*Cardinalis cardinalis*), antibodies were completely undetectable in adult and juvenile birds two years post-exposure.

POSSIBLE COMPLICATIONS
Death and chronic debilitation can occur.

EXPECTED COURSE AND PROGNOSIS
Even with intense management, the short term and long term prognosis for the individual bird should be considered guarded.

MISCELLANEOUS

ASSOCIATED CONDITIONS
N/A

AGE-RELATED FACTORS
Juvenile birds seem more susceptible; however, naïve geriatric birds may also be at risk.

ZOONOTIC POTENTIAL
Generally not considered a direct zoonotic disease because of the mosquito involvement; however, risk during handling tissues and fluids, inhalation, mucous membrane contact, open cuts and puncture wounds from a needle stick or contaminated equipment. Gloves should be worn when handling suspect animals and bedding.

FERTILITY/BREEDING
In mammals, there are reported effects on reproductive organs, but not reported in avian species. The level of mortality from the disease can have significant effects on the populations/genetic diversity of endangered species and sensitive species such as corvids and raptors.

SYNONYMS
N/A

SEE ALSO
Atherosclerosis
Heavy metal toxicity
Liver disease
Neurologic conditions
Nutritional deficiencies
Ocular lesions
Seizures
Trauma
Tumors
Viral disease

ABBREVIATIONS
WNV—West Nile virus
DMSO—dimethyl sulfoxide
NSAIDs—nonsteroidal anti-inflammatory drugs

INTERNET RESOURCES
http://www.cdc.gov/westnile/
http://diseasemaps.usgs.gov/mapviewer/
http://c.ymcdn.com/sites/www.aazv.org/resource/resmgr/IDM/IDM_West_Nile_Virus_2013.pdf
http://en.wikipedia.org/wiki/West_Nile_virus
https://ahdc.vet.cornell.edu/test/detail.aspx?testcode=WNVSN

Suggested Reading

Rizzoli, A., Jiménez-Clavero, M.A., Barzon, L., et al. (2015). The challenge of West Nile virus in Europe: knowledge gaps and research priorities. *Euro Surveill*. 20(20): pii=21135. Available online: http://www.eurosurveillance.org/ViewArticle.aspx?ArticleId=21135

Angenvoort, J., Fischer, D., Fast, C., et al. (2014). Limited efficacy of West Nile virus vaccines in large falcons (*Falco* spp.). *Veterinary Research*, **45**:41.

Young, J.A., Jefferies, W. (2013). Towards the conservation of endangered avian species: A recombinant West Nile virus vaccine results in increased humoral and cellular immune responses in Japanese quail (*Coturnix japonica*). *PLoS ONE*, **8**(6):e67137.

Gamino, V., Höfle, U. (2013). Pathology and tissue tropism of natural West Nile virus infection in birds: A review. *Veterinary Research*, **44**:39.

McKee, E., Walker, E., Anderson, T., et al. (2015). West Nile virus antibody decay rate in free ranging birds. *Journal of Wildlife Diseases*, **51**(3):601–608.

Pérez-Ramírez, E., Llorente, F., Jiménez-Clavero, M.A. (2014). Experimental infections of wild birds with West Nile virus. *Viruses*, **6**(2):752–781.

Plante, J.A., Burkhalter, K.L., Mann, B.R., et al. (2014). Co-circulation of West Nile virus variants, Arizona, USA, 2010. *Emerging Infectious Diseases*, **20**(2):272–275.

VanDalen, K.K., Hall, J.S., Clark, L., et al. (2013). West Nile virus infection in American Robins: New insights on dose response. *PLoS ONE*, **8**(7):e68537.

Author Eric Klaphake, DVM, DABVP (Avian, Reptile/Amphibian), DACZM

Acknowledgement Zubi and Kunye bustards, Jenyva Turner, Shelly Cook, Allison Rossing; and Drs. Matthew Johnston, Liza Dadone, Genevieve Vega Weaver

Xanthomas

BASICS

DEFINITION
Xanthomas are benign dermal masses or thickenings, often yellow, orange, to white in color. The subcutaneous or deeper tissues may also be affected.

PATHOPHYSIOLOGY
• The exact pathophysiology is unclear; however, there does appear to be an association with high-fat diets and hypovitaminosis A, as well as prior trauma to the affected area. • Chronic steatitis has also been suggested as a causative factor.

SYSTEMS AFFECTED
• Skin. • Musculoskeletal.

GENETICS
None identified.

GEOGRAPHIC DISTRIBUTION
Reported worldwide.

SIGNALMENT
• **Species:** Any species may develop a xanthoma; however, cockatiels and budgerigars are over-represented. • **Mean age and range:** Older birds are more commonly affected, though a xanthoma may develop at any age. • **Predominant sex:** Females may be more prone to developing xanthomas.

SIGNS
Historical Findings
• Mass, often on the wing or sternum.
• Bleeding. • Self-mutilation.

Physical Examination Findings
• Lesions are often found on the wing or sternum, though other locations are reported. • Discrete yellow dermal mass, may be broad-based or pedunculated. • May also present as a diffuse thickening of the skin with yellow discoloration. • Overlying skin is featherless, friable, and yellow to yellow-tan in color. • Masses may be ulcerated or hemorrhaging. • Masses are often found on the wings, but may be present elsewhere. • The bird may be otherwise healthy on examination, or may be obese.

RISK FACTORS
• Obesity. • Hypovitaminosis A. • Elevated serum cholesterol. • Prior trauma at the location of the xanthoma.

DIAGNOSIS

DIFFERENTIAL DIAGNOSIS
• Lipoma. • Articular gout may mimic the gross appearance of lesions on joints.
• Granuloma. • Other dermal or connective tissue neoplasm. • Histopathology can aid differentiation, though gross appearance is often highly suggestive.

CBC/BIOCHEMISTRY/URINALYSIS
• CBC and biochemistry profile results may be within reference ranges.
• Hypercholesterolemia and hyperlipidemia may be noted. • Inflammatory leukogram may be noted if there is trauma or inflammation of the mass.

OTHER LABORATORY TESTS
N/A

IMAGING
Radiography: While this is not directly useful in the diagnosis of a xanthoma, it can be used to evaluate underlying structures for evidence of infiltration (e.g., body wall herniation).

DIAGNOSTIC PROCEDURES
The gross appearance of a xanthoma is often pathognomonic. However, confirmation by biopsy and histopathologic examination is the gold standard.

PATHOLOGIC FINDINGS
• The mass or thickened portion of skin is soft or firm, yellow to yellow-tan in color, and may be friable. • Histologically, there are aggregates of macrophages with foamy-appearing cytoplasm and lipid vacuoles, along with focal or diffuse aggregates of giant cells around or containing cholesterol clefts.

TREATMENT

APPROPRIATE HEALTH CARE
• Surgical removal of the xanthoma is preferred. • Dietary management may be successful with some individuals.

NURSING CARE
• Temporary bandaging may be required to prevent trauma to the site. • An Elizabethan collar may be indicated if the bird is self-mutilating. • Antibiotics and analgesics should be used as indicated if the mass is traumatized.

ACTIVITY
If the size or presence of the mass impede the bird's ability to ambulate or fly, these activities should be avoided (flight) or modified (perching, ambulating) to limit the risk of falling. This may entail removing higher perches from the cage, padding the cage bottom in case of falls, and keeping food and water within easy reach of the cage bottom.

DIET
Converting the bird to a low-fat pelleted diet may help prevent further xanthomatous tissue from developing, but may or may not induce regression of a xanthoma already present. Less severe lesions may respond to dietary improvement as a sole therapy.

CLIENT EDUCATION
While diet may not cause the lesion to completely regress, it is an important step in preventing recurrence and can improve the bird's overall health.

SURGICAL CONSIDERATIONS
• Care should be taken to completely excise the xanthoma with a margin of healthy tissue; xanthomatous tissues are prone to dehiscence postoperatively. • Consider the use of radiosurgery to limit blood loss, as xanthomas tend to be vascular. • A larger mass may require second intention healing of the site. Infiltrative masses may necessitate amputation (distal wing) or resection of deeper tissues. • As obesity is a risk factor for the development of a xanthoma, it is important to ensure adequate ventilation of the anesthetized patient. • Cryosurgery may also be used.

MEDICATIONS

DRUG(S) OF CHOICE
N/A

CONTRAINDICATIONS
N/A

PRECAUTIONS
N/A

POSSIBLE INTERACTIONS
N/A

ALTERNATIVE DRUGS
N/A

FOLLOW-UP

PATIENT MONITORING
• The surgical site should be monitored daily for evidence of dehiscence. • Surgical incisions left to heal by second intention should be monitored at the surgeon's discretion. • Routine weighing of the patient should be performed when dietary management is employed, to ensure there are no rapid drops in weight and that the patient achieves a healthy weight.

PREVENTION/AVOIDANCE
• Limit the bird's fat intake to prevent development of lesions. • Avoid trauma to the distal limbs.

POSSIBLE COMPLICATIONS
• Dehiscence of the surgical site may be seen, especially if complete excision was not possible. This may necessitate a second surgery or the wound may be allowed to heal by second intention. • If left *in situ*, the mass may become abraded or otherwise traumatized by the bird's normal activities (flight, grooming, play). This can lead to infection of the lesion, bleeding, and pain. A large mass may also impede movement.

XANTHOMAS (CONTINUED)

EXPECTED COURSE AND PROGNOSIS
- With complete excision, the prognosis for that particular lesion is good. This cannot always be accomplished, however, as xanthomas can be extensive and infiltrative.
- A sizable xanthoma located on the distal wing may require amputation of the distal wing for complete excision. • Without addressing the diet, there is a risk for reoccurrence, in the same area or elsewhere.

 MISCELLANEOUS

ASSOCIATED CONDITIONS
- Obesity. • Lipomas.

AGE-RELATED FACTORS
As older birds tend to be more commonly affected, there may be other concurrent age-related disease present. This may affect the decision to perform surgery as opposed to palliative care.

ZOONOTIC POTENTIAL
None.

FERTILITY/BREEDING
N/A

SYNONYMS
N/A

SEE ALSO
Atherosclerosis
Dyslipidemia
Hyperuricemia
Lipoma
Obesity
Renal failure
Squamous cell carcinoma
Trauma
Tumors

INTERNET RESOURCES
http://www.vcahospitals.com/main/pet-health-information/article/animal-health/tumors-xanthomas-in-birds/965

Suggested Reading
Lightfoot, T.L. (2010). Geriatric psittacine medicine. *The Veterinary Clinics of North America. Exotic Animal Practice*, **13**(1):27–49.
Pass, D.A. (2007). The pathology of the avian integument: a review. *Avian Pathology*, **18**:1–72.

Author Kristin M. Sinclair, DVM, DABVP (Avian)

APPENDIX I

COMMON DOSAGES FOR BIRDS

Drug	Dosage	Indication/Comments
Acetic acid (vinegar)	16 mL/L drinking water	May be useful for noninvasive gastrointestinal yeast infections
Acyclovir (Zovirax)	10–40 mg/kg IM q12–24h 29–330 mg/kg PO q8–12h	Antiviral agent; active against herpesvirus and cytomegalovirus
Albendazole (Valbazen)	5.2–50 mg/kg PO q12–24h	Broad spectrum anthelmintic; toxic in some species
Albuterol	2.5 mg in 3 cc saline q4–6h for nebulization	Bronchodilator
Allopurinol	10–30 mg/kg PO q4–24h	Xanthine oxidase inhibitor; use in treatment of gout controversial
Aloe vera	Topical	Anti-inflammatory, antithromboxane activity
Aluminum hydroxide	30–90 mg/kg PO q12h	Phosphate binder; antacid
Amikacin	5–6 mg/mL sterile water or saline for nebulization	Discontinue if renal disease or polyuria develops
Aminophylline	4–10 mg/kg PO, IV 3 mg/mL sterile water or saline for nebulization	Bronchial and pulmonary vasculature smooth muscle relaxation, pulmonary edema
Amitriptyline (Flavil)	1–5 mg/kg PO q12–24h	Tricyclic antidepressant; inhibits serotonin reuptake; antihistamine
Amoxicillin/clavulanate (Clavamox)	125 mg/kg PO q12h	Beta-lactamase inhibitor
Amphotericin B	1.5 mg/kg IV q8h × 3–7 days 1 mg/kg intratracheal q8–12h, dilute to 1 mL with sterile water 100 mg/kg PO q12–24h × 10–30 days for Macrorhabdus 0.1–7 mg/mL saline q12h for nebulization	Fungicidal; intratracheal administration may cause tracheitis; nephrotoxic, lipid-based product less toxic
Amprolium (Corid, Amprol Plus)	2.2–30 mg/kg PO q24h 50–575 mg/mL drinking water	Pyridimine derivative coccidiostat; some coccidial organisms show resistance
Asparaginase (Elspar)	400–1650 U/kg SC, IM	Lymphosarcoma; premedicate with diphenhydramine; higher dosage associated with bone marrow suppression in some species
Aspirin (acetylsalicylic acid)	5–150 mg/kg PO 325 mg/250 mL drinking water	Contraindicated with tetracycline, insulin, or allopurinol therapy
Atropine sulfate	0.01–0.5 mg/kg SC, IM, IV	Anticholinergic agent; antidote for organophosphate toxicosis
Azithromycin (Zithromax)	40 mg/kg PO q24–48h	Macrolide antibiotic
Bismuth sulfate (Bismusal)	1–2 mL/kg PO	Weak adsorbent; may be useful for toxin removal
Bupivicaine HCl	2–10 mg/kg SC, perineurally, into incision site, intra-articular 50:50 mixture with DMSO for topical application	Local anesthetic agent; minimize dose to limit toxic effects
Buprenorphine HCl	0.05–6 mg/kg IM, IV	Opiod agonist–antagonist
Busprione HCl (Buspar)	0.5 mg/kg PO q12h	Anxiolytic
Butorphanol tartrate	0.05–6 mg/kg IM, IV q 1–4h	Opiod agonist–antagonist; PO biovailability <10%
Calcitonin	4 U/kg IM q12h	Hypercalcemia reduction caused by cholecalciferol rodenticide toxicity
Calcium EDTA (edetate calcium disodium)	10–50 mg/kg IM, IV q12h	Preferred initial chelator for lead and zinc toxicosis; maintain hydration

Common Dosages for Birds (continued)

Drug	Dosage	Indication/Comments
Calcium glubionate	23–150 mg/kg PO q12–24h 750 mg/mL drinking water	Hypocalcemia, calcium supplementation
Calcium gluconate (10%)	5–500 mg/kg SC, IM, IV 1 mL/30 mL (3300 mg/L) drinking water	Hypocalcemia; dilute 1:1 with saline or sterile water for IM or IV injections
Carboplatin	5 mg/kg IV, IO, intralesional 125 mg/m^2 IV	Osteosarcoma, carcinoma, other sarcomas
Carnidazole (Spartrix)	5–10 mg/bird PO 12.5–50 mg/kg PO once, repeat in 10–14 days	May be effective against *Trichomonas*, *Hexamita*, *Histomonas*
Carprofen	1–40 mg/kg PO, SC, IM, IV, in feed	Nonsteroidal anti-inflammatory; use caution in some species
Celecoxib (Celebrex)	10 mg/kg PO q24h	Psittacines/clinical proventricular dilatation disease
Cephalexin	40–100 mg/kg PO, IM q6–8h	First-generation cephalosporin
Charcoal, activated	52–8000 mg/kg PO	Adsorbs toxins from GI tract
Chlorambucil (Leukeran)	1 mg/bird PO twice weekly, ducks 2 mg/kg PO twice weekly	Lymphocytic leukemia or lymphosarcoma
Chloroquine phosphate	5–60 mg/kg PO repeated q12h–7days	Used with primaquine for *Plasmodium*, *Haemoproteus*, *Leucocytozoon*
Cimetidine	3–10 mg/kg PO, IM, IV q8–12h	Histamine-2 blocker
Ciprofloxacin (Cipro)	15–20 mg/kg PO, IM q12h	Broad spectrum quinolone; do not use in food animals
Ciprofloxacin HCl 0.3% (Ciloxan)	Apply topically	Antibiotic; corneal ulcers, conjunctivitis
Cisapride (Propulsid)	0.25–1.5 mg/kg PO q8–12h	Gastrointestinal prokinetic agent; not commercially available, can be compounded
Cisplatin	1 mg/kg IV over 1 h	Osteosarcoma, carcinoma, other sarcomas
Clindamycin	25–100 mg/kg PO q8–24h	Lincosamide antibiotic; indicated for bone/joint infections
Clomipramine (Clomicalm)	0.5–8 mg/kg PO q12–24h	Tricyclic antidepressant; antihistamine; extrapyramidal signs and death reported in some species
Colchicine	0.01–0.2 mg/kg PO q12–24h	Anti-inflammatory used to treat gout or hepatic fibrosis/cirrhosis; may potentiate gout formation
Cyclophospamide	200–300 mg/m^2 PO, IO	Lymphosarcoma, sarcoma
Deferoxamine mesylate (Desferal)	20–100 mg/kg PO, SC, IM	Preferred iron chelator for hemochromatosis
Deslorelin (Suprelorin)	4.7 mg or 9.4 mg implant placed SC intrascapularly	GnRH agonist available as long-term implant
Dexamethasone	0.2–8 mg/kg SC, IM, IV	Steroidal anti-inflammatory; use with caution
Dextrose (50%)	50–1000 mg/kg IV (slow bolus)	Hypoglycemia; can dilute with fluids
Diazepam	0.05–15.6 mg/kg PO, IM, IV, intranasally	Benzodiazepine; IM administration may cause muscle irritation; reversal with flumazenil
Dimercaprol (BAL in oil)	2.5–35 mg/kg PO, IM	Heavy metal toxicosis; arsenical compound toxicosis
Dimercaptosuccinic acid (DMSA or succimer)	25–40 mg/kg PO q12h	Oral chelator for lead or zinc; may be effective for mercury toxicosis
Diphenhydramine	2–4 mg/kg PO, IM	Antihistamine
Doxorubicin	2 mg/kg IV 30–60 mg/m^2 IO, IV	Osteosarcoma, mesenchymal and epithelial tumors, and lymphoproliferative disease

Common Dosages for Birds (continued)

Drug	Dosage	Indication/Comments
Doxycycline (Vibramycin)	25–50 mg/kg PO q12–24h	Antibiotic of choice for *Chlamydia* and *Mycoplasma*; may cause regurgitation in some species; wide variation in dosage recommendations for food/water treatment
Doxycycline (Vibravenos)	25–100 mg/kg IM q5–7 days	Not available in the USA without FDA permission; injectable doxycycline of choice for use in birds
Enalapril	0.2–5 mg/kg PO q8–24h	ACE inhibitor; heart failure
Enrofloxacin (Baytril)	5–20 mg/kg PO, SC, IM q12–24h	Broad spectrum quinolone; IM formulation extremely painful for repeated injection; best to avoid IV use in birds
Epinephrine (1:1000)	0.5–1 mL/kg IM, IO, IV, intratracheal	CPR; bradycardia
Fatty acids (omega-3, omega-6)	0.1–0.2 mL/kg of flaxseed oil to corn oil mixed at a ratio of 1:4 PO or added to food	Glomerular disease; pancreatitis; use to reduce thromboxane A_2 synthesis
Fenbendazole (Panacur)	1.5–50 mg/kg PO q24h	Effective against cestodes, nematodes, trematodes, *Giardia*, acanthocephalans; toxic in many species; can cause feather abnormalities
Fipronil (Frontline)	7.5 mg/kg; spray on skin once, repeat in 30 days	Ectoparasite treatment; apply via pad to base of neck, tail base, under wings; avoid plumage during application to minimize feather damage
Fluids	10–25 mL/kg IO, IV 50–90 mL/kg SC, IO, IV	Fluid therapy; dehydration; hypovolemic shock
Fluconazole	2–15 mg/kg PO q12–24h	Fungistatic; penetrates CNS and eyes; may be ineffective against aspergillosis; can cause death in some species at higher doses
Flumazenil	0.013–0.31 mg/kg IM, IV, intranasally	Benzodiazepine antagonist
Fluoxetine (Prozac)	0.4–4 mg/kg PO q12–24h	Selective serotonin reuptake inhibitor; antidepressant
Furosemide	0.1–10 mg/kg PO, SC, IM, IV	Diuretic; some species extremely sensitive
Gabapentin	3–11 mg/kg PO q12–24h	GABA analogue; indicated for neuropathic pain
Gemfibrozil	30 mg/kg PO q8h	Lipid regulating agent; yolk emboli; give with niacin
Gentamicin sulfate	1 drop topical q4–8h	Antibiotic; corneal ulcers; causes irritation
Glipizide	0.5–1.25 mg/kg PO q12–24h	Sulfonylurea antidiabetic; diabetes mellitus
Glucosamine/Chondroitin Sulfate	20 mg/kg PO (of the chondroitin component) q12h	Nutraceutical used as an adjunctive treatment for osteoarthritis or other painful conditions
Glycopyrrolate	0.01–0.04 mg/kg IM, IV	Anticholinergic agent; slower onset than atropine
Guafenisin	0.8 mg/kg PO q12h	Expectorant; bronchodilation
Haloperidol (Haldol)	0.1–2 mg/kg PO q12–24h	Butyrophenone dopamine antagonist tranquilizer; extrapyramidal effects and death reported in some species
Hetastarch	10–15 mL/kg IV slowly	Hypovolemia; hypoproteinemia; may be associated with coagulopathy or acute renal disease in mammals
Hyaluronidase	5 U/kg IV q12h 75–150 U/L fluids	Egg-yolk related disease; increase absorption rate of fluids
Hydroxyzine (Atarax)	2–2.2 mg/kg PO q8–12h 30–40 mg/L drinking water	Antihistamine with mild sedative effects
Insulin	0.5–3 U/kg IM q12–48h	NPH insulin
Iodine (Lugol's iodine)	0.2 mL/L drinking water daily 3 drops into 100 mL drinking water	Thyroid hyperplasia

Common Dosages for Birds (continued)

Drug	Dosage	Indication/Comments
Iron dextran	10 mg/kg IM, repeat in 7–10 days	Iron deficiency anemia
Isoflurane	0.5–4% (usually 1.5–2%)	Inhalant anesthetic of choice in birds
Isoxsuprine	5–10 mg/kg PO q24h	Peripheral vasodilator; wing tip edema; atherosclerosis
Itraconazole (Sporanox)	2.5–10 mg/kg PO q12–24h	Fungistatic; effective for systemic and superficial mycosis; use caution in African grey parrots
Ivermectin (Ivomec)	0.2–2 mg/kg PO, SC, topical on skin, IM once, repeat in 7–14 days	Effective against nematodes, acanthocephalans, leeches, most ectoparasites; toxicity reported
Kaolin/pectin	2–15 mL/kg PO q6–12h	Intestinal protectant; antidiarrheal
L-carnitine	1000 mg/kg feed	Budgerigars; lipomas
Lactulose	0.3 mL/kg PO q12h 150–650 mg/kg PO q8–12h	Prophylactic laxative
Leuprolide acetate (Lupron Depot)	100–1250 ug/kg IM	Synthetic GnRH agonist depot drug
Levetiracetam (Keppra)	20–100 mg/kg PO q8h	Anticonvulsant; measure drug levels since metabolism may vary between species
Levothyroxine (l-thyroxine)	0.02 mg/kg PO 1–1000 μg/kg PO	May induce molt; use with caution
Lidocaine	1–3 mg/kg 15–20 mg/kg perineurally	Local anesthetic agent; toxic doses reported at 2.7–3.3 mg/kg in some species
Lorazepam (Ativran)	0.1 mg/kg PO q12h	Benzodiazepine with anxiolytic and sedative effects
Magnesium sulfate (Epsom salts)	500–1000 mg/kg PO q12–24h	Cathartic used in lead toxicosis to reduce lead absorbtion; give 30 min after activated charcoal
Mannitol	0.2–2 mg/kg IV (slow)	Cerebral edema; anuric renal failure
Meloxicam (Metacam)	0.1–2 mg/kg PO, IM, IV	Nonsteroidal anti-inflammatory
Methocarbamol	32.5–50 mg/kg PO, IV	Capture myopathy; muscle relaxation
Metoclopramide	0.1–2 mg/kg PO, IM, IV	Gastrointestinal motility disorders
Metronidazole	10–50 mg/kg PO,IM q12–24 hours	Effective against most anaerobic bacteria; antiprotozoal (*Giardia, Histomonas, Spironucleus, Trichomonas*)
Miconazole (Monistat)	Topical	Antifungal; IV formulation and cream can be used topically
Midazolam HCl	0.1–15.6 mg/kg SC, IM, IV, intranasally	Benzodiazepine with shorter duration than diazepam
Mineral oil	5–15 mg/kg via gavage	Cathartic; foreign body passage; administer directly into crop to avoid aspiration
Monensin (Coban 45)	53–108 mg/kg in feed × 8–10 weeks	Ionophore antibiotic anticoccidial feed additive
Moxidectin (ProHeart)	0.2–1 mg/kg PO, IM 1 mg/bird topically in budgerigars	Treatment for *Serratospiculim, Capillaria*, acanthocephalans, *Paraspiralatus, Physaloptera*
N-acetyl-L-cysteine 10–20% (Mucomyst)	22 mg/mL sterile water for nebulization	Mucolytic agent; tracheal irritation and bronchoconstriction reported in mammals
Naloxone HCl (Narcan)	2 mg/kg IV	Opiod antagonist
Naltrexone HCl	1.5 mg/kg PO q8–12h	Opiod antagonist; feather picking; self-mutilation
Neomycin/polymyxin B/gramicidin	1 drop topical q2–8h	Antibiotic; corneal ulcers, conjunctivitis
Niacin (nicotinic acid)	50 mg/kg PO q8h	Yolk emboli; give with gemfibrozil
Nystatin	100,000–600,000 U/kg PO q8–12h	Candidiasis; not systemically absorbed; lesions must be treated with direct contact

Common Dosages for Birds (continued)

Drug	Dosage	Indication/Comments
Oxytocin	0.5–10 U/kg IM	For egg binding and dystocia; use should be preceded by calcium administration
Pancreatic enzyme powder (Viokase-V Powder)	2–5 g/kg 1/8 tsp/kg feed	Exocrine pancreatic insufficiency; maldigestion; mix with food and let stand 30 min
Paramomycin (Humatin)	100 mg/kg PO q12h × 7 days 1000 mg/kg food	*Cryptosporidium*; may cause secondary bacterial or fungal infection
Paroxetine (Paxil)	1–3 mg/kg PO q24h	Selective serotonin reuptake inhibitor
Penicillamine (Cuprimine)	30–55 mg/kg PO q12h	Preferred chelator for copper toxicosis; may be used for lead, zinc, and mercury toxicosis
Permethrin (Adams)	Dust plumage lightly	Lice, fleas
Phenobarbital sodium	1–7 mg/kg PO, IV 50–80 mg/L drinking water	Barbiturate anticonvulsant; may not reach therapeutic levels with oral dosing in African grey parrots
Pimobendan	0.25–0.35 mg/kg PO q12h	Inodilator for management of CHF
Piperacillin/tazobactam (Zosyn)	100 mg/kg IM, IV q6–12h	Extended-spectrum penicillin
Policosanol	0.3–2 mg PO	Hyperlipidemia
Polysulfated aminoglycosaminoglycan (Adequan)	N/A	Hemorrhagic diathesis reported in multiple avian species
Potassium bromide	25–80 mg/kg PO q24h	Long-term seizure management
Pralidoxime (2-PAM)	10–100 mg/kg IM q24–48h	Administer within 24–36 h of organophospate intoxication; use lower dose in combination with atropine
Praziquantel (Droncit)	1–10 mg/kg PO, SC, IM, repeat in 10–14 days	Cestodes, trematodes; toxicity and death in some species
Prednisolone (prednisone)	0.5–4 mg/kg PO, IM, IV	Steroidal anti-inflammatory; use with caution
Prednisolone sodium succinate (Solu-Delta-Cortef)	0.5–30 mg/kg IM, IV	Steroidal anti-inflammatory; use with caution
Primaquine	0.03–1.25 mg/kg PO q24h	Used with chloroquine for *Plasmodium, Haemoproteus, Leucocytozoon*
Propanolol	0.04–0.2 mg/kg IM, IV	Supraventricular arrhythmia; atrial flutter; fibrillation
Propofol	1–15 mg/kg IV	IV sedative–hypnotic agent; intubation, ventilation, and supplemental oxygen recommended
Prostaglandin E_2 (dinoprostone)	0.02–0.1 mg/kg applied topically to uterovaginal sphincter 1 mL/kg applied topically to uterovaginal sphincter	Dystocia; relaxes uterovaginal sphincter
Psyllium (Metamucil)	0.5 tsp/60 mL hand feeding formula or gruel 1 Tbs/60 mL water q24h	Bulk diet; delay absorption of ingested toxin
Pyrantel pamoate	4.5–70 mg/kg PO	Intestinal nematodes
Pyrethrins (0.15%)	Dust plumage lightly to moderately prn	Ectoparasites
Pyrimethamine (Fansidar)	0.25–1 mg/kg q12h	*Toxoplasma, Atoxoplasma, Sarcocystis*; may be effective for *Leucocytozoon*
Ronidazole (Ronivet-S)	2.5–20 mg/kg PO 100–1000 mg/L drinking water × 5–7 days	*Trichomonas, Cochlosoma*
Selamectin (Revolution)	23 mg/kg topically; repeat in 3–4 weeks	Ectoparasites including *Knemidokoptes*
Selenium	0.05–0.1 mg Se/kg IM q14d	Neuromuscular diseases
Silver sulfadiazine	Topical q12–24h	Useful for treatment of burns and ulcers

Common Dosages for Birds (continued)

Drug	Dosage	Indication/Comments
Silymarin (milk thistle)	100–150 mg/kg PO divided q8–12h	Hepatic antioxidant; used in patients with liver disease and as ancillary to chemotherapy
Sucralfate	25 mg/kg PO q8h	Esophageal, gastric, duodenal ulcers; give one hour before food or other drugs
Sulfachlorpyridazine (Vetisulid)	150–500 mg/L drinking water	Coccidostat
Sulfadimethoxine (Albon)	25–55 mg/kg PO q24h 250–500 mg/L drinking water	Coccidiostat
Tamoxifen citrate	2 mg/kg PO 40 mg/kg IM to induce molt	Nonsteroidal antiestrogen; may cause leukopenia
Tea (black tea leaves, decaffeinated)	8 g/kg diet	Add to food to decrease iron absorption
Terbinafine	10–30 mg/kg PO q12–24h 1 mg/mL solution via nebulization	Fungicidal; questionable efficacy for treatment of aspergillosis; may be more effective at higher doses or in combination with itraconazole
Terbutaline	0.01–0.1 mg/kg PO, IM 0.01 mg/kg with 9 mL saline, nebulize	Bronchodilation
Testosterone	2–8.5 mg/kg IM	Anabolic steroid; contraindicated with hepatic or renal disease; stimulate sexual behavior
Theophylline	2 mg/kg PO q12h	Bronchodilation
Thyroid stimulating hormone (thyrotropin; TSH)	0.1 U IM in cockatiels 0.2–2 U/kg IM	Obtain blood at 0 hours, then 4–6 h after TSH stimulation
Toltrazuril	7–35 mg/kg PO q24h 2–125 mg/L drinking water	Coccidiocidal
Tramadol HCl	5–30 mg/kg PO, IV	Synthetic analogue of codeine with opiod, alpha-adrenergic, and serotonergic receptor activity
Trimethoprim/sulfamethoxazole (Bactrim)	10–100 mg/kg PO, IM q12–24h	Broad spectrum sulfonamide; may cause regurgitation
Urate oxidase (Uricozyme)	100–200 U/kg IM	Lowers plasma uric acid
Vincristine sulfate	0.1 mg/kg IV q7–14d 0.5–0.75 mg/m^2 IV q7d	Lymphoma and other lymphoid tumors, mast cell tumors, some sarcomas
Vitamin A (Aquasol)	200–50,000 U/kg IM	Hypovitaminosis A
Vitamin B_1 (thiamine)	1–50 mg/kg PO, IM	Thiamine deficiency
Vitamin B_{12} (cyanocobalamin)	0.25–0.5 mg/kg IM 2–5 mg/bird SC, pigeons	Vitamin B_{12} deficiency; anemia
Vitamin C (ascorbic acid)	20–150 mg/kg PO, IM	Nutritional support; supplemental therapy for pox infection
Vitamin D_3 (Vital E-A+D)	3300–6600 U/kg IM	Hypovitaminosis D_3
Vitamin E (Bo-SE)	0.06–400 mg/kg PO, IM	Hypovitaminosis E
Vitamin K_1	0.2–2.5 mg/kg PO, SC, IM	Rodenticide anticoagulant toxicosis
Voriconazole (Vfend)	10–40 mg/kg PO, IV q8–24h	Most active drug against aspergillosis
Zonisamide (Zonegran)	20 mg/kg PO q12h	Anticonvulsant; can be used in combination with levetiracetam

Adapted from: Hawkins, M., Barron, H., Speer, B., Polloc, K. C., Carpenter, J. (2013). Birds. In: Carpenter, J. (ed.), *Exotic Animal Formulary*, 4th edn. St. Louis, MO: Elsevier, pp. 183–437.

Note: Dosing recommendations vary drastically between species; see above reference for most complete information.

Appendix II

Avian Hematology Reference Values

Table II–A

Hematological and serum biochemical reference ranges for select parrots

Parameter	African grey parrot (Psittacus erithacus)	Budgerigar parakeet (Melopsittacus undulates)	Cockatiel (Nymphicus hollandicus)	Amazon parrots (Amazona spp)	Cockatoos (Cacatua spp)	Conures
WBC ($10^3/\mu L$)	5–15[a] 5–11[b]	3–8.5[a,b]	5–13[a] 5–10[b]	6–17[a] 6–11[b]	5–13[a] 5–11[b]	4–13[a] 4–11[b]
Heterophils (%)	45–75[a] 55–75[b]	40–75[a] 50–75[b]	40–70[a] 55–80[b]	30–80[a] 55–80[b]	15–64[a] 55–80[b]	40–70[a] 55–75[b]
Lymphocytes (%)	20–50[a] 25–45[b]	20–45[a] 25–45[b]	25–55[a] 20–45[b]	20–65[a] 20–45[b]	29–83[a] 20–45[b]	20–50[a] 25–45[b]
Monocytes (%)	0–3[a,b]	0–2[a,b]	0–2[a,b]	0–3[a,b]	0–9[a] 0–1[b]	0–3[a] 0–2[b]
Basophils (%)	0–5[a] 0–1[b]	0–1[a,b]	0–6[a] 0–2[b]	0–5[a] 0–1[b]	0–3[a] 0–1[b]	0–5[a] 0–1[b]
Eosinophils (%)	0–2[a,b]	0–1[a] 0–2[b]	0–2[a,b]	0–1[a,b]	0[a] 0–2[b]	0–3[a] 0–2[b]
PCV (%)	40–55[a] 42–50[b]	44–58[a] 42–53[b]	45–54[a] 45–57[b]	40–55[a] 44–49[b]	42–54[a] 38–48[b]	42–54[a] 42–49[b]
Total protein (g/dL)	3–5[a] 3–4.6[b]	2–3[a] 2.5–4.5[b]	2.4–4.1[a,b]	3–5[a,b]	3–5[a,b]	2.5–4.5[a] 3–4.2[b]
Albumin (g/dL)	1.57–3.23[a,b]	0.79–1.35[b]	0.7–1.8[a,b]	1.9–3.5[a,b]	1–1.6[a] 1.8–3.1[b]	1.9–2.6[a,b]
Uric acid (mg/dL)	4–10[a] 4.5–9.5[b]	3–8.6[a] 4.5–14[b]	3.5–11[a] 3.5–10.5[b]	2–10[a] 2.3–10[b]	2–8.5[a] 3.5–10.5[b]	2.5–10.5[a] 2.5–11[b]
Bile acid ($\mu mol/L$)	18–71[a] 13–90[b]	20–65[a] 15–70[b]	25–85[a] 20–85[b]	19–144[a] 18–60[b]	20–70[a] 25–87[b]	20–45[a] 15–55[b]
RIA Colorimetric	12–96[a]	32–117[a]	15–139[a]	33–154[a]	34–112[a]	32–105[a]
AST (U/L)	100–350[a] 100–365[b]	55–154[a] 145–350[b]	100–396[a] 95–345[b]	130–350[a,b]	120–360[a] 145–355[b]	125–378[a] 125–345[b]
GGT (U/L)	1–10[a,b]	1–10[a,b]	0–5[a] 1–30[b]	NA[a] 1–12[b]	0–4[a] 1–45[b]	1–15[a,b]
CK (U/L)	123–875[a] 165–412[b]	54–252[a] 90–300[b]	30–245[a,b]	45–265[a] 55–345[b]	140–410[a] 95–305[b]	35–355[a,b]
Ca (mg/dL)	8–13[a] 8.5–13[b]	6.4–11.2[a] 6.5–11[b]	8.5–13[a] 8–13[b]	8–13[a] 8.5–14[b]	8–11[a] 8–13[b]	8–15[a] 7–15[b]
P (mg/dL)	3.2–5.4[a,b]	3–5.2[a,b]	3.2–4.8[a,b]	3.1–5.5[a,b]	3.5–6.5[a] 2.5–5.5[b]	2–10[a,b]

[a]Carpenter, J. (ed.) (2013). *Exotic Animal Formulary*, 4th edn. St. Louis, MO: Elsevier
[b]Harrison, G.J., Lightfoot, T.L. (eds) (2006) *Clinical Avian Medicine*, Vol. 2. Palm Beach, FL: Spix Publishing, Inc.

AVIAN HEMATOLOGY REFERENCE VALUES (CONTINUED)

Table II–B

Hematological and serum biochemical reference ranges for select avian species

Parameter	Canary (Serinus canaria)	Chicken (Gallus gallus)	Turkey (Meleagridis gallopavo)	Wood duck (Aix sponsa)	Red tailed hawk (Buteo jamaicensis)	Great horned owl (Bubo virginianus)
WBC ($10^3/\mu L$)	4–9[a,b]	9–32[a]	16–25.5[a]	19.9–31.3[a]	19.1–33.4[a]	6–8[a]
Heterophils (%)	50–80[a,b]	15–50[a]	29–52[a]		24–46[a]	36–58[a]
Lymphocytes (%)	20–45[a,b]	29–84[a]	35–48[a]		35–53[a]	20–34[a]
Monocytes (%)	0–1[a,b]	0.1–7[a]	3–10[a]		3–9[a]	5.4–12.6[a]
Basophils (%)	0–1[a,b]	0–8	1–9		Rare	Rare
Eosinophils (%)	0–2[a,b]	0–16	0–5		9–17	0–2.2
PCV (%)	37–49[a]	23–55[a]	30.4–45.6[a]	42.1–48.9[a]	31–43[a]	40–46[a]
Total protein (g/dL)	2.8–4.5[a,b]	3.3–5.5[a]	4.9–7.6[a]	2.1–3.3[a]	3.9–6.7[a]	4.3[a]
Albumin (g/dL)	0.81–1.23[b]	1.3–2.8[a]	3–5.9[a]	1.5–2.1[a]		1.3[a]
Uric acid (mg/dL)	4–12[a,b]	2.5–8.1[a]	3.4–5.2[a]	2.5–12.9[a]	8.1–16.8[a]	13.7[a]
Bile acid (umol/L) RIA	23–90[a,b]			22–60[a]	8.4–10.2[a]	
AST (U/L)	145–345[a,b]			45–123[a]	76–492[a]	287[a]
GGT (U/L)	1–14[a,b]			0–2.9[a]	0–20[a]	
CK (U/L)	55–350[a,b]			110–480[a]		
Ca (mg/dL)	5.5–13.5[a,b]	13.2–23.7[a]	11.7–38.7[a]	7.6–10.4[a]	10–12.8[a]	10.2[a]
P (mg/dL)	2.9–4.9[a,b]	6.2–7.9[a]	5.4–7.1[a]	1.8–4.1[a]	1.9–4[a]	4.3[a]

[a]Carpenter, J. (ed.) (2013). *Exotic Animal Formulary*, 4th edn. St. Louis, MO: Elsevier
[b]Harrison, G.J., Lightfoot, T.L. (eds) (2006) *Clinical Avian Medicine*, Vol. 2. Palm Beach, FL: Spix Publishing, Inc.

AUTHORS
Laura Kleinschmidt, BS DVM,
J. Jill Heatley, DVM MS DABVP (Avian, Reptile & Amphibian) DACZM

Appendix III

Laboratory Testing (USA)

Test	Laboratory
Adenovirus	Infectious Diseases Laboratory 110 Riverbend Rd. Riverbend North, Rm 150 University of Georgia, Athens, GA 30602 706 542 8092 http://www.vet.uga.edu/idl/
	National Veterinary Services Laboratories United States Department of Agriculture USDA-APHIS-VS-NVSL 1920 Dayton Ave. Ames, IA 50010 515 337 7266 http://www.aphis.usda.gov/wps/portal/aphis/ourfocus/animalhealth
	Zoologix Inc. 9811 Owensmouth Avenue Suite 4 Chatsworth, CA 91311-3800 818 717 8880 http://zoologix.com/
	Charles River Avian Vaccine Services Franklin Commons, 106 Route 32 North Franklin, CT 06254 800 772 3271 http://www.criver.com/products-services/avian-vaccine-services
Anticoagulants (blood or tissue)	Louisiana Animal Disease Diagnostic Laboratory Louisiana State University River Road Room 1043 Baton Rouge, LA 70803 225 578 9777 http://www1.vetmed.lsu.edu/laddl/index.html
	Diagnostic Center for Population and Animal Health Michigan State University 4125 Beaumont Road Lansing, MI 48910-8104 517 353 1683 http://www.animalhealth.msu.edu/
Avian encephalomyelitis virus	National Veterinary Services Laboratories United States Department of Agriculture USDA-APHIS-VS-NVSL 1920 Dayton Ave. Ames, IA 50010 515 337 7266 http://www.aphis.usda.gov/wps/portal/aphis/ourfocus/animalhealth

Laboratory Testing (USA) (continued)

Test	Laboratory
Avian influenza virus	National Veterinary Services Laboratories United States Department of Agriculture USDA-APHIS-VS-NVSL 1920 Dayton Ave. Ames, IA 50010 515 337 7266 http://www.aphis.usda.gov/wps/portal/aphis/ourfocus/animalhealth
	Mississippi State Veterinary Research and Diagnostic Laboratory System College of Veterinary Medicine Mississippi State University 3137 Highway 468, West Pearl, MS 39208 601 420 4700 http://www.cvm.msstate.edu/contact?id=71
	Diagnostic Services and Teaching Laboratory Poultry Diagnostic & Research Center 953 College Station Road Athens, GA 30605 706 542 1904 http://vet.uga.edu/avian/
Avian leucosis/sarcoma virus	Diagnostic Services and Teaching Laboratory Poultry Diagnostic & Research Center 953 College Station Road Athens, GA 30605 706 542 1904 http://vet.uga.edu/avian/
Avian nephritis virus	National Veterinary Services Laboratories United States Department of Agriculture USDA-APHIS-VS-NVSL 1920 Dayton Ave. Ames, IA 50010 515 337 7266 http://www.aphis.usda.gov/wps/portal/aphis/ourfocus/animalhealth
Bacterial culture and sensitivity (Aerobic)	Most microbiology laboratories
Bacterial culture and sensitivity (Anaerobic)	Clinical Laboratory UC Davis Veterinary Medicine Teaching Hospital Central Laboratory Receiving, Room 1033 1 Garrod Drive Davis, CA 95616-8747 530 752 8684 http://www.vetmed.ucdavis.edu/vmth/small_animal/laboratory/index.cfm

Laboratory Testing (USA) (continued)

Test	Laboratory
Bordetella	Kansas State Veterinary Diagnostic Laboratory Kansas State University 1800 Denison Avenue Manhattan, KS 66506 785 532 5650 http://www.ksvdl.org/
	Zoologix Inc. 9811 Owensmouth Avenue Suite 4 Chatsworth, CA 91311-3800 818 717 8880 http://zoologix.com/
Bornavirus	Infectious Diseases Laboratory 110 Riverbend Rd. Riverbend North, Rm 150 University of Georgia, Athens, GA 30602 706 542 8092 http://www.vet.uga.edu/idl/
	Department of Veterinary Pathobiology Avian BornaVirus PCR Testing Services Texas AgriLife Research Texas A&M University College of Veterinary Medicine and Biomedical Sciences College Station, TX 77843-4467 979 862 2327 http://vetmed.tamu.edu/files/vetmed/schubot/services/Schubot-ABV-PCR-Testing-2013.pdf
Botulism	National Botulism Reference Laboratory School of Veterinary Medicine University of Pennsylvania New Bolton Center 382 West Street Road Kennett Square, PA 19348 610 925 6383 http://www.vet.upenn.edu/veterinary-hospitals/NBC-hospital/diagnostic-laboratories/national-botulism-reference-laboratory
Campylobacter	Diagnostic Services and Teaching Laboratory Poultry Diagnostic & Research Center 953 College Station Road Athens, GA 30605 706 542 1904 http://vet.uga.edu/avian/
	Zoologix Inc. 9811 Owensmouth Avenue Suite 4 Chatsworth CA 91311-3800 818 717 8880 http://zoologix.com/
Candida (cytology)	Most clinical pathology laboratories

Laboratory Testing (USA) (continued)

Test	Laboratory
Candida (identification)	The Fungus Testing Laboratory Department of Pathology Room 329E. Mail Code 7750 The University of Texas Health Science Center at San Antonio San Antonio, Texas 78229-3900 210 567 4131 http://strl.uthscsa.edu/fungus/
Chicken anemia virus	Diagnostic Services and Teaching Laboratory Poultry Diagnostic & Research Center 953 College Station Road Athens, GA 30605 706 542 1904 http://vet.uga.edu/avian/
Chlamydia psittaci	Infectious Diseases Laboratory 110 Riverbend Rd. Riverbend North, Rm 150 University of Georgia, Athens, GA 30602 706 542 8092 http://www.vet.uga.edu/idl/
Cholinesterase	Louisiana Animal Disease Diagnostic Laboratory Louisiana State University River Road Room 1043 Baton Rouge, LA 70803 225 578 9777 http://www1.vetmed.lsu.edu/laddl/index.html
Circovirus	Infectious Diseases Laboratory 110 Riverbend Rd. Riverbend North, Rm 150 University of Georgia, Athens, GA 30602 706 542 8092 http://www.vet.uga.edu/idl/
Clostridium	North Carolina Veterinary Diagnostic Laboratory System North Carolina State Laboratory of Public Health 4312 District Drive Raleigh, N.C. 27611 919 733 3986 http://www.ncagr.gov/vet/ncvdl/ Zoologix Inc. 9811 Owensmouth Avenue Suite 4 Chatsworth CA 91311-3800 818 717 8880 http://zoologix.com/
Cryptosporidium (feces)	Diagnostic Center for Population and Animal Health Michigan State University 4125 Beaumont Road Lansing, MI 48910-8104 517 353 1683 http://www.animalhealth.msu.edu/

Laboratory Testing (USA) (continued)

Test	Laboratory
Cytology	Most clinical pathology laboratories
Drug levels	
Amikacin, Benzodiazepines, Bromide, Carvedilol, Cyclosporine, Digoxin, Enrofloxacin/Ciprofloxacin, Gabapentin, Gentamicin, Itraconazole, Leflunomide, Levetiracetam, Lidocaine, Phenobarbital, Phenytoin, Procainamide, Theophylline, Valproic Acid, Vancomycin, Voriconazole, Zonisamide	Clinical Pharmacology Laboratory College of Veterinary Medicine Auburn University 1500 Wire Rd, 214 SRRC Auburn University, AL 36849 334 844 7187 http://www.vetmed.auburn.edu/veterinarians/clinical-labs/
Voriconazole, posaconazole, itraconazole, fluconazole, amphotericin B, micafungin, caspofungin, anidulafungin	The Fungus Testing Laboratory Department of Pathology Room 329E. Mail Code 7750 The University of Texas Health Science Center at San Antonio San Antonio, Texas 78229-3900 210 567 4131 http://strl.uthscsa.edu/fungus/
Eastern equine encephalitis virus	National Veterinary Services Laboratories United States Department of Agriculture USDA-APHIS-VS-NVSL 1920 Dayton Ave. Ames, IA 50010 515 337 7266 http://www.aphis.usda.gov/wps/portal/aphis/ourfocus/animalhealth
Fungal culture	Most mycology laboratories
Fungal identification	The Fungus Testing Laboratory Department of Pathology Room 329E. Mail Code 7750 The University of Texas Health Science Center at San Antonio San Antonio, Texas 78229-3900 210 567 4131 http://strl.uthscsa.edu/fungus/
Heavy metals	Louisiana Animal Disease Diagnostic Laboratory Louisiana State University River Road Room 1043 Baton Rouge, LA 70803 225 578 9777 http://www1.vetmed.lsu.edu/laddl/index.html
Herpesvirus	Infectious Diseases Laboratory 110 Riverbend Rd. Riverbend North, Rm 150 University of Georgia, Athens, GA 30602 706 542 9092 http://www.vet.uga.edu/idl/

Laboratory Testing (USA) (continued)

Test	Laboratory
Infectious bronchitis virus	Animal Health Diagnostic Center Cornell University College of Veterinary Medicine 240 Farrier Road Ithaca, NY 14853 607 253 3943 https://ahdc.vet.cornell.edu/
	National Veterinary Services Laboratories United States Department of Agriculture USDA-APHIS-VS-NVSL 1920 Dayton Ave. Ames, IA 50010 515 337 7266 http://www.aphis.usda.gov/wps/portal/aphis/ourfocus/animalhealth
	Diagnostic Services and Teaching Laboratory Poultry Diagnostic & Research Center 953 College Station Road Athens, GA 30605 706 542 1904 http://vet.uga.edu/avian/
Infectious bursal disease virus	Animal Health Diagnostic Center Cornell University College of Veterinary Medicine 240 Farrier Road Ithaca, NY 14853 607 253 3943 https://ahdc.vet.cornell.edu/
	Diagnostic Services and Teaching Laboratory Poultry Diagnostic & Research Center 953 College Station Road Athens, GA 30605 706 542 1904 http://vet.uga.edu/avian/
Infectious laryngotracheitis virus	Diagnostic Services and Teaching Laboratory Poultry Diagnostic & Research Center 953 College Station Road Athens, GA 30605 706 542 1904 http://vet.uga.edu/avian/
Insecticide (crop/gastric content)	Louisiana Animal Disease Diagnostic Laboratory Louisiana State University River Road Room 1043 Baton Rouge, LA 70803 225 578 9777 http://www1.vetmed.lsu.edu/laddl/index.html
Macrorhabdus ornithogaster (cytology)	Most clinical pathology laboratories

Laboratory Testing (USA) (continued)

Test	Laboratory
Marek's disease virus	National Veterinary Services Laboratories United States Department of Agriculture USDA-APHIS-VS-NVSL 1920 Dayton Ave. Ames, IA 50010 515 337 7266 http://www.aphis.usda.gov/wps/portal/aphis/ourfocus/animalhealth
	Diagnostic Services and Teaching Laboratory Poultry Diagnostic & Research Center 953 College Station Road Athens, GA 30605 706 542 1904 http://vet.uga.edu/avian/
Metapneumovirus	National Veterinary Services Laboratories United States Department of Agriculture USDA-APHIS-VS-NVSL 1920 Dayton Ave. Ames, IA 50010 515 337 7266 http://www.aphis.usda.gov/wps/portal/aphis/ourfocus/animalhealth
Mycobacterium	National Jewish Health Mycobacteriology Laboratory 1400 Jackson Street, Room K422 Denver, CO 80206 800 550 6227 http://www.nationaljewish.org/professionals/clinical-services/diagnostics/adx/about-us/lab-expertise/mycobacteriology
Mycoplasma	Mississippi State Veterinary Research and Diagnostic Laboratory System College of Veterinary Medicine Mississippi State University 3137 Highway 468, West Pearl, MS 39208 601 420 4700 http://www.cvm.msstate.edu/contact?id=71
	Animal Health Diagnostic Center Cornell University College of Veterinary Medicine 240 Farrier Road Ithaca, NY 14853 607 253 3943 https://ahdc.vet.cornell.edu/
	Diagnostic Services and Teaching Laboratory Poultry Diagnostic & Research Center 953 College Station Road Athens, GA 30605 706 542 1904 http://vet.uga.edu/avian/

Laboratory Testing (USA) (continued)

Test	Laboratory
Mycotoxins (feed)	Department of Agricultural Chemistry Louisiana State University AgCenter Room 102 Ag Chemistry Building 110 LSU Union Square Baton Rouge, LA 70803 225 342 5812 http://www.lsuagcenter.com/en/our_offices/departments/Ag_Chemistry/forms/index.htm
Mycotoxins quantification (feed and gastric content)	Diagnostic Center for Population and Animal Health Michigan State University 4125 Beaumont Road Lansing, MI 48910-8104 517 353 1683 http://www.animalhealth.msu.edu/
Newcastle disease virus (Paramyxovirus-1)	Animal Health Diagnostic Center Cornell University College of Veterinary Medicine 240 Farrier Road Ithaca, NY 14853 607 253 3943 https://ahdc.vet.cornell.edu/
Organophosphate/carbamate	Louisiana Animal Disease Diagnostic Laboratory Louisiana State University River Road Room 1043 Baton Rouge, LA 70803 225 578 9777 http://www1.vetmed.lsu.edu/laddl/index.html
Other hormones	Diagnostic Center for Population and Animal Health Michigan State University 4125 Beaumont Road Lansing, MI 48910-8104 517 353 1683 http://www.animalhealth.msu.edu/
	Endocrinology Service Biomedical and Diagnostic Sciences University of Tennessee College of Veterinary Medicine 2407 River Drive, Room A105 Knoxville, TN 37996-4543 865 974 5638 https://vetmed.tennessee.edu/vmc/dls/endocrinology/Pages/default.aspx
Paramyxovirus	National Veterinary Services Laboratories United States Department of Agriculture USDA-APHIS-VS-NVSL 1920 Dayton Ave. Ames, IA 50010 515 337 7266 http://www.aphis.usda.gov/wps/portal/aphis/ourfocus/animalhealth

Laboratory Testing (USA) (Continued)

Test	Laboratory
Parathyroid hormone	Diagnostic Center for Population and Animal Health Michigan State University 4125 Beaumont Road Lansing, MI 48910-8104 517 353 1683 http://www.animalhealth.msu.edu/
Pasteurella (Serotyping)	Diagnostic Services and Teaching Laboratory Poultry Diagnostic & Research Center 953 College Station Road Athens, GA 30605 706 542 1904 http://vet.uga.edu/avian/
Polyomavirus	Infectious Diseases Laboratory 110 Riverbend Rd. Riverbend North, Rm 150 University of Georgia, Athens, GA 30602 706 542 8092 http://www.vet.uga.edu/idl/
Poxvirus (Fowl pox)	Diagnostic Services and Teaching Laboratory Poultry Diagnostic & Research Center 953 College Station Road Athens, GA 30605 706 542 1904 http://vet.uga.edu/avian/
Reovirus	National Veterinary Services Laboratories United States Department of Agriculture USDA-APHIS-VS-NVSL 1920 Dayton Ave. Ames, IA 50010 515 337 7266 http://www.aphis.usda.gov/wps/portal/aphis/ourfocus/animalhealth
Rotavirus	National Veterinary Services Laboratories United States Department of Agriculture USDA-APHIS-VS-NVSL 1920 Dayton Ave. Ames, IA 50010 515 337 7266 http://www.aphis.usda.gov/wps/portal/aphis/ourfocus/animalhealth
Salmonella (culture and serogrouping)	Diagnostic Services and Teaching Laboratory Poultry Diagnostic & Research Center 953 College Station Road Athens, GA 30605 706 542 1904 http://vet.uga.edu/avian/

Laboratory Testing (USA) (continued)

Test	Laboratory
Thyroid hormones	Diagnostic Center for Population and Animal Health Michigan State University 4125 Beaumont Road Lansing, MI 48910-8104 517 353 1683 http://www.animalhealth.msu.edu/
	Endocrinology Service Biomedical and Diagnostic Sciences University of Tennessee College of Veterinary Medicine 2407 River Drive, Room A105 Knoxville, TN 37996-4543 865 974 5638 https://vetmed.tennessee.edu/vmc/dls/endocrinology/Pages/default.aspx
Tremorgenic mycotoxins (bromethalin, penitrem A, roquefortine, and strychnine)	Diagnostic Center for Population and Animal Health Michigan State University 4125 Beaumont Road Lansing, MI 48910-8104 517 353 1683 http://www.animalhealth.msu.edu/
Trichomonas	Zoologix Inc. 9811 Owensmouth Avenue Suite 4 Chatsworth, CA 91311-3800 818 717 8880 http://zoologix.com/
Vitamin D	Diagnostic Center for Population and Animal Health Michigan State University 4125 Beaumont Road Lansing, MI 48910-8104 517 353 1683 http://www.animalhealth.msu.edu/
West Nile virus	Animal Health Diagnostic Center Cornell University College of Veterinary Medicine 240 Farrier Road Ithaca, NY 14853 607 253 3943 https://ahdc.vet.cornell.edu/

Author João Brandão, LMV, MS

APPENDIX IV

VIRAL DISEASES OF CONCERN

Virus	Characteristics	Species	Lesions: clinical and gross/HP	Transmission/testing	AKA Disease
Adenovirus	Aviadenoviruses nonenveloped dsDNA	Falcons	Acute death; hepatic, splenic necrosis; large IN basophilic to amphophilic IB	Horizontally – fecal oral; virus neutralizing assay, PCR	Falcon adenovirus
Adenovirus	Aviadenoviruses nonenveloped dsDNA	Hawks	Acute death; hepatic, splenic, proventricular and ventricular necrosis; large IN basophilic to amphophilic IB	Horizontally – fecal oral; PCR	Raptor adenovirus
Adenovirus	Aviadenoviruses nonenveloped dsDNA	Chicken	Sudden mortality; nephritis; necrotic focal hepatic lesions with basophilic IN IB	Vertically Horizontally – fecal oral; ELISA, IIF, HI, PCR	Fowl adenoviruses
Adenovirus	Aviadenoviruses nonenveloped dsDNA	Duck	Asymptomatic infection; may cause mortality in ducklings, hepatitis, tracheitis with eosinophilic IN inclusions	Vertically Horizontally-fecal oral; PCR	Duck hepatitis
Adenovirus	Aviadenoviruses nonenveloped dsDNA	Chicken	Decreased/abnormal egg production; severe inflammation of shell gland with IN IB in the epithelial cells; epithelial sloughing in GI, duodenum lesions most severe	Vertically Horizontally – fecal oral; ELISA, HI, virus isolation, PCR	Egg drop syndrome
Adenovirus	Aviadenoviruses nonenveloped dsDNA	Ostrich	High mortality in chicks under one month, gray chalky stools, ascites, proventricular impaction, chalky or wrinkled eggs	Vertically Horizontally – fecal oral; PCR	Ostrich adenovirus
Adenovirus	Aviadenoviruses nonenveloped dsDNA Two clinical disease presentations	Pigeon	IN viral inclusions: hepatocytes, renal tubular epithelium, enterocytes of the small intestine	Horizontally – fecal oral; PCR	Inclusion body hepatitis and viral enteritis
Adenovirus	Aviadenoviruses nonenveloped dsDNA	Poultry	Myocardial and hepatic necrosis; lymphoid depletion of spleen, thymus and bursa of Fabricius; basophilic inclusions in hepatocytes	Vertically Horizontally – fecal oral; ELISA, HI, PCR	Hydropericardium syndrome
Adenovirus	Aviadenoviruses nonenveloped dsDNA	Psittacine	Basophilic IN IB in hepatocytes, enterocytes, pancreas, renal tubular epithelium. Maybe no disease in some species.	Horizontally – fecal oral; PCR	Psittacine adenovirus
Adenovirus	Aviadenoviruses nonenveloped dsDNA	Quail	Exudate in nasal passages, trachea, mainstem bronchi; necrosis of tracheal epithelium w/ IN IB; multifocal hepatic necrosis	Horizontally – fecal oral; ELISA, PCR	Quail bronchitis
Adenovirus	Group 2 Aviadenoviruses nonenveloped dsDNA	Turkey	Acute death; hepatomegaly, splenomegaly; lesions similar to chicken but do not involve GI	Horizontally – fecal oral; ELISA, PCR	Hemorrhagic enteritis virus of turkeys

Viral Diseases of Concern (continued)

Virus	Characteristics	Species	Lesions: clinical and gross/HP	Transmission/testing	AKA Disease
Adenovirus	Group 2 Aviadenoviruses nonenveloped dsDNA	Pheasant	Pulmonary edema, enlarged mottled spleens; lesions resemble those of seen in the chicken but do not involve GI	Horizontally – fecal oral; ELISA, PCR	Marble spleen disease
Adenovirus	Siadenovirus Aviadenoviruses nonenveloped dsDNA	Gouldian finch	Large clear to basophilic IN IB in renal tubular epithelial cells, maybe no disease	Suspected to be oral or respiratory transmission; PCR	Gouldian finch adenovirus 1
Avian bornavirus	Borna disease virus, enveloped ssRNA	Psittacines (variable susceptibility)	Lymphoplasmacytic ganglioneuritis and leiomyositis, no viral inclusions; anorexia, passage of whole seed in stool, maldigested excrement, abnormal frequency of defecation or regurgitation	Horizontally – fecal oral; western blot, ELISA, IF, antiganglioside antibody, PCR	Proventricular dilatation disease (PDD)
Avian encephalomyelitis (AE)	Picornavirus nonenveloped ssRNA	Chickens, pheasant, turkeys	Lenticular cataract, neurologic signs (muscular tremors in chicks), drop in egg production	Horizontally – fecal oral, vertically; serologic testing of paired serum samples, using virus neutralization or ELISA tests	Tremovirus
Avian influenza	Orthomyxoviridae, enveloped, RNA. Classified by subtypes based on hemagglutinin and neuraminidase and pathogenicity to domestic chickens (High or low pathogenicity)	All birds	No viral inclusions, lesions vary	Horizontally – fecal oral route, aerosol; Specific RT-PCR tests	Common reservoirs: waterfowl and shorebirds.
Avian paramyxovirus-1	Enveloped ssRNA virus	Pigeon	No clinical disease to polydipsia, ataxia, poor balance, torticollis, head tremors, inability to fly, and diarrhea: Interstitial nephritis, lymphoplasmacytic hepatitis and pancreatitis, with nonsuppurative encephalitis.	Horizontally – fecal oral, aerosol; PCR	Pigeon paramyxovirus type I (PPMV-I)
Avian paramyxovirus-1	Enveloped ssRNA virus	All birds	Variable depending on system affected	Horizontally – fecal oral, aerosol; ELISA, DNA	Newcastle disease - NDV/VVND
Avian retrovirus	Alpharetrovirus enveloped	Chickens, ubiquitous in chicken flocks	Lymphoid neoplasia, bursa of Fabricious enlargement	Horizontally and vertically; ELISA, PCR	Lymphoid leukosis (LL)

Viral Diseases of Concern (continued)

Virus	Characteristics	Species	Lesions: clinical and gross/HP	Transmission/testing	AKA Disease
Avian retrovirus	Gammaretrovirus enveloped	Chickens	Lymphoid neoplasia, immunosuppression	Horizontally and vertically; ELISA, PCR	Reticuloendotheliosis (RE)
Birnavirus	Birnavirus nonenveloped dsRNA	Chickens	Immunosuppression, cloacitis, bursal necrosis	Horizontally – fecal oral, aerosol; AGID	Infectious bursal disease (Gumboro disease)
Circovirus	Nonenveloped circular ssDNA	Psittacines, Rare in Neotropical parrots	Intracytoplasmic botryoid IB in growing feathers and the bursa of Fabricius *African grey*: anorexia, weight loss, vomiting, weakness, and crop stasis *Cockatoos*: chronic form, retained feather sheaths, blood in shafts, short clubbed feathers, deformed curled feathers, stress lines in vanes, circumferential constrictions Beak; progressive elongation, transverse or longitudinal fractures, palatine necrosis, oral ulceration	Horizontally – fecal oral, vertical; PCR, HI	Psittacine beak and feather disease (PBFDv)
Circovirus	Nonenveloped, circular ssDNA	Chicken	Intracytoplasmic botryoid inclusion bodies in bone marrow and lymphoid tissue	Horizontally – fecal oral, vertically; PCR, ELISA	Chicken anemia virus (CAV)
Circovirus	Nonenveloped, circular ssDNA	Pigeon	Intracytoplasmic botryoid inclusion bodies in the bursa of Fabricius	Horizontally – fecal oral, vertically; PCR	Pigeon circovirus (PiCV)
Circovirus	Nonenveloped, circular ssDNA	Finch, canary	Intracytoplasmic botryoid inclusion bodies in monocytes	Horizontally – fecal oral, vertically; PCR	"Blackspot disease"
Circovirus	Nonenveloped, circular ssDNA	Goose	Intracytoplasmic botryoid inclusion bodies in the bursa of Fabricius and lymphohistiocytic depletion, slow growth and some feather deformities	Horizontally – fecal oral, vertically; PCR	Goose circovirus (GoCV)
Duck viral hepatitis	Picornavirus nonenveloped ssRNA	Ducklings, not pathogenic for goslings	High mortality, enlarged liver with hemorrhagic foci	Horizontally – fecal oral; IFA, RT-PCR	Duck viral hepatitis
Herpesvirus	Alphaherpesvirus enveloped dsDNA	Ducks, geese, and swans of all ages	High mortality, blood stained vents and nares, penile prolapse, necrotizing and hemorrhagic enteritis	Horizontally – fecal oral; virus neutralization serum, PCR	Duck virus enteritis (DVE), duck plague

Viral Diseases of Concern (continued)

Virus	Characteristics	Species	Lesions: clinical and gross/HP	Transmission/testing	AKA Disease
Eastern equine encephalitis	Alphavirus enveloped ssRNA	Emus, pheasant, whooping crane	High mortality, depression, anorexia, hemorrhagic gastroenteritis	Mechanical – mosquito transmission; HI, CF, ELISA, VN, PCR	EEE, Triple E, sleeping sickness
Herpesvirus	Alphaherpesvirus enveloped dsDNA	New World psittacines	Mucosal papillomas, no viral inclusions	Horizontally – fecal oral; PCR	Mucosal papillomatosis
Herpesvirus	Alphaherpesvirus, dsDNA	Chickens over age 3–4 weeks, ubiquitous in chicken flocks	Lymphoid neoplasia, enlarged peripheral nerves, diffuse or nodular lymphoid tumors in various organs and enlarged feather follicles, no viral inclusions	Horizontally – fecal oral; PCR	Marek's disease (MD)
Herpesvirus	Alphaherpesvirus dsDNA	Chickens, pheasants	Hemorrhagic, necrotizing tracheitis, IN IB in tracheal epithelium	Horizontally – fecal oral, respiratory; ELISA, PCR	Infectious laryngotracheitis (ILT)
Herpesvirus	Psittacid Herpesvirus-2, Alphaherpesvirus, dsDNA	African grey parrots	Cutaneous papillomas, no viral inclusions	Horizontally – fecal oral; PCR	Cutaneous papillomatosis
Herpesvirus in hawks	Herpesvirus, Columbid Herpesvirus-1 (Previously falconid and strigid herpesvirus-1), dsDNA	Hawks	Splenic and hepatic necrosis with eosinophilic IN inclusions, high mortality	Horizontally – transmitted to raptors from rock pigeons; PCR	Inclusion body disease
Infectious bronchitis virus	Coronavirus, enveloped ssRNA	Chickens	Abnormal eggs, sinusitis, nephritis, air sacculitis	Horizontally – fecal oral, respiratory; Virus isolation, HI, ELISA, RNA qRT-PCR	Infectious bronchitis virus
Papillomavirus, suspected	Papillomavirus, dsDNA	Finches, chaffinches	Cutaneous papillomas, no viral inclusions	Horizontal transmission, PCR	Cutaneous papillomatosis
Parvovirus	Parvovirus nonenveloped ssDNA	Goslings and Muscovy ducklings	Hemorrhagic nephritis and enteritis, perihepatitis, pericarditis, Cowdry type-A IN IB	Horizontally – fecal oral, vertical; Clinical signs, histology, virus isolation, PCR, serology	Goose parvovirus (Derzsy disease), goose hepatitis, goose plague
Polyomavirus	Polyomavirus nonenveloped dsDNA	Finches	Acute mortality in 2–3 day olds, fledglings, and adults and survivors have poor feather development, long tubular misshapen beaks	Horizontally – fecal oral, vertical transmission suspected; Histopathology, PCR	APV

Viral Diseases of Concern (continued)

Virus	Characteristics	Species	Lesions: clinical and gross/HP	Transmission/testing	AKA Disease
Polyomavirus	Polyomavirus nonenveloped dsDNA	Psittacines	Peracute death in nestlings and adult Eclectus, Painted conure, White-bellied caiques. *Clinical Signs*: depression, anorexia, weight loss, delayed crop emptying, regurgitation, diarrhea, dehydration, SQ hemorrhages, dyspnea, posterior paresis and paralysis, polyuria	Horizontally – fecal oral, vertical transmission suspected; Histopathology, PCR	Budgerigar Fledgling disease, APV
Poxvirus	Avipoxvirus enveloped dsDNA	16 have been described and named after the species they infect	Hyperplastic, ballooning degenerative epithelium with IC IB (Bollinger bodies). Three forms: cutaneous (dry form), diphtheroid (wet or mucosal form), and septicemia or systemic poxvirus	Mechanically with insect vectors, direct contact with fomites; histopathology, PCR, serology	Psittacinepox, canarypox, pigeonpox, etc.
Poxvirus	*Chordopoxvirinae* subfamily, Large, dsDNA, enveloped	Chickens, turkeys, waterfowl	Proliferative skin lesions, laryngotracheitis, eosinophilic cytoplasmic IB	Mechanically with insect vectors, direct contact with fomites; histopathology, FA, IHC, PCR	Fowlpox
Reovirus	Nonenveloped dsRNA	Chicken	Arthritis, tenosynovitis, runting-stunting syndrome, atrophy of spleen and bursa of Fabricius	Horizontally – fecal oral, respiratory and vertical transmission; serology, PCR	Viral arthritis
West Nile virus	Flaviviridae, arthropod borne ssRNA virus	All birds susceptible, most cases subclinical as birds are a natural reservoir host	Heterophilic to lymphoplasmacytic myocarditis, encephalitis, ganglionitis, hepatitis, and nephritis, no viral inclusions	Mechanically – arthropod borne; PCR, IHC	WNV (high mortality in corvids and select raptors)
Western equine encephalitis	Alphavirus enveloped ssRNA	Emus and rheas	High morbidity, low mortality, anorexia, weight loss, weakness, depression, drowsiness, stupor, ataxia, incoordination	Mechanically – mosquito transmission, may be transmitted by ticks; HI, CF, ELISA, VN, PCR	WEE (inapparent infection in native birds as virus cycles between mosquitoes and passerines)

Authors Drury R. Reavill, DVM, DABVP (Avian/Reptile & Amphibian), DACVP
Jennifer Graham, DVM, DABVP (Avian/ECM), DACZM

Appendix V

Selected Zoonotic Diseases of Concern and Personal Protection*

Infectious agent	Mode of transmission	Disease in humans
Viruses		
Avian influenza	Aerosol	Respiratory
Avian paramyxovirus – 1 (Newcastle disease)	Direct contact	Mild, self-resolving conjunctivitis
Arboviruses: West Nile virus, Eastern equine encephalitis virus, Western equine encephalitis virus	Arthropod borne (mosquitoes)	Fever, flu-like symptoms, encephalitis
Bacteria		
Campylobacter spp.	Fecal–Oral, food-borne	Gastrointestinal, Guillain–Barre syndrome
Chlamydia psittaci	Aerosol	Respiratory
E. coli	Fecal–Oral, food-borne	Gastrointestinal, hemolytic uremic syndrome
Mycobacterium spp.	Aerosol	Possibly respiratory but no confirmed cases
Salmonella spp.	Fecal–Oral, food-borne	Gastrointestinal
Fungi		
Cryptococcus neoformans	Passed in feces and then aerosolized	Respiratory, meningoencephalitis
Histoplasma capsulatum	Passed in feces and then aerosolized; most commonly associated with pigeons	Respiratory
Parasite		
Fowl mites: *Ornithonyssus sylviarum, Dermanyssus gallinae*	Direct contact	Pruritus, papules

STEPS TO REDUCE ZOONOTIC DISEASE TRANSMISSION

A more extensive description of preventive measures can be found in the *Compendium of Veterinary Standard Precautions for Zoonotic Disease Prevention in Veterinary Personnel*, which is written and distributed by the National Association of State Public Health Veterinarians and can be found at: http://www.nasphv.org/Documents/VeterinaryPrecautions.pdf

Hand Hygiene
This practice can greatly reduce the risk of disease transmission and should be performed after handling each animal and prior to taking breaks where food or drink will be consumed. Warm water and soap are necessary to clean hands if organic material is present, but alcohol-based hand sanitizers are effective against many bacteria and viruses.

Gloves and Sleeves
These items provide a protective barrier and should always be worn with handling feces, vomitus, exudates or if breaks in one's skin are present. Gloves should also be worn when cleaning cages or handling dirty laundry.

Facial Protection
This practice protects the mucous membranes of the eyes, nose and mouth and can reduce the likelihood of inhalation of infectious agents. Facial protection should be worn whenever splashes or sprays are likely to occur, such as flushing wounds or abscesses, or performing a necropsy.

Respiratory Protection
Masks and respirators can be worn to reduce or eliminate the possibility of inhaling an infectious agent. Masks typically worn in surgery are not adequate protection against inhaled pathogens. Respiratory protection must be fitted to the individual and filters in most masks must be replaced periodically.

Protective Outerwear
Garments such as laboratory coats, smocks, aprons and coveralls can be worn to reduce contamination but are typically not fluid-proof; these items should be laundered daily. Footwear that can be easily disinfected, such as rubber boots, can be useful, especially if the veterinarian is making house or farm calls.

*Additional information on zoonoses associated with birds can be found in: Souza, M.J. (2011). Zoonoses, public health, and the exotic animal practitioner. *Veterinary Clinics of North America. Exotic Animal Practice*, **14**(3):xi-xii.

SELECTED ZOONOTIC DISEASES OF CONCERN AND PERSONAL PROTECTION (CONTINUED)

Animal Handling
Although injuries from birds are not as common as those associated with dogs, proper handling and communication between the veterinarian and assistant can reduce the likelihood of trauma. It is important to know what part of the bird is "dangerous". For example, the beak of parrots is of most concern, but the feet and talons of raptors are of most concern. Any wounds from trauma inflicted by an animal should be cleaned immediately. Depending on severity, a physician should be contacted.

Disinfection
Appropriate disinfection of areas where animals have been housed or examined will reduce contamination. Most bacteria and viruses are killed with commonly used disinfectants, but some organisms, such as *Mycobacterium spp.*, may need special disinfectants to be effective.

Education
Staff training is essential to eliminate disease transmission. Established protocols should outline specific steps to be taken for disinfection and disease prevention.

AUTHOR
Marcy J. Souza, DVM, MPH, DABVP (Avian), DACVPM

Appendix VI

Common Avian Toxins and their Clinical Signs

Plants	Lesions	Clinical signs	Treatment/Antidotes
Avocado (*Persea americana*)	Myocardial degeneration with subcutaneous edema of the neck, chest, and pulmonary system	Listless, ruffled feathers, cessation of perching, dyspnea	Anti-inflammatory agents, oxygen, and antimicrobials
Azalea/Rhododendron (*Rhododendron spp*)	Edema of lungs, renal tubular and hepatocellular necrosis	Weakness, dyspnea, weight loss	Supportive care
Black Locust (*Robinia pseudoacacia*)	Hemorrhagic inflammation of GI tract	Diarrhea, regurgitation/vomiting, dyspnea	Supportive care
Castor bean (*Ricinus communis*)	GI tract inflammation, edema, necrosis	Weakness, severe diarrhea, hypotension, collapse	Aggressive fluid replacement replacing electrolytes and glucose as indicated
Dieffenbachia (*Dieffenbachia maculata*)	Lingual inflammation, laryngeal inflammation	Anorexia, hypersalivation	Oral fluids as tolerated, IV fluids
English ivy (*Hedera helix*)	GI tract irritation/inflammation	Anorexia, regurgitation/vomiting, diarrhea	Fluid replacement, supportive care
Foxglove (*Digitalis purpurea*)	Myocardial degeneration, fibrosis and necrosis	Marked weakness, crop stasis, anorexia, arrhythmias, seizures	IV fluids, atropine and/or propranolol depending on arrhythmias, supportive care
Lily of the valley (*Convallaria majalis*)	GI hemorrhages and inflammation, myocardial degeneration	Weakness, anorexia, arrhythmias	IV fluids, atropine for arrhythmias
Marijuana (*Cannabis sativa*)	None	Behavior changes, disorientation, hyperactivity, tremors	Supportive care
Mistletoe (*Phoradendron leucarpum*)	GI tract irritation	Anorexia, regurgitation/vomiting, diarrhea	Correct fluid and electrolyte imbalances
Oleander (*Nerium oleander*)	GI hemorrhages and inflammation, myocardial degeneration, fibrosis and necrosis	Marked weakness, crop stasis, anorexia, arrhythmias, seizures	IV fluids, atropine and/or propranolol depending on arrhythmias, supportive care
Pothos (*Epipremnum aureum*)	Oral mucosa, crop, and upper GI tract irritation/inflammation	Anorexia, dysphonia, regurgitation, hypersalivation	Oral fluid gavage, milk, supportive care
Peace lily (*Spathiphyllum spp*)	Oral mucosa, crop, and upper GI tract irritation/inflammation	Anorexia, dysphonia, regurgitation, hypersalivation	Oral fluid gavage, milk, supportive care
Philodendron (*Philodendron spp*)	Oral mucosa, crop, and upper GI tract irritation/inflammation	Anorexia, dysphonia, regurgitation, hypersalivation	Oral fluid gavage, milk, supportive care
Poinsettia (*Euphorbia pulcherrima*)	Localized irritation of oral mucosa	Anorexia	Supportive care
Pokeweed (*Phytolacca americana*)	GI tract inflammation	Diarrhea	Supportive care
Virginia creeper (*Pathenocissus quinquefolia*)	GI tract irritation	Regurgitation/Vomiting	Supportive care
Yew (*Taxus spp*)	Pulmonary congestion and edema	Bradycardia, weakness, sudden death	Atropine, fluids, supportive care

Common Avian Toxins and their Clinical Signs (continued)

Common drugs at toxic doses	Clinical signs	Treatment/Antidotes
Acetaminophen	Vomiting, weakness, anorexia	Fluids and supportive care; possibly N-acetylcysteine as antidote
NSAIDs	Anorexia, dullness, ruffled feathers, lethargy, depression, recumbence, sunken eyes, watery droppings; elevations in serum uric acid and creatinine	Fluids, GI protectant, and supportive care
Topical Vitamin D analogs and oral Vitamin D3 in excess	Increased thirst, dystrophic mineralization, renal failure	Fluids, phosphate binders, calcitonin
Ivermectin	Seizures	Steroids, fluids, supportive care
Aminoglycosides	Polydipsia, weakness, increased concentrations of uric acid and creatinine	Fluids, supportive care
Vitamin A in excess	Osteodystrophy	Supportive care

Heavy metals	Clinical signs	Treatment/Antidotes
Arsenic	Fluffed feathers, pruritis, drooped eyelid, weight loss, feather picking, loss of righting reflex, immobility, seizures	Vitamin D3, chelation (dimercaprol), fluids and supportive care
Iron	Lethargy, emaciation, anorexia	Chelation (deferoxamine or calcium disodium EDTA), restrict sources, supportive care
Lead	Depression, weakness, crop stasis, regurgitation, polydipsia, seizures, blindness, hemoglobinuria, diarrhea	Removal of source, fluids, nutritional support, chelation (calcium disodium EDTA), supportive care
Mercury	Depression, hematuria, weakness, collapse	Fluids, chelation (dimercaprol), supportive care
Zinc	Feather loss/picking, depression, polydipsia, GI tract irritation, crop stasis, regurgitation, weight loss, weakness, anemia, seizures	Removal of source, fluids, nutritional support, chelation (calcium disodium EDTA), supportive care

Inhaled toxins	Clinical signs	Treatment/Antidotes
Ammonia	Respiratory tract irritation, immunosuppression with secondary infections	Fresh air, oxygen, supportive care, antimicrobials for infections
Carbon monoxide	Respiratory depression, weakness, collapse, seizures	Fresh air, oxygen, supportive care
Polytetrafluoroethylene (PTFE)	Respiratory failure, dyspnea, hemorrhage, collapse and death	Fresh air, oxygen, bronchodilators, NSAIDs, supportive care
Smoke (Marijuana)	Depression or hyperactivity, behavior changes	Fresh air, oxygen, supportive care
Smoke (Tobacco)	Dyspnea, coughing, tachycardia, pulmonary irritation, ocular irritation	Fresh air, oxygen, supportive care
Smoke (Wood)	Dyspnea, coughing, sneezing, pulmonary and ocular irritation	Fresh air, oxygen, supportive care

Common Avian Toxins and their Clinical Signs (continued)

Household	Clinical signs	Treatment/Antidotes
Alcohol	Lethary, regurgitation, ataxia, impaired motor coordination, vomiting	Fluids with dextrose, supportive care
Aluminum chloride	Oral irritation, GI tract inflammation	Oral gavage/fluids, supportive care
Ammonia	Respiratory tract irritation	Fresh air, oxygen, supportive care
Chlorine (bleach)	Respiratory tract irritation, photophobia, coughing, sneezing, hyperventilation, GI irritation if ingested	Fluids, milk, GI protectant, supportive care
Cigarettes/cigars/tobacco	Vomiting, depression, dyspnea, tachycardia, seizures	Fluids, supportive care
Methylene chloride	Respiratory depression, weakness, collapse, seizures	Fresh air, oxygen, supportive care
Petroleum products	Depression, respiratory tract irritation, pneumonia	Fluids, oxygen, supportive care
Salt (table salt & salty foods)	GI tract irritation, polydipsia, dehydration, weakness and depression	Fluids replaced slowly, supportive care
Soaps/detergents	GI tract irritation, vomiting/regurgitation, diarrhea	Fluids, supportive care
Yeast dough	Crop stasis, depression	Crop gavage, fluids, supportive care

Pesticides	Clinical signs	Treatment/Antidotes
Anticoagulant rodenticides	Hemorrhage, weakness, collapse, dyspnea	Vitamin K1, supportive care
Bromethalin	Depression, seizures	Mannitol, fluids, supportive care
Carbamates/organophosphorus insecticides	Ataxia, spastic nictitans, seizures	Fluids, atropine, pralidoxime, steroids, supportive care
Pyrethrins/pyrethroids	Incoordination, weakness, muscle tremors, diarrhea	Fluids, supportive care

Mycotoxins	Clinical signs	Treatment/Antidotes
Aflatoxin	Hemorrhage, icterus, anemia, depression, ruffled feathers, anorexia and other signs of severe hepatic damage, immunosuppression	Removal of suspected feed source, increase high quality protein, supportive care
Trichothecenes (i.e., diacetoxyscirpenol)	Enteritis, dehydration, weight loss, malformed feathers, immunosuppression and secondary infections	Removal of suspected feed source, nutritional support, antimicrobials, supportive care
Ochratoxin	Anorexia, polydipsia, depression, elevations in serum uric acid and creatinine	Removal of suspected feed source, fluids, supportive care

AUTHOR
John H Tegzes, MA, VMD, Dipl. ABVT

APPENDIX VII

CLINICAL ALGORITHMS

Algorithm 1 Sick Bird Syndrome

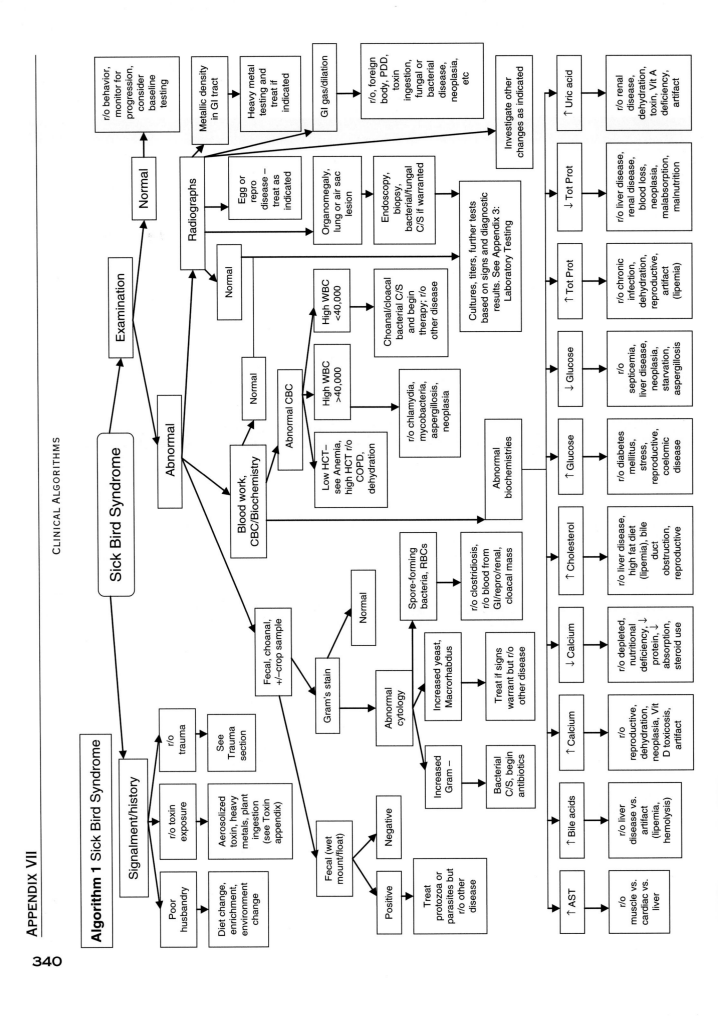

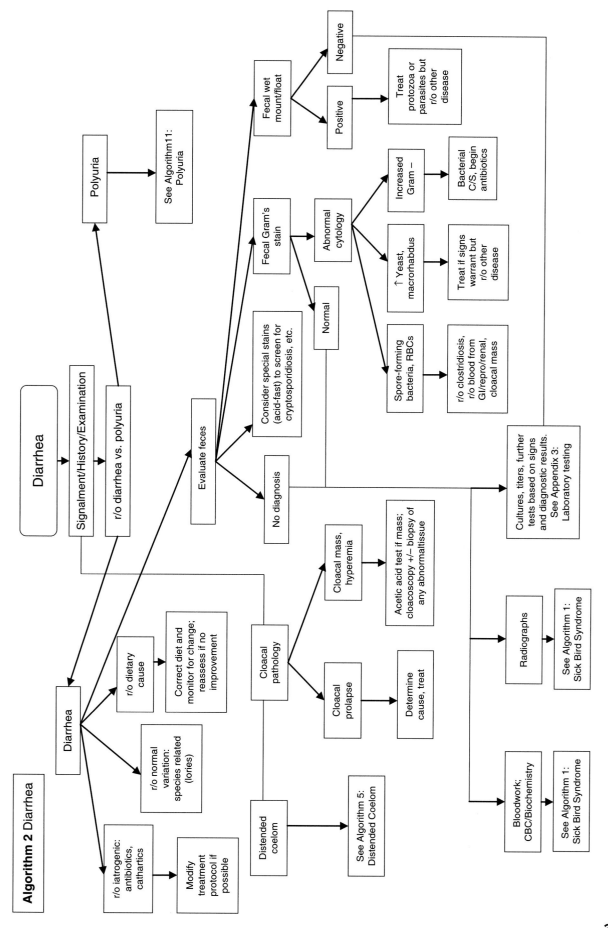

Algorithm 3
Regurgitation/Vomiting

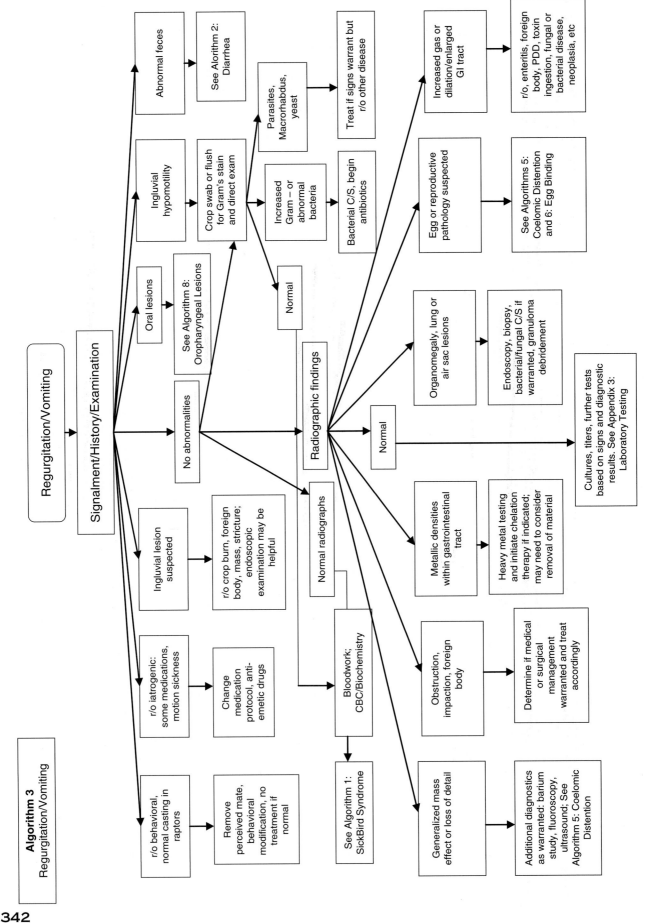

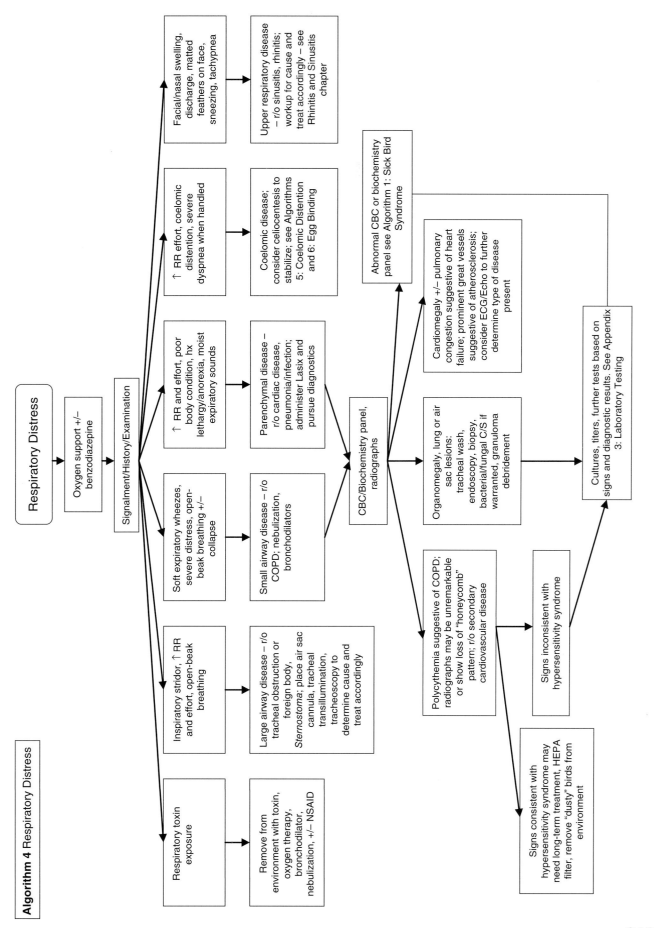

Algorithm 4 Respiratory Distress

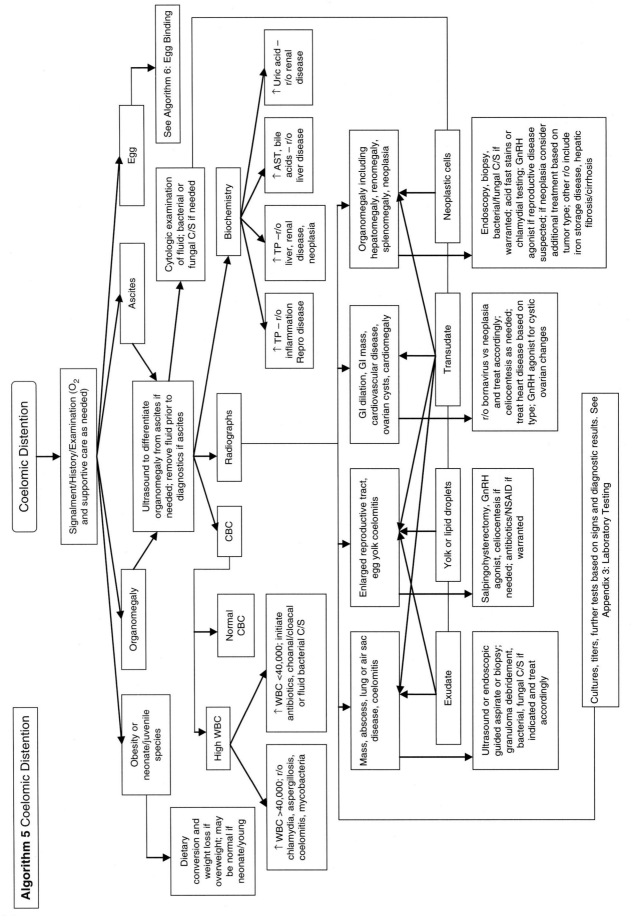

Algorithm 5 Coelomic Distention

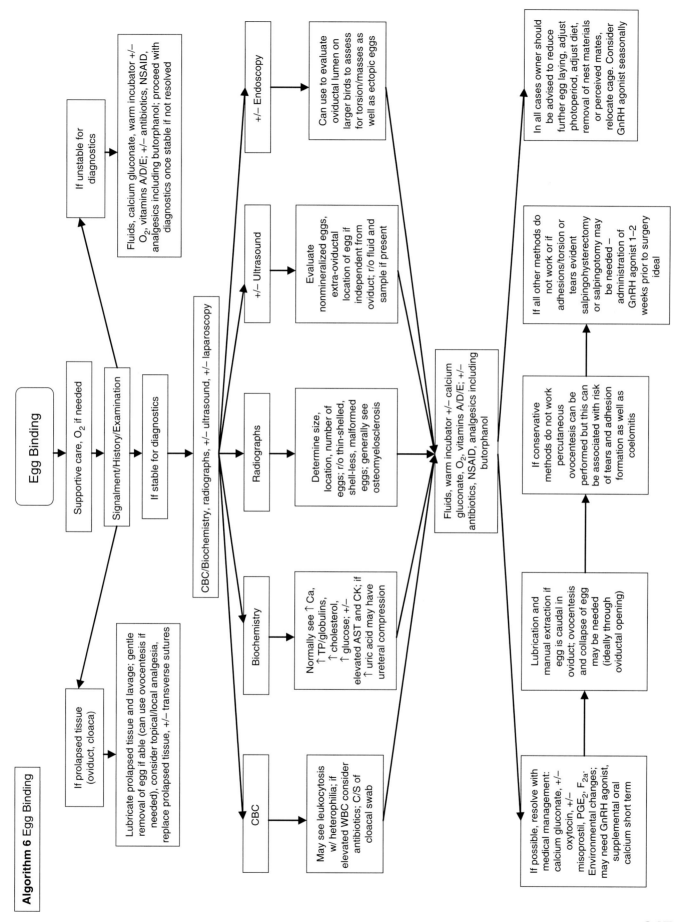

Algorithm 6 Egg Binding

Algorithm 7 Anemia

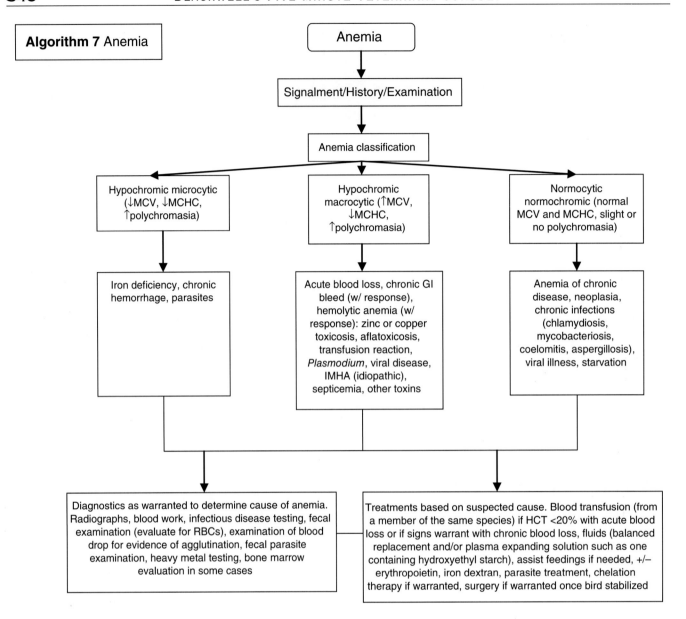

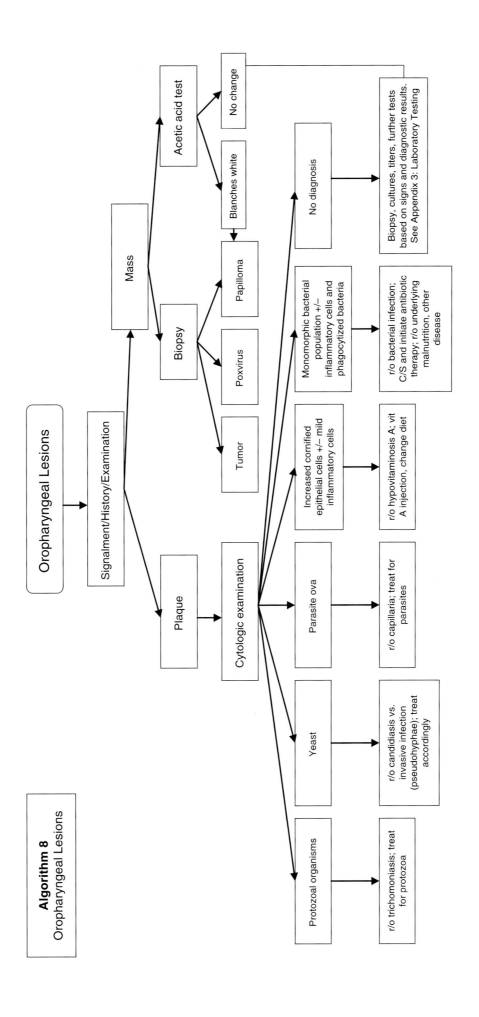

Algorithm 8
Oropharyngeal Lesions

348 BLACKWELL'S FIVE-MINUTE VETERINARY CONSULT

Algorithm 9 Neurologic Signs

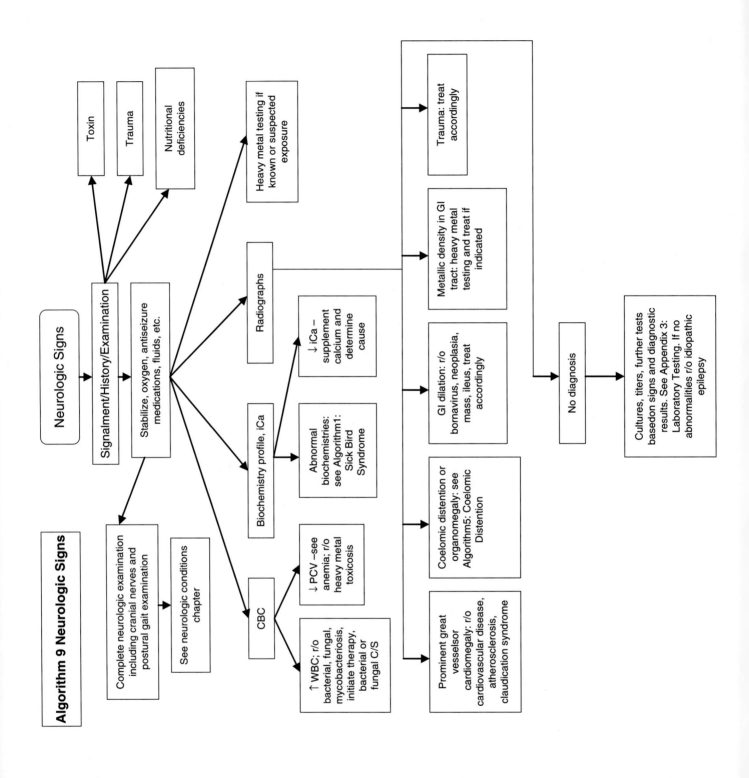

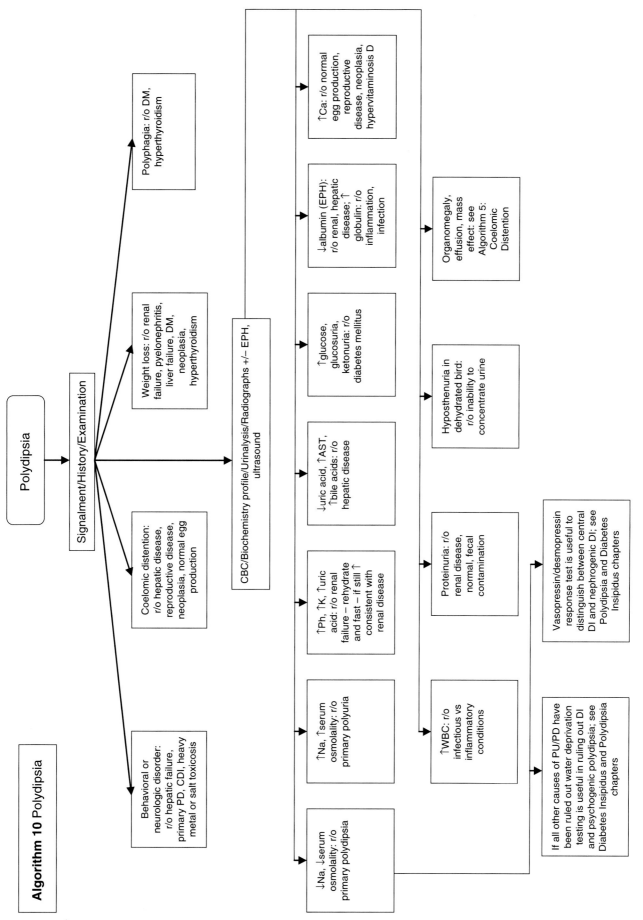

Algorithm 11 Polyuria

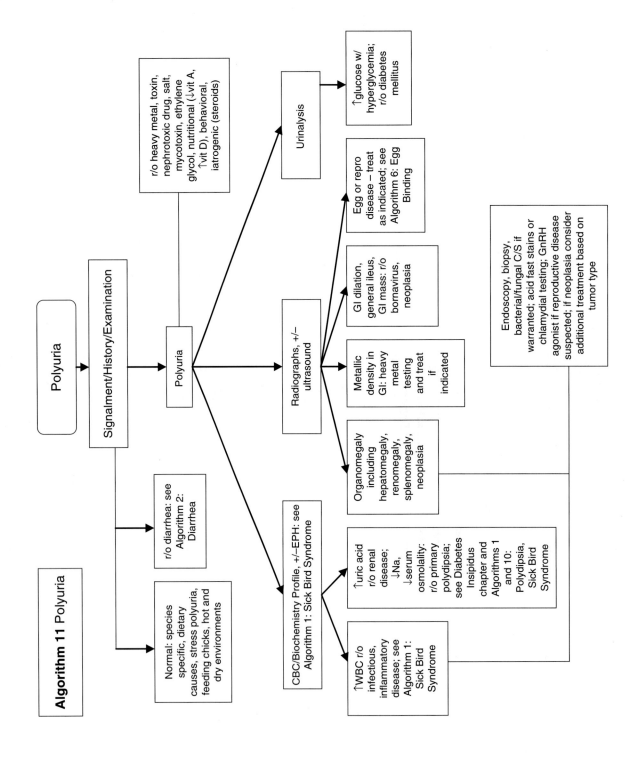

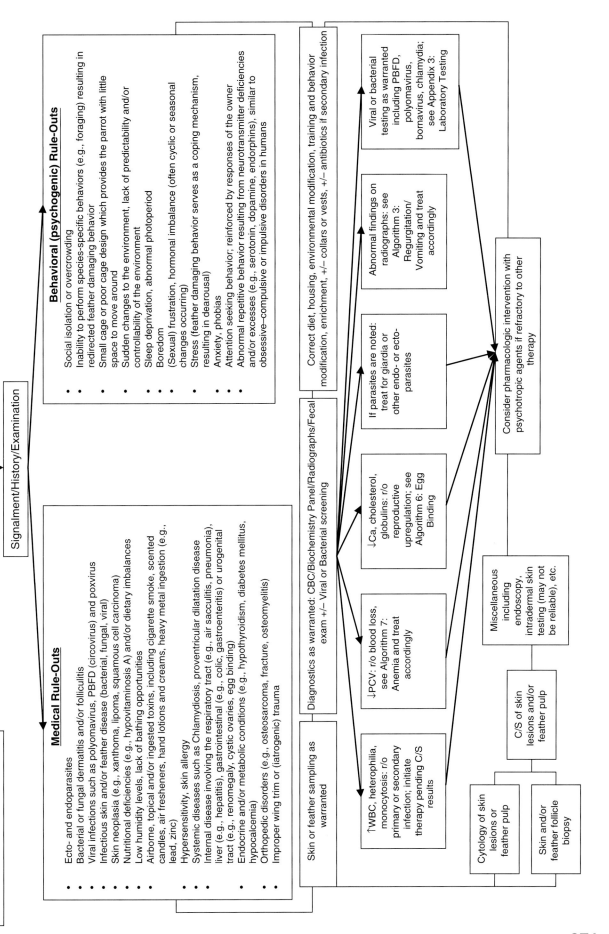

Algorithm 13 Lameness

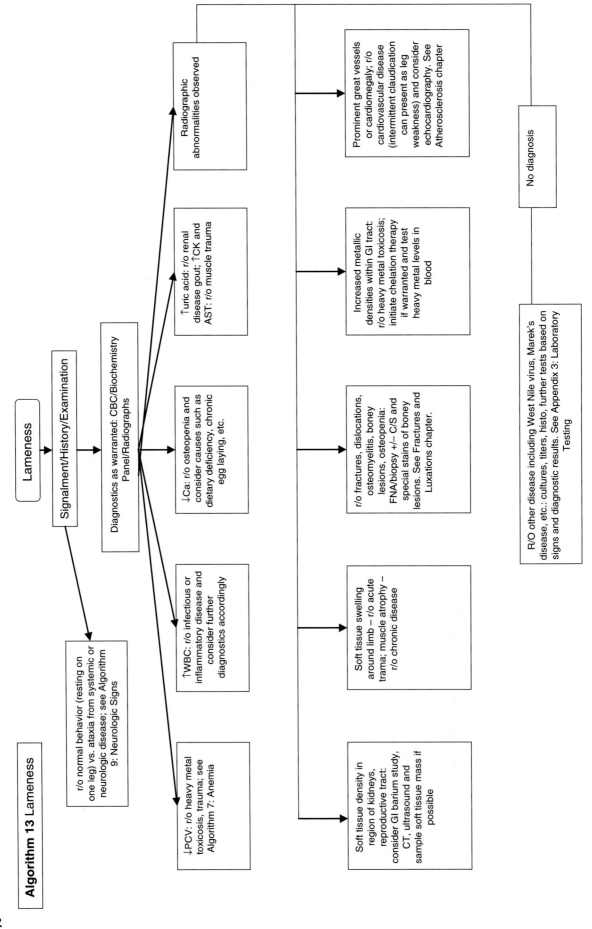

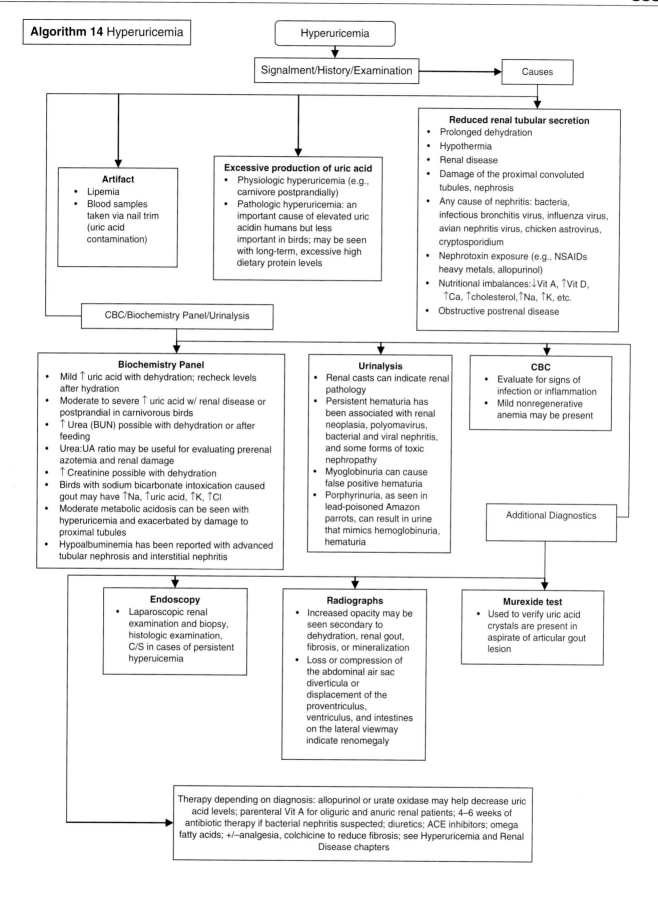

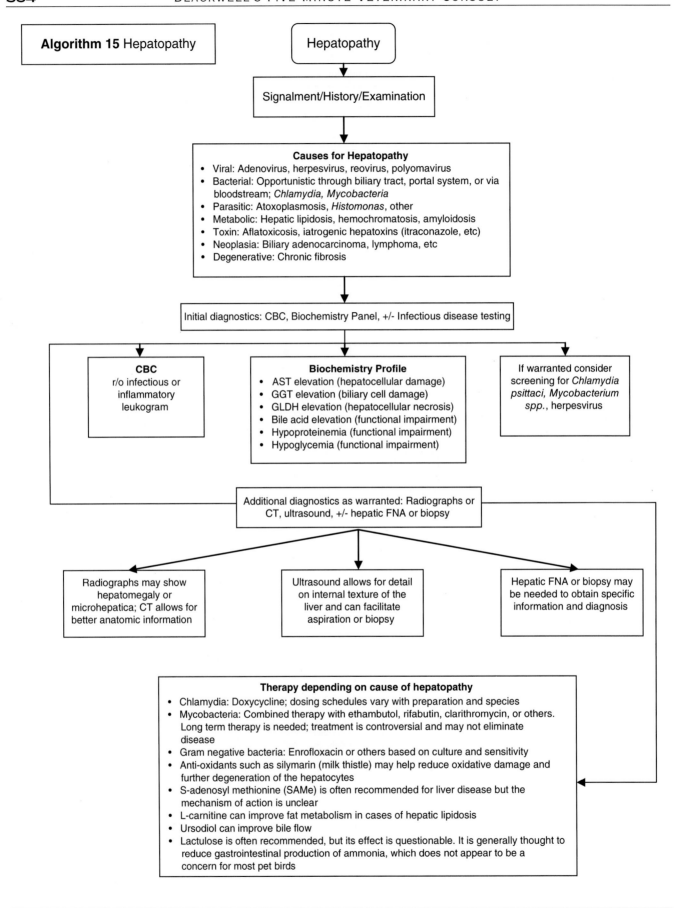

Index

Page numbers in *italics* refer to pages on which figures/tables appear. Page numbers in **bold** refer to the main topics of this book.
vs denotes differential diagnosis
Abbreviations, PDD - proventricular dilatation disease

A

"ABCs," in bite wounds, 41
abdominal distension *see* coelomic distension
abdominal mass, lipomas, 161
abscess, 42, 79, 82
acanthocephalans *see* parasitic infection
acaricide *see* antiparasitic therapy
ACE (angiotensin-converting enzyme) inhibitor, 22, 31, 53, 80, 227, 253
acetaminophen, 338
acetic acid (vinegar), 205, 312
N-acetyl-beta-D-glucosaminidase (NAG), 252
acetylcholinesterase (AChE), 200
N-acetylcysteine (NAC), 138, 315
acid-fast stains, 85, 174, 175, 249
acidosis, 105
ACT (Activated Clotting Time), 72
activated charcoal, 201, 226, 313
activated partial thromboplastin time (APTT), 72
acupuncture, 100, 159, 300
acute death *see* sudden death
acute mycotoxicosis, 179
acute tubular necrosis, 129
acyclovir, 141, 198, 312
S-adenosyl methionine (SAMe), 138, 164
adenovirus(es), **1–2**, 299, 320, 330–331
liver disease due to, 1, 163
Adequan (polysulfated glycosaminoglycan), 19, 135, 159, 316
adhesions, 80, 98, 99, 104
adsorbents, 110
Aegyptianella, 131, 132
aflatoxicosis/aflatoxins, 72, 79, 163, 179, 180, 339
aggression, 59, **244–246**, *245*, 284
airborne toxins, **3–4**, 338
feather damaging/self-injurious behavior, 113
pneumonia and, 228, 229
rhinitis and sinusitis due to, 257
tracheal/syringeal disease and, 281

air sac(s), 7, 228
aspiration, 27
compression, egg yolk/reproductive coelomitis, 103, 104
in pneumonia, 228, 229
rupture, **7–8**, 33, 153, 284
air sac cannula/cannulation, 27, 229, 255, 282
air sacculitis, 228
in aspergillosis, 24
in bordetellosis, 44
in chlamydiosis, 56
in mycobacteriosis, 174
in mycoplasmosis, 177
air sac mites, **5–6**, 229, 254, 281
airway disease, 254
airway obstruction, 27
alanine aminotransferase (ALT), 132, 134
albumin, 318, 319
albuterol *see* bronchodilator(s)
Aldar™ (imiquimod), 275, 276
alkaloids, 225, *225*, 226
allometric scaling, 106
allopurinol, 19, 146, 306, 312
alopecia *see* feather(s), loss
ALT (alanine aminotransferase), 132, 134
amantadine, 247, 248, 300
Amazon foot necrosis, 158
Amazon tracheitis (ILT), **140–142**, 281, 325
ambulation *see* gait
aminoglycosides, 204, 205, 338
aminophylline *see* bronchodilator(s)
amoxicillin-clavulanic acid, 42, 312
amphotericin B, 312
aspergillosis, 25
candidiasis, 50
macrorhabdosis, 169, 170, 296
tracheal/syringeal disease, 282
see also antifungal therapy
amprolium *see* antiparasitic therapy
amputation
in arthritis, 19
in beak injuries, 37
in phallus prolapse, 224

of pygostyle, 111
of uropygium, 297
amylase, 93, 143, 213, 215, 239
amyloidosis, 137, 213, 233
in arthritis, 18
hyperglycemia/diabetes mellitus due to, 143
in renal disease, 251, 252
amylorrhea, 109
anabolic steroids, in hepatic lipidosis, 138
analgesia
in airborne toxin sensitivity, 4
in air sac rupture, 8
in arthritis, 19
in beak injury, 37
in beak malocclusion, 39
in bite wounds, 42
in bordetellosis, 44, 45
in cloacal diseases, 64
in dystocia and egg binding, 99
in egg yolk/reproductive coelomitis, 104
in fractures and luxations, 122
in gastrointestinal foreign bodies, 124
in hemorrhage, 135
in ingluvial hypomotility/crop stasis/ileus, 154
in lameness, 159
in lymphoid neoplasia, 167
in metabolic bone disease, 171
in oral plaques, 198
in pancreatic disease, 214
in phallus prolapse, 224
in pododermatitis, 232
in renal disease, 253
in respiratory distress, 255
in squamous cell carcinoma, 275
in trauma, 285
in tumors, 290
anaphylaxis, 16, 73, 308
anatid herpesvirus, **140–142**
anemia, **9–10**, 15
in anticoagulant rodenticide toxicosis, 15, 16
in bite wounds, 42
in botulism, 46
in chlamydiosis, 56
in circovirus infections, 61

355

anemia (*Continued*)
  in cloacal disease, 64
  in clostridiosis, 66
  in coagulopathies, 72
  in coccidiosis, 75, 76
  diagnostic algorithm, *346*
  in diarrhea, 93, 94
  ectoparasites and, 101
  in feather cyst, 111
  in gastrointestinal foreign bodies, 124
  in gastrointestinal helminthiasis, 126
  in heavy metal toxicity, 129
  hemolytic *see* hemolytic anemia
  in hemoparasite infection, 131, 132
  in hemorrhage, 134, 135
  in hepatic lipidosis, 137
  in iron storage disease, 156
  in lymphoid neoplasia, 166
  in macrorhabdosis, 169
  in mycobacteriosis, 174, 175
  in mycoplasmosis, 177
  in mycotoxicosis, 179, 180
  in neurologic conditions, 183
  nonregenerative, 9, 56, 166, 185, 262
    in mycoplasmosis, 177
    in renal disease, 252
    in thyroid disease, 277
    in tumors, 289
  in nutritional deficiencies, 185
  in oil exposure, 193, 194, 195
  oral plaques and, 197–198
  in PDD, 247
  in pneumonia, 228
  regenerative, 9, 295
  in salpingitis, oviductal/uterine disorders, 262
  in thyroid disease, 277
  in trauma, 284, 285
  undigested food in droppings and, 293
  in urate and fecal discoloration, 295
anesthesia, 41, 42, 159
aneurysm, ruptured, 29, 30
angel wing, **11–12**
angiography, in pododermatitis, 231
anorexia, **13–14**
  in arthritis, 18
  in beak injury, 37
  in bordetellosis, 44
  in candidiasis, 50
  in chlamydiosis, 56
  in circovirus infection, 61
  in clostridiosis, 66
  in coelomic distension, 79
  in colibacillosis, 82
  in dehydration, 87
  in dystocia and egg binding, 98
  in egg yolk/reproductive coelomitis, 103
  in emaciation, 105
  in enteritis and gastritis, 108, 109
  in flagellate enteritis, 119
  in gastrointestinal foreign bodies, 124
  in gastrointestinal helminthiasis, 126
  in heavy metal toxicity, 129
  in hemorrhage, 134
  in hepatic lipidosis, 137
  in herpesvirus infections, 140
  in hypocalcemia and hypomagnesemia, 147
  in ingluvial hypomotility/crop stasis/ileus, 153
  in liver disease, 163, 164
  in lymphoid neoplasia, 166
  in macrorhabdosis, 169
  in mycotoxicosis, 179
  in neurological conditions, 182
  in obesity, 188
  in ocular lesions, 190
  oral plaques and, 197, 198
  in otitis, 204
  in ovarian cysts and neoplasia, 207
  in pancreatic diseases, 213
  in pneumonia, 228
  in polyostotic hyperostosis, 202
  in poxvirus infections, 241
  in renal disease, 251
  in sick bird syndrome, 270
  in squamous cell carcinoma, 274, 275
  in toxoplasmosis, 279
  in tracheal/syringeal disease, 281 282
  in trichomoniasis, 287
  undigested food in droppings and, 292
  in urate and fecal discoloration, 295
  in viral disease, 299
  in vitamin D toxicosis, 305
antecedent-behavior-consequence (ABC), 244, 245
antianxiety analgesics, 4
antibiotic therapy
  air sac rupture, 8
  arthritis, 19
  ascites, 22
  aspiration, 28
  beak injury, 37
  bite wounds, 42
  bordetellosis, 44
  botulism, 46
  campylobacteriosis, 48
  chlamydiosis, 57, 183
  cloacal diseases, 64
  clostridiosis, 66
  coelomic distension, 80
  colibacillosis, 83–84
  crop burns and frostbite, 54
  cryptosporidiosis, 86
  dystocia and egg binding, 99
  egg yolk/reproductive coelomitis, 104
  enteritis and gastritis, 110
  fractures and luxations, 122
  gastrointestinal foreign bodies, 124
  heavy metal toxicity, 130
  hemorrhage, 135
  hepatic lipidosis and, 138
  infertility and, 151
  ingluvial hypomotility/crop stasis/ileus, 154
  liver disease, 164
  lymphoid neoplasia, 167
  mycobacteriosis, 175
  mycoplasmosis, 177
  ocular lesions, 191
  oil exposed birds, 195
  oral plaques, 198
  otitis, 205
  pancreatic disease, 214
  pasteurellosis, 222
  phallus prolapse, 224
  pneumonia, 229, 230
  pododermatitis, 232
  renal disease and, 252, 253
  respiratory distress, 255
  rhinitis and sinusitis, 258
  salmonellosis, 260
  secondary infections viral disease, 300
  sick bird syndrome, 271
  tracheal and syringeal disease, 282
  trauma, 285
  undigested food droppings and, 293
  uropygial gland disease, 297
anticoagulant(s), 135, 320
anticoagulant rodenticide toxicosis, **15–17**, 72, 339
anticonvulsant drugs, 268, 269
antidepressants, 268
antidiuretic hormone (ADH), avian *see* arginine vasotocin (AVT)
antiemetic drugs, 109, 138, 214, 250, 306
antifungal therapy
  aspergillosis, 25
  in beak injury, 37
  candidiasis, 51
  in fractures and luxations, 122
  infertility and, 151
  in ingluvial hypomotility/crop stasis/ileus, 154
  Macrorhabdus ornithogaster infection, 170
  in oil exposed birds, 195
  oral plaques, 198
  pneumonia, 229
  respiratory distress, 255
  rhinitis and sinusitis, 258
  sick bird syndrome, 271
  tracheal and syringeal disease, 282
  undigested food in droppings and, 293
antiganglioside antibody, 247

antigenic shift/drift, 33
antimicrobial drugs *see* antibiotic therapy; antifungal therapy; antiparasitic therapy
antimycobacterial drugs, 175
antioxidant drugs/agents, 138, 157, 164, 227
antiparasitic therapy
   air sac mite infection, 5, 6
   coccidiosis, 75, 77
   cryptosporidiosis, 86
   ectoparasites, 101–102
   enteritis and gastritis, 110
   flagellate enteritis, 120
   gastrointestinal helminthiasis, 127
   hemoparasite infections, 133
   oral plaques, 198
   respiratory distress, 255
   Sarcocystis infections, 266
   toxoplasmosis, 280
   trichomoniasis, 288
   undigested food in droppings and, 293
antiseptics, in bite wounds, 42
antithrombotic drugs, 31
antitoxins, in botulism treatment, 47
antiviral agents, 198, 300
   acyclovir, 141, 198, 312
   amantadine, 247, 248, 300
anuria, 251
anxiolytics, 4, 115
   see also midazolam
appetite, loss *see* anorexia
applied behavior analysis (ABA), 244
aqueous flare, 190
arboviruses, 307, 335
arginine vasotocin (AVT), 91, 92, 234, 235
arrhenoblastomas, 207
arrhythmias, 21, 186, 305
   atherosclerosis and, 29, 30
   cardiac disease, 52, 53
   plant and avocado toxicosis, 226
arsenic, 338
arterial blood gas, in tracheal/syringeal disease, 281
arthritis, **18–20**
   in colibacillosis, 83
   in infertility, 150
   lameness and, 158, 159
   in obesity, 189
   in pododermatitis, 233
   in salmonellosis, 260
arthrocentesis, 19
arthrodesis, in arthritis, 19
articular gout *see* gout
ascarids, 108, **126–128**
ascites, **21–23**
   in cardiac disease, 30, 31, 52
   in coelomic distension, 21, 79, 80

in egg yolk/reproductive coelomitis, 103, 104
in hepatic lipidosis, 137, 138
in iron storage disease, 156
in liver disease, 163, 164
in lymphoid neoplasia, 166–167
in ovarian cysts and neoplasia, 22, 206, 207
in tumors, 290
aspartate aminotransferase (AST), 318, 319
   in aspergillosis, 24
   in campylobacteriosis, 48
   in chlamydiosis, 56
   in dystocia and egg binding, 99
   in enteritis and gastritis, 109
   in hemoparasites, 132
   in hemorrhage, 134
   in hepatic lipidosis, 137
   in herpesvirus infections, 141
   in inglovial hypomotility/crop stasis/ileus, 153
   in iron storage disease, 156
   in lameness, 158
   in liver disease, 164
   in lymphoid neoplasia, 166
   in *Macrorhabdus* infection, 169
   in mycobacteriosis, 174
   in neurological conditions, 183
   in ovarian cysts and neoplasia, 206
   in polydipsia, 234
   in *Sarcocystis* infections, 265
   seizures and, 267
   in toxoplasmosis, 279
   in tumors, 289, 290
aspergillosis, **24–26**
   after oil exposure, 194
   ascites due to, 21
   beak malocclusion and, 38
   coelomic distension and, 79
   granulomas, 24, 281, 282, 283
   oil exposure and, 194
   overgrown beak/nails, 210
   pneumonia, 228
   respiratory distress in, 24, 254
   tracheal/syringeal disease, 24, 281, 282, 283
aspiration, **27–28**
   air sac rupture, 7–8
   ascites, 22
   bone marrow, 167, 290
   coelomic distension, 80
   in emaciation, 105
   feather cyst, 111
   fine needle *see* fine needle aspiration
   nasal, rhinitis and sinusitis, 258
   respiratory, **27–28**
aspiration cytology, in lipomas, 161

aspiration pneumonia, 27, 105, 230, 254
   in enteritis and gastritis, 109
   inglovial hypomotility/crop stasis/ileus/ileus, 153
   in plant and avocado toxicosis, 227
   regurgitation and vomiting, 249, 250
AST *see* aspartate aminotransferase (AST)
Atadenovirus see adenovirus
ataxia, 68, 182, 200, 279
   hypocalcemia and hypomagnesemia, 147
   plant and avocado toxicosis, 226
   see also neurologic conditions
atherosclerosis, **29–32**, 52, 53, 188
   in dyslipidemia/hyperlipidemia, 95, 96
   neurological conditions due to, 182
   in obesity, 188, 189
atherothrombosis, 29
atomic absorption spectrophotometry (AAS), in iron storage disease, 156
atoxoplasmosis *(Atoxoplasma)*, 76, 131, 132, 133, 163
atropine, 201
attention seeking behavior, 113
atypical mycobacteria, 174
autoimmune thyroiditis, 277
automutilation *see* self-mutilation
avascular necrosis, 37
Aviadenovirus see adenovirus
avian bornavirus (ABV), **247–248**, 322, 331
   CNS tumors *vs*, 68
   enteritis and gastritis due to, 108
   neurological conditions, 183
   regurgitation and vomiting due to, 249, 250
   see also proventricular dilatation disease (PDD)
avian cholera (pasteurellosis), **221–222**
avian encephalitis virus, 320
avian encephalomyelitis (AE), 331
avian gastric yeast *see* macrorhabdosis
avian influenza, **33–35**, 72, 299, 300, 321, 331, 335
   H5N1, 33, 301
   in pneumonia, 228
avian leucosis virus, 68, 166, 321
avian metapneumoviruses (aMPV), **218–220**
avian nephritis virus, 321
avian paramyxoviruses (APMV), **218–220**, 331, 335
avian pathogenic *E. coli* (APEC), 82
avian polyomavirus (APV) infection, **237–238**
avian poxvirus (APV), 197, 198, **241–243**
avian retrovirus, 331, 332

avocado toxicosis, *225*, **225–227**, 337
   cardiac disease due to, 52
azithromycin *see* antibiotic therapy
azole drugs, 50, 282, 283

B

Babesia, 131
bacterial culture *see* culture
bacterial infection(s)
   in air sac mite infection, 5
   in arthritis, 18
   in bite wounds, 41
   cloacal, 63
   conjunctivitis due to, 190
   diarrhea due to, 93
   in egg yolk/reproductive coelomitis, 103
   enteritis and gastritis due to, 108, 110
   feather damaging/self-injurious behavior, 113
   in flagellate enteritis, 120
   infertility and, 150
   ingluvial hypomotility/crop stasis/ileus, 153, 154
   in liver disease, 163, 164
   in ocular lesions, 190, 191
   oral plaques and, 197, 198
   otitis due to, 204
   overgrown beak due to, 210
   pancreatic disease due to, 213
   pneumonia, 228, 229, 230
   in pododermatitis, 231
   regurgitation and vomiting, 249
   in renal disease, 251
   respiratory distress due to, 254
   rhinitis and sinusitis due to, 257
   sick bird syndrome due to, 270
   squamous cell carcinoma and, 275
   tracheal/syringeal disease due to, 281
   undigested food in droppings and, 292
   in uropygial gland disease, 297, 298
bacteriophage, 84
balance, loss, 132, 158, 204
bandage/bandaging
   bite wounds, 42
   in dermatitis and constricted toe, 89
   figure-of-eight, 11, 12, 111, 121, 159
   in fractures and luxations, 121
   in lameness, 159
   in metabolic bone disease, 171
   in pododermatitis, 232, 233
   in slipped tendon, 272, 273
   in xanthomas, 310
barium contrast radiography, 207
basal metabolic rate (BMR), 106
Baylisascaris procyonis, 68, 127

beak, **36–40**
   abnormal formation, 61
   fracture, 36–37
   injury/trauma, **36–37**, 210, 211
   malocclusion, **38–40**, 210
   mass, squamous cell carcinoma, 210, 274
   overgrowth, **210–212**
      circovirus infection, 61
      hepatic lipidosis, 137
      liver disease, 163
   rubbing, air sac mite infection, 5
   scissors deformity, **38–40**, 210, 211
   soft/bent, in metabolic bone disease, 171
   trimming, 211
behavioral changes, **244–246**, *245*
   in arthritis, 18
   in aspergillosis, 24
   in beak injury, 36
   in beak malocclusion, 38
   in chronic egg laying, 59
   in clostridiosis, 66
   in CNS tumors, 68
   in coccidiosis (systemic), 76
   in enteritis and gastritis, 108
   in feather cyst, 111
   feather damaging *see* feather(s), damaging behavior
   in heavy metal toxicity, 129
   in hemoparasite infection, 131
   in hyperglycemia and diabetes mellitus, 143
   in ingluvial hypomotility/crop stasis/ileus, 153
   in lipomas, 161
   in liver disease, 163
   in mycotoxicosis, 179
   in nutritional deficiencies, 185
   in oil exposure, 193
   in ovarian cysts and neoplasia, 206
   in overgrown beak/nails, 210
   in pancreatic disease, 213
   in phallus prolapse, 223
   in polydipsia, 234
   in polyostotic hyperostosis, 202
   regurgitation and vomiting due to, 249
   reproductive, 95, 150
   self-injurious behavior *see* self-mutilation
   in viral disease, 299, 307
benazepril, 22, 31, 80
benzodiazepines, 149, 183, 255, 268
   see also midazolam
beta-blocker, 31
bile acid(s), 318, 319
   in ascites, 21
   in aspergillosis, 24
   in atherosclerosis, 30

   in beak injury, 36
   in campylobacteriosis, 48
   in chlamydiosis, 56
   in coelomic distension, 79, 81
   in hemorrhage, 134
   in hepatic lipidosis, 137
   in liver disease, 164
   in lymphoid neoplasia, 166
   in mycobacteriosis, 174
   in mycotoxicosis, 180
   in neurological conditions, 183
   seizures and, 267
   in toxoplasmosis, 279
   in tumors, 289
   undigested food in droppings and, 293
   urate and fecal discoloration and, 295
bile ducts
   carcinoma, 141, 163
   in mycotoxicosis, 180
bile pigments, 295
biliverdin, 132, 137, 163, 295
biliverdinuria, 163, 265, 292, 295
   see also urate, discoloration
biopsy
   in arthritis, 19
   in chlamydiosis, 56
   in coccidiosis (intestinal), 74
   crop, 109, 247, 249, 293
   in cutaneous papillomas, 216
   in enteritis and gastritis, 109
   in feather damaging/self-injurious behavior, 114
   in feather disorders, 117
   in gastrointestinal helminthiasis, 126–127
   in hepatic lipidosis, 137
   in herpesvirus infections, 141
   in hyperglycemia and diabetes mellitus, 143
   infertility and, 151
   in ingluvial hypomotility/crop stasis/ileus, 154
   in iron storage disease, 156
   in lameness, 158
   in lipomas, 161
   liver *see* liver biopsy
   in nutritional deficiencies, 186
   in otitis, 204
   in pancreatic disease, 214
   in pneumonia, 228, 229
   in pododermatitis, 231
   renal, 252
   in salpingitis, oviductal/uterine disorders, 262
   skin, nutritional deficiencies, 186
   in squamous cell carcinoma, 274
   in thyroid disease, 277
   in tracheal/syringeal disease, 281

in tumors, 274, 290
in viral disease, 300
biosecurity measures
avian influenza, 34
colibacillosis, 83, 84
paramyxovirus infections, 219, 220
poxvirus infections, 242
viral neoplasms, 303
biotin deficiency, 138, 185, 272
birnaviruses, 332
bite wounds, **41–43**
beak injuries, 36
dermatitis due to, 89
hemorrhage due to, 134
biting, **244–246**, *245*
black spot disease, 133
bleeding *see* hemorrhage
bleeding diathesis/disorders, 71–73, 134, 135, 295
blepharospasm, 190
blindness, 129, 182, 192, 279, 302
blood, in droppings, 74, 140, 295
see also hematochezia; melena
blood count *see* CBC (complete blood count); leukocytosis
blood feather, broken, 134, 284
blood loss
anemia, 9–10
anticoagulant rodenticide toxicity, 15
hyperuricemia and, 145
trauma, 15, 284
see also hemorrhage
blood pressure, 30, 52, 134, 135, 252
blood transfusion
in anemia, 10
in anticoagulant rodenticide toxicity, 16
in bite wounds, 42
in coagulopathies, 72
in hemorrhage, 135
in plant and avocado toxicosis, 226
in trauma, 285
blood urea nitrogen (BUN), 145, 252
"blowing bubbles", 287
BMR (basal metabolic rate), 106
Bollinger bodies, 198, 242, 334
bone, 147, 158, 171
callus, 122
cortices, 122
cyst, 18
deformity, metabolic bone disease, 171, 172
fixation, 122
fracture *see* fracture(s)
healing, 121, 122
metabolic disease *see* metabolic bone disease
mineral density decrease, 171
opacity, increased, 202
pneumatic, 122

bone marrow
aspiration/biopsy, 167, 290
in lymphoid neoplasia, 166, 167
ossification, 202
suppression, in mycotoxicosis, 179
bordetellosis *(Bordetella)*, **44–45**, 190, 281, 322
boredom, 113, 114, 115, 244
boric acid, in otitis, 205
bornavirus *see* avian bornavirus (ABV); proventricular dilatation disease (PDD)
Borrelia, 131
botulism, **46–47**, 182, 322
bradycardia, 134, 200, 307
brain tumors, **68–70**, 166
breathing, **254–256**
open-beaked/mouthed, 153, 241, 254
pneumonia, 228
rhinitis and sinusitis, 257
tracheal and syringeal disease, 281
see also respiratory distress
brodifacoum, 16, 71
broken blood feather, 134, 284
bromethalin, 339
bronchial lavage, in mycoplasmosis, 177
bronchodilator(s), 28, 229, 266
in respiratory distress, 255
toxins and, 4, 227
bronchopneumonia, 24, 76
broody behavior, 59, 98
brown hypertrophy of the cere, 54
bruising, 41, 121, 134, 158, 284
see also hemorrhage
bumblefoot *see* pododermatitis
burns
crop, **54–55**
regurgitation and vomiting, 249
Bursa of Fabricius, 61
butorphanol *see* analgesia

C

cabergoline, in ovarian cysts, 208
caEDTA *see* calcium EDTA
calcitonin, 147, 306, 312
calcitriol, 147
calcium, 147, 171, 318, 319
in chronic egg laying, 59, 60
deficiency, 185, 186, 234
dystocia and egg binding, 99
splay leg and slipped tendon, 273
deposits, vitamin D toxicosis, 305
homeostasis, 147, 171, 305
ionized, 147, 148, 171, 183
in metabolic bone disease, 172
metabolism, 187
in polyostotic hyperostosis, 202, 207
storage, osteomyelosclerosis, 202

supplementation, 80, 99, 148, 149, 172, 186
in hemorrhage, 135
see also hypercalcemia; hypocalcemia
calcium EDTA, 125, 130, 159, 312
in ingluvial hypomotility/crop stasis/ileus, 154
in renal disease, 253
in sick bird syndrome, 271
calcium oxalate, 227
calcium:phosphorus ratio, 172
callus, 159
campylobacteriosis *(Campylobacter)*, **48–49**, 108, 322, 335
candidiasis *(Candida* spp.), **50–51**, 197, 249, 322–323
enteritis and gastritis due to, 108
ingluvial hypomotility/crop stasis/ileus, 153
oral plaques, 197, 198
regurgitation and vomiting, 249
undigested food in droppings and, 292, 293
cannibalism, 134
capillariasis *(Capillaria* spp.), 108, 126, 127, 197
capnometry, in tracheal/syringeal disease, 282
capture myopathy, 252
carbamate toxicity, **200–201**, 327, 339
carbaryl powder, 102
carbon monoxide toxicity, 3, 338
carcinoma *see specific carcinomas*
carcinomatosis, 80
ascites, 21, 22
ovarian neoplasms, 207
cardiac disease, 21, **52–53**
in atherosclerosis, 29–31
in coelomic distension, 80
in colibacillosis, 82
in PDD, 247
in polyomavirus infection, 237
in *Sarcocystis* infections, 265, 266
in vitamin D toxicosis, 305, 306
see also cardiovascular disease
cardiac glycosides, 225, *225*, 226
cardiomegaly, 21, 22, 30, 80
cardiomyopathy, 21, 52
see also cardiac disease
cardiovascular disease
in anemia, 9
in organophosphate/carbamate toxicity, 200
in pasteurellosis, 221
in PDD, 247
in plant and avocado toxicosis, 225
in polydipsia, 234
in polyomavirus infection, 237
in viral disease, 299
see also cardiac disease

carnidazole *see* antiparasitic therapy
L-carnitine, 80, 138, 164, 189, 315
carpal valgus (angel wing), 11–12
carprofen *see* NSAIDs
cartilage, 18
casting behavior, 249
cataracts, 190, 191
cathartics, 94, 226
CBC (complete blood count)
   in air sac mite infection, 5
   in anemia, 9, 10
   in anorexia, 13
   in bite wounds, 41
   in clostridiosis, 66
   infertility and, 151
   in mycobacteriosis, 174
   in ocular lesions, 191
   in otitis, 204
   in ovarian cysts and neoplasia, 206
   in overgrown beak/nails, 210
   in pododermatitis, 231
   in sick bird syndrome, 270
   see also leukocytosis
celiocentesis, 22, 80, 167
   in egg yolk/reproductive coelomitis, 103
   in respiratory distress, 255
celioscopy, 91, 156, 229
celiotomy, 76
cellulitis, 83, 231, 232
cere and skin, 54
   circovirus infection, 61
   color changes, **54–55**
   dermatitis *see* dermatitis
   ectoparasites and, 101
   in nutritional deficiencies, 185
   see also skin
cerebellar lesions, 182
cerebral edema, 183, 284
cerebrospinal fluid (CSF) analysis, 268
cervical esophageal perforation, 153
cestodes, 126
chelation therapy, 183
   in gastrointestinal foreign bodies, 124, 125
   heavy metal toxicity, 130, 268
   iron storage disease, 22, 80, 157, 314
   in pancreatic disease, 214
   in renal disease, 253
   in urate and fecal discoloration, 296
chemical toxins, **339**
chemotherapy, 290
   lymphoid neoplasia, 167–168
   pancreatic neoplasia, 214
   renal disease, 253
   squamous cell carcinoma, 274, 275, 276
   uropygial gland disease, 297, 298
CHF *see* congestive heart failure (CHF)
chicken anemia virus (CAV), **61–62**, 323

chlamydiosis *(Chlamydia psittaci)*, **56–58**, 72, 183, 323, 335
   atherosclerosis and, 30
   feather damaging behavior due to, 113
   in hyperglycemia and diabetes mellitus, 143
   infertility and, 150, 151
   liver disease due to, 163, 164
   neurological conditions due to, 182
   in pneumonia, 229, 230
   seizures, 267
   tracheal/syringeal disease and, 281, 282, 283
choanal atresia, 257, 258
choanal papillae, blunting, 197
cholangiohepatitis, 66
cholecalciferol, 147
cholestasis, 137
cholesterol
   in atherosclerosis, 30
   in chronic egg laying, 59
   in dyslipidemia/hyperlipidemia, 95
   in hepatic lipidosis, 137
   in obesity, 188
   see also hypercholesterolemia
cholesterol absorption inhibitors, 96
cholestyramine, 110
choline, deficiency, 272
cholinesterase, 200, 323
cholinesterase inhibitors, 127, 200
choroid plexus tumors, 68, 69
chronic egg laying, **59–60**, 98, 103
   lameness, 158
   in ovarian cysts and neoplasia and, 206
   in polyostotic hyperostosis, 202
   salpingitis, oviductal/uterine disorders, 262
chronic respiratory disease (CRD), 177
cilostazol, 31
Cimex lectularius, 101
circoviruses, **61–62**, 72, 299, 323, 332
   feather disorders, 117, 118
   overgrown beak/nails, 210
cirrhosis, 21, 22, 79, 80
cisplatin, 275, 290, 291, 313
CK *see* creatine kinase (creatine phosphokinase; CK)
clicking *see* respiratory distress
cloaca, *63*, 251, 262
cloacal disease, **63–65**
   in botulism, 46
   in clostridiosis, 66
   in coccidiosis (systemic), 76
   in coelomic distension, 79
   in dystocia and egg binding, 98, 99
   in gastrointestinal helminthiasis, 126
   in herpesvirus infections, 140

   in hypocalcemia and hypomagnesemia, 147
   in lipomas, 161
   in phallus prolapse, 223, 224
cloacal obstruction, 63
cloacal papilloma, **140–142**, 163
cloacal prolapse, 63, 66, 147, 223
cloacitis, 63, 64, 66, 134
cloacolith, 63, 64
cloacopexy, 64
cloacoscopy, 296
cloacotomy, 64
clomipramine, 115, 313
clostridiosis (*Clostridium* spp.), **66–67**, 323
   botulism, **46–47**, 182, 322
   in enteritis and gastritis, 108
   in neurologic conditions, 182
   undigested food in droppings and, 293
Clostridium botulinum, 46–47, 66
   botulism, **46–47**, 182, 322
Clostridium perfringens, 66
clotrimazole, in aspergillosis, 25
clotting factors, 71, *71*
CNS tumors, **68–70**, 166
coagulation, 15–16, *71*, **71–73**, 134, 156
coagulopathies, **71–73**
   in anticoagulant rodenticide toxicosis, 15
   hemorrhage due to, 134
   in hepatic lipidosis, 137
   in liver disease, 163
   in pasteurellosis, 221
coaptation, external, 121, 159, 171
coccidiosis (disseminated visceral), 76
coccidiosis (intestinal), 66, **74–75**, 108, 292
coccidiosis (systemic), **76–78**
   renal, 251, 252–253
Cochlosoma, in flagellate enteritis, 119, 120
coelioscopy, 91, 156, 229
coeliotomy, 76
coelomic disease, respiratory distress in, 254
coelomic distension, **79–81**
   ascites *vs*, 21
   in chronic egg laying, 59
   in coccidiosis (systemic), 76
   diagnostic algorithm, *344*
   dystocia and egg binding *vs*, 98
   in egg yolk/reproductive coelomitis, 103
   in hepatic lipidosis, 137
   in hypocalcemia and hypomagnesemia, 147
   in iron storage disease, 156
   in lameness, 158
   in lipomas, 161

in liver disease, 163
in lymphoid neoplasia, 166, 167
in mycobacteriosis, 174, 175
in obesity, 188
in ovarian cysts and neoplasia, 206, 207, 208
in pancreatic diseases, 213, 214
pneumonia *vs*, 228
polydipsia and, 234, 235
polyostotic hyperostosis *vs*, 202
in renal disease, 251, 252
in respiratory distress, 254, 255
in salpingitis, oviductal/uterine disorders, 262
in tumors, 289, 290
undigested food in droppings and, 292
in viral neoplasms, 302
coelomic effusion, 167, 207
coelomic fluid analysis, 290
coelomitis, 59, **103–104**, 213
see also egg yolk coelomitis
coelomocentesis *see* celiocentesis
colchicine, 19, 22, 80, 146, 253, 313
colibacillosis, 44, **82–84**
coliform infections, fecal discoloration, 295
colisepticemia, 83
collisions, 121
colopexy, 64
columbid herpesvirus, **140–142**
coma, in plant and avocado toxicosis, 225
computed tomography (CT)
 in air sac rupture, 7
 in arthritis, 19
 in aspiration, 27
 in atherosclerosis, 30
 in CNS tumors, 69
 in coelomic distension, 80
 in diabetes insipidus, 91
 in egg yolk/reproductive coelomitis, 103
 in enteritis and gastritis, 109
 in fractures and luxations, 121
 in ingluvial hypomotility/crop stasis/ileus, 154
 in liver disease, 164
 in metabolic bone disease, 171
 in neurological conditions, 183
 in ocular lesions, 191
 in ovarian cysts and neoplasia, 207
 in overgrown beak/nails, 211
 in pododermatitis, 232
 in respiratory distress, 254
 in rhinitis and sinusitis, 257–258
 in salpingitis, oviductal/uterine disorders, 262
 in *Sarcocystis* infections, 265
 in squamous cell carcinoma, 274

in trauma, 284
in tumors, 290
congestive heart failure (CHF), 53, 79, 80
 ascites and, 21, 22, 23
 atherosclerosis and, 30
 in dyslipidemia/hyperlipidemia, 95
conjunctivitis, 33, 190, 191
 in bordetellosis, 44
 in cryptosporidiosis, 85
 in herpesvirus infections, 140
 in mycoplasmosis, 177, 190
 in otitis, 204
 in toxoplasmosis, 279
constricted toe, 89, 90
contrast radiography *see* radiography
Conure bleeding syndrome, 9, 71
convulsions *see* seizures
COPD, 3
copper toxicity, 129, 130
coprodeum *see* cloaca
corneal pathology, 190, 193
cor pulmonale, 29
corticosteroid(s)
 in airborne toxin exposure, 4
 contraindication in seizures, 269
 hyperglycemia/diabetes mellitus due to, 143
 in lymphoid neoplasia, 167, 168
 in neurologic disease, 184
 in pododermatitis, 233
 in renal disease, 253
 risks in ocular lesions, 191
 squamous cell carcinoma and, 275
 in tracheal/syringeal disease, 282, 283
 in tumors, 291
coryza, 44
cough, 281
 in airborne toxin exposure, 3
 in air sac mite infection, 5
 in aspiration, 27
 in avian influenza, 33, 34
 in bordetellosis, 44
 in candidiasis, 50
 in cryptosporidiosis, 85
 in pneumonia, 228
 in respiratory distress, 254
 in tracheal lesions/disease, 197, 281
 see also respiratory distress
Cowdry bodies, 198
COX-2 inhibitors, 250, 291, 293
cranial nerve(s), examination, 182
creatine kinase (creatine phosphokinase; CK), 318, 319
 in atherosclerosis, 30
 in dystocia and egg binding, 99
 in enteritis and gastritis, 109
 in feather damaging/self-injurious behavior, 114

in ingluvial hypomotility/crop stasis/ileus, 153
in iron storage disease, 156
in lameness, 158
in lymphoid neoplasia, 166
in ovarian cysts and neoplasia, 206
in PDD, 247
in pneumonia, 228
in *Sarcocystis* infections, 265
seizures and, 267
in tumors, 290
in vitamin D toxicosis, 305
creatinine, 145, 252
crepitus, 18, 158
critically ill birds, 270
crop, 153
 biopsy, 109, 247, 249, 293
 burn, **54–55**, 105
 delayed emptying, polyomavirus infection, 237
 distension
  ingluvial hypomotility/crop stasis/ileus, 153
  undigested food in droppings and, 292
 foreign bodies, 249
 lavage, in plant and avocado toxicosis, 226
 motility, 153
 ulcer, 54
crop disorders, regurgitation/vomiting, 249
crop stasis, **153–155**
 in bite wounds, 41
 in candidiasis, 50
 in flagellate enteritis, 119
 in gastrointestinal foreign bodies, 124
 in heavy metal toxicity, 129
 in Marek's disease, 302
 in organophosphate/carbamate toxicity, 200
 in renal disease, 251
 in thyroid disease, 277
 in vitamin D toxicosis, 305
crop suspension bandages, 154
crushing injury, 121
crusts, 89
cryotherapy, squamous cell carcinoma, 275
Cryptococcus, 38, 335
cryptosporidiosis (*Cryptosporidium*), **85–86**, 145, 323
 regurgitation and vomiting, 249
 undigested food in droppings, 292, 293, 294
CT *see* computed tomography (CT)
culture (bacterial/fungal), 321
 in ascites, 21
 in bite wounds, 41
 in bordetellosis, 44

culture (bacterial/fungal) (*Continued*)
   in campylobacteriosis, 48
   in candidiasis, 50
   in chlamydiosis, 56
   in cloacal diseases, 64
   in clostridiosis, 66–67
   in coelomic distension, 79
   in colibacillosis, 82–83
   in cryptosporidiosis, 85
   in dystocia and egg binding, 99
   in egg yolk/reproductive coelomitis, 103
   in enteritis and gastritis, 109
   in feather cyst, 111
   in feather damaging/self-injurious behavior, 114
   in ingluvial hypomotility/crop stasis/ileus, 153
   in mycobacteriosis, 174
   in otitis, 204
   in overgrown beak/nails, 211
   in pasteurellosis, 221
   in pneumonia, 228
   in renal disease, 252
   in respiratory distress, 254
   in tracheal/syringeal disease, 281
   urate and fecal discoloration and, 296
cutaneous papillomas, **216–217**
cutaneous pox, 241, 242
cyanosis, 33, 281
cyclosporine, in PDD, 247
cyst(s)
   bone, 18
   feather, **111–112**
   ovarian *see* ovarian cysts
   pancreatic, 213
cystatin C, urinary, 252
cystic ovaries *see* ovarian cysts
Cytodites nudus (air sac mite), **5–6**, 229, 254, 281
cytology, 324
   in ascites, 21
   in bordetellosis, 44
   in candidiasis, 50
   in cloacal diseases, 64
   in coccidiosis (systemic), 76
   in coelomic distension, 79
   in cryptosporidiosis, 85
   in diarrhea, 93
   in egg yolk/reproductive coelomitis, 103
   in feather cyst, 111
   in feather damaging/self-injurious behavior, 114
   in flagellate enteritis, 119–120
   in hemoparasites, 132
   in ingluvial hypomotility/crop stasis/ileus, 153
   in lipomas, 161
   in oral plaques, 198

   in otitis, 204
   in pneumonia, 228
   in pododermatitis, 231
   in renal disease, 252
   in respiratory distress, 254
   in rhinitis and sinusitis, 257
   in salpingitis, oviductal/uterine disorders, 262
   in tumors, 290
   undigested food in droppings and, 293
   urate and fecal discoloration and, 296
   in uropygial gland disease, 297
   in viral disease, 300

D

defecation difficulty/failure, 98, 105
   in dystocia and egg binding, 98
   in lipomas, 161
deferoxamine, 22, 80, 157, 314
degenerative joint disease, 122, 231
dehydration, **87–88**
   in anorexia, 13
   in bite wounds, 42
   in botulism, 46
   in campylobacteriosis, 48
   in cardiac disease, 52
   in chlamydiosis, 56
   in coccidiosis, 74, 76
   in colibacillosis, 82, 83
   in cryptosporidiosis, 85
   in diabetes insipidus, 91, 92
   in diarrhea, 93, 94
   in emaciation, 105
   in enteritis and gastritis, 108, 109, 110
   in flagellate enteritis, 119
   in gastrointestinal foreign bodies, 124
   in hepatic lipidosis, 137, 138
   in hyperglycemia and diabetes mellitus, 143
   in hyperuricemia, 145, 146
   in ingluvial hypomotility/crop stasis/ileus, 153
   in *Macrorhabdus* infection, 169
   in oil exposure, 193–194
   oral plaques and, 197, 198
   in pancreatic disease, 213, 214
   in plant and avocado toxicosis, 226, 227
   in pneumonia, 228, 229
   in polydipsia, 235
   in polyuria, 239, 240
   in regurgitation and vomiting, 249, 250
   in renal disease, 251, 252, 253
   in respiratory distress, 255
   in sick bird syndrome, 270
   in squamous cell carcinoma, 274

   in toxoplasmosis, 279
   in trauma, 285
   in tumors, 290
demyelination, 183
Depo-Provera, 208
dermal mass, xanthomas, 310
Dermanyssus gallinae, 101, 335
dermatitis, **89–90**
   in colibacillosis, 82
   ectoparasites and, 101
   feather damaging/self-injurious behavior, 113, 114
   pododermatitis, **231–233**
deslorelin, 22, 60, 80, 104, 313
   ovarian cysts, 208
   polyostotic hyperostosis, 203
   salpingitis, oviductal/uterine disorders, 263
   see also GnRH (gonadotropin-releasing hormone) agonist
deworming, 127
dexamethasone *see* corticosteroid(s)
diabetes insipidus, **91–92**, 234, 252
diabetes mellitus, **143–144**
   dyslipidemia/hyperlipidemia, 95
   iron storage disease and, 156, 157
   pancreatic disease and, 213, 214, 215
   polydipsia, 234
diarrhea, **93–94**, 292
   in adenovirus infection, 1
   in avian influenza, 33, 34
   in campylobacteriosis, 48
   in chlamydiosis, 56
   in clostridiosis, 66
   in coccidiosis (intestinal), 74–75
   in coccidiosis (systemic), 76
   in colibacillosis, 82
   in cryptosporidiosis, 85
   diagnostic algorithm, *341*
   in enteritis, 108
   in flagellate enteritis, 119
   in gastrointestinal helminthiasis, 126
   green, 287
   in heavy metal toxicity, 129
   in hepatic lipidosis, 137
   in herpesvirus infections, 140
   in liver disease, 163
   in lymphoid neoplasia, 166
   in *Macrorhabdus* infection, 169
   in mycotoxicosis, 179
   in oil exposure, 193–194, 195
   in pancreatic diseases, 213
   in paramyxovirus infections, 218, 219
   in PDD, 247
   in phallus prolapse, 223
   in plant and avocado toxicosis, 226, 227
   in polyomavirus infection, 237

polyuria *vs*, 239
in renal disease, 251
in salmonellosis, 260
in sick bird syndrome, 270
in trichomoniasis, 287
undigested food in droppings and, 292
urate and fecal discoloration and, 287, 295
diatomaceous earth, 102, 127
diclofenac *see* NSAIDs
diet
ideal, 188–189
iron storage disease due to, 156
mycotoxicosis and, 179
obesity associated, 188
dietary management
angel wing, 11, 12
anorexia, 14
anticoagulant rodenticide toxicosis, 16
arthritis, 19
aspiration, 27–28
atherosclerosis, 31
beak injury, 36
bordetellosis, 45
cardiac disease, 53
chlamydiosis, 57
chronic egg laying, 59
coagulopathy, 72
coccidiosis (systemic), 77
diabetes mellitus and hyperglycemia, 143
diarrhea, 93
dyslipidemia/hyperlipidemia, 96
dystocia and egg binding, 99
egg yolk and reproductive coelomitis, 104
emaciation and starvation, 105–106
enteritis and gastritis, 109
feather cyst, 112
feather damaging/self-injurious behavior, 115
feather disorders, 117, 118
gastrointestinal foreign bodies, 124
hemorrhage, 135
hepatic lipidosis, 138
hypocalcemia and hypomagnesemia, 148
infertility and, 151
ingluvial hypomotility/crop stasis/ileus, 154
iron storage disease, 157
lameness, 159
lipomas, 161
liver disease, 164
lymphoid neoplasia, 167
metabolic bone disease, 172
mycobacteriosis, 175
neurological conditions, 183
nutritional deficiencies, 186

obesity, 188–189
oil exposed birds, 194, 195
oral plaques, 198
overgrown beak, 211
pancreatic disease, 214
plant and avocado toxicosis, 226
pododermatitis, 232
problem behaviors and, 245
proventricular dilatation disease, 247
respiratory distress, 255
salpingitis, oviductal/uterine disorders, 263
seizures, 268
splay leg and slipped tendon, 272
thyroid disease, 277
tumors, 291
viral disease, 300
vitamin D toxicosis, 305
xanthomas, 310
see also nutritional support
differential selection of alternative (DRA) behavior, 245
differential selection of incompatible (DRI) behavior, 245
digoxin, 53
dimercaprol, 130, 313
dimetridazole, 120, 293
diphtheroid (wet) pox, 241, 242
"disease for all seasons" (pasteurellosis), 221
disinfection, 336
in avian influenza, 34
bite wounds, 41
in gastrointestinal helminthiasis, 127
in paramyxovirus infections, 219
in polyomavirus infection, 238
in poxvirus infections, 242
Dispharynx spp., 126
disseminated intravascular coagulation (DIC), 15, 71, 72, 73
distension *see* coelomic distension
diuresis, 159, 305
diuretics
in cardiac disease, 53
in coelomic distension, 80
contraindicated in polyuria, 240
in hepatic lipidosis, 138
in plant and avocado toxicosis, 227
in renal disease, 253
in respiratory distress, 255
in vitamin D toxicosis, 305, 306
DMSA (dimercaptosuccinic acid), 125, 130, 159, 313
see also chelation therapy
DMSO (dimethyl sulfoxide), in phallus prolapse, 224
DNA, viral, identification, 219
donut-shaped pads, 232

dopamine D2-receptor antagonist, 154
dopamine receptor agonists, 208
doxycycline, 57, 314
see also antibiotic therapy
Dremel, 211
drugs, 324
common dosages, **313–318**
toxic, **338**
drying, in oil exposed birds, 195
duck plague, **140–142**
duck viral enteritis, 182, 183
duck viral hepatitis, 332
duodenostomy tube, 109
dust bath, 102
dysbiosis (sour crop), 153
dyschondroplasia, 159
dyslipidemia, **95–97**, 182, 184
see also hyperlipidemia
dysphagia, 126, 140, 197, 198
dyspnea
in air sac mite infection, 5
in ascites, 21
in aspiration, 27
in bordetellosis, 44
in coccidiosis (systemic), 76
in cryptosporidiosis, 85
in dystocia and egg binding, 98
in egg yolk/reproductive coelomitis, 103
in hepatic lipidosis, 137
in herpesvirus infections, 140
in hypocalcemia and hypomagnesemia, 147
in ingluvial hypomotility/crop stasis/ileus, 153
in iron storage disease, 156
in lipomas, 161
in lymphoid neoplasia, 166
in mycobacteriosis, 174
in mycoplasmosis, 177
in obesity, 188
oral plaques and, 197
in ovarian cysts and neoplasia, 206, 207
in paramyxovirus infections, 219
in plant and avocado toxicosis, 226
in renal disease, 145251
in *Sarcocystis* infections, 265, 266
in toxoplasmosis, 279
in tracheal/syringeal disease, 281, 282
see also respiratory distress
dystocia, **98–100**
in lipoma, 161
in nutritional imbalances, 185, 186, 187
in obesity, 188, 189
respiratory distress in, 254
see also egg binding

E

ear discharge, in otitis, 204
Eastern Equine Encephalitis (EEE) virus, 68, 134, 299, 324, 333, 335
echocardiography, 30, 80, 81
    in iron storage disease, 156
    lameness and, 158
    in respiratory distress, 254
    seizures and, 268
E-collar *see* Elizabethan collar
ectoparasites, **101–102**
    dermatitis due to, 89
    feather damaging/self-injurious behavior, 113
    in uropygial gland disease, 297
edema
    in avian influenza, 33, 34
    in cardiac disease, 52
    dermatitis due to, 89
    pulmonary, 2, 29, 265, 266, 279, 280
    trauma causing, 284
egg(s)
    abnormal/misshapen, 59, 98, 150, 151, 218, 262
    ectopic, 99, 100
    necropsy, 151
egg binding, **98–100**
    in chronic egg laying, 59, 60
    in coelomic distension, 80, 81
    diagnostic algorithm, *345*
    infertility and, 150
    lameness and, 159
    in nutritional imbalances, 185, 187
    in ovarian cysts and neoplasia, 206
    in polyostotic hyperostosis, 202
    in renal disease, 252
    in salpingitis, oviductal/uterine disorders, 262
    see also dystocia
egg drop syndrome *see* adenovirus(es)
egg laying, 98
    chronic *see* chronic egg laying
    in coelomic distension, 79, 80
    in dystocia and egg binding, 98
    in egg yolk/reproductive coelomitis, 103
    in hepatic lipidosis, 138
    in hypocalcemia and hypomagnesemia, 147
    infertility and, 150
    in metabolic bone disease, 172
    in polyostotic hyperostosis, 202
    in salpingitis, oviductal/uterine disorders, 262
egg production
    in avian influenza, 33, 34
    in coccidiosis (intestinal), 74
    drop/reduced
        ectoparasites causing, 101
        mycoplasmosis, 177, 178
        mycotoxicosis, 179
        paramyxovirus infections, 218, 219
        viral neoplasms, 302
egg shell, abnormal, 1, 98, 262
egg yolk coelomitis, **103–104**, 182
    ascites and, 21, 23
    in chronic egg laying, 59
    in coelomic distension, 79
    dystocia and egg binding *vs*, 98
    hyperglycemia/diabetes mellitus due to, 143
egg yolk peritonitis, 100, 103
Eimeria infection, 66, 74, 76
    see also coccidiosis
electric shock, 284
electrocardiography, 30, 52, 53, 226
electrocautery, 135
electroencephalography (EEG), 268
electrolyte imbalance, 59, 91, 145, 226
electromyography, 158, 183
electron microscopy, 216, 300
electrophoresis *see* protein electrophoresis (EPH)
elementary bodies, 56
ELISA (enzyme-linked immunosorbent assay)
    in aspergillosis, 24
    in avian influenza, 34
    in botulism, 46
    in chlamydiosis, 56
    in flagellate enteritis, 119
    in oil exposure, 194
    in paramyxovirus infections, 219
    in PDD, 247
    in viral disease, 299
    in viral neoplasms, 302
    in West Nile virus infection, 307
Elizabethan collar
    in air sac rupture, 7
    in bite wounds, 42
    in feather cyst, 111
    in feather damaging/self-injurious behavior, 114
    in lameness, 159
emaciation, **105–107**
    in campylobacteriosis, 48
    in coccidiosis (intestinal), 74
    in coccidiosis (systemic), 76
    in flagellate enteritis, 119
    in gastrointestinal helminthiasis, 126
    in iron storage disease, 156
    in *Macrorhabdus* infection, 169
    in PDD, 247
embryonal tumors, 68, 69
embryonic deaths, 150, 151
emergency management
    anticoagulant rodenticide toxicosis, 16
    bite wounds, 41, 42
    coccidiosis, 74, 77
    hyperglycemia, diabetes mellitus, 143
    plant and avocado toxicity, 226
    respiratory distress, 254, 255
    seizures, 268, 269
    tracheal/syringeal granulomas, 25
    trauma, 284, 285
    see also intensive care
emphysema, subcutaneous, 7, 8, 191
enalapril *see* ACE (angiotensin-converting enzyme) inhibitor
encephalomalacia, 68
endocarditis, 52
endocrine disease, 95
    cystic ovarian disease, 206
    tumors, 289
endoparasites *see* gastrointestinal helminthiasis
endoscope-assisted gastric lavage (EAGL), 109, 249
endoscopy
    in airborne toxin exposure, 3
    in air sac mite infection, 5
    in aspergillosis, 24
    in atherosclerosis, 30
    in candidiasis, 50
    in cloacal diseases, 64
    in coccidiosis (intestinal), 74
    in diabetes insipidus, 91
    in diarrhea, 93
    in dystocia and egg binding, 99
    in egg yolk/reproductive coelomitis, 103
    in feather damaging/self-injurious behavior, 114
    in gastrointestinal foreign bodies, 124
    in hemorrhage, 134, 135
    in hepatic lipidosis, 137
    infertility and, 151
    in oil exposure, 194
    in ovarian cysts and neoplasia, 207
    in respiratory distress, 254, 255
    in rhinitis and sinusitis, 258
    in salpingitis, oviductal/uterine disorders, 262
    in thyroid disease, 277
    in tracheal/syringeal disease, 282
    in trichomoniasis, 287
    undigested food in droppings and, 293
endotracheal wash, in pneumonia, 229
energy metabolism, 185, 197
enrichment, environmental, 114
enrofloxacin, 57, 314
enteral feeding *see* gavage feeding
enteritis, **108–110**, 119
    in adenovirus infection, 1–2
    in campylobacteriosis, 48
    in chlamydiosis, 56
    in clostridiosis, 66–67
    in coccidiosis, 74, 76

in colibacillosis, 82, 83
in cryptosporidiosis, 85
diarrhea and, 93
flagellate *see* flagellate enteritis
in gastrointestinal helminthiasis, 126, 127
hemorrhagic, 74
in herpesvirus infections, 141
in *Macrorhabdus* infection, 169
in salmonellosis, 260, 261
undigested food in droppings and, 2–293
Enterococcus, 44
enzyme supplements, plant-based organic, 214
eosinophilia, 119, 126, 293, 318, 319
EPH (electrophoresis) *see* protein electrophoresis (EPH)
epilepsy, 147–148, 182, 267, 268
epistaxis, 71, 134
ergotism, 179, 180
erythrocytes, 9, 134
erythrophagocytosis, 111
eschar, 231
Escherichia coli, **82–84**, 335
in bordetellosis, 45
in enteritis and gastritis, 108
in pododermatitis, 231, 232
esophageal disorders
regurgitation and vomiting, 249
squamous cell carcinoma, 274
esophagostomy, 154
essential oils, 84
Eustrongylides ignotus, 126
exercise, 30, 96
lack, obesity due to, 188
in lipomas, 161
in obesity management, 188
in pododermatitis, 232
exercise intolerance *see* lethargy
external skeletal fixator (ESF), 122
exudate, in ascites, 21
eye(s)
lubrication, 268
movements, 182
"sunken," in anorexia, 13
see also ocular lesions
eyelid margins, in color changes, 54

F

facial nerve deficits, 182
falling, 18, 29, 124, 134, 147, 158
Fanconi syndrome, 129
fat atrophy, in lymphoid neoplasia, 167
fatigue, in starvation, 105
fatty acids *see* omega-3 fatty acids
fatty liver hemorrhagic syndrome, 138
feather(s), 113

abnormal development, reticuloendotheliosis, 302
abnormal growth, 117
blood, broken, 134, 284
in circovirus infections, 61
color, 117
cyst, **111–112**
damaging behavior, **113–116**
chronic egg laying, 59
circovirus infection *vs*, 61
clostridiosis, 66
coelomic distension, 79
dermatitis due to, 89
diagnostic algorithm, *351*
dyslipidemia/hyperlipidemia, 95
dystocia and egg binding, 98
ectoparasites, 101
feather cyst, 111
flagellate enteritis, 119, 120
hepatic lipidosis, 137
pancreatic diseases, 213
renal disease, 251
squamous cell carcinoma and, 274, 275, 276
in trauma, 284
undigested food in droppings, 292
discoloration, **117–118**, 137
disorders, 101, **117–118**
dystrophy, 61–62, **117–118**, 237
in ear opening, in otitis, 204
enlarged follicles, in Marek's disease, 302
fluffed/fluffing
air sac mite infection, 5
chlamydiosis, 56
circovirus infection, 61
coccidiosis, 74, 76
colibacillosis, 82
egg yolk/reproductive coelomitis, 103
flagellate enteritis, 119
hypocalcemia and hypomagnesemia, 147
obesity, 188
pneumonia, 228
renal disease, 251
sick bird syndrome, 270
follicle cyst, 89
food matting, regurgitation/vomiting, 249
loss, 61
feather damaging/self-injurious behavior, 113
pancreatic disease, 213
in trauma, 284
in obesity, 188
in oil exposure, 193, 194
picking *see* feather(s), damaging behavior

poor condition, in hepatic lipidosis, 137
stress bars, 137, 185
in thyroid disease, 277
trimming, 113
waterproofing, 297
feather duster disease, 117, 118
fecal acid fast stain, 85, 174, 175, 249
fecal blood testing, 134
fecal caking, in phallus prolapse, 223
fecal direct smear
in coccidiosis, 75, 76
in cryptosporidiosis, 85
in diarrhea, 93
in hemoparasite infections, 132
undigested food in droppings and, 293
urate and fecal discoloration and, 296
see also fecal wet mount
fecal discoloration, **295–296**
in anemia, 9
in botulism, 46
in campylobacteriosis, 48
in chlamydiosis, 56
in coccidiosis, 74, 76
in diarrhea, 93
in flagellate enteritis, 119
in heavy metal toxicity, 129
in hemoparasite infection, 132
in hepatic lipidosis, 137
in herpesvirus infections, 140
in pasteurellosis, 221
in plant and avocado toxins, 226
in renal disease, 251
fecal float, 75, 76, 78, 85
in diarrhea, 93
in feather damaging/self-injurious behavior, 114
in gastrointestinal helminthiasis, 126
in hemoparasite infections, 132
in regurgitation and vomiting, 249
fecal Gram stain *see* Gram stain
fecaliths, 81
fecal malodor, 66, 108
in flagellate enteritis, 119, 120
in lipomas, 161
in pancreatic disease, 213
undigested food in droppings and, 292
fecal material adhered to vent, 98, 108, 119
fecal production
in hypocalcemia and hypomagnesemia, 147
in ingluvial hypomotility/crop stasis/ileus, 153
scant, 98, 105
see also defecation difficulty/failure

fecal retention
    in chronic egg laying, 59
    in clostridiosis, 66
    in enteritis and gastritis, 108
    in polyostotic hyperostosis, 202
fecal trichrome stain, 119
fecal wet mount
    in air sac mite infection, 5
    in coccidiosis (intestinal), 74
    in coccidiosis (systemic), 76, 78
    in ectoparasites, 101
    in feather damaging/self-injurious behavior, 114
    in flagellate enteritis, 119
    in gastrointestinal helminthiasis, 126
    in *Macrorhabdus* infection, 169
    in mycobacteriosis, 174
    in regurgitation and vomiting, 249
    in trichomoniasis, 287
    see also fecal direct smear
fecal zinc sulfate centrifugation, 119
feed analysis, for mycotoxins, 179–180
fenbendazole, 120, 127, 198, 314
    see also antiparasitic therapy
ferritin, serum, in iron storage disease, 156
fertility *see* infertility
FGAR (first-generation anticoagulant rodenticides), 15–16
fibrates, 31, 96
fibrinogen, 72
fibrinolysis, 71, *71*
fibroatheromatous lesion, 29
figure-of-eight bandage, 11, 12, 111, 121, 159
filariasis, subcutaneous, 18
fine needle aspiration (FNA)
    in coccidiosis (systemic), 76
    in coelomic distension, 80
    in feather cysts, 111
    in hepatic lipidosis, 137
    in lameness, 158
    in liver disease, 164
    in mycoplasmosis, 177
    in pododermatitis, 231
    in salmonellosis, 260
fipronil *see* antiparasitic therapy
fistulograms, in pododermatitis, 231
flagellate enteritis, 109, **119–120**, 198
    trichomoniasis, 287
    undigested food in droppings, 292, 293, 294
flight
    difficulties, 158, 251
    erratic, pasteurellosis, 221
    heavy metal toxicosis affecting, 129
    in hemoparasite infections, 132
    inability, fractures/luxations, 121
    loss, in oil exposure, 193
    trauma due to, 284

flight testing, 122
fluconazole *see* antifungal therapy
fluid overload, 251
fluid therapy, 314
    in anemia, 10
    in anticoagulant rodenticide toxicosis, 16
    in arthritis, 19
    in ascites, 22
    in aspergillosis, 25
    in aspiration, 27
    in beak injury, 36
    in bite wounds, 42
    in coccidiosis (intestinal), 74
    in coccidiosis (systemic), 77
    in colibacillosis, 83
    in dehydration, 87
    in diarrhea, 93
    in egg yolk/reproductive coelomitis, 103–104
    in emaciation, 105, 106
    in enteritis and gastritis, 109
    in feather cysts, 111
    in gastrointestinal foreign bodies, 124
    in heavy metal toxicosis, 129
    in hemorrhage, 135
    in hepatic lipidosis, 138
    in hyperuricemia, 146
    in ingluvial hypomotility/crop stasis/ileus, 154
    in lameness, 159
    in lipomas, 161
    in liver disease, 164
    in lymphoid neoplasia, 167
    in *Macrorhabdus* infection, 169
    in neurological conditions, 183
    in oil exposure, 194
    in overgrown beak, 211
    in pancreatic disease, 214
    in plant and avocado toxicosis, 226
    in pneumonia, 229
    in polydipsia, 235
    in polyuria, 239
    in respiratory distress, 255
    in *Sarcocystis* infections, 266
    in seizures, 268
    in sick bird syndrome, 271
    in thyroid disease, 277
    in toxoplasmosis, 279
    in trauma, 284, 285
    in trichomoniasis, 287
    in tumors, 290
    in viral disease, 300
    in vitamin D toxicosis, 305
    in West Nile virus infection, 308
fluorescein stain, 190, 191, 194
fluoroscopy, 109, 124, 247, 254
flush, nasal, 254, 255
FNA *see* fine needle aspiration (FNA)

folate, deficiency, 185
foods, novel, in obesity management, 188–189
foraging, 114, 188
foreign bodies, 13
    crop, regurgitation and vomiting, 249
    ear canal, otitis, 204
    gastrointestinal, **124–125**, 249, 250, 292
    in granulomas, 62
    heavy metal, diarrhea, 93
    ingluvial, 153, 154
    inhalation, 27, 229
    lead, 158, 183
    metallic, 124
    mineral, 124
    ocular, 191
    PDD and, 247
    regurgitation and vomiting, 249
    rhinitis and sinusitis due to, 257
    tracheal, 27, 254, 255, 281, 282
"fowl glioma", 68
"fowl paralysis", 302
fracture(s), **121–123**
    in air sac rupture, 7
    beak injury, 36–37
    in chronic egg laying, 59–60
    in circovirus infections, 61
    in dystocia and egg binding, 98
    in hypocalcemia and hypomagnesemia, 147, 148
    lameness and, 158, 159
    liver, 164, 221
    in metabolic bone disease, 171, 172
    in neurologic conditions, 182, 183
    in overgrown beak, 210
    pathological, 101, 202
    in trauma, 284, 285
fracture stabilization, 159
free fatty acids, 143
frostbite, 54
fructosamine, 143
fungal infection(s)
    in beak injury, 37
    conjunctivitis due to, 190
    feather damaging/self-injurious behavior, 113
    infertility and, 150
    ingluvial hypomotility/crop stasis/ileus, 153
    neurological conditions due to, 182
    oral plaques and, 197, 198
    in otitis, 204
    overgrown beak due to, 210
    in pneumonia, 228, 229, 230
    in pododermatitis, 231
    renal disease due to, 251
    rhinitis and sinusitis due to, 257
    sick bird syndrome due to, 270

squamous cell carcinoma and, 275
tracheal and syringeal disease due to,
  281
fungi, 179, 324
furosemide, 22, 53, 80, 138, 266, 306,
  314
  see also diuretics

G

gabapentin, 19, 184, 268, 314
  see also analgesia
gait, 79, 177, 182
  abnormal, lameness, 158
galactomannan, 24, 282
gamma-glutamyl transferase (GGT),
  141, 318, 319
  in liver disease, 164
  in lymphoid neoplasia, 166
  in tumors, 289
ganglioneuritis, 247
gas exchange, 228
gas gangrene, 67
gastric acidity, 170
gastric adenocarcinoma, 170
gastric emptying *see* gastrointestinal
  emptying
gastric lavage, regurgitation/vomiting,
  249
gastric mucus secretion, increased, 169
gastric ulcer, 124
gastric wash, 109
gastritis, **108–110**
gastrointestinal dilatation *see*
  proventricular dilatation disease
  (PDD)
gastrointestinal emptying, delayed, 50,
  200, 223
gastrointestinal foreign bodies, **124–125**,
  249, 250, 292
gastrointestinal helminthiasis, **126–128**
  undigested food in droppings and,
  292, 293
gastrointestinal hemorrhage, 134, 183
gastrointestinal motility, 200, 213
  hypomotility/ileus, 46, 79, 124,
  153–155
gastrointestinal motility stimulants, 214
gastrointestinal obstruction, 79, 80, 124,
  127, 249
gastrointestinal perforation, 124
gastrointestinal protectants, 110, 125
  sucralfate, 110, 125, 227, 250, 317
gastrointestinal stasis, 292
gastrointestinal tract
  in botulism, 46–47
  in candidiasis, 50
  in cryptosporidiosis, 85, 86
  in nutritional deficiencies, 185
  in oil exposure effect, 193

in plant and avocado toxicosis, 225
regurgitation/vomiting, 249
in trichomoniasis, 287
tumors, 249, 289
undigested food in droppings, 292
in viral disease, 299
in West Nile virus infection, 307
gavage feeding
  in coelomic distension, 80
  in emaciation, 105–106
  in gastrointestinal foreign bodies, 124
  in hepatic lipidosis, 138
  in liver disease, 164
  in *Macrorhabdus ornithogaster*
    infection, 169
  in metabolic bone disease, 171
  in neurological conditions, 183
  in ovarian cysts and neoplasia, 207
  in plant and avocado toxicosis, 226
  in regurgitation and vomiting, 250
  see also nutritional support
geriatric disease
  arthritis, 18, 158, 159
  atherosclerosis, 29
  congestive heart failure, 21
  lameness, 158, 159
  nutritional deficiencies, 187
germ cell tumors, 69
GGT *see* gamma-glutamyl transferase
  (GGT)
"ghost rods", 198
GHV-1 (Gallid herpesvirus 1), **140–142**
giardiasis *(Giardia)*, 119
  enteritis and gastritis due to, 108
  in flagellate enteritis, 119
  undigested food in droppings, 292,
  293
gliomas, 68, 69
glipizide, 144, 157
glomerular filtration rate (GFR), 251,
  252
glomerular lipidosis, 251
glossitis, pustular, 197
glossopharyngeal nerve deficits, 182
glucagon, 105, 106, 143, 144
gluconeogenesis, 105
glucosamine/chondroitin sulfate, 19,
  159, 314
glucose, 141
  in starvation, 105
  see also hyperglycemia; hypoglycemia
glucose curve, 143, 144
glucosuria, in polydipsia, 234
glutamate dehydrogenase (GLDH), 137,
  164
gnathotheca *see* beak
GnRH (gonadotropin-releasing
  hormone) agonist
  in ascites, 22
  in atherosclerosis, 31

in chronic egg laying, 59
in coelomic distension, 80, 81
in dyslipidemia/hyperlipidemia, 96
in dystocia and egg binding, 99
in egg yolk/reproductive coelomitis,
  104
in feather damaging/self-injurious
  behavior, 115
in hepatic lipidosis, 138
in ovarian cysts, 208
in polyostotic hyperostosis, 202
problem behaviors and, 245
salpingitis, oviductal and uterine
  disorders, 263
in tumors, 291
see also deslorelin
goiter, 27, 185, 186, 277, 278, 281
gonads, abnormal development,
  infertility, 150
gout, 18, 19, 145, 146, 158, 159
  after/in renal disease, 239, 253
  articular, 18, 158, 159
    hyperuricemia, 145, 146
    pododermatitis *vs*, 231
    vitamin D toxicosis, 305, 306
    xanthomas *vs*, 310
  visceral, 146, 253, 305
  see also urate
Gram stain
  in candidiasis, 50
  in cloacal diseases, 64
  in clostridiosis, 66
  in diarrhea, 93
  in flagellate enteritis, 119
  in *Macrorhabdus ornithogaster*
    infection, 169
  urate and fecal discoloration and, 296
granulomas, 174
  aspergillosis, 25, 281, 282, 283
  in coccidiosis (systemic), 77
  foreign body, 62
  oral plaques, 197
  pulmonary, pneumonia, 229
  xanthomas, 310
granulomatous disease, in colibacillosis,
  83
granulomatous inflammation, 174
granulosa cell tumors, 207
grayanotoxin, 225, *225*, 226
growth rate, reduced, 177
gunshot injuries, 121, 284

H

Haemoproteus, 131, 132, 133
halitosis, 197, 274
haloperidol, 13, 314
HDL-C (high-density lipoprotein
  cholesterol), 95
head down posture, 129

head flicking, 287
head-shaking, 5, 226
head-sweating, 249
head tilt, 182, 204, 205
    in avian influenza, 33
    in Marek's disease, 302
    in paramyxovirus infections, 219
    in toxoplasmosis, 279
    see also neurologic conditions
head trauma, 284, 285
hearing problems, in otitis, 204
heart disease see cardiac disease
heart failure, 29, 52, 95, 189, 228
    congestive see congestive heart failure (CHF)
heart murmur, 21, 29, 53
heavy metal toxicity, **129–130**, 324, **338**
    anemia due to, 9
    aspiration due to, 27
    diarrhea due to, 93, 94
    enteritis and gastritis due to, 108
    feather damaging/self-injurious behavior, 113, 114
    gastrointestinal foreign bodies, 124, 125
    hemorrhage due to, 134, 135
    hypocalcemia (and hypomagnesemia), 147–148
    ingluvial hypomotility/crop stasis/ileus/ileus, 153
    lameness and, 158
    neurological conditions due to, 182
    polydipsia due to, 234
    polyuria due to, 239
    regurgitation and vomiting, 249
    renal disease due to, 251–252
    seizures due to, 267, 268
    sick bird syndrome, 270
    undigested food in droppings and, 292, 293
    urate and fecal discoloration, 295, 296
Heinz bodies, 226
helminthiasis, **126–128**
hemagglutination inhibition assays, 62, 299
hematochezia, 66, 71, 74–75, 98
    in enteritis and gastritis, 108
    tumors causing, 289
    undigested food in droppings and, 292
hematocrit, in cardiac disease, 52
hematology reference values, **318–319**
hematoma, 71, 72, 153
hematuria, 72, 295
    in hemorrhage, 134
    in hyperuricemia, 145
    in neurological conditions, 183
    in renal disease, 251, 252
hemochromatosis, 52, 156, 163, 166
    see also iron storage disease

hemoconcentration, 76, 226
hemoglobinuria, 124, 129, 132, 225, 295
    plant and avocado toxicosis, 225, 226
    undigested food in droppings and, 292
hemolytic anemia, 9, 10
    in hemoparasite infection, 131
    in oil exposure, 193
    in plant and avocado toxicosis, 225, 226
hemoparasites, **131–133**
hemorrhage, **134–136**
    in adenovirus infection, 2
    in airborne toxin exposure, 3
    anemia due to, 9
    in anticoagulant rodenticide toxicosis, 15–16
    in bite wounds, 41
    in coagulopathies, 71–72
    in colibacillosis, 83
    in enteritis and gastritis, 108, 109, 110
    in feather cyst, 111
    in feather damaging/self-injurious behavior, 113, 114
    gastrointestinal, 134, 183
    in gastrointestinal helminthiasis, 126, 127
    in herpesvirus infections, 141
    in lameness, 158
    in lipomas, 161
    in liver disease, 164
    in neurologic conditions, 182, 183
    overgrown beak/nails, 211
    in pancreatic diseases, 214
    in pasteurellosis, 221
    in phallus prolapse, 223, 224
    in pneumonia, 229
    in polyomavirus infection, 237
    in regurgitation and vomiting, 249
    renal, 251, 252
    in *Sarcocystis* infections, 265, 266
    in trauma, 284, 285
    urate and fecal discoloration and, 295
    in uropygial gland disease, 297
hemorrhagic enteritis, 1
hemosiderin, 164
hemosiderosis, 143
hemostasis, 161, 162, 253
heparin, in overheparinization, 72
hepatic disease see liver disease
hepatic encephalopathy, 138, 163, 182, 267
hepatic failure, 234
hepatic fibrosis, 22, 79, 163, 164, 179, 180
    chronic, 163

hepatic lipidosis, **137–139**, 163, 185, 188
    in anorexia, 13
    in coelomic distension, 79, 80
    in dyslipidemia/hyperlipidemia, 95, 96
    in lipomas, 161
    in liver disease, 163, 164
    in nutritional imbalances, 137, 185, 186
    in obesity, 188, 189
    in ovarian cysts, 208
hepatic lymphosarcoma, 166
hepatic necrosis, 1, 61, 164, 180, 237, 279
    in hemoparasite infections, 132
    in herpesvirus infections, 140, 141
hepatitis, 76, 79
    in campylobacteriosis, 48
    in chlamydiosis, 56
    in colibacillosis, 82
    hepatic lipidosis *vs*, 137
hepatoencephalopathy, in polydipsia, 234
hepatomegaly, 164
    in adenovirus infection, 1
    in campylobacteriosis, 48
    in cardiac disease, 52
    in chlamydiosis, 56
    in coccidiosis (systemic), 77
    in coelomic distension, 79
    in hemoparasite infections, 132
    in hepatic lipidosis, 137
    in iron storage disease, 156
    in liver disease, 56, 137, 163
    in mycobacteriosis, 174
    in toxoplasmosis, 279
    in trichomoniasis, 287
    urate and fecal discoloration and, 295, 296
hepatopathy see liver disease
hepatosplenomegaly, in lymphoid neoplasia, 166
hepatotoxins, 163
Hepatozoon, 131
hepcidin assay, in iron storage disease, 156
herbal therapy, in viral disease, 300
"herd health" basis, 194
heritable conditions, feather disorders due to, 117
hermaphroditism, 150
hernia
    in chronic egg laying, 59
    in coelomic distension, 79, 81
    in egg yolk/reproductive coelomitis, 103
    in lipomas, 161
    in ovarian cysts and neoplasia, 206

herpesviruses, **140–142**, 197, 299, 324, 332, 333
    in cloacal diseases, 63
    in cutaneous papillomas, 216
    liver disease due to, 163
    in lymphoid neoplasia, 166
    neurologic conditions due to, 182, 183
    oral plaques and, 197, 198
    ovarian neoplasia due to, 206
    in pancreatic diseases, 213
    in pneumonia, 228, 229
    tracheal and syringeal disease and, 281
    tumors due to, 290
    undigested food in droppings and, 292
    viral neoplasms and, 302
hetastarch, 22, 138, 194, 314
Heterakis gallinarum, 108, 127
heterophilia, 56, 318, 319
    in clostridiosis, 66
    in egg yolk/reproductive coelomitis, 103
    in feather damaging/self-injurious behavior, 114
    in gastrointestinal helminthiasis, 126
    in mycobacteriosis, 175
    in mycoplasmosis, 177
    in otitis, 204
    in pneumonia, 228
    in polydipsia, 234
    in tracheal/syringeal disease, 281
hexamitiasis *(Hexamita)*, 108, 119, 120
hind limb weakness, 98
hippoboscid flies, 101, 131
histamine-2 (H2) blockers, 110, 250
Histoplasma capsulatum, 335
homeopathy, 100
honeycomb proliferative lesion, 210
hormonal lipogenesis, 137
hormone assays, 327
housing, oil exposed birds, 194–195
HPAI viruses (avian influenza), 33, 34
human chorionic gonadotropin (HCG), in ovarian cysts, 208
humidification, 255, 282, 283
husbandry, poor, 228, 231, 232, 251, 260
hydrochlorothiazide, 92
hydrostatic pressure, 21, 79
hydrotherapy, in bite wounds, 42
hydroxyapatite, 147
25-hydroxycholecalciferol, 147, 148, 171, 305
hygiene, 335
    in flagellate enteritis, 120
    in gastrointestinal helminthiasis, 127
    poor, pododermatitis due to, 231, 233
    viral neoplasms and, 303

hygromycin B, 127
hypercalcemia, 59, 305
    in coelomic distension, 79
    in dystocia and egg binding, 98
    in egg yolk/reproductive coelomitis, 103
    in lymphoid neoplasia, 166
    in polydipsia, 234
    in polyostotic hyperostosis, 202
    in salpingitis, oviductal and uterine disorders, 262
    in tumors, 290
    in vitamin D toxicosis, 305
hypercholesterolemia, 59, 99
    in neurological conditions, 183
    in obesity, 188, 189
    in polyostotic hyperostosis, 202
    in thyroid disease, 277
    in xanthomas, 310
    see also hyperlipidemia
hyperestrogenism, 182
hypergammaglobulinemia, 265
hyperglobulinemia, 59, 99
    in egg yolk/reproductive coelomitis, 103
    in hemoparasite infections, 132
    in lymphoid neoplasia, 166
    in salpingitis, oviductal/uterine disorders, 262
hyperglycemia, **143–144**, 184
    in diabetes insipidus, 91
    in pancreatic diseases, 143, 144, 213
    in polydipsia, 234
hyperkalemia, in polydipsia, 234
hyperkeratosis, 101
hyperlipidemia, 29–31, **95–97**
    atherosclerosis and, 29, 30, 31
    in chronic egg laying, 59
    in coelomic distension, 79
    in dystocia and egg binding, 99
    in egg yolk/reproductive coelomitis, 103
    in hepatic lipidosis, 137
    in hyperglycemia and diabetes mellitus, 143
    in lipomas, 161
    in neurologic conditions, 183, 184
    in pancreatic diseases, 214
    in xanthomas, 310
    see also hypercholesterolemia
hypermetabolic state, 105
hypernatremia, 145
hyperostosis, 206, 207
    see also polyostotic hyperostosis
hyperparathyroidism, nutritional secondary, 148, 234
hyperphosphatemia, 171, 234, 262, 305
hyperproteinemia, 109, 132
hypersensitivity syndrome, 3, 89–90, 254, 255

feather damaging/self-injurious behavior, 113
    pneumonia vs, 228
hypertension, 30, 31, 134
hyperthermia, in oil exposure, 193
hyperthyroidism, 278
hypertonic sugar solution, 224
hypertriglyceridemia, 59, 213
    see also hyperlipidemia
hyperuricemia, 18, **145–146**
    in arthritis, 18, 19
    in chronic egg laying, 59
    in coccidiosis (systemic), 76
    in coelomic distension, 79
    diagnostic algorithm, *353*
    in dystocia and egg binding, 98, 99
    in egg yolk/reproductive coelomitis, 103
    in enteritis and gastritis, 109
    in lameness, 158
    in lymphoid neoplasia, 166
    in nutritional imbalances, 186
    oral plaques and, 198
    in PDD, 247
    in polydipsia, 234
    in polyostotic hyperostosis, 202
    in renal disease, 252, 253
    in tumors, 289–290
    in vitamin D toxicosis, 305, 306
hypervitaminosis A, 143, 145, 147, 338
hypervitaminosis D, 186, **305–306**, 338
hyphema, 190
hypoalbuminemia, 79, 137
    in ascites, 21
    in hemoparasite infection, 132
    in iron storage disease, 156
    in mycobacteriosis, 174
    in polydipsia, 234
    in renal disease, 252
    in trichomoniasis, 287
hypocalcemia, **147–149**, 184, 185, 186, 262
    in chronic egg laying, 59, 60
    in dystocia and egg binding, 98
    in feather damaging/self-injurious behavior, 113
    infertility due to, 150
    in lameness, 158, 159
    in metabolic bone disease, 171
    in neurological conditions, 182, 183
    seizures and, 267, 268
hypochloremia, 145, 169
hypocholesterolemia, 169
hypoglossal nerve deficits, 182
hypoglycemia, 105, 183
    hemorrhage and, 135
    in hepatic lipidosis, 137
    in iron storage disease, 156
    in liver disease, 164
    in *Macrorhabdus* infection, 169

hypoglycemia (*Continued*)
   oral plaques and, 198
   seizures and, 267, 268
hypokalemia, 105, 137, 145
hypomagnesemia, 105, **147–149**, 185, 186, 267
hypomotility/ileus, 46, 79, 124, **153–155**
hyponatremia, 169, 234
hypophosphatemia, 105, 169, 171
hypoproteinemia, 71
   in ascites, 21
   in campylobacteriosis, 48
   in coelomic distension, 79
   in enteritis and gastritis, 109
   in flagellate enteritis, 119, 120
   in gastrointestinal helminthiasis, 126
   in hepatic lipidosis, 137, 138
   in iron storage disease, 156
   in liver disease, 163, 164
   in *Macrorhabdus* infection, 169
   in PDD, 247
   undigested food in droppings and, 293
   see also hypoalbuminemia
hyposthenuria, 234
hypotension, 134, 234, 284
hypothermia, 119, 145, 179, 193
hypothyroidism, 61, 95, 113, 188, **277–278**
   lipomas due to, 161
   obesity and, 188
hypouricemia, 137
hypovitaminosis A
   air sac rupture and, 7
   in feather cyst, 111
   feather damaging/self-injurious behavior due to, 113
   ingluvial hypomotility/crop stasis/ileus, 153, 154
   oral plaques and, 197, 198
   in pneumonia, 228
   in pododermatitis, 231
   polydipsia due to, 234
   in renal disease, 251
   squamous cell carcinoma and, 274
   tracheal/syringeal disease due to, 281
   uropygial gland disease due to, 297
   see also vitamin A, deficiency
hypovolemia, 108
   see also hypotension
hypovolemic shock, 71, 134, 135, 221
   management (phases), 285
   in trauma, 284, 285

I

ileus, 46, 79, 124, **153–155**
imaging *see specific imaging modalities*
immune-mediated arthritis, 18
immune mediated hemolytic anemia, 9
immune-mediated injuries, beak, 36
immune modulators, 185
immunofluorescence assay (IFA), 56, 299
immunohistochemistry, 167, 279, 299
immunomodulators, in squamous cell carcinoma, 275, 276
immunosuppression, 56, 57, 61, 132
   in clostridiosis, 66
   in oil exposure, 193
   in reticuloendotheliosis, 302
   in tumors, 291
   in West Nile virus infection, 307
immunotherapy, 290
impaction, 124
   uropygial gland, 297, 298
inclusion bodies
   in duck viral enteritis, 183
   intracytoplasmic, 242
      circovirus infection, 62
      oral plaques and, 198
   intranuclear
      adenovirus infection, 1, 2
      herpesvirus infection, 141
      oral plaques and, 198
      polyomavirus infection, 237
   in poxvirus infections, 242
infection(s)
   in bite wounds, 41
   cardiac disease and, 52
   in cloacal diseases, 63
   conjunctivitis due to, 190
   in dystocia and egg binding, 98, 99
   emaciation due to, 105
   in feather cysts, 111
   in fractures and luxations, 122
   hypocalcemia and hypomagnesemia *vs*, 147–148
   ingluvial hypomotility/crop stasis/ileus due to, 153
   lameness and, 158
   liver disease due to, 163
   neurological conditions due to, 182
   otitis due to, 204
   overgrown beak due to, 210
   in PDD, 248
   pneumonia, **228–230**
   polyuria due to, 239
   regurgitation and vomiting, 249
   renal disease due to, 251
   rhinitis and sinusitis due to, 257
   salpingitis, oviductal/uterine disorders, 262
   tracheal and syringeal disease due to, 281
   ulcerative dermatitis due to, 89
   see also bacterial infection(s); fungal infection(s); parasitic infection(s); viral disease
infectious bronchitis virus, 145, 251, 325, 333
infectious bursal disease, 325
infectious laryngotracheitis (ILT), **140–142**, 281, 325
infertility, **150–152**
   in air sac mite infection, 6
   in anorexia, 14
   in cloacal diseases, 65
   in cryptosporidiosis, 86
   dehydration and, 88
   in ectoparasite infestation, 102
   in emaciation, 106
   in heavy metal toxicity, 130
   in herpesvirus infections, 140, 142
   in hypocalcemia and hypomagnesemia, 149
   lipomas and, 161, 162
   in metabolic bone disease, 172
   in mycotoxicosis, 179
   in nutritional deficiencies, 187
   obesity and, 188, 189
   in organophosphate/carbamate toxicity, 201
   in ovarian cysts and neoplasia, 206, 207
   in phallus prolapse, 224
   viral neoplasms and, 304
inflammation
   in arthritis, 18
   in aspergillosis, 24
   in atherosclerosis, 29
   in bite wounds, 41
   in bordetellosis, 44
   in cloacal disease, 63, 64
   in clostridiosis, 66
   in coccidiosis (systemic), 77
   in dystocia and egg binding, 98
   ear (otitis), **204–205**
   in egg yolk/reproductive coelomitis, 103
   emaciation due to, 105
   in feather damaging/self-injurious behavior, 114
   in gastritis and enteritis, 108, 109, 110
   in gastrointestinal helminthiasis, 126–127
   in lameness, 158
   in liver disease, 164
   in mycobacteriosis, 174
   in mycoplasmosis, 177
   in pneumonia, 228
   polydipsia due to, 234
   pseudomembranous, in mycotoxicosis, 179
   in salpingitis, 262
   squamous cell carcinoma and, 274
influenza *see* avian influenza
ingluvial flush, 154

ingluvial gavage *see* gavage feeding
ingluvial hypomotility, **153–155**
   see also crop stasis
ingluvioliths, 249
ingluviostomy tube, 154
ingluviotomy, 125, 154, 183
ingluvitis, 82, 83
injuries *see* trauma
insect control, 127, 133
insecticides, 325
in situ hybridization (ISH), 300
in situ lesions, 274
insulin, 143, 144, 157, 214, 314
intensive care
   in heavy metal toxicity, 130
   in plant and avocado toxicosis, 226
   in respiratory distress, 255
   in viral diseases, 300, 308
   see also emergency management
interferon α, 300
intermittent claudication, 18, 29
internal papillomatosis, **140–142**, 292
intestinal diseases, 295
intestinal helminthiasis, **126–128**
intestinal obstruction, 79, 80, 81, 124, 127, 249
intestinal scraping, 74
intracranial pressure, elevated, 268
intradermal skin test, 113
intranuclear inclusion body *see* inclusion bodies
intraosseous catheter, 16, 77, 87, 99
intravenous catheter, 16
intubation, tracheal stenosis after, 281, 283
intussusception, 63, 249
iodine, 185, 186, 314
   deficiency, 277, 281
   in flagellate enteritis, 119
iodine therapy, radioactive, 278
ipronidazole *see* antiparasitic therapy
iron, 156, 157
   deficiency, 185
   excess, 156
   toxicity, 338
iron chelators, 22, 80, 157, 314
iron storage disease, **156–157**
   ascites, 22
   cardiac disease and, 52
   coelomic distension, 79, 80
   liver disease and, 163, 166
isoflurane, 145, 315
Isospora, 76, 132
   see also coccidiosis (systemic)
isthmus gland, 169
itraconazole *see* antifungal therapy
ivermectin, 101–102, 127, 211, 229, 315, 338
   see also antiparasitic therapy

J

jaundice, 132
joint(s)
   decreased range of motion, 122
   effusion, pododermatitis, 231
   in lameness, 158
   luxation, **121–123**
   pain, in vitamin D toxicosis, 305
   swelling, 177, 260

K

keel movement, exaggerated, 228
keel wraps, 194
keratin, 36, 111, 198, 210
"keratin pearls", 274
ketoacidosis, 143
ketones, 143
ketonuria, in polydipsia, 234
ketoprofen *see* NSAIDs
kidney
   compression, in coelomic distension, 79, 81
   milky white, 146
   see also renal disease
Knemidokoptes, 38, 89, 101–102, 113
   overgrown beak/nails due to, 210, 211
knuckling over, 182
koilin, 124, 169
koilin dysplasia, 249

L

laboratory testing, **320–330**
laceration, 121, 284
lactate, in polyuria, 239
lactate dehydrogenase (LDH), 24, 156, 265
Lactated Ringers Solution (LRS), 36, 42, 77, 87, 211, 226, 255
lactulose, 138, 164, 315
Lafora body neuropathy, 183
lagophthalmos, 191
lameness, **158–160**
   in arthritis, 18, 19
   in colibacillosis, 82
   diagnostic algorithm, *352*
   in fractures and luxations, 121
   in hemoparasite infections, 132
   in Marek's disease, 302
   in mycoplasmosis, 177
   in ovarian cysts and neoplasia, 206
   in pododermatitis, 231
   in renal disease, 251
   in splay leg and slipped tendon, 272
laparoscopy
   in coelomic distension, 80
   in hyperuricemia, 146
   in lameness, 158
   in oil exposure, 194
   in salpingitis, oviductal/uterine disorders, 262
   in trichomoniasis, 287
   undigested food in droppings and, 293
   urate and fecal discoloration and, 296
laxatives, 125
LDH (lactate dehydrogenase), 24, 156, 265
LDL-C (low-density lipoprotein cholesterol), 95
lead
   gastrointestinal foreign bodies, 124
   normal levels, in diarrhea, 93
lead toxicity/toxicosis, 24, 124, **129–130**, 295, 338
   CNS tumors *vs*, 68
   emaciation due to, 105
   lameness and, 158, 159
   neurologic conditions due to, 182, 183, 184
   seizures and, 267, 268
   undigested food in droppings and, 292, 293, 294
   urate and fecal discoloration, 295, 296
leg weakness, 76
lens opacities, 190
lethargy, 200
   in arthritis, 18
   in aspergillosis, 24
   in avian influenza, 33
   in campylobacteriosis, 48
   in chlamydiosis, 56
   in clostridiosis, 66
   in coagulopathies and coagulation, 71
   in coccidiosis, 74, 76
   in coelomic distension, 79
   in colibacillosis, 82
   in dyslipidemia/hyperlipidemia, 95
   in dystocia and egg binding, 98
   in egg yolk/reproductive coelomitis, 103
   in emaciation, 105
   in flagellate enteritis, 119
   in gastrointestinal helminthiasis, 126
   in heavy metal toxicity, 129
   in hemoparasites infection, 131, 132
   in hemorrhage, 134
   in hepatic lipidosis, 137
   in herpesvirus infections, 140
   in hyperglycemia and diabetes mellitus, 143
   in hypocalcemia and hypomagnesemia, 147
   in liver disease, 163
   in *Macrorhabdus* infection, 169
   in mycobacteriosis, 174
   in ovarian cysts and neoplasia, 206, 207
   in pancreatic disease, 213

lethargy (*Continued*)
   in paramyxovirus infections, 218
   in pasteurellosis, 221
   in plant and avocado toxicosis, 226
   in pneumonia, 228
   in polyostotic hyperostosis, 202
   in poxvirus infections, 241
   in renal disease, 251
   in salmonellosis, 260
   in *Sarcocystis* infections, 265
   in sick bird syndrome, 270
   in toxoplasmosis, 279
   in tracheal/syringeal disease, 281
   in trichomoniasis, 287
   in tumors, 289
   in viral neoplasms, 302
Leucocytozoon, 131, 132, 133
leukemia, **166–168**
leukocytosis, 318, 319
   in arthritis, 18
   in ascites, 21
   in aspergillosis, 24
   in chlamydiosis, 56
   in clostridiosis, 66
   in coelomic distension, 79
   in diarrhea, 93
   in dystocia and egg binding, 98–99
   in egg yolk/reproductive coelomitis, 103
   in enteritis and gastritis, 108
   in feather damaging/self-injurious behavior, 114
   in gastrointestinal helminthiasis, 126
   in hemoparasite infection, 131, 132
   in hepatic lipidosis, 137
   in lymphoid neoplasia, 166
   in neurological conditions, 183
   oral plaques and, 197
   in otitis, 204
   in pancreatic disease, 213
   in PDD, 247
   in pneumonia, 228
   in polyomavirus infection, 237
   in rhinitis and sinusitis, 257
   in salpingitis, oviductal/uterine disorders, 262
   in *Sarcocystis* infections, 265
   in tracheal/syringeal disease, 281
   in trichomoniasis, 287
   in tumors, 289
   undigested food in droppings and, 293
   in viral disease, 299
leukopenia, 66, 137, 141, 289
leuprolide acetate, 22, 64, 99, 100, 315
   in chronic egg laying, 60
   in egg yolk/reproductive coelomitis, 104
   in hepatic lipidosis, 138
   in ovarian cysts, 208

   in polyostotic hyperostosis, 202, 203
   in salpingitis, oviductal/uterine disorders, 263
   see also GnRH (gonadotropin-releasing hormone) agonist
levetiracetam, 268, 315
levothyroxin, 162, 315
Libyostrongylus douglassi, 126
lice *see* ectoparasites
limb deformity, 272
limb disease, in mycoplasmosis, 177
lipase, 143, 213, 215
lipemia, 95, 137, 145, 188
lipid(s), metabolism, 95, 137, 143
lipid-lowering agents, 31, 96, 184
lipomas, 80, 81, **161–162**, 188, 189
   infertility and, 150
   xanthomas *vs*, 310
lipoproteins, 29, 30, 95
liver biopsy, 164
   in hepatic lipidosis, 137, 138
   in iron storage disease, 156
   in lymphoid neoplasia, 166
   in nutritional deficiencies, 186
liver disease/conditions, **163–165**, 295
   in adenovirus infection, 1
   in anorexia, 13
   in ascites, 21
   in atherosclerosis, 29
   in campylobacteriosis, 48
   in cardiac disease, 52
   in chlamydiosis, 56
   in clostridiosis, 66, 67
   in coccidiosis (systemic), 76, 77
   in colibacillosis, 82
   diagnostic algorithm, *354*
   in dyslipidemia/hyperlipidemia, 95
   in egg yolk/reproductive coelomitis, 103
   in enteritis and gastritis, 108, 110
   feather disorders due to, 117
   in gastrointestinal helminthiasis, 126
   in heavy metal toxicity, 129
   in hemoparasite infection, 131, 132
   hemorrhage due to, 134, 135
   in hepatic lipidosis, 137
   in hepatomegaly, 56, 137, 163
   in herpesvirus infections, 141
   in iron storage disease, 156
   in mycobacteriosis, 174, 175
   in mycotoxicosis, 179
   necrosis *see* hepatic necrosis
   in oil exposure, 193
   in ovarian cysts and neoplasia, 206
   overgrown beak and, 210
   in pancreatic disease, 213, 215
   in polydipsia, 234
   in polyomavirus infection, 237
   in polyuria, 239

   in poxvirus infections, 242
   in *Sarcocystis* infections, 265
   in toxoplasmosis, 279
   in trichomoniasis, 287
   undigested food in droppings, 292
   urate and fecal discoloration, 295, 296
   in viral disease, 299
liver fracture, 164, 221
liver tumors, 289
lockjaw syndrome, 44
   see also bordetellosis
LPAI viruses (avian influenza), 33, 34
lumbosacral plexus, compression, 251
lung(s)
   biopsy, 229
   in coccidiosis (systemic), 76, 77
   fluid accumulation, OP/CA intoxication, 200
luxations, 18, **121–123**, 284
   in air sac rupture, 7
   lameness and, 158
   lens, 190
lymphocytes, 166, 167, 318, 319
lymphocytosis, 131, 132, 166
   see also leukocytosis
lymphoid leucosis (LL), **302–304**
lymphoid neoplasia, **166–168**
lymphoma, 163, **166–168**, 182
   ovarian, 206, 207, 208
lymphoplasmocytic ganglioneuritis, 247
lymphoproliferative diseases, **166–168**
lymphosarcoma, 167

M

macrorhabdosis (*Macrorhabdus ornithogaster*), **169–170**, 325
   enteritis/gastritis due to, 108
   regurgitation/vomiting, 249, 250
   undigested food in droppings and, 292, 293
   urate and fecal discoloration and, 295, 296
magnesium, 147
   deficiency, 105, **147–149**, 185, 186, 267
   supplements, 149
magnesium sulfate, 125, 186
magnetic resonance imaging (MRI)
   in arthritis, 19
   in atherosclerosis, 30
   in CNS tumors, 69
   in diabetes insipidus, 91
   inglon uvial hypomotility/crop stasis/ileus, 154
   in iron storage disease, 156
   in neurological conditions, 183
   in overgrown beak/nails, 211
   in respiratory distress, 254

in rhinitis and sinusitis, 257–258
in tumors, 290
maintenance requirements, nutritional, 105–106
malabsorption, 119, 147, 169, 185
malaria, 133
maldigestion, 93, 109
malnutrition
    beak injury due to, 36
    in beak malocclusion, 38
    in candidiasis, 50
    emaciation and, 105, 106
    in oil exposure, 193
    in pneumonia risk, 228
    in pododermatitis, 231
    respiratory distress in, 254
    see also nutritional imbalances
malodorous feces *see* fecal malodor
malunion of fracture, 122, 159
mandibular prognathism, 210
mandibular ramp prosthesis, 39
manganese, deficiency, 272
mannitol, 183, 315
marble spleen disease *see* adenovirus(es)
Marek's disease, 183, **302–304**, 326
    ovarian lymphoma with, 206, 207, 208
    ovarian neoplasia, 206
    vaccination, 192, 208, 303
Marek's disease virus (MDV), 158, 166, 182, 183
MEC (daily minimum energy cost), 106
medicated water, in candidiasis, 50
megacloaca, 46, 66, 67
megacolon, in clostridiosis, 66, 67
melena, 295
    in candidiasis, 50
    in coagulopathies and coagulation, 71
    in coccidiosis (intestinal), 74
    in enteritis and gastritis, 108
    in gastrointestinal foreign bodies, 124
    in hemorrhage, 134
    in hepatic lipidosis, 137
    in *Macrorhabdus* infection, 169
    tumors causing, 289
    see also fecal discoloration
meloxicam *see* NSAIDs
meningeal tumors, 68, 69
meningitis/meningoencephalitis, 83
mercury toxicity, 129, 130, 295, 338
metabolic acidosis, 145
metabolic bone disease, **171–173**
    beak injury due to, 36
    beak malocclusion, 38
    chronic egg laying, 59
    polyostotic hyperostosis *vs*, 202
metallic foreign body, 124
metapneumoviruses, 218, 326
metastases, 291
methionine deficiency, 185, 272

metoclopramide, 154, 315
metritis, 98, 99, 150
metronidazole, 120, 198, 315
metyrapone, 303
midazolam, 315
    in airborne toxin exposure, 4
    in ascites, 22
    in heavy metal toxicity, 130
    in hypocalcemia and hypomagnesemia, 149
    in neurological conditions, 183
    in tracheal/syringeal disease, 282
milk thistle (silymarin), 22, 296, 317
mineralization, 30
mineral supplements, 186
misoprostil, 100
mites
    air sacs, **5–6**, 229, 254, 281
    feather damaging/self-injurious behavior, 113
    see also ectoparasites
molt failure, 61
molting difficulties, 119
monocytosis, 56, 66, 114, 281, 318, 319
mosquito control, 308
motion sickness, 249
moxidectin *see* antiparasitic therapy
MRI *see* magnetic resonance imaging (MRI)
mucolytics, 282
mucous discharge, in pasteurellosis, 221
mucous membranes
    in hemoparasite infections, 132
    hemorrhage, 134
    pale, in lymphoid neoplasia, 166
    in plant and avocado toxicosis, 226
    in polyuria, 239
    in poxvirus infections, 241, 242
    see also oral mucosa
multiorgan failure, hemorrhage and, 134
murexide test, 145–146, 252
muscarinic signs, 200
muscle
    atrophy, 167, 193
    catabolism, 105
    damage in dystocia and egg binding, 98, 99
    disuse atrophy, lameness, 158
    fasciculations, 182, 201
    loss, in liver disease, 163
    pectoral mass loss, ovarian cysts/neoplasia, 206
    wasting, undigested food in droppings, 292
muscle tetanus, 171
musculoskeletal system tumors, 289
mycobacteriosis (*Mycobacterium* infections), **174–176**, 326, 335
    ascites and, 21, 22, 23
    enteritis and gastritis due to, 108, 109

    liver disease due to, 163, 164
    oral plaques and, 197, 198
    pneumonia, 229
    regurgitation and vomiting, 249
    tracheal/syringeal disease and, 281
    undigested food in droppings, 292, 294
Mycobacterium tuberculosis, 174, 175
mycoplasmosis (*Mycoplasma* spp.), **177–178**, 326
    infertility and, 150
    ocular lesions, 177, 190
    pneumonia, 229
    tracheal/syringeal disease and, 281, 283
mycotoxicosis and mycotoxins, **179–181**, 327, 329, **339**
mydriasis, 191
myocardial ischemia, 29
myocarditis, 76, 279
myoglobinuria, 145, 252
myositis, in toxoplasmosis, 279

N

NAC (N-acetylcysteine), 138, 315
nail(s)
    in hepatic lipidosis, 137
    overgrowth, **210–212**
    trimming, 211
nasal aspiration, 258
nasal cavity, 255
nasal discharge, 287
    in air sac mite infection, 5
    in avian influenza, 33, 34
    in bordetellosis, 44
    in candidiasis, 50
    in cryptosporidiosis, 85
    in herpesvirus infections, 140
    in mycoplasmosis, 177
    in pasteurellosis, 221
    in rhinitis and sinusitis, 257
nasal flush, respiratory distress, 254, 255
nasal lavage, rhinitis and sinusitis, 258
nasal swab, 258
nebulization, 25, 27, 229, 230, 255, 258, 282
necropsy, 179, 300
necrosis, 237
    in pododermatitis, 232
    in poxvirus infections, 242
    in trauma, 284
    see also specific organs
necrotizing enterocolitis, 66, 67
nematodes, 126, 127
neoplasia, **289–291**
    arthritis, 18
    ascites and, 21
    beak injury due to, 36
    beak malocclusion and, 38

neoplasia (*Continued*)
   in cloacal disease, 63–64
   coelomic distension and, 79, 80
   dystocia and egg binding due to, 98
   in egg yolk/reproductive coelomitis, 103
   feather damaging/self-injurious behavior due to, 113
   hemorrhage due to, 134
   in herpesvirus infections, 140, 141
   hypocalcemia and hypomagnesemia *vs*, 147–148
   ingluvial hypomotility/crop stasis/ileus due to, 153
   in lameness, 158
   liver disease due to, 163
   in mycotoxicosis, 179
   neurological conditions due to, 182
   osteomyelosclerosis/polyostotic hyperostosis *vs*, 202
   otitis due to, 204
   rhinitis and sinusitis due to, 257
   tracheal and syringeal disease and, 281
   viral-induced, **302–304**
   see also specific tumors
nephritis, 76, 251, 252
   infectious, 239, 240, 251, 252
nephrocalcinosis, 305, 306
nephrogenic diabetes insipidus, **91–92**
nephrotoxicity, 145, 146, 252, 253, 306
nerve conduction tests, 158
nervous system tumors, 289
neurologic conditions/features, **182–184**
   in anticoagulant rodenticide toxicosis, 15
   in arthritis, 18
   in atherosclerosis, 29
   in avian influenza, 33
   in bite wounds, 41
   in botulism, 46, 182
   in cardiac disease, 52
   in CNS tumors, 68
   in coagulopathies, 71
   in coccidiosis (systemic), 76
   in coelomic distension, 79
   in colibacillosis, 82
   diagnostic algorithm, *348*
   in dyslipidemia/hyperlipidemia, 95
   in dystocia and egg binding, 98
   in fractures and luxations, 121
   gastrointestinal foreign bodies causing, 124
   in heavy metal toxicity, 129
   in hemoparasite infection, 131, 132
   in herpesvirus infections, 140
   in hypocalcemia and hypomagnesemia, 147, 148, 149
   lameness and, 159
   in mycotoxicosis, 179
   in nutritional deficiencies, 185
   in oil exposure, 193
   in overgrown beak/nails, 210
   in paramyxovirus infections, 218, 219
   in PDD, 247
   in plant and avocado toxicosis, 225
   in *Sarcocystis* infections, 265
   seizures *see* seizures
   in toxoplasmosis, 279, 280
   in trauma, 284
   undigested food in droppings and, 292
   in viral disease, 299
   in West Nile virus infection, 307
neuromuscular changes, 153, 171, 185, 200, 221, 284
neurotransmitters, 114, 200
Newcastle disease, 33, 183, **218–220**, 327
nicotinic signs, 200
nitroimidazoles, 288
nitrotoxin, 225, *225*, 226
nocardiosis, in pneumonia, 229
nonregenerative anemia *see* anemia
nonsteroidal anti-inflammatories *see* NSAIDs
nonunion of fractures, 122, 159
notifiable diseases *see* reportable diseases
NSAIDs, 338
   in airborne toxin exposure, 4
   in arthritis, 19
   in bite wounds, 42
   in bordetellosis, 44
   coagulopathies and, 72
   in dystocia and egg binding, 99
   in egg yolk/reproductive coelomitis, 104
   in enteritis and gastritis, 110
   in fractures and luxations, 122
   in gastrointestinal foreign bodies, 124
   in hyperuricemia, 146
   infertility and, 151
   in ingluvial hypomotility/crop stasis/ileus, 154
   in ocular lesions, 191
   in otitis, 205
   in pancreatic disease, 214
   in PDD, 183–184, 247, 248
   in phallus prolapse, 224
   in plant and avocado toxicosis, 227
   in pododermatitis, 232, 233
   polyuria due to, 240
   regurgitation and vomiting, 250
   in renal disease, 252, 253
   in respiratory distress, 255
   in salpingitis, oviductal/uterine disorders, 263
   in splay leg and slipped tendon, 273
   in tumors, 290
   in viral disease, 300, 308
nutritional deficiencies, **185–187**
   oral plaques due to, 197
   splay leg and slipped tendon, 272
   tracheal/syringeal disease, 281
   see also nutritional imbalances
nutritional imbalances, **185–187**, 228
   in angel wing, 11
   in aspergillosis, 24, 25
   in atherosclerosis, 30, 31
   in beak injury, 36, 37
   in beak malocclusion, 38
   in chronic egg laying, 59
   circovirus infection *vs*, 61
   in clostridiosis, 66
   in coelomic distension, 79
   in colibacillosis, 82
   diabetes insipidus *vs*, 91
   in dyslipidemia/hyperlipidemia, 95
   in dystocia and egg binding, 98
   in emaciation, 105
   in feather cyst, 111
   feather damaging/self-injurious behavior due to, 113
   in feather disorders, 117
   hemorrhage and, 134, 135, 136
   in hepatic lipidosis, 137, 185, 186
   in hyperuricemia, 145
   infertility due to, 150, 187
   iron storage disease due to, 156
   in lipomas, 161
   in liver disease, 163
   in metabolic bone disease, 171
   neurological conditions due to, 182
   in obesity in, 188
   in overgrown beak/nails, 210
   in polyostotic hyperostosis, 202
   in polyuria, 239
   in renal disease, 251
   in thyroid disease, 277
   in vitamin D toxicosis, 305
   see also malnutrition; nutritional deficiencies
nutritional secondary hyperparathyroidism (NSHP), 148, 234
nutritional support
   in bite wounds, 42
   in coelomic distension, 80
   in hemorrhage, 135
   in metabolic bone disease, 172
   in neurological conditions, 183
   in regurgitation and vomiting, 250
   in sick bird syndrome, 271
   in West Nile virus infection, 308
   see also dietary management; gavage feeding
nystatin, 50, 169, 198, 315
   see also antifungal therapy

O

obesity, **188–189**
    in arthritis, 18, 19
    atherosclerosis and, 30
    coelomic distension and, 79, 81
    in dyslipidemia/hyperlipidemia, 95
    in egg yolk/reproductive coelomitis, 103
    in hepatic lipidosis, 137
    in hyperglycemia and diabetes mellitus, 143
    in hypothyroidism, 277
    lameness and, 158
    lipomas and, 161, 162
    pancreatic disease and, 213
    xanthomas risk, 310
ochratoxicosis, and ochratoxin, 179, 180, 339
ocular discharge, 140, 219, 257
ocular lesions/signs, **190–192**
    in avian influenza, 33
    in bite wounds, 41
    in bordetellosis, 44, 190
    in cere/skin color changes, 54
    in chlamydiosis, 56
    in CNS tumors, 68
    in colibacillosis, 82, 83
    in cryptosporidiosis, 85
    in dehydration, 87
    in lymphoid neoplasia, 166
    in Marek's disease, 302
    in mycobacteriosis, 174, 175
    in mycoplasmosis, 177, 190
    in nutritional deficiencies, 185, 190
    in oil exposure, 193, 194
    oral plaques and, 197
    in rhinitis and sinusitis, 257
    in toxoplasmosis, 279
    in West Nile virus infection, 307
    see also conjunctivitis
ocular tumors, 289
oculomotor dysfunction, 182
oil exposure, **193–195, 193–196**
oil spills, management, 193, 194, 195
oliguria, 251
omega-3 fatty acids
    in arthritis, 19
    in atherosclerosis, 30, 31
    in dyslipidemia/hyperlipidemia, 96
    in pancreatic disease, 214
    in pneumonia, 230
    in renal disease, 253
omega-6 fatty acids, 214
oophoritis, infertility and, 150
open mouth breathing *see* breathing, open-beaked/mouthed
ophthalmic conditions *see* ocular lesions/signs
ophthalmic irrigation solution, 194, 195

ophthalmic medications, oil exposed birds, 195
ophthalmologic examination, 191
opioids *see* analgesia
oral bleeding, 71
oral cavity/lesions, 197
    diagnostic algorithm, *347*
    in diphtheroid pox, 241
    in mycotoxicosis, 179, 180
    in nutritional deficiencies, 185
    squamous cell carcinoma, 274
oral mucosa
    candidiasis, 50
    hyperkeratosis, 54
    in repeated tonguing, 197
    squamous metaplasia, 154, 197, 274
oral plaques, 50, **197–199**
orchitis, in infertility, 150
organomegaly
    in ascites, 22
    cardiomegaly, 21, 22
    in coelomic distension, 79, 80
    in egg yolk/reproductive coelomitis, 103
    in lymphoid neoplasia, 166, 168
    in pancreatic disease, 213
    renomegaly, 146, 166, 179, 251, 252
    see also hepatomegaly; splenomegaly
organophosphate toxicity, 182, **200–201**, 327, 339
Ornithonyssus sylvarium, 101, 335
Ornithostrongylus spp., 126, 127
oropharyngeal lesions, diagnostic algorithm, *347*
orthomyxoviridae *see* avian influenza
osmoregulatory dysfunction, 193
osteoarthritis, 18–20, 83
    see also arthritis
osteodystrophy, 147
osteomalacia, 147, 171
osteomyelitis, 158, 159, 205, 261
    fractures and luxation complication, 122
osteomyelosclerosis, 59, 99, **202–203**
osteopenia, 99, 147, 251
    lameness and, 158, 159
osteopetrosis, 202
osteophyte, 18
osteoporosis, in chronic egg laying, 59, 60
otitis, **204–205**
otoscopy, 191
ovarian carcinomas, 207
ovarian cysts, **206–209**
    ascites, 22
    coelomic distension, 79
    in egg yolk/reproductive coelomitis, 103
    infertility and, 150
ovarian follicles, 207

ovarian neoplasia, **206–209**, 291
    ascites, 22
    lameness and, 159
    lymphoma, 206, 207, 208
overgrown beak *see* beak, overgrowth
overgrown nails, **210–211**
overhydration, 234, 239
overpreening *see* preening
oviduct, 262
    adenocarcinoma, 263
    persistent, 98
oviductal disease, 59, 98, **262–264**
oviductal tissue, prolapsed, 63
ovocentesis, 99
oxalate crystals, 225, *225*, 226
oxygen therapy
    in airborne toxin exposure, 3
    in anticoagulant rodenticide toxicosis, 16
    in ascites, 22
    in aspergillosis, 25
    in aspiration, 27
    in bite wounds, 42
    in coelomic distension, 80
    in egg yolk/reproductive coelomitis, 104
    in lymphoid neoplasia, 167
    in plant and avocado toxicosis, 226
    in pneumonia, 228, 229
    in respiratory distress, 255
    in *Sarcocystis* infections, 266
    in sick bird syndrome, 271
    in thyroid disease, 277
    in toxoplasmosis, 279
    in trauma, 284, 285
    in tumors, 290
Oxyglobin, 135, 285
oxytocin, 99, 316

P

Pacheco's disease, 63, **140–142**, 163
packed cell volume (PCV), 16, 318, 319
    in dehydration, 87
    in hemorrhage, 134, 135
    in iron storage disease, 157
    in ovarian cysts and neoplasia, 206
pain
    arthritis, 18, 19
    beak injury, 36
    lameness, 158, 159
    lymphoid neoplasia, 167
    overgrown nails, 210
    pancreatic disease, 213
pain management *see* analgesia
pallor, in circovirus infection, 61
palpebral reflex, 182
pancreas
    atrophy, 213, 214
    biopsy, 214

pancreas (*Continued*)
  cystic, 213
  necrosis, 213, 214, 215
  neoplasia, 213, 214
pancreatic diseases, **213–215**, 291, 292
  in herpesvirus infections, 140, 141
  undigested food in droppings and, 292, 293
pancreatic duct carcinoma, 141
pancreatic enzymes, 214
pancreatic insufficiency, exocrine, 213, 214
pancreatitis, 76, 95, 213, 214, 215
  hyperglycemia/diabetes mellitus due to, 143
  in polyuria, 239
pancytopenia, 61, 166
panophthalmitis, 83
papillomas, cutaneous, **216–217**
papillomatosis, 134, **140–142**, 289, 299
  see also herpesviruses
papillomaviruses, 216, 333
paramyxoviruses, 68, **218–220**, 327
  hemorrhage and, 134
  neurological conditions, 183
  pneumonia, 230
parasitic infection(s)
  cloacal diseases, 63, 64
  conjunctivitis due to, 190
  diarrhea due to, 93
  flagellate enteritis, 119–120
  gastrointestinal helminthiasis, **126–128**
  liver disease due to, 163
  oral plaques and, 197, 198
  otitis due to, 204
  in pneumonia, 228
  regurgitation and vomiting, 249
  respiratory distress due to, 254
  sick bird syndrome due to, 270
  tracheal/syringeal disease due to, 281, 282, 283
  undigested food in droppings and, 292, 293
parathyroid gland, 147, 148, 150
parathyroid hormone (PTH), 147, 148, 171, 328
parenchymal disease, lung, 254
parenteral nutrition, 109
paresis/paralysis, 98, 158, 182, 226, 251, 302
  see also neurologic conditions
parrot fever *see* chlamydiosis
parvoviruses, 333
pasteurellosis (*Pasteurella multocida*), **221–222**, 328
PBFD (psittacine beak and feather disease), 36, 38, **61–62**, 210
  hypothyroidism *vs*, 277
PCR *see* polymerase chain reaction (PCR)

PCV *see* packed cell volume (PCV)
PDD *see* proventricular dilatation disease (PDD)
pedicle advancement flaps, 232
D-penicillamine, 94, 125, 130, 316
pentoxifylline, 31
pericardial effusion, 237
pericardiocentesis, 53, 226
periocular swelling, 190
peristalsis, 153
peritonitis
  egg yolk, 100, 103
  see also coelomitis
permethrin *see* antiparasitic therapy
persin, 225, *225*, 226
personal protective equipment (PPE), 194, 222, 237, 335
petechiation, 71, 134
petroleum exposure, 193, 194, 195
phallectomy, 224
phallus
  intromittent (protruding), 223
  nonintromittent, 223
  prolapse, **223–224**
pharyngitis, in oral plaques, 197
phenobarbital, 184, 268, 269, 316
phlebotomy, in iron storage disease, 157
phorbol ester, 225, *225*, 226
phosphorus, 171, 318, 319
  deficiency, 171, 185, 186
  dietary, 147, 172
photodynamic therapy, 290
photoperiod, 99, 100, 104, 113, 203
photophobia, 190
photosensitization, 89, 90
phototherapy, in squamous cell carcinoma, 275
physiotherapy, in metabolic bone disease, 172
pigeon circovirus (PiCV), **61–62**
pigeon paramyxovirus 1 (PPMV-1), 218, 219, 220
pimobendan, 22, 53, 80, 316
pineal tumors, 68, 69
pin sites, cleaning, 122
piperacillin, in bite wounds, 42, 316
pituitary tumors, **68–70**, 91, 182
plant toxins, *225*, **225–227**, 337
plaques *see* oral plaques
plasma osmolality, 91, 234, 235
plasma protein electrophoresis *see* protein electrophoresis (EPH)
plasma transfusion, in pancreatic disease, 214
Plasmodium, **131–133**
pleurodesis, 8
PMV3 *see* paramyxoviruses
pneumonia, **228–230**
  in air sac mite infection, 5
  aspiration *see* aspiration pneumonia

  in bordetellosis, 44
  in candidiasis, 50
  in coccidiosis (systemic), 76
  in mycoplasmosis, 177
  in toxoplasmosis, 279
pododermatitis (bumblefoot), 52, 54, 121, 149, **231–233**
  in arthritis, 19
  chronic, 233
  in colibacillosis, 82
  in hypocalcemia and hypomagnesemia, 148
  lameness and, 158, 159
  in metabolic bone disease, 172
  in obesity, 189
  ulcerative, 89
poisoning
  anticoagulant rodenticides, **15–17**, 72, 339
  plants and avocado, *225*, **225–227**, 337
  see also heavy metal toxicity
polychromasia, 134, 284, 285
polycythemia, 3, 52, 281
polydipsia (PD), **234–236**, 239
  in aspergillosis, 24
  in diabetes insipidus, 91
  diagnostic algorithm, *349*
  extreme, regurgitation, 249
  in heavy metal toxicity, 129
  in hepatic lipidosis, 137
  in hyperglycemia and diabetes mellitus, 143
  in pancreatic disease, 213
  psychogenic, 91, 234, 235, 239, 252
  in vitamin D toxicosis, 305
polymerase chain reaction (PCR)
  in chlamydiosis, 56
  in circovirus infections, 62
  in clostridiosis, 66–67
  in cryptosporidiosis, 85
  in enteritis and gastritis, 109
  in feather damaging/self-injurious behavior, 114
  in flagellate enteritis, 119
  in hemoparasite infections, 132
  in herpesvirus infections, 141
  in hyperglycemia and diabetes mellitus, 143
  in lameness, 158
  in mycobacteriosis, 174
  in neurological conditions, 183
  in paramyxovirus infections, 219
  in PDD, 247
  in polydipsia, 234
  in poxvirus infections, 242
  reverse transcriptase (RT-PCR), 247.219, 307
  in rhinitis and sinusitis, 257
  seizure causes, 268

in tracheal/syringeal disease, 281
in tumors, 290
in viral disease, 299–300
in West Nile virus infection, 307
polyomavirus, 72, **237–238**, 299, 328, 333, 334
   ascites and, 22
   cere and skin color changes, 54
   circovirus infection *vs*, 61
   feather disorders, 117, 118
   hemorrhage and, 134
   in hyperglycemia and diabetes mellitus, 143
   in liver disease, 163
   in pneumonia, 229
   in thyroid disease, 277
polyostotic hyperostosis, **202–203**
   in chronic egg laying, 59
   in dystocia and egg binding, 99
   in egg yolk/reproductive coelomitis, 103
   in ovarian cysts and neoplasia, 206, 207
polyphagia, 108, 143, 188, 234
polyserositis, 83
polysulfated glycosaminoglycan (PSGAG), 19
polytetrafluoroethylene (PTFE), 3, 228, 338
polyuria (PU), **239–240**
   in aspergillosis, 24
   in diabetes insipidus, 91
   diagnostic algorithm, *350*
   in diarrhea, 93
   in heavy metal toxicity, 129
   in hepatic lipidosis, 137
   in hyperglycemia and diabetes mellitus, 143
   in ingluvial hypomotility/crop stasis/ileus, 153
   in pancreatic disease, 213
   in PDD, 247
   in phallus prolapse, 223
   in polydipsia response, 234
   in renal disease, 251
   in trichomoniasis, 287
   in vitamin D toxicosis, 305
   in West Nile virus infection, 307
ponazuril *see* antiparasitic therapy
porphyrinuria, 129, 145, 252
portal system, 275
positron emission tomography (PET), in tumors, 290
posture, abnormal, 158, 182
potassium bromide, 268
poxvirus, 89, 229, **241–243**, 328, 334
pralidoxime iodide, 201, 316
praziquantel *see* antiparasitic therapy
preening, increased, 113
   in otitis, 204

in uropygial gland disease, 297
pressure sores, 158, 159, 193, 194, 201
primitive neuroectodermal tumors (PNETs), 68, 69
probiotics, 84, 110, 247
problem behaviors, **244–246**, *245*
   see also behavioral changes
proctodeum *see* cloaca
prolapse
   cloacal, 63, 66, 147, 223
   in dystocia and egg binding, 98
   intestinal, 63
   oviductal tissue, 63
   phallus, **223–224**
prostaglandin E_2, 99, 263, 316
prostaglandin $F_{2\alpha}$, 99, 263
protein, dietary
   angel wing and, 11
   deficiency, 185
   in renal disease, 251
protein electrophoresis (EPH)
   in chronic egg laying, 59
   in dyslipidemia/hyperlipidemia, 96
   in dystocia and egg binding, 99
   in egg yolk/reproductive coelomitis, 103
   in mycobacteriosis, 174
   in pancreatic disease, 214
   in pneumonia, 228
   in polydipsia, 234
   in polyostotic hyperostosis, 202
   in renal disease, 252
   in respiratory distress, 254
   in rhinitis and sinusitis, 257
   in *Sarcocystis* infections, 265
protein-losing enteropathy, 110
protein-losing nephropathy, 79
proteinuria, 145, 234, 252
prothrombin time (PT), 72, 134
protoanemonin, 225, *225*, 226
proventricular adenocarcinoma, 169
proventricular dilatation disease (PDD), 183–184, **247–248**
   aspiration in, 27
   enteritis and gastritis, 108
   feather damaging/self-injurious behavior, 113
   gastrointestinal foreign bodies and, 124, 249
   ingluvial hypomotility/crop stasis/ileus, 153, 154
   Macrorhabdus infection *vs*, 169
   neurologic conditions in, 108, 182, 183–184, 267
   regurgitation and vomiting due to, 182, 183–184, 249, 299
   undigested food in droppings and, 292, 293, 294, 299
proventricular diseases, 249
proventriculitis, 169

pruritus, 89, 101, 119, 120
pseudohyphae, 198
Pseudomonas, in pododermatitis, 231
psHV-1 (Psittacid-herpesvirus 1), 89, **140–142**
psHV-2 (Psittacid-herpesvirus 2), **140–142**, 216
psittacine beak and feather disease *see* PBFD
psittacosis, 57, 150, 151
   see also chlamydiosis
psychogenic polydipsia, 91, 234, 235, 239, 252
psychotropic medication, 115
PTH (parathyroid hormone), 147, 148, 171, 328
Ptilonyssus spp. (air sac mite), **5–6**, 229, 254, 281
ptyalism, 200, 226
pudendal nerve, 182
pulmonary edema, 2, 29, 265, 266, 279, 280
pulmonary hemorrhage, in coagulopathies, 71, 72
pulmonary hypersensitivity, 228
pulmonary neoplasia, pneumonia *vs*, 228
pulsed field gel electrophoresis, 260
pulse oximetry, in tracheal/syringeal disease, 282
puncture wounds, 41
   see also wound management
puzzle-feeders, 115
pyrethrins/pyrethroids, 339
pyridoxine (vitamin B_6), deficiency, 185, 272

Q

quarantine, 57, 84
   trichomoniasis, 288
   viral diseases, 300

R

radiography
   in air sac rupture, 7
   in angel wing, 11
   in anorexia, 13
   in anticoagulant rodenticide toxicosis, 16
   in arthritis, 18–19
   in ascites, 21
   in aspergillosis, 24
   in aspiration, 27
   in atherosclerosis, 30
   in beak injury, 36
   in beak malocclusion, 38
   in campylobacteriosis, 48
   in candidiasis, 50
   in cardiac disease, 52

radiography (Continued)
   in chlamydiosis, 56
   in chronic egg laying, 59
   in cloacal diseases, 64
   in clostridiosis, 67
   in CNS tumors, 69
   in coccidiosis (systemic), 76
   in coelomic distension, 79
   contrast
      ascites, 21
      cloacal diseases, 64
      coelomic distension, 79
      diarrhea, 93
      egg yolk/reproductive coelomitis, 103
      enteritis and gastritis, 109
      gastrointestinal foreign bodies, 124
      herpesvirus infection, 141
      ingluvial hypomotility/crop stasis, 153
      macrorhabdosis, 169
      mycobacteriosis, 174
      oral plaques, 198
      ovarian cysts and neoplasia, 207
      pododermatitis, 231, 232
      polyuria, 239
      renal disease, 252
      respiratory distress, 254
      rhinitis and sinusitis, 257
   in diabetes insipidus, 91
   in diarrhea, 93
   in dystocia and egg binding, 99
   in egg yolk/reproductive coelomitis, 103
   in enteritis and gastritis, 109
   in feather damaging/self-injurious behavior, 114
   in fractures and luxations, 121
   in gastrointestinal foreign bodies, 124
   in heavy metal toxicity, 129
   in hepatic lipidosis, 137
   in herpesvirus infections, 141
   in hypocalcemia and hypomagnesemia, 148
   infertility and, 151
   in ingluvial hypomotility/crop stasis/ileus, 153
   in iron storage disease, 156
   in lameness, 158
   in lipomas, 161
   in liver disease, 164
   in *Macrorhabdus* infection, 169
   in metabolic bone disease, 171
   in mycobacteriosis, 174
   in mycoplasmosis, 177
   in neurological conditions, 183
   in nutritional deficiencies, 186
   in obesity, 188
   in ocular lesions, 191
   in oil exposure, 194
   in oral plaques, 198
   in otitis, 204
   in ovarian cysts and neoplasia, 206
   in overgrown beak/nails, 211
   in PDD, 247
   in plant and avocado toxicosis, 226
   in pneumonia, 229, 230
   in pododermatitis, 231, 233
   in polydipsia, 234–235
   in polyostotic hyperostosis, 202
   in polyuria, 239
   in renal disease, 252
   in respiratory distress, 254
   in rhinitis and sinusitis, 257
   in salmonellosis, 260
   in salpingitis, oviductal/uterine disorders, 262
   in *Sarcocystis* infections, 265
   in seizures, 268
   in sick bird syndrome, 270
   in splay leg and slipped tendon, 272
   in squamous cell carcinoma, 274
   in thyroid disease, 277
   in tracheal/syringeal disease, 282
   in trauma, 284
   in trichomoniasis, 287
   in tumors, 290
   undigested food in droppings and, 293
   in vitamin D toxicosis, 305
   in xanthomas, 310
radiotherapy, in tumors, 290
   in lymphoid neoplasia, 167
   in squamous cell carcinoma, 275, 276
rales *see* respiratory distress
range of motion, decreased, 122, 158
refeeding syndrome, 106
reflexes, 182, 251
regurgitation, **249–250**
   in candidiasis, 50
   in clostridiosis, 66
   diagnostic algorithm, *342*
   in enteritis and gastritis, 108
   in heavy metal toxicity, 129
   in hepatic lipidosis, 137
   in herpesvirus infections, 140
   in ingluvial hypomotility/crop stasis/ileus, 153
   in *Macrorhabdus* infection, 169
   in Marek's disease, 302
   in oil exposure, 193–194
   oral plaques and, 197
   in pancreatic disease, 213
   in plant and avocado toxicosis, 226
   in polyomavirus infection, 237
   in renal disease, 251
   in trichomoniasis, 287
   tumors causing, 289
   undigested food in droppings and, 292
reinforcement of behaviors, 244
releaves (holistic supplement), 208
renal adenoma, 252
renal agenesis, 251
renal biopsy, 252
renal carcinoma, 251, 252
renal disease, **251–253**
   in botulism, 46
   in heavy metal toxicity, 129
   in hyperuricemia, 145, 146
   in plant and avocado toxins, 225, 226
   in polyuria, 239
   undigested food in droppings, 292
renal failure, 234, 251
   in gastrointestinal foreign bodies, 125
   in plant and avocado toxicosis, 225
   in vitamin D toxicosis, 305, 306
renal hemorrhage, 251, 252
renal ischemia, 234
renal tubular constipation, 153
renal tubules, squamous metaplasia, 185, 197
renal tumors, 251, 253, 289
renomegaly, 146, 166, 179, 251, 252
reovirus, 72, 328, 334
reportable diseases, 57, 175, 184, 220
reproductive behavior, 95, 150
reproductive coelomitis, **103–104**
reproductive disease
   ascites due to, 21
   in colibacillosis, 82
   in mycotoxicosis, 179
   in ocular lesions with, 190
   in ovarian cysts and neoplasia, 206
   in phallus prolapse, 223
reproductive tract
   enlarged, 79
   in infertility, 150, 151
   masses, egg binding and, 98
   in oil exposure, 193
   in polyostotic hyperostosis, 202
   tumors, 289
   undigested food in droppings, 292
respiratory disease/signs
   in coccidiosis (systemic), 76, 77
   in colibacillosis, 82
   in cryptosporidiosis, 85
   in mycobacteriosis, 174
   in mycoplasmosis, 177
   in nutritional deficiencies, 185
   in ocular lesions with, 190
   in otitis, 204
   in overgrown beak/nails, 210
   in pancreatic disease, 213
   in paramyxovirus infections, 218, 219
   in plant and avocado toxicosis, 225
   in poxvirus infections, 241
   in rhinitis and sinusitis, 257
   in *Sarcocystis* infections, 265
   in thyroid disease, 277

in toxoplasmosis, 279
in tracheal/syringeal disease, 281
in trichomoniasis, 287
respiratory distress, **254–256**
    in air sac mite infection, 5
    in anticoagulant rodenticide toxicosis, 15
    in ascites, 21
    in aspergillosis, 24
    in aspiration, 27
    in atherosclerosis, 29
    in avian influenza, 33, 34
    in bite wounds, 41
    in bordetellosis, 44
    in botulism, 46
    in cardiac disease, 52
    in chlamydiosis, 56
    in coelomic distension, 79
    diagnostic algorithm, *343*
    in dystocia and egg binding, 98
    in egg yolk/reproductive coelomitis, 103
    in hemoparasite infections, 131
    in hepatic lipidosis, 137
    in herpesvirus infections, 140
    in hypocalcemia and hypomagnesemia, 147
    in ingluvial hypomotility/crop stasis/ileus, 153
    in lipomas, 161
    in liver disease, 163
    in lymphoid neoplasia, 166
    in Marek's disease, 302
    in mycobacteriosis, 174
    in nutritional deficiencies, 186
    in oil exposure, 193
    in organophosphate/carbamate toxicity, 200
    in ovarian cysts and neoplasia, 206
    in pancreatic disease, 213
    in pasteurellosis, 221
    periodic, 254
    in pneumonia, 228
    in *Sarcocystis* infections, 265
    in tracheal/syringeal disease, 281
    in trichomoniasis, 287
    see also dyspnea; tachypnea
respiratory epithelium, in squamous metaplasia, 197
respiratory failure, 200
respiratory rate/effort, 228, 241, 254
respiratory system, 3, 255
respiratory tract tumors, 289
response-to-injury hypothesis, 29
reticulocytes, in anemia, 9
reticuloendotheliosis (RE), **302–304**
reticuloendotheliosis virus (REV), 166
retinal detachment, 190
retinal disease, 190
retinal hemorrhage, 134

retropulsion, 68
retroviruses, 166, 213, 302, 331, 332
rhabdomyolysis, 96
rhamphotheca *see* rhinotheca
rhinitis, 44, 61, 250, **257–259**
rhinoliths, 257, 258
rhinotheca, 210
    overgrown, hepatic lipidosis, 137
    see also beak
riboflavin, deficiency, 185
rickets, 148, 158
rodenticides, 15, 252
    anticoagulant, toxicosis, **15–17**, 72, 339
Romanowsky stain, 169
ronidazole *see* antiparasitic therapy
rotavirus, 328
roundworm *see* parasitic infection(s)
Russell Viper Venom Time (RVV), 72

S

salmonellosis *(Salmonella)*, 108, **260–261**, 328, 335
salpingitis, **262–264**
    in colibacillosis, 83
    in dystocia and egg binding, 99
    infertility and, 150
    in polyuria, 239
salpingohysterectomy, 60, 81, 99, 100, 103, 104, 263
salpingotomy, 99
salt glands, 234
SAMe (S-adenosyl-methionine), 138, 164
sanitation, 34, 108, 109
Sarcocystis, 68, 229, **265–266**
scaly face and leg mite *see* ectoparasites
scarring, in poxvirus infections, 241, 242
schistosomiasis encephalitis, 68
Schroeder–Thomas splint, 159
scintigraphy, 239, 290
scissors beak, **38–40**, 210, 211
screaming, **244–246**, *245*
sedation, 41, 42, 154, 159, 282
seizures, 183, 184, 185, 200, **267–269**
    gastrointestinal foreign bodies and, 124, 125
    in heavy metal toxicosis, 129
    in hypocalcemia and hypomagnesemia, 147, 148, 149
    nutritional deficiencies causing, 182
    in pasteurellosis, 221
    in PDD, 247
    in plant and avocado toxicosis, 226
    in toxoplasmosis, 279
    undigested food in droppings and, 292
    urate and fecal discoloration and, 295

    see also neurologic conditions
selenium, 59
    deficiency, 52, 98, 182, 213, 265, 272, 279
self-mutilation (self-injurious behavior), 89, 90, **113–116**, 284
    in feather cysts, 111, 112
    hemorrhage due to, 134
    in phallus prolapse, 223
    in renal disease, 251
    squamous cell carcinoma *vs*, 274
sepsis, 82, 103, 104, 239
septic arthritis, 18, 233
septicemia, in pasteurellosis, 221
septicemic pox, 241, 242
serology
    mycoplasmosis, 177
    paramyxovirus infections, 219
    in PDD, 247
    in pneumonia, 228
    in rhinitis and sinusitis, 257
    salmonellosis, 260
    in tracheal/syringeal disease, 281
SGAR (second-generation anticoagulant rodenticides), 15–16
shell gland, 98
shock
    in fractures and luxations, 121
    in hemorrhage, 134, 135
    hypovolemic *see* hypovolemic shock
    in pasteurellosis, 221
Siadenovirus see adenovirus(es)
sialolithiasis, 197
sick bird syndrome, 228, **270–271**
    avian influenza, 33
    diagnostic algorithm, *340*
    flagellate enteritis, 119–120
    gastrointestinal foreign bodies, 124
    hepatic lipidosis, 137
    hypocalcemia and hypomagnesemia, 147
    liver disease, 163
    mycotoxicosis, 179
    pneumonia, 228
sildenafil citrate, in ascites, 22
silver sulfadiazine cream, 233, 316
silymarin (milk thistle), 22, 296, 317
sinus aspiration/lavage, 258
sinusitis, 44, 177, **257–259**
sinusotomy, 258
sinus trephination, 258
SIRS (systemic inflammatory response syndrome), 41
skin
    biopsy, in nutritional deficiencies, 186
    bite wounds, 41
    color changes, **54–55**, 188, 310
    ectoparasite infestation, 101–102
    in feather damaging/self-injurious behavior, 113

skin (*Continued*)
   in feather disorders, 117
   in fractures and luxations, 121
   in liver disease, 163
   in lymphoid neoplasia, 166, 167
   mass, squamous cell carcinoma, 274
   in mycobacteriosis, 174
   necrosis, in mycotoxicosis, 179
   in oil exposure, 193
   papillomas (warts), **216–217**
   in pododermatitis, **231–233**
   in poxvirus infections, 241
   scrapings, 101, 191, 204, 211
   tenting, in dehydration, 87
   in thyroid disease, 277
   tumors, 289
   see also cere and skin
sleep deprivation, 113
slipped tendon, **272–273**
small airway disease, 254
small intestinal disease, 292
smoke, second hand, 30, 95, 230
smoke inhalation, 3, 338
sneezing *see* respiratory distress
sodium benzoate, 170
sodium toxicity, 68
soft tissue swelling, 18, 121, 158, 177
somatostatins, 144
sour crop *see* candidiasis
Spironucleus, in flagellate enteritis, 119, 120
splay leg, 159, **272–273**
spleen, in lymphoid neoplasia, 166
splenitis, 76, 77
splenomegaly, 1, 52, 56, 79
   in hemoparasite infections, 132
   in herpesvirus infections, 141
   in lymphoid neoplasia, 166
   in mycobacteriosis, 174
   in toxoplasmosis, 279
splints, 159, 273
split keel, 89
squamous cell carcinoma (SCC), **274–276**
   beak, 36, 38, 210
squamous metaplasia, 185, 274
   crop mucosa, 154
   oral, 197, 274
Staphylococcus, in feather damaging/self-injurious behavior, 113
Staphylococcus aureus, in pododermatitis, 231, 232
starvation, 105, 106, 134
   see also emaciation
statins, 31, 96, 184
steatorrhea, 109, 163
stents, 7, 282
stereotypy *see* behavioral changes

Sternostoma tracheacolum (air sac mite), **5–6**, 229, 254, 281
steroids *see* corticosteroid(s)
St John's Wort, 89
stomatitis, 50
stools, 295
   in diarrhea, 93
   mucus-laden, 108
   see also entries beginning fecal
straining, 98, 99, 147, 262
stress, 113
stress bars, 137, 185
stress hyperglycemia, 143
strontium-90, 275
styptic powder, 211
subcutaneous emphysema, 7, 8, 191
subluxation, 121
sucralfate, 110, 125, 227, 250, 317
suction, in respiratory distress, 255
sudden death, 140, 237
   in atherosclerosis, 29
   in chronic egg laying, 59
   in dystocia and egg binding, 98
   in egg yolk/reproductive coelomitis, 103
   in heavy metal toxicity, 129
   in herpesvirus infections, 140
   in ingluvial hypomotility/crop stasis/ileus, 153
   in iron storage disease, 156
   in pasteurellosis, 221
   in polyomavirus infection, 237, 238
   in renal disease, 251
sulfachlorpyridazine, 133, 317
sulfa drugs, coagulopathy due to, 71
sulfur-containing alkaloids, 225, *225*, 226
"sunken sinus syndrome", 257
sunlight exposure, 172
surgery
   in angel wing, 11–12
   in arthritis, 19
   in ascites, 22
   in aspergillosis, 25
   in beak injuries, 36–37
   in beak malocclusion, 39
   in bite wounds, 42
   cervicocephalic-clavicular air sac shunt, 8
   in chronic egg laying, 60
   in cloacal diseases, 64
   in coelomic distension, 80
   in cutaneous papillomas, 216
   debridement, in pododermatitis, 232
   in dystocia and egg binding, 99, 100
   in egg yolk/reproductive coelomitis, 103, 104
   in feather cyst, 111, 112
   fistula, in air sac rupture, 7–8
   in fractures and luxations, 122

   in gastrointestinal foreign bodies, 125, 250
   in hemorrhage, 134, 135
   in hepatic lipidosis, 138
   in herpesvirus infections, 141
   in lameness, 159
   in lipomas, 162, 189
   in lymphoid neoplasia, 167
   in mycobacteriosis, 175
   in ocular lesions, 191
   in ovarian cysts and neoplasia, 208
   in pancreatic disease, 214
   in phallus prolapse, 223–224
   in pododermatitis, 232
   in renal disease, 253
   in respiratory distress, 255
   in rhinitis and sinusitis, 258
   in salpingitis, oviductal/uterine disorders, 263
   in splay leg and slipped tendon, 272, 273
   in squamous cell carcinoma, 275, 276
   in thyroid disease, 278
   in tracheal/syringeal disease, 282
   in tumors, 290
   in uropygial gland disease, 297
   in xanthomas, 310, 311
sutures, 64, 223–224
swab(s)
   in herpesvirus infections, 141
   in hyperglycemia and diabetes mellitus, 143
   in rhinitis and sinusitis, 258
swelling
   in fractures and luxations, 121
   soft tissue, 18, 121, 158, 177
swimming in circles, 221
swollen head syndrome, 83
syncope, 267
syringeal disease, **281–283**
syrinx, 281, 282

T

tachycardia, 134
tachypnea
   in arthritis, 18
   in ascites, 21
   in candidiasis, 50
   in hemorrhage, 134
   in ingluvial hypomotility/crop stasis/ileus, 153
   in plant and avocado toxicosis, 226
   tumors causing, 289
   see also respiratory distress
tail bob, 79, 98, 228, 254, 281
   see also respiratory distress
tannins, 157, 164
tarsorrhaphy, 191
tassle foot *see* ectoparasites

tattoo placement site, masses, 174
teletherapy, in lymphoid neoplasia, 167
tendons, slipped, 177
tenesmus, in herpesvirus infections, 140
terbinafine *see* antifungal therapy
terbutaline *see* bronchodilator(s)
testicular neoplasia, in lameness, 159
tetracycline *see* antibiotic therapy
Tetrameres spp., 126, 127
Tetratrichomona gallinarum, 68
theophylline *see* bronchodilator(s)
thermal injury, 54
thermal support *see* warmth
thermography, in pododermatitis, 232
thermoregulation, 42, 113, 115, 300
    in organophosphate/carbamate toxicity, 200, 201
thiamin (thiamine, vitamin B_1), 106, 195
    deficiency, 185, 186, 267
    supplements, 186, 268
thiaminase, 185
thinness *see* emaciation
thrombocyte, 71
thrombocytopenia, 72, 134, 135, 285
thromboelastography (TEG), 72, 134
thrombosis, in pododermatitis, 232
thrush *see* candidiasis
thyroid diseases, 137, 161, **277–278**
    see also hypothyroidism
thyroid gland, infertility and, 150
thyroid hormones, 185, 188, 329
thyroid hyperplasia, 277, 278
thyroid neoplasia, 277, 278
thyroid stimulating hormone (TSH), 277
thyroxine, 189, 277, 278
ticks *see* ectoparasites
tissue plasminogen activator (TPA), 191
toes
    bleeding, 211
    constricted, 89, 90
    discolored, in mycotoxicosis, 179
    skin, color changes, 54
    swelling/redness *see* pododermatitis (bumblefoot)
toltrazuril *see* antiparasitic therapy
tonometry, in ocular lesions, 191
tophi, 145
torticollis, 68, 204, 205, 279
total calcium, 147
total iron binding capacity (TIBC), 156
total solids (TS), in dehydration, 87
toxicities/toxicosis
    anticoagulant rodenticide, **15–17**, 72, 339
    botulism, **46–47**
    carbamate, **200–201**
    diabetes insipidus *vs*, 91
    hepatic lipidosis *vs*, 137
    hypocalcemia/hypomagnesemia *vs*, 147–148
    lameness in, 158
    neurologic conditions due to, 182
    organophosphate, **200–201**
    polydipsia due to, 234
toxins, **337–339**
    airborne *see* airborne toxins
    avian pathogenic *E. coli* (APEC), 82
    avocado *see* avocado toxicosis
    in cardiac disease pathophysiology, 52
    clostridial, 46, 66
    diarrhea due to, 93
    feather disorders due to, 117
    gastrointestinal foreign bodies, 124
    infertility due to, 150
    liver disease due to, 163
    pancreatic disease due to, 213
    plant, *225*, **225–227**, 337
    renal disease due to, 251–252
    respiratory distress due to, 254
    tracheal and syringeal disease and, 281
toxoplasmosis (*Toxoplasma gondii*), 68, **279–280**
trachea
    foreign bodies, 27, 254, 255, 281, 282
    resection and anastomosis, 282
tracheal disease, 5, 141, **281–283**
tracheal displacement, in thyroid disease, 277
tracheal lavage, in mycoplasmosis, 177
tracheal obstruction, 254
tracheal stenosis, 281, 282, 283
tracheal stenting, 282
tracheitis, 44, 141, 177
tracheoscopy, 5, 282
tracheostomy, 282
tracheotomy, 282
tramadol *see* analgesia
transillumination, 5, 27
trans-sinus pin and tension band procedure, 39
transudate, 21
trauma, **284–286**
    in arthritis, 18
    beak injury, **36–37**, 210, 211
    beak malocclusion, 38
    bite wounds, 41–43
    dermatitis due to, 89
    emaciation after, 105
    to feather cyst, 111
    feather damaging/self-injurious behavior, 114
    feather disorders due to, 117
    fractures and luxations, 121
    hemorrhage, 134
    ingluvial hypomotility/crop stasis/ileus due to, 153, 154
    lameness due to, 158
    nervous system, 182
    ocular lesions due to, 190, 192
    overgrown beak/nails, 210
    phallus prolapse due to, 223
    pneumonia *vs*, 228
    pododermatitis due to, 231
    renal disease due to, 252
    respiratory distress due to, 254
    seizures due to, 267, 268
    splay leg and slipped tendon, 272
    tracheal and syringeal disease due to, 281
    uropygial gland, 298m297
    uveitis due to, 190
trematodes, 126, 251
triage, in oil exposure, 194
trichomoniasis (*Trichomonas* spp.), 197, 249, **287–288**, 329
    in oral plaques, 197, 198
trichothecene mycotoxicosis, 179, 180, 339
trigeminal nerve dysfunction, 182
triglycerides, 95, 137, 143, 213, 214
trimethoprim/sulfamethoxazole, 75, 83, 317
trimming
    beak and nail, 211
    wing, 113
trochlear dysfunction, 182
Trypanosoma, 131, 132, 133
trypsin, 213
trypsin-like immunoreactivity (TLI), 213–214
tube feeding *see* gavage feeding
tumors *see* neoplasia
"Turkish Towel", 50
typhlitis, 108

U

ulcer(s)
    crop, 54
    in dermatitis, 89
    in flagellate enteritis, 120
    hemorrhage due to, 134
    in herpesvirus infections, 141
    regurgitation and vomiting, 249
    in renal disease, 253
    uropygial gland, 297, 298
ulcerative dermatitis, 89
ultracentrifugation, in dyslipidemia/hyperlipidemia, 95–96
ultrasonography, 252
    in arthritis, 19
    in ascites, 21–22
    in cardiac disease, 52
    in chlamydiosis, 56
    in chronic egg laying, 59
    in cloacal diseases, 64
    in coccidiosis (systemic), 76
    in coelomic distension, 79–80
    in diabetes insipidus, 91

ultrasonography (*Continued*)
  in dystocia and egg binding, 99
  in egg yolk/reproductive coelomitis, 103
  in enteritis and gastritis, 109
  in feather damaging/self-injurious behavior, 114
  in hepatic lipidosis, 137
  in herpesvirus infections, 141
  infertility and, 151
  in ingluvial hypomotility/crop stasis/ileus, 154
  in iron storage disease, 156
  in lameness, 158
  in lipomas, 161
  in liver disease, 164
  in *Macrorhabdus* infection, 169
  in mycobacteriosis, 174
  in nutritional deficiencies, 186
  in ocular lesions, 191
  in ovarian cysts and neoplasia, 207
  in pancreatic disease, 214
  in pododermatitis, 231
  in respiratory distress, 254
  in tumors, 290
  in viral disease, 300
umbilical swelling, 287
undigested food in droppings, **292–294**
  in cryptosporidiosis, 85
  in diarrhea, 93
  in emaciation, 105
  in enteritis and gastritis, 108
  in gastrointestinal helminthiasis, 126
  in herpesvirus infections, 140
  in *Macrorhabdus* infection, 169
  in pancreatic disease, 213
upper motor neuron lesion, 182
upper respiratory tract infections, 228, 254, 255
urate oxidase, 146
urate/uric acid, 145, 295, 318, 319
  adhered to vent, in enteritis/gastritis, 108
  caking, phallus prolapse, 223
  crystals, in arthritis, 19
  decreased, 156, 234
  discoloration, **295–296**
    in ascites, 21
    in campylobacteriosis, 48
    in chlamydiosis, 56
    in circovirus infection, 61
    in coelomic distension, 79
    in enteritis and gastritis, 108
    in hemoparasites, 132
    in hepatic lipidosis, 137
  in gout, 18, 145
  in iron storage disease, 156
  in polydipsia, 234
  in renal disease, 252
  in toxoplasmosis, 279

vent stained, 98
  see also hyperuricemia
urea:creatinine ratio, 252
urea:UA ratio, 145, 252
ureterolith, 18
uric acid *see* urate/uric acid
urinalysis
  in hyperglycemia and diabetes mellitus, 143
  in hyperuricemia, 145
  in polydipsia, 234
  in renal disease, 252
urine, discolored, 226, 295
urine osmolality, 91, 234, 239
urine output, 239, 251
urine specific gravity (USG), 91, 234
urodeum *see* cloaca
uropygial gland, ulcer, 297, 298
uropygial gland disease, 185, **297–298**
  tumors, 274, 289
ursodeoxycholic acid, 138
ursodiol, 164
uterine disorders, **262–264**
uterine tumors, 262
UVB, 147, 148
UV damage, 274
uveitis, 190

V

vaccination
  avian influenza, 34
  botulism, 47
  coccidiosis (intestinal), 75
  colibacillosis, 84
  hemoparasites, 133
  infectious laryngotracheitis, 141
  lymphoid leukosis, 303
  Marek's disease, 208, 303
  mycoplasmosis, 178
  ocular lesions and, 192
  paramyxovirus infections, 220
  pasteurellosis, 222
  polyomavirus infection, 238
  poxvirus infections, 242
  reticuloendotheliosis, 303
  viral diseases, 300
  West Nile virus, 308
vacuum-assisted closure (VAC), in pododermatitis, 232
vagal nerve deficits, 182
vascular congestion, 137
vascular disease, 18, 52, 279
vascular injury, in hemorrhage, 134
vasculitis, in pododermatitis, 232
vasoconstriction, in mycotoxicosis, 179
vasodilators, 31, 53, 54
vasopressin/desmopressin response test, 235
vent (and vent feathers), 63

  dorsal displacement, ovarian cysts and neoplasia, 206
  dysfunction, salpingitis, oviductal/uterine disorders, 262
  flaccid, egg yolk and reproductive coelomitis, 103
  matted
    enteritis and gastritis, 108
    West Nile virus infection, 307
  soiled, 66, 93, 98
    pasteurellosis, 221
    phallus prolapse, 223
ventplasty, 64
ventricular diseases
  dilation *see* proventricular dilatation disease (PDD)
  regurgitation and vomiting, 249
  undigested food in droppings due to, 292
ventricular feeding, 183
ventriculitis, in *Macrorhabdus* infection, 169
vestibular lesions, 182, 204, 205, 267
vestibulocochlear nerve deficits, 182
"Vinegar test", 141
viral antibodies, 302–303
viral disease, **299–301**
  CNS tumors, 68
  conjunctivitis due to, 190
  cutaneous papillomas and, 216
  diarrhea due to, 93
  feather cyst due to, 111
  feather damaging/self-injurious behavior due to, 113
  infection types, 299
  infertility and, 150
  ingluvial hypomotility/crop stasis/ileus, 154
  liver disease due to, 163
  in lymphoid neoplasia, 166
  oral plaques due to, 197, 198
  otitis due to, 204
  pancreatic disease due to, 213
  pneumonia, 228
  renal disease due to, 251, 253
  respiratory distress due to, 254
  rhinitis and sinusitis due to, 257
  tracheal and syringeal disease due to, 281
  see also specific viruses
viral diseases of concern, **330–334**
viral neoplasms, **302–304**
viricides, 308
virus isolation (VI), 34, 219, 268, 300, 303, 307
visceral gout, 146, 253, 305
vitamin A, 187
  deficiency, 185, 186
    aspergillosis, 24
    candidiasis, 50

cere and skin color changes, 54
hyperuricemia, 145, 146
ocular lesions, 190, 191
see also hypovitaminosis A
excess, 143, 145, 147, 338
supplementation, 186, 191, 232, 253, 317
vitamin B, 106
deficiencies, 186
in hepatic lipidosis, 138
vitamin B_1 *see* thiamin (thiamine, vitamin B_1)
vitamin B_6, deficiency, 185, 272
vitamin B_{12}, 317
deficiency, 272
vitamin C, in iron storage disease, 157
vitamin D, 149, 187, 329
in chronic egg laying, 59
deficiency, 59, 171, 185, 186
excess, 186, **305–306**, 338
in hypocalcemia and hypomagnesemia, 147
in metabolic bone disease, 171, 172
toxicosis, **305–306**
vitamin D_3, 147, 148, 172, 186, 305, 317
deficiency, in splay leg and slipped tendon, 273
in hemorrhage, 135
vitamin E, 59, 84, 317
deficiency, 98, 120, 158, 185, 272
excess, 147
in hepatic lipidosis, 138
in hypocalcemia and hypomagnesemia, 147, 148, 149
vitamin K, 72, 134, 135
deficiency, 71, 134
vitamin K_1, 317
in anticoagulant rodenticide toxicosis, 16
in coagulopathies, 72
in hemorrhage, 135
in hepatic lipidosis, 138
Vitamin K_3 (menadione), 73
vitamin supplements, 186, 195
vitellogenesis, in dyslipidemia/hyperlipidemia, 95
VLDL-C (very low-density lipoprotein cholesterol), 95
vocalization changes
in air sac mite infection, 5
in aspergillosis, 24
in bordetellosis, 44
in egg yolk/reproductive coelomitis, 103
in emaciation, 105

in hypocalcemia and hypomagnesemia, 147
in nutritional deficiencies, 185
in pneumonia, 228
in squamous cell carcinoma, 274
in tracheal/syringeal disease, 281, 283
vomiting, **249–250**
in cryptosporidiosis, 85
diagnostic algorithm, *342*
in enteritis and gastritis, 108
in gastrointestinal helminthiasis, 126
in hepatic lipidosis, 137
oral plaques and, 197
in pancreatic disease, 213
in PDD, 247
in plant and avocado toxicosis, 226, 227
in polydipsia, 235
in renal disease, 251
undigested food in droppings and, 292
in vitamin D toxicosis, 305
voriconazole, 25, 317

W

warfarin, 15
warmth
in anticoagulant rodenticide toxicity, 16
in bite wounds, 42
in coccidiosis, 74, 77
diarrhea management, 93
in dystocia and egg binding, 99
in egg yolk/reproductive coelomitis, 103
in hemorrhage, 135
in organophosphate/carbamate toxicity, 201
in plant and avocado toxicosis, 226
in respiratory distress, 255
in sick bird syndrome, 271
in West Nile virus infection, 308
warts (cutaneous papillomas), **216–217**
washing, oil exposed birds, 195
water consumption, 234, 235, 239
excessive, 239
see also polydipsia
water deprivation test, 91, 235, 252
weakness *see* neurologic conditions
weight gain, 59, 202
weight loss
in air sac mite infestation, 5
in anorexia, 13
in aspergillosis, 24
in aspiration, 27
in bordetellosis, 44
in campylobacteriosis, 48
in clostridiosis, 66

in CNS tumors, 68
in coccidiosis, 74, 76
in cryptosporidiosis, 85
in emaciation, 105
in enteritis and gastritis, 108, 110
in flagellate enteritis, 119
in gastrointestinal foreign bodies, 124
in gastrointestinal helminthiasis, 126
in heavy metal toxicity, 129
in hemoparasite infection, 132
in herpesvirus infections, 140
in hyperglycemia and diabetes mellitus, 143
in ingluvial hypomotility/crop stasis/ileus, 153
in lipomas, 161
in liver disease, 163
in *Macrorhabdus* infection, 169
in mycobacteriosis, 174
in mycoplasmosis, 177
in mycotoxicosis, 179
in obesity management, 188
in oil exposure, 193
oral plaques and, 197
in overgrown beak, 211
in pancreatic disease, 213
in PDD, 247
in pneumonia, 228
in regurgitation and vomiting, 249, 250
in renal disease, 251
in toxoplasmosis, 279
in trichomoniasis, 287
in vitamin D toxicosis, 305
weight monitoring, 106, 239, 248
Western Equine Encephalitis (WEE) virus, 68, 134, 299, 334, 335
West Nile virus (WNV), 299, 300, 301, **307–309**, 329, 334, 335
CNS tumors *vs*, 68
lameness due to, 158
seizures, 267
wheezing, 140
white blood cell (WBC) count, 318, 319
mycoplasmosis, 177
squamous cell carcinoma, 274, 275
see also leukocytosis
wide-based stance, 98, 103, 188
wing droop, 11, 111, 121, 126, 158, 307
wing trim, 113
wound(s), **284–286**
bite, **41–43**
closure, 42, 285
dressings, in bite wounds, 42
healing, phases, 284
wound management, 285
in bite wounds, 41–42
in fractures and luxations, 121
in uropygial gland disease, 297
Wright's stain, 85

X

xanthine dehydrogenase, 145
xanthine oxidase, 145
xanthoma, 95, 161, 188, **310–311**

Y

yolk coelomitis *see* egg yolk coelomitis
yolk emboli, 182

Z

zinc, normal blood levels, 93
zinc sulfate centrifugation technique, 119
zinc toxicity/toxicosis, 66, 124, **129–130**, 338
anemia associated, 9
gastrointestinal foreign bodies, 124
neurological conditions due to, 182
pancreatic disease and, 213, 214
undigested food in droppings and, 292, 293, 294
urate and fecal discoloration and, 295, 296
zonisamide, 184, 268
zoonotic disease, **335–336**
avian influenza, 33–34
bornavirus infections, 248
campylobacteriosis, 48–49, 335
candidiasis, 50
chlamydiosis, 57, 165, 335
colibacillosis, 84, 335
cryptosporidiosis, 86, 294
ectoparasites, 102
flagellate enteritis, 120
mycobacteriosis, 164, 165, 175, 335
in ocular disease, 191, 192
oil spills and, 195
pancreatic disease and, 215
paramyxovirus infections, 220
pasteurellosis, 222
pneumonia causes, 230
salmonellosis, 260, 261, 335
undigested food in droppings and, 294
viral diseases, 301
West Nile virus infection, 309, 335